11–19 PROGRESSION

Edexcel

Mathematics

Foundation

Student Book

Confidence • Fluency • Problem-solving • Reasoning

Series Editors:
Dr Naomi Norman • Katherine Pate

ALWAYS LEARNING

PEARSON

Published by Pearson Education Limited, 80 Strand, London WC2R 0RL.

Registered in England 872828

www.pearsonschoolsandfecolleges.co.uk

Copies of official specifications for all Edexcel qualifications may be found on the website: www.edexcel.com

Text © Pearson Education Limited 2014
Typeset by Tech-set Ltd, Gateshead
Original illustrations © Pearson Education Limited 2014

The rights of Chris Baston, Ian Bettison, Ian Boote, Tony Cushen, Tara Doyle, Kath Hipkiss, Su Nicholson, Katherine Pate, Jenny Roach, Carol Roberts, Peter Sherran, Catherine Murphy, Naomi Norman and Robert Ward-Penny to be identified as authors of this work have been asserted by them in accordance with the Copyright, Designs and Patents Act 1988.

First published 2015

18 17
10 9 8 7 6 5

British Library Cataloguing in Publication Data
A catalogue record for this book is available from the British Library

ISBN 978 1 447 98019 3

Printed in Slovakia by Neografia

Acknowledgements
We would like to thank Glyn Payne for his work on this book and Amanda Hill, Kath Hipkiss, Mel Muldowney, Pietro Tozzi and Narsh Srikanthapalan for their feedback on the ordering of this book.

The publisher would like to thank the following for their kind permission to reproduce their photographs:

(Key: b-bottom; c-centre; l-left; r-right; t-top)

123RF.com: digieye 277, Olena Makovey 562; **Alamy Images:** imageBROKER / Alexander Pöschel 151; **Getty Images:** Blend Images / Jon Feingersh 32, Claus Lunau 539, NiseriN 594, Oli Scarff 612; **Imagestate Media:** John Foxx Collection 209; **Peter Evans:** 344; **Shutterstock.com:** Albert Ziganshin 243, DavidEwingPhotography 296, Denis Cristo 79, diversepixel 552, Gunter Nezhoda 427, iofoto 139, Jason Stitt 398, JeremyCulpDesign 58, Joe Belanger 379, joingate 1, mama_mia 92, Marek Szumlas 265bl, mark higgins 265, Patryk Kosmider 439, Somchai Buddha 504, spaxiax 526, swinner 197, topora 149, wavebreakmedia 121; **TopFoto:** PA Photos 181; **Veer / Corbis:** motorolka 411, unkreatives 479

Cover images: Front: Created by Fusako, Photography by NanaAkua

All other images © Pearson Education

A note from the publisher
In order to ensure that this resource offers high-quality support for the associated Edexcel qualification, it has been through a review process by the awarding organisation to confirm that it fully covers the teaching and learning content of the specification or part of a specification at which it is aimed, and demonstrates an appropriate balance between the development of subject skills, knowledge and understanding, in addition to preparation for assessment.

While the publishers have made every attempt to ensure that advice on the qualification and its assessment is accurate, the official specification and associated assessment guidance materials are the only authoritative source of information and should always be referred to for definitive guidance.

Edexcel examiners have not contributed to any sections in this resource relevant to examination papers for which they have responsibility.

No material from an endorsed resource will be used verbatim in any assessment set by Edexcel.

Endorsement of a resource does not mean that the resource is required to achieve this qualification, nor does it mean that it is the only suitable material available to support the qualification, and any resource lists produced by the awarding organisation shall include this and other appropriate resources.

Contents

Welcome to Edexcel GCSE (9-1) Mathematics Foundation Student Book

This Student Book is packed full of features to help you enjoy and feel confident in maths as well as preparing you for your GCSE.

Choose only the topics in *Strengthen* that you need a bit more practice with. You'll find more hints here to lead you through specific questions. Then move on to *Extend*.

At the end of the *Master* lessons, take a *Check up* test to help you to decide whether to *Strengthen* or *Extend* your learning.

Extend helps you to apply the maths you know to some different situations.

Unit Openers put the maths you are about to learn into a real-life context. Have a go at the question – it uses maths you have already learnt so you should be able to answer it at the start of the unit.

When you have finished the whole unit, a *Unit test* helps you see how much progress you are making.

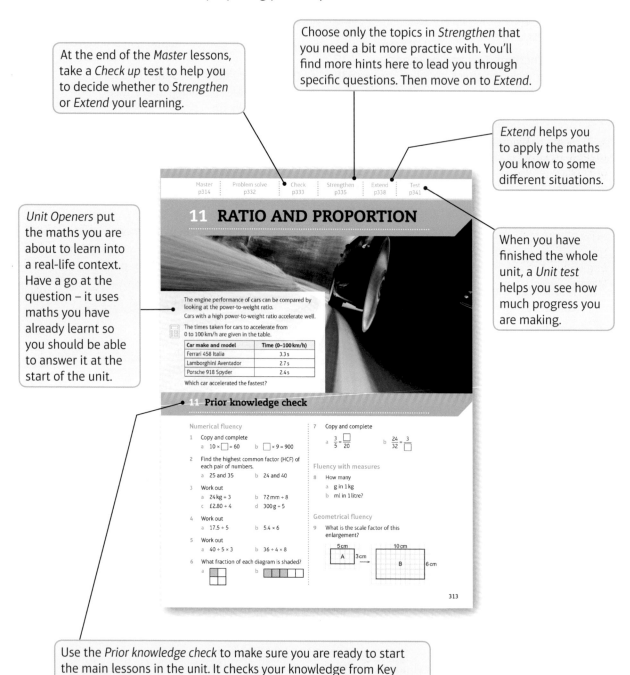

| Master p314 | Problem solve p332 | Check p333 | Strengthen p335 | Extend p338 | Test p341 |

11 RATIO AND PROPORTION

The engine performance of cars can be compared by looking at the power-to-weight ratio.
Cars with a high power-to-weight ratio accelerate well.

The times taken for cars to accelerate from 0 to 100 km/h are given in the table.

Car make and model	Time (0–100 km/h)
Ferrari 458 Italia	3.3 s
Lamborghini Aventador	2.7 s
Porsche 918 Spyder	2.4 s

Which car accelerated the fastest?

11 Prior knowledge check

Numerical fluency

1 Copy and complete
   a $10 \times \square = 60$
   b $\square \times 9 = 900$

2 Find the highest common factor (HCF) of each pair of numbers.
   a 25 and 35
   b 24 and 40

3 Work out
   a $24\,kg \div 3$
   b $72\,mm \div 8$
   c $£2.80 \div 4$
   d $300\,g \div 5$

4 Work out
   a $17.5 \div 5$
   b 5.4×6

5 Work out
   a $40 \div 5 \times 3$
   b $36 \div 4 \times 8$

6 What fraction of each diagram is shaded?
   a   b

7 Copy and complete
   a $\frac{3}{5} = \frac{\square}{20}$
   b $\frac{24}{32} = \frac{3}{\square}$

Fluency with measures

8 How many
   a g in 1 kg
   b ml in 1 litre?

Geometrical fluency

9 What is the scale factor of this enlargement?

5 cm
A   3 cm
10 cm
B   6 cm

313

Use the *Prior knowledge check* to make sure you are ready to start the main lessons in the unit. It checks your knowledge from Key Stage 3 and from earlier in the GCSE course. Your teacher has access to worksheets if you need to recap anything. If you find you often struggle with these, you may want to work through our Access to Foundation Workbooks and then come back to this Student Book.

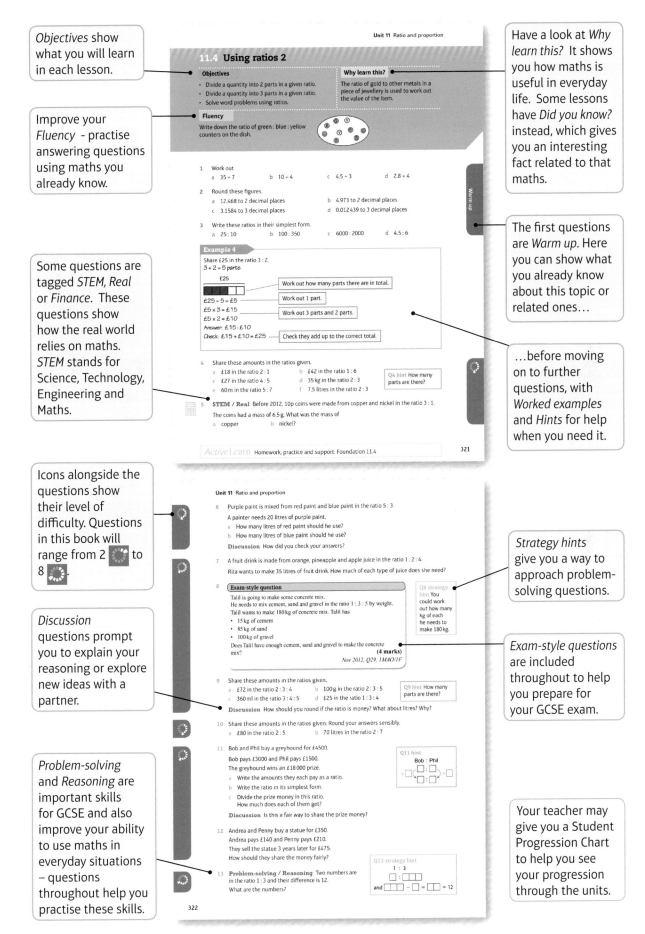

Objectives show what you will learn in each lesson.

Improve your *Fluency* - practise answering questions using maths you already know.

Some questions are tagged *STEM, Real* or *Finance*. These questions show how the real world relies on maths. *STEM* stands for Science, Technology, Engineering and Maths.

Icons alongside the questions show their level of difficulty. Questions in this book will range from 2 to 8.

Discussion questions prompt you to explain your reasoning or explore new ideas with a partner.

Problem-solving and *Reasoning* are important skills for GCSE and also improve your ability to use maths in everyday situations – questions throughout help you practise these skills.

Have a look at *Why learn this?* It shows you how maths is useful in everyday life. Some lessons have *Did you know?* instead, which gives you an interesting fact related to that maths.

The first questions are *Warm up*. Here you can show what you already know about this topic or related ones…

…before moving on to further questions, with *Worked examples* and *Hints* for help when you need it.

Strategy hints give you a way to approach problem-solving questions.

Exam-style questions are included throughout to help you prepare for your GCSE exam.

Your teacher may give you a Student Progression Chart to help you see your progression through the units.

Unit 11 Ratio and proportion

11.4 Using ratios 2

Objectives
- Divide a quantity into 2 parts in a given ratio.
- Divide a quantity into 3 parts in a given ratio.
- Solve word problems using ratios.

Why learn this?
The ratio of gold to other metals in a piece of jewellery is used to work out the value of the item.

Fluency
Write down the ratio of green : blue : yellow counters on the dish.

1 Work out
   a $35 \div 7$ b $10 \div 4$ c $4.5 \div 3$ d $2.8 \div 4$

2 Round these figures.
   a 12.468 to 2 decimal places   b 4.973 to 2 decimal places
   c 3.1584 to 3 decimal places   d 0.012439 to 3 decimal places

3 Write these ratios in their simplest form.
   a $25 : 10$ b $100 : 350$ c $6000 : 2000$ d $4.5 : 6$

Example 4
Share £25 in the ratio 3 : 2.
3 + 2 = 5 parts

£25 ÷ 5 = £5 — Work out how many parts there are in total.
£5 × 3 = £15 — Work out 1 part.
£5 × 2 = £10 — Work out 3 parts and 2 parts.
Answer: £15 : £10
Check: £15 + £10 = £25 — Check they add up to the correct total.

4 Share these amounts in the ratios given.
   a £18 in the ratio 2 : 1   b £42 in the ratio 1 : 6
   c £27 in the ratio 4 : 5   d 35 kg in the ratio 2 : 3
   e 60 m in the ratio 5 : 7   f 7.5 litres in the ratio 2 : 3

   Q4 hint How many parts are there?

5 **STEM / Real** Before 2012, 10p coins were made from copper and nickel in the ratio 3 : 1. The coins had a mass of 6.5 g. What was the mass of
   a copper   b nickel?

*Active*Learn Homework, practice and support: Foundation 11.4   321

Unit 11 Ratio and proportion

6 Purple paint is mixed from red paint and blue paint in the ratio 5 : 3.
   A painter needs 20 litres of purple paint.
   a How many litres of red paint should he use?
   b How many litres of blue paint should he use?
   Discussion How did you check your answers?

7 A fruit drink is made from orange, pineapple and apple juice in the ratio 1 : 2 : 4.
   Rita wants to make 35 litres of fruit drink. How much of each type of juice does she need?

8 **Exam-style question**
   Talil is going to make some concrete mix.
   He needs to mix cement, sand and gravel in the ratio 1 : 3 : 5 by weight.
   Talil wants to make 180 kg of concrete mix. Talil has
   • 15 kg of cement
   • 85 kg of sand
   • 100 kg of gravel
   Does Talil have enough cement, sand and gravel to make the concrete mix?
   (4 marks)
   Nov 2012, Q29, 1MAO/1F

   Q8 strategy hint You could work out how many kg of each he needs to make 180 kg.

9 Share these amounts in the ratios given.
   a £72 in the ratio 2 : 3 : 4   b 100 g in the ratio 2 : 3 : 5
   c 360 ml in the ratio 3 : 4 : 5   d £25 in the ratio 1 : 3 : 4
   Discussion How should you round if the ratio is money? What about litres? Why?

   Q9 hint How many parts are there?

10 Share these amounts in the ratios given. Round your answers sensibly.
    a £80 in the ratio 2 : 5   b 70 litres in the ratio 2 : 7

11 Bob and Phil buy a greyhound for £4500.
    Bob pays £3000 and Phil pays £1500.
    The greyhound wins an £18 000 prize.
    a Write the amounts they each pay as a ratio.
    b Write the ratio in its simplest form.
    c Divide the prize money in this ratio. How much does each of them get?
    Discussion Is this a fair way to share the prize money?

    Q11 hint
    Bob : Phil
    □ : □
    □ : □

12 Andrea and Penny buy a statue for £350.
    Andrea pays £140 and Penny pays £210.
    They sell the statue 3 years later for £475.
    How should they share the money fairly?

13 **Problem-solving / Reasoning** Two numbers are in the ratio 1 : 3 and their difference is 12.
    What are the numbers?

    Q13 strategy hint
    1 : 3
    □ : □ □
    and □□ – □ = □ = 12

322

vii

Problem-solving lessons

As well as problem-solving and reasoning throughout, this book includes a problem-solving lesson in every unit. There are two types:

- Some, such as the Unit 11 one below, give you strategies to approach problem-solving questions, for example using bar models. You are given a worked example which talks you through answering a question using the strategy and then a number of questions to practise on.

- Others, such as the Unit 7 one below, give you problem-solving questions in a real-life context to help you see how mathematical problem-solving is a part of many real-life activities.

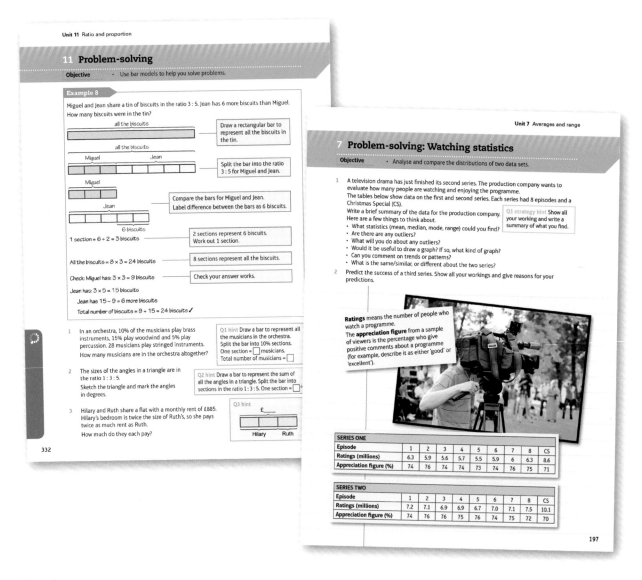

Further support

You can easily access extra resources that tie in to each lesson – look for the *ActiveLearn Homework, practice and support* references on the first page of each lesson. This is online practice that is clearly mapped to the lessons and provides interactive exercises with lots of extra support for when you are working independently.

The Practice, Problem-solving and Reasoning Books are full of extra practice for key questions and will help you reinforce your learning and track your own progress.

1 NUMBER

The Fibonacci sequence of numbers starts with 1, 1. The next number in the sequence is found by adding the two previous numbers together: 1, 1, 2, 3, 5, 8, 13, 21, ... Fibonacci spirals can be created using the Fibonacci sequence of numbers. They are found in many places in nature, for example the nautilus shell.

You can draw a Fibonacci spiral by joining the opposite corners of squares with sides of length 1 cm, 1 cm, 2 cm, 3 cm, 5 cm, 8 cm, ...

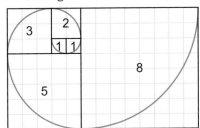

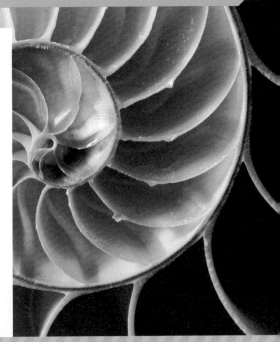

Continue the Fibonacci sequence to the first 20 terms.
1, 1, 2, 3, 5, 8, 13, 21, …

1 Prior knowledge check

Numerical fluency

1 Write down the value of the digit 2 in these numbers.
   a 125
   b 2300
   c 628374
   d 47.28
   e 0.372
   f 8.021

2 Write these decimal numbers in order of size. Start with the largest.
   4.3, 4.12, 4.61, 3.12, 3.09

3 Round
   a 2385 to the nearest 1000
   b 872 to the nearest 100
   c 59 to the nearest 10
   d 84.19 to the nearest integer.

4 Work out
   a 25 × 100
   b 14 ÷ 10
   c 3.5 × 1000
   d 138.6 ÷ 100
   e 4.7 × 1000000
   f 20 × 30

5 Work out
   a 25 + 32
   b 153 + 19
   c 87 − 29
   d 476 − 38

6 Work out
   a 25 × 30
   b 125 × 234

7 Look at this set of numbers.
   4, 8, 12, 15, 27, 29, 36, 49
   Which are
   a odd numbers
   b divisible by 2
   c multiples of 3
   d square numbers
   e cube numbers?

8 Write the prime numbers between 4 and 10.

9 List the factors of 24.

10 Work out
   a $8\overline{)816}$
   b $6\overline{)282}$
   c $5\overline{)67}$
   d 252 ÷ 14
   e 1278 ÷ 12

1

11 Work out
   a the square of 9
   b the square root of 64
   c the cube of 4
   d $\sqrt{16}$

12 Work out
   a 5 + −3   b 6 − 8
   c −7 + 2   d −4 − −3
   e −3 + 10   f −1 − −4

13 What number is
   a 20 more than −40
   b 30 less than −50?

14 Work out
   a −5 × 2   b 5 × −2   c −5 × −2
   d 12 ÷ −3 e −12 ÷ 3 f −12 ÷ −3

15 Work out
   a 3.1 + 5.27   b 16.4 − 9.18

16 Work out
   a 0.2 × 4   b 5 × 0.7
   c 1.8 ÷ 3   d 2.4 ÷ 6

17 Copy and complete.
   a $5^{\square} = 25$ b $3^{\square} = 27$

18 Work out
   a 20^2 b 6^3

19 Work out
   a (8 − 5) × 3   b 12 ÷ (7 − 4)

20 Put the correct sign (< or >) between each pair of numbers to make a true statement.
   a 1 ☐ 7   b 0.8 ☐ 0.2
   c −4 ☐ 3   d −1 ☐ −9

Fluency with measures

21 Find the area of the square.

←7 cm→
7 cm

22 Sienna buys a cup of coffee for £2.60 and a sandwich for £5.95. How much change should she get from a £20 note?

23 Ji has four parcels to post. They weigh 580 g, 825 g, 673 g and 742 g. What is the total mass in kg?

> **Q23 hint**
> 1000 g = 1 kg

24 What is the temperature on this thermometer in
   a °C
   b °F?

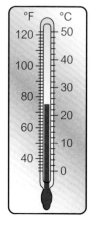

25 What is the difference between
   a 5 °C and 12 °C
   b −7 °C and 8 °C
   c −3 °C and −11 °C?

26 The daily average maximum temperatures in Moscow range from −9 °C in January to 23 °C in July. What is the difference between these temperatures?

27 How many 240 ml glasses can be completely filled with a 2 litre bottle of water?

> **Q27 hint**
> 1000 ml = 1 litre

28 Abi needs 18.4 m of skirting board. Skirting board is sold in packs of four lengths. Each length is 2400 mm. How many packs does Abi need to buy?

> **Q28 hint**
> 1000 mm = 1 m

✱ Challenge

29 A birthday algorithm.
   Step 1 Start with the number for the month of your birthday. For example, for December use 12.
   Step 2 Multiply this by 5.
   Step 3 Add 7.
   Step 4 Multiply by 4.
   Step 5 Add 13.
   Step 6 Multiply by 5.
   Step 7 Add the date of your birthday. For example, for 18 January add 18.
   Step 8 Subtract 205.

   If you have done this correctly you will be left with the month and day of your birthday!

1.1 **Calculations**

Objectives

- Use priority of operations with positive and negative numbers.
- Simplify calculations by cancelling.
- Use inverse operations.

Did you know?

The Romans used small pebbles on a counting machine called an abacus to help them with calculations. The word 'calculate' comes from the Latin word 'calculus', which means 'small pebble'.

Fluency

- What is 5 less than 3? What is $\frac{1}{3}$ of 12? What is 2^2? What is 5^2?
- What is the total of 18, 7 and 2?
- Is 2×3 the same as 3×2?

1 Simplify

   a $\frac{25}{10} = \frac{\square}{2} =$ b $14 \div 4 = \frac{14}{4} =$ c $45 \div 18 =$

2 Write the missing number facts. The first one has been started for you.

   a $3 \times -5 = -15$, so $-15 \div 3 = -5$ and $-15 \div -5 = \square$

   b $-8 \times 4 = -32$ so $-32 \div -8 = \square$ and $-32 \div 4 = \square$

   c $-7 \times -6 = 42$, so $42 \div -7 = \square$ and $42 \div -6 = \square$

3 Work out

   a $2 \times \frac{1}{8}$ b $2 \times \frac{3}{12}$ c $3 \times \frac{2}{5}$

> **Q3a hint** $2 \times \frac{1}{8} = 2$ eighths $= \frac{2}{\square}$

> Questions in this unit are targeted at the steps indicated.

4 Copy and complete these rules for multiplying and dividing. The first one has been done for you.

   a $+ \times + = +$ b $+ \times - = \square$ c $- \times + = \square$ d $- \times - = \square$

   e $+ \div + = \square$ f $+ \div - = \square$ g $- \div + = \square$ h $- \div - = \square$

Discussion What is an easy way to remember the rules in **Q4**?

Key point 1

The priority of operations is: Brackets, Indices, Division and Multiplication, Addition and Subtraction.

> **Communication hint**
> Indices means 'powers'. It is the plural of index (power).

5 Work out

   a $(4 - 1)^2 + 2 = \square^2 + 2 =$ b $12 \div (6 - 4)^2$

   c $(4 + 8) \div 2^2$ d $(4 \times 3) \div (5 - 9)$

Check your answers using a calculator.

6 Use a calculator to work out

   a $(7 - 2)^2 \times (3 - 5)$ b $(2 - 5) \div (7 - 8)^3$

   c $(4 + 1)^2 \times \sqrt{9}$ d $(5 - 3)^3 \div (9 - 7)^2$

> **Q6 hint** Use the x^2, x^3, $\sqrt{}$ and brackets keys on your calculator.

Warm up

7 **Problem-solving** Here are five calculations.

A $12 - 3 \times 2$ **B** $12 \times 3 - 2$ **C** $(12 - 3) \times 2$

D $\dfrac{12}{3 \times 2}$ **E** $12 \div 3 + 2$

Which of these calculations have the same answer?

Key point 2

The symbol $\neq$ means 'not equal to'.

---

8 Write $=$ or $\neq$ for each pair of calculations.

a $8 - 4 \times 2 \;\square\; (8 - 4) \times 2$ b $(5 + 2) \times 4 \;\square\; 5 \times 4 + 2 \times 4$

c $(3 + 4)^2 \;\square\; 3^2 + 4^2$ d $\dfrac{24}{4 + 8} \;\square\; \dfrac{24}{4} + \dfrac{24}{8}$

e $(8 - 3) - 4 \;\square\; 8 - (3 - 4)$

Q8 hint Work out the calculation on both sides. Do you get the same answer?

Example 1

Work out $(4 \times 5) \div (2 \times 30)$

$(4 \times 5) \div (2 \times 30) = \dfrac{4 \times 5}{2 \times 30}$ — Write as a fraction.

$= \dfrac{4}{2} \times \dfrac{5}{30}$ — Split into two fractions.

$= 2 \times \dfrac{1}{6} = \dfrac{2}{6}$ — Simplify.

$= \dfrac{1}{3}$ — Write the fraction in its simplest form.

---

9 Work out

a $\dfrac{3 \times 8}{6} = \dfrac{3}{6} \times 8 = \square \times 8 = \square$ b $\dfrac{30 \times 20}{40} = 30 \times \dfrac{20}{40} = 30 \times \square = \square$

c $(6 \times 7) \div (12 \times 21)$ d $\dfrac{3 \times 10}{2 \times 9}$

e $(5 \times 3) \div (6 \times 10)$ f $\dfrac{9 \times 4}{12 \times 18}$

Q9d strategy hint

$\dfrac{3 \times 10}{2 \times 9}$ is the same as $\dfrac{10 \times 3}{2 \times 9}$.

Key point 3

A **function** is a rule. The function $+2$ adds 2 to a number.

$7 \longrightarrow \boxed{+2} \longrightarrow 9$

$7 \longleftarrow \boxed{-2} \longleftarrow 9$

The **inverse** function is -2 because it *reverses* the effect of the function $+2$.

---

10 Write down the inverse for each function machine.

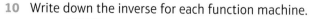

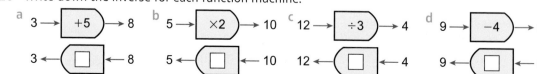

11 Write down a calculation to check each of these.

   a 387 + 579 = 966   b 687 − 598 = 89

   c 506 ÷ 46 = 11   d 264 × 12 = 3168

Q11a hint

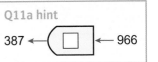

387 ←— ☐ ←— 966

12 Which is the correct calculation to check 96 ÷ 6 + 4 = 20?

   A 20 − 4 × 6   **B** 20 × 6 − 4

   C (20 − 4) × 6   **D** 20 × (6 − 4)

   Discussion How did you use the priority of operations?

Q12 hint Draw the function machine.

 13 Complete these calculations. Check the answers using inverse operations.

   a 15 × 5 − 3 = ☐   b (18 + 7) ÷ 5 = ☐   c 30 × 6 ÷ 9 = ☐

> **Key point 4**
>
> Finding the **square root** is the inverse of finding the square.
> Finding the **cube root** is the inverse of finding the cube.

 14 Complete these calculations. Check your answers using an inverse operation.

   a $\sqrt{625}$ = ☐   b 5^3 = ☐   c 45^2 = ☐   d $\sqrt[3]{1331}$ = ☐

15 **Exam-style question**

   Vincent uses this rule to convert from euros (€) to British pounds (£).
   Multiply the number of euros by 4 and then divide by 5.

   a Use the rule to convert 80 euros to British pounds.   **(2 marks)**

   Vincent uses the rule to convert an amount in euros to British
   pounds. The result is £100.

   b What amount in euros did Vincent convert to British pounds?

   (3 marks)

Q15 strategy hint
Start by drawing the function machine for Vincent's rule. Use it to work out the inverse function in part **b**.

1.2 Decimal numbers

Objectives

- Round to a given number of decimal places.
- Multiply and divide decimal numbers.

Why learn this?

Decimals are used commonly in everyday life, for example in money and measurement.

Fluency

- What is the value of the digit 5 in the number 3.152?
- What is the value of the digit 6 in the number 0.506?

1 Copy and complete.

   a 1 cm = ☐ mm   b 1 m = ☐ cm   c 1 km = ☐ m

2 Work out

   a 5)395   b 2)343   c 4)211

3 Copy and complete these equivalent fractions.

   a $\dfrac{3}{5} = \dfrac{☐}{50}$ b $\dfrac{7}{10} = \dfrac{☐}{100}$

Warm up

> **Key point 5**
>
> To round a number to 1 decimal place (1 d.p.), look at the digit in the 2nd decimal place. If it is 5 or more, round up. For example, 35.2**3** is 35.2 (1 d.p.) and 35.2**7** is 35.3 (1 d.p.).

4 Round these numbers to 1 decimal place.
   a 3.462   b 0.539   c 12.082   d 8.973

> **Key point 6**
>
> To round a number to 2 decimal places (2 d.p.), look at the digit in the 3rd decimal place. To round a number to 3 decimal places (3 d.p.), look at the digit in the 4th decimal place.

5 Round these numbers to 2 decimal places.
   a 4.0258   b 16.1723   c 0.1349   d 11.896

6 Round these numbers to 3 decimal places.
   a 8.4621   b 22.8057   c 9.1063

7 Round

Q7a hint
3.53 cm = 35.3 mm rounds to ☐ mm.

   a 3.53 cm to the nearest mm   b 9.4615 m to the nearest mm
   c 6.846 m to the nearest cm   d 25.3254 km to the nearest m.

8 **Problem-solving** Jamie cuts a 3.5 m length of wood into 6 equal pieces. What is the length of each piece of wood correct to the nearest mm?

9 **Problem-solving / Finance** Jasmine pays £52.40 for 40 litres of petrol. What is the cost per litre?

Q9 communication hint Cost per litre means the cost for 1 litre.

   Discussion How many decimal places should you round money answers to?

10 **Problem-solving / Reasoning** Four friends share a £35.90 taxi fare equally between them. How much should they each pay? Check your answer.

11 Work out
   a 0.4 × 20 = 0.4 × 2 × 10   b 0.6 × 30   c 0.3 × 500
              = 0.8 × ☐
              = ☐

12 Work out
   a 3 × 2   b 3 × 0.2   c 0.3 × 2   d 0.3 × 0.2
   Discussion What do you notice?

> **Example 2**
>
> Work out 3.4 × 5.6
>
> Estimate 3 × 6 = 18 ──── Estimate your answer by rounding.
>
> $$\begin{array}{r} 3\ 4 \\ \times\ \ 5\ 6 \\ \hline 2\ 0\ _2 4 \\ 1\ 7\ _2 0\ 0 \\ \hline 1\ 9\ 0\ 4 \end{array}$$ ──── Use a standard method to work out 34 × 56.
>
> 3.4 × 5.6 = 19.04 ──── Use your estimated answer to see where to put the decimal point.

13 Work out
  a 8.5 × 3   b 0.62 × 8   c 71 × 0.3   d 162 × 1.81

14 **Exam-style question**

  The table shows the cost of two different diaries.

Small	£2.40
Large	£3.50

  Eleanor buys 10 small diaries and 5 large diaries. She pays with a
  £50 note. Work out how much change she should get.   **(3 marks)**

Exam hint
Show your working
by writing down
every calculation
you need to do.
Make sure you
clearly state your
final answer.

15 **Real** Freya buys a 25 kg sack of potatoes. The potatoes cost £1.19 per kg.
  How much does she pay?

16 Work out 0.8 × 0.08

17 **Problem-solving** Kelly multiplies two decimals and gets the answer 0.06
  What two decimals did she multiply? Is there more than one answer?

18 Work out
  a $\dfrac{8}{4}$ b $\dfrac{0.8}{4}$ c $\dfrac{8}{0.4}$ d $\dfrac{0.8}{0.4}$
  Discussion What do you notice?

19 Work out
  a 24.5 ÷ 5   b 29.4 ÷ 3   c 34.3 ÷ 7

  Q19a hint Write as $5\overline{)24.5}$

  Discussion How could you estimate the answer to use as a check?

Key point 7

To divide by a decimal, multiply both numbers by a power of 10 (10, 100, …) until you have a
whole number to divide by. Then work out the division.

Example 3

Work out 35.1 ÷ 1.5

$35.1 \div 1.5 = \dfrac{35.1}{1.5}$

$\dfrac{35.1}{1.5} = \dfrac{351}{15}$ (×10)

1.5 has 1 decimal place, so
multiply both numbers by 10.

$15\overline{)351.0}^{\,23.4}$ — Divide.

Check using an inverse operation and estimation.

Check: 15 × 23.4 ≈ 20 × 20 = 400

Communication hint ≈ means 'approximately equal to'.

20 Work out
  a 6.24 ÷ 0.4   b 22.5 ÷ 0.15   c 280 ÷ 0.07   d 234 ÷ 1.8

  Q20b hint 0.15
  has 2 decimal
  places, so
  multiply by 100.

21 **Reasoning** Lex pays £5.60 for some new pens. The pens cost 35p each.
  How many did he buy?

22 **Reflect** Look at these statements.
  • Multiplying a number by a number less than 1 gives you an answer smaller than the first
    number.
  • Dividing a number by a number less than 1 gives you an answer larger than the first number.
    Are these statements true?

1.3 **Place value**

Objectives

- Write decimal numbers of millions.
- Round to a given number of significant figures.
- Estimate answers to calculations.
- Use one calculation to find the answer to another.

Did you know?

Numbers are often rounded to the degree of accuracy required. For example, when measuring a length of wood you may only need to measure to the nearest cm.

Fluency

What is
- 115 to the nearest 100
- 217 to the nearest 10
- 46.3 to the nearest whole number?

1 Work out
   a 1.3 × 1000
   b 2.6 × 1000000
   c 34500000 ÷ 1000000

2 Simplify
   a $\dfrac{8}{0.5}$
   b $\dfrac{15 \times 20}{40}$

3 Copy and complete.
   7.6 × 2.1 = ☐, so ☐ ÷ 7.6 = 2.1 and ☐ ÷ 2.1 = 7.6

4 **Reasoning** Sarah multiplies 14.3 by 0.96 on her calculator. Her answer is 14.688.
   Without finding the exact value of 14.3 × 0.96, explain why her answer must be wrong.

5 Write these numbers in figures.
   a 4.6 million
   b 10.1 million
   c 2.45 million
   d 3.125 million

   > **Q5a hint** 4.6 million in figures is 4.6 × 1000000.

6 Write these numbers in millions.
   a 34100000
   b 4250000
   c 58420000
   d 16325000

   > **Q6a hint** 34100000 in millions is 34100000 ÷ 1000000.

7 Work out these. Write your answers in figures.
   a 5.3 million + 6.25 million
   b 0.5 million + 10.25 million
   c 14.2 million − 1.34 million
   d 11.3 million − 0.75 million

   > **Q7 hint** Add the number of millions. Then convert to figures.

Key point 8

You can round numbers to a number of significant figures (s.f.). The 1st significant figure is the one with the highest place value. It is the first non-zero digit in the number, counting from the left.

8 Round
   a 561.837 to 4 s.f.

   > **Q8a hint** The 4th significant figure of 561.837 is 8. The next digit is 3, so leave the 8 as it is.

   b 0.003 468 to 3 s.f.

   > **Q8b hint** The 3rd significant figure is 6.

   c 48725 to 2 s.f.

   > **Q8c hint** 48 725
   > You will need to write in zeros to keep the place value.

 *Active*Learn Homework, practice and support: Foundation 1.3

> **Key point 9**
>
> Rounded numbers must have the same place value as the original number. For numbers greater than zero this means you may need to put in zeros as 'place fillers'.

9   Round these to the number of significant figures shown.
   a   76 432 (3 s.f.)   b   0.0578 (1 s.f.)   c   342 510 (2 s.f)   d   47.376 (4 s.f.)

10   Work out
   a   700 × 400
   b   2000 ÷ 500
   c   8000 × 300
   d   60 000 ÷ 30

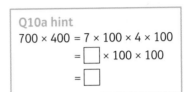

Q10a hint
700 × 400 = 7 × 100 × 4 × 100
= ☐ × 100 × 100
= ☐

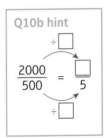

Q10b hint

$$\frac{2000}{500} = \frac{☐}{5}$$

> **Key point 10**
>
> To estimate the answer to a calculation, you can round every number to 1 s.f.

11   Estimate an answer for each calculation.
   a   275 × 421 ≈ 300 × ☐ =
   b $\dfrac{876}{29}$
   c $\dfrac{41 \times 482}{1182}$
   d $\dfrac{675 \times 2346}{374}$
   e $\dfrac{284 \times 10.34}{0.52}$
   f $\dfrac{5.21 \times 3.84}{6.72}$
   g $\dfrac{9.83 \times 3.24}{7.65}$

Q11f & g hint You may need to estimate the final division, for example 21 ÷ 4 ≈ 5

12   Check your answers to **Q11** using a calculator.

Q12 hint Use the fraction key on your calculator.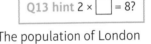

13   **Real / Problem-solving**
   a X is 8 million. Y is 2 million. How many times larger is X than Y?
   b   At the end of 2013, the population of Beijing was 21.15 million. The population of London was 8.308 million. Estimate the number of times greater the population of Beijing was than the population of London.

Q13 hint 2 × ☐ = 8?

> **Example 4**
>
> Use the information that 282 × 56 = 15 792 to work out the value of
> a   28.2 × 5.6
> b   15.792 ÷ 5.6
>
> a   282 × 56 = 15 792
>
> The digits in the questions are the same.
>
>   Estimate: 28.2 × 5.6 ≈ 30 × 6 = 180
>   28.2 × 5.6 = 157.92
>
> Use the estimate to decide where to put the decimal point.
>
> b   15 792 ÷ 56 = 282
>
> Write the related division.
>
>   Estimate: 15.792 ÷ 5.6 ≈ 20 ÷ 6 ≈ 3
>   15.792 ÷ 5.6 = 2.82
>
> Estimate the answer.
>
> Use the estimate to decide where to put the decimal point.

14 **Reasoning** Use the information that 45 × 127 = 5715 to work out the value of
   a   4.5 × 1.27   b   57.15 ÷ 12.7   c   5.715 ÷ 0.127

15 **Reasoning** Use the information that 148 × 39 = 5772 to work out the value of
   a   148 × 40   b   147 × 39

16 Calculate the value of $\frac{23.5 - 4.43}{18.40 - 3.22}$

   a   Write down all the figures on your calculator display.
   b   Give your answer correct to 2 decimal places.

   | Q16 hint Use the fraction key on your calculator. |

   Discussion How can you decide how much to round an answer, if the question doesn't tell you?

17 **Exam-style question**

   Calculate the value

   of $\frac{12.79 \times (18.3 - 10.43)}{(47.1 + 19.6) \times 4.1}$.   **(3 marks)**

   a   Write down all the figures on your calculator display.
   b   Give your answer correct to 2 decimal places.

   Exam hint
   Key the calculations into the top and bottom parts of the fraction on your calculation.

1.4 **Factors and multiples**

Objectives

· Recognise 2-digit prime numbers.
· Find factors and multiples of numbers.
· Find common factors and common multiples of two numbers.
· Find the HCF and LCM of two numbers by listing.

Did you know?

Prime numbers are used in sophisticated encryption codes. They were used by Alan Turing as a key to crack the Enigma code used in World War II.

Fluency

3, 7, 11, 12, 13
Which of these numbers is a
· prime number   · factor of 24   · multiple of 4   · product of 3 and 4?

1   What is the smallest prime number?

2   List the first 8 multiples of 7.

3   Copy and complete each sentence using the correct word from the cloud.
   a   4 is a .... of 36.
   b   21 is a .... of 7 and 3.
   c   18 is the .... of 6 and 3.
   d   11 is a ....

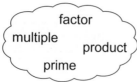

factor
multiple
product
prime

Key point 11

A prime number has exactly two factors, itself and 1.

10

4　List the prime numbers between 20 and 30.

> Q4 hint Write out the numbers from 20 to 30. Cross out the ones that are divisible by 2, divisible by 3, divisible by 5, and so on.

5　List the prime numbers between 50 and 70.

6　**Reasoning** The product of two prime numbers is always even. Is this true or false?
　Discussion How did you decide?

7　List all the factors of 36.

> Q7 hint List all of its factor pairs.
> $1 \times 36 = 36$, $2 \times \square = 36$, and so on.

8　List all the factors of
　a　28　　　　　b　32　　　　　c　40

Key point 12

The **highest common factor (HCF)** of two numbers is the largest number that is a factor of both numbers.

Example 5

Find the HCF of 18 and 24.

1×18　　　　1×24

2×9　　　　2×12　　　—　　Work out the factors.

3×6　　　　3×8

　　　　　　　4×6

18:　①,　②,　③,　⑥,　9,　18
24:　①,　②,　③,　4,　⑥,　8,　12,　24

Ring the common factors.

The HCF is 6.

9　Find the HCF of
　a　16 and 40　　　　b　35 and 63　　　　c　24 and 84
　d　30 and 75　　　　e　16, 24 and 40

Key point 13

The **lowest common multiple (LCM)** of two numbers is the smallest number that is a multiple of both numbers.

10　a　Write out the first 10 multiples of
　　　　i　3　　　ii　4
　b　Ring the common multiples of 3 and 4.
　c　Which is the LCM of 3 and 4?

11　Find the LCM of
　a　7 and 5　　　　b　4 and 12　　　　c　8 and 6
　d　12 and 30　　　e　5, 8 and 10
　Discussion How many multiples do you need to write for each pair of numbers?

12 Find the HCF and LCM of

  a 12 and 18   b 40 and 60

13 **Reasoning** Write two numbers with a HCF of 16.

14 **Reasoning** Write a pair of numbers with a LCM of 40.

15 **Problem-solving** Matt prepares lucky dip bags for a pet shop. He wants each to be the same with no items left over.

Matt has 18 chews and 27 toys. What is the greatest number of lucky dips he can prepare?

> **Q15 hint** Find the HCF of 18 and 27.
> 18 = ☐ × 2   27 = ☐ × 3
> Use a diagram approach to check your answer.

16 **Problem-solving** One lighthouse flashes its lights every 20 seconds. A second lighthouse flashes its lights every 25 seconds. They both flash together at 6 pm. After how long will they next flash together?

> **Q16 hint** Find the LCM of 20 and 25.

17 (**Exam-style question**

Veena bought some food for a barbecue.

She is going to make some hot dogs.

She needs a bread roll and a sausage for each hot dog.

There are 40 bread rolls in a pack.

There are 24 sausages in a pack.

Veena bought exactly the same number of bread rolls and sausages.

  a How many packs of bread rolls and packs of sausages did she buy?

  .... packs of bread rolls

  .... packs of sausages

  b How many hot dogs can she make?   .... hot dogs

(5 marks)

March 2011, Q4, 5MB2H/01

> **Exam hint**
> When using the LCM to find the answer to a question, show all multiples clearly.

18 **Problem-solving** Three sets of Christmas tree lights flash every 0.4, 0.6 and 0.8 seconds. They all flash together at midnight. How long does it take before they next flash together?

> **Q18 hint** For small numbers you may find it easier to list the multiples to find the LCM.

1.5 Squares, cubes and roots

Objectives

- Find square roots and cube roots.
- Recognise powers of 2, 3, 4 and 5.
- Understand surd notation on a calculator.

Did you know?

The symbol we use today for square root ($\sqrt{\ }$) first appeared in the 16th century.

Fluency

- What is the area of this square?
- What are the first 6 square numbers?

6 cm

6 cm

1   Round these numbers to the number of decimal places given.
     a   13.27 (1 d.p.)   b   4.632 (2 d.p.)   c   0.8477 (3 d.p.)

2   Round these numbers to the number of significant figures given.
     a   52.36 (3 s.f.)   b   112.3 (2 s.f.)   c   482 (1 s.f.)

3   Work out
     a 3×-4 b -3×4 c -3×-4

4   Work out
     a 5.4^2 b $\sqrt{79}$ to 1 d.p.

     c 6.1^3 d $\sqrt[3]{112}$ to 3 s.f.

     Q4 hint Use the power and root keys on your calculator.

5   Work out
     a 6^2 and $(-6)^2$ b $(-9)^2$ and 9^2

     Discussion What do you notice about these answers?

6   Use your answers to **Q5** to work out
     a $\pm\sqrt{36}$ b $\pm\sqrt{81}$

     Q6 hint The symbol ± shows that you are being asked for the positive **and** the negative square root.

7   Work out
     a 5^3 b $(-5)^3$

     Discussion What can you say about the cube of a negative number?

8   Use your answers to **Q7** to work out
     a $\sqrt[3]{125}$ b $\sqrt[3]{-125}$

9   Work out
     a $(-4.2)^2$ b $(3.7)^3$ to 3 s.f.   c $\sqrt{481}$ to 1 d.p.   d $\sqrt[3]{-564}$ to 3 s.f.

10   Find the positive square root of 2.56.

     Q10 hint Use the key on your calculator.

11   **Problem-solving** A square rug is 3.6 m by 3.6 m. What is the area of the rug?

12   **Problem-solving** The area of another square rug is 49 m².
     What is the length of one of the sides?

     Discussion Do you use the positive or negative square root? Why?

13   Find the pairs.

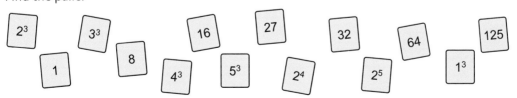

14   Evaluate
     a $2^2 \times 3$ b $2^2 + 3^2$ c $3^3 - \sqrt{64}$
     d $2^3 - 5^2$ e $2^2 + 3^2 - 4^2$ f $2^3 \times 5 \times 10$

     Q14 communication hint Evaluate means 'work out' the value.

15 Work out

    a $2.7^2 + 1.5^3$ b $6.3^3 - 7.2^2$ c $\sqrt{8^2 - 3.4^2}$ to 3 s.f.   d $\sqrt[3]{\dfrac{9.4}{5.2}}$ to 3 s.f.

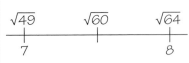

Example 6

Between which two numbers does $\sqrt{60}$ lie?

60 is between 49 and 64. ——————— | Find the two square numbers 60 lies between.

$$\sqrt{49} \qquad \sqrt{60} \qquad \sqrt{64}$$
$$7 \qquad\qquad\qquad\qquad 8$$

So $\sqrt{60}$ lies between 7 and 8. ——— | Work out their square roots.

16 **Reasoning** Between which two numbers does $\sqrt{90}$ lie?

> **Q16 strategy hint**
> Go through the square numbers in order.

17 **Reasoning** Work out $\sqrt{4} \times \sqrt{16}$ and $\sqrt{4 \times 16}$
    What do you notice about your answers?

18 **Reasoning** $9 \times 25 = 225$
    Use this fact to work out $\sqrt{225}$

19 Work out

    a $\sqrt{3} \times \sqrt{3}$ b $\sqrt{5} \times \sqrt{5}$

    What do you notice about your answers?

> **Q19 hint** $\sqrt{3} \times \sqrt{3} = \sqrt{3 \times 3}$

20 **Exam-style question**

    Use your calculator to work out the value of $2.58 \times \sqrt{2}$.

    a Write down all the figures on your calculator display.   **(1 mark)**

    b Write your answer to part **a** correct to 1 decimal place.   **(1 mark)**

    Nov 2008, Q7, 5543F/10A

21 **Exam-style question**

    Use a calculator to work out $\sqrt{\dfrac{21.6 \times 15.8}{3.8}}$
    Write down all the figures on your calculator display.   **(2 marks)**

    Nov 2008, Q2, 5544H/15H

> **Exam hint**
> Do not just write the answer as the calculation is worth 2 marks. Show part calculations too.

22 Give your answer to **Q21** correct to 3 significant figures.

Key point 14

Expressions with square roots like $3\sqrt{2}$ are in **surd** form. $3\sqrt{2}$ means $3 \times \sqrt{2}$.
An answer in surd form is exact.

23 Work out $\sqrt{12}$. Give your answer
    a in surd form
    b as a decimal.

> **Q23 hint** Use the $\boxed{S \Leftrightarrow D}$ button on your calculator to switch your answer between decimal form and surd or fraction form.

24 Work out these. Give your answers in surd form.

    a $\sqrt{6^2 - 5^2}$ b $\sqrt{\dfrac{5 \times 16}{25 - 15}}$ c $\sqrt{3^2 + 6^2}$ d $\sqrt{\dfrac{4 \times 12}{6}}$ e $\sqrt{10^2 - 5^2}$

1.6 Index notation

Objectives

- Use index notation for powers of 10.
- Use index notation in calculations.
- Use the laws of indices.

Why learn this?

Index notation is a short way of writing a number that is multiplied by itself several times.

For example, the number of stars in the Milky Way is about 100 billion or 100 000 000 000, which can be written as 10^{11}.

Fluency

Work out
- 10^2
- 10^3
- What is the index number in each?

1 Work out

   a $2 + -3$ b $2 - 5$ c $3 - -1$ d $\dfrac{2 \times 2 \times 2}{2}$

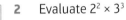

2 Evaluate $2^2 \times 3^3$

3 Work out

   a 14^2 b 11^3

4 Convert

   a 65 mm to cm   b 275 cm to m.

Key point 15

In index notation, the number that is being multiplied by itself is called the **base**.

The number written above the base is called the **index** or the **power**.

The index tells you the number of times that the base must be multiplied by itself.

Index or power

Base $\longrightarrow 10^{11} = 10 \times 10 \times 10 \times 10 \times 10 \times 10 \times 10 \times 10 \times 10 \times 10 \times 10$

5 Copy and complete the pattern.

$10^1 = 10$

$10^2 = 10 \times 10 = \square$

$10^3 = 10 \times 10 \times 10 = \square$

$10^4 = 10 \times 10 \times 10 \times 10 = 10\ 000$

$10^5 = \ldots$

$10^6 = \ldots$

6 Copy and complete.

   a $3 \times 3 \times 3 \times 3 \times 3 \times 3 = 3^{\square}$ b $5 \times 5 \times 5 = 5^{\square}$ c $4 \times 4 \times 4 \times 4 \times 4 = 4^{\square}$

7 Write as a product.

   a 2^6 b 5^4 c 7^5

> **Q7 Communication**
> hint A **product** is a multiplication.

8 Write each product using powers.

   a $4 \times 4 \times 4 \times 4 \times 4$ b $6 \times 6 \times 6$

   c $2 \times 2 \times 2 \times 3 \times 3$ d $2 \times 2 \times 5 \times 5 \times 5$

9 Copy and complete using = or ≠.

a $20^2 \square (2 \times 10)^2$

b $30^2 \square 3^2 \times 10^2$

c $2^2 \times 5^3 \square 2^3 \times 5^2$

d $2^2 \times 2^3 \square 2 \times 2 \times 2 \times 2 \times 2 \times 2$

> **Q9 hint** Work out the calculation on both sides. Do you get the same answer?

Key point 16

To multiply powers of the same number, add the **indices**.

> **Communication hint** Indices is the plural of index.

10 Find pairs with the same value.

| $2^4 \times 2^3$ | | 2^5 | | $2^4 \times 2^2$ | | 2^7 | | $2^2 \times 2^3$ | | 2^6 |

Discussion How can you write $2^3 \times 2^5$ as a single power?

11 Write these expressions as a single power.

a $3^3 \times 3^4$ b $5^2 \times 5^3$ c 7×7^4 d $8^4 \times 8^5$

> **Q11c hint** $7 = 7^1$

12 a Work out $\dfrac{3 \times 3 \times 3 \times 3 \times 3 \times 3 \times 3}{3 \times 3 \times 3 \times 3}$ by cancelling.

b Write your answer to part **a** as a power of 3.

c Copy and complete.

$$\frac{3 \times 3 \times 3 \times 3 \times 3 \times 3}{3 \times 3 \times 3} = \frac{3^{\square}}{3^{\square}} = 3^{\square}$$

d Copy and complete.

$$3^5 \div 3^3 = \frac{3^5}{3^3} = \frac{\square \times \square \times \square \times \square \times \square}{\square \times \square \times \square} = 3^{\square}$$

Discussion How can you work out divisions using the indices?

Key point 17

To divide powers of the same number, subtract the indices.

13 Write as a single power.

a $4^6 \div 4^4$ b $5^4 \div 5$ c $7^5 \div 7^4$ d $2^7 \div 2^2$

14 Evaluate

a $\dfrac{2^4 \times 2^2}{2^3}$ b $\dfrac{4^3 \times 4^5}{4^6}$ c $\dfrac{5^4 \times 5}{5^3}$ d $\dfrac{3^2 \times 3^3 \times 3^4}{3^6}$

Example 7

Evaluate $(2^2)^3$

$(2^2)^3 = 2^2 \times 2^2 \times 2^2$ ——— Write out in full.

$\qquad = 2^6$ ——— Add the indices.

Discussion Can you see a quicker way to evaluate expressions like this?

15 Write as a single power.

a $(3^2)^5$ b $(4^3)^4$ c $(5^4)^2$ d $(6^5)^3$

16 (**Exam-style question**

Write as a power of 5

a $5^4 \times 5^2$ **b** $5^9 \div 5^6$ **c** $\dfrac{5^2 \times 5^3}{5}$ **d** $(5^3)^3$

(4 marks)

> **Q16 strategy hint**
> Write the expressions out in full if it helps.

17 Copy and complete the pattern.

$10^3 = 1000$

$10^2 = 100$

$10^1 = 10$

$10^\square = 1$

$10^{-1} = \dfrac{1}{10} = 0.1$

$10^\square = \dfrac{1}{100} = \dfrac{1}{10^2} = 0.01$

$10^\square = \square = \square = \square$

18 Write as a single power.

  a $10^3 \times 10^{-1}$ b $10^2 \times 10^{-3}$

  c $\dfrac{10^2}{10^5}$ d $\dfrac{10^3}{10^{-1}}$

> **Q18a hint** $3 + -1 = \square$

19 Copy and complete.

  a   1 million = $\square$ = $10^\square$

  b   1 billion = $\square$ = $10^\square$

  c   1 trillion = $\square$ = $10^\square$

> **Q19 communication hint** A billion is a thousand million. A trillion is a thousand billion.

Key point 18

Some powers of 10 have a name called a **prefix**. Each prefix is represented by a letter.

---

20 **STEM** Copy and complete the table of prefixes.

Prefix	Letter	Power	Number
tera	T	10^{12}	1 000 000 000 000
giga	G		1 000 000 000
mega	M	10^6	
kilo	k		1000
deci	d	10^{-1}	
centi	c		0.01
milli	m	10^{-3}	
micro	μ		0.000 001
nano	n	10^{-9}	
pico	p		0.000 000 000 001

21 **STEM** Copy and complete.

  a   1 kilogram (kg) = $\square$ g   b   1 megabyte (MB) = $\square$ B

  c   1 microsecond (μs) = $\square$ s   d   1 picometre (pm) = $\square$ m

> **Q21 hint** Refer to your completed table in **Q20**.

22 **STEM** What is 10^{12} bytes in gigabytes?

1.7 **Prime factors**

Objectives

- Write a number as the product of its prime factors.
- Use prime factor decomposition and Venn diagrams to find the HCF and LCM.

Did you know?

You can use the LCM to work out when two or more events will happen at the same time, for example when orbiting planets will line up.

Fluency

- What are the factors of 15, 12, 18?
- Which of these factors are prime?

Warm up

1   Write as a product of powers.
$2 \times 2 \times 2 \times 3 \times 3 \times 3 \times 3 = 2^{\square} \times 3^{\square}$

2   Evaluate $2^2 \times 3 \times 5$.

3   a   Find the HCF of 16 and 36.   b   Find the LCM of 4 and 9.

Example 8

Write 180 as a product of its prime factors.

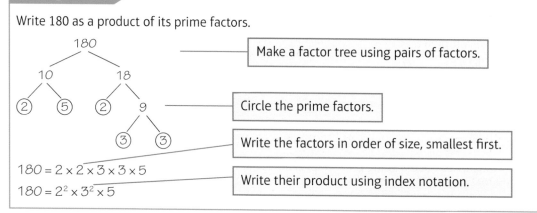

Make a factor tree using pairs of factors.

Circle the prime factors.

Write the factors in order of size, smallest first.

Write their product using index notation.

$180 = 2 \times 2 \times 3 \times 3 \times 5$
$180 = 2^2 \times 3^2 \times 5$

---

4   a   Complete these factor trees for 24.

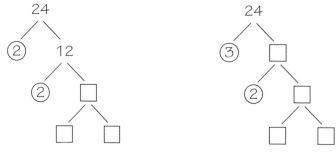

b   Write 24 as a product of its prime factors.

Discussion Does it matter which two factors you choose first?

Key point 19

All numbers can be written as a product of prime factors. This is called **prime factor decomposition**.

5   Write these numbers as products of their prime factors.
    a   18   b   30   c   56   d   72

Active Learn   Homework, practice and support: Foundation 1.7

6   **Reasoning** Flo says, 'Prime numbers cannot be written as a product of two numbers.'
Is this true or false? Show examples to explain.

Discussion Are there any numbers that cannot be written as a product of two numbers?

Example 9

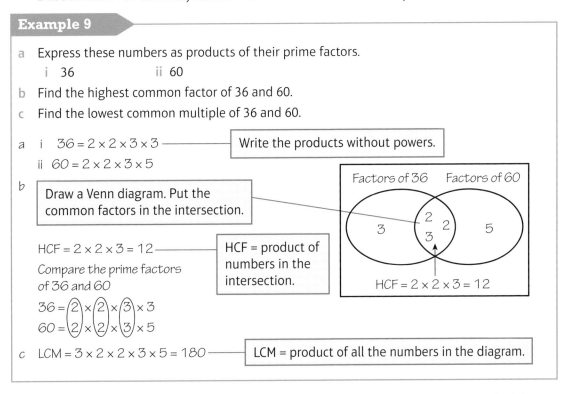

a   Express these numbers as products of their prime factors.
   i 36   ii 60
b   Find the highest common factor of 36 and 60.
c   Find the lowest common multiple of 36 and 60.

a i $36 = 2 \times 2 \times 3 \times 3$ ——— Write the products without powers.
   ii $60 = 2 \times 2 \times 3 \times 5$

b   Draw a Venn diagram. Put the
common factors in the intersection.

Factors of 36   Factors of 60

HCF $= 2 \times 2 \times 3 = 12$ ——— HCF = product of
numbers in the
intersection.

Compare the prime factors
of 36 and 60

$36 = 2 \times 2 \times 3 \times 3$
$60 = 2 \times 2 \times 3 \times 5$

HCF $= 2 \times 2 \times 3 = 12$

c   LCM $= 3 \times 2 \times 2 \times 3 \times 5 = 180$ ——— LCM = product of all the numbers in the diagram.

7   Express these numbers as products of their prime factors. Draw Venn diagrams to find the
HCF and LCM.
   a   60 and 96   b   24 and 108   c   120 and 150

Discussion Compare finding the HCF and LCM by prime factor decomposition and by
listing factors and multiples. Which method do you prefer? Do you think one method is easier
for large numbers?

8   What is 300 as a product of its prime factors?
   A 3×100 **B** $2^2 \times 3 \times 25$ **C** $4 \times 3 \times 25$ **D** $2^2 \times 3 \times 5^2$ **E** $4 \times 3 \times 5^2$

9   $2^4 \times 3 \times 5$ is the prime factor decomposition of
   A   48   **B**   80   **C**   120   **D**   240   **E**   480

10   **Reasoning** Use the information that
$13 \times 17 = 221$ to find the LCM of
   a   39 and 17
   b   13 and 34

> **Q10 hint** What type of numbers are 13
> and 17? What is the LCM of 13 and 17?
> $39 = 13 \times \boxed{}$ $34 = 17 \times \boxed{}$

11   **Reasoning** The number 360 can be written as a product of its prime factors.
$360 = 2^3 \times 3^2 \times 5$
Are these numbers factors of 360? Explain your answers.
   a   24   b   50   c   90

> **Q11a hint** Write 24 as a product
> of its prime factors. Does this
> divide into $2^3 \times 3^2 \times 5$?

12   **Reasoning / Problem-solving** Lea makes 225 milk chocolates, 165 dark chocolates and
180 white chocolates. The chocolates cannot be mixed.

She wants to buy the largest possible box for the chocolates that can be filled leaving no
spaces. How many chocolates should each box hold?

13 ┌──────────────────────────────┐
   │ **Exam-style question** │
   └──────────────────────────────┘

The number 324 can be written as a product of its prime factors:

$$324 = 2^2 \times 3^4$$

Write $\sqrt{324}$ as a product of its prime factors. **(2 marks)**

Exam hint

$$\sqrt{324} \times \sqrt{324} = 324$$
$$(2^{\square} \times 3^{\square}) \times (2^{\square} \times 3^{\square}) = 324$$

1 Problem-solving

Objective
• Use pictures to help you solve problems.

┌─ **Example 10** ──┐

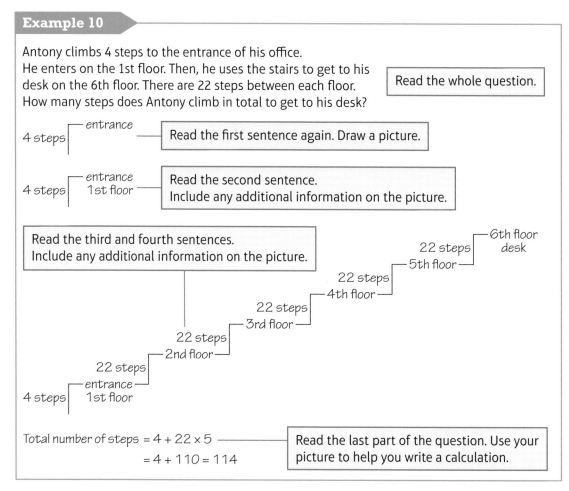

Antony climbs 4 steps to the entrance of his office.
He enters on the 1st floor. Then, he uses the stairs to get to his desk on the 6th floor. There are 22 steps between each floor. How many steps does Antony climb in total to get to his desk?

Read the whole question.

4 steps — entrance

Read the first sentence again. Draw a picture.

4 steps — entrance, 1st floor

Read the second sentence.
Include any additional information on the picture.

Read the third and fourth sentences.
Include any additional information on the picture.

4 steps — entrance, 1st floor — 22 steps — 2nd floor — 22 steps — 3rd floor — 22 steps — 4th floor — 22 steps — 5th floor — 22 steps — 6th floor desk

Total number of steps = 4 + 22 × 5
= 4 + 110 = 114

Read the last part of the question. Use your picture to help you write a calculation.

└──┘

1   A circus clown climbs a 12 m slippery pole. When the music plays he climbs up 2.2 m. When the music stops, he slides down 1 m. How many times must the music play for the clown to reach the top of the slippery pole?

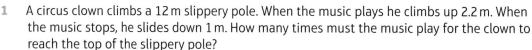

Q1 hint Read the sentences, one at a time, and include any information on a picture. (There is not just one 'correct' picture.)

2   Jacqui's rectangular chicken run is 7.2 m wide and 8.4 m long. She plans to enclose it with a fence, using a supporting post every 1.2 m. How many posts does Jacqui need?

Q2 hint Draw a picture.

3 When two tent poles are placed end to end, their total length is 6.3 metres. When the two poles are placed side by side, one is 3.7 metres longer than the other. What lengths are the poles?

> **Q3 hint** Draw one picture to represent each sentence.

4 **Finance** In a café, tea costs £1.40 and hot chocolate costs £1.80.
   A group of friends bought some drinks and paid £10.
   How many of each kind of drink did they buy?

> **Q4 hint** Draw a picture.
>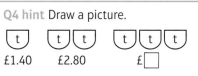
> £1.40   £2.80   £☐
> Look for a total number of teas and hot chocolates that together cost £10.

5 **Finance** At a sale, sheep cost £95 and lambs cost £125. A farmer buys some sheep and lambs. He pays a total of £1190.
   How many sheep and lambs does the farmer buy?

6 Abbie has 3 hats and 5 scarves. How many different combinations of hat and scarf can she wear?

> **Q6 hint**
> h – hat; s – scarf
>

7 An ice-cream seller offers vanilla, toffee, banana or mint ice-cream. He also offers a choice of nuts, chocolate chips or sprinkles for toppings. How many different combinations of a single flavour ice-cream with one topping can be ordered?

8 A tour company starts tours at 9 am each day. The castle tours leave every 15 minutes and the museum tours leave every 20 minutes. When do the castle and museum tours next leave at the same time?

9 **Reflect** How can you solve problems **6**, **7** and **8** without drawing a picture?

   Discussion Does it matter how you solve a maths problem?

1 Check up

Log how you did on your Student Progression Chart.

Calculations

1 Work out
   a $(8 - 3)^2 + 4$ b $8 \div 2 \times 3 + 1$ c $20 \div (2 - 6)$
   d $-6 \times (2 - 5)^2$ e $\dfrac{5 \times 9}{3}$ f $(15 \times 4) \div (16 \times 5)$

2 Write down a calculation you could do to check each of these.
   a $840 \div 24 = 35$ b $13 \times 3 - 7 = 32$

3 Work out
   a 4.2×3.8 b $3.5 \div 0.7$

4 Calculate an estimate for
   a $\dfrac{491}{52.1}$ b $\dfrac{764 \times 96}{38}$

5 Use the information that $436 \times 178 = 77\,608$ to work out the value of
   a 43.6×1.78 b $776.08 \div 4.36$

6 Work out
   a 0.7×0.05 b $8 \div 0.25$

7   Use a calculator to work out $\sqrt{\dfrac{6700 - 2.38^2}{3.6^2 + 5.71}}$

   a   Write down all the figures on your calculator display.

   b   Give your answer to part **a** correct to 3 significant figures.

8   Use a calculator to work out $2.6^3 - \sqrt[3]{5.4}$

   Give your answer to 3 decimal places.

Powers and roots

9   Copy and complete.

   a $5^{\square} = 125$ b $3^{\square} = 81$ c $\sqrt[3]{64} = \square$

   d $6 \times 6 \times 6 \times 6 \times 6 = 6^{\square}$ e $10^{\square} = 1000$

10   Evaluate

   a $2^3 \times 3^2$ b $\sqrt{81} - 2^2$ c $\sqrt{4} \times \sqrt{4}$

11   Write as a single power of 7

   a $7^3 \times 7^4$ b $7^5 \div 7^3$ c $(7^4)^2$ d 7×7^3

Factors, multiples and primes

12   Write the prime numbers between 30 and 40.

13   a   Find the HCF of 12 and 30.   b   Find the LCM of 8 and 10.

14   a   Write 72 and 96 as products of their prime factors.

   b   Find the highest common factor of 72 and 96.

   c   Find the lowest common multiple of 72 and 96.

15   How sure are you of your answers? Were you mostly

   Just guessing   Feeling doubtful   Confident

   What next? Use your results to decide whether to strengthen or extend your learning.

✱ Challenge

16   You can use Euclid's algorithm to find the highest common factor (HCF) of two positive integers, M and N.

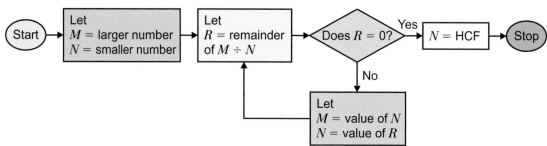

   The table shows what happens if you input the numbers 36 and 8.

M	N	R	
36	8	4	$36 \div 8 = 4$ remainder 4 ($R \neq 0$)
8	4	0	$8 \div 4 = 2$ remainder 0
8	4	0	HCF = $N = 4$

   Use Euclid's algorithm to find the HCF of

   a   45 and 18   b   1620 and 228.

1 **Strengthen**

Calculations

1 Round these numbers to 1 decimal place.
a 6.32   b 15.07
c 0.438   d 11.972

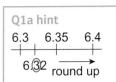

Q1a hint

2 Round these numbers to 2 decimal places.
a 11.257   b 9.072
c 0.6352   d 28.983

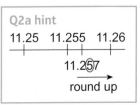

Q2a hint

3 Round these numbers to 3 decimal places.
a 8.0462   b 14.1732
c 0.0568   d 21.8139

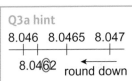

Q3a hint

4 Copy these numbers. Circle the first significant figure. Write its value.
a 47.823   b 0.00572
c 432 650   d 0.6718

Q4b hint The first significant figure is the first non-zero digit starting from the left.
0.00⑤72

5 Round these numbers to 1 significant figure (1 s.f.).
a 51.3   b 487.2
c 6234   d 8753

Q5a hint Circle the first significant figure. It's in the tens column, so you are rounding to the nearest 10.

6 Round these numbers to the number of significant figures shown.
a 14.08 (3 s.f.)   b 7.192 (2 s.f.)
c 0.04318 (3 s.f.)   d 0.006 052 (2 s.f.)

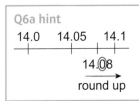

Q6a hint

7 Use priority of operations to work out
a $10 \div 5 \times 6 - 1$ b $(6 - 2) \times (5 + 2)$
c $3^2 + 2 \times 4$ d $(10 - 4)^2 - 5$
e $\sqrt{100} - 2^3$ f $4^3 + \sqrt{16}$

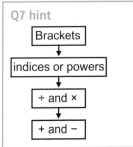

Q7 hint

8 Split into two fractions. Then simplify.
a $\dfrac{7 \times 12}{4}$ b $\dfrac{15 \times 3}{5}$

c $\dfrac{7 \times 6}{12}$ d $\dfrac{5 \times 6}{25 \times 2}$

e $\dfrac{8 \times 9}{3 \times 24}$ f $(16 \times 7) \div (14 \times 4)$

Q8a hint
$\dfrac{7 \times 12}{4} = \dfrac{7}{4} \times 12$ or $= 7 \times \square = \square$
does not simplify

9   Work out

    a -6×2 b 4×-8

    c -3×-4 d $-12 \div 4$

    e $(3 - 7) \div -2$ f $-9 \times (2 - 6)$

    g $(6 - 2)^2 \div -8$ h $10 \times (3 - 5)^2$

> **Q9 hint**
> For multiplying and dividing:
> • same signs give a positive answer
> • different signs give a negative answer.

10 Draw a function machine for each calculation. Work backwards through it. Write down the calculation to check it.

    a $792 - 456 = 336$

    b $321 + 245 = 566$

    c $851 \div 23 = 37$

    d $48 \times 62 = 2976$

    e $22 \div 2 + 5 = 16$

    f $25 \div 5 - 3 = 2$

    g $13 \times 4 - 2 = 50$

    h $32 \div 4 \times 3 = 24$

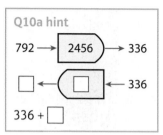

> **Q10a hint**

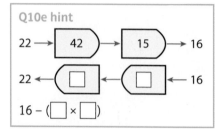

> **Q10e hint**

11 Work out

    a 6.4×5.8

    b 2.3×5.7

    c 7.1×0.32

    d 2.34×3.16

> **Q11a hint** Work out 64×58.
> $6.4 \times 5.8 = \square\square.\square\square$
> 1 d.p. + 1 d.p. = 2 d.p.
> Check your answer by estimating, $6 \times 6 = 36$.

12 Work out

    a $28 \div 1.4$ b $4.44 \div 2.4$

    c $18.9 \div 1.2$ d $7.2 \div 0.06$

    e $8.1 \div 0.09$ f $71.5 \div 0.11$

> **Q12a hint** It is easier to divide by a whole number.
> $\times 10 \big(\dfrac{28 \div 1.4}{280 \div 14} \big) \times 10$

13 Rewrite each calculation with the numbers rounded to 1 significant figure to work out an estimated answer.

    a $546 \times 372 \approx 500 \times \square$

    b $\dfrac{618}{34.6}$

    c $\dfrac{291 \times 42}{59.3}$

    d $\dfrac{45.3 \times 217.8}{0.4}$

14 **Reasoning** $\boxed{4.2 \times 5.4 = 22.68}$

    Use this fact to write down the value of

    a 42×5.4 b 42×54

    c 0.42×5.4 d 0.42×0.54

> **Q14a hint**
> $\times 10 \big(\dfrac{4.2 \times 5.4 = 22.68}{42 \times 5.4 = \square} \big) \times 10$

15 **Reasoning** $4.5 \times 3.7 = 16.65$

    a   Copy and complete.

        i $\dfrac{16.65}{3.7} = \square$ ii $\dfrac{16.65}{4.5} = \square$

    b   Work out

        i $\dfrac{166.5}{3.7}$ ii $\dfrac{16.65}{45}$ iii $\dfrac{1.665}{4.5}$

> **Q15bi hint** Use an estimate.
> $\dfrac{166.5}{3.7} \approx \dfrac{200}{4} = 50$

16 Use your calculator to work out the value of $\sqrt{(4.5^2 - 0.5^3)}$

    a   Write down all the figures on your calculator display.

    b   Write your answer correct to 2 decimal places (2 d.p.).

Powers and roots

1 Copy and complete.
   a $2^3 = 2 \times \square \times \square = \square$ b $\square^\square = 4 \times 4 \times 4 = \square$
   c $2 \times 2 \times 2 \times 2 = 2^\square$ d $2^2 \times 3^3 = 2 \times 2 \times 3 \times \ldots$

> **Q1 hint** $5^{4\leftarrow}$ This number tells you how many fives are multiplied together.
> $5^4 = 5 \times 5 \times 5 \times 5$

2 a Copy and complete.
      i $\sqrt[3]{64} = 4$ because $4^3 = 4 \times 4 \times 4 = \square$.
      ii $\sqrt[3]{1000} = \square$ because $10^3 = \square \times \square \times \square = 1000$.
      iii $\sqrt[3]{\square} = 2$ because $2^3 = \square$.

> **Q2a iii hint** Work out 2^3 first.

      iv The cube root of 27 is $\square$.
   b Write the value of each root.
      i $\sqrt[3]{125}$ ii $\sqrt{81}$ iii $\sqrt[3]{1}$ iv $\sqrt{144}$

3 a Work out
      i 4^2 ii $(-4)^2$

> **Q3a ii hint** $(-4)^2 = -4 \times -4$

   b Copy and complete.
      The two square roots of 16 are $\square$ and $\square$.
   c Write the two square roots of
      i 9   ii 25   iii 144

4 Write each product as a single power.
   a $2^3 \times 2^4 = 2^{\square + \square} = 2^\square$
   b $4^2 \times 4^3 = 4^{2\square 3} =$
   c $3^4 \times 3^5$
   d $5^4 \times 5$
   e $6^2 \times 6^3 \times 6^4 = 6^{\square + \square + \square} = 6^\square$
   f $10^5 \times 10^4 \times 10$

> **Q4a hint**
>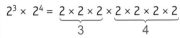
> How many 2s are multiplied together?

> **Q4d hint** Write 5 as 5^1.

5 Write each division as a single power.
   a $7^6 \div 7^3 = 7^{\square - \square} = 7^\square$
   b $4^8 \div 4^2 = 4^{8\square 2} =$
   c $3^5 \div 3^2$
   d $6^4 \div 6$

> **Q5a hint**
> $7^6 \div 7^3 = \dfrac{7^6}{7^3} = \dfrac{\overbrace{\cancel{7} \times \cancel{7} \times \cancel{7} \times 7 \times 7 \times 7}^{6}}{\underbrace{\cancel{7} \times \cancel{7} \times \cancel{7}}_{3}}$
> How many 7s are left after cancelling 3 of them?

6 Work out
   a $(6^2)^3 = 6^2 \times \square^\square \times \square^\square = 6^\square$
   b $(2^5)^2 = \square^\square \times \square^\square =$
   c $(3^4)^3$

7 Work out
   a $\sqrt{2} \times \sqrt{2}$ b $\sqrt{11} \times \sqrt{11}$ c $\sqrt{7} \times \sqrt{7}$

8 Write down the value of
   a $\sqrt{4} \times \sqrt{4}$ b $\sqrt{6} \times \sqrt{6}$ c $\sqrt{57} \times \sqrt{57}$

> **Q8b hint** Use your answers to **Q7** to help.

Factors, multiples and primes

1 Copy and complete.
   a Factors of 20: 1, 2, ...
   b Factors of 30: 1, 2, ...
   c *Common* factors of 20 and 30: 1, ...
   d *Highest* common factor of 20 and 30: ...
   e Highest common factor of 28 and 42: ...

> **Q1c hint** Circle the numbers that appear in both lists. These are the common factors.

2   Copy and complete.
    a   First 10 multiples of 6: 6, 12, …   b   First 10 multiples of 7: 7, 14, …
    c   *Lowest common multiple* of 6 and 7: …   d   Lowest common multiple of 5 and 8: ….

3   a   Copy and complete the factor tree for 360.

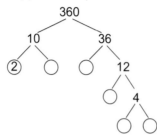

> **Q3a hint** The numbers in the circles are the prime factors.

    b   Write 360 as the product of its prime factors.
    c   Write each number as the product of its prime factors.
        i   144   ii   396   iii   450   iv   72   v   84

> **Q3b hint**
> $2^{\square} \times 3^{\square} \times \square$

4   Write 72 and 84 as products of their prime factors.
    a   Copy and complete the Venn diagram of their prime factors.

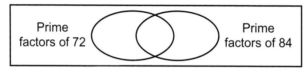

Prime factors of 72   Prime factors of 84

> **Q4b hint**
>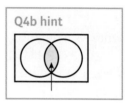

    b   Multiply the *common* prime factors to find the HCF.
    c   Multiply all the prime factors to find the LCM.

1   Extend

1   Put brackets in the expression so that its value is 45.024.
    $1.6 + 3.8 \times 2.4 \times 4.2$

2   Use your calculator to work out the value of $3.67 \times (\sqrt{2})^3$.
    a   Write down all the figures on your calculator display.
    b   Write your answer to part **a** correct to 1 decimal place.

3   **Exam-style question**

Comp Parts and Z Parts both sell memory sticks.

Comp parts	Z Parts
Memory sticks £4 each	Memory sticks
	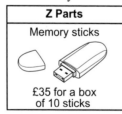
1 free stick for every 10 sticks bought	£35 for a box of 10 sticks

There are 150 students in Year 10 in a school.
A teacher needs to buy a memory stick for each student.
At which of the shops should he buy the memory sticks?
You must show all your working.   **(5 marks)**

March 2011, Q13, 5MB2F/01

Exam hint
Show your final answer in a clear written statement which includes the reason why it should be bought at the shop.

4   Round these numbers to the number of significant figures (s.f.) stated.
   a   486 000 000 (2 s.f.)   b   3 872 000 (3 s.f.)
   c   16 370 000 (2 s.f.)   d   352 840 000 (3 s.f.)

5   **Problem-solving / Finance** Mrs Baker took a group of children to the theatre.
   Adult ticket price   £13.20
   Child ticket price   £8.30
   The total cost of *one* adult ticket and *all* the child tickets was £146.
   Work out the number of children Mrs Baker took to the theatre.

6   Work out
   a   12 ÷ 0.3   b   15 ÷ 0.5   c   20 ÷ 0.4   d   18 ÷ 0.6
   Discussion What do you notice?

7   **Real / problem-solving** Trains from Manchester to Stockport leave every 15 minutes.
   Trains from Manchester to London leave every 20 minutes.
   A train to Stockport and a train to London leave together at 12 noon. What is the next time a train will leave for Stockport and London at the same time?

8   Copy and complete the pattern.

$2^3 = 2 \times 2 \times 2 = \Box$

$2^2 = 2 \times \Box = \Box$

$2^1 = \Box$

$2^0 = 1$

$2^{-1} = \dfrac{1}{2^1} = \Box$

$2^{-2} = \dfrac{1}{2^2} = \Box$

$2^{-3} = \Box = \Box$

9   Rhona wants to make up some party bags with identical contents. She has 24 balloons and 30 party poppers.
   What is the greatest number of party bags she can make up with no items left over?

10 **Problem-solving / Finance** An estate agent sells a house for £2.4 million. On the same day she sells another house for £750 000.
   How many times smaller is £750 000 than £2.4 million?

11 **Reasoning** Rebecca divides 14.314 by 0.17 on her calculator.
   Her answer is 8.42.
   Without finding the exact value of 14.314 ÷ 0.17, explain why her answer must be wrong.

> **Q11 hint**
> Estimate the answer.

12 ┌─ **Exam-style question** ────────────────

   Use your calculator to work out $\dfrac{4.6 + 3.85}{3.2^2 - 6.51}$.

   a   Write down all the numbers on your calculator display.   **(2 marks)**
   b   Give your answer to part **a** correct to 1 significant figure.   **(1 mark)**

   June 2009, Q27, 1380/2F

> **Exam hint**
> Put brackets around the numerator and denominator before dividing.

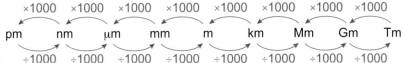

Key point 20

←── To convert bigger units to smaller units, multiply

×1000 ×1000 ×1000 ×1000 ×1000 ×1000 ×1000 ×1000

pm   nm   μm   mm   m   km   Mm   Gm   Tm

÷1000 ÷1000 ÷1000 ÷1000 ÷1000 ÷1000 ÷1000 ÷1000

To convert smaller units to bigger units, divide ──→

13 Convert the units.

    a   0.65 Gm to m   b   0.024 μm to pm   c   0.000 006 Mm to mm

14 **Reasoning** Decide whether each statement is true or false. If it is false, give an example.

    a   The product of two prime numbers is always a prime number.

    b   The sum of two prime numbers is always a prime number.

    c   The difference between two consecutive numbers is never 2.

15 **Reasoning** Given that $\dfrac{14.4}{(0.8)^2} = 22.5$, work out the value of $\dfrac{1.44}{(0.08)^2}$

> **Q15 hint** $0.08 = 0.8 \div \square$
> $(0.08)^2 = 0.08 \times \square$

16 Write down the whole number that is closest in value to $\sqrt{40}$.

> **Q16 hint** First find the two whole numbers $\sqrt{40}$ lies between.

17 $2160 = 2^2 \times 3^3 \times 5$

Copy and complete.

    a $2160 = 2^2 \times \square$

> **Q17a hint** Work out $3^3 \times 5$.

    b $2160 = 3^3 \times \square$

    c $2160 = 3^{\square} \times 240$

18 The number 96 can be written in the form $2^n \times 6$. Find the value of n.

19 **Exam-style question**

Work out the value of

    a $(2^2)^3$

    b $(\sqrt{3})^2$

    c $\sqrt{2^4 \times 9}$ **(4 marks)**

> **Exam hint**
> First work out what is under the square root, writing this answer clearly.

20 $A = 2^3 \times 3^4 \times 5$ $B = 2^4 \times 3^2 \times 5^2$

Write down as a product of its prime factors

    a   the highest common factor (HCF) of A and B

    b   the lowest common multiple (LCM) of A and B.

> **Q20 strategy hint**
> Draw a Venn diagram

21 **Reasoning**

    a   Given that $2304 = 36 \times 64$, work out $\sqrt{2304}$.

    b   Given that $216 = 8 \times 27$, work out $\sqrt[3]{216}$.

> **Q21b hint**
> $\sqrt{2304} = \sqrt{36 \times 64} = \sqrt{36} \times \sqrt{64}$

1 **Knowledge check**

⊙ The priority of operations is Brackets, Indices (powers and roots), Division and Multiplication, Addition and Subtraction. .... *Mastery lesson 1.1*

⊙ ≠ means 'not equal to'. .... *Mastery lesson 1.1*

⊙ ≈ means 'approximately equal to'.. *Mastery lesson 1.1*

⊙ A **function** is a rule that acts on a number (the input) to give an output number. .... *Mastery lesson 1.1*

⊙ The **inverse** function reverses the effect of the original function. ..... *Mastery lesson 1.1*

- The inverse of add (+) is subtract (−). .... *Mastery lesson 1.1*

- The inverse of multiply (×) is divide (÷). .... *Mastery lesson 1.1*

- Finding the **square root** is the inverse of squaring. .... *Mastery lesson 1.1*

- Finding the **cube root** is the inverse of cubing. .... *Mastery lesson 1.1*

- To round a number to 1 decimal place (1 d.p.), look at the digit in the second decimal place. If it is 5 or more round up.
  For example, to 1 d.p. 35.23 is 35.2 and 35.27 is 35.3. .... *Mastery lesson 1.2*

- To round a number to 2 decimal places, look at the digit in the third decimal place. To round a number to 3 decimal places, look at the digit in the fourth decimal place. .... *Mastery lesson 1.2*

- To divide by a decimal, first multiply both numbers by a power of 10 (10, 100, …) until you have a whole number to divide by. Then work out the division. .... *Mastery lesson 1.2*

- For any number, the first significant figure is the one with the highest place value. It is the first non-zero digit in the number, counting from the left. .... *Mastery lesson 1.3*

- To estimate a calculation, round each number to 1 significant figure (1 s.f.) .... *Mastery lesson 1.3*

- The **highest common factor (HCF)** is the highest factor that is common to two or more numbers. .... *Mastery lesson 1.4*

- The **lowest common multiple (LCM)** is the lowest multiple that is common to two or more numbers. .... *Mastery lesson 1.4*

- To multiply powers of the same number, add the indices.
  To divide powers of the same number, subtract the indices. .... *Mastery lesson 1.6*

- When using a Venn diagram
  - to find the HCF, multiply the *common* prime factors
  - to find the LCM, multiply *all* the prime factors. .... *Mastery lesson 1.7*

Reflect For each statement A, B and C, choose a score:

1 – strongly disagree;   2 – disagree;   3 – agree;   4 – strongly agree

A I always try hard in mathematics.

B Doing mathematics never makes me worried.

C I am good at mathematics.

For any statement you scored less than 3, write down two things you could do to help you agree more strongly in the future.

Reflect

1 Unit test

Log how you did on your Student Progression Chart.

1 Gavin buys 36 packets of raisins. Each packet of raisins costs £0.35.
Work out the total cost of all the packets that Gavin buys. *(2 marks)*

2 Exam-style question

The table shows temperatures at midnight and midday on one day in five cities.

City	Midnight temperature	Midday temperature
Belfast	−3°C	4°C
Cambridge	−1°C	4°C
Edinburgh	−7°C	−1°C
Leeds	−6°C	3°C
London	−2°C	6°C

a Which city had the lowest midnight temperature?

b How many degrees higher was the midnight temperature in Cambridge than the midnight temperature in Leeds?

c Which city had the greatest rise in temperature from midnight to midday? **(3 marks)**

May 2009, Q6, 5384F/11F

3 **Problem-solving** Three friends share the cost of a pizza equally between them.
The pizza costs £12.35. How much should they each pay? *(3 marks)*

4 Work out
a $9 - (3 - 7)$
b $(8 - 2)^2 \div \sqrt{9}$ *(3 marks)*

5 **Problem-solving** Abdul has a '5p off per litre' voucher to use at his local petrol station.
The petrol normally costs 130.9p per litre.
Abdul fills his tank with 42 litres of petrol. How much does he pay? *(3 marks)*

6 Work out $\dfrac{15 \times 28}{7 \times 20}$ *(2 marks)*

7 Estimate the value of $\dfrac{70.1 \times 5.92}{0.19}$ *(3 marks)*

8 **Reasoning** In the 2011 census the population of England was 53.0 million. The population of York was 198 051.
Work out an estimate for the number of times greater the population of England was than the population of York. *(3 marks)*

9 Work out
a 25^2
b 0.5^3
c $\sqrt{0.16}$
d $\sqrt[3]{64}$ *(4 marks)*

10 Given that $32 \times 14 = 448$, write down the value of
a 3.2×1.4
b $44.8 \div 1.4$
c $4.48 \div 320$ *(3 marks)*

Active Learn   Homework, practice and support: Foundation 1 Unit test

11 **Reasoning** A lighthouse flashes its light every 4 seconds. A second lighthouse flashes its light every 6 seconds.

They flash together at 6 am. How many times in the next minute will they flash together?

(3 marks)

12 Write each as a power of 9.

a $9^3 \times 9^5$ b $9^6 \div 9^2$ c $\dfrac{9 \times 9^3}{9^2}$ d $(9^3)^5$ *(4 marks)*

13 a Express 75 and 90 as products of their prime factors.   *(2 marks)*

   b For the numbers 75 and 90

     i find the highest common factor   *(2 marks)*

     ii find the lowest common multiple.   *(2 marks)*

14 Work out the value of $\sqrt[3]{\dfrac{(3.3^2 + 4.2)}{5.1 - 2.02}}$

   a Write down all the figures on your calculator display.   *(2 marks)*

   b Give your answer to 2 significant figures.   *(1 mark)*

Sample student answers

Which student gives the better answer and why?

Exam-style question

Miss Phillips needs to decide when to have the school sports day.

The table shows the number of students who will be at the sports day on each of 4 days.

It also shows the number of teachers who can help on each of the 4 days.

	Tuesday	Wednesday	Thursday	Friday
Number of students	179	162	170	143
Number of teachers	15	13	14	12

For every 12 students at the sports day there must be at least 1 teacher to help.

On which of these days will there be enough teachers to help at the sports day?

You must show all your working.   **(3 marks)**

June 2014, Q24, 1MA0/2F

Student A

1 teacher for every 12 students.

Tuesday $179 \div 12 = 14.9\ldots$ 15 teachers needed ✓

Wednesday $162 \div 12 = 13.5$ 14 teachers needed ✗

Thursday $170 \div 12 = 14.1\ldots$ 15 teachers needed ✗

Friday $143 \div 12 = 11.9\ldots$ 12 teachers needed ✓

Answer: Enough teachers on Tuesday and Friday.

Student B

Max number of students

Tuesday $15 \times 12 = 180 > 179$ ✓

Wednesday $13 \times 12 = 156 < 162$ ✗

Thursday $14 \times 12 = 168 < 170$ ✗

Friday $12 \times 12 = 144 > 143$ ✓

2 ALGEBRA

Average speed = $\dfrac{\text{distance}}{\text{time}}$.

This racing car took just 12 seconds to cover 900 metres.
What was its average speed?

2 Prior knowledge check

Numerical fluency

1 Work out
   a −6 + 1
   b −4 − 5
   c 3 × −4
   d 10 − 21 + 9
   e −2 × −10
   f 8 ÷ −4
   g −15 ÷ −3
   h −2 × −5 × −6
   i 3(5 + 2)
   j $5^2 − 2 \times 3 \times 4$

2 Work out
   a 10 squared
   b $2^2 + 3^3$
   c $(−1)^2$
   d 3×2^2
   e $\sqrt{81}$
   f $\sqrt{36}$

3 Write each expression as a single power.
   a $3^2 \times 3^3$
   b $7^3 \times 7$
   c $4^6 \div 4^3$
   d $9^5 \div 9^2$

4 Work out the HCF of
   a 8 and 12
   b 3 and 6
   c 18 and 24

Algebraic fluency

5 Simplify
   a $3t + 5t + 7t$
   b $5m + 10m − 3m$
   c $10x − 12x$
   d $a − 12a + 6a$

6 To work out his pay, Harry uses this word formula.
   pay = hourly rate × number of hours
   His hourly rate is £5. Work out his pay when he works 4 hours.

7 This word formula can be used to work out the perimeter of a rectangle.
   perimeter = 2 × (base + height)
   Work out the perimeter of a rectangle with base 7 cm and height 5 cm.

* Challenge

8 Write four algebraic expressions that simplify to give $25x$.

2.1 Algebraic expressions

Objectives

- Use correct algebraic notation.
- Write and simplify expressions.

Did you know?

- Algebra is a universal language that has been used for centuries in countries all over the world.

Fluency

Simplify
- $a + a$
- $h + h + h$
- $8x − 7x$
- $4s − 7s$

ActiveLearn Homework, practice and support: Foundation 2.1

1   Simplify
    a $5a + 3a - 2a$ b $9r + 4r - 6r$ c $8x - 5x - 7x$ d $10s - 7s - 4s$

2   What is the perimeter of the rectangle?

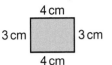

> **Q2 hint**
> perimeter = distance all round the outside

Key point 1

A **term** is a number, a letter, or a number and a letter multiplied together.

Like terms contain the same letter to the same power (or do not contain a letter). You can simplify an expression by collecting like terms.

$3x$ $7x$ These are 'like terms' as the **letters** are the same.

$3x$ $7y$ $2x^2$ These are not 'like terms' as the letters are different or the powers are different.

expression

$3x + 1$

terms

Example 1

Simplify these expressions by collecting like terms.

a $2a + 3 + a + 4$

b $2x^2 - 2x + 7x^2 + 4x$

a $2a + 3 + a + 4 = 3a + 7$ ———— Add the letter terms: $2a + a$. Add the numbers: $3 + 4$

b $2x^2 - 2x + 7x^2 + 4x = 9x^2 + 2x$ ———— x^2 and x are not like terms.

> Questions in this unit are targeted at the steps indicated.

3   Simplify by collecting like terms
    a $2y + 4 + 7y + 8$ b $3a + 6b - a - 9b$ **Q3b hint** $1a = a$
    c $8m^2 + 5 - 7m^2 - 3$ d $r^2 + 6r + 8r^2 - 5r$

4   **Reasoning** Sam and Ben simplify $4a + 7b - a$. Who is correct?

    Sam's answer: $5a + 7b$ Ben's answer: $3a + 7b$

    Discussion What mistake did the other student make?

5   a   Add these terms together, moving clockwise round the set.

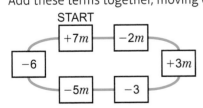

    b   Now add the terms together moving *anti*clockwise. Do you get the same result?
    Reflect Does it matter what order you add them in?

Key point 2

Terms can be simplified when multiplying or dividing, even when they are not like terms.

$a \times b = ab$ $x \div y = \dfrac{x}{y}$

When multiplying:
- write letters in alphabetical order
- write numbers before letters

Example 2

Simplify

a $6 \times y$ b $5 \times 2p$ c $c \times b$ d $t \div 5$

a $6y$ — 6 lots of y

b $10p$ — Multiply the numbers first: $(5 \times 2) \times (p)$

c bc — Write letters in alphabetical order.

d $\dfrac{t}{5}$

---

6 Simplify

    a $3 \times m$ b $c \times 40$ c $4t \times 2$

    d $d \times c$ e $h \div 3$ f $a \div b$

7 **Exam-style question**

    a Simplify $f + f + f + f - f$ **(1 mark)**

    b Simplify $2m \times 3$ **(1 mark)**

    c Simplify $3a + 3h + a + 3h$ **(2 marks)**

    Nov 2012, Q13, 1MA0/1F

Exam hint
Write your answer using lowercase letters (e.g. f not F); as in the question.

Key point 3

You write an algebraic expression by using letters to stand for numbers. The letter is called a **variable** because its value can change or **vary**.

8 Write an expression for these.

    a 6 more than x

    b 7 less than x

    c 12 multiplied by y

    d 3 lots of m

    e y divided by 2

    f d halved

Q8a hint

x 6

$x + \square$

9 **Reasoning** Avinash is y years old.

    a His brother is 2 years younger. Write an expression in y for Avinash's brother's age.

    b His grandmother is 5 times as old as Avinash. Write an expression in y for Avinash's grandmother's age.

    c His cousin is 3 years older. Write an expression in y for Avinash's cousin's age.

    d Write and simplify an expression for the combined ages of all four.

Q9 communication hint
An 'expression in y' is an expression that contains the letter y.

10 Each bag holds n sweets. Write an expression for the number of sweets in

    a 2 bags   b 4 bags

    c 10 bags   d x bags

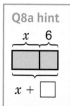

n sweets

Q10a hint $2 \times n =$

Q10d hint $\square \times n =$

11 Write an expression for the perimeter P of the rectangle.

h cm b cm h cm

b cm

2.2 Simplifying expressions

Objectives

- Use the index laws.
- Multiply and divide expressions.

Did you know?

- Simplifying makes algebra less complicated and easier to use.

Fluency

Work out
- 3×4
- $15 \div 5$
- $a \times b$

Warm up

1 Simplify
   a $3 \times y$ b $l \times m$ c $h \div 4$ d $2n \times 4$

2 Write using index notation
   a $2 \times 2 \times 2$ b $3 \times 3 \times 4 \times 4 \times 4$
   c $\dfrac{7 \times 7 \times 7}{7}$ d $\dfrac{5 \times 5 \times 5 \times 5 \times 5}{5 \times 5}$

3 Work out
   a -4×-5 b $-2 \times 4 \times 3$ c $12 \div -6$ d $-8 \div 2$

4 Copy and complete.
   a $3 \times 3 = 3^{\square}$ b $x \times x = x^{\square}$ c $4 \times 4 \times 4 = 4^{\square}$ d $y \times y \times y = y^{\square}$

5 Copy and complete.
   a $2^2 \times 2^4 = 2^{\square}$ b $x^2 \times x^4 = x^{\square}$ c $y^6 \times y^4 \times y = y^{\square}$

6 Write two terms that multiply together to give these answers.
   a $\square \times \square = y^2$ b x^3 c x^7

   Discussion Is there more than one answer to part **c**?

Key point 4

To multiply powers of the same letter, add the indices.

7 Copy and complete.
   a $5^7 \div 5^4 = 5^{\square}$ b $y^7 \div y^4 = y^{\square}$ c $9^8 \div 9^3 = 9^{\square}$ d $x^8 \div x^3 = x^{\square}$

8 Simplify
   a $\dfrac{a^6}{a^2} = a^6 \div a^2 = \square$ b $\dfrac{z^7}{z^3}$ c $\dfrac{x^4}{x}$ d $\dfrac{g^{10}}{g^5}$

Key point 5

To divide powers of the same letter, subtract the indices.

Example 3

Simplify $2a \times 3b$

$2a \times 3b = 2 \times 3 \times a \times b$ — Multiply the numbers first: 2×3. Then multiply the letters: $a \times b$

$ = 6ab$ — Put the number first, then the letters in alphabetical order.

9 Simplify
   a $3a \times 4b$
   b $2r \times 7s$
   c $6a \times 3b \times 2c$
   d $9x \times 2y \times z$
   e $-6m \times 7n$
   f $a \times 4a$
   g $8s \times 3r \times 2s$
   h $4a \times -2a \times 2b$

10 **Reasoning** Dave and Alesha work out the answer to $2a \times 4b \times 3c$
   Dave's answer: $24abc$
   Alesha's answer: $cba24$
   Who wrote their answer better?
   Discussion What did the other student do wrong?

11 Find the product of each pair of expressions connected by the six lines.

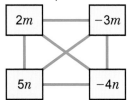

Key point 6

To divide algebraic terms, divide the numbers first and then the letters.
$$\frac{10x}{2} = \frac{10}{2} \times x = 5x$$

12 Simplify
   a $\dfrac{9b}{3} = \dfrac{\square}{\square} \times b = \square \times b =$
   b $\dfrac{-40a}{10}$
   c $\dfrac{26z}{13}$
   d $\dfrac{2m}{4} = \dfrac{2}{4} \times m = \dfrac{\square}{\square} \times m = \dfrac{m}{\square}$
   e $\dfrac{10p}{30}$
   f $-\dfrac{8e}{4}$
   g $\dfrac{12t}{16} = \dfrac{12}{16} \times t = \dfrac{\square}{\square} \times t = \dfrac{\square t}{\square}$
   h $\dfrac{6f}{9}$
   i $\dfrac{-6d}{-12}$

Key point 7

$\dfrac{1}{2}x = \dfrac{x}{2}$ These fractions both mean 'half of x'.

13 Lyn says $\dfrac{x}{x} = 0$. Jessie says $\dfrac{x}{x} = 1$. Who is correct?

> **Q13 hint** What is $\dfrac{4}{4}$? ... $\dfrac{6}{6}$?

14 Simplify
   a $\dfrac{3a^2}{a} = 3 \times \dfrac{a^2}{a} = 3 \times \square =$
   b $\dfrac{8c^4}{2c^2}$
   c $\dfrac{5x}{20x}$
   d $\dfrac{-10z^2}{2z}$
   e $\dfrac{6p^2}{-36p}$
   f $\dfrac{-15m^3}{-20m}$

Discussion Why is there no 'x' in the answer to **Q14c**?

15 **Reasoning** Insert the missing term in each question.

Choose from: $2x$ x^2 x

a $\dfrac{2x}{2} = \square$

b $\dfrac{\square}{x} = 2$

c $\dfrac{6x^2y}{30\square} = \dfrac{y}{5}$

d $\dfrac{16x^2}{\square} = 8x$

e $\dfrac{3x^3y}{3xy} = \square$

2.3 Substitution

Objective

- Substitute numbers into expressions.

Did you know?

- In Algebra 'substitution' means putting numbers in place of letters.

Fluency

- Is 3×5 the same as 5×3? Is $y \times x$ the same as $x \times y$?
- Does the order you multiply in matter?

1 Work out

a $-9 + 11$

b -6×-7

c $21 \div -3$

d $4 \div -16$

2 Match each statement to an expression.

$x - 2 \qquad \dfrac{x}{2} \qquad x + 2 \qquad 2x$

a x divided by 2

b double x

c x subtract 2

d x add 2

3 Work out

a $3 \times 4 + 2$

b $2 \times 5 + 7 \times 3$

c $6 \times 2 \div 3$

d $2 + 3 \times 2^2$

4 Write an expression for these statements.
Use n to represent the starting number.

a Tom thinks of a number and subtracts 20

b Suzanne thinks of a number and multiplies it by 10

c Ahmed thinks of a number and divides it by 5

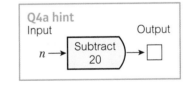

Q4a hint

Input → Output

$n \rightarrow$ Subtract 20 $\rightarrow \square$

5 Write an expression for these statements.
Use n to represent the starting number.

a Christina thinks of a number, multiplies it by 4 then adds 5

b Javed thinks of a number, doubles it then subtracts 6

c Louisa thinks of a number, multiplies it by 3 then divides by 4

d Matty thinks of a number, divides it by 2 then adds 4

e Lisa thinks of a number, adds 4 then divides by 2

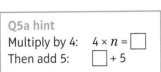

Q5a hint

Multiply by 4: $4 \times n = \square$

Then add 5: $\square + 5$

Discussion Matty and Lisa both have the same starting number.
Will their answers be the same? If not, why not?

Example 4

When $x = 2$ and $y = 5$ work out the value of

a $x + y$ b xy c $\dfrac{5x}{y}$ d $4x + 3y$

a $2 + 5 = 7$

b $2 \times 5 = 10$ | Replace x and y with the values given. |

c $5 \times 2 \div 5 = 10 \div 5 = 2$

d $4 \times 2 + 3 \times 5 = 8 + 15 = 23$ | Use the priority of operations. |

6   Work out the value of these expressions when $a = 5$ and $b = 2$

     a $a + b$ b $a - 2$ c ab d $\dfrac{8a}{b}$

     e $4a - 2b$ f $5ab$ g $\dfrac{-16}{b}$ h $\dfrac{-8a}{b^2}$

> **Q6h hint**
> Work out the index (power) first.

7   Find the value of each expression when $a = 3$, $b = -5$ and $c = 2$

     a $10a + 10b$ b $a + 2b$ c $\dfrac{10a}{b}$ d $\dfrac{ab}{b}$

     e $2b^2$ f $a + c^2$ g $b^2 - a$ h $b + a^2c$

8   **Exam-style question**

     $f = 8$

     a Work out the value of $2f + 7$ **(2 marks)**

     $T = 3g + 5h$

     $g = -2$

     $h = 4$

     b Work out the value of T.   **(2 marks)**

     March 2013, Q18, 1MA0/2F

> **Exam hint**
> $2f$ means $2 \times f$

9   A plate of biscuits has c chocolate biscuits, and p plain biscuits.

     a Write an expression in c and p for the total number biscuits on the plate.

     b Use your expression to work out the total number of biscuits on the plate when $c = 5$ and $p = 3$.

> **Q9b hint**
> Substitute the values into your expression from part **a**.

10   **Reasoning** Anna buys n cupcakes.

     a She gives one to her mum. Write an expression in n for the number of cupcakes Anna has left.

     b She gives half of the remaining cupcakes to her brother. Write an expression in n for the number of cupcakes Anna gives to her brother.

     c Use your answer to part **b** to work out how many cupcakes Anna gives to her brother when $n = 5$.

11   a **Reasoning** A pencil costs x pence. Write an expression for the cost of 3 pencils.

     b A ruler costs y pence. Write an expression for the cost of 2 rulers.

     c Josef buys 3 pencils and 2 rulers. Write an expression in x and y for the cost.

     d Use your answer to part **c** to work out much Josef spends when $x = 35$ pence and $y = 90$ pence. Give your answer in pounds (£).

12   **Problem-solving** An airline charges passengers £55 for up to 20 kg of luggage, plus £4.50 for each kilogram over this limit.

     a The expression used to work out how much each passenger will pay is 55 + £4.50k. Explain what k represents.

     b Copy and complete the table.

kg over limit	$55 + 4.50k$
1	
3	
7	
10	

2.4 **Formulae**

Warm up

Objectives

- Recognise the difference between a formula and an expression.
- Substitute numbers into a simple formula.

Did you know?

- In Science, formulae are regularly used for calculations, such as finding the mass of an object by using the formula mass = density × volume.

Fluency

- Work out **a** $-1 \times 4b$ **b** $(-6)^2$ **c** $5 + (-4)$ **d** $20 - (-5)$
- $a = 2$ and $b = -3$.

Work out **a** a^2 **b** b^2

1 Work out
   a 3^2
   b -2^2
   c $5^2 + 4^2$

2 Work out the value of these expressions when $a = 3$, $b = 4$, $c = 5$
   a $a + 4b$
   b $\dfrac{12c}{b}$
   c $c^2 \times b$
   d $a^2 + c^2$

3 A red sweet costs r pence. A yellow sweet costs 4 pence more than a red sweet.
   Write an expression for the cost of a yellow sweet.

Key point 8

A **formula** is a general rule that shows a relationship between variables.
For example, speed = distance ÷ time, which we can write as $s = \dfrac{d}{t}$
Speed, distance and time are variables. Although their values can vary, the rule stays true.

4 **Reasoning** A packet of red hair bands costs n pence.
   a Write an expression in n for a gold packet that costs 10 pence more.
   b Write a formula for the cost, C, of the gold packet.

   Q4b hint To make the expression in part **a** into a formula, write '$C =$' in front. This shows the relationship between C and n.

5 **Reasoning** There are 12 pencils in a box.
   a Write an expression for the number of pencils in x boxes.
   b Using your expression, write a formula for the number n of pencils in x boxes.

6 **Reasoning** A packet of sweets costs x pence.
   Write formulae for the cost, C,
   of another packet which costs
   a 15 pence less
   b 3 times more
   c half as much.

   Q6 communication hint The plural of formula is formulae.

7 **STEM** The formula to work out the force on an object is $F = ma$, where
   F = force (in newtons, N), m = mass (in kilograms, kg) and a = acceleration (in metres per second per second, m/s²).
   Use the formula $F = ma$ to work out the force in newtons on an object when
   a mass = 10 kg and acceleration = 2 m/s²
   b mass = 15 kg and acceleration = 3 m/s²

   Q7 hint Write the units with your answer.

8   A formula to calculate a rough estimate for the area of a circle is $A = 3r^2$, where r is the radius.
    Calculate an estimate for the area of a circle in cm² when
    a $r = 3$ cm   b $r = 5$ cm   c $r = 10$ cm

> **Q8 hint** The radius is half the diameter of a circle.
>

9   **STEM / Modelling** The formula for the time taken for a journey is $t = \dfrac{d}{s}$, where t = time, d = distance and s = speed.
    Work out the time (t) in hours a car takes to travel
    a   180 km at 60 km/h
    b   280 km at 70 km/h
    c   60 km at 30 km/h

> **Q9a hint** $d = 180$, $s = 60$

10 **STEM / Modelling** The formula for speed is $s = \dfrac{d}{t}$, where s = speed, d = distance and t = time.
    Work out the speed, s, in miles per hour when
    a d = 200 miles and t = 8 hours
    b d = 70 miles and t = 2 hours
    c   a car travels a distance of 240 miles in 4 hours
    d   a plane flies a distance of 1750 miles in 5 hours.

11 **Exam-style question**

    $y = 4x + c$
    $x = 7.5$
    $c = 5.4$
    Work out the value of y.   **(2 marks)**

    Nov 2012, Q18a, 1MA0/2F

> **Exam hint**
> Write down what you input into your calculator for the method mark.

12 Which of these are formulae and which are expressions?
    a $2a + 3$ b $F = ma$ c $m^2 - 3pq$
    d $\dfrac{t^3}{3}$ e $P = 2w + 2l$ f $A = bh$

> **Q12 hint** A formula always has an equals sign.

Example 5

Sarah is a hairdresser. She works h hours per week at an hourly rate of £m.

Write a formula to work out Sarah's total pay, P, using the number of hours worked, h, and her hourly rate of pay, m.

Hourly rate of pay, m → $\boxed{\times h \text{ hours worked}}$ → total pay, P ── Use a function machine.

$P = hm$

13 **Reasoning** Darren is a tour guide. He works n hours per day and is paid a rate of £r per hour.
    a   Write a formula for Darren's total pay per day, D, in terms of n and r
    b   Use your formula to work out D when $n = 8$ and $r = 7$
    c   At the end of each day Darren and a colleague share equally the tips, t, they have earned together. Write a formula for S, the amount Darren receives from the tips, in terms of t.
    d   Use your formula for part **c** to work out Darren's share of the tips when t = £30.
    e   Use your answers to parts **b** and **d** to work out Darren's total earnings for the day.

    Reflect Would it matter if your formula used the letters x, y and z instead of D, n and r?

14 **Reasoning** A taxi firm charges £3 fixed charge plus £4 per kilometre.
  a Write a formula for the cost of a journey, C, of k kilometres.
  b Use your formula to work out the cost of travelling 5 km.

15 In a right-angled triangle $c^2 = a^2 + b^2$, where c is the hypotenuse.

> **Q15 hint** Substitute the amounts to find c^2. Square root to get c.

Find the length, c, of the hypotenuse when
  a $a = 3, b = 4$ b $a = 7, b = 24$ c $a = 2.8, b = 3.9$ d $a = 4.25, b = 11.5$
Round your answers to 1 d.p. if necessary.

2.5 Expanding brackets

Objectives

- Expand brackets.
- Simplify expressions with brackets.
- Substitute numbers into expressions with brackets and powers.

Did you know?

- Brackets are part of the order of operations. Brackets in an expression show you which part to work out first.

Fluency

Work out
- -3×4 • -7×-6 • $3 \times b$ • $4 \times 3a$
- $a \times 2a$ • $10c \times 3d$ • $-2y \times 8xy$

1 Work out
  a $2 \times 10 + 5^2$ b $12 - (8 - 2)^2$ c $4^2 \div 2 - 6$ d $3 \times (4 + 4) \div 2^2$

2 Simplify by collecting like terms
  a $3y - 6y$ b $2a + 4b - a - 7b$ c $12m^2 + 1 - 9m^2 - 10$

3 An electrician charges a call-out fee of £35 plus £20 per hour, h.
  a Write a formula for the cost, C, that a customer pays for h hours of work.
  b Use your formula to work out the cost of 3 hours of work.

4 Find the value of each expression when $a = 2$ and $b = 3$
  a $2a(b - 1)$ b $9(a + b)$ c $(3b + a)^2$
  d $a(b - 7)$ e $(5a)^2$

> **Q4 hint** Substitute the values of a and b. Work out the brackets first.

5 Work out
  a $3(5 + 4)$ b $3 \times 5 + 3 \times 4$ c $2(6 + 1)$ d $2 \times 6 + 2 \times 1$
  Discussion What do you notice about your answers?

Example 6

Expand $4(3a + 2)$

$4(3a + 2) = 12a + 8$ ——— Multiply each term in the bracket by the term on the outside.

6 Expand the brackets

 a $3(x + 2)$ b $2(a + 7)$ c $10(t - 6)$
 d $5(3 - m)$ e $3(2w + 1)$ f $-2(a + 5)$
 g $-3(2b + 6)$ h $-4(a - 6)$ i $-(x + 4)$

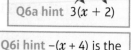

Q6a hint $3(x + 2)$

Q6i hint $-(x + 4)$ is the same as $-1(x + 4)$

7 **Reasoning** Hannah and Benton expand $3(2c - 2d)$.
  Hannah's answer is $6c + 6d$. Benton's answer is $6c - 6d$
  Who is correct?
  Discussion What mistake has the other person made?

8 Expand

 a $x(x + 1)$ b $r(r + 4)$ c $g(3g - 2)$

Q8a hint $x(x + 1)$

9 Write an expression for each statement. Use n to represent the starting number.

 a I think of a number, add 3 and then double it.
 b I think of a number, add 1 and multiply it by 4.
 c I think of a number, add 5 and multiply it by 10.

Q9a hint Use brackets.
$(n + 3) \times 2 = \boxed{}(n + 3)$

10 **Reasoning** A can of oil holds $3n + 1$ litres.

 a Write and simplify an expression for the oil in 5 cans.
 b When $n = 20$, how many litres are there in 5 cans?

11 Expand and collect like terms

 a $3(t + 4) + 2$ b $5(m - 2) + 6$ c $a(a - 8) + 2a$
 d $-2(m + 3) + 7$ e $20 - 2(3 - 5x)$ f $17e - (e + 2f)$

Q11a hint Expand the bracket then collect like terms.

12 Expand and simplify

 a $2(x + 1) + 3(x + 2)$ b $4(2a + 3) - 3(a + 2)$ c $2(2d - 3) + 3(d - 4)$

13 Work out the value of these expressions.

 a $(5e)^2$ when $e = -2$
 b $(2e - 2f)^2$ when $e = 9$ and $f = -1$
 c $\dfrac{e(f - 2)^2}{ef}$ when $e = -4$ and $f = 1$
 d $\dfrac{(10e + 6f)^2}{(f)^2}$ when $e = 4$ and $f = -6$

14 **Reasoning** Brad is x years old and his sister is 2 years older. Their cousin is half the age of Brad's sister.

 a Write an expression for the age of Brad's sister.
 b Write an expression for the age of Brad's cousin.
 c Use your answer to part **b** to work out the age of Brad's cousin when Brad is 14.
 Discussion Did you all get the same answer? How did you do the calculation?

15 **Reasoning** An entertainer charges £25, plus £12 per child, for birthday parties.

 a There are n children at a party. Write an expression in n for the cost of the party.
 b Write a formula for the cost C, in terms of n.
 c The entertainer puts on three parties. The same number of children, n, attend each one. Write a formula in n for her earnings, E.
 d When n is 30, how much does the entertainer earn for three parties?

Q15c hint Use brackets.

16 **Reasoning** The flow rate of water from a shower is $x + 2$ litres per minute.
  In one shower, 70 litres of water are used, and then the shower is run for another 5 minutes.
  The expression $70 + 5(x + 2)$ represents the total amount of water used.

 a Write what the terms in the expression represent.
 b $x = 6$. How many litres in total were used during the shower?
 Reflect Brackets are used in text for information that isn't essential to the meaning of a sentence. For example: this is a GCSE maths book (for use in schools). Write a short paragraph explaining how brackets are used in maths.

2.6 Factorising

Objectives

- Recognise factors of algebraic terms.
- Factorise algebraic expressions.
- Use the identity symbol ≡ and the not equals symbol ≠

Did you know?

- Factorising is the reverse of expanding.

Fluency

- Give the HCF of these pairs of numbers.   **a** 2 and 8   **b** 9 and 21   **c** 20 and 30

Warm up

1   Work out the HCF of
    a   16 and 20   b   6 and 15

2   Expand
    a $3(x + 4)$ b $5(x - 6)$ c $x(x + 1)$ d $2x(x - 1)$

3   Find the HCF of
    a   −12 and 4   b   6 and −9

> **Q3a hint** Factors of −12 are −12, −6, −4, −3, −2, −1, 1, 2, 3, 4, 6, 12

4   Write the missing terms
    a $2(a + \boxed{}) = 2a + 6b$
    b $\boxed{}(a - 4) = 4a - 16$
    c $10a(\boxed{} + \boxed{}) = 10a^2 + 10ab$
    d $\boxed{}(7a - 1) = 84a - 12$

> **Key point 9**
>
> The factors of a term are all the numbers and letters that divide exactly into it.
> A **common factor** is a factor of two or more terms.

5   a   Copy and complete to find the **highest common factor** (HCF) of $10t$ and 20.
        i   The factor pairs for 20 are 1×20, $2 \times \boxed{}$, $\boxed{} \times 5$
        ii The factor pairs for $10t$ are $1 \times 10t$, $\boxed{} \times 5t$, $5 \times \boxed{}$, $10 \times \boxed{}$
        iii The highest common factor is $\boxed{}$
    b   What is the highest common factor of $24r$ and 42?
        i   Work out the factor pairs for each term.
        ii By looking at all of these factors, work out the HCF.
    c   Find the HCF of
        i $9x$ and 3   ii $16y$ and 12

> **Example 7**
>
> Factorise $10y + 25$
> *The highest common factor of 10y and 25 is 5.*
> $10y + 25 = 5(2y + 5)$
> $5(2y + 5) = 10y + 25$
>
> Write the HCF of both terms outside the bracket. Work out the terms inside the bracket by dividing each term in the expression by the HCF.
>
> Check your answer by expanding.

6 Copy and complete. Check your answers by expanding the brackets.

   a $8y + 16 = 8(y + \boxed{})$ b $10m - 25 = 5(\boxed{} - 5)$

   c $6y + 24 = \boxed{}(y + 4)$ d $7m - 21 = \boxed{}(m - \boxed{})$

7 Emily and Phil factorise $8x + 12$

Emily says the answer is $2(4x + 6)$. Phil says the answer is $4(2x + 3)$.

Who is correct?

Discussion What mistake has the other student made?

8 Factorise completely

   a $9x + 18$ b $3w - 12$ c $15a + 10$ d $12 - 21t$

9 **Reasoning** Charlotte and Zhir factorise $10x - 5$

Charlotte's answer is $5(2x - 5)$. Zhir's answer is $5(2x - 1)$

Who is correct? What mistake has the other student made?

10 Work out the HCFs of

   a cd and d | Q10a hint $cd = c \times d$ |

   b a^2 and a

   c bc and ab

   d $6xy$ and $36y$ | Q10d hint Find the HCF of the numbers, then the letters. |

   e $10a^2$ and $-50a$

   f $6xy$ and $10x^2$

Example 8

Factorise

a $y^2 + y$

b $2ef + 4f$

a $y^2 + y = y(y + 1)$ The HCF is y

b $2ef + 4f = 2f(e + 2)$ The HCF is $2f$

---

11 Factorise

   a $n^2 + n$ b $4x^2 + 3x$ c $2st + 4t$ d $5ab - 3b$

12 Choose the correct factorisation for each expression.

   a $a^2 + 7a$ **A** $a(a + 7)$ **B** $a(a + 7a)$

   b $2ab - 3b$ **A** $b(2a - 3b)$ **B** $b(2a - 3)$

   c $y^3 + y^2$ **A** $y(y^2 + y)$ **B** $y^2(y + 1)$

   d $5d - d^2$ **A** $d(5 - d)$ **B** $5d(1 - d)$

13 Factorise these completely. Expand your answers to check them.

   a $x^2 + 5x$ b $7x^2 - 21x$ c $9x + 12y$ d $6y^2 - 2y$

Discussion When do you know that your answer has been completely factorised?

14 **Exam-style question**

   a Factorise $4x + 10y$ **(1 mark)**

   b Factorise $x^2 + 7x$ **(1 mark)**

Nov 2012, Q26, 1MA0/2F

| Q14 strategy hint |
| Expand your answers to check them. |

> **Key point 10**
>
> The ≡ symbol shows an identity.
> An identity is an equation that is true for all values of the variable.
> $5(x + 1) \equiv 5x + 5$ is an identity.
> $5(x + 1)$ has the same value as $5x + 5$ for all values of x.

15 Which of these are true for *all* values of t rather than just *some* values of t?
   Rewrite any identities that you find, replacing = with ≡

   a $t + 2 = 6$ b $2t + 4 = 2(t + 2)$

   c $t^2 = 6t$ d $5t + 7 = 7 + 5t$

   > **Q15 hint** Substitute numbers for t.

16 Use ≡ to write an identity for each of these expressions.

   a $4a + a$ b $0.5a$ c $3(a + 4)$ d $a + 2$

17 State whether each of these is an expression, formula or identity.

   a $F = ma$ b $x^2 + 4x = x(x + 4)$ c $9x - 3x^2 + 4$

   d $4x^2 = (2x)^2$ e $y^2 + 2y$ f $E = mc^2$

> **Key point 11**
>
> The ≠ symbol is used to show that two expressions are not always equal.
> For example, $5x + 12 \neq 5(x + 6)$

18 Put the sign ≠ or ≡ in each box.

   a $0.5x \square \dfrac{x}{2} - 1$ b $4(x + 1) \square 4x$

   c $16t + 4 \square 4(4t + 1)$ d $p^2 - p \square -p(1 - p)$

2.7 Using expressions and formulae

Objectives

- Write expressions and simple formulae to solve problems.
- Use maths and science formulae.

Did you know?

- STEM means Science, Technology, Engineering and Maths. STEM formulae are regularly used in industry.

Fluency

- Using x, write expressions for

   a 4 less than x b half of x c twice x d x divided by 3   e 10 lots of x

1 Work out the value of each expression when $a = 2$, $b = -3$, $c = 4$

   a $b^2 - a$ b $10a + 10b$ c $a - b$ d ab

   e $\dfrac{b^3}{b}$ f $\frac{1}{2}ac^2$ g $c + ab$ h $ab + \frac{1}{2}c^2$

2 A stapler costs r pence.
   a Write an expression for the cost of 3 staplers.
   b A hole-punch costs 20 pence more than a stapler. Write an expression for the cost of a hole-punch.
   c Beth buys 3 staplers and a hole-punch. Using your expressions, write a formula for the cost, C, simplifying your answer.

> **Q2c hint** A formula has an equals sign.

3 Write a formula for P, the perimeter of each rectangle.

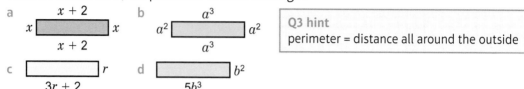

> **Q3 hint**
> perimeter = distance all around the outside

4 Write an expression using n as the unknown starting number.
   a I think of a number, multiply it by 5 then add 2.
   b I think of a number, subtract 1 and then multiply by 2.
   c I think of a number, multiply it by 3 and divide by 2.
   d I think of a number and multiply it by itself.
   e I think of a number, multiply it by itself and then multiply it by itself again.
   f I think of a number and square root it.

> **Q4 hint** Use a function machine.

5 **Reasoning** Cakes are sold in boxes or packets. A packet contains x cakes and a box contains y cakes.
   a Write an expression for the number of cakes in 3 packets.
   b Write an expression for the number of cakes in 4 boxes.
   c Josh buys 3 packets and 4 boxes of cakes.
     Write an expression in terms of x and y for the number of cakes he buys.
   d Josh buys p packets and b boxes of cakes.
     Write an expression in terms of x, y, p and b for the number of cakes he buys.
   e When $x = 2$ and $y = 4$, use your answer to part **c** to work out how many cakes Josh buys.

6 **Reasoning** Alice scores m marks in her physics exam.
   a Write an expression in terms of m for
     i her biology marks, 20 marks less than in physics
     ii her chemistry marks, twice as much as the marks you calculated for biology.
   b Write and simplify an expression for Alice's total marks in physics, biology and chemistry.
   c Alice scored 60 marks in physics. Use your answer to part **b** to work out her total marks for all three science papers.

> **Q6a ii hint** 'Twice as much' means multiply the **whole** expression by 2. Use brackets.

7 **STEM** The formula to work out the mass, m, of an object is
$m = dv$ where d = density and v = volume.
Work out m, in kg, when $d = 3\ \text{kg/m}^3$ and $v = 2\text{m}^3$.

Example 9

The formula $v = u + at$ gives the final velocity (speed in a particular direction) of an object, where v = final velocity (in m/s), u = initial velocity (in m/s), a = acceleration (in m/s²) and t = time (in s).

Work out the final velocity of a car when $u = 0$, $a = 2$ and $t = 11$

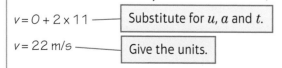

$v = 0 + 2 \times 11$ —— Substitute for u, a and t.

$v = 22$ m/s —— Give the units.

Warm up

8 **STEM** Use the formula $v = u + at$ to work out v when
  a $u = 0\,\text{m/s}$, $a = 4\,\text{m/s}^2$, $t = 15\,\text{s}$
  b $u = 10\,\text{m/s}$, $a = 5\,\text{m/s}^2$, $t = 30\,\text{s}$

9 **STEM** The formula $s = ut + \frac{1}{2}at^2$ gives the height of a ball thrown upwards, where
  s = distance (m), u = initial velocity (m/s), a = acceleration (m/s^2) and t = time (s).
  Work out the value of s when
  a $u = 20\,\text{m/s}$, $t = 3\,\text{s}$, $a = -10\,\text{m/s}^2$
  b $u = 30\,\text{m/s}$, $t = 4\,\text{s}$, $a = -10\,\text{m/s}^2$

10 **STEM** Use the formula $v^2 = u^2 + 2as$ to work out
   the value of v when $u = 0\,\text{m/s}$, $a = 5\,\text{m/s}^2$ and $s = 10\,\text{s}$

> **Q10 hint** Substitute the values to work out v^2. Square root v^2 to get v

11 **Exam-style question**

A temperature in F (°Fahrenheit) can be converted to C
(°Celsius) using the formula $F = \dfrac{9C}{5} + 32$.

  a Work out F when $C = 15°$ **(2 marks)**
  b On Monday the temperature was $25°$ Celsius and on
    Tuesday it was $75°$Fahrenheit. Which day was hotter?
    Give reasons for your answer. **(2 marks)**

> **Q11b strategy hint**
> First convert 25°C to
> Farenheit.

12 **Problem-solving** To cook a chicken takes 40 minutes per kg plus an extra 20 minutes.
   a How long does it take to cook a 2.5 kg chicken?
   b Write a formula for the number of minutes, M, it takes to
     cook a chicken that weighs w kg.
   c Use your formula to find M when $w = 4$.
   d What time should you put a 2 kg chicken in the oven to be ready for 7 pm?

> **Q12 communication hint**
> 'per kg' means 'for
> each kg'.

2 Problem-solving

Objective · Use smaller numbers to help you solve problems.

Example 10

Rebecca buys 12 packets of biscuits. Each packet contains 18 biscuits. She puts the same
number of biscuits on each of 9 plates.
a How many biscuits are on one plate?
b Write an expression for the number of biscuits on one plate when there are p packets
  containing q biscuits, put on n plates.

> Replace the numbers in the
> question with smaller numbers.

Using smaller numbers:
1 packet of biscuits, 5 biscuits per pack, 2 plates.

1 packet of biscuits   2 plates

 ← contains
        5 biscuits

> Draw a picture.

Total number of biscuits: $1 \times 5 = 5$

> Use your picture to help you calculate
> how many biscuits on one plate.

Number of biscuits on each plate: $5 \div 2 = 2.5$

a   Total number of biscuits: $12 \times 18 = 216$

    Number of biscuits on one plate: $216 \div 9 = 24$

> Replace the smaller numbers in your calculation with the numbers in the question.

b   Total number of biscuits: $p \times q = pq$

    Number of biscuits on one plate: $pq \div n = \dfrac{pq}{n}$

> Replace the numbers in your calculation with the letters.

1   Every day Luke does 11 puzzles in a puzzle book. After 18 days, he still has 7 puzzles left to do. How many puzzles are in his book?

> **Q1 hint** Read each sentence, one at a time. Replace each number with a smaller number and draw a picture to represent it. This will show you what calculation you need to do.

2   Adam makes kites. Last week he made 16 kites. He put a 22.5 metre string on 9 kites, and a 25 metre string on the rest.

    a   What is the total length of string Adam used last week?

    b   Write an expression for the length of string Adam needs per week if he makes b kites with a 22.5 metre string and c kites with a 25 metre string.

3   A teacher has 15 boxes of pens. There are 8 pens in each box. She gives 1 pen to each student in a class. There are 31 students in the class.

    a   How many pens are left?

    b   Write an expression for the number of pens left when there are r boxes containing s pens, and t students in a class.

4   There is a group of 8 friends. Each friend sends one text to every other member of the group.

    a   How many texts are sent altogether?

    b   Write an expression for the number of texts sent by q friends.

> **Q4 hint** Sometimes it is helpful to try a series of smaller numbers to look for a pattern.
>
> | 2 friends | 3 friends | 4 friends | 8 friends | q friends |
> |---|---|---|---|---|
> | A B | A B C | | | |
> | texts ↓ ↓ | ⋀ ⋀ ⋀ | | | |
> | B A | B C A C A B | | | |
>
> number of texts: 1×2 2×3

5   An artist puts paint on the bottom of his shoes and creates a picture of footprints. He does 5 hops then 9 jumps, 5 hops then 9 jumps and so on.

    a   How many footprints are in the picture after he has done this 20 times?

    b   Write an expression for the number of footprints in the picture after he does 5 hops then 9 jumps p times.

> **Q5 hint** How many footprints are made when the painter hops? What about when he jumps? Try a series of smaller numbers to look for a pattern.

6   a   Write down the 37th odd number.

    b   Write an expression for the nth odd number.

> **Q6 hint** Write down the 1st, 2nd, 3rd odd number. How do you find the 37th odd number?

7   **Reflect** Did using smaller numbers help you?

    Is this a strategy you would use again to solve problems?

    What other strategy helped you to solve these problems?

2 Check up

Expressions and substitution

1 Simplify
  a $2w + 3w - 4w$
  b $6a + 4b - a - 9b$
  c $10m^2 - 2m - 9m^2 - 6m$
  d $5m \times 4n$
  e $4a \times a \times b$
  f $\dfrac{64x^2}{8x}$
  g $\dfrac{15ab}{b}$

2 Work out the value of each expression when $x = 4$, $y = -2$, $z = 10$
  a $2x + z$
  b $z - 2x$
  c y^2
  d $\dfrac{z}{y}$

3 Use x as the starting number to write expressions.
  a I think of a number and add 5
  b I think of a number, multiply it by 4 and then divide by 5

Expanding and factorising

4 Expand and simplify
  a $2(a + 1)$
  b $5f(3f + 2)$
  c $y(6y - 2)$
  d $-2(3a + 5)$

5 Factorise completely. Check your answers.
  a $36x + 12$
  b $4x^2 + 16x$
  c $9x + 21y$
  d $15xy - 5y$

6 Choose the correct sign, $\neq$ or $\equiv$
  a $10a + 20ab \ \boxed{}\ 10a(a + 2b)$
  b $4x(x + 1) \ \boxed{}\ 4x^2 + 4x$
  c $36t + 6 \ \boxed{}\ 6(6t + 1)$
  d $0.75x \ \boxed{}\ \dfrac{x}{4}$

Writing and using formulae

7 State whether each of these is an expression, formula or identity.
  a $y = mx + c$
  b $0.5x = \dfrac{x}{2}$
  c $2x + 4$

8 **STEM** a Use the formula $s = \dfrac{d}{t}$ to work out the value of s when $d = 40$ km and $t = 5$ hours
  b Use the formula $A = (a + b)\dfrac{h}{2}$ to work out the value of A when $a = 4$, $b = 6$ and $h = 5$

9 **Reasoning** There are b blue sweets and p pink sweets in a box.
  a Write a formula for the total amount, T, of sweets in the box using b and p.
  b Use your formula to work out T when $b = 20$ and $p = 15$

10 **STEM** Use the formula $v^2 = u^2 + 2as$ to work out the value of v when $u = 0$, $a = 2$ and $s = 16$

11 How sure are you of your answers? Were you mostly

Just guessing 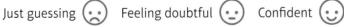   Feeling doubtful   Confident

What next? Use your results to decide whether to strengthen or extend your learning.

✳ Challenge

12 a Multiply together the four pairs of connected terms and expand your answers.
  b Add together your answers and simplify the result.
  c Factorise your simplified expression.

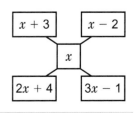

2 **Strengthen**

Expressions and substitution

1   Simplify
    a $3x + 4x$ b $8b + 4b + 2b$
    c $7h - 5h$ d $6y - 3y + 8y$

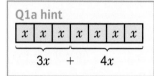

Q1a hint

| x | x | x | x | x | x | x |

$\underbrace{\qquad}_{3x} \quad + \quad \underbrace{\qquad}_{4x}$

2   Simplify by collecting like terms
    a $2d + 4d + 7e + 3e$
    b $5x + 2y + 10x + 8y$
    c $7r + 6s - 5r + 2s$
    d $3p - 4q + p - 5q$
    e $6v - 10w - 9v + 4$
    f $8 + 3g + 4h - g + 2$
    g $2x^2 + 3x^2 + 5x^2$
    h $7a^2 + 3a + a^2 - a$

Q2a hint $2d + 4d = \square d$
$7e + 3e = \square e$

Q2c hint The sign before a term belongs with the term.
$7r - 5r = \square r$

Q2h hint a^2 and a are not like terms.

3   Simplify
    a $a \times 6$
    b $4 \times n$
    c $y \times -2$
    d $-4 \times k$
    e $b \times a$
    f $g \times f$

Q3a hint

| a | a | a | a | a | a |

Q3c hint The negative sign goes in front of the term: $-\square y$

Q3e hint Write letters in alphabetical order.

4   Simplify
    a $5 \times 2a$
    b $6y \times 3$
    c $-4s \times 10t$
    d $2p \times -6q$
    e $\dfrac{20a}{10}$
    f $18b \div -3$
    g $16x \div 4x$

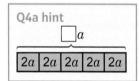

Q4a hint

$\overbrace{}^{\square a}$

| $2a$ | $2a$ | $2a$ | $2a$ | $2a$ |

Q4e hint $\dfrac{20a}{10} = \dfrac{20}{10} \times a = \square a$

Q4c hint $-4 \times 10 = \square$
$s \times t = \square$

Q4g hint $\dfrac{16}{4} = \square$
$\dfrac{x}{x} = \square$

5   Write an expression for these statements. Use n to represent the unknown starting number.
    a   Lucy thinks of a number and adds 4
    b   Adam thinks of a number and subtracts 6
    c   Keisha thinks of a number and multiplies it by 2
    d   Bill thinks of a number and divides it by 3

Q5a hint
... a number and adds 4
         ↑   ↑
         n $+4$

6   Simplify
    a $a \times a$
    b $2a \times a$
    c $a^3 \times a^2$
    d $q^4 \times q$

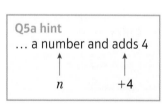

Q6c hint

$\underbrace{a^3}_{a \times a \times a} \quad \times \quad \underbrace{a^2}_{a \times a}$

$a^\square$

*Active*Learn Homework, practice and support: Foundation 2 Strengthen

7   Work out

   a $\dfrac{4 \times 4}{4} = \dfrac{\cancel{4} \times 4}{\cancel{4}} = \square$
   b $\dfrac{a \times a}{a} = \square$
   c $\dfrac{s \times s \times t}{s} = \dfrac{\cancel{s} \times s \times t}{\cancel{s}} = \square$

8   Simplify

   a $\dfrac{x^5}{x^2} = \dfrac{x \times x \times x \times x \times x}{x \times x} = \square$
   b $\dfrac{a^7}{a^3}$

   c $\dfrac{6x^5}{2x}$
   d $\dfrac{30f^6}{10f^2}$

> **Q8c hint** Divide the numbers and use the laws of indices to simpify the letters.

9   Copy and complete to find the value of the expressions when $a = 2$ and $b = 5$

   a $2a = 2 \times \square = \square$
   b $4b = 4 \times \square = \square$
   c $2a + 4b = \square + \square = \square$
   d $2b - a = 2 \times \square - \square = \square - \square = \square$
   e $11a - 2b = \square$
   f $3b^2 = 3 \times \square \times \square = \square$
   g $25a^2 = \square$
   h $(b - a)^2 = (\square - \square)^2 = \square^2 = \square$
   i $\dfrac{5a}{b} = \dfrac{5 \times \square}{\square} = \dfrac{\square}{\square} = \square$

10 Work out the value of these expressions when $a = 3$ and $b = 4$

   a $a + b$
   b $3a - b$
   c $7(a + b)$
   d $\dfrac{8a^2}{b}$

11 Find the value of each expression when $f = -2$ and $g = 4$

   a $f + g$
   b $g - f$
   c $3g + 2f$
   d $4f^2$
   e $\dfrac{2g}{f}$
   f $2(f - g)$

Expanding and factorising

1   You can use a grid to expand brackets.

×	**3a**	**+2**
a	$3a^2$	$+2a$

$a(3a + 2) = 3a^2 + 2a$

> **Q1c hint** Don't forget negative signs.
>
> | × | **2d** | **−5** |
> |---|---|---|
> | **d** | | |

   Expand these using a grid.

   a $b(3b + 4)$
   b $5(t + 2)$
   c $d(2d - 5)$
   d $5(2f - 1)$

2   Expand

   a $p(4p + 1) + 3p$
   b $x(2x + 3) - 5x$
   c $2r + 7(r - 6)$
   d $5b + 2b(b - 2)$
   e $3(2a + 7) + 2(3a + 4)$
   f $r(r + 2) - 3r(r + 1)$

> **Q2a hint** Expand the bracket then collect like terms and simplify.

> **Q2c hint** Don't forget negative signs.

3   What is the highest common factor (HCF) of

   a   5 and 10
   b   21 and 49
   c   12 and 16
   d   18 and 27
   e a^2 and a^3
   f a^5 and a^2
   g a and a^2
   h ab and ab^2

> **Q3e hint** $a^2 = @ \times @$
> $a^3 = @ \times @ \times a$

> **Q3h hint** $ab = @ \times \textcircled{b}$
> $ab^2 = @ \times \textcircled{b} \times b$

4   Factorise each expression completely by taking out the HCF and putting it in front of the brackets. Check your answers by expanding the brackets.

a $3a^2 - 9a = \boxed{}(a - 3)$

b $16x^2 + 12x = \boxed{}(4x + 3)$

c $5a^2 + 15ab$

d $2q^3 + 8q$

e $84a - 12 = \boxed{}(\boxed{} - \boxed{})$

f $5a^2 + ab - ac = \boxed{}(\boxed{} + \boxed{} - \boxed{})$

g $y^3 + y^2 = \boxed{}(\boxed{} + \boxed{})$

> **Q4a hint** What is the HCF of 3 and 9? ...a^2 and a?

5   Fay writes $8x^2 + 16x = 2x(4x + 8)$. Has she factorised the expression completely?

6   Choose the correct sign, $\neq$ or $\equiv$

a $10a^2 + 20a \boxed{} 10a(a + 2)$

b $4x(x + 1) \boxed{} 4x^2 + 5x$

c $81t + 9 \boxed{} 9(9t + 1)$

d $m^2 - m \boxed{} m(1 - m)$

> **Q6 hint** Use the identity sign $\equiv$ if the two sides of the equation are the same when the brackets are completely expanded.
> Use the not equals sign $\neq$ if the two sides of the equation are *not* always the same when the brackets are completely expanded.

Writing and using formulae

1   **STEM** You can use the formula $F = ma$ to work out the force acting on an object, where

F = force (newtons)

m = mass (kg)

a = acceleration (m/s²)

Work out F when

a $m = 60$ and $a = 2$

b $m = 20$ and $a = 1.5$

c $m = 100$ and $a = -5$

> **Q1a hint** $ma = m \times a$
> $= 60 \times 2 = \boxed{}$ newtons

2   **Reasoning** The rectangle has base b cm and height h cm.

b cm

h cm $\boxed{}$ h cm

b cm

> **Q2 hint** Write a word formula first, then replace the words with letters.

a   Write a formula for P, the perimeter of the rectangle, using b and h. Simplify your answer.

b   Use your formula to work out the perimeter of the rectangle when $b = 6$ cm and $h = 7$ cm.

3   **Reasoning / Real** A film company pays extras an amount per day. The amount paid depends on the acting role.

a   Work out the total amount paid to an extra who works for

  i   2 days at £50 per day

  ii 10 days at £35 per day.

b   Write a formula for T, the total amount paid, at £a, for d days.

c   Use your formula to work out the total amount paid to an extra who works for 3 days at £25.50 per day.

> **Q3b hint** A formula will always have an equals sign.
> $T = d$ lots of £$a = \boxed{}$

4   **Reasoning / Real** Amy has two part-time waitressing jobs. She is paid according to how many hours she works. She also gets tips.

a   Work out Amy's pay when she works for 4 hours at £5 per hour and earns £10 in tips.

b   Write an expression for Amy's total earnings when she works h hours at $£x$ per hour and gets $£t$ in tips.

c   Write a formula for Amy's total earnings E in terms of x, h and t.

d   Use your formula to work out how much Amy earns in her second part-time job, when h = 10 hours, x = £5 and t = £20.

> Q4a hint
> earnings per hour   tips
> | £5 | £5 | £5 | £5 | + | £10 |

> Q4b hint
> An expression doesn't have an equals sign.

2 Extend

1   **Problem-solving / Reasoning** Tallulah thinks of a number, n, adds 3 and then doubles the answer.

a   Write an expression in terms of n for Tallulah's result. Simplify your answer.

b   Work out another way for Tallulah to get the same result.

2   Collect like terms and simplify

a $3x + 4x^2 + 4 + 6x^2 + x + 10$

b $5xy - 6x + 7xz - 10xy - 4xz$

3   Collect like terms and simplify

a $0.5x + x = \boxed{}x$

b $2.5x^2 + 0.5x^2 + 3x$

c $0.75x - 0.5x$

> Q3a hint $x = 1x$

4   Simplify

a $a^6 \times a^3$

b $8a^{10} \times 2a^4$

c $a^m \times a^n = a^{\square + \square} =$

d $4a^y \times 6a^z$

e $\dfrac{a^7}{a^4}$

f $\dfrac{15a^8}{3a^6}$

g $\dfrac{a^5}{a^{10}}$

h $\dfrac{a^m}{a^n} = a^{\square - \square} =$

5   Simplify

a $\dfrac{ab^2}{b} = a \times \dfrac{b^2}{b} =$

b $\dfrac{c^4d}{c^2} = \dfrac{c^4}{c^2} \times \boxed{} =$

c $\dfrac{xy^5}{y^3}$

d $\dfrac{m^7p^6}{m^4p^2}$

6   **STEM** You can use this formula to work out s, the distance an object has travelled in metres.

$s = ut + \frac{1}{2}at^2$

where

u = starting velocity (m/s)

t = time (s)

a = acceleration (m/s²)

Work out the value of s when

a u = 0 m/s, t = 30 s, a = 5 m/s²

b u = 7 m/s, t = 10 s, a = 2.5 m/s²

7   **STEM / Problem-solving** You can use this formula to work out the density, d, of an object in g/cm³

$d = \dfrac{m}{v}$

where m = mass (g) and v = volume (cm³)

Work out d when

a m = 16 g and v = 4 cm³

b m = 30 g and v = 20 cm³

8
Exam-style question

There are $300\,ml$ (millilitres) of medicine in a bottle. Mary has to take two $5\,ml$ spoons full of medicine twice a day. Mary has to take the medicine until the bottle is empty.

Take two $5\,ml$ spoons full twice a day

$300\,ml$

a How many days does Mary have to take the medicine for? **(3 marks)**

> **Exam hint**
> You must show how you worked out your answer so that the examiner can see your method.

You can work out the amount of medicine, $c\;ml$, to give to a child using the formula

$$c = \frac{ma}{150}$$

m is the age of the child, in months
a is an adult dose, in ml.

A child is 30 months old.
An adult's dose is $40\,ml$.

b Work out the amount of medicine you can give to the child. **(2 marks)**

> **Exam hint**
> ma means $m \times a$

June 2012, Q26, 1MA0/1F

9 Expand
  a $9(x + 2y - z)$　　　　**b** $-2(r - 7t - 3s)$　　　　**c** $3c(3a - 3b)$
  d $-p(r - s - t)$　　　　**e** $5x(x + 8y - 9z)$

10 Factorise
  a $-7t - 21$　　　　**b** $-a^2 - 5a$　　　　**c** $-6f^2 - 12f$　　　　**d** $-2r - 4r^2$

11
Exam-style question

a Expand $5(2c + 3d)$ **(1 mark)**

b Here are two straight lines, $ABCDE$ and PQ.

Diagrams NOT accurately drawn

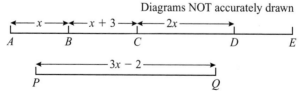

> **Exam hint**
> First work out the length of AE in terms of x.

In the diagrams all the lengths are in cm.
$AE = 2PQ$
Find an expression, in terms of x, for the length of DE.
Give your answer in its simplest form. **(4 marks)**

Nov 2013, Q15, 5MB2F/01

12 Reasoning Alex says that when you raise a negative number to an even power the answer is always positive, and when you raise a negative number to an odd power the answer is always negative.
Is he right? Give reasons for your answer.

13 Work out

   a $ab^2 - (ab)^2$ when $a = -5$ and $b = -2$ b $\dfrac{a^2 - b^2}{a + b}$ when $a = 7$ and $b = -2$

   c $\sqrt{a^2 + b^2}$ when $a = -4$ and $b = -3$ d $\dfrac{a^2 b}{\sqrt[3]{c}}$ when $a = -3$, $b = 6$ and $c = 1000$

14 Use the values $q = 8$, $p = 3$, $r = -6$, $s = -2$ and no other numbers to write three expressions that will give each of the answers in the box. At least one of your expressions should involve a square or square root.

> | 12 | 6 | 28 | 27 |

15 Expand and simplify

   a $9a - 5a(3b - 2a) + 5ab$ b $2p(3p^2 + q) - 5p^2(2p - q)$ c $x(a + b) - a(x - b) + b(a - x)$

2 Knowledge check

⊙ A **term** is a number, a letter, or a number and a letter multiplied together. When a term has numbers and letters, numbers are written first, then letters in alphabetical order. .... *Mastery lesson 2.1*

⊙ An expression is a collection of terms. .... *Mastery lesson 2.1*

⊙ Like terms contain the same letter or power of a letter (or are just numbers). Simplifying an expression involving adding and subtracting is called '**collecting like terms**'. .... *Mastery lesson 2.1*

⊙ To multiply powers of the same letter, add the indices. To divide powers of the same letter, subtract the indices. .... *Mastery lesson 2.2*

⊙ When multiplying algebraic terms, multiply the numbers first, then the letters. $2a \times 3b = 6ab$ .... *Mastery lesson 2.2*

⊙ In algebra $x \div y$ is written as $\dfrac{x}{y}$ .... *Mastery lesson 2.2*

⊙ Dividing algebraic terms is similar to cancelling numbers in fractions. Cancel the numbers first, then the letters. When there is more than one letter, cancel identical letters in turn. .... *Mastery lesson 2.2*

⊙ Substitution means putting in numbers in place of letters. .... *Mastery lesson 2.3*

⊙ A **formula** is a general rule that shows the relationship between variables. A formula always has an equals sign. .... *Mastery lesson 2.4*

⊙ To expand brackets, multiply each term inside the bracket by the term outside the bracket. $3(x + 2)$ .... *Mastery lesson 2.5*

⊙ The highest common factor (HCF) is the largest factor of two or more terms. HCFs can contain both letters and numbers. .... *Mastery lesson 2.6*

⊙ Expanding removes brackets from an expression. Factorising inserts brackets into an expression. To factorise, write the common factor of its terms outside the brackets. This is called 'taking out the common factor'. .... *Mastery lesson 2.6*

⊙ The ≡ (identity) symbol is used to show that two expressions are *always* equal whatever their values. The ≠ (not equals) symbol is used to show that two expressions are *not* equal. .... *Mastery lesson 2.6*

Reflect

Choose A, B or C to complete each statement about algebra.

In this unit, I did…	**A** well	**B** OK	**C** not very well
I think algebra is…	**A** easy	**B** OK	**C** hard
When I think about doing algebra I feel…	**A** confident	**B** OK	**C** unsure

Did you answer mostly As and Bs? Are you surprised by how you feel about algebra? Why?

Did you answer mostly Cs? Find the three questions in this unit that you found the hardest. Ask someone to explain them to you. Then complete the statements above again.

2 Unit test

Log how you did on your Student Progression Chart.

1 Simplify
   a $7e - e + 3e$
   b $4m \times 5m$
   c $\dfrac{16b}{4}$ *(3 marks)*

2 Simplify
   a $10x^2 + x + 4x^2 + 3x$
   b $2a^2 \times 3a^3$ *(4 marks)*

3 Expand and simplify
   a $4(p + 8)$
   b $m(n + 1) - n(m - 1)$ *(4 marks)*

4 Factorise completely
   a $84a - 12$
   b $3a^2 - 9a$
   c $a^3 + a^2$ *(3 marks)*

5 Work out the value of each expression when $a = 4$ and $b = -2$
   a $a - b$
   b $\dfrac{3a}{b^2}$
   c $5a + (b + a)^2$ *(4 marks)*

6 Choose the correct sign, ≠ or ≡
   a $\dfrac{a}{a} \square 0$
   b $2x - 4 \square 2(x - 2)$ *(2 marks)*

7 Write an expression using n as the unknown starting number.
   a I think of a number and add 9
   b I think of a number, add 5 and double the result *(2 marks)*

8 Tickets for a funfair are £a per adult and £c per child. Write an expression for the cost in terms of a and c for 2 adults and 3 children. *(2 marks)*

9 State whether each of these is an expression, formula or identity.
   a $v = u + at$
   b $u^2 - 2as$
   c $4x(x - 2) = 4x^2 - 8x$ *(3 marks)*

10 Use the formula $E = \dfrac{kx^2}{2l}$ to work out the value of E when $k = 50$, $x = 4$ and $l = 10$ *(2 marks)*

11 **Reasoning** The fuse size I amps of an electrical appliance can be found by dividing the power rating, P watts by the voltage, V volts.
   a Write a formula for I in terms of P and V. *(2 marks)*
   b Elsa needs to replace the fuse in her hairdryer, which has a power rating of 1100 watts. Use your formula to work out the size of fuse, I, she needs to buy when $P = 1100$ watts and $V = 220$ volts. *(2 marks)*

 *Active*Learn Homework, practice and support: Foundation 2 Unit test

12 **Reasoning** A paint-ball party company charges a fixed amount per guest plus an amount per hour for the venue hire.

    a   Work out how much it would cost for a party of 10 guests at £25 each plus venue hire at £12 per hour for 2 hours. *(2 marks)*

    b   Write an expression for the cost for a party of g guests at £p pounds per person plus venue costs for h hours at £v per hour. *(2 marks)*

    c   Write a formula for the total cost, C, in terms of g, p, h and v. *(1 mark)*

    d   Use your formula to work out C when $g = 15$ guests, $p = £20$ per person, $v = £8$ per hour, and $h = 4$ hours. *(2 marks)*

Sample student answer

Look at the student answer.

a   What is good about this layout?

b   Why is it a good idea to use brackets for multiplying?

c   Why is it a good idea to write out the whole sum, not just the answer?

Exam-style question

Emma has x number of books.

Isabelle has 5 more books than Emma.

Laura has 3 fewer books than Emma.

Scarlett has twice as many books as Isabelle.

Write an expression in terms of x for the number of books the girls have altogether. Simplify your answer. **(3 marks)**

Student answer

Emma	x
Isabelle	$x + 5$
Laura	$x - 3$
Scarlett	2 times $x + 5$
	$= 2(x + 5)$
	$= 2x + 10$
Altogether	$x + x + 5 + x - 3 + 2x + 10$
	$= 5x + 12$

3 GRAPHS, TABLES AND CHARTS

Use the graph to decide which of these statements are true, false or unable to tell.

A Healthcare spending per person increased more between 2002 and 2006 than between 2008 and 2012.

B Healthcare spending per person in 2012 was more than double the amount for 2002.

C Healthcare spending in 2014 will be more than in 2012.

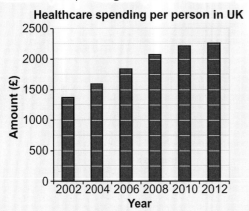

Healthcare spending per person in UK

Source: ONS

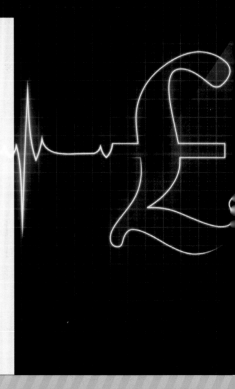

3 Prior knowledge check

Numerical fluency

1 Work out

   a 180 − 72   b 360 −120 − 60

   c 360 ÷ 30   d 360 ÷ 36

   e $\frac{1}{2}$ of 360   f 10% of 180

Fluency with data

2 Jason conducted a survey to find out which flavour of sweet people like the best.

Here are his results.
lemon, lime, lime, orange, blackberry, orange, blackberry, lime, lime, lemon, blackberry, lime, orange, lime, lime, lime, blackberry, lemon, blackberry, lime

   a Construct and fill in a frequency table.

   b How many people were surveyed?

   c How many people like orange sweets the best?

3 Natasha asked some of her class how many hours they spend online every day. Her results are shown in the table.

Hours online	2	3	4	5	6
Frequency	1	6	8	4	2

What is the modal number of hours online each day?

4 The table shows some information about five students.

Name	Gender	Age	Favourite subject
Ella	Female	16	Science
Liam	Male	15	French
Neil	Male	12	History
Penny	Female	15	Maths
Rashida	Female	14	English

   a What is Liam's favourite subject?

   b Write the name of the oldest student.

   c Write the name of the female student who is 15 years old.

5 Dima collected data on pets from students in her class.

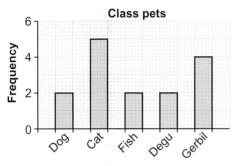

Class pets

a How many pets do Dima's classmates own in total?

b Which pet is the mode?

6 The composite bar chart shows how Jess and Phoebe spent their allowance last month.

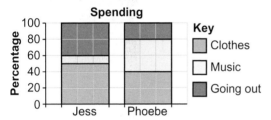

Spending

Key
- Clothes
- Music
- Going out

a What did Jess spend most on?

b Who spent the least percentage on music?

c What percentage of her allowance did Phoebe spend on music?

7 The graph shows the number of bird nests on an island.

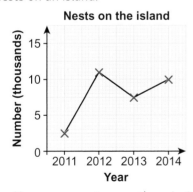

Nests on the island

a How many nests were there in 2013?

b How many nests were there in 2014?

c Between which two years did nest numbers decrease?

d Between which two years was the biggest increase?

8 A class of 28 students were asked if they liked school dinners. The results are displayed in the pictogram.

a How many students said no?

b Draw the symbols for the 10 students who 'don't know'.

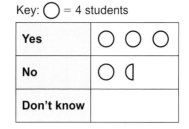

Key: ◯ = 4 students

Yes	◯ ◯ ◯
No	◯ ◖
Don't know	

9 The chart shows the times taken by some students to complete a task.

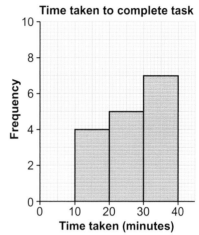

Time taken to complete task

a How many students took between 20 and 30 minutes?

b How many students took 30 minutes or less?

c How many students completed the task?

10 Charlie bought a new car in 2011 for £10 500. The value of the car changed each year.

Year	2011	2012	2013	2014
Value (£)	10 500	9600	8200	7000

Draw a bar chart to display this data.

11 Nav recorded the number of letters he got each day for a period of time.

Number of letters	1	2	3	4
Number of days	6	10	7	3

Draw a vertical bar chart to display this data.

Geometrical fluency

12 Draw a circle of radius 3 cm.

13 Draw an arc of radius 4 cm with an angle of 120°.

14 Match the labels to the diagrams. Each diagram can be matched to *two* labels.

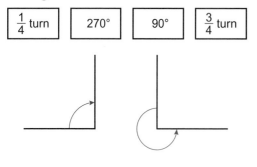

$\frac{1}{4}$ turn   270°   90° $\frac{3}{4}$ turn

✱ Challenge

15 Draw a circle of radius 5 cm.
Draw these angles at the centre of the circle:
60°   140°   30°.
Measure the remaining angle.

Hint

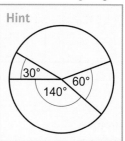

3.1 Frequency tables

Objectives

- Designing tables and data collection sheets.
- Reading data from tables.

Why learn this?

Frequency tables are a useful and clear way of displaying data.

Fluency

- Work out 45 + 62 + 39
- How many tallies?   a 卌||| b 卌卌|

Warm up

1 **Real** Eloise conducted a survey on what time Year 11 students get up on a Sunday morning.
   a How many Year 11 students get up at 8 am or later?
   b How many students did Eloise survey?

Time	Number of students
Before 7 am	6
7 to 7.59 am	12
8 to 8.59 am	35
9 to 9.59 am	25
10 am or later	62

2 A sports coach recorded the length of time (in seconds) John took to complete 15 races.
13.4, 13.0, 13.4, 13.1, 13.2, 13.2, 12.9, 13.3, 12.9, 13.1, 13.3, 13.4, 13.0, 12.9, 12.9
Draw a tally chart for this data.

> Questions in this unit are targeted at the steps indicated.

3 Chris works in a café. During lunch he records the number of customers sitting at each table.
   a Work out the total number of tables in the café.
   b Work out the total number of customers in the café.

Number of customers at table	Number of tables
0	4
1	5
2	10
3	7

Key point 1

A **grouped frequency table** contains sorted data in groups called classes.

4   A shop records the shoe sizes of 20 customers.
    3, 5, 11, 8, 9, 5, 6, 5, 2, 6, 6, 6, 8, 4, 4, 6, 7, 5, 9, 3
    Copy and complete the grouped frequency table.

Shoe size	Tally	Frequency
2–4		
5–7		

Key point 2

An **inequality** is a mathematical sentence. $30 < y \leqslant 31$ means that a number (y) is greater than 30 but less than or equal to 31.

5   For each inequality, list the numbers in the cloud that belong to it.
    a $25 < y \leqslant 26$ b $26 \leqslant y \leqslant 27$

25   25.5
26   27
26.5

Key point 3

Discrete data can only have particular values. For example, shoe sizes are usually whole numbers. For discrete data you can write groups like 1–5, 6–10.
Continuous data is measured and can have any values, for example length and time. Write inequalities for the groups with no gaps between them.

6   **Real** A college records the ages of 22 people taking a night class.
    22, 22, 28, 23, 26, 26, 18, 27, 19, 30, 26, 29, 28, 17, 25, 32, 34, 24, 17, 23, 20, 21
    a   Copy and complete the grouped frequency table for this data.
    b   What is the least common age group?
    c   How many people are aged 30 years or less?

Age (years)	Tally	Frequency
$15 < y \leqslant 20$		
$20 < y \leqslant 25$		
$25 < y \leqslant 30$		
$30 < y \leqslant 35$		

Key point 4

A suitable number of classes for a grouped frequency table is four to six. The classes should be of equal width.

Example 1

Edward recorded the time, in seconds, it took some Year 11 students to complete a task.
13, 14, 18, 21, 13, 18, 19, 13, 21, 20, 15, 15, 18, 13, 14
Design a suitable grouped frequency table for his data.

The smallest value is 13 seconds.   ——— Find the smallest and largest values.
The largest value is 21 seconds.

There are 8 seconds between the smallest and largest values.

A suitable number of classes is four. ———   Decide on a sensible number of classes. Four works here because the difference between the smallest and largest values is not very big.

Time (seconds)	Tally	Frequency
$13 \leqslant t < 15$		
$15 \leqslant t < 17$		
$17 \leqslant t < 19$		
$19 \leqslant t < 22$		

Use inequalities because the data is continuous.

7   Aric measured the lengths, in cm, of 20 books.
    13.2, 16, 18.5, 12, 16.2, 15, 19.4, 20, 15, 19.8, 14, 16.3, 14, 17, 12.5, 18, 16, 15.6, 19, 16.1
    a   Is this data discrete or continuous?
    b   Design and complete a suitable grouped frequency table.

8 Ffion asked some students which country they would like to visit and recorded her results.

France, Spain, Spain, Greece, Peru, Spain, Ghana, Greece, France, France, Greece, Spain, Greece, India, Spain, France

Design and complete a data collection sheet for Ffion's data.

Q8 communication hint A **data collection sheet** is a table to record data.

Q8 hint

Country	Tally	Frequency
France		

Peru, Ghana and India are only mentioned once. To stop the table getting too big, put these countries in an 'Other' row.

9 Ben records the number of leaves on some tomato seedlings.

2, 3, 3, 5, 3, 4, 3, 5, 3, 2, 4, 3, 4, 5, 5

Design and complete a data collection sheet for Ben's data.

10 An elite cyclist recorded the number of hours she trained each week for 15 weeks.

15, 12, 11.5, 26, 23, 21, 23, 27, 15.5, 14, 21, 24, 19, 22, 28

Design and complete a data collection sheet for this data.

Q10 hint Group the data.

3.2 Two-way tables

Objectives

- Use data from tables.
- Design and use two-way tables.

Why learn this?

You can use a two-way table to show how data falls into two different categories, for example gender and favourite subject at school.

Fluency

- How many minutes in an hour? How many hours in a day?
- Change 4.25 pm to 24-hour time. Change 19.50 to 12-hour time.

1 Work out the time difference between
  a 11.15 am and 1.25 pm
  b 9.40 am and 11.45 am
  c 09.15 and 20.30

2 a Alice works from 10.30 am to 2.45 pm.
    For how long is she at work?
  b A plane takes $2\frac{1}{2}$ hours to get to Spain. It arrives at 15.45 UK time. What time did it leave?

3 **Real / Communication** Here is part of a train timetable.

Welwyn	0953	0959	1022	1029	1052
Hatfield	0957	1004	1026	1034	1056
Welham	—	1007	—	1037	—
Brookmans	—	1010	—	1040	—
Kings Cross	1020	1041	1049	1111	1119

Q3 hint All five trains start at Welwyn. Some trains do not stop at Welham and Brookmans.

  a How long does the first train from Hatfield take to travel to Kings Cross?
  b Beth wants to travel from Welham to Kings Cross. She wants to arrive after 11.00. What train should she catch?
  c Luca says the trains from Welwyn take the same time to travel to Brookmans. Is he correct? Explain how you know.

Key point 5

A **distance chart** is a convenient way of showing the distances between several places.

4 The distance chart shows distances in miles between four cities.

Leeds

70	Lincoln
42	87	Manchester
35	45	42	Sheffield

> **Q4a hint** Move your finger down the 'Leeds' column until you reach the 'Manchester' row. The number in that square is the number of miles between the two cities.

How far is it from

a Leeds to Manchester   b Manchester to Sheffield   c Sheffield to Lincoln?

5 **Exam-style question**

The chart shows the shortest distances, in kilometres, between cities.

London
196	Nottingham
300	101	Manchester
325	158	56	Liverpool
639	446	346	348	Glasgow

a Write down the distance between Nottingham and Liverpool. **(1 mark)**

Daniel drives from London to Manchester by the shortest route. He drives 137 km and stops for a rest.

b Work out how many more kilometres he must drive. **(2 marks)**

c Write down the names of the two cities which are the least distance apart. **(1 mark)**

Nov 2007 specimen paper, Q7, 5384F/11F

> **Q5a strategy hint** See which number in the table is below Nottingham and also to the left of Liverpool.

Key point 6

A **two-way table** divides data into groups in rows across the table and in columns down the table. You can calculate the totals across and down.

6 The table shows the numbers of medals won by a team in the Junior and Senior Games.
How many more medals, in total, did the team win in the Senior Games than in the Junior Games?

	Gold	Silver	Bronze
Junior Games	29	17	19
Senior Games	34	43	43

7 Use the two-way table to work out how many females swim.

	Male	Female	Total
Swim	9		34
Run	24	12	36

8 **Reasoning** Martin wants to know if Rita, Sveta and Ali can teach squash, badminton, football and tennis.
Design a form to collect the information.

> **Q8 hint** Use a two-way table. Put sports down the left-hand side and names across the top.
>
> | | Rita | | |
> |--|------|--|--|
> | **Squash** | | | |

9 **Reasoning** A teacher collects data on how late students are to school. She wants a table to record this information for each year group from Year 7 to Year 11.
Design a two-way table to record the data.

10 A factory makes three sizes of bookcase – small, medium and large.
Each bookcase can be made from pine, oak or yew.
The two-way table shows some information about the number of bookcases the factory makes in a week.

	Small	Medium	Large	Total
Pine	7			23
Oak		16		34
Yew	3	8	2	13
Total	20		14	

> **Q10a hint** Use the 'small' column. The total is 20.

> **Q10c hint** Look for rows or columns that only have one missing value.

a How many small oak bookcases does the factory make?
b How many large oak bookcases does the factory make?
c Copy and complete the two-way table.

Example 2

50 people chose one activity from swimming, squash or going to the gym.
21 of the people were female.
6 of the 8 people who played squash were male.
18 of the people went to the gym.
9 males went swimming.
a Put the data into a two-way table.
b Use the table to find the number of females who went to the gym.

> Put the activity down the left-hand side and male/female across the top. Include a 'total' column and row.

a

	Male	Female	Total
Swimming	9	$24 - 9 = 15$	24
Squash	6	$8 - 6 = 2$	8
Gym	$29 - 9 - 6 = 14$	$21 - 15 - 2 = 4$	18
Total	$50 - 21 = 29$	21	50

> Complete your table by working out the missing values.
> Start with a row or column that only has one missing value:
> Total swimming = 50 – 18 – 8
>   = 24

> Put the data you know into the table.

b 4 females went to the gym ——— Read the value from your completed table.

11 **Problem-solving** There are 40 people at a meeting.
Each person travelled to the meeting by car or by train.
   13 of the people are male.
   10 females travelled by train.
   8 males travelled by car.
Work out the total number of people who travelled by car.

> **Q11 hint** Draw a two-way table.

12 **Problem-solving** Nadine asked 50 people which subject they like best from maths, English and science.
Here is some information about her results.
   19 out of the 25 males said they like science best.
   5 females said they like English best.
   Of the 7 people who said they like maths best, 4 were female.
Work out the number of people who like science best.

3.3 **Representing data**

Objectives

- Draw and interpret comparative and composite bar charts.
- Interpret and compare data shown in bar charts, line graphs and histograms.

Why learn this?

Displaying data in a graph makes it easier to interpret. For example, if a bar chart is used to display the hours of sunshine per day in a number of holiday destinations, you can see at a glance which destination is the sunniest.

Fluency

What is missing from this graph?

1   The bar charts show the temperature at midday in three different cities.

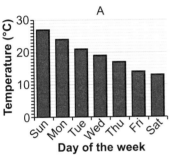
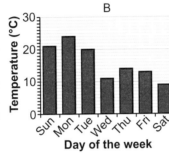
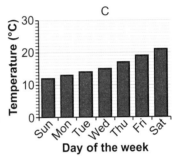

a   Which bar chart shows midday temperature
  i   decreasing each day
  ii increasing each day?
b   What was the highest midday temperature for city B?

Example 3

Kitty and George sell cars.

The table shows the number of cars sold by Kitty and George in the first four months of 2014.

	January	February	March	April
Kitty	2	5	13	10
George	4	7	9	10

Show this information in a suitable chart.

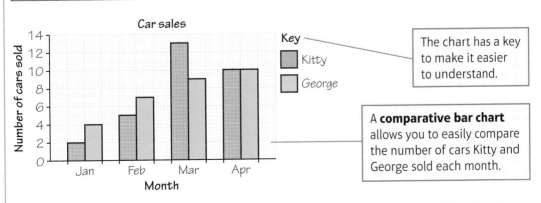

The chart has a key to make it easier to understand.

A **comparative bar chart** allows you to easily compare the number of cars Kitty and George sold each month.

---

2 **Real / Communication** Kirsten records the number of mugs sold in a table. She divides the year into 'quarters' (periods of three months).

	Jan–Mar	Apr–Jun	Jul–Sep	Oct–Dec
2013	470	420	510	630
2014	490	540	770	820

a Draw a comparative bar chart for the data.

b Which quarter of which year did she sell the most?

c Compare the sales of mugs in 2014 and 2013.

> **Q2a communication hint**
> Comparative bar charts are also known as dual bar charts.

> ### Key point 7
> A multiple or **composite bar chart** compares features within a single bar.

3 **Exam-style question**

The table shows the number of gold, silver and bronze medals won by a team in 2008 and 2012.

	Gold	Silver	Bronze	Total
2008	15	17	15	47
2012	29	17	19	65

a Copy and complete the composite bar chart. **(4 marks)**

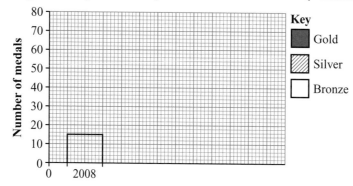

> **Q3a communication hint** A composite bar chart is also known as a compound bar chart.

b Compare the performance of the team in 2008 and 2012.

(2 marks)

> ### Key point 8
> A **histogram** is a type of frequency diagram used for grouped continuous data. There are no gaps between the bars.

4 Jacob asked a group of people how many minutes it took them to get to work. His results are shown in the table.

Time taken, t (minutes)	Frequency
$0 < t \leqslant 10$	4
$10 < t \leqslant 20$	7
$20 < t \leqslant 30$	9
$30 < t \leqslant 40$	5

> **Q4 hint**
> There should not be gaps between your bars.

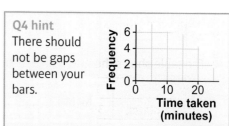

Draw a histogram to display this data.

5 **Reasoning / Real** Here is a nursery's income for the first three months of 2014.

	January	February	March
Morning	£8500	£10500	£9000
Afternoon	£12000	£16750	£14000

Draw a chart for this data.

Discussion Is there another type of graph that could be used to display the data?

> **Q5 hint** Your graph must enable you to compare the data for morning and afternoon. Look back at the graphs you have drawn so far in this lesson. Which one could you use to display this type of data?

> **Key point 9**
>
> A **line graph** is useful for identifying trends in data. The **trend** is the general direction of change.

6 **Real** An online book shop recorded the numbers of books and ebooks sold over 5 years.

Year	1	2	3	4	5
Books (1000s)	6	4.75	5.25	4.5	4.6
Ebooks (1000s)	3.5	3.75	3.5	4	4.8

a Draw a set of axes. Put Year on the horizontal axis and Number sold on the vertical axis.

> **Q6a hint** Label your axes properly – the values are in thousands.

b Draw a line graph for the books sold. Use a different colour to draw a line graph for the ebooks sold.

c Describe the trend in the number of books sold. Describe the trend in the number of ebooks sold.

> **Q7 hint** Draw two lines, one for morning and one for afternoon.

7 Draw a line graph of the data in **Q5**.

8 **Reflect** Look back at all of the graphs you have seen in this lesson. Which type of graph would you use to represent each of these sets of data?

a Percentage of males and females gaining GCSE grades in 2012.

Grade	No grades	D–G only	1–4 at A*–C	5+ at A*–C
Males (%)	6	23	25	46
Females (%)	4	15	25	56

b Salaries in a company.

Salary (£000)	1–10	11–31	32–52	53+
Frequency	14	23	8	2

3.4 Time series

Objectives

- Plot and interpret time series graphs.
- Use trends to predict what might happen in the future.

Why learn this?

A time series graph is very useful for showing how something changes over time, for example a hospital patient's temperature.

Fluency

Write   1.6 million as a number   160000 in millions.

1 The table shows the amount of rainfall (in mm) each day for a week.

Day	1	2	3	4	5	6	7
Rainfall (mm)	35	42	30	25	28	10	21

Plot a line graph for the data.

ActiveLearn Homework, practice and support: Foundation 3.4

Warm up

2 The graph shows the height of a balloon at different times during a flight.

  a What is the height of the balloon at 35 seconds?

  b Estimate how long it took the balloon to reach a height of 40 m.

  Discussion Did the balloon gain height steadily? How do you know?

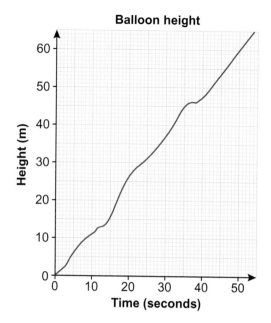

Balloon height

3 **Communication** Mandy's sales figures for the last three months are shown in the table.

Month	Oct	Nov	Dec
Sales figures (£)	12 340	14 500	20 090

Mandy says, 'My sales figures are increasing at a steady rate.' Is Mandy correct? Explain how you know.

> **Q3 communication hint** Steady means going up or down by the same amount each time.

> **Q3 hint** Draw a graph of the data.

Key point 10

A **time series** graph is a line graph with time plotted on the horizontal axis.

4 The table shows the temperature every two hours during one morning in January.

Time	4 am	6 am	8 am	10 am
Temperature (°C)	−2	2	6	9

Draw a time series graph to represent the data.

> **Q4 hint** Plot time along the horizontal axis and temperature on the vertical axis. The vertical axis needs to go down to −2. Join the points with straight lines.

5 **Real** Jess is an electrician. She records the money she receives from her customers.

  March: £280 from Mrs Jenkins, £1500 from Greens Garage, £4020 from Brants

  April: £1340 from Fox Services, £1260 from Mr Cox

  May: £4500 from ASA International

  a Construct a time series table for Jess.

  b Represent the data as a time series graph.

> **Q5a hint** Find the total amount of money Jess receives each month.

6 **Communication** The table shows the number of overseas visitors to the UK in 2007 and 2008.

	Q1	Q2	Q3	Q4
2007	860 000	1 300 000	1 580 000	1 100 000
2008	790 000	1 210 000	1 420 000	770 000

> **Q6 communication hint** Q1 means the first quarter of the year (January to March).

  a Draw a time series graph for each year's data on the same axes.

  b Describe how the number of visitors changes during 2007.

7   **Reasoning / Communication** The table shows the number of students taking A Level Physics and French in each of the years 2007 to 2012.

Year	2007	2008	2009	2010	2011	2012
No. taking Physics	27 000	28 000	30 000	33 000	35 000	37 000
No. taking French	15 000	13 000	13 000	11 000	12 000	11 000

a Represent the data as two time series on a single graph.
b Describe the difference in the number of students taking Physics and French between 2007 and 2012.
c Sally predicts the number of students taking Physics A Level is likely to increase in 2013 and 2014. Is she correct? Explain your answer.
d What is likely to happen to the number of students taking French A Level in 2013 and 2014?

3.5 Stem and leaf diagrams

Objectives

- Construct and interpret stem and leaf and back-to-back stem and leaf diagrams.

Why learn this?

A stem and leaf diagram gives you a detailed overview of a set of data at a glance.

Fluency

Put these numbers in size order, starting with the smallest.
7, 15, 9, 24, 17, 31, 6, 25

1   Put these numbers in size order, starting with the smallest.
a 132, 123, 125, 135, 143, 146, 125, 123
b 3.2, 6.7, 3.9, 4.1, 7.5, 1.4, 8.4, 3.9

Key point 11

A **stem and leaf diagram** shows numerical data split into a 'stem' and 'leaves'. The leaf is usually the last digit and the stem is the other digits.
In a stem and leaf diagram the numbers are placed in order.

2   **Real** The stem and leaf diagram shows the length, in metres, of some bridges.

```
1 | 4  5  5  6  7  7      Key
2 | 0  4  5  5  5  9      1 | 4 represents 14 metres
3 | 1  1  4
```

a How many bridges were under 20 m long?
b What is the length of the longest bridge?

3   **STEM** Jason recorded the height (in mm) of some plants. The information is shown in the stem and leaf diagram.

a How many plants are there?
b How many plants are more than 3 cm tall?
c What is the difference between the tallest and the shortest height?

```
1 | 1  2  3  3
2 | 3  3  5  9  9      Key
3 | 0  2  2  6  6  7   4 | 8 means 48 mm
4 | 1  1  4  8
```

Example 4

Here are the heights of some students (in cm).

169, 163, 153, 173, 166, 178, 177

Construct a stem and leaf diagram for this data.

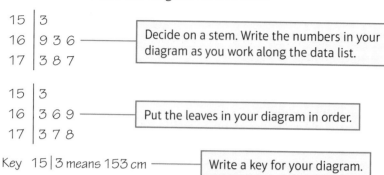

```
15 | 3
16 | 9 3 6 ————————— Decide on a stem. Write the numbers in your
17 | 3 8 7                diagram as you work along the data list.

15 | 3
16 | 3 6 9 ————————— Put the leaves in your diagram in order.
17 | 3 7 8
```

Key   15|3 means 153 cm ————— Write a key for your diagram.

4 Here are the times (in minutes) it took 21 teachers at a school to get to work.

13, 18, 20, 35, 45, 34, 44, 23, 33, 12, 46, 21, 22, 17, 22, 31, 23, 8, 15, 22, 10

Construct a stem and leaf diagram to show this information.

5 **Exam-style question**

Here are the ages, in years, of 15 students.

19 18 20 25 37 33 21 17 29 20 42 18 23 37 22

Show this information in an ordered stem and leaf diagram.   **(3 marks)**

March 2013, Q1, 1MA0/2H

Exam hint
Remember to include a key.

6 Here are the times some students took to complete a task (in seconds).

220, 238, 220, 230, 235, 238, 205, 198, 238, 192

Draw a stem and leaf diagram to display the data.

Q6 hint Use the 'hundreds' and 'tens' digits as the stem.

7 Daley measured the distance (in metres) that some Year 9 students jumped in a long jump competition.

4.6, 4.8, 5.1, 4.7, 3.9, 3.9, 3.4, 3.9, 4.7, 4.9

Draw a stem and leaf diagram to display this data.

Q7 hint For decimals, use the whole number part as the stem.

Key point 12

A back-to-back stem and leaf diagram compares two sets of data.

8 **STEM / Communication**

Some students watched a film. At the end of the film, the students' heart rates were recorded in beats per minute (bpm).

a What is the lowest female heart rate?

b What is the highest male heart rate?

c What is the difference between the lowest male heart rate and the highest female heart rate?

d Did more males than females have a heart rate of more than 100 bpm? Explain your answer.

Females					Males				
		8	5	7	6	7	9		
7	5	4	3	0	8	8	3	5	
	9	8	6	1	9	9	2	3	5 7 8
				10	1	3	7		

Key

For females   For males

5|7 means 75 bpm   7|6 means 76 bpm

Discussion What does the shape of a back-to-back stem and leaf diagram show you?

9 **Real / Communication** A hotel chain records the age (in years) of the guests at two of its hotels.

Abbey Hotel: 2, 11, 15, 28, 32, 33, 19, 40, 45, 58, 39, 33, 35, 17, 21, 36, 23, 29, 36, 47, 47, 49, 39, 37, 39, 39, 48

Balmoral Hotel: 40, 45, 34, 37, 62, 64, 71, 63, 65, 50, 50, 50, 53, 56, 46, 26, 49, 40, 34, 51, 45, 63, 50, 75, 57, 67, 70, 56

a Draw a back-to-back stem and leaf diagram to display the data.

b Bronte says, 'The residents of the Abbey Hotel are younger than the residents of the Balmoral Hotel.' Is Bronte correct? Explain your answer.

10 Phoebe measures the height of some Year 8 students to the nearest centimetre.

Boys	149	153	155	156	163	165	165	165	170	172
Girls	146	148	151	151	152	155	156	157	164	169

Draw a back-to-back stem and leaf diagram to display the data.

> **Q10 hint** Write the 'hundreds' and 'tens' digits down the central stem of your diagram.
>
> | Boys | | Girls | |
> |---:|:---:|:---|:---|
> | 9 | 14 | 6 | 8 |
> | 5   3 | 15 | 1 | 1 |

11 **STEM / Communication** Alex records the height (in cm) of tomato seedlings grown in the dark and in the light.

Grown in dark (cm)	1.6	1.9	1.5	1.8	2.2	2.0	1.7	1.9
Grown in light (cm)	2.8	2.7	3.5	4.4	3.8	4.1	4.5	4.3

> **Q11a hint** Use the whole number part as the stem.

a Draw a back-to-back stem and leaf diagram to display Alex's data.

b Did the tomato seedlings grow better in the dark or the light? Explain your answer.

3.6 Pie charts

Objectives

- Draw and interpret pie charts.

Why learn this?

A theme park could use a pie chart to show which of its rides was most popular.

Fluency

- What fraction or percentage of each of these circles is shaded?

- How many degrees are there at the centre of a circle?

1 a Draw a circle of radius 4 cm.

  b Draw an angle of 60° at the centre of the circle.

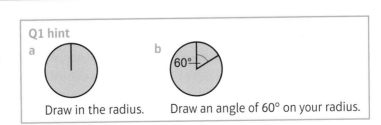

Q1 hint

a   Draw in the radius.

b   60°   Draw an angle of 60° on your radius.

> **Key point 13**
>
> A **pie chart** is a circle divided into sectors. Each sector represents a set of data.

2 **Real / Communication** A council surveyed householders for their opinion on how well waste is collected.

The pie charts show the results.

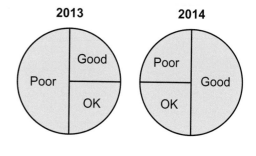

a What fraction of householders thought waste collection was good in 2014?

b Compare the percentage of householders that thought the waste collection was poor in 2013 and 2014.

c Do you think household waste collection is improving? Explain your answer.

3 **Communication** A group of Year 11 students were asked if they wanted more after school sports clubs.

The results are shown in the pie chart.

a What percentage of Year 11 students said 'Yes'?

b Measure the 'No' section with a protractor.

c Bethan says, 'More than twice the number of Year 11s said 'Don't know' rather than 'Yes'.' Explain why Bethan is correct.

4 **Reasoning / Communication** The pie charts show the match results of a school's two netball teams.

Team A played 20 games and Team B played 28 games.

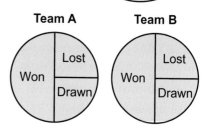

a Calculate the number of games that Team A lost.

b The head teacher says, 'Well done both teams, you won the same number of games.' Is he correct? Explain your answer.

Example 5

The table shows the match results of a football team. Draw a pie chart to represent the data.

Result	Won	Drawn	Lost
Frequency	28	12	20

Total number of games = 28 + 12 + 20 = 60

$\div 60 \left(\begin{array}{c} 60 \text{ games} : 360° \\ 1 \text{ game} : 6° \end{array} \right) \div 60$

> The total number of games is the total frequency.

1 game = 360 ÷ 60 = 6°

> Work out the angle for one game.

Won: 28 × 6° = 168°

Drawn: 12 × 6° = 72°

> Work out the angle for each result.

Lost: 20 × 6° = 120°

Check: 168 + 72 + 120 = 360

> Check that your angles total 360°.

Team results

> Draw the pie chart. Give it a title and label each section, or make a key.

5   **Real** A café owner records the drinks sold in his café
    on one day. The information is shown in the table.

    a   Work out the total frequency.

    b   Work out the angle for one drink on a pie chart.

    c   Work out the angles for each type of drink on a pie chart.

    d   Draw a pie chart to show the information.

Drink	Frequency
Hot chocolate	20
Milkshake	15
Coffee	25
Tea	30

6   **Exam-style question**

    40 students went on holiday abroad.

    The table shows the number of students who visited
    each country.

    Draw an accurate pie chart to show the information
    shown in the table.   **(4 marks)**

    Exam hint
    Remember to label the bar chart with the countries.

Country	Number of students
France	16
Spain	12
Germany	5
Italy	7

7   30 people used a sports centre one evening. They each took part in one of four activities.

    gym, swimming, squash, swimming, aerobics, swimming,
    aerobics, aerobics, aerobics, gym, aerobics, gym, gym, gym,
    squash, squash, gym, squash, gym, gym, gym, aerobics,
    aerobics, squash, gym, gym, aerobics, squash, gym, aerobics

    Draw a pie chart to show this information.

    Q7 hint Put the data into a frequency table first.

8   **Reasoning** The pie chart shows the percentage of cake
    sales in a cake shop over one week.
    The cake shop sold 30 chocolate cakes.

    a   How many cakes did it sell in total?

    b   How many
        i   lemon   ii   banana cakes did it sell?

Cake sales

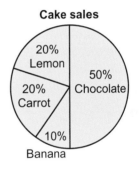

9   Margaret asked the students in a Year 8 English class their
    favourite type of book.
    She drew a pie chart of her results.

    a   Margaret knows that eight students said 'Romance'.
        How many people are in the English class?

    b   What is the modal type of book?

    Q9a hint Measure the angle
    in the pie chart for 'Romance'.

    Q9b hint The mode will
    be the most popular one.

Favourite book type

10 **Reasoning** Kato sells umbrellas. Here are his sales figures for last week.

Umbrella type	Spotty	Single colour	Multicolour
Number sold	24	16	32

    Q10 hint Think of
    the differences
    between a pie chart
    and a bar chart.

    a   Draw a bar chart for the data.   b   Now draw a pie chart.

    Discussion Which chart best shows

    •   the different numbers of each type sold?

    •   the fractions of umbrellas sold that were spotty?

    •   the mode?

3.7 Scatter graphs

Objectives

- Plot and interpret scatter graphs.
- Determine whether or not there is a relationship between sets of data.

Why learn this?

A scatter graph allows you to see the relationship between two sets of data, for example how house prices change as you get closer to a city centre.

Fluency

Is it generally true that that the older the car, the less it is worth?

1 The table shows the number of bedrooms and number of bathrooms in four houses.

Number of bedrooms	2	4	5	5
Number of bathrooms	1	2	3	5

a Draw x- and y-axes on graph paper from 0 to 6.
b Plot the data, using crosses.

> **Q1a hint** Plot 'Number of bedrooms' on the x-axis and 'Number of bathrooms' on the y-axis.

Key point 14

A **scatter graph** shows the relationship between two sets of data. Plot the points with crosses. Do not join them up.

2 Max recorded the sale price of nine cars.

Age of car (years)	3	7	8	2	1	3	5	4	6
Sale price (£thousands)	7.5	1.5	1.2	8.0	9.2	7.4	3.6	6.3	2.1

a Draw a scatter graph to display the data.
b Copy and complete the sentence.
As the age of the car increases, the value of the car ....

> **Q2a hint** Put age of car on the horizontal axis and sale price on the vertical axis.

3 Beatrice recorded the age of nine people and the time it took them to run 200 m.

Age (years)	35	17	28	23	19	37	51	43	60
Time (seconds)	32	26	29	27	25	33	36	34	38

a Draw a scatter graph to display the data.
b Copy and complete the sentence.
As a person's age increases, their time to run 200 m ....

Key point 15

The relationship between the sets of data is called **correlation**. The sets of data are called **variables**.

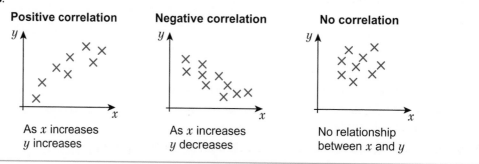

Positive correlation	Negative correlation	No correlation
As x increases y increases	As x increases y decreases	No relationship between x and y

4   Look back at your graphs for **Q1–3**. Describe the type of correlation in each one.

5   Match each real-life example to the correct type of correlation.
    a   The relationship between height and arm span.   **A**   No correlation
    b   The relationship between car engine size and fuel economy.   **B**   Positive correlation
    c   The relationship between eye colour and intelligence.   **C**   Negative correlation

6   ⬭ Exam-style question ⬭

The tread on a car tyre and the distance travelled by that tyre were recorded for a sample of cars. The data is shown in the scatter graph.

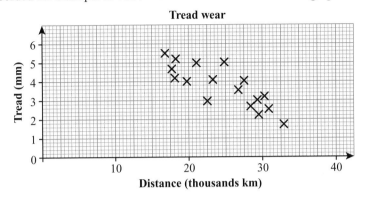

Tread wear

    a   Describe the relationship between the tread on a car tyre and the
        distance travelled by that tyre.   **(1 mark)**
    b   A tyre with less than 1.6 mm of tread is illegal. If the law changed to
        less than 2.5 mm, how many tyres in *this data set* would be illegal?
                                                                          (1 mark)

Exam hint
This means write
the type of
correlation so,

_____ correlation.

7   The table lists the masses of 12 books and the number of pages in each one.

Number of pages	80	155	100	125	145	90	140	160	135	100	115	165
Mass (g)	160	330	200	260	320	180	290	330	260	180	230	350

    a   Copy and complete the scatter graph
        to show the information in the table.
    b   Copy and complete the sentences.
        As the number of pages in a book
        increases, the mass ....
        This is .... correlation.

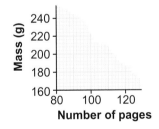

Q7a hint You
do not need
to start the
vertical or
horizontal
axis at 0.

Key point 16

An **outlier** is a value that does not fit the pattern of the data.

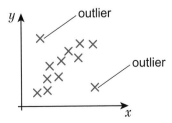

An outlier can be ignored if it is due to a measuring or recording error.

8 The heights and lengths of seven sheep are given in the table.

Height (cm)	65	80	52	78	65	62	84
Length (cm)	100	110	86	106	80	95	115

Q8a hint Plot height on the horizontal axis and length on the vertical axis.

a Plot the information on a scatter graph.

b One of the points is an outlier. Use your scatter graph to identify this outlier.

c Describe the correlation between heights and lengths of the sheep.

9 The table shows the number of divorces and the amount of margarine eaten.

Number of divorces in a population of 1000	5.0	4.7	4.6	4.4	4.3	4.1	4.2	4.2	4.2	4.1
Amount of margarine eaten (lbs)	8.2	7.0	6.5	5.3	5.2	4.0	4.6	4.5	4.2	3.7

Plot the information on a scatter graph.

Discussion Does 'Number of divorces' correlate with 'Amount of margarine eaten'? Does eating more margarine cause the number of divorces to increase?

Key point 17

Correlation shows that there may be a link between two events. Correlation does *not* show that one event caused the other.

10 Copy and complete the table.
Tick the correct type of correlation for each set of data and whether you think one causes the other.

	Positive correlation	Negative correlation	No correlation	Possible correlation
Arm length and leg length				
Exercise and weight				
Size of garden and running speed				
Hours of TV watched and shoe size				

3.8 Line of best fit

Objectives

- Draw a line of best fit on a scatter graph.
- Use the line of best fit to predict values.

Why learn this?

A café could use a line of best fit to predict the number of sales of ice cream sundaes as the temperature increases.

Fluency

What is a real-life example of negative correlation?

1 a What is the value of x when y is 4?
  b What is the value of y when x is 3?

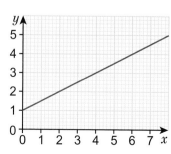

2 Copy and complete the statements.
  a The more I work, the more I get paid. This is .... correlation.
  b The more I watch TV, the less I revise for my exam. This is .... correlation.

ActiveLearn Homework, practice and support: Foundation 3.8

Key point 18

A **line of best fit** is a straight line drawn through the middle of the points on a scatter graph. It should pass as near to as many points as possible and represent the trend of the points.

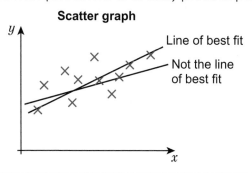

Scatter graph

3 **STEM** A scientist recorded the density of air and the speed of sound in each of six experiments. The table shows her results.

Speed of sound (m/s)	355	340	337	320	317	340
Density of air (kg/m³)	1.1	1.2	1.3	1.2	1.4	1.3

> **Q3b hint** A line of best fit is often drawn by eye. There should be roughly the same number of points either side of the line.

a   Draw a scatter graph for the data.
b   Draw a line of best fit on your graph.

4 **Communication** A school records the marks achieved by a group of students in a chemistry exam and a music exam.

Chemistry mark	22	56	34	28	40	26	72	68
Music mark	46	62	82	30	70	62	38	88

a   Draw a scatter graph for the data.
b   Describe the correlation.
c   Can you draw a line of best fit? Explain your answer.

Example 6

Jayden records the time he spends watching TV (x) and the time he spends revising (y) for one week.

Time spent watching TV (hours)	0.75	2.00	2.75	3.75	4.25	5.00	5.75
Time spent revising (hours)	5.25	4.75	3.75	3.25	3.00	2.00	1.75

a   Show Jayden's results on a scatter graph. Draw a line of best fit.
b   Jayden spends four hours watching TV. Estimate how much time he will spend revising.

a

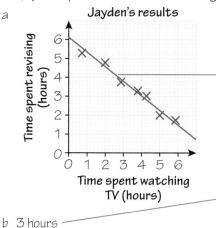

Jayden's results

Position a transparent ruler over the scatter graph so it follows the overall trend. Move it slightly so you have roughly the same number of points above and below the line.

Start at 4 on the horizontal axis, go up to the line of best fit and read off the answer on the vertical axis.

b   3 hours

5   The table gives the lengths and
    widths of seven fossils.

Length (cm)	1.7	2.3	2.4	5.8	6.3	7.8	9.0
Width (cm)	0.9	2.3	3.3	6.0	6.2	7.0	8.5

   a   Plot the points on a scatter graph.
   b   Draw a line of best fit on the scatter graph.
   c   Another fossil is 4.0 cm in length. Use your line of best fit to estimate its width.

6   **Reasoning / Communication** Some students took a French test and a Spanish test.
    The table gives the marks for seven of the students.

French mark	20	25	27	50	64	80	90
Spanish mark	20	30	52	58	75	86	87

   a   Plot the points on a scatter graph.
   b   Are there any outliers? Explain your answer.
   c   Describe the relationship between the French and Spanish marks.
   d   Cherie's French mark is 60. Estimate her Spanish mark.

> **Q6b hint** An outlier is a value that does not fit the pattern of the data.

7   **Real / Modelling** Toni sells used cars. The table gives the price and mileage of five cars sold.

Mileage	28 000	32 000	45 000	50 000	56 000
Price (£)	9200	8400	7500	6500	6200

> **Q7a hint** Plot mileage on the horizontal axis, from 20 000 to 70 000 miles. Plot price on the vertical axis, from £4000 to £10 000.

   a   Plot the data on a scatter graph.
   b   Describe the correlation between mileage and price.
   c   Draw a line of best fit.
   d   Use your line of best fit to estimate the cost of a car with a mileage of 65 000 miles.
   Discussion Is your estimate more or less reliable than your estimate for **Q6**?
   Give reasons for your answer.

Key point 19

Using a line of best fit to predict data values within the range of the data given is called **interpolation** and is usually reasonably accurate.
Using a line of best fit to predict data values outside the range of the data given is called **extrapolation** and may not be accurate.

---

8   **Exam-style question**

    The scatter graph shows information about the height and arm length of each of 8 students in Year 11.

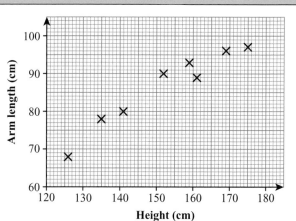

    Exam hint
    Use a line of best fit to answer part **b**.

   a   What type of correlation does this scatter graph show?   **(1 mark)**
   A different student in Year 11 has a height of 148 cm.
   b   Estimate the arm length of this student.   **(2 marks)**

                                                                    Nov 2012, Q2, 1MA0/1H

3 Problem-solving: Worldwide data

Objective	• Use statistics to test statements about real-world data.

Government agencies and charities collect data to make comparisons between countries. It helps them decide how to distribute money and aid.

1 Jenny says, 'Richer countries have a lower infant mortality rate.'
 a Look at the table below. Which column measures the wealth of a country?
 b Plot a graph using two appropriate columns from the table to test Jenny's statement.
 c Do you think Jenny's statement is true or false? Why?
 Discussion The table only has a sample of countries of the world. Does this affect your answer? How?

2 Rupal says, 'Countries with less unemployment have a higher life expectancy.'
 a Choose two columns from the table to plot a graph to test Rupal's statement.
 b Do you think Rupal's statement is true or false? Why?
 Discussion When there is correlation between two variables, does it mean one causes the other?

3 Write your own statement, and test it using the data in the table. You could find out information about more countries, to increase the size of the sample.

> Communication hint **GDP** means 'Gross Domestic Product'. It measures the wealth of a country.

Country	Infant Mortality Deaths/1000 Births	GDP Per Person	Life Expectancy Years	Unemployment % of Population	Health Expenditure % of GDP	Adult Literacy % of Population
Australia	4.43	$43,000	82.07	5.7	9.0	96.0
Brazil	19.21	$12,100	73.28	5.7	8.9	91.3
India	43.19	$4,000	67.80	8.8	3.9	74.4
Japan	2.13	$37,100	84.46	4.1	9.3	99.0
Morocco	24.52	$5,500	76.51	9.5	6.0	67.1
Norway	2.48	$55,400	81.60	3.6	9.1	100.0
South Africa	41.61	$11,500	49.56	24.9	8.5	93.0
Thailand	9.86	$9,900	74.18	0.7	4.1	93.5
Tonga	12.36	$8,200	75.82	13.0	5.3	98.9
Turkey	21.43	$15,300	73.29	9.3	6.7	94.1
United Kingdom	4.44	$37,300	80.42	7.2	9.3	99.0
United States	6.17	$52,800	79.56	7.3	17.9	99.0

3 Check up

Tables

1 Mandip measures the height of some students in metres.
1.35, 1.28, 1.36, 1.42, 1.33, 1.29, 1.23, 1.41, 1.40, 1.34, 1.34, 1.46
Use the information to copy and complete the table.

Height (m)	Tally	Frequency
$1.2 \leqslant h < 1.3$		
$1.3 \leqslant h < 1.4$		
$1.4 \leqslant h < 1.5$		

2 Here is part of a train timetable from Peterborough to London.
   a Which station should the train leave at 09:01?
   b The train arrives in Sandy at 09:12.
     How many minutes should it wait there?
   c The train should take 41 minutes to travel from Arlesey to London.
     What time should the train arrive in London?

Station	Time of leaving
Peterborough	08:44
Huntingdon	09:01
St Neots	09:08
Sandy	09:15
Biggleswade	09:19
Arlesey	09:24

3 The two-way table shows numbers of tickets sold at a theatre.
   a Write the number of adults who chose luxury seats.
   b Copy and complete the two-way table.

	Budget seats	Standard seats	Luxury seats	Total
Adult		17	19	
Child	24		30	
Total	39			130

Graphs and charts

4 The table gives information about the numbers of fish in a lake.
Draw an accurate pie chart to display this information.

Fish	Frequency
Perch	10
Bream	23
Carp	39

5 The table shows the number of sunny days and windy days in a four-month period.

	May	Jun	Jul	Aug
Sunny days	14	22	18	23
Windy days	16	13	8	5

   a Draw a comparative bar chart to display the data.
   b Write a statement comparing the number of sunny and windy days during the four-month period.

6 **Reasoning** Caroline and Marc are in a darts team.
The pie charts show the number of games Caroline and Marc won and lost last year.
Caroline played 52 games.
Marc played 160 games.
How many more games did Marc win than Caroline?

Caroline

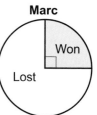

Marc

*Active*Learn Homework, practice and support: Foundation 3 Check up

7 Amrita recorded the heart rate
(in beats per minute) of 15 people.
She asked them to walk up some stairs
and recorded their heart rates again.
She showed her results in a back-to-back
stem and leaf diagram.

Did more people have a higher heart rate
after walking up the stairs than before?
Explain how you know.

Before						After			
		9	8	5					
7 6 6 4	1	0	6	5 8 8 9					
9 8	6	3	2	7	2 4 7 8				
		4	1	8	5 6 8				
			9	1 3 7					
			10	2					

Key

Before

8 | 5 means 58 bpm

After

6 | 5 means 65 bpm

Time series and scatter graphs

8 The graph shows the height of a candle as it burns.
a What is the height of the candle after 2 hours?
b How long does the candle take to burn down
from 15 cm to 4 cm?

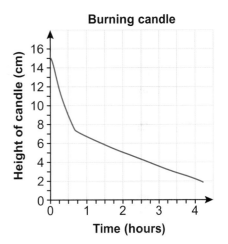

Burning candle

9 A beach café sells ice creams. Describe the relationship between the number of hours of
sunshine and the number of ice creams sold.

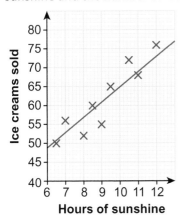

10 How sure are you of your answers? Were you mostly

Just guessing   Feeling doubtful   Confident

What next? Use your results to decide whether to strengthen or extend your learning.

✳ Challenge

11 Design a data collection sheet for a traffic survey to be carried out near a primary school.
Decide what the purpose of the survey will be and the types of data you need to collect.

3 Strengthen

Tables

1   Erin measures the length of some runner beans from her garden.
    14.7, 12.0, 13.8, 8.9, 14.9, 12.4, 9.0, 13.7, 7.0, 9.7, 11.5, 11.9, 14.2, 11.0, 12.5
    Use the information to copy and complete the frequency table.

Length, l (cm)	Tally	Frequency
$7 \leqslant l < 9$		
$9 \leqslant l < 11$		
$11 \leqslant l < 13$		
$13 \leqslant l < 15$		

> **Q1 hint** The first class includes all lengths up to, but not including, 9.0 cm. Which class should include the length 9.0 cm? Add tally marks to your table as you read through the data set. Total your tallies to complete your frequency column.

2   Jack carries out a survey of the type of pets students in his class have.
    Design a data collection sheet for Jack.

> **Q2 hint** What pets do you expect people to have? Include an 'Other' category for any more unusual pets. You will need a column for tallies and for one frequency.

3   **Real** Here is part of a train timetable from Dundee to London.

Dundee	5:56 am	6:30 am	7:28 am
Peterborough	11:30 am	12:20 pm	1:20 pm
London	12:35 pm	1:35 pm	2:40 pm

> **Q3 hint** Count on in hours from 6:30 am until you reach 1:30 pm. Add on the minutes to get the total amount of time.

How long does it take the 6:30 am train from Dundee to get to London?

4   **Real** Here is part of a railway timetable.

New Street	10 13	10 30	10 33
Marston Green	10 26	↓	10 41
Birmingham International	10 29	10 39	10 45
Hampton-in-Arden	10 32	↓	10 48
Tile Hill	10 40	↓	10 55
Coventry	10 47	10 49	11 00

> **Q4a hint** Count on in minutes.

> **Q4b hint** A train leaves Birmingham International at 10:29, 10:39 and 10:45. Read down each column until you reach the row for Tile Hill. Which of the trains get into Tile Hill by 10:50?

a   Work out how long the 10:13 train takes to travel from New Street to Coventry.

b   Harry is at Birmingham International.
    He needs to be at Tile Hill by 10:50.
    What time is the latest train from Birmingham International he can catch?

5   The diagram shows some shapes.
    Copy and complete the two-way table to show the number of shapes in each category.

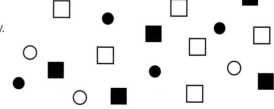

	White	Black
Circle		
Square		

6 A group of students were asked their favourite zoo animal.

	Monkey	Lion	Giraffe	Elephant	Zebra
Girls	2	3	6	4	8
Boys	4	7	3	9	4

a How many more boys than girls chose elephants?

b How many more girls than boys chose zebras?

c Which animal was chosen by the greatest number of students?

> **Q6c hint** Find the total number of students who chose each animal.

7 The two-way table shows how 100 students travelled to school on one day.

	Walk	Car	Bicycle	Total
Boys	15		14	54
Girls		8	16	
Total	37			100

a Work out the number of girls who walked to school.

b Work out the number of boys who travelled to school by car.

c Copy and complete the rest of the table.

> **Q7c hint** Total the 'Girls' row and the 'Car' and 'Bicycle' columns.

8 80 students went on a school trip. They went to either London or York.

23 boys and 19 girls went to London.

14 boys went to York.

Use this information to complete the two-way table.

> **Q8 hint** Fill in the information you already know. (Make sure you include the total number of students.) Find a column or row with just one value missing. Work out the value using the method from **Q7**.

	London	York	Total
Boys			
Girls			
Total			

Graphs and charts

1 The table shows the number of computer games and DVDs Ted bought in 2013 and 2014.

	Games	DVDs
2013	7	9
2014	10	6

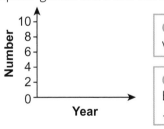

a Draw a comparative bar chart.

b Write a statement comparing the numbers of games and DVDs he bought each year.

> **Q1a hint** What is the highest value you need to plot?

> **Q1b hint** In .... he bought more .... than ....

2 The table shows the average daily hours of sunshine in Majorca and Crete over a five-month period.

	April	May	June	July	August
Majorca	9	9	11	11	10
Crete	6	8	11	13	12

a Display the data in a comparative bar chart.

b Write a statement comparing the hours of sunshine in Majorca and Crete during the five-month period.

> **Q2a hint** Plot the month on the horizontal axis and the hours on the vertical axis.

3 Dexter has a tube of sweets. The table shows the number of sweets of each colour in the tube.

Colour	Number of sweets	Angle
Red	10	
Green	5	
Yellow	6	
Orange	9	
Total		360°

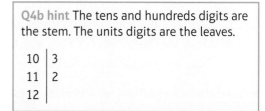

Q3b hint

÷ ☐ ↰ ☐ sweets is 360° ÷ ☐
1 sweet is ☐°

a Work out the total number of sweets.
b Work out the angle for one sweet in a pie chart.
c Work out the angle for the 10 red sweets.
d Work out the angle for the green, yellow and orange sweets.
e Draw the pie chart. Give it a title and label the sectors (or make a key).

4 **STEM** In a chemistry experiment, Juan recorded the mass of chemical produced by a reaction (in grams).
He repeated the experiment 15 times.
Here are his results.

Q4b hint The tens and hundreds digits are the stem. The units digits are the leaves.

```
10 | 3
11 | 2
12 |
```

105, 112, 117, 127, 123, 103, 110, 125,
121, 108, 113, 125, 114, 119, 125

a Write the data in order, starting with the smallest.
b Draw an ordered stem and leaf diagram to show this information.

Time series and scatter graphs

1 The table shows the temperature in a school greenhouse every 2 hours.

Time	11 am	1 pm	3 pm	5 pm
Temperature (°C)	12	20	29	25

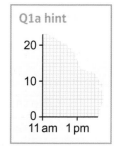

Q1a hint

a Draw a time series graph for the data.
b What is the time when the temperature is 25 °C?
c Estimate the temperature when the time is 4 pm.
d Estimate the difference between the temperature at midday and the temperature at 5 pm.

2 For each graph, decide whether it shows positive correlation, negative correlation or no correlation.

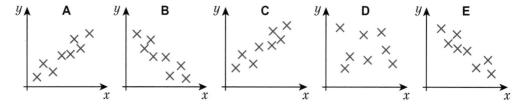

Q2 hint **Positive correlation** – looking from (0, 0), the points go 'uphill': the values are increasing.
Negative correlation – looking from (0, 0), the points go 'downhill': the values are decreasing.
No correlation – the points are not close to a straight line, uphill or downhill.

3 Extend

1   The pictogram shows the numbers of text messages Katherine sent.

Write an expression in terms of m for the number of text messages Katherine sent on

a   Monday

b   Tuesday

c   Wednesday.

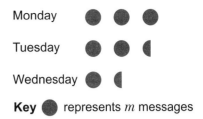

Monday

Tuesday

Wednesday

Key represents m messages

2   **Problem-solving** Students in Year 11 were asked whether they walk or cycle, or walk *and* cycle to school.

The bar chart shows some of the results.

All 203 Year 11 students took part in the survey. How many both walk *and* cycle to school?

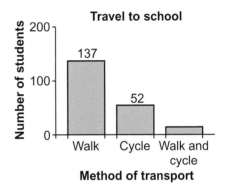

Travel to school

3   **Reasoning** The diagram gives information about the height of some buildings.

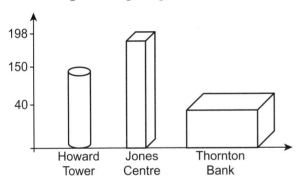

> **Q3 hint** What do you notice about the scale on the vertical axis? What about the shape of the bars?

Write *two* things that could be wrong or misleading in the diagram.

4   **Real / Communication** The graph shows the cost of the gas Charlie has used over the last 12 months.

Charlie wants to make monthly payments for his gas. He thinks his gas is going to cost the same for the next 12 months.

Charlie works out he should pay exactly £40 per month to cover the cost over 12 months.

Is he correct? Explain your answer.

> **Q4 communication hint**
> Quarterly means every quarter (three months) of the year.

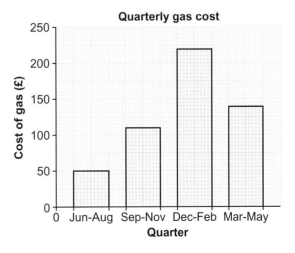

Quarterly gas cost

5 The table shows the sports chosen by 65 students.
Draw a pie chart to show this data.

Q5 hint: Round angles to the nearest degree.

Sport	Tennis	Football	Swimming	Basketball
Frequency	16	26	8	15

6 ⟨ **Exam-style question** ⟩

The bar chart shows the proportion of different types of mice found in samples from homes near woodland and away from woodland.

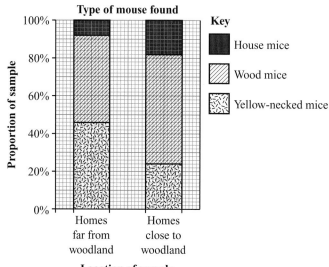

Exam hint
Read values from the graph as accurately as you can.

a Approximately what percentage of mice in homes close to woodland are wood mice?

b Approximately what percentage of mice in homes far from woodland are not wood mice?

c One of the black bars is taller than the other. Does that mean there are more house mice in homes far from woodland than in homes close to woodland?
Explain how you know.

(4 marks)

7 The graph shows at what time the sun sets and rises.

a Which month is the longest day in?

b Which month is the shortest day in?

c How many hours of sunlight were there on the longest day?

Q7a hint The longest day is the one with the most hours of sunlight.

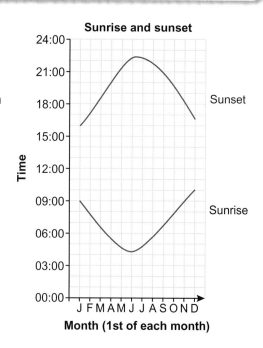

8 **Real** This is a record of the money an electricity supplier receives in February.

Date	Amount (£)	Date	Amount (£)
5th	153 000	13th	468 000
7th	543 000	16th	38 000
9th	223 000	18th	540 000

Draw a graph or chart to represent this information. Choose a sensible scale.

9 **Real / Problem solving** Here is part of a bus timetable from Harrow Lane to Cartbridge Street.

Harrow Lane to Cartbridge Street

Harrow Lane	08 02	09 04	10 12	11 02	12 04	12 12
Elm Drive	08 19	09 21	10 29	11 19	12 21	12 29
Hamden Road	08 32	09 34	10 42	11 32	12 34	12 42
Swipe Crescent	08 41	09 43	10 51	11 41	12 43	12 51
Cartbridge Street	08 50	09 52	11 01	11 50	12 52	13 01

Cartbridge Street to Harrow Lane

Cartbridge Street	13 11	14 14	15 07	16 11	17 14	18 07
Swipe Crescent	13 20	14 24	15 16	16 20	17 24	18 16
Hamden Road	13 29	14 33	15 25	16 29	17 33	18 25
Elm Drive	13 43	14 47	15 39	16 43	17 47	18 39
Harrow Lane	13 53	14 57	15 49	16 53	17 57	18 49

Ganesh lives in Harrow Lane. His grandmother lives in Swipe Crescent.

Ganesh travels by bus to visit his grandmother. He wants to spend at least 3 hours with her.

He needs to be back at Harrow Lane by 16:00.

Plan Ganesh's journey to visit his grandmother and get back to Harrow Lane.

You must include the times of the buses.

3 Knowledge check

- A **grouped frequency table** has data in classes. The classes should be of equal width. . . . . *Mastery lesson 3.1*

- **Discrete data** can only take particular values, for example shoe sizes are usually whole numbers. **Continuous data** is measured and can take any value, for example length and time. *Mastery lesson 3.1*

- For discrete data use groups like 0–4, 5–8. For continuous data use inequalities like $0 < t \leqslant 4$. . . . . *Mastery lesson 3.1*

- A **distance chart** shows the distances between several places. . . . . *Mastery lesson 3.2*

- A **two-way table** divides data into groups in rows across the table and in columns down the table. You can calculate the totals across and down. . . . . *Mastery lesson 3.2*

⊙ A **comparative bar chart** compares two or more sets of data on the same chart. **A composite bar chart** compares features within a single bar. .... *Mastery lesson 3.3*

⊙ A **histogram** shows grouped continuous data. .... *Mastery lesson 3.3*

⊙ A **line graph** shows trends in data. .... *Mastery lesson 3.3*

⊙ A **time series graph** is a line graph with time plotted on the horizontal axis. .... *Mastery lesson 3.4*

⊙ A **stem and leaf diagram** shows ordered numerical data split into a 'stem' and 'leaves'. The leaf is usually the last digit and the stem is the other digits. A back-to-back stem and leaf diagram compares two sets of results. .... *Mastery lesson 3.5*

⊙ A **pie chart** is a circle divided into sectors. Each sector represents a set of data. .... *Mastery lesson 3.6*

⊙ A **scatter graph** shows the relationship between two sets of data. Plot the points with crosses. Do not join them up. .... *Mastery lesson 3.7*

⊙ An **outlier** is a value in a data set that is much larger or smaller than the other numbers in the set. .... *Mastery lesson 3.7*

⊙ The relationship between the sets of data is called **correlation**. The sets of data are called **variables**. .... *Mastery lesson 3.7*

⊙ A **line of best fit** is a straight line that passes as close as possible to points on a scatter graph and follows the trend of the data. .... *Mastery lesson 3.8*

⊙ Using a line of best fit to predict data values within the range of the data given is called **interpolation** and is usually reasonably accurate. Using a line of best fit to predict data values outside the range of the data given is called **extrapolation** and may not be accurate. .... *Mastery lesson 3.8*

The mathematics studied in this unit is called 'statistics'.
Choose A, B or C to complete each statement about statistics.

In this unit, I did…	A well	B OK	C not very well
I think statistics is…	A easy	B OK	C hard
When I think about doing statistics I feel…	A confident	B OK	C unsure

Did you answer mostly As and Bs? Are you surprised by how you feel about statistics? Why?

Did you answer mostly Cs? Find the three questions in this unit that you found the hardest. Ask someone to explain them to you. Then complete the statements above again.

Reflect

3 Unit test

Log how you did on your Student Progression Chart.

1 This frequency table records the lengths of grubs found in a garden.

Length l (mm)	$0 < l \leq 5$	$5 < l \leq 10$	$10 < l \leq 15$	$15 < l \leq 20$	$20 < l \leq 25$
Frequency	3	7	14	4	1

a How many grubs were found in total?

b How many were greater than 15 mm in length? *(2 marks)*

*Active*Learn Homework, practice and support: Foundation 3 Unit test

2   The bar chart shows how many students go to a French club.

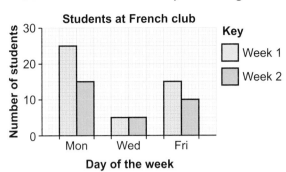

a   How many students attended the French club during the two weeks?   *(1 mark)*
b   Tom said, 'The same students attended the French club on both Wednesdays.'
    Is he right? Explain your answer.   *(2 marks)*

3   The chart shows the distances, in kilometres, between four cities.

Caen

692	Lyon		
573	457	Metz	
295	662	705	Nantes

How far is it from
a   Caen to Metz
b   Lyon to Nantes
c   Nantes to Caen?   *(3 marks)*

4   The table gives the match results of a football team.

Result	Frequency
Won	10
Drawn	6
Lost	8

Draw an accurate pie chart to show this information.   *(4 marks)*

5   **Reasoning** There are 25 students in a Year 10 class.
    The table shows if they travel to school by bus.

	Travel by bus	Do not travel by bus
Boys	6	7
Girls	4	8

a   How many students do not travel by bus?   *(1 mark)*
b   Copy and complete the table.
    Another Year 10 class contains seven girls and 13 boys.
    No girls travel by bus.
    Four boys do not travel by bus.

	Travel by bus	Do not travel by bus
Boys		
Girls		

(3 marks)

6 **Reasoning** The incomplete table and composite bar chart show the percentage population in three age ranges, for two countries.

	Under 15	15 to 64	65 and over
UK			
Mexico	36%	60%	4%

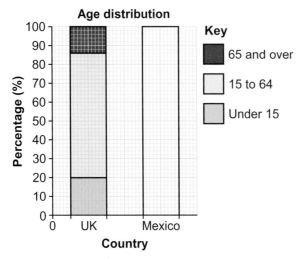

Age distribution

Key

■ 65 and over

□ 15 to 64

▨ Under 15

a Use the information from the composite bar chart to complete the table. *(3 marks)*

b Use the information from the table to complete the Mexico bar on the chart. *(3 marks)*

7 A patient's temperature is taken every hour for eight hours.

Hour	1	2	3	4	5	6	7	8
Temperature (°C)	37.4	37.8	38.3	38.0	38.2	37.4	37.3	37.2

Draw a time series graph to represent this data. *(5 marks)*

8 Monica measures the body length and foot length of eight different rodents. The data is displayed in the table.

Body length (mm)	25	100	75	75	175	105	250	125
Foot length (mm)	14	25	18	24	35	19	43	28

a Draw the scatter graph and a line of best fit. *(2 marks)*

b What is the type of correlation between body length and foot length? *(1 mark)*

c An animal has a body length of 230 mm and a foot length of 20 mm.
Is this animal likely to be one of these types of rodent?
Explain your answer. *(2 marks)*

9 **Reasoning** 120 students went on a school activities day.
They could go bowling, skating or to the cinema.

66 of the students were girls.

28 of the girls went bowling.

36 students went to the cinema.

20 of the students who went to the cinema were girls.

15 boys went skating.

Work out the number of students who went bowling. *(4 marks)*

10   Joan records her dressage test scores for her two horses, Jigsaw and Percy.

Jigsaw	57	63	51	68	50	49	66	70
Percy	55	62	60	81	59	59	74	73

a   Draw a back-to-back stem and leaf diagram for this data.   *(3 marks)*
b   Which horse performed better at dressage?   *(1 mark)*

Sample student answer

a   Which is the best type of graph to use for this question? Explain why.
b   Why would the student who used the best graph still not get full marks?

> **Exam-style question**
>
> The table shows information about some students' favourite football teams.
>
> | | United | City | Rovers | Latics | Wanderers |
> |-------|--------|------|--------|--------|-----------|
> | **Boys** | 12 | 5 | 3 | 7 | 6 |
> | **Girls** | 4 | 9 | 5 | 3 | 10 |
>
> Represent this information in a suitable diagram or chart.   **(4 marks)**
>
> *Practice Paper Set C, Q7, 1MA0/2F*

Student A

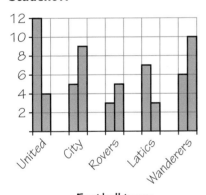

Football team

KEY ☐ Boys ☐ Girls

Student B

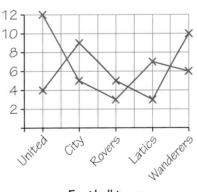

Football team

KEY — Boys — Girls

Master
p.93

Problem-solve
p.111

Check
p.112

Strengthen
p.113

Extend
p.116

Test
p.119

4 FRACTIONS AND PERCENTAGES

A survey found 2 in every 7 people thought the customer service in the shop was excellent.
How many people would you expect to think the customer service in the shop was excellent in a survey of

70 people   35 people   350 people?

4 Prior knowledge check

Numerical fluency

1 What fraction is shaded in this diagram?

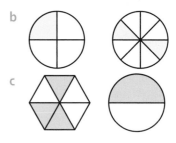

2 Which diagrams have $\frac{1}{2}$ shaded?

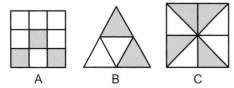

A   B   C

3 For each pair of diagrams, write down the equivalent fractions. The first one has been started for you.

a
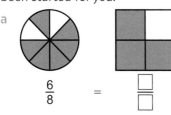

$$\frac{6}{8} = \frac{\square}{\square}$$

b

c

4 Work out

a $\frac{3}{5} + \frac{1}{5}$ b $\frac{1}{4} + \frac{1}{4}$ c $\frac{3}{5} - \frac{1}{5}$ d $\frac{3}{4} - \frac{1}{4}$

5 Write these fractions in order of size. Start with the smallest.
$\frac{2}{9}, \frac{5}{9}, \frac{1}{9}, \frac{8}{9}$

6 Copy and complete

a $\frac{3}{4} = \frac{6}{\square}$ b $\frac{1}{5} = \frac{\square}{20}$ c $\frac{6}{10} = \frac{\square}{5}$

7 Write each fraction in its simplest form.

a $\frac{3}{9}$ b $\frac{6}{8}$ c $\frac{12}{15}$

8 Write $\frac{12}{7}$ as a mixed number.

9 Write these mixed numbers as improper fractions.

a $4\frac{1}{2}$ b $6\frac{3}{4}$

10 Write these fractions as decimals.

a $\frac{1}{2}$ b $\frac{3}{4}$ c $\frac{7}{10}$ d $\frac{2}{3}$

11 Write these fractions as percentages.

a $\frac{1}{2}$ b $\frac{1}{5}$ c $\frac{3}{4}$ d $\frac{1}{3}$

12 Write these percentages as fractions.

a 25%   b 40%   c 90%

13 Write these decimals as percentages.

a 0.2   b 0.5   c 0.75

14 a Write in order of size, starting with the smallest.

50%, 0.75, $\frac{3}{10}$

b Write in order of size, starting with the largest.

$\frac{4}{5}$, 1, 0.5, 60%

15 Write

a 5 as a fraction of 12

b 4 as a fraction of 16.

16 Work out

a 25% of 8   b 75% of 200.

Fluency with measures

17 Work out

a $\frac{1}{5}$ of £90

b 10% of £60

c $\frac{1}{2}$ of 32 cm.

18 Write as a fraction of an hour

a 30 minutes   b 15 minutes

Fluency with probability

19 What is the probability of getting black on this as a fraction.

* Challenge

20 Write these fractions in order of size. Start with the smallest.

$\frac{17}{3}$, $3\frac{1}{4}$, $5\frac{2}{3}$, $6\frac{3}{4}$, $\frac{17}{5}$

4.1 Working with fractions

Objectives

- Compare fractions.
- Add and subtract fractions.
- Use fractions to solve problems.

Did you know?

From as early as 1800 BC, the Egyptians were working with fractions.

Fluency

- Are $\frac{1}{2}$ and $\frac{2}{4}$ equivalent fractions?
- What is the denominator of $\frac{2}{9}$?
- What is the numerator of $\frac{5}{6}$?

1 Find the LCM (lowest common multiple) of these numbers.

a 3 and 4   b 6 and 8   c 2, 3 and 6

2 Write < or > between each pair of fractions.

a $\frac{3}{16} \square \frac{1}{16}$ b $\frac{4}{7} \square \frac{6}{7}$

3 Copy and complete these equivalent fractions.

a $\frac{2}{3} = \frac{\square}{6}$ b $\frac{4}{9} = \frac{\square}{27}$ c $\frac{2}{5} = \frac{8}{\square}$ d $\frac{2}{3} = \frac{\square}{6} = \frac{8}{\square}$

Q3 hint

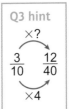

4 Write each fraction in its simplest form.

a $\frac{6}{8}$ b $\frac{12}{18}$ c $\frac{9}{15}$ d $\frac{16}{24}$

Active Learn   Homework, practice and support: Foundation 4.1

Warm up

Questions in this unit are targeted at the steps indicated.

5 Write each pair of fractions with a common **denominator**.

a $\frac{1}{2}$ and $\frac{3}{4}$

b $\frac{7}{10}$ and $\frac{3}{5}$

Q5a hint Use the LCM of the denominators as the common denominator.

The LCM of 2 and 4 is 4. $\frac{1}{2} = \frac{\square}{4}$

Key point 1

To compare fractions, write them with a common denominator.

6 **Problem-solving** Marcia has two spinners, A and B.

For spinner A, P(red) = $\frac{5}{8}$.

For spinner B, P(red) = $\frac{7}{12}$.

Which spinner is more likely to land on red?

Q6 hint Write the fractions with a common denominator. Which is larger?

7 **Reasoning / Communication** Is $\frac{4}{9} > \frac{1}{3}$?

Show your working to explain your answer.

8

Exam-style question

Here are two fractions $\frac{3}{5}$ and $\frac{2}{3}$.

Explain which is the larger fraction.

You may use the grids to help with your explanation. **(3 marks)**

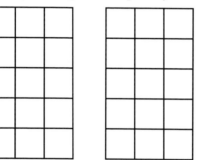

2008 Practice paper B, Q19, 4444/12

Exam hint

Make a statement after all your working to say which is the larger fraction.

Example 1

Write $\frac{4}{5}$, $\frac{5}{8}$ and $\frac{3}{4}$ in order of size. Start with the smallest.

The LCM of 5, 8 and 4 is 40.

$\frac{4}{5} = \frac{32}{40}$ $\frac{5}{8} = \frac{25}{40}$ $\frac{3}{4} = \frac{30}{40}$

Answer: $\frac{5}{8}$, $\frac{3}{4}$, $\frac{4}{5}$

Write the fractions with a common denominator.

Write the original fractions in order.

9 a Write in order of size, starting with the smallest. $\frac{2}{3}$, $\frac{3}{5}$, $\frac{3}{4}$

b Write in order of size, starting with the largest. $\frac{7}{12}$, $\frac{5}{6}$, $\frac{2}{3}$

10 **Reasoning** a Predict which of these fractions is smallest.

$\frac{3}{10}$, $\frac{3}{4}$, $\frac{5}{8}$

b Work out which fraction is smallest.

Reflect Was your prediction correct?

11 **Reasoning / Communication**
    a Write two fractions with the same numerator.
       Explain how you can tell which one is smaller.
    b Explain which is larger, $\frac{7}{8}$ or $\frac{8}{9}$.

> **Q11 communication hint**
> 'Explain' means show your working.

Key point 2

To add or subtract fractions, write them with a common denominator.

Example 2

Work out $\frac{2}{3} + \frac{1}{9}$.

$\frac{2}{3} + \frac{1}{9} = \frac{6}{9} + \frac{1}{9} = \frac{7}{9}$

> The LCM of 3 and 9 is 9. Write the fractions with denominator 9 and then add.

---

12 Work out these calculations. Give each answer in its simplest form.
    a $\frac{1}{2} + \frac{3}{8}$ b $\frac{3}{5} + \frac{2}{15}$ c $\frac{3}{4} - \frac{1}{2}$ d $\frac{2}{3} - \frac{1}{9}$
    e $\frac{5}{8} - \frac{2}{16}$ f $\frac{1}{3} + \frac{3}{9}$ g $\frac{2}{5} - \frac{1}{15}$ h $\frac{17}{20} - \frac{1}{4}$

13 Work out
    a $\frac{1}{5} + \frac{1}{3}$ b $\frac{1}{2} - \frac{1}{3}$ c $\frac{1}{2} + \frac{1}{5} - \frac{1}{7}$
    Discussion When do you have to change all the denominators?

> **Q13a hint** Change both fractions so they have the same denominator. Use the LCM.

14 **Reasoning / Real** John is planning a gymnastics event.
    He needs $\frac{1}{3}$ hour to work out the results and $\frac{1}{2}$ hour to give out the medals.
    What fraction of an hour will this take?

15 **Reasoning** A group of students went to a restaurant.
    $\frac{1}{5}$ of them bought a chicken burger and $\frac{1}{2}$ of them bought a beef burger.
    The remainder only bought a drink.
    What fraction of the group bought a type of burger?

Key point 3

A **unit fraction** has **numerator** 1.

---

16 **Problem-solving** The ancient Egyptians
    only used unit fractions.
    For $\frac{3}{4}$ they wrote $\frac{1}{2} + \frac{1}{4}$.
    a For what fraction did they write $\frac{1}{3} + \frac{1}{5}$?
    b How did they write $\frac{7}{12}$?
       Check your answer using a calculator.

> **Q16 hint** Use the fraction button on your calculator to enter a fraction.

> **Q16b hint** Try different unit fractions.
>

17 Work these out. Give each answer in its simplest form.
    a $\frac{5}{12} + \frac{1}{8}$ b $\frac{3}{4} + \frac{1}{6}$ c $\frac{5}{6} - \frac{1}{4}$
    d $\frac{9}{10} - \frac{3}{4}$ e $\frac{3}{4} + \frac{1}{10}$ f $\frac{3}{4} - \frac{1}{16}$
    Check your answers using a calculator.

> **Q17 hint** You may need to change both fractions so they have a common denominator.

18 Work out
    a $1 - \frac{1}{3}$ b $1 - \frac{3}{5}$ c $2 - \frac{1}{4}$

> **Q18a hint** $1 = \frac{3}{3}$

> **Q18c hint** $2 = \frac{\square}{4}$

4.2 Operations with fractions

Objectives

- Find a fraction of a quantity or measurement.
- Use fractions to solve problems.

Did you know?

Fractions can be found in ancient Indian writings. The fractions $\frac{1}{2}$ and $\frac{3}{4}$ have been found in a manuscript dating back to around 1000 BC and a mixed number, $3\frac{3}{8}$, can be found in a manuscript dating back to 400 AD.

Fluency

Work out $\frac{5}{6} - \frac{1}{6}$ $2 - \frac{1}{3}$ $\frac{5}{6} = \frac{\square}{18}$ $\frac{1}{4}$ of 20 $\frac{1}{5}$ of 30

1 Convert
   a 5 m to cm   b 350 cm to m.

2 Work out
   a $\frac{1}{8} + \frac{1}{8}$ b $\frac{5}{12} - \frac{1}{8}$

3 Change these improper fractions to mixed numbers.
   a $\frac{7}{2}$ b $\frac{11}{8}$ c $\frac{50}{6}$ d $\frac{25}{12}$

4 Change these mixed numbers into improper fractions.
   a $2\frac{1}{2}$ b $4\frac{3}{4}$ c $7\frac{1}{6}$ d $10\frac{1}{2}$

Example 3

Work out $\frac{3}{5}$ of 40. ——— In mathematics, 'of' means multiply.

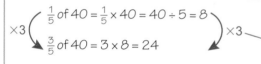

$\times 3 \left(\begin{array}{l} \frac{1}{5} \text{ of } 40 = \frac{1}{5} \times 40 = 40 \div 5 = 8 \\ \frac{3}{5} \text{ of } 40 = 3 \times 8 = 24 \end{array} \right) \times 3$

Multiply by 3 to find $\frac{3}{5}$

5 Work out
   a $\frac{2}{9}$ of 18   b $\frac{7}{8}$ of 32   c $\frac{4}{5}$ of 80.

   Q5a hint First work out $\frac{1}{9}$ by dividing by $\square$.

6 Find
   a $\frac{3}{4}$ of 200 kg   b $\frac{4}{5}$ of 90 cm.

7 A test has 60 marks. Monty gets $\frac{3}{4}$ of the marks. How many marks does he get?

8 **Problem-solving / Real** Alicia and Gaby sell CDs on an internet auction site. They split the money so that Alicia gets $\frac{2}{3}$ and Gaby gets the rest. They get £90 in total. How much does Gaby get?

9 **Real / Communication** Rory wants to fit 7 shelves along a wall of his garage. Each shelf is 150 cm long. The wall is 10 m long. Show that Rory cannot fit 7 shelves along the wall.

10 **Exam-style question**

   There are 700 students in a college.
   All of the students are 16 years old, 17 years old, or 18 years old.
   $\frac{1}{10}$ of the students are 16 years old.
   $\frac{1}{5}$ of the students are 18 years old.
   Work out how many of the students are 17 years old.   **(4 marks)**

   March 2013, Q13, 5MB1F/01

   Exam hint
   Check your answer gives a correct total number of students.

11 **Reasoning / Communication** The frequency table shows the numbers of animals on a farm.

Animal	Frequency
Hens	30
Sheep	15
Cows	5
Goats	10

Leandra says, 'In a pie chart, the goats will be represented by $\frac{1}{4}$ of its area.'
Explain why Leandra is wrong.

12 **Problem-solving** 600 children were asked their favourite sport.
This pie chart shows the results.

Favourite sports

a How many children chose swimming?
b Estimate the number of children who chose football.

> **Q12b hint** Estimate the fraction for football.

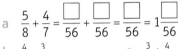

13 Work these out. Give your answers as mixed numbers.

a $\frac{5}{8} + \frac{4}{7} = \frac{\square}{56} + \frac{\square}{56} = \frac{\square}{56} = 1\frac{\square}{56}$

b $\frac{4}{5} + \frac{3}{8}$ c $\frac{3}{5} + \frac{4}{7}$ d $\frac{5}{8} + \frac{5}{6}$

Check your answers using a calculator.
Discussion How does your calculator show a mixed number?

Key point 4

To add or subtract mixed numbers, convert to improper fractions first.

Example 4

Work out $1\frac{1}{2} + 2\frac{3}{4}$.   ⟶   Convert to improper fractions.

$1\frac{1}{2} + 2\frac{3}{4} = \frac{3}{2} + \frac{11}{4}$

$= \frac{6}{4} + \frac{11}{4} = \frac{17}{4}$ ⟶   Add the fractions.

$= 4\frac{1}{4}$ ⟶   Convert back to a mixed number.

---

14 Work out

a $2\frac{1}{4} + 3\frac{1}{2}$ b $7\frac{1}{2} + 5\frac{3}{4}$ c $8\frac{1}{2} + 3\frac{1}{4}$

d $12\frac{1}{2} + 3\frac{3}{4}$ e $1\frac{1}{10} + 2\frac{1}{2}$ f $4\frac{1}{2} + 2\frac{1}{6}$

15 Work out

a $3\frac{3}{4} - 2\frac{1}{3}$ b $10\frac{1}{2} - 5\frac{3}{4}$ c $11\frac{1}{2} - 8\frac{2}{5}$

d $7 - 3\frac{3}{4} - 2\frac{1}{2}$ e $6\frac{1}{10} - 4\frac{3}{5}$ f $8\frac{2}{3} - 6\frac{1}{6}$

> **Q15 hint** Convert to improper fractions and then subtract.

Discussion Do you always need to convert to improper fractions before you add and subtract?

16 **Problem-solving** In this diagram, the number in each box is the sum of the two numbers below it.

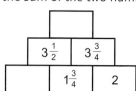

> **Q16 communication hint**
> Sum means add.

Find the missing numbers.

4.3 Multiplying fractions

Objectives

- Multiply whole numbers, fractions and mixed numbers.
- Simplify calculations by cancelling.

Why learn this?

Recipes may contain fractional amounts, like $\frac{1}{2}$ teaspoon. When scaling a recipe up, you need to multiply fractions.

Fluency

- Work out $\frac{1}{3}$ of 18.
- How many grams in 1 kg?
- Is 3 × 2 the same as 2 × 3?

1 Work out

   a $\frac{5 \times 1}{6}$

   b $\frac{2 \times 3}{7}$

   c $\frac{5 \times 2}{15}$

   d $\frac{7 \times 2}{16}$

> **Q1 hint** Write answers in their simplest form.

2 Convert

   a $\frac{15}{2}$ to a mixed number

   b $3\frac{3}{5}$ to an improper fraction.

3 Copy and complete.

   a $7 \times \frac{1}{2} = \frac{7 \times 1}{2} = \frac{\square}{2} = \square\frac{1}{2}$

   b $7 \times \frac{5}{2} = \frac{\square}{2} = \square\frac{1}{2}$

4 Work these out. Write your answers as mixed numbers.

   a $2 \times \frac{5}{2}$
   b $3 \times \frac{5}{2}$
   c $7 \times \frac{12}{5}$
   d $\frac{5}{3} \times 6$
   e $\frac{13}{4} \times 3$

5 **Communication** Scott has to complete 5 tasks. Each task takes $\frac{3}{4}$ of an hour to complete. Show that Scott will take $3\frac{3}{4}$ hours to complete the 5 tasks.

6 **Reasoning / Finance** Ali is selling his bike for £460. He reduces the price by $\frac{1}{4}$. What is the new price?

   Discussion How did you reduce the price by $\frac{1}{4}$?

ActiveLearn Homework, practice and support: Foundation 4.3

Warm up

7 Substitute $x = 3$ and $y = 6$ into

a $\frac{1}{2}(x + 5)$

b $\frac{1}{4}(y + 10)$

c $\frac{1}{3}(x + y)$

d $\frac{1}{2}(y - x)$

> **Q7a hint** Work out brackets first, then multiply by $\frac{1}{2}$.

8 A cake recipe uses $\frac{3}{4}$ kg of flour for 1 cake. How much flour is needed for 8 cakes?

Key point 5

To multiply fractions together, multiply the numerators together and the denominators together.

Example 5

Work out $\frac{2}{3} \times \frac{1}{5}$.

> Multiply the numerators and the denominators.

$$\frac{2}{3} \times \frac{1}{5} = \frac{2 \times 1}{3 \times 5} = \frac{2}{15}$$

---

9 Work out

a $\frac{1}{2} \times \frac{1}{4}$

b $\frac{1}{5} \times \frac{1}{3}$

c $\frac{1}{2} \times \frac{3}{4}$

d $\frac{1}{7} \times \frac{6}{7}$

e $\frac{4}{5} \times \frac{3}{7}$

f $\frac{2}{7} \times \frac{4}{9}$

g $\frac{3}{10} \times \frac{2}{3}$

h $\frac{5}{6} \times \frac{3}{5}$

> **Q9g hint** Write answers in their simplest form.

Discussion Does the rule 'multiply the numerators and multiply the denominators' work for $7 \times \frac{3}{2}$ and $\frac{2}{5} \times 6$?

10 **Problem-solving / Finance** A company spends $\frac{1}{4}$ of its profit on new machinery. The company makes a profit of £$\frac{3}{4}$ million.

How much money does the company spend on new machinery?

Reflect What operation do you use to work out $\frac{1}{4}$ of a quantity?

Key point 6

Numerators can be cancelled with denominators if they are divisible by the same number.

Example 6

Work out $\frac{5}{6} \times \frac{2}{7}$.

> Look for numbers in the numerator and denominator with a common factor. 2 is a factor of 2 and 6.

$$\frac{5}{6} \times \frac{2}{7} = \frac{5 \times 2}{6 \times 7}$$

$$= \frac{\overset{1}{2} \times 5}{\underset{3}{6} \times 7} = \frac{1 \times 5}{3 \times 7}$$

> Change the order to give $\frac{2}{6}$, which simplifies to $\frac{1}{3}$.

$$= \frac{5}{21}$$

---

11 Work out

a $\frac{4}{5} \times \frac{1}{2}$

b $\frac{7}{9} \times \frac{3}{10}$

c $\frac{5}{12} \times \frac{6}{7}$

d $\frac{8}{9} \times \frac{1}{4}$

e $\frac{10}{21} \times \frac{7}{15}$

f $\frac{13}{16} \times \frac{4}{7}$

Discussion Which is easier – rearrange and simplify or simplify at the end?

12 In a cycling club, $\frac{1}{4}$ of the members are female. Of these, $\frac{2}{3}$ are under 16.

What fraction of the cycling club are female *and* under 16?

> **Q12 hint** $\frac{1}{4}$ of $\frac{2}{3} = \frac{1}{4} \,\square\, \frac{2}{3}$

13 Garcia buys $\frac{3}{4}$ kg of cheese. He gives his friend $\frac{1}{3}$ of it.

Garcia needs 600 g of cheese for a recipe. Does he have enough?

14 **Exam-style question**

   A full petrol tank holds 52 litres of petrol.
   The fuel gauge shows that the tank is three-quarters full.
   Work out how much petrol (in litres) is in the petrol tank.   **(2 marks)**

> **Exam hint**
> Check your answer is sensible.

15 Work out
   a $3 \times 1\frac{3}{4}$ b $6 \times 2\frac{1}{7}$
   c $2\frac{1}{2} \times 7$ d $3\frac{1}{5} \times 11$

> **Q15a hint** Write $1\frac{3}{4}$ as an improper fraction first, then multiply by 3.

16 Uzma's paper round takes her $1\frac{1}{2}$ hours each day.
   How long does she spend on her paper round each week (Monday to Friday)?

17 Nathalie needs 6 sections of copper piping $1\frac{3}{4}$ m in length. She has 10 m.
   Does she have enough piping?

18 **Problem-solving / Finance**   Riccardo's parents need a mortgage to buy a house.
   They are offered two options.

> **Q18 strategy hint** Work out the amount for each option.

   Option 1   Three and a half times their joint salary.

   Option 2   Six times the larger salary plus one and a half times the smaller salary.
   Riccardo's dad's salary is £22 000. His mum's salary is £37 000.
   Riccardo's parents need a mortgage greater than $\frac{1}{4}$ million pounds.
   Show that only one of the options is suitable.

4.4 Dividing fractions

Objectives

- Divide a whole number by a fraction.
- Divide a fraction by a whole number or a fraction.

Why learn this?

Dividing by a fraction helps you work out how many $\frac{1}{3}$-litre glasses you can fill from a 2-litre bottle.

Fluency

How many   twos in 8   threes in 12   halves in 1 whole   thirds in 1 whole?

1   Copy and complete.
    a $\frac{1}{2}$ of 8 = 8 ÷ ☐   b $\frac{1}{4}$ of 48 = 48 ÷ ☐

2   Write $3\frac{3}{4}$ as an improper fraction.

3   Write $\frac{19}{3}$ as a mixed number.

4   Work out
    a $\frac{2}{3} \times 5$ b $\frac{2}{3} \times \frac{9}{11}$

Key point 7

The **reciprocal** of a fraction is the 'upside down' fraction. The reciprocal of 2 (or $\frac{2}{1}$) is $\frac{1}{2}$.

     ActiveLearn   Homework, practice and support: Foundation 4.4

5 Write down the reciprocal of

a $\frac{3}{4}$ b $\frac{1}{5}$ c 4

Reflect Multiply $\frac{3}{4}$, $\frac{1}{5}$ and 4 by their reciprocals.
Copy and complete this sentence to explain what you notice.

Multiplying a number by its reciprocal .....

Key point 8

Dividing by a number is the same as multiplying by its reciprocal.
$12 \div 2$ means $\frac{1}{2}$ of 12 or $12 \times \frac{1}{2}$.

Example 7

Work out $6 \div \frac{2}{3}$.

$6 \div \frac{2}{3} = 6 \times \frac{3}{2}$ Change to multiplication by the reciprocal.

$= \frac{6 \times 3}{1 \times 2} = \frac{3\cancel{6} \times 3}{1\cancel{2} \times 1} = 9$ Give the answer in its simplest form.

6 Work out

a $4 \div \frac{1}{3} = 4 \times \square = \square$ b $5 \div \frac{1}{5}$ c $12 \div \frac{1}{8}$ d $4 \div \frac{3}{4}$

7 **Real** How many $\frac{3}{4}$-litre cups can you fill from a 2-litre jug?

8 Work out these divisions. Write your answer as a mixed number if necessary.

a $5 \div \frac{3}{2} = 5 \times \frac{2}{\square} =$ b $12 \div \frac{4}{3}$ c $14 \div \frac{18}{5}$

9 Work out these divisions.
Write your answer as a mixed number if necessary.

a $7 \div 2\frac{1}{2} = 7 \div \frac{\square}{2} =$ b $10 \div 2\frac{2}{5}$ c $8 \div 5\frac{1}{3}$

> **Q9 hint** Write mixed numbers as improper fractions first.

10 Work out

a $\frac{1}{3} \div 2$ b $\frac{2}{3} \div 6$

c $\frac{6}{5} \div 6$ d $3\frac{3}{4} \div 8$

> **Q10a hint**
> $\frac{1}{3} \div 2 = \frac{1}{2}$ of $\frac{1}{3} = \frac{1}{2} \times \frac{1}{3}$
>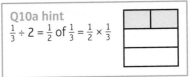

11 Kirsten has to walk $2\frac{3}{4}$ km to school.
How far has she walked when she is halfway?
Give your answer in metres.
Check your answer using an inverse operation.

> **Q11 hint**
>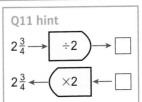

12 Sharnia takes $2\frac{2}{5}$ of an hour to do 3 tasks.
Each task takes the same time.
What fraction of an hour does 1 task take?

13 **Exam-style question**

Gunnar has 20 m of fabric. He needs $\frac{4}{5}$ of a metre of fabric to make a cushion.
How many cushions can he make from this fabric?   **(2 marks)**

> **Exam hint**
> Check your answer using the inverse operation.

14 Work out these divisions. Write your answer as a mixed number if necessary.

a $\frac{1}{2} \div \frac{1}{3} = \frac{1}{2} \times 3 = \frac{\square}{\square} = 1\frac{\square}{\square}$

b $\frac{3}{4} \div \frac{1}{5}$ c $\frac{7}{10} \div \frac{1}{2}$

4.5 **Fractions and decimals**

Warm up

Objectives

- Convert fractions to decimals and vice versa.
- Use decimals to find quantities.
- Write one number as a fraction of another.

Why learn this?

You can compare fractions with different denominators by converting them to decimals.

Fluency

What is the value of the digit 6 in   0.6   0.56   0.246?

1   Write **a** $\frac{7}{10}$ as a decimal   **b** 0.4 as a fraction.

2   Work out **a** $4\overline{)5.00}$ **b** $6\overline{)25.00}$ as a decimal.

3   Write **a** 7 as a fraction of 12   **b** 6 as a fraction of 9.

Key point 9

You can use short or long division to convert fractions to decimals. For example, $\frac{3}{5}$ means $3 \div 5$.

Example 8

Write $\frac{7}{8}$ as a decimal.

$$\frac{7}{8} = 8\overline{)7} = 8\overline{)7.7^06^04^0} \quad \begin{array}{c} 0.\ 8\ 7\ 5 \end{array}$$

$\frac{7}{8} = 0.875$

4   **Exam-style question**

Write $\frac{7}{25}$ as a decimal.   **(1 mark)**

Exam hint
Estimate the answer to check your working out.

5   Copy and complete this table.

Fraction	$\frac{1}{1000}$	$\frac{1}{100}$	$\frac{1}{10}$	$\frac{1}{8}$	$\frac{1}{5}$	$\frac{1}{4}$	$\frac{1}{2}$
Decimal			0.1			0.25	

6   Write these fractions as decimals. Use your table from **Q5** to help you.
    a $\frac{3}{8} = 3 \times \frac{1}{8} = 3 \times \square =$ **b** $\frac{2}{5}$ **c** $\frac{5}{4}$ **d** $\frac{7}{2}$

7   Work out
    a $\frac{1}{4}$ of 10 = 0.25 × 10 =   **b** $\frac{3}{5}$ of 20
    c $\frac{5}{2}$ of 10   **d** $\frac{3}{10}$ of 500 kg.

Q7 hint Converting a fraction to a decimal can make a calculation easier.

8   Which of these fractions is closest to $\frac{1}{4}$?
    $\frac{7}{20}, \ \frac{4}{10}, \ \frac{11}{4}, \ \frac{5}{12}$

Q8 strategy hint
Convert fractions to decimals.

Example 9

Write 0.723 as a fraction.

$0.723 = \frac{723}{1000}$ The smallest place value is 3 thousandths, so use denominator 1000.

9   Write these decimals as fractions in their simplest form.

a   0.61   b   0.78   c   0.229

d   0.450   e   0.096

Q9b hint
Simplify your answer.

10   Write these in order of size, starting with the smallest.

a $\frac{2}{3}$, 0.6, $\frac{5}{8}$, 0.628

b $\frac{3}{5}$, 0.605, $\frac{1}{2}$, 0.51

c $\frac{5}{2}$, $-\frac{17}{4}$, −4.5, 2.8

Q10 strategy hint Convert the fractions into decimals to compare. Write the original values in order.

11   **Communication** James says that $\frac{1}{7}$ of £250 is £35.71 rounded to the nearest penny.
Show that James is correct.

12   **Finance / Reasoning** Ed and Sam share a £200 car repair bill.
Ed pays $\frac{2}{3}$ and Sam pays $\frac{1}{3}$.

a   How much does each pay?

b   Will the amounts in your answers to part **a** pay the whole bill?

Reflect Should you round up or down when sharing a bill?

13   **Reasoning** Randell uses this formula to work out the time to cycle to his aunt's house.

Time (hours) = $\dfrac{\text{distance (km)}}{18}$

The distance is 42 km.

Work out the time. Give your answer in hours and minutes.

14   **Communication** Zule draws a pie chart to display the meals bought in the school canteen.

School meals bought

Show that the fraction of lasagne meals is $\frac{5}{12}$.

15   Sally's mum pays rent of £200 a week. The rent goes up to £240.
Write the new rent as a fraction of the old rent.
Give your answer in its simplest form.

Q15 hint

$\dfrac{240}{200} = \dfrac{\square}{20} = \dfrac{\square}{\square}$

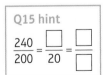

16   Craig buys a ring for £500. He sells it for £750.
Write the selling price as a fraction of the cost price in its simplest form.

17   Write 18 minutes as a fraction of an hour in its simplest form.

4.6 Fractions and percentages

Objectives

- Convert percentages to fractions and vice versa.
- Write one number as a percentage of another.

Did you know?

Per cent comes from the Latin 'per centum', which means out of a hundred.

Fluency

Convert to a percentage.

$\frac{1}{4}$, $\frac{10}{20}$, 0.5, 0.2, 0.68

Warm up

1. Convert to a decimal.

    a $\frac{7}{8}$ b $\frac{6}{15}$

2. Write these decimals as percentages.

    a 0.2   b 0.35

    c 0.8   d 0.55

3. Write 20 minutes as a fraction of 1 hour.

4. Write as a fraction in its simplest form

    a 8%   b 24%   c 65%

    d 64%   e 96%

    > **Q4a hint** 8% means '8 out of 100' or $\frac{8}{100}$. Simplify your answer.

5. 35% of students are driven to school.
   Write this percentage as a fraction in its simplest form.

6. Show that 15% is larger than $\frac{1}{10}$.

    > **Q6 hint** Write both as percentages.

Example 10

There are 20 students in a class. 6 are male. What percentage of the class is male?

Method A: $\frac{6}{20} \times 100\% = 6 \times \frac{\overset{5}{\cancel{100}}}{\underset{1}{\cancel{20}}}\%$

$= 30\%$

Method B: $\overset{\times 5}{\frac{6}{20}} = \frac{30}{100} = 30\%$ $\underset{\times 5}{}$

Convert to a fraction with denominator 100.

7. A class of 25 primary school children chose a musical instrument to learn.
   10 chose the violin. The rest chose the recorder.
   What percentage chose

   a the violin

   b the recorder?

   Reflect Add together your answers to **a** and **b**. What do you notice? Write a calculation to work out the percentage that chose the recorder, using the percentage that chose the violin.

ActiveLearn Homework, practice and support: Foundation 4.6

8 **Communication / Reasoning** A cinema wants to give a reduction of between 20% and 30% on a film price.
Which of these fractions could they use?
$\frac{1}{2}, \frac{3}{8}, \frac{1}{4}$
Show your working to explain your answer.

9 **Communication / Reasoning** Vakita says that 30% is $\frac{1}{3}$. Show that Vakita is wrong.

10 **Communication** Lucy is on work experience at a garage.
She draws a bar chart to show the number of cars serviced.

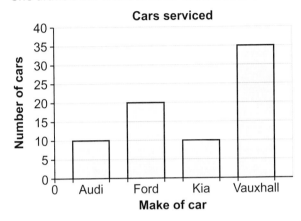

a What percentage of the cars serviced were Vauxhalls?
Lucy says there were half the number of Kias and Audis serviced than Vauxhalls.
b Show that Lucy is incorrect.

11 Write
a 15 as a percentage of 50
b 350 as a percentage of 750
c 70p as a percentage of £3.50
d 500 ml as a percentage of 8 litres.

> **Q11c, d hint** Write both amounts in the same units.

12 **Reasoning / Real** Harry got 70 out of 80 in a recent test. Jill scored 84% in the same test.
Who achieved the higher score? Explain your answer.

13 **Reasoning / Real** Elliot took a maths test and an English test.
He scored 35 out of 40 for maths and 50 out of 60 for English.
In which test did Elliot score the lower percentage?

14 There are 240 passengers on a train.
85 are getting off at the next station.
What percentage of passengers are getting off at the next station?
Round your answer to the nearest whole number.

> **Q14 hint** Use your calculator to convert $\frac{85}{240}$ to a decimal.

15 **Exam-style question**

A factory line produces 260 chocolate bunnies in an hour. 17 are rejected.
What percentage of the choclate bunnies are rejected?
Give your answer to 1 decimal place. **(3 marks)**

Exam hint
Write down the full calculator answer before you round it.

4.7 **Calculating percentages 1**

Objectives

- Convert percentages to decimals and vice versa.
- Find a percentage of a quantity.
- Use percentages to solve problems.
- Calculate simple interest.

Why learn this?

Consumer goods have reductions given as percentages.
You need to check if the percentage reductions are correct.

Fluency

- What is   50% of 26   25% of 200   10% of 70?
- Write $\frac{30}{100}$ as a decimal.

Warm up

1   Write these percentages as fractions.
   a   30%   b   60%   c   33.3̇%

2   Write these fractions as decimals.
   a $\frac{1}{5}$ b $\frac{4}{1000}$ c $\frac{9}{10}$

3   Write these percentages as decimals.
   a   65%   b   42%

> **Q3 hint** 65% = $\frac{65}{100}$ = 0.☐

4   The diagram shows a fraction–decimal–percentage triangle.

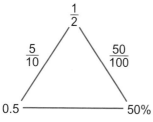

   Draw fraction–decimal–percentage triangles for
   a $\frac{1}{4}$ b   80%   c   0.6   d $\frac{2}{3}$

5   Write in order of size, starting with the largest.
   a   0.6, 65%, $\frac{1}{2}$, $\frac{11}{20}$ b   0.5%, $\frac{1}{100}$, 0.5 , $\frac{26}{100}$

> **Q5 hint** Convert to decimals.

6   A student loan has an interest rate of 3.3%.
   Write this percentage as a decimal.

> **Q6 hint**
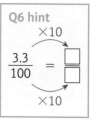

7   a   Write 20% as a fraction.
   b   Work out 20% of 1500.

8   Work out
   a   90% of 400   b   70% of 300   c   30% of 900.

9   a   Write 20% as a decimal.

      b   Work out 20% of 300.

 10   Work out these percentages by converting to a decimal first.

      a   60% of 375   b   35% of 600   c   5% of 280

 11   Jo buys a £25 coat reduced by 35% in a sale. How much money does Jo save?

12   Josh wants to buy a pair of jeans costing £40.50.

     He has two vouchers. Explain why Josh chose the 15% off voucher.

13   On Friday a shop sells 140 loaves of bread.

     25% of the loaves are wholemeal bread. How many of the loaves sold are not wholemeal?

     Reflect   Here are two ways of starting to solve this problem.

     1)   25% of 140

     2)   100% – 25% = 75%

     How did you start? Work out the answer in another way.

14   **Problem-solving / Real**   Ava's grandmother's annual council tax bill is £1050.60.

     She only pays 75% of this as she lives alone.

     The council says Ava's grandmother must pay £87.50 each month for a year.

     Check that the council are charging correctly. Show your working.

15   **Problem-solving / Real**   Ashley sells a chair on an internet auction site.

     He pays £6 to put the chair in the auction and 5% of the selling price.

     The chair sells for £140. How much does Ashley have to pay?

> **Key point 10**
>
> Percentages can be bigger than 100%. The cost of a loaf of bread has increased by more than 100% since 1975.

16   Write as a decimal

     a   130%   b   120%   c   350%

> **Q16a hint** $130\% = \frac{130}{100} = \boxed{}.\boxed{}$

17   A loaf of bread costs $1\frac{1}{2}$ times as much today as it did 10 years ago. What percentage has it increased by?

> **Q17 hint** $1\frac{1}{2} = 1.5 = \boxed{}\%$

 18   Work out

     a   150% of £80   b   125% of 200 ml   c   250% of £700

19   **Finance**   A used car dealer buys a car for £8000. She sells it at 130% of this cost.

     How much did the car dealer sell the car for?

20   **Finance**   A supermarket buys chicken for £1.35 per kg and sells it at 250% of the cost price.

     How much does the supermarket sell 750 g of chicken for? Write your answer to the nearest penny.

Key point 11

Simple interest is interest paid out each year by banks and building societies.

Example 11

Find the simple interest when £5000 is invested at 2.75% per annum over 2 years.

2.75% = 0.0275 ——— | Convert the percentage to a decimal multiplier. |

5000 × 0.0275 = £137.50 ——— | This is the interest earned over 1 year. |

£137.50 × 2 = £275

| Multiply your answer by 2. |

| Communication hint
Per annum or p.a. means 'each year'. |

21 **Finance** Find the simple interest when
  a   £3000 is invested at 2.75% per annum (p.a.) over 1 year
  b   £250 is invested at 3.25% p.a. over 3 years
  c   £4000 is invested at 2.2% p.a. over 18 months.

22 **Finance / Real** John's grandmother gives him £2000. He saves the money in a bank account with a simple interest rate of 6.2% per annum.
How much money will John have in the bank account after 30 months?

23 **Problem-solving / Finance** The cost of living increased by 30% from 2004 to 2014.
In 2004, Sharnia's wage was £240 a week.
In 2014, her wage was £300 a week.
Explain if Sharnia's wage has increased more or less than the cost of living.

> **Q23 hint** Increase £240 by 30%. Compare the answer with £300.

24 **Finance** Harry buys a car for £9800.
The value of the car **depreciates** by 15% each year.
Work out the value of the car at the end of the year.

> **Q24 communication hint**
> **Depreciates** means that the value of the car decreases.

25 **Exam-style question**

Jessica's annual income is £12 000
She pays 10% of the £12 000 in rent.
She spends $\frac{1}{4}$ of the £12 000 on clothes.
Work out how much of the £12 000 Jessica has left. **(2 marks)**

> **Exam hint**
> Make sure all your calculations are linked with a label, for example, Rent = …

4.8 Calculating percentages 2

Objectives

- Calculate percentage increases and decreases.
- Use percentages in real-life situations.
- Calculate VAT (value added tax).

Did you know?

Inflation is a percentage increase.
Deflation is a percentage decrease.
Inflation increases the price of goods and deflation decreases the price of goods.

Fluency

- Work out
  100% + 10%   100% – 40%   100% – 70%
- Change to a decimal
  20%   350%

1   Work out
    a   10% of 40
    b   110% of 120
    c   250% of 70.

Key point 12

To increase a number by a percentage, work out the increase and add this to the original number.

Example 12

Increase 30 by 10%.

10% of 30 = 3 ——— | Work out 10%. |

30 + 3 = 33 ——— | Add your answer to the original number. |

2   Increase
    a   33 by 20%
    b   400 by 33%
    c   2000 by 0.1%.

3   **Reasoning** Daisy works part time for £40 a week. Her employer agrees to raise her weekly wage by 3%. What is Daisy's new weekly wage?

Key point 13

To decrease a number by a percentage, work out the decrease and subtract this from the original number.

Example 13

Decrease 60 by 30%.

30% of 60 = 0.3 × 60 = 18 ——— | Convert the percentage to a decimal multiplier. |

60 − 18 = 42 ——— | Subtract your answer from the original number. |

4   Decrease
    a   48 by 25%
    b   600 by 45%
    c   5000 by 0.6%.

5   **Problem-solving** A carpet company reduces its prices by 40%.
    A carpet was £300. What is the new price of the carpet?

6   **Problem-solving** Marcia took 5% off her 5000 m running time during school sports day. Her time was previously 25 minutes. What is her new time?

7   Train fares are increasing by 25%. A train fare was originally £18.
    a   Work out 25% of £18.
    b   Work out the new price of the train fare.
    c   Work out 125% of £18.
    d   What do you notice about your answers to parts **b** and **c**? Explain.

8   Write the decimal multiplier you can use to work out an increase of
    a   20%
    b   75%
    c $87\frac{1}{4}$%.

    | Q8a hint |
    | 100% + 20% = ☐% = ☐.☐ |

9   Increase
    a   120 by 20%
    b   500 by 5%
    c   5000 by 87.5%.
    d   1.5 by 140%

10 Write the decimal multiplier you can use to work out a decrease of
    a   40%
    b   35%
    c   6%
    d   1.2%.

    | Q10a hint |
    | 100% − 40% = 60% |
    | = ☐.☐ |

11 Decrease

a 200 by 40%   b 500 by 35%   c 130 by 6%   d 4000 by 1.2%.

Discussion Explain why you cannot decrease a number by more than 100%.

12 **Problem-solving / Real** A 5% service charge is added to the cost of a meal.
A meal costs £52. What is the total charge?

13 | **Exam-style question**

Ashley wants to buy some tins of paint.

He finds out the costs of paint at two shops.

Paint R Us	**Deco Mart**
Normal price £2.19 a tin	Normal price £1.80 a tin
Special Offer	Special Offer
Buy 2 tins at the normal price and get the 3rd tin free	10% off the normal price

Ashley needs 9 tins of paint.
Ashley wants to get all the tins of paint from the same shop.
He wants to pay the cheapest possible total price.
Which of the two shops should Ashley buy the paint from?   **(6 marks)**

Nov 2012, Q16, 1MA0/2F

Exam hint
Make sure you make a statement at the end that includes the cost of paint for each shop (with units), from which shop Ashley should buy the paint, and why.

14 **Finance / Reasoning** Orlav is self-employed. Last year, he earned £18 940.
He does not pay income tax on the first £10 000 he earned.
He pays tax of 20% for each pound he earned above £10 000.
How much tax must he pay?

15 **Problem-solving / Real** Debbie's mum gets two quotes for plumbing work.
Quote 1   £660 including VAT
Quote 2   £500 excluding VAT
Which is the more expensive quote? Show your working.

Q15 hint Add VAT at 20% to Quote 2.

16 **Problem-solving / Real** Ben buys two tickets for a football match.
The cost of one ticket is £299 plus VAT.
VAT is 20%. Work out the cost of the tickets.

17 **Problem-solving** A company plans a competition for its employees.
Each winner gets a prize of two tickets to a sporting event.
The cost of two tickets is £120 plus VAT at 20%.
The company wants to spend up to £1000. How many winners can there be?

18 **Problem-solving / Real** Mr Elliot and his five children are going to London by train.
An adult ticket costs £24. A child ticket costs £12.
Mr Elliot has a family rail card which gives $\frac{1}{3}$ off adult tickets and 60% off child tickets.
Work out the cost of the tickets.

19 **Reasoning** Luke's head teacher predicts there will be 15% more Year 7 students next year.
This year there are 220 students in Year 7.
How many Year 7 students does Luke's head teacher predict for next year?

4 Problem-solving

Objective • Use bar models to help you solve problems.

Example 14

Sophie spends 40% of her birthday money on a necklace, and $\frac{1}{2}$ of the remainder on earrings.
She is left with £16.50. How much birthday money did Sophie receive?

Draw a rectangular bar to represent all the birthday money.

necklace

| 10% | 10% | 10% | 10% | 10% | 10% | 10% | 10% | 10% | 10% |

Split the bar into 10% sections. Label 40% for the necklace.

necklace   earrings

Label $\frac{1}{2}$ of the remainder for earrings.

necklace   earrings   £16.50

Label the £16.50 she has left.

1 section = £16.50 ÷ 3 = £5.50

3 sections = £16.50. Work out 1 section.

Total birthday money = £5.50 × 10 = £55

The bar represents all the birthday money.

Check: Sophie spends

Check your answer.

40% of £55 on a necklace: 40% of £55 = 4 × 10% of £55 = 4 × £5.50 = £22

$\frac{1}{2}$ of the remainder on earrings: remainder = £55 − £22 = £33; $\frac{1}{2}$ of £33 = £16.50

She has £16.50 left.

Total = £22 + £16.50 + £16.50 = £55 ✓

---

1   At a pantomime, 60% of the audience are children.
    $\frac{1}{4}$ of the remaining audience are men.
    The rest of the audience are women.
    There are 45 women in the audience.
    How many people are at the pantomime?

> **Q1 hint** Draw a bar to represent all the audience. Split the bar into 10% sections. Work out what 1 section represents. Then work out the total number of people in the audience.

2   A small business has 16 employees. They are working on
    two different projects, project A and project B.
    4 employees are working only on project A.
    6 employees are working on both project A and project B.
    2 employees are not working on either project.
    a   How many employees are working on project B only?
    b   How many employees, in total, are working on project B?

> **Q2a hint** Draw a bar to represent all the employees working for the business.
> How many sections should you split the bar into?

> **Q2b hint** Total number of employees working on project B = employees working on both project A and project B + employees working on project B only.

3   $\frac{1}{6}$ of stalls at a market sell food.
    $\frac{2}{3}$ of food stalls sell fruit and vegetables.
    What fraction of the market stalls sell fruit and vegetables?
    Give your answer as a fraction in its simplest form.

4   **Finance** Amy spent $\frac{1}{3}$ of her savings on a car.
    She spent $\frac{5}{12}$ of the remainder on a holiday.
    The car cost £200 more than the holiday.
    How much of her savings does she have left?
    Discussion How did you decide how to split the bar?

> **Q4 hint** Draw a bar to represent all of Amy's savings.
> Compare the sections for 'car' and 'holiday'. Label the difference £200.

5   Toby, Ed and Isy throw balls into a basketball hoop.
    Toby gets 5 more balls in the net than Ed.
    Isy gets 3 more balls in the net than Toby.
    Write an expression for the total number of balls
    they all got in the net, using x for the number of balls
    Ed gets in.

> **Q5 hint** Start by drawing a bar for Ed. Label the bar x.
> Draw two more bars beneath, one for Toby and one for Isy.

6   At a skateboarding competition, each skateboarder performs a trick.
    $\frac{3}{8}$ perform a flip, 40% of the remaining skaterboarders perform an aeriel, $\frac{1}{3}$ of the
    skateborders who have not performed either a flip or an aeriel perform a grind.
    The rest perform a slide.
    3 skateborders perform a grind.
    Draw a bar chart to show the number of
    skateboarders that perform each trick.

> **Q6 hint** Start by drawing a bar model to work out the number of skateboarders that perform each trick.

7   **Reflect**
    a   How do you feel when you can't solve a
        problem straight away?
    b   What do you do?

> **Q7b hint** Do you read the hint for that question? Is it OK to move on to the next problem, and then come back to the one you were stuck on?

4 Check up

Log how you did on your Student Progression Chart.

Operations with fractions

1   Work these out. Simplify your answers where needed.
    a $\frac{1}{5} + \frac{1}{3}$ b $\frac{3}{8} + \frac{1}{4}$ c $\frac{3}{4} - \frac{1}{2}$ d $\frac{11}{12} - \frac{1}{8}$

2   Work these out. Simplify if necessary.
    a $\frac{1}{2} \times \frac{1}{3}$ b $\frac{2}{5} \times \frac{3}{5}$ c $\frac{2}{7} \times \frac{14}{5}$

3   Work out
    a $4 \div \frac{3}{8}$ b $\frac{3}{4} \div 8$ c $\frac{3}{4} \div \frac{1}{2}$

4   Work these out. Write your answer as a mixed number.
    a $\frac{7}{8} + \frac{1}{6}$ b $3\frac{3}{4} + 2\frac{1}{8}$ c $5\frac{3}{4} - 3\frac{1}{2}$ d $4\frac{1}{2} - 2\frac{3}{4}$

Percentages, decimals and fractions

5   Write $\frac{3}{8}$ as a decimal.

6   Write these decimals as fractions in their simplest form.
    a   0.007   b   0.325

7   Work out $\frac{3}{8} \times 100$, giving your answer as a decimal.

8   Write as a fraction in its simplest form

   a   5%   b   28%   c   150%

9   Write as a percentage

   a $\frac{3}{25}$ b $\frac{3}{8}$

10 Put these in order of size, smallest first.

   $\frac{3}{5}$, 0.62, 58%, $\frac{2}{3}$

11 **Reasoning** a A 450 g pot of yoghurt contains 60 g of fruit.
       What percentage of the total is fruit? Give your answer to the nearest whole number.
   b   A 300 g pot of yogurt contains 45 g of fruit. Does this have a greater or smaller
       percentage of fruit?

Calculating percentages

12 Work out

   a   12% of 150 cm   b   8% of 225 km   c   130% of £82.

13 Lindy buys a bag priced at £22. She gets 15% off. How much does she save?

14 A price of £65 is increased by 20%. Work out the new price.

15 £500 is reduced by 35%. Work out the new price.

16 Building work costs £8500 + VAT. VAT is 20%. Work out the total cost.

17 Find the simple interest when £2500 is invested for 1 year at 1.75%.

18 How sure are you of your answers? Were you mostly

   Just guessing   Feeling doubtful   Confident ?

   What next? Use your results to decide whether to strengthen or extend your learning.

✷ Challenge

19 Ole says $2\frac{1}{5}$ is 2.5.
   Explain why Ole is wrong.

---

4 Strengthen

Operations with fractions

1   Work these out.

   a $\frac{1}{4} + \frac{1}{2}$ b $\frac{3}{8} + \frac{1}{4}$

   c $\frac{2}{9} + \frac{1}{3}$ d $\frac{1}{3} - \frac{1}{6}$

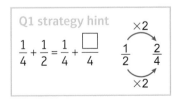

Q1 strategy hint

$\frac{1}{4} + \frac{1}{2} = \frac{1}{4} + \frac{\square}{4}$

$\frac{1}{2} \xrightarrow{\times 2} \frac{2}{4}$ (×2)

2   Work out

   a $\frac{1}{8} + \frac{1}{6}$ b $\frac{1}{5} + \frac{1}{8}$

   c $\frac{1}{2} - \frac{1}{3}$ d $\frac{1}{3} - \frac{1}{10}$

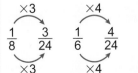

Q2a hint The lowest number that 6 and 8 'go into' is 24.

$\frac{1}{8} \xrightarrow{\times 3} \frac{3}{24}$ $\frac{1}{6} \xrightarrow{\times 4} \frac{4}{24}$

$\frac{1}{8} + \frac{1}{6} = \frac{\square}{24} + \frac{\square}{24}$

3 Work these out. Simplify if needed.

   a $\frac{5}{6} - \frac{1}{4}$ b $\frac{3}{4} + \frac{1}{5}$

   c $\frac{7}{12} + \frac{1}{8}$ d $\frac{2}{5} - \frac{1}{4}$

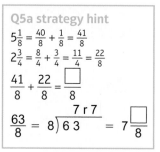

Q3a hint $\frac{5}{6} - \frac{1}{4} = \frac{\square}{12} - \frac{\square}{12}$ $\frac{5}{6} = \frac{10}{12}$ ($\times 2$) $\frac{1}{4} = \frac{3}{12}$ ($\times 3$)

4 Work these out. Give each answer as a mixed number.

   a $\frac{3}{4} + \frac{1}{2}$ b $\frac{2}{3} + \frac{5}{9}$

   c $\frac{4}{5} + \frac{1}{4}$ d $\frac{6}{7} + \frac{1}{2}$

Q4a hint

$\frac{3}{4} + \frac{1}{2} = \frac{3}{4} + \frac{2}{4} = \frac{\square}{4} = 1\frac{\square}{4}$

5 Work out

   a $5\frac{1}{8} + 2\frac{3}{4}$ b $4\frac{3}{4} + 3\frac{1}{2}$

   c $3\frac{1}{2} + 4\frac{7}{8}$ d $3\frac{3}{4} + 2\frac{5}{8}$

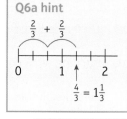

Q5a strategy hint

$5\frac{1}{8} = \frac{40}{8} + \frac{1}{8} = \frac{41}{8}$

$2\frac{3}{4} = \frac{8}{4} + \frac{3}{4} = \frac{11}{4} = \frac{22}{8}$

$\frac{41}{8} + \frac{22}{8} = \frac{\square}{8}$

$\frac{63}{8} = 8\overline{)63}^{\,7\ r\ 7} = 7\frac{\square}{8}$

6 Work these out. The first one has been worked out for you.

   a $2 \times \frac{2}{3} = \frac{2 \times 2}{3} = \frac{4}{3} = 1\frac{1}{3}$

   b $2 \times \frac{3}{4}$ c $5 \times \frac{4}{7}$ d $15 \times \frac{3}{5}$

Q6a hint

$\frac{2}{3} + \frac{2}{3}$

$\frac{4}{3} = 1\frac{1}{3}$

7 Work out

   a $5\frac{3}{4} - 2\frac{1}{2}$ b $4\frac{5}{7} - 1\frac{2}{7}$ c $3\frac{1}{4} - 1\frac{1}{2}$

8 Work these out.

   a $\frac{1}{2} \times \frac{1}{3} = \frac{1 \times 1}{2 \times 3} = \frac{1}{\square}$

   b $\frac{1}{3} \times \frac{1}{5}$ c $\frac{2}{5} \times \frac{2}{3}$ d $\frac{4}{5} \times \frac{1}{2}$

   e $\frac{3}{4} \times \frac{3}{7}$ f $\frac{7}{11} \times \frac{4}{5}$ g $\frac{4}{5} \times \frac{4}{5}$

Q8a hint

$\frac{1}{6}$ $\frac{1}{3}$

9 Work these out. The first one has been worked out for you.

   a $\frac{2}{5} \times \frac{15}{4} = \frac{15 \times 2}{5 \times 4} = \frac{15}{5} \times \frac{2}{4} = 3 \times \frac{1}{2} = 1\frac{1}{2}$

   b $\frac{1}{2} \times \frac{4}{7}$ c $\frac{3}{4} \times \frac{8}{10}$ d $\frac{4}{5} \times \frac{5}{7}$

   e $\frac{2}{3} \times \frac{9}{12}$ f $\frac{3}{4} \times \frac{12}{15}$ g $\frac{7}{8} \times \frac{12}{21}$

Q9 hint Change the order so you can cancel common factors.

10 Work out

   a $3 \div \frac{2}{3} = 3 \times \frac{3}{2} = \frac{9}{2} = 4\frac{\square}{2}$

   b $4 \div \frac{1}{4}$ c $4 \div \frac{2}{3}$ d $7 \div \frac{2}{3}$

Q10a hint

3 $\div$ $\frac{2}{3}$

stick   switch   flip

$\downarrow$ $\downarrow$ $\downarrow$

3 $\times$ $\frac{3}{2}$

11 Work these out. The first one has been done for you.

   a $\frac{1}{2} \div 3 = \frac{1}{2} \times \frac{1}{3} = \frac{1}{6}$

   b $\frac{1}{3} \div 4$ c $\frac{2}{5} \div 6$ d $\frac{3}{7} \div 6$

Q11a hint

$\frac{1}{2}$

$\frac{1}{6}$

12 Work out these out. Write your answer as a mixed number if required.

   a $\frac{1}{4} \div \frac{2}{15}$ b $\frac{7}{12} \div \frac{1}{3}$

   c $\frac{2}{5} \div \frac{3}{4}$ d $\frac{3}{7} \div \frac{6}{21}$

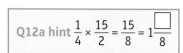

Q12a hint $\frac{1}{4} \times \frac{15}{2} = \frac{15}{8} = 1\frac{\square}{8}$

Percentages, decimals and fractions

1 Write these fractions as decimals. The first one has been started for you.

a $\frac{5}{8}$

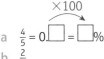

$\frac{5}{8} = 8\overline{)5.^5 0^2 0^4 0}$ with $0.\,6\square\square$

b $\frac{3}{8}$ c $\frac{3}{5}$

2 Write these decimals as fractions. Simplify if necessary.

a $0.6 = \frac{6}{10} = \frac{\square}{5}$ b $0.03 = \frac{\square}{100}$ c 0.25

d $0.005 = \frac{\square}{1000}$ e 0.033 f 0.125

3 Write as a fraction in its simplest form

a $2\% = \frac{2}{100} = \frac{1}{\square}$ b $8\% =$ c 48% d 78% e $160\% = \frac{160}{100} =$ f 180%

4 Write as a percentage

a $\frac{14}{50} = \frac{28}{100} = \square\%$ b $\frac{41}{50}$ c $\frac{3}{25}$

5 Write as a decimal, then a percentage

$\overset{\times 100}{\frown}$

a $\frac{4}{5} = 0.\square = \square\%$

b $\frac{2}{5}$ c $\frac{3}{8}$

> **Q5 hint**
> $5\overline{)4.0}$ with $0.\square$

6 Place these in order, smallest first.

a $2\frac{3}{4}$, 2.35, 223%

b $\frac{3}{8}$, 0.357, 0.335, 35%

c 12.5%, 0.127, $\frac{5}{8}$, 55%

d $\frac{3}{4}$, 73.2%, 0.07, 72.3%

> **Q6 hint** Convert to decimals. Write the decimals in order. Write the original numbers in order.

7 a Write 15 as a fraction of 40.
   b Write 34 as a fraction of 50.

> **Q7a hint**
> $\frac{15}{40} = \frac{\square}{\square}$ with $\div 5$ above and $\div 5$ below

8 a Write 24 as a percentage of 50.
   b Write 8 as a percentage 50.
   c Write 12 as a percentage of 25.
   d Write 10 as a percentage of 40.

> **Q8a hint**
> $\frac{24}{50} = \frac{\square}{100} = \square\%$ with $\times 2$ above and $\times 2$ below

9 Write these percentages as decimals.

a $40\% = \frac{40}{100} = \frac{4}{10} = 0.\square$ with $\div 10$ above and $\div 10$ below

b 70%   c $2\% = \frac{2}{100} = 0.\square\square$ d 2.5%

> **Q9d hint**
> $2\% = 0.\square\square$
> $2.5\% = 0.\square\square\square$

Calculating percentages

1 Work out

a 30% of 80   b 20% of 70   c 60% of 220

d 5% of 60   e 25% of 200   f 75% of 1250.

> **Q1 hint** 30% of 80 = 0.3 × 80
> or 30% of 80 = $\frac{3}{10}$ × 80

2 Work out
  a   12% of 46
  b   23% of 150
  c   47% of 700
  d   124% of 90

3 a   Increase 40 by 10%.
   b   Increase 30 by 10%.
   c   Increase 20 by 50%.
   d   Increase 180 by 80%.
   e   Increase £3000 by 5%.

> **Q3 hint**
> 100% + 10% = 110% = 1.1
> 1.1 × 40 = ☐

4 a   Decrease 70 by 10%.
   b   Decrease 60 by 10%.
   c   Decrease 90 by 20%.
   d   Decrease 40 by 60%.
   e   Decrease £1000 by 5%.

> **Q4 hint**
> 100% − 10% = 90% = 0.9
> 70 × 0.9 = ☐

5 **Finance / Real** VAT is 20%. Add VAT to
   a   £750
   b   £1200
   c   £32

6 **Finance / Real**
   a   Find the simple interest when £1500 is invested for 1 year at 1.75%.
   b   Find the simple interest when £10 000 is invested for 2 years at 1.75%.
   c   Find the simple interest when £1.5 million is invested for 2 years at 3.75%.

> **Q6b hint**
> Simple interest for 2 years = 2 × simple interest for 1 year.

4 Extend

1 **Real / Problem-solving** Scotland has an area of 30 000 square miles.
   $\frac{1}{6}$ of the area is woodland.
   $\frac{3}{4}$ of the area of woodland is covered with pine trees.
   Work out the area of woodland covered with pine trees.

2 **Reasoning** There are 21 questions in a test on biology, chemistry and physics.
   There are 7 biology questions and 8 chemistry questions.
   a   What fraction of the questions are on biology?
   b   What percentage of the questions are on physics?

3 **Reasoning / Finance** Last year Sasha was paid £3400 for her weekend job.
   This year Sasha's pay increased by 2%.
   How much pay does Sasha get this year?

4 **Real / Problem-solving** Cherry wants to buy a pair of designer jeans.
   In Jeans R Us the original price for the jeans was £120.
   This price has now been reduced by 25%.
   The original price for the same pair of jeans at Designer Bargains was £110 but has been reduced by 20%.
   At which shop would Cherry pay less for the jeans?

> **Q4 strategy hint** Show your working for Jeans R Us and Designer Bargains separately.

5 Millie is collecting small coins in a jar.
   She empties the jar and counts 120 coins.
   $\frac{2}{5}$ are 1p coins.
   25% are 2p coins.
   The rest are 5p coins.
   What percentage of the coins are 5p coins?

> **Q5 strategy hint** Work out the percentage of 1p coins first.

6 **Problem-solving** Anwar and Bethany each earn the same weekly wage.

Each week, Anwar saves 12% of his wage and spends the rest.

Each week, Bethany spends $\frac{7}{8}$ of her wage and saves the rest.

Who saves more money each week?

You must show each stage of your working.

7 **Exam-style question**

Kylie wants to invest £1000 for one year.

She considers two investments, Investment A and Investment B.

Investment A	**Investment B**
£1000	£1000
Earns £2.39 per month	Earns 3.29%
plus	interest per annum
£4.50 bonus for each complete year	

Kylie wants to get the greatest return on her investment.

Which of these investments should she choose?

8 **Reasoning** Last year, Jake spent 20% of his salary on rent, $\frac{2}{5}$ of his salary on entertainment and $\frac{1}{4}$ of his salary on living expenses. He saved the rest of his salary.

Jake spent £3600 on living expenses.

Work out how much money he saved.

9 **Exam-style question**

Jennie's council has a target for households to recycle $\frac{1}{5}$ of their waste.

In January, Jennie recycled $\frac{1}{10}$ of her household waste.

In February, she recycled 15 kg of her 120 kg of household waste.

Her result for March was 13% recycled out of 112 kg of household waste.

Has Jennie met the council's target?

Which was her best month for recycling?

Exam hint

For each month write the amount recycled in the same form, for example all as fractions or all as percentages.

10 **Reasoning** There are 2 adults and 2 children in the Smith family.

The family wants a holiday for 7 nights, starting on 1st August.

	Park Palace		Dubai Grand	
Date holiday starts	5 nights	extra night	5 nights	extra night
16 Jul – 20 Aug	£810	£80	£854	£53
21 Aug – 10 Dec	£810	£80	£869	£94
Discount for each child	$\frac{1}{5}$ off		15% off	

One hotel will be cheaper for them than the other hotel.

Work out the cost of the cheaper holiday.

11 **Finance** Des buys two tyres with valves and balancing and has to pay VAT at 20%.

Work out the total amount Des pays for the tyres.

Dunlap tyres: £62 each

Valves: 50p per tyre

Balancing: £1 per tyre

117

12 **Finance / Reasoning** Nadine receives an electric bill for £214 plus VAT.
VAT on fuel is set at 5%. How much must Nadine pay?

13 **Finance / Reasoning** A new television normally costs £800. This price has been reduced by 10%.
For loyalty card members, the price is further reduced by 5%.
a   How much will the television cost without a loyalty card?
b   Jamelia says that if she uses her loyalty card she will get 15% off the normal price.
Is she correct? Explain.

14 **Reasoning / Real** The value of a car depreciates by 20% each year. The price when new was £7500.
Work out the value of a two-year-old car.

> **Q14 hint** Work out the value after 1 year, then decrease that by another 20%.

15 **Reasoning** Gunnar is doing a project for his health and social care course.
He finds the following information.

Year	Percentage of patients unable to secure a GP or nurse appointment
2013	8.9
2014	10.3

> **Q15 hint** Increase 8.9% by 10% and compare the values.

Gunnar writes, 'The percentage of patients unable to secure an appointment has increased by more than 10%.'
Is Gunnar correct? Show your working.

4   Knowledge check

- ⊙ To compare, add or subtract fractions, write them with a common **denominator**. .... *Mastery lesson 4.1*

- ⊙ To add or subtract **mixed numbers**, it often helps to convert to **improper fractions** first.
For example, write $2\frac{2}{3}$ as $\frac{8}{3}$. .... *Mastery lesson 4.2*

- ⊙ To multiply fractions together, multiply the **numerators** together and the denominators together. .... *Mastery lesson 4.3*

- ⊙ The **reciprocal** of a fraction is the original fraction turned upside down. The reciprocal of 2 (or $\frac{2}{1}$) is $\frac{1}{2}$. .... *Mastery lesson 4.4*

- ⊙ Dividing by a number is the same as multiplying by its reciprocal.
For example, $12 \div 2$ means $\frac{1}{2}$ of 12 or $12 \times \frac{1}{2}$. .... *Mastery lesson 4.4*

- ⊙ To convert a fraction to a decimal, divide the numerator by the denominator. .... *Mastery lesson 4.5*

- ⊙ **Percentage** means 'out of 100'. For example $24\% = \frac{24}{100}$. .... *Mastery lesson 4.6*

- ⊙ You multiply a fraction by 100 to change it to a percentage.
For example $\frac{3}{20} = \frac{3 \times 100}{20}\% = \frac{300}{20}\% = 15\%$. .... *Mastery lesson 4.6*

- ⊙ **Simple interest** is interest paid out each year by banks and building societies. .... *Mastery lesson 4.7*

⊙ To increase a number by a percentage, work out the increase and add this to the original number.
To decrease a number by a percentage, work out the decrease and subtract this from the original number. .... *Mastery lesson 4.8*

Look back at the topics in this unit.

a Which one are you most confident that you have mastered?
What makes you feel confident?

b Which one are you least confident that you have mastered?
What makes you least confident?

c Discuss a question you feel least confident about with a classmate.
How does discussing it make you feel?

4 Unit test

Log how you did on your Student Progression Chart.

1 Samuel gets 24 out of 50 in a science test.
Write 24 out of 50 as a percentage. *(2 marks)*

2 **Reasoning** Wayne took two tests.
He got $\frac{12}{25}$ in the first test and $\frac{56}{100}$ in the second test.
Wayne says he did better in the first test. Is he correct? Show why you think this. *(2 marks)*

3 Put these in order of size, smallest first.
0.88, $\frac{4}{5}$, $\frac{8}{9}$, $\frac{43}{50}$, 0.82 *(2 marks)*

4 a Increase £75 by 5%. *(2 marks)*
  b Decrease £150 by 12%. *(2 marks)*

5 The rate of simple interest is 3% per year.
Work out the simple interest paid on £8000 in one year. *(2 marks)*

6 Work out $2\frac{2}{3} \times 3$. *(3 marks)*

7 Work out $\frac{2}{3} + \frac{1}{5}$. *(2 marks)*

8 A tiling bill is £360 plus VAT at 20%.
Work out the total amount. *(3 marks)*

9 **Reasoning** Here are two fractions: $\frac{3}{4}$ and $\frac{4}{5}$.
Which is the larger fraction?
You must show your working to explain your answer. *(2 marks)*
You may use the grids to help with your explanation.

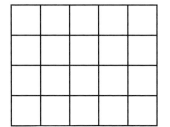 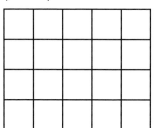

10 **Reasoning** A gift shop buys ribbon in rolls.
Each roll has 30 m of ribbon.
Eloise uses a machine to cut ribbon into lengths of $\frac{2}{5}$ m.
She needs 400 lengths. How many rolls of ribbon does she need? *(4 marks)*

11 Rosemary has £18 200 in a bank account that pays simple interest of 1.2%.
How much interest does she receive each year? *(2 marks)*

12 **Reasoning** William needs a new mobile phone contract for 1 year.
He can get 25% off the £150 contract cost or pay £24.61 for 1 month and
£7.99 for the next 11 months.
Show that whatever method of payment he chooses, William pays the same amount. *(4 marks)*

13 Work out $5\frac{2}{3} - 2\frac{3}{4}$. *(3 marks)*

14 **Reasoning** Karim is mixing some ingredients for a cake.
He needs $1\frac{1}{3}$ cups of sugar, $1\frac{1}{3}$ cups of flour and $2\frac{1}{2}$ cups of raisins.
His mixing bowl holds 5 cups. Will the bowl be large enough? *(3 marks)*

Sample student answer

The student has shown all the correct working out, but will only get 4 out of the 5 marks.
Explain why.

Exam-style question

Samantha wants to buy a new pair of trainers.

There are 3 shops that sell the trainers she wants.

Sports '4' All Trainers	Edexcel Sports Trainers	Keef's Sports Trainers
£5 plus 12 payments of £4.50	$\frac{1}{5}$ off usual price of £70	£50 plus VAT at 20%

From which shop should Samantha buy her trainers to get the best deal?

You must show all of your working. **(5 marks)**

Practice Paper Set A, Q22, 1MA0/1F

Student answer

12 × 4.50 = 54 $\frac{1}{5}$ of 70 =   20% of 50
54 + 5 = 59   70 ÷ 5 = 14   10% = 5
          70 − 14 = 56   20% = 10
                    50 + 10 = 60

5 EQUATIONS, INEQUALITIES AND SEQUENCES

Patient safety depends upon a nurse's ability to use precise calculations.
Nurses use a formula to calculate doses.

They use this formula: dose $= \dfrac{DV}{H}$

What is the dose when $D = 5000$ units, $H = 10\,000$ units and $V = 1$ ml?

5 Prior knowledge check

Numerical fluency

1 Write $>$ or $<$ between each pair of numbers.
  a 3 … 5
  b 7 … 6
  c 9 … −7
  d −6 … 2

2 Work out
  a $3 \times 2 + 1$
  b $5 \times 3 - 6$
  c $\dfrac{9}{3}$
  d $\dfrac{15}{5}$

3 Find the missing number.
  a $10 \div 2 = \square$
  b $\square \times 6 = 18$
  c $\square \times 7 = -14$
  d $-16 \div 2 = \square$
  e $-81 \times -1 = \square$
  f $63 \div \square = -7$
  g $8 \times -4 = \square$
  h $-56 \div \square = -7$

Algebraic fluency

4 a Draw pattern 4.

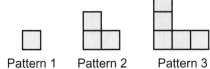

  Pattern 1   Pattern 2   Pattern 3

  b How many squares would be in pattern 6?
  c What is the term-to-term rule?

5 Write the next two terms and the term-to-term rule for each sequence.
  a 2, 6, 10, 14, 18, $\square$, $\square$
  b 32, 28, 24, 20, 16, $\square$, $\square$
  c 5, 7, 9, 11, 13, $\square$, $\square$
  d 32, 27, 22, 17, 12, $\square$, $\square$

6 Find the missing two terms in each sequence.
  a 2, 4, 6, $\square$, $\square$, 12, 14
  b 1, 4, 7, $\square$, $\square$, 16, 19
  c 15, 13, 11, $\square$, $\square$, 5, 3
  d 44, 39, 34, $\square$, $\square$, 19, 14

7 Write down the 10th term in each sequence.
  a 7, 14, 21, 28, 35, …
  b 4, 7, 10, 13, 16, …

8 Find the next two terms in each sequence.
  a 1, 2, 4, 8, 16, $\square$, $\square$
  b 10, 100, 1000, 10 000, $\square$, $\square$
  c 2, 6, 18, 54, 162, $\square$, $\square$
  d 400, 200, 100, $\square$, $\square$
  Write down the term-to-term rule for each sequence.

121

9 Write down the outputs for each function machine.

a Input   Output

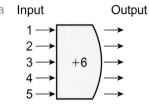

b Input   Output

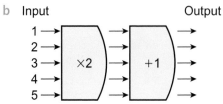

10 Simplify

a $2x + 3x$ b $2a + a - 10$

c $3b + 10 + 2b + 6$ d $4x + 3 + 2x - 5$

11 Write an expression for the perimeter of this shape.

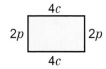

12 Solve

a $a + 3 = 4$ b $c - 6 = 4$

c $15 = g + 4$ d $21 + h = 23$

e $11 = k - 6$ f $l - 7 = 14$

13 Solve

a $4h = 40$ b $3m = 15$

14 Expand

a $2(a + 8)$ b $4(11 + c)$

c $5(9 - f)$ d $2(5b + 3)$

e $8(2 - 2x)$ f $-2(x + 3)$

15 Substitute $a = 3$, $b = 4$ and $h = 6$ to find the values of these expressions.

a $a + b$ b $a \times b$ c bh d $3a + b$

16 Substitute $u = 3$, $t = 5$ and $a = 2$ to find the values of these formulae.

a $v = u + at$ b $s = ut + \frac{1}{2}at^2$

* Challenge

17 When you add the expressions in each row, column or diagonal in this grid, the total is always the same.

$2a$	$7a$	$6a$
$9a$	$5a$	a
$4a$	$3a$	$8a$

Write each expression below into one of the cells in a similar grid so that when each row, column or diagonal is added, the simplified expression is the same.

$8a + 3b$	$2a + b$	$7a + 6b$
$5a + 2b$	$a + 4b$	$4a + 5b$
$6a + 9b$	$3a + 8b$	$7b$

5.1 Solving equations 1

Objectives

- Understand and use inverse operations.
- Rearrange simple linear equations.
- Solve simple linear equations.

Why learn this?

You can use an equation to work out the density of an object when you know its mass and volume.

Fluency

- What is the inverse of each function machine?
- Work out each output when the input is 2.

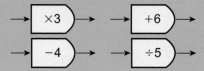

1 Solve these equations.

a $a + 3 = 9$ b $b - 4 = 6$

c $7 + c = 10$ d $11 - d = 9$

e $8 = f - 5$ f $4h = 12$

Q1a hint

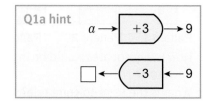

Key point 1

An **equation** contains an unknown number (a letter) and an '=' sign.
When you **solve** an equation you work out the value of the unknown number.

Questions in this unit are targeted at the steps indicated.

2   Copy and complete to solve this equation.

$\frac{y}{7} = 8$

$y = 8 \times \square$

$y = \square$

Q2 hint

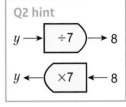

3   Solve

a $\frac{a}{10} = 9$ b $\frac{b}{3} = 4$ c $\frac{c}{5} = 4$ d $\frac{d}{4} = 3$ e $\frac{e}{8} = 8$ f $\frac{f}{6} = 2$

Key point 2

In an equation, the expressions on both sides of the = sign have the same value.
You can visualise them on balanced scales.

The scales stay balanced if you complete the
same operation to both sides. You can use
this **balancing method** to solve equations.

Example 1

Solve the equation $x - 3 = 7$

Visualise the equation as balanced scales.

The inverse of −3 is + 3. Do this to both sides
to keep the equation balanced.

$x = 7 + 3$

$x = 10$

Check: $x - 3 = 10 - 3 = 7$ ✓

4   Copy and complete to solve these equations using a balancing method.

a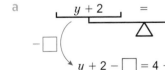

$y = \square$

b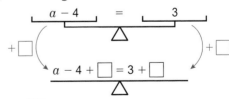

$a = \square$

c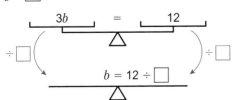

$b = \square$

d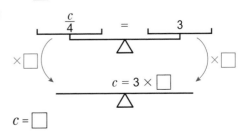

$c = \square$

Discussion How do you decide which operation to use on both sides?

123

5 Use a balancing method to solve

   a $5 + x = 7$ b $y - 7 = 11$ c $x - 4 = 1$ d $y + 2 = 6$

6 Simplify the left-hand side, then solve.

   a $2a + 5a = 28$ b $8p + 3p = 33$ c $4m - 2m = 6$ d $9q - q = 32$

7 **Reasoning** a Write an equation for the sum of these angles.

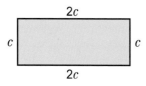

> **Q7a hint**
>
> $\square + \square + \square = 180°$

   b   Solve your equation to find the value of a.

8 **Reasoning** The perimeter of this rectangle is 30 cm.

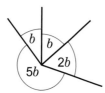

   a   Write an equation for the perimeter.
   b   Work out the value of c.
   c   Work out the length of the longer side.

> **Q8c hint**
>
> $2c = \square$

9 **Reasoning** Calculate the size of the largest angle.

> **Q9 hint** What do the angles around a point add up to?

10 ┌─────────────────────────────
    Exam-style question

    The perimeter of this rectangle is 60 cm.
    Work out the length and width.

    $4v$

    v **(4 marks)**

> **Exam hint**
>
> When you have found v, remember to find $4v$.

11 **Real** A T-shirt costs £s. The cost of 4 T-shirts is £36.
   a   Write an equation involving s.
   b   Solve your equation to find the cost of a T-shirt.

12 **Real** Alex has sold 11 games from his collection of n games. He now has 18 games.
   a   Write an equation involving n.
   b   Solve your equation to find the oiginal number of games in his collection.

13 **Reflect** The example used scales to help visualise balancing an equation.
   Did this help you? How?

5.2 Solving equations 2

Objectives

- Solve two-step equations.

Why learn this?

Knowing how to solve an equation helps you solve other problems, such as finding one mass when given another.

Fluency

What are the values of a, b, c and d?

- $a + 3 = 4$
- $7 - b = 4$
- $\dfrac{c}{3} = 2$
- $4d = 24$

1   Copy and complete the table for each function machine below.

a
Input $\longrightarrow$ ×4 $\longrightarrow$ +2 $\longrightarrow$ Output

b
Input $\longrightarrow$ ×3 $\longrightarrow$ −4 $\longrightarrow$ Output

Input	Output
1	
2	
3	
4	
5	

2   Simplify
a $8d - 4d$
b $5f + 3f$
c $2p + 3 + p + 9$
d $3 - 2b - 4 + 3b$

3   Copy and complete these inverse function machines.

a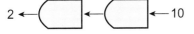
2 $\longrightarrow$ ×3 $\longrightarrow$ +4 $\longrightarrow$ 10

2 $\longleftarrow$ $\longleftarrow$ $\longleftarrow$ 10

b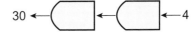
30 $\longrightarrow$ ÷5 $\longrightarrow$ −2 $\longrightarrow$ 4

30 $\longleftarrow$ $\longleftarrow$ $\longleftarrow$ 4

4   a   Solve the equation $2x + 5 = 9$ using this function machine.

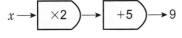
$x \longrightarrow$ ×2 $\longrightarrow$ +5 $\longrightarrow$ 9

☐ $\longleftarrow$ $\longleftarrow$ $\longleftarrow$ 9

b   Solve these equations using function machines.
i $5k - 7 = 8$ ii $4w + 8 = 28$ iii $6w - 3 = 24$

Example 2

Solve the equation $3a + 7 = 13$

$3a + 7 - 7 = 13 - 7$

$3a = 6$

$\dfrac{3a}{3} = \dfrac{6}{3}$

$a = 2$

Check: $3a + 7 = 3 \times 2 + 7 = 13$ ✓

Subtract 7 from both sides.

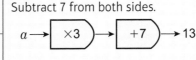
$a \longrightarrow$ ×3 $\longrightarrow$ +7 $\longrightarrow$ 13

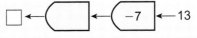
☐ $\longleftarrow$ $\longleftarrow$ −7 $\longleftarrow$ 13

Divide both sides by 3.

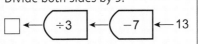
☐ $\longleftarrow$ ÷3 $\longleftarrow$ −7 $\longleftarrow$ 13

Warm up

5 Copy and complete.

a $2a + 3 = 23$

$2a + 3 - \square = 23 - \square$

$2a = \square$

$a = \square$

b $2c - 4 = 14$

$2c - 4 + \square = 14 + \square$

$2c = \square$

$c = \square$

6 Solve these equations.

a $2a + 1 = 5$

b $2a - 1 = 5$

c $3a + 2 = 8$

d $3a + 5 = 4$

e $7f - 12 = 9$

f $-5c + 12 = 2$

g $3a + 1 = 8$

h $2p - 4 = -5$

i $8t + 2 = -3$

Q6g hint

$3a = 7$

$a = \dfrac{\square}{\square}$

7 **Exam-style question**

a Solve $y + 5 = 12$ **(1 mark)**

b Solve $\dfrac{x}{4} = 3$ **(1 mark)**

c Solve $5w - 6 = 10$ **(2 marks)**

November 2012, Q17, 1MA0/1F

Exam hint

Write your solutions clearly.

$y = \square$

$x = \square$

$w = \square$

8 **Reasoning** I think of a number. I multiply it by 6 and subtract 8. The result is 46. Find the number.

Q8 hint Use n for the number.

9 **Reasoning** a Write an equation for the diagram.

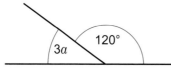

b Solve your equation to find the value of a.

10 **Reasoning** The sizes of the angles on a straight line are $a + 25°$, $2a + 40°$ and $55° - a$.

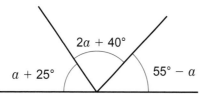

Find the value of a.

11 **Reasoning** The length of each side of a square is $3y - 3$ centimetres. The perimeter of the square is 36 cm. What is the value of y?

Q11 strategy hint Draw a diagram. Write an expression for the perimeter.

12 **Reasoning** The length of a rectangle is 3 cm greater than its width. The perimeter of the rectangle is 54 cm. Find its length.

Q12 hint width = w, length = $\square$ + 3

13 Solve these equations.

a $\dfrac{4c}{5} = 4$

b $\dfrac{5d}{12} = 10$

c $\dfrac{7e}{2} = 21$

d $-\dfrac{2f}{3} = 2$

Q13a hint

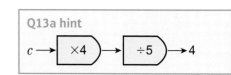

14 Solve these equations.

a $\dfrac{a}{6} + 2 = 4$ b $\dfrac{b}{3} - 8 = 2$ c $\dfrac{c}{4} + 6 = 9$

d $\dfrac{d}{2} - 4 = 2$ e $\dfrac{e}{4} - 8 = -4$ f $\dfrac{f}{9} + 2 = 5$

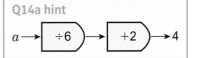

Q14a hint

$a \rightarrow \boxed{\div 6} \rightarrow \boxed{+2} \rightarrow 4$

Reflect How can you check your solutions to equations?

15 Solve

a $\frac{1}{2}x = 8$ b $\dfrac{x}{6} = 20$ c $\frac{3}{4}x = 9$

d $\frac{2}{5}x = 12$ e $\dfrac{2x}{3} = 3$ f $\dfrac{x}{2} = \dfrac{5}{7}$

Q15 hint Remember to use inverse operations.

5.3 Solving equations with brackets

Objective

- Solve linear equations with brackets.
- Solve equations with unknowns on both sides.

Why learn this?

Engineers building large structures have to solve many equations to make sure that the structure is strong enough and will not fall down.

Fluency

Expand

- $3(a + 7)$ • $5(b - 6)$

Warm up

1 Expand

a $2(2a + 3)$ b $5(6p - 4)$

c $3(6 - 7c)$ d $-4(3 - 2x)$

2 Solve

a $4e + 5 = 29$ b $4a + 4 = 0$

c $-3b + 5 = -4$ d $-7g - 4 = 12$

e $\dfrac{x}{2} = 12$ f $\dfrac{x}{3} + 5 = 2$

Q2 hint
Use inverse operations.

Key point 3

In an equation with **brackets**, expand the brackets first.

3 Expand and solve

a $5(a - 5) = 70$ b $6(b + 5) = 30$

c $3(d - 5) = 15$ d $3(2d - 5) = 27$

e $4(m - 4) = 12$ f $9(b - 11) = 9$

g $7(4 - c) = 35$ h $-2(e + 2) = -10$

i $-3(7 - f) = -3$

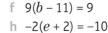

Q3a hint

$5(a - 5) = \boxed{} - \boxed{} = 70$

4 Solve

a $\dfrac{3c + 4}{3} = 2$

b $\dfrac{4g - 5}{5} = 3$

c $\dfrac{5g + 7}{4} = 6$

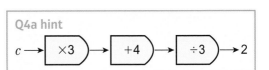

Q4a hint

$c \rightarrow \boxed{\times 3} \rightarrow \boxed{+4} \rightarrow \boxed{\div 3} \rightarrow 2$

Example 3

Solve $4d + 17 = 8d - 3$

$$\underbrace{4d + 17} \quad = \quad \underbrace{8d - 3}$$

$-\boxed{4d}$ ⟨△⟩ $-\boxed{4d}$

> Visualise the equation as balanced scales. Subtract $4d$ from both sides.

$$17 = 4d - 3$$
△

$17 + 3 = 4d - 3 + 3$ ← The inverse of -3 is $+3$. Do this to both sides.

$20 = 4d$ ← Divide both sides by 4.

$5 = d$

$d = 5$

Check: LHS $= 4d + 17 = 4 \times 5 + 17 = 37$
RHS $= 8d - 3 = 8 \times 5 - 3 = 37$ ✓

---

5 Solve these equations.

a $5k = 2k + 3$ b $2a = a + 14$

c $3c - 1 = c + 9$ d $5p - 7 = 2p + 11$

Discussion How do you decide which term to subtract first?

> **Q5a hint**
>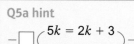

Key point 4

Whatever you do to one side of an equation, you must do to the other side.

6 Solve these equations.

a $2a + 9 = a + 5$ b $8b + 9 = 3b + 14$ c $3d + 7 = 2d + 19$

d $6v - 7 = 3v + 7$ e $3e = 7e - 18$ f $2h + 7 = 8h - 1$

7 Solve these equations.

a $40 - 3x = 1$ b $9 - 5x = 3x + 1$ c $1 - 6x = 9 - 7x$

d $8 + 3x = 1 - 4x$ e $13 - 2x = 3 - 7x$ f $3 - 9x = 5 - 6x$

> **Q7 hint** Be careful with the negative signs.

8 **Exam-style question**

Solve $4(x - 5) = 2x - 8$ **(4 marks)**

> **Exam hint**
> Substitute your value for x back into the equation to check it works.

9 **Reasoning** Steve is 30 years older than his son Jenson. He is also 11 times as old as Jenson. How old is each of them?

> **Q9 hint** Write an equation and solve it. Use x for Jenson's age.

10 Expand the brackets on both sides, then solve.

a $3(s + 4) = 4(s - 4)$ b $4(f + 5) = 6(f + 4)$ c $3(x - 2) = 6(x + 3)$

d $8(m - 2) = 4(m + 9)$ e $6(2y - 7) = 3(5y + 6)$ f $4(7t - 5) = 2(9t + 5)$

11 a Work out the length and width of the rectangle.

b What is the perimeter?

> **Q11 hint** Write equations for the equal sides.

$9 - 7x$ cm

$1 - 6y$ cm [ ] $3y + 2$ cm

$10 - x$ cm

12 **Problem-solving** The diagram shows the plan of a room. Lengths are in metres.

The area of the room is 45 m².

a Write an equation and solve it to find b.

b Josh buys new skirting board for the room. What length of skirting board does he need?

$b + 5$

5

5.4 **Introducing inequalities**

Objectives

- Use correct notation to show inclusive and exclusive inequalities.
- Solve simple linear inequalities.
- Write down whole numbers which satisfy an inequality.
- Represent inequalities on a number line.

Why learn this?

Inequalities can be used to compare quantities.

Fluency

Which numbers in the cloud belong to this inequality?
$3 < x \leqslant 7$

6.9
1.9   3   6.2
4.5   7

1   Put the correct sign ($<$, $>$ or $=$) between each pair of numbers to make a true statement.
   a   4 … 6
   b   5 … 2
   c   15 … 8
   d   6 … 0.7
   e   4.8 … 4.79
   f   4.5 … 4.5

Key point 5

You can show solutions to inequalities on a number line.
An empty circle ○ shows the value is *not* included.
A filled circle ● shows the value is included.
An arrow ○——→ shows that the solution continues towards infinity.

Example 4

Use a number line to show the values that satisfy each inequality.

a $x < 3$

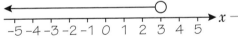

> This includes all the numbers less than 3 (not including 3).

b $2 < x \leqslant 6$

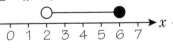

> This includes all the numbers greater than 2 (not including 2) and less than or equal to 6 (including 6).

2   Draw six number lines from −5 to +5. Show these inequalities.
   a $x > 1$
   b $x < 4$
   c $x < -5$
   d $x \geqslant 3\frac{1}{2}$
   e $x \leqslant 3$
   f $x \geqslant -2$

3   Draw six number lines from −5 to +5. Show these inequalities.
   a $-3 \leqslant x < 4$
   b $-2 < x < 5$
   c $0 < x < \frac{4}{5}$
   d $-2 \leqslant x < 1$
   e $-4 \leqslant x < 1$
   f $0 \leqslant x \leqslant 3$

4 Write down the inequalities represented on these number lines.

a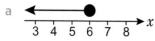

b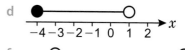

c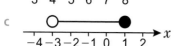

d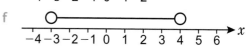

e

f

Reflect Did you check your answers? How?

5 Phil's teacher tells him that his mark in his maths test, m, is at least 55 and at most 63. Show this on a number line.

Key point 6

An **integer** is a positive or negative whole number or zero. The integer values that satisfy $2 < x < 3$ are −1, 0, 1 and 2.

6 Write down the integer values of x that satisfy each of these inequalities.

a $4 < x < 6$

b $3 < x < 8$

c $0 \leqslant x < 4$

d $-1 < x < 5$

e $-3 \leqslant x \leqslant 3$

f $-2 \leqslant x \leqslant 6$

Q6a hint

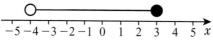

7 $n < 3$

Write an inequality for

a $2n$

b $3n$

c $5n$

d $3n + 1$

Key point 7

You can solve inequalities in the same way as linear equations.

8 Solve the inequalities. Show each solution on a number line.

a $x + 3 < 7$

b $x + 5 \geqslant 1$

c $2x \leqslant 12$

d $\dfrac{x}{3} > 2$

e $5x - 7 < 3$

f $3x + 7 \geqslant 1$

g $3(x + 4) \leqslant 9$

h $2(2x + 3) \geqslant 22$

Q8 hint Use the balancing method – do the same to both sides.

9 **Exam-style question**

a n is an integer.

$-1 \leqslant n < 4$

List the possible values of n. **(2 marks)**

b Write down the inequality shown in the diagram.

(2 marks)

c Solve $3y - 2 > 5$ **(2 marks)**

Nov 2012, Q25, 1MA0/2F

Exam hint

Make sure you do not use = signs in this question, only inequality symbols.

10 Solve

a $5x + 3 \geqslant 2x + 9$

b $9x - 7 < 5x + 3$

c $7 - 2x \geqslant 3x + 2$

d $6 - 5x \leqslant 2 - 3x$

e $3 - 5x \geqslant 4 - 7x$

f $10 - 3x > 2x - 1$

11

The sum of a number and 2 more than the number is less than 20.
What could the number be? **(4 marks)**

Exam hint
Write an inequality.
Use n for the number.

12 **Reasoning / Problem-solving** A bus timetable shows that the buses arrive at
20 minutes past the hour and at 10 minutes to the hour. The buses can be up to
4 minutes early and up to 8 minutes late.
Sam arrives at the bus stop at $14:10$.
Write an inequality to show how long he could wait for a bus to arrive.

13 **Modelling / Problem-solving** I think of a number, add 2 and multiply by 4.
My answer is greater than when I multiply the number by 3 and add 2.
Write an inequality.
Find three possible values for my number.

5.5 More inequalities

Objectives

- Solve two-sided inequalities.

Why learn this?

An inequality remains true when the same
number is added to or subtracted from both
sides.

Fluency

What are the possible integer values of x?
$-5 \leqslant x < 4$

1 Solve

   a $x + 4 < 2$

   b $x + 2 < -3$

   c $3x + 4 < 16$

2 Solve

   a $\dfrac{x}{4} > -3$

   b $2(x - 3) \geqslant -4$

   c $5(x + 2) > 15$

   d $\dfrac{1}{3}x \leqslant 2$

   e $4(x - 3) < 24$

   f $\dfrac{x}{2} + 4 \leqslant 6$

3 Solve

   a $4 \leqslant 2x \leqslant 6$

   b $-9 \leqslant 3x \leqslant 9$

   c $-16 < 4x \leqslant 0$

   d $-7 < 5x \leqslant 20$

   e $-5 < 2x < 2$

   f $-10 < 3x < 3$

Q3a hint

Key point 8

You can solve **two-sided inequalities** using a balancing method.

Example 5

Solve $7 < 2x - 1 < 13$

$7 + 1 < 2x - 1 + 1 < 13 + 1$ ——— Add 1 to all the parts.

$\div 2 \left(\begin{array}{c} 8 < 2x < 14 \\ 4 < x < 7 \end{array} \right) \div 2$ ——— Divide by 2.

4   Solve these double inequalities.

   a $5 < 3x - 1 < 14$ b $2 < 3x + 5 \leqslant 17$ c $1 \leqslant 4x - 4 < 11$

   d $-1 < 2x + 1 < 5$ e $-7 \leqslant 3x - 1 < 2$ f $-6 \leqslant 4x - 6 \leqslant 15$

5   Solve each two-sided inequality and show its solution set on a number line.

   a $14 < 2x < 20$ b $5 \leqslant 3x - 1 \leqslant 8$ c $-2 \leqslant 3x - 4 \leqslant 5$

   d $0.12 \leqslant 4x - 1 \leqslant 1.8$ e $5 \leqslant 4(x + 3) < 15$ f $10 < 2(3x + 4) \leqslant 24$

6   On a number line you can see that $-3 < -2$.

   $$\overline{-3 \quad -2 \quad -1 \quad 0 \quad 1 \quad 2 \quad 3}$$

   a Multiply both sides of the inequality by -1. Is the inequality still true?

   Discussion What happens to the inequality sign when you multiply both sides
   of an inequality by -1?

   b $-x < -2$. Write an inequality for x.

> ### Key point 9
>
> When you multiply or divide both sides of an inequality by a negative number you reverse the
> inequality sign.

7   Solve these double inequalities and show each solution set on a number line.

   a $-4 < -2x + 6 < 4$ b $4 < 3(-2x + 1) \leqslant 7$ c $5 \leqslant 3(5 - 2x) < 12$

8   Solve the inequality $11 - 2x < 2 - 5x$
   What is the smallest integer that satisfies it?

9   **Problem-solving** Harry has three parcels. The first parcel has a mass of x kg.
   The second parcel has a mass twice that of the first parcel.
   The third parcel has a mass 3 kg less than the first parcel.
   The total mass of the parcel is less than or equal to 22 kg.
   What is the largest possible mass of the smallest parcel?

10 **Problem-solving** The perimeter of the square P is more than the perimeter of the
   rectangle Q.
   Work out the range of values for the side length of the square.

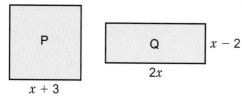

11 For each pair find the integer values of x which
   satisfy both inequalities.

   a $x \geqslant 2$ and $x \leqslant 5$

   b $x > -3$ and $x < 2$

   c $x \leqslant 4$ and $x < -1$

   d $x > -1$ and $x > 3$

> **Q11c hint** $x > 2$ and $x > 4$ are both
> satisfied by $x > \boxed{}$.
>
> $$\xrightarrow{} x$$
> $$0 \quad 1 \quad 2 \quad 3 \quad 4 \quad 5 \quad 6 \quad 7$$

12 **Exam-style question**

   Find the integer value of x that satisfies both the inequalities
   $x + 4 > 8$ and $2x - 7 < 5$ **(2 marks)**

   Exam hint
   Solve each inequality and then
   show your solutions on a number
   line to get the final answer.

5.6 Using formulae

Objectives

- Substitute values into formulae and solve equations.
- Change the subject of a formula.
- Know the difference between an expression, an equation and a formula.

Why learn this?

You can use a formula to work out how much interest your savings will earn over a period of time.

Fluency

What is the inverse of each operation?

$+2$　　-10　　$\div 3$　　$\times 8$

1 $T = pq + m$
   Work out the value of T when
   a $m = 5$, $p = 6$ and $q = 7$　　b $m = -3$, $p = -2$ and $q = 4$　　c $m = 7$, $p = -5$ and $q = -3$

2 $s = \frac{1}{2}at^2$
   Work out the value of s when
   a $a = 3$ and $t = 8$　　b $a = 4$ and $t = -5$　　c $a = -7$ and $t = -4$

3 $A = bh$
   a   Find the value of h when   i $A = 36$, $b = 4$　　ii $A = 42$, $b = 7$
   b   Find the value of b when   i $A = 65$, $h = 5$　　ii $A = 144$, $h = 6$

> Q3a i hint $A = b \times h$
> $36 = 4 \times h$
> $\boxed{} = h$

4 $y = 4x - 5$
   Work out the value of x when
   a $y = 3$　　　　　　b $y = -31$
   c $y = 75$　　　　　d $y = -6$

5 $V = lwh$
   a   Work out the value of h when $V = 100$, $l = 10$ and $w = 2$
   b   Work out the value of l when $V = 40$, $h = \frac{1}{2}$ and $w = 4$
   c   Work out the value of w when $V = 150$, $h = 5$ and $l = 3$

6 $y = \dfrac{x}{7}$
   Work out the value of x when
   a $y = 3$　　　　　　b $y = 8$
   c $y = -5$　　　　　d $y = 4.3$
   e $y = \frac{1}{7}$　　　　　f $y = \frac{3}{4}$

7 $P = a + 2b$
   a   Work out the value of a when   i $P = 11$ and $b = 4$　　ii $P = 7$ and $b = 5.2$
   b   Work out the value of b when   i $P = 9$ and $a = 14$　　ii $P = -23$ and $a = 6$

8 **STEM / Modelling** The formula to work out the speed of an object is $s = \dfrac{d}{t}$
   where d = distance in miles, t = time in hours and s = speed in mph.
   a   Work out the distance when
   　　i $s = 50$ mph and $t = 4$ hours　　ii $s = 65$ mph and $t = 5.5$ hours.
   b   Work out the time when
   　　i   the distance is 220 miles and the speed is 55 mph
   　　ii   a plane travels a distance of 1800 miles at 450 mph.

9 **Exam-style question**

$y = 4x + c$ $x = 7.5$ $c = 5.4$

a Work out the value of y. **(2 marks)**

$y = 4x + c$ $y = 18.8$ $c = -2.4$

b Work out the value of x. **(2 marks)**

Nov 2012, Q18, 1MA0/2F

Exam hint
First substitute and then rearrange the equation.

10 **STEM / Modelling** In the formula $v^2 = u^2 + 2as$, v = final velocity (m/s), s = distance (m), u = initial velocity (m/s), a = acceleration (m/s²) and t = time (s).
Use $v^2 = u^2 + 2as$ to work out s when $u = 10$ m/s, $a = 5$ m/s² and $v = 20$ m/s.

11 **STEM / Modelling** You can use this formula to work out s, the distance an object has travelled in metres: $s = ut + \frac{1}{2}at^2$, where u = starting velocity (m/s), t = time (s) and a = acceleration (m/s²).
Work out the value of a when $s = 200$ m, $u = 5$ m/s and $t = 8$ s.

12 **STEM / Modelling** You can use this formula to work out the density of an object:
$d = \dfrac{m}{V}$ where d = density (g/cm³), m = mass (g) and V = volume (cm³).

Work out the mass when the density is 2.7 g/cm³ and the volume is 24 cm³.

Key point 10

A **formula** shows the relationship between two or more variables (letters).
You can use **substitution** to find an unknown value.
An **equation** contains an unknown number (a letter) and an = sign. You can solve it to find the value of the letter.

13 State whether each of these is an expression, an equation or a formula.

a $x + 3 = 3(2x - 4)$

b $V = IR$

c $q(p - 8)$

d $y = 4x + 5$

e $10(x + 2) = 3x^2$

f $s = \dfrac{d}{t}$

Reflect What is the same about equations and formulae?
What is the difference between equations and formulae?

Example 6

Rearrange $y = 2x + 5$ to make x the **subject** of the formula.

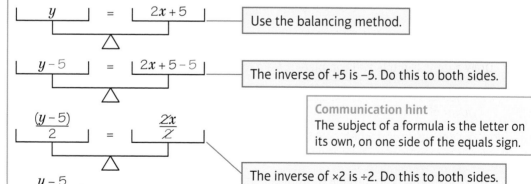

| y | $=$ | $2x + 5$ |

Use the balancing method.

| $y - 5$ | $=$ | $2x + 5 - 5$ |

The inverse of +5 is −5. Do this to both sides.

Communication hint
The subject of a formula is the letter on its own, on one side of the equals sign.

| $\dfrac{(y - 5)}{2}$ | $=$ | $\dfrac{2x}{2}$ |

The inverse of ×2 is ÷2. Do this to both sides.

$x = \dfrac{y - 5}{2}$ ✓

14 Rearrange each formula to make the letter in the square brackets the subject.

a $y = x + 4$ $[x]$

b $y = x - 7$ $[x]$

c $P = IV$ $[I]$

d $P = 5d$ $[d]$

e $y = 5x + 3$ $[x]$

f $y = 4x - 3$ $[x]$

g $M = 7N - 5$ $[N]$

h $V = \dfrac{W}{3}$ $[W]$

i $T = \dfrac{D}{V}$ $[V]$

5.7 Generating sequences

Objective

- Recognise and extend sequences.

Why learn this?

You can use sequences to perform a piece of music or do a kata in martial arts.

1 Write down the next two terms in each sequence.
   a 41, 37, 33, 29, ☐, ☐
   b 17, 14, 11, 8, ☐, ☐

2 Write down the 10th term in each sequence.
   a 1, 5, 9, 13, ☐
   b 3, 7, 11, 15, ☐
   c 4, 8, 12, 16, ☐
   What is the term-to-term rule for each sequence?

> **Key point 11**
>
> A **sequence** is a pattern of numbers or shapes that follow a rule. The numbers in a sequence are called **terms**. The **term-to-term rule** describes how to get from one term to the next.

3 Write down the next two terms in each sequence.
   a 1.5, 2, 2.5, 3, ☐, ☐
   b $-\frac{2}{3}, -\frac{1}{3}, 0, \frac{1}{3}$, ☐, ☐
   c 3.5, 2.7, 1.9, 1.1, ☐, ☐
   d −1.5, −2.5, −3.5, −4.5, ☐, ☐
   e $\frac{3}{5}, -\frac{1}{5}, -1, -1\frac{4}{5}$, ☐, ☐
   f −10.6, −9.9, −9.2, −8.5, ☐, ☐

4 Use the first term and the term-to-term rule to generate the first five terms of each sequence.
   a start at 3 and add 0.4
   b start at 10 and subtract 0.2
   c start at 7 and add 3
   d start at 7 and add 2
   e start at −3 and add 2
   f start at −7 and subtract 5

5 In a Fibonacci sequence, the term-to-term rule is 'add the two previous terms to get the next one'. Write the next 3 terms in each Fibonacci sequence.
   a 1, 1, 2, 3, 5, …
   b 3, 3, 6, 9, 15, …
   c 5, 5, 10, 15, 25, …

6 Each sequence is made up of a pattern of sticks. For each sequence
   i Draw the next two patterns.
   ii Draw a table like this and complete it for each pattern.

Pattern number	1	2
Number of sticks		

   iii Write down the rule to continue the pattern.
   iv Work out the number of sticks needed for the 10th pattern.

   a
   b
   c
   d

7 Write down the next two terms in each sequence.

a 5, 50, 500, 5000, ☐, ☐ b 2, 1, $\frac{1}{2}$, $\frac{1}{4}$, ☐, ☐

c 100, 10, 1, 0.1, ☐, ☐ d 2, 8, 32, 128, ☐, ☐

8 Find the term-to-term rule for each sequence.

a 2, 4, 8, 16, 32, 64, … b 100, 50, 25, 12.5, 6.25, …

c 1, 3, 9, 27, 81, 243, … d 5, −10, 20, −40, 80, −160, …

e 1, 10, 100, 1000, … f 18, 6, 2, $\frac{2}{3}$, …

9 This sequence made from counters shows the first three triangular numbers.

a Draw the next two patterns in the sequence.

b Write down the number of counters in each pattern.

c Work out the differences between the numbers of counters.

d Follow the sequence to work out the number of counters in the 10th pattern.

Q9c and d hint

1, 3, 6,

+2 +3 +☐

10 This sequence made from counters shows the first three square numbers.

a Draw the next two patterns in the sequence.

b Work out the number of counters in the 8th pattern.

Discussion The first three cube numbers are 1, 8, 27. What are the next two terms in the sequence?

11 **Exam-style question**

Here are some patterns made from white centimetre squares and grey centimetre squares.

a Draw pattern 4.

b Find the number of grey squares in Pattern 6.

A pattern has 20 grey squares.

c Work out how many white squares there are in this pattern.

Pattern 1   Pattern 2   Pattern 3   **(4 marks)**

Nov 2013, Q10, 5MB2F/01

5.8 Using the nth term of a sequence

Objectives

- Use the nth term to generate terms of a sequence.
- Find the nth term of an arithmetic sequence.

Why learn this?

Many things around us follow a pattern. Knowing how a pattern continues can help us know what to expect, for example where the planets are going to be in the night sky.

Fluency

Substitute n = 1, 2, 3, 4, 5 into $4n$.

1 a Work out $5x$ when $x = 10$. b Work out $3p + 2$ when $p = 3$.

2 Solve
   a $2x + 3 = 7$ b $2x - 7 = 15$ c $4x + 2 = 20$

Key point 12

The nth term of a sequence tells you how to work out the term at position n (any position). It is also called the **general term** of the sequence.

---

3 A sequence has nth term $2n + 1$. Copy and complete the table to work out the first 5 terms of the sequence.

n	1	2	3	4	5
Term	$2 \times \text{①} + 1 = \square$	$2 \times \text{②} + 1 = \square$			

4 Write the first five terms of the sequences with these nth terms.
   a $3n$ b $7n - 4$ c $5n + 1$ d $21 - 2n$
   e $14 - 3n$ f $\frac{1}{2}n + 2$ g $30 - 4n$ h $-4n + 3$

   Discussion Look at the common difference for each sequence and its nth term. What do you notice?

Example 7

a Work out the nth term of the sequence 7, 11, 15, 19, 23, …
b Is 33 a term of the sequence?

a $4n$ 4, 8, 12, 16, 20, … $\Big) +3$
   7, 11, 15, 19, 23, …
   $+4 \ +4$

   The common difference is 4. Write out the first five terms of the sequence for $4n$, the multiples of 4. Work out how to get from each term in $4n$ to the term in the sequence.

   The nth term is $4n + 3$
b 33 $= 4n + 3$

   Write an equation using the nth term and solve it.

   $33 - 3 = 4n + 3 - 3$
   $30 = 4n$
   $7.5 = n$

   33 cannot be in the sequence because 7.5 is not an integer.

---

5 Find the nth term for each sequence.
   a 2, 5, 8, 11, 14, 17, … b 2, 6, 10, 14, 18, 22, … c 2, 7, 12, 17, 22, 27, …
   d 5, 7, 9, 11, 13, 15, … e 19, 17, 15, 13, 11, 9, … f 20, 18, 16, 14, 12, 10, …

6 For each sequence, explain whether each number in the brackets is a term in the sequence or not.
   a 2, 5, 8, 11, 14, … (50, 66) b 5, 8, 11, 14, 17, … (50, 62)
   c 1, 5, 9, 13, 17, … (101, 150) d 4, 9, 14, 19, 24, … (168, 169)
   e 40, 35, 30, 25, 20, … (85, 4) f 5, 11, 17, 23, 29, … (119, 72)

   Q6a hint Work out the nth term
   $\square n - \square = 50$
   $n = \square$

7 Using the nth term given, find the 20th term.
   a $2n$ b $3n + 1$ c $11 - 3n$

   Q7 hint Use a function machine to help you visualise.

8 Find the nth term for each sequence. Use it to work out the 10th term.
   a 1, 3, 5, 7, … b 3, 6, 9, 12, … c 10, 8, 6, 4, … d 3, 7, 11, 15, …

9 Find the first term over 100 for each sequence.
   a 9, 18, 27, 36, 45, … b 7, 10, 13, 16, 19, …
   c 4, 9, 14, 19, 24, … d 10, 15, 20, 25, 30, …

   Q9 hint Solve nth term = 100

10 Here are the first four terms in a number sequence.

124, 122, 120, 118

a Write down the next term in this number sequence.

b Work out the nth term for this number sequence.

c Can 9 be a term in this sequence? Give a reason for your answer.

11 Write down the first five terms of the sequence with nth term

a n^2 b $3n^2$ c $n^2 - 1$

d $\frac{1}{4}n^2$ e $n^2 + 4$ f $65 - n^2$

12 **Exam-style question**

Here is a pattern made from dots.

4   7   10

a Draw the next pattern in the sequence.

b Copy and complete this table for the numbers of dots used to make the patterns.

Pattern number	1	2	3	4	5	6
Number of dots						

c Write, in terms of n, the number of dots needed for pattern n.

d How many dots are needed for pattern 30?   **(5 marks)**

13 **Modelling / Problem-solving** Sam makes a pattern sequence with tiles.

a Draw the next pattern in the sequence.

b Sam has 35 tiles. Does he have enough to make
   i the 10th pattern   ii the 20th pattern?

> **Q13b hint** Work out the nth term of the sequence.

c Work out the pattern number of the biggest pattern he can make.

14 **Modelling / Problem-solving**
Here is a pattern sequence of blue and
white tiles.

a Copy and complete the table for the numbers
   of white tiles and the numbers of blue tiles.

Pattern number	1	2	3	4	5
Number of white tiles	4	5			
Number of blue tiles	2	4			

b Write down the nth term for the sequence of the numbers of blue tiles.

c Write down the nth term for the sequence of the numbers of white tiles.

d How many blue tiles are there in the 20th pattern?

e How many white tiles are there in the 30th pattern?

f Alex has 50 blue tiles and 45 white tiles.
   Which is the largest complete pattern she can make?

5 Problem-solving: Paediatric medicine

Objectives
- Use formulae by substituting values.
- Rearrange formulae.

1 **Real** Six children are prescribed the same medicine.
The adult dose is 300 mg per day.
For each child in Table 1 work out the dose using
a Clark's rule
b Young's rule.
Discussion For which children do you get two very different doses? Why do you think this is?

> **Q1 communication hint**
> 1000 mg = 1 gram

2 **Real** A GP prescribes medicine for four children.
The adult dose is 500 mg per day.
a Estimate the age of each child.
Show your working.
b **Reflect** What information did you use
to answer this question? How did you
decide on the information to use?
Discussion What maths skills did you use to answer this question?
Try to list them all.

Jon	120 mg
Fiona	165 mg
Jill	270 mg
Wayne	235 mg

> **Q2a hint**
> Use Young's rule.

Clark's rule and Young's rule

Clark's rule and Young's rule are ways to calculate medicine doses for children.
These methods can be used when the packaging only gives the adult dose.

Clark's rule: $\text{child's dose} = \text{adult dose} \times \left(\dfrac{\text{child's weight in pounds}}{150} \right)$

Young's rule: $\text{child's dose} = \text{adult dose} \times \left(\dfrac{\text{age of child}}{\text{age of child} + 12} \right)$

Table 1

Name	Age (years)	Weight (pounds)
Alex	8	58
Brian	4	45
Clara	3	32
Dilip	7	52
Emily	2	24
Farhana	10	58

Table 2
Average weight for girls and boys

Age (years)	Average weight (pounds)	
	Girls	Boys
2	26.5	27.5
3	31.5	31.0
4	34.0	36.0
5	39.5	40.5
6	44.0	45.5
7	49.5	50.5
8	57.0	56.5
9	62.0	63.0
10	70.5	70.5
11	81.5	78.5

5 Check up

Equations and formulae

1 Solve
   a $c - 2 = -1$
   b $3f + 7 = 1$
   c $\frac{x}{2} + 5 = 12$

2 Solve
   $3a + 8 = 6a + 1$

3 Expand and solve
   a $4(x + 3) = 24$
   b $2(x - 2) = -14$

4 $V = IR$
   Work out I when $V = 20$ and $R = 40$

5 $P = 2w + 2l$
   Work out l when $P = 25$ and $w = 3$

6 Rearrange each formula to make the letter in the square brackets the subject.
   a $P = \frac{M}{4}$ [M]
   b $y = 3x - 5$ [x]

7 Solve
   a $3(a + 5) = a + 21$
   b $2(3t + 4) = 5(2t - 1)$

8 An isosceles triangle has side lengths $3x$ cm, $3x$ cm and $2x - 20$ cm
   Write an expression for its perimeter.
   The perimeter is 68 cm. What is the length of each side?

Inequalities

9 Show each inequality on a number line.
   a $x > 2$
   b $x < -\frac{1}{2}$
   c $-1 < x \leqslant 4$

10 Write down the inequalities represented on these number lines.

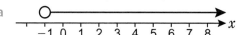

11 For each inequality, list the integers which satisfy it.
   a $-3 \leqslant x < 4$
   b $-6 < x \leqslant 1$

12 Solve each inequality.
   a $x - 6 > 4$
   b $6x \leqslant 36$
   c $2x - 5 > 15$
   d $2(x - 3) \geqslant 14$

13 y is an integer. What is the smallest possible value of y?
   $-3 < y \leqslant 1$

14 Solve
   a $-10 < 2x \leqslant 4$
   b $-4 \leqslant x - 7 < 3$

Sequences

15 Here are the first four terms of a sequence.
   98, 96, 94, 92
   a Write down the next term in the sequence.
   b Write down the 7th term in the sequence.
   c Write a formula for the nth term of this sequence.
   d Can −3 be a term in this sequence? Explain how you know.

16 Write down the next two terms in each sequence.
   a 1, 2, 4, 8, 16, ☐ , ☐
   b 100, –50, 25, –12.5, ☐ , ☐

17 Here are the first five terms of a number sequence.
   2, 6, 10, 14, 18
   a Work out the nth term of this sequence.
   b Work out the 10th term.

18 How sure are you of your answers? Were you mostly

   Just guessing   Feeling doubtful   Confident 🙂

   What next? Use your results to decide whether to strengthen or extend your learning.

✳ Challenge

19 The answer to a question is $x = 3$.
   Write three questions with this answer.

> **Q19 hint** One possible question would be 'Expand and solve $2(3x + 1) = 20$'

5 Strengthen

Equations and formulae

1 What is the inverse of each operation?
   a + 7   b – 10   c × 2

> **Q2a hint** You need to get only unknowns (x) on one side of the '=' sign, and only numbers on the other side.

2 Copy and complete to solve these equations.
   a $2x + 7 = 13$ b $3x – 10 = 2$

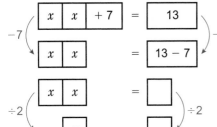
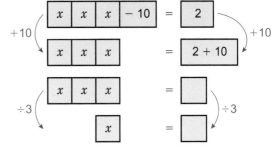

3 Solve these equations.
   a $3x + 4 = 19$
   b $2y – 7 = 13$
   c $5s – 12 = 18$
   d $27 = 3 + 4p$
   e $50 = 9s + 14$

> **Q3 hint** Use the method in **Q2**.

> **Q3d hint**
>
> | 27 | | = | 3 + | p | p | p | p |

4 $F = ma$
   Work out
   a F when $m = 6$ and $a = 5$ b m when $F = 30$ and $a = 10$

5 $v = u + at$
   Work out
   a v, when $u = 12$, $a = 4$ and $t = 2$
   b t, when $v = 20$, $u = 10$ and $a = 2$
   c a, when $v = 15$, $u = 7$ and $t = 4$

> **Q5a hint** Substitute all the values you know into the formula.
>
> $v = u + a \quad t$
>
> $v = 12 + 4 \times 2$

6　Solve these equations.

　a　$3(2x + 5) = 45$

　b　$6(x - 7) = 18$

　c　$5(4x - 4) = 100$

> **Q6a hint** Expand the brackets: $6x +$ ▢ $= 45$
> Then use the method in **Q2**.

7　Copy and complete to solve this equation.

　$2a + 5 = 4a - 7$

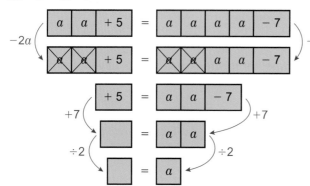

> **Q7 hint** Subtract the same number of unknowns (a) from both sides so that the equals is still true. Do this until only one side has unknowns.

8　Solve these equations.

　a　$4a + 1 = 5a - 3$　　b　$7p - 3 = 5p + 1$

　c　$5y + 2 = 3y + 12$　　d　$2b + 9 = 4b + 13$

> **Q8 hint** Use the method in **Q7**.

9　Solve these equations.

　a　$2(5x + 7) = x - 13$

　b　$4(q - 5) = 2q - 1$

　c　$4(3a + 2) = 27a + 3$

　d　$3(4z + 6) = 4(2z + 5)$

> **Q9a hint** Expand the brackets: $10x +$ ▢ $= x - 13$
> Then use the method in **Q7**.

> **Q9b hint** The solution may be a fraction.

10　What is the inverse of each of these?

　a　$+x$　　b　$-y$　　c　$\times a$　　d　$\div p$

> **Q10 hint** The inverse of $+ 7$ is ▢.
> So the inverse of $+ x$ is ▢.

11　What is the subject of each formula?

　A　$y = x - z$　　**B**　$A = x + h$　　**C**　$q = rx$

　D　$F = x - ab$　　**E**　$m = \dfrac{x}{2}$

> **Q11 hint** Which letter is on its own on one side of the '=' sign?

12　a　Copy and complete to make x the subject of $M = x - y$

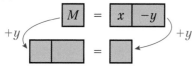

　b　Make x the subject of each formula in Q11.

Inequalities

1　The cloud contains ten integers.

　Choose at least two integers that could be x, when

　a　x is greater than 6

　b　x is less than or equal to 1

　c　$x \geqslant 7$

　d　$x > 4$

　e　$x \leqslant -5$

　f　$-2 > x$

> **Q1c hint** Read the inequality aloud: x is _____ than or equal to 7.

> **Q1f hint** Read the inequality aloud: x is _____ ▢.

2   Follow these instructions to list the integers that satisfy $-3 \leqslant x < 1$

   a   Cover the last part of the inequality with your finger, so you can
      see only one inequality. $-3 \leqslant x$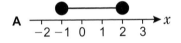

      List ten integers that could be x, starting with the lowest and
      counting up.

   b   Cover the first part of the inequality with your finger, so you can see only the other
      inequality. 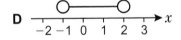 $x < 1$

      Circle all the integers in your list from part a that satisfy this inequality.

   c   Write down only the integers you circled.
      These are the integers that satisfy $-3 \leqslant x < 1$.

> **Q2 Communication hint** 'Satisfy' means 'that would make the inequality true'.

3   List the integers that satisfy each inequality.

   a $2 < x \leqslant 7$

   b $-8 < x < -3$

   c $1 \geqslant x < -4$

   d $3 < x < 5$

> **Q3 hint** Use the method in **Q2**.

> **Q3c hint** List ten integers that satisfy $1 \geqslant x$, starting with the highest and counting down.

4   Match each inequality to the number line that represents it.

   a $x < 2$ b $x > 2$ c $x \geqslant 2$

   d $-1 < x < 2$ e $-1 \leqslant x < 2$ f $-1 \leqslant x \leqslant 2$

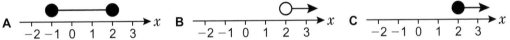

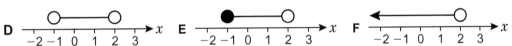

5   Follow these instructions to draw a number line that shows the inequality $n < -1$.

   a   Draw a number line with -1 in the middle, three integers
      above and three integers below.

   b   Draw a circle above -1.

   c   Should your circle be empty or filled?

   d   Draw an arrow from your circle.

> **Q5 hint**
>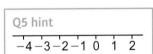

6   Show each inequality on a number line.

   a $n \geqslant 3$

   b $x < 8$

   c $y \leqslant -5$

   d $x > \frac{1}{2}$

   e $17 < c \leqslant 20$

   f $-7 \leqslant m \leqslant -2$

> **Q6d hint** Draw a number line with three integers above, and three integers below $\frac{1}{2}$. Mark $\frac{1}{2}$ on your number line.

7   Solve each inequality and show your answer on a number line.

   a $x + 4 > 9$ b $3x - 10 \geqslant 8$ c $6(y + 3) < 66$

      $x > \Box$ $6y + \Box < 66$

> **Q7 strategy hint** Use the same method as for solving an equation.

8 Solve these inequalities.

a $9 < 2x + 1 \leqslant 15$

b $5 < 4x - 3 < 13$

c $-1 \leqslant 3x + 2 < -4$

> **Q8a hint** Cover part of the inequality.
> $9 < 2x + 1$ 🔲 Solve to find ☐ $< x$.
>
> Cover the other part of the inequality.
> 🔲 $2x + 1 \leqslant 15$ Solve to find $x \leqslant$ ☐.
>
> Write your solution as ☐ $< x \leqslant$ ☐.

Sequences

1 Write down the next two terms in each sequence.

a 3, 7, 11, 15, ☐, ☐

b 5, 1, −3, −7, ☐, ☐

c 3, 8, 13, 18, ☐, ☐

d 1, 3, 9, ☐, ☐

e 120, 60, 30, ☐, ☐

f 1, −2, 4, −8, ☐, ☐

> **Q1 hint** What do you have to do to get to the next term? Do you add, subtract, multiply or divide by the same number each time?

2 Write down the nth term of each sequence.

a 2, 4, 6, 8, …

b 4, 8, 12, 16, …

c 5, 10, 15, 20, …

d −4, −8, −12, −16, …

> **Q2a hint**
>
> $\times$ ☐ (
>
> | n | 1 | 2 | 3 | 4 |
> |------|---|---|---|---|
> | Term | 2 | 4 | 6 | 8 |
>
>) ☐ $\times$

> **Q2d hint** The nth term for 2, 4, 6, 8… is $2n$.
> The nth term for −2, −4, −6, −8… is $-2n$.

3 For sequence **a** in **Q1**:

a What is the term-to-term rule?

b Compare the terms of the sequence with the sequence $4n$.

Term	3	7	11	15
4n	4	8	12	16

What do you notice?

c Write down the nth term of the sequence 3, 7, 11, 15, …

d Check your nth term works.

   i Substitute $n = 1$ into your answer for part **c**. What number do you get?

   ii Substitute $n = 2$ into your answer for part **c**. What number do you get?

e What is the 10th term of the sequence?

f Is 49 a term in this sequence? If so, which term is it?

g Is 100 a term in this sequence? If so, which term is it?

> **Q3a hint**
>
> $+$☐ $+$☐ $+$☐ $+$☐
> 3 → 7 → 11 → 15 →

> **Q3c hint** $4n -$ ☐

> **Q3d hint**
> $4n -$ ☐ $= 4 \times 1 -$ ☐ $=$ ☐

> **Q3e hint** Substitute $n = 10$ into your answer for part **c**. What number do you get?

> **Q3f hint** Write your nth term = 49. Solve the equation to find n. Is n a whole number?

4 For each of the sequences **b** and **c** in **Q1**:

a Work out the nth term.

b Find the 10th term.

c Show whether 49 is a term in the sequence.

> **Q4a hint** Use the method in **Q3** parts **a**, **b** and **c**. What sequence do you think you need to compare each sequences to? Look back at **Q1** to help you.

5 Extend

1 Solve these equations.

a $1.7 + 2f = 8.4$

b $3.8 + 3g = 11.7$

c $4h - 0.22 = 1.1$

2 Solve these equations.

a $\dfrac{a}{4} = 0.5$

b $\dfrac{b}{3} + 3 = 3.\dot{3}$

c $\dfrac{c}{8} - 2 = 0.1$

d $\dfrac{9d - 4}{2} = 1$

e $\dfrac{18e + 12}{3} = -1$

> **Q2 hint** The solutions to these equations could be fractions, decimals or whole numbers.

3 **Exam-style question**

Here are the first six terms of a number sequence.

5   9   13   17   21   25

a Write down the next term of the sequence. **(1 mark)**

b i Work out the 11th term of the sequence.

  ii Explain how you found your answer. **(2 marks)**

Nov 2012, Q8, 5MB2F/01

> **Exam hint**
> 'Explain' means that you should show how you worked out the nth term.

4 **Exam-style question**

Here are the first four patterns in a sequence.

Each pattern is made from squares and circles.

      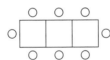   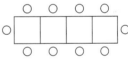

Pattern number 1   Pattern number 2   Pattern number 3   Pattern number 4

a How many squares are there in the pattern with exactly 14 circles? **(2 marks)**

b How many circles are needed for Pattern number 50? **(2 marks)**

Nov 2011, Q11, 5MB2F/01

5 Find the term-to-term rule for each sequence.

a $x + 1, x + 2, x + 3, x + 4, \ldots$

b $x - 2, x - 4, x - 6, x - 8, \ldots$

c $2x, 2x^2, 2x^3, 2x^4, \ldots$

d $x + 3, x + 7, x + 11, x + 15, \ldots$

e $x, x, 2x, 3x, 5x, \ldots$

f $2x, 4x, 8x, 16x, \ldots$

6 **Problem-solving** The first term in a sequence is p.

The term-to-term rule is $\times 3 + 2$

The third term is 17.

What is p?

> **Q6 hint** Write the 1st three terms in terms of p.
> $p, \square, \square$

7 Here is a sequence: 120, 113, 106 …

a Work out the nth term.

b Which term is the first negative term? State the term and its position in the sequence.

8

The diagram shows a garden in the shape of a rectangle.

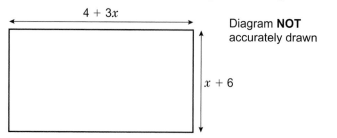

Diagram **NOT** accurately drawn

All measurements are in metres.

The perimeter of the garden is 32 metres.

Work out the value of x. **(4 marks)**

June 2013, Q28, 1MA0/1F

Exam hint

Write an equation that links the lengths of the sides to the given perimeter.

9 **Reasoning** Work out the size of the largest angle.

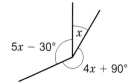

10 Solve these equations.

a $8 + 3x = 10 - x$

b $1 - 9x = 3x - 3$

c $14 - 5x = 10x + 5$

11 **Problem-solving** Jug A contains w litres of water.

Jug B contains $w + 1.5$ litres of water.

Jim pours 0.5 litres out of jug A and into a glass. Jug B now has twice as much water as jug A.

Work out the value of w.

12 The perimeter of shape A is equal to the perimeter of shape B.

a Write an equation using the perimeters.

b Solve the equation to find the perimeter.

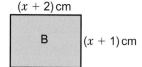

13 Find the integer values of x that satisfy each inequality.

a $\frac{7}{8} < \frac{1}{4}x \leqslant 2$

b $8 \leqslant 4.5x - 1 \leqslant 11$

c $-6 \leqslant \frac{2}{3}x + 4 < 0$

d $-1 < \frac{1}{3}(2x - 5) < 2$

e $-1 < -6x + 11 < 1$

14 **Problem-solving** Ben chooses three different whole numbers.

The first number is a prime number.

The second number is 7 times the first number.

The third number is 10 more than the second number.

The sum of the three numbers is greater than 100 and less than 160.

a Write an expression for the sum of the three numbers.

b What are the three numbers?

Q14a hint Use x for the first number.

Q14b hint Write an inequality using your answer to part **a** and the information in the question.

15 Here are the first five terms of an arithmetic sequence.

10, 16, 22, 28, 34

Peter says that 555 is a term in this sequence.

Explain why Peter is wrong.

5 Knowledge check

- In an **equation**, a letter or symbol represents an unknown number. Equations have terms in one letter and an equals sign. You can solve an equation to find the value of the unknown. .... *Mastery lesson 5.1*
- You can solve an equation by using the **balancing method** and inverse operations to get the unknown on one side of the = and a number on the other. .... *Mastery lesson 5.1*
- An integer is a positive or negative whole number or zero. .... *Mastery lesson 5.4*
- Solutions to equations can be integers, fractions, decimals or negatives. .... *Mastery lesson 5.2*
- In an equation with **brackets**, expand the brackets first. .... *Mastery lesson 5.3*
- You can solve problems in mathematics and other subjects by writing and then solving equations. .... *Mastery lesson 5.1*
- A **formula** shows the relationship between two or more variables (letters). You can substitute values to find an unknown value. .... *Mastery lesson 5.6*
- Inequalities use these symbols: $<, >, \leq, \geq$ .... *Mastery lesson 5.4*
- You can show inequalities on a number line. An empty circle shows that a number is not included. A filled circle shows that a number is included. .... *Mastery lesson 5.4*
- You can solve an inequality the same way as an equation. .... *Mastery lesson 5.4*
- When you multiply or divide both sides of an inequality by a negative number you reverse the inequality sign. .... *Mastery lesson 5.5*
- The numbers in a sequence are called **terms**. The **term-to-term** rule describes how to get from one term to the next. .... *Mastery lesson 5.5*
- The nth term of a sequence tells you how to work out the term at position n. .... *Mastery lesson 5.8*

Look back at this unit.
Which lesson made you think the hardest?
Write a sentence to explain why.

Hint Begin your sentence with: Lesson ☐ made me think the hardest because ☐.

Reflect

5 Unit test

Log how you did on your Student Progression Chart.

1   Solve these equations.

   a $\dfrac{a}{5} = 3$ b $b - 8 = 12$ c $6p + 3 = 33$ d $-2q + 10 = 2$ *(4 marks)*

2   $a = \dfrac{b}{c}$

   Work out the value of b when

   a $a = 2$ and $c = 3$ b $a = 8$ and $c = -4$ c $a = \frac{1}{4}$ and $c = 12$ *(3 marks)*

3   Make the letter in square brackets the subject of each formula.

   a $d = 5t$ $[t]$ b $y = x - 6$ $[x]$ c $q = 3p + 8$ $[p]$ d $m = \dfrac{n}{4}$ $[n]$ *(4 marks)*

4   Solve these equations.

   a $60 - 4x = 5x + 6$ b $x + 22 = 5(x + 4)$ *(4 marks)*

5 **Reasoning** The diagram shows a rectangle.
The perimeter of the rectangle is 146 cm.
Find the value of x. $(2x + 3)$ cm *(2 marks)*

$(3x - 6)$ cm

6 **Reasoning** Anna has x sweets.
Isabel has $x + 15$ sweets.
Anna gives 5 sweets to Meena. Anna now has half as many sweets as Isabel.
How many sweets did Isabel have to start with? *(2 marks)*

7 Draw four number lines from −5 to +5. Show these inequalities.
a $-4 \leqslant x \leqslant 4$ b $x \leqslant 3$ c $-3 < x < 0$ d $-2 < x$ *(4 marks)*

8 Find the integer solutions that satisfy each inequality.
a $5 < x + 6 < 11$ b $2 < 2x + 10 \leqslant 8$ c $7 \geqslant 1 - x \geqslant 5$ *(6 marks)*

9 **Reasoning** Penny chooses three different whole numbers.
The second number is 5 times the first number.
The third number is 20 less than the second number.
The sum of the three numbers is greater than 40 and less than 50.
a Write an expression for the sum of the three numbers. *(2 marks)*
b What are the three numbers? *(2 marks)*

10 Here are the first five terms of a sequence.
8, 13, 18, 23, 28
a Write the next two terms in the sequence. *(1 mark)*
b Write an expression for the nth term. *(2 marks)*

11 **Reasoning** Here are the first five terms of an arithmetic sequence.
8, 15, 22, 29, 36
Peter says that 163 is a term in this sequence.
Explain why Peter is wrong. *(2 marks)*

Sample student answer

Is '$x = 11$' the correct final answer? Explain why?

Exam-style question

Here is a rectangle.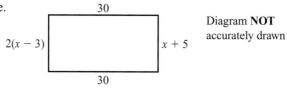

30

Diagram **NOT**
accurately drawn

$2(x - 3)$

$x + 5$

30

All measurements are given in centimetres.
Find the perimeter of the rectangle.

_____ cm **(4 marks)**

Practice Paper Set C, Q7, 1MA0/4H

Student answer
$2(x - 3) = x + 5$
$2x - 6 = x + 5$
$\quad -x \qquad -x$
$\quad x - 6 = 5$
$\quad +6 \qquad +6$
$\qquad x = 11$

6 ANGLES

Engineers and architects use geometrical relationships with angles and shapes to build unusual buildings and bridges. What shapes can you see in this bridge?

6 Prior knowledge check

Numerical fluency

1   Work out
  a   90 − 70
  b   15 + 45 + 70
  c   180 − 120
  d   2 × 45
  e   5 × 108
  f   1440 ÷ 8
  g   180 − 110 − 45
  h   360 − 280
  i   360 − 110 − 155

2   Work out the missing numbers
  a $15 + \square = 90$
  b $10 + 70 + \square = 180$
  c $110 + 175 + \square = 360$

3   Work out
  a   (180 − 28) ÷ 2
  b   360 − (54 × 2)
  c $\frac{1}{2}(90 - 18)$
  d   180(5 − 2)

Algebraic fluency

4   a   Simplify
     i $2x + 90 - x$
     ii $180 - 2x + 90 + x$
  b   Find the value of each expression in part **a** when $x = 40$.

5   Solve
  a $6x = 180$ b $x + 45 = 3x - 135$

Geometrical fluency

6   a   Identify pairs of lines that are parallel.

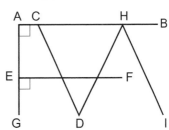

  b   Copy and complete the statement.
     Lines AE and AB are _____

7 Is each angle acute, right, obtuse or reflex?

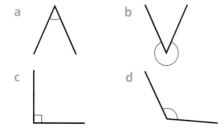

a   b

c   d

8 Is each angle acute, obtuse or reflex?
   a 153°   b 28°
   c 294°   d 79°
   e 120°   f 190°

9 a Estimate the sizes of angles a and b.

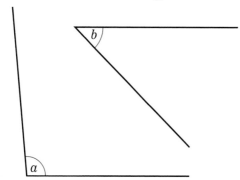

   b Measure angles a and b.

10 Work out the sizes of angles a and b.

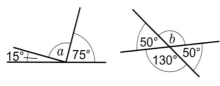

   For each angle, choose a reason for your answer from the box.

   > Vertically opposite angles are equal.
   > Angles on a straight line add up to 180°.
   > Angles around a point add up to 360°.

11 Look at triangle XYZ.

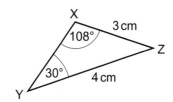

   a Copy and complete the statement.
     Side _____ is 4 cm in length.

b Which of these are correct ways to name the 30° angle?

   Angle XYZ   Angle YXZ   Angle XZY
   Angle Y   Angle YZ   Angle XY
   Angle XZ   XŶZ   ∠XYZ

12 a Label each triangle. Choose a label from the box.

   > scalene   isosceles
   > right-angled   equilateral

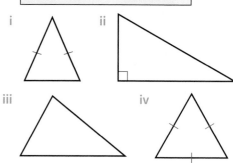

   i   ii

   iii   iv

   b Copy the triangles in part **a** and draw on all the lines of symmetry.
   c What is the order of rotational symmetry of each triangle in part **a**?

13 A   B

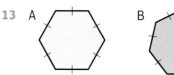

   a How many lines of symmetry does each shape have?
   b What is the order of rotational symmetry of each shape?

14 Name each shape. Choose a name from the box.

   > rhombus   rectangle   trapezium
   > kite   parallelogram

   a   b

   c   d

15 Copy the shapes in **Q14**.
   a Draw on any lines of symmetry.
   b State the order of rotational symmetry of each shape.

16 Copy and complete the diagram.

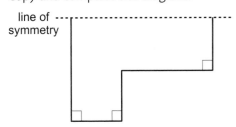

17 a What do the angles in a quadrilateral add up to?

b What size is angle c in this quadrilateral?

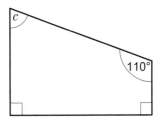

Graphical fluency

18 Draw a coordinate grid with axes labelled from −5 to +5.
   a Plot the points A(−4, −5), B(−3, 1) and C(4, 3).
   b Plot point D so that ABCD is a parallelogram.

* Challenge

19 Find sizes for the angles at A, B, C and D and side lengths AB, BC, CD and DA, to change the quadrilateral ABCD to
   a a rhombus   b a trapezium.

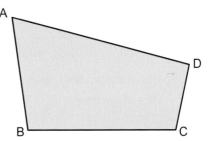

6.1 **Properties of shapes**

Objectives

- Solve geometric problems using side and angle properties of quadrilaterals.
- Identify congruent shapes.

Did you know?

Artists use properties of quadrilaterals to create optical illusions like this one.

Fluency

How many lines of symmetry does each shape have? What is its order of rotational symmetry?

1 Draw an angle of   a 65°   b 128°

2 a What do the four angles of each of these quadrilaterals add up to?

i
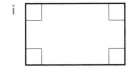

ii
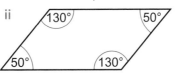

iii
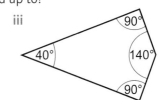

b Copy and complete this rule.
   The four angles of any quadrilateral add up to ☐°.

Key point 1

Two shapes are **congruent** when they are exactly the same shape *and* size. Two shapes are **similar** when they are the same shape. Similar shapes may be different sizes.

Questions in this unit are targeted at the steps indicated.

3   Write down the letters of three pairs of shapes that are congruent.

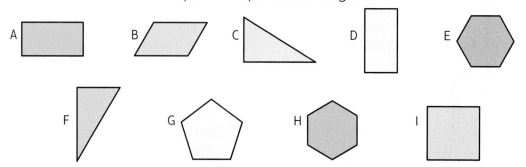

Q3 hint Congruent shapes are exactly the same size and shape – if you cut them out, one shape fits on top of the other exactly. It does not matter if you turn them over or turn them round.

Reflect In your own words, write mathematical definitions for congruent and similar.

4   These parts of shapes are drawn on a centimetre square grid.

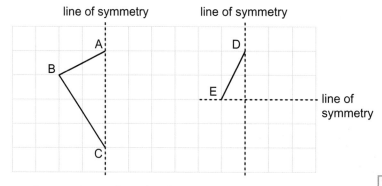

a   Copy and complete the diagrams.
b   Name each shape and measure and label all its angles.

Q4b hint Use a protractor to measure the angles.

Discussion What can you say about the opposite angles in a rhombus and a kite?

5   a   Name each shape. Which of these shapes have
    i   two pairs of equal opposite angles
    ii diagonals that bisect at 90°
    iii two pairs of parallel sides?

Q5 communication hint
A **diagonal** joins opposite vertices of a shape. '**Bisect**' means cut in half.

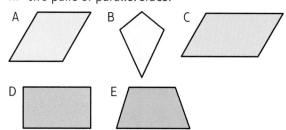

6   **Reasoning** Name each quadrilateral being described.
    a   I have four equal sides and my opposite angles are equal.
    b   All my angles are 90° and my diagonals bisect at 90°.
    c   My diagonals bisect at 90° but are not the same length.
    d   I have one pair of parallel sides and one pair of equal sides.
    Discussion Is there more than one answer for each part?

7   **Problem-solving** Draw a coordinate grid with axes labelled from −5 to 5.
    Plot these points.
    A(−5, 1), B(−2, −1), C(3, −2), D(5, 2), E(3, 3), F(2, 1), G(1, 2), H(−2, 5), I(−4, 3) J(−5, 3)
    Which four points can you join to make
    a a parallelogram   b a kite   c an isosceles trapezium?

8   Work out the sizes of the missing angles in each quadrilateral.
    a   b   c

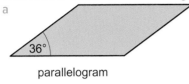

parallelogram

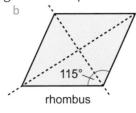

115°

rhombus

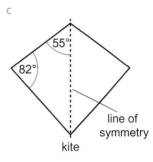

line of
symmetry

kite

9   **STEM / Reasoning** Bronwen
    is an engineer.
    She draws a design for a bridge.
    Bronwen uses these rules in her
    design.
    • Angle *g* is 10% less than
      angle *e*.
    • Angle *e* is 10% less than
      angle *c*, rounded to the
      nearest degree.
    • Angle *c* is 10% less than
      angle *a*, rounded to the
      nearest degree.
    Work out the sizes of the angles *a* to *g*.

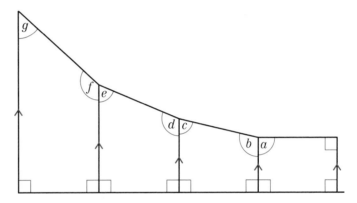

10 **Exam-style question**

    PQRS is a quadrilateral. Find the size of angle PQR.
    Explain each stage of your working.

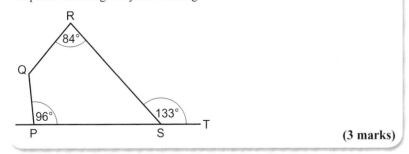

                                    (3 marks)

Exam hint
Line PST is
straight. What
does the straight
line tell you about
the angles at
S? Make sure
you explain your
reasoning at each
stage.

11 **Problem-solving** Draw a coordinate grid with axes labelled from −5 to 5.
    a Here are some coordinates for three shapes ABCD. Plot the points on your grid.
        i   A(−1, 2), B(1, 4), C(5, 0)
        ii A(−2, 0), B(−4, 1), C(−2, 2)
        iii A(−1, −1), B(−3, 2), C(−1, 3)
    b Join the points A to B and B to C in each shape.
    c Draw a point D for each shape so that it has rotation symmetry of order 2.
    d Name each shape.

12 **Problem-solving / Communication**
The diagram shows a pattern made from four identical parallelograms.
Show that angles a, b, c and d add up to 360°.

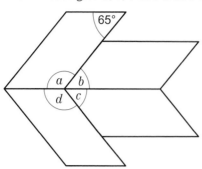

Q12 hint Work out all the angles in one parallelogram first.

13 **Problem-solving** The diagram shows a kite.
Find the size of angle a.

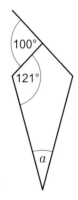

6.2 Angles in parallel lines

Objectives

• Understand and use the angle properties of parallel lines.
• Find missing angles using corresponding and alternate angles.

Why learn this?

Engineers need to know about angle properties when designing buildings like The Shard in London.

Fluency

• Find two parallel lines.
• Find two perpendicular lines.

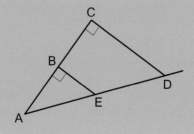

1 Which of the angles marked with letters are obtuse? Which of them are acute?

  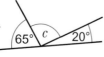

ActiveLearn Homework, practice and support: Foundation 6.2

Key point 2

Parallel lines are shown with arrows.

When a line crosses two parallel lines it creates a 'Z' shape.

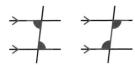

Inside the Z shape are **alternate angles**.
Alternate angles are equal.
Alternate angles are on different or alternate sides of the line.

2 The diagram shows a line crossing two parallel lines
   and angles labelled p, q, r and s.
   Write down two pairs of alternate angles.

3 Find the sizes of the angles marked with letters.

   a 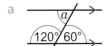 b 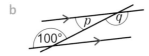 c

   Q3 hint Angles on a
   straight line add up
   to 180°

Key point 3

When a line crosses two parallel lines it creates an 'F' shape.
Inside the F shape are **corresponding** angles.
Corresponding angles are equal.

4 Find the sizes of the angles marked with letters.

   a 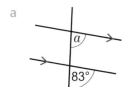 b 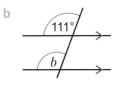 c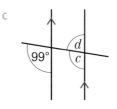

5 Find the sizes of angles s to w. Give reasons for your answers.
   Choose your reasons from the box.

   Angles on a straight line add up to 180°.
   Angles ☐ and ☐ are alternate.
   Angles ☐ and ☐ are corresponding.

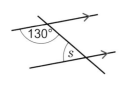 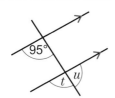 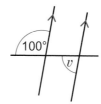 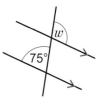

Discussion How did you find angle s? Did everyone use the same method?

6 **Communication** Which angles in the diagram
are equal?
Give reasons for your answers.

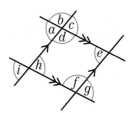

7 **Communication** Angles inside two parallel lines
are called **co-interior angles**. Use this diagram to
explain why co-interior angles add to 180°.

 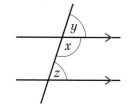

8 | **Exam-style question**

ABC and DEF are straight lines.
ABED is a parallelogram.
Find the size of the angle marked x.
Give reasons to explain your answer.

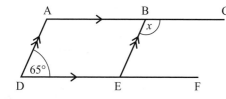

Diagram NOT
accurately drawn

(4 marks)

9 Find the sizes of the angles a to j. Give reasons for your answers.

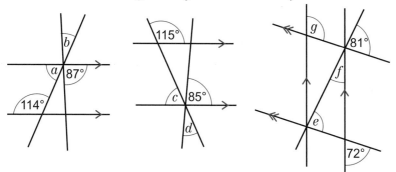

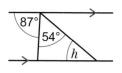

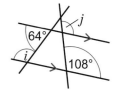

10 | **Exam-style question**

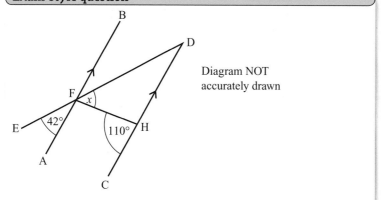

Diagram NOT
accurately drawn

AFB and CHD are parallel lines.
EFD is a straight line.
Work out the size of the angle marked x.
(3 marks)
Nov 2010, Q4, 5MB2H/01

Exam hint
Write on the
diagram any angles
you work out.
Make sure you use
the fact that AFB
is parallel to CHD.

11 ABCD is a trapezium.
Work out the sizes of angles x and y.

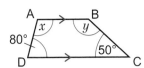

6.3 **Angles in triangles**

Objectives

- Solve angle problems in triangles.
- Understand angle proofs about triangles.

Did you know?

Knowing the angles in a triangle helps pinpoint the exact position of aircraft.

Fluency

Match each triangle to a name from the box.

A   B   C   D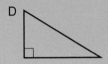

| right-angled triangle |
| scalene triangle |
| equilateral triangle |
| isosceles triangle |

1 Copy and complete the sentence.
The three angles in a triangle always add up to ☐°.

2 Work out
   a 180 − 100 − 50   b 180 − 57 − 57   c $\dfrac{180 - 35}{2}$

3 Calculate the sizes of the angles marked with letters.

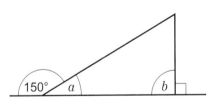

 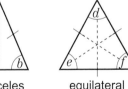

Choose a reason from the box to explain each answer.

| Reason 1 | Angles about a point add up to 360°. |
| Reason 2 | Angles on a straight line add up to 180°. |

4 **Reasoning** The diagram shows an isosceles triangle and an equilateral triangle.
   a An isosceles triangle has a line of symmetry.
     What does this tell you about the two angles a and b?
   b An equilateral triangle has three lines of symmetry.
     What does this tell you about angles d, e and f?

isosceles   equilateral

Discussion Are these facts true for all isosceles/equilateral triangles?

5 Find the sizes of the missing angles.

a 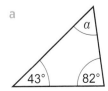   b 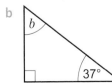   c 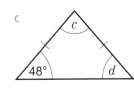   d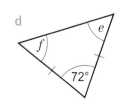

6 **Real / Communication** For safety, a ladder should make an angle of 15° with the vertical.
   a How many degrees is this with the horizontal?
   b Is each of these ladders safe? Explain your answer.

Q6 communication hint
vertical |
horizontal —

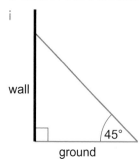
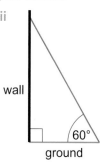

7 **Real** An architect draws a roof extension.

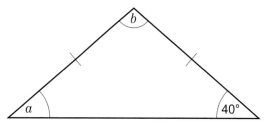

What sizes are angles a and b?

8 **Reasoning** Sketch a copy of this diagram.
Copy and complete these statements to prove that the sum of the angles in a triangle is 180°.

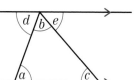

Statement	Reason
$d + b + e = \square$	Angles d, b and e lie on a straight line.
$d = \square$	Angles d and $\square$ are alternate.
$e = \square$	Angles e and $\square$ are alternate.

So $d + b + e$
  $= a + b + c$
  $= \square$

This proves that the angles in a triangle sum to $\square$.
Discussion Why is this a proof?

9 **Problem-solving** Work out the sizes of the angles marked with letters.

a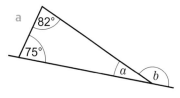

b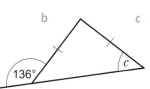

c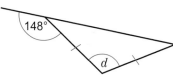

10 **Reasoning** Copy the diagram.
Copy and complete this proof to show that $w = x + y$.
$x + y = 180° - \square$ because the angles in a triangle add up to $\square°$.
$w = 180° - \square$ because the angles on a _____ line add up to 180°.
So $x + y = w$

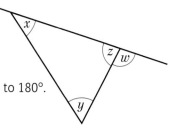

11 Reasoning Which statement is true about this triangle?

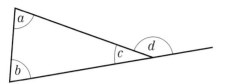

Q11 hint Choose numbers for a, b and c, and work out d. Which statement is true?

A Angle d = angle b
B Angle d = angle a + angle c
C Angle d = angle a + angle b
D Angle d = angle b + angle c

12 Reasoning What sizes could angles a, b and c be? Choose possible angles from the cloud.

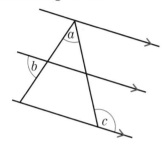

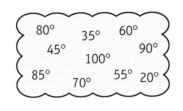

Q12 hint Mark the angles equal to b and c on your diagram.

13 Problem-solving Work out the sizes of the angles marked with a letter.

a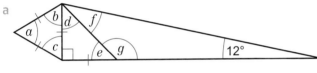

Q13a hint What types of triangles can you see?

b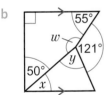

Q13b hint What do the angles of a quadrilateral add up to? How can you use properties of parallel lines?

14 (**Exam-style question**

AB is parallel to CD.
EG = FG
$A\widehat{E}G = 110°$
Calculate the size of $D\widehat{G}H$.

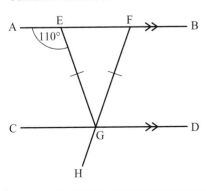

Exam hint
Work out the angles in triangle EFG first. Then use the fact that AB is parallel to CD.

(3 marks)

6.4 Exterior and interior angles

Objective

- Calculate the interior and exterior angles of regular polygons.

Did you know?

The 50p and 20p coins are both regular heptagons with slightly curved sides. They were designed this way to make them easier to identify and to avoid confusion with other similar sized coins.

Fluency

Match the name of the shape with the number of sides.
Triangle, pentagon, octagon, decagon, hexagon, quadrilateral, nonagon, heptagon

3 8 5 6
9 7 10 4

Warm up

1 Work out
   a $360 \div 10$ b $360 \div 12$ c $360 \div 18$

2 What size are the angles inside
   a a square   b an equilateral triangle?

3 Work out
   a $180 - 77$ b $180 - 137$ c $360 - 273$ d $180 - 37$ e $360 - 193$

Key point 4

You can draw an **exterior angle** of a shape by extending one of its edges.
The exterior angle is between the extended line and the next side of the shape.

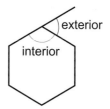

exterior

interior

4 Copy the polygons. Draw one **interior angle** and one exterior angle for each shape.

a

b

c

d

> **Q4 communication hint**
> A **polygon** is a 2-dimensional shape bounded by straight edges.

Key point 5

A **regular polygon** has all equal side lengths and all equal angles. An **irregular polygon** has unequal side lengths and unequal angles.

*Active*Learn Homework, practice and support: Foundation 6.4

5   Write down whether each polygon is regular or irregular.

a   b   c   d   e

6   a   Work out the size of one exterior angle of each regular polygon.

i    ii 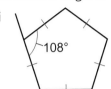   iii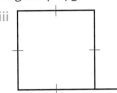

> **Q6 hint**
> In a regular polygon, all the exterior angles are equal.

b   Work out the sum of the exterior angles of each regular polygon. What do you notice?

Discussion If you walked around the outside of a polygon, what size turn would you make?

> **Key point 6**
>
> The sum of the exterior angles of a regular polygon is always 360°.

7   A regular polygon has 15 sides.
   a   How many equal exterior angles does it have?
   b   Work out the size of *one* exterior angle.

> Q7b hint 360° ÷ ☐ = ☐

8   The exterior angles of some regular polygons are
   a   30°   b   45°   c   18°
   Work out the number of sides of each regular polygon.

> Q8a hint 30° × ☐ = 360°

9   The diagram shows an irregular pentagon.
   What do the exterior angles add up to?
   Discussion What is the sum of the
   exterior angles of any irregular polygon?

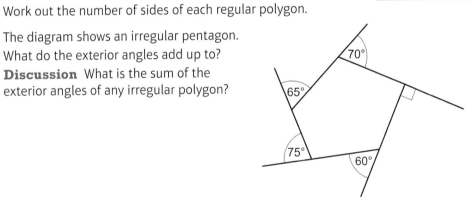

10   Work out the sizes of the missing exterior angles for each polygon.

a 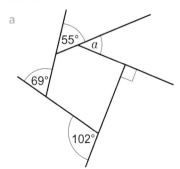   b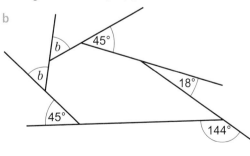

11   **Reasoning / Communication** Cerys says,
   'I can work out the size of the exterior angles of an irregular hexagon by dividing 360° by 6.'
   Is Cerys correct? Explain.

12 **Reasoning / Communication** The diagram shows an exterior and an interior angle of a regular polygon.

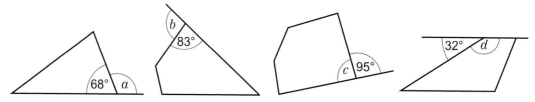

a Work out the number of sides of the polygon.

b What do the interior and exterior angles add up to?

c Copy and complete this rule.

The interior and exterior angles always sum to ☐°.

d Explain why this rule is true.

13 Work out the sizes of the angles marked with letters.

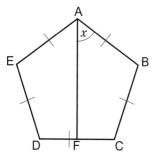

14 Copy and complete this table.

Regular polygon	Exterior angle	Interior angle
pentagon		
hexagon		
decagon		

> **Q14 hint**
> Work out the exterior angles first.

15 A regular polygon has an interior angle of 144°.
Find the size of the exterior angle and work out how many sides the polygon has.

16 **Problem-solving** Point F lies on the midpoint of CD. Find the size of angle x.

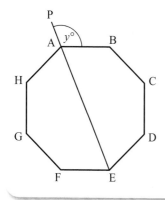

> **Q16 hint** AF is a line of symmetry so it bisects the interior angle EAB.

17 **Exam-style question**

P is at the top. ABCDEFGH is a regular octagon.
PAE is a straight line.
Angle PAB = $y°$
Work out the value of y.

(4 marks)

Nov 2012, Q9, 5MB2H/01

> **Exam hint**
> Start by working out the size of each exterior angle and put this on the diagram. Note that y is *not* an exterior angle.

6.5 More exterior and interior angles

Objectives

- Calculate the interior and exterior angles of polygons.
- Explain why some polygons fit together and others do not.

Did you know?

The Greek word for seven is hepta.
The heptathlon has 7 events and a heptagon has 7 sides.

Fluency

How many interior angles does each of these polygons have?
Quadrilateral, pentagon, heptagon, decagon

1 Work out the value of $180 \times (n - 2)$ when
  a $n = 6$ b $n = 3$ c $n = 10$

2 a Is each lettered angle interior, exterior or neither?

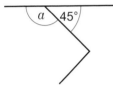 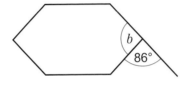 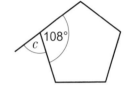

  b Work out the sizes of the lettered angles.

Key point 7

For shapes to fit together (**tessellate**), all the angles where the shapes meet must add up to 360°.

3 Copy this parallelogram onto squared paper.
Continue your diagram to show how parallelograms fit together.
Show that parallelograms fit together by considering the interior angles.

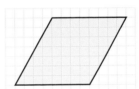

Q3 hint Sketch four parallelograms meeting at a point. What do the angles add up to?

4 Do regular pentagons fit together? Explain your answer by considering the interior angles.

Q4 hint Sketch three pentagons meeting at a point.

5 Copy this trapezium onto squared paper.
Show how trapeziums fit together by drawing five more on your diagram.
Discussion 'Trapeziums will fit together no matter what their interior angles are.'
Is this statement always, sometimes or never true?

6 **Reasoning** Copy and complete these statements.
This shape has ☐ sides.
It is made of ☐ triangles.
The interior angles in a triangle add up to ☐°.
So the sum of the interior angles in this shape is ☐ × ☐° = ☐°

7 **Reasoning**
   a What is the sum of the interior angles of this polygon?

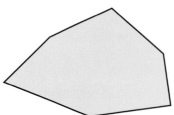

> **Q7a hint** Draw diagonals from the top vertex to divide the polygon into triangles.

> **Q7 communication hint** **Vertex** means corner.

> **Q7b hint** Try with 4 and 7-sided. What do you notice?

   b How many triangles can you divide an n-sided polygon into?.
   c Write a formula to work out the sum of the interior angles, S, for a polygon with n sides.

8 For each irregular polygon, work out
   i the sum of the interior angles
   ii the size of the angle marked with a letter.

> **Q8 i hint** Use your formula from **Q7c**.

a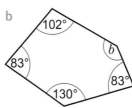
b
c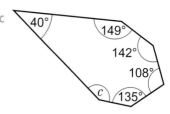

Discussion Does the formula work for regular polygons?

9 For each shape, work out the size of
   i the angle sum   ii the interior angle.
   a regular hexagon   b regular pentagon   c 18-sided regular polygon
   Reflect Write step-by-step instructions for finding the sum of the interior angles of any regular polygon. Is there more than one method you could use? Which do you prefer?

10 For each polygon, work out the number of sides from the sum of its interior angles.
   a 1620°   b 2160°   c 2700°   d 3960°

11 The diagrams show regular polygons with some angles given.
   Work out the sizes of the angles marked with letters.
   Give reasons for your answers.
   Choose reasons from the box.

> Vertically opposite angles are equal.
> The triangle is isosceles.
> Exterior and interior angles add up to 180°.
> Exterior angle of an octagon.

> **Q11c hint** Work out angle w first.

a

b

c

Example 1

AB, BC and CD are three sides of a regular octagon.

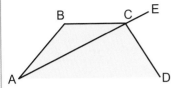

Find the size of angle BAC. Give reasons for your answer.

| Use the facts you know about interior and exterior angles. |

Angle ABC = interior angle

Exterior angle of a regular octagon = $360° \div 8 = 45°$

Interior angle = $180 - 45 = 135°$ ─── | Exterior angle + interior angle = $180°$ |

Angle BAC = angle BCA (triangle ABC is isosceles)

Angle BAC = $(180 - 135) \div 2 = 22.5°$ (angles in a triangle)

| Subtract the known angle from the sum of the interior angles of a triangle. Divide by 2 as angles BAC and BCA are identical. |

12 **Problem-solving** ABCDEF is a regular hexagon.
Find the size of angle AFB.

| Q12 hint Use the method in **Example 1**. |

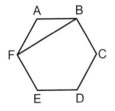

13 **Exam-style question**

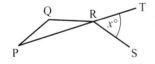

Diagram NOT accurately drawn

PQ, QR and RS are three sides of a regular decagon.
PRT is a straight line.
Angle TRS = $x°$
Work out the value of x.

(5 marks)

Exam hint
Calculate the angles in the order shown here.

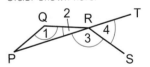

6.6 Geometrical problems

Objectives

- Solve angle problems using equations.
- Solve geometrical problems showing reasoning.

Did you know?

Geometrical shapes occur everywhere in real life. For example, a football has polygons on its surface that tessellate on a curved surface.

Fluency

Find the sizes of angles a, b, c and d.

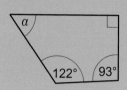 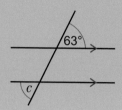 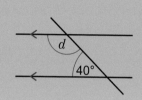

Warm up

1 Solve these equations.

  a $5x = 180$
  b $3x + 30 = 180$
  c $6x - 270 = 360$
  d $60 + 8x = 10x - 105$

2 An angle is x. Another angle is 60° more than x.

  a Write an expression for the larger angle in terms of x.
  b Write an expression for the total of the two angles. Simplify your expression.

3 The diagram shows a straight line.

  a Write an equation in terms of x.
  b Solve your equation to find the value of x.
  c Write down the sizes of the two angles.

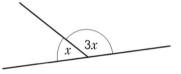

Example 2

Work out the size of angle x.
Explain your answer.

> Use the angle facts you know to find the sizes of other angles in the diagram.

Angle EFB = x (corresponding angles)

$x + 2x = 180°$ (angles x and $2x$ lie on a straight line)

$3x = 180°$

> Write an equation and solve it to find x.

$x = 60°$

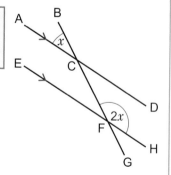

4 Work out the value of x in each diagram.

  a   b   c   d

  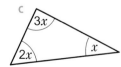

Q4a hint Write an equation in terms of x.

Q4d hint Copy the diagram. Mark the angles on the diagram as you find them.

5 **Reasoning** Look at the diagram.
   a i Explain why $x + 40° = 2x − 10°$
      ii Work out the value of x.
      iii Work out the sizes of the angles.
   b Work out the sizes of the angles.

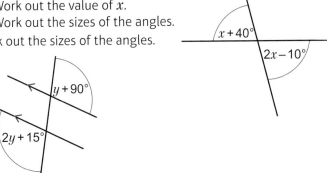

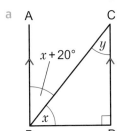

> **Q5a iii hint**
> Substitute the value
> of x into $x + 40°$.

6 **Problem-solving** Work out the values of x and y.

   a   b

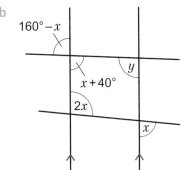

> **Q6b strategy hint**
> Start with the
> quadrilateral.

7 Look at the diagram.

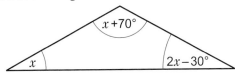

   a Write an equation in terms of x.
   b Solve this equation for x.
   c Write down the sizes of the three angles of the triangle.

8 Write an equation in terms of x and use it to find the three angles of the triangle.

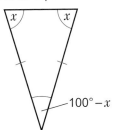

9 a **Communication** The left-hand column of the
       table shows the angles in three different triangles.
       The right-hand column shows equations
       that have been written about the triangles,
       but not in order.
       Match each set of angles with the
       correct equation.

Angles	Equations
20°, 20°, 140°	$2a = 90°$
60°, 60°, 60°	$2a + b = 180°$
30°, 50°, 100°	$3a = 180°$
90°, 45°, 45°	$a + b + c = 180°$

   b Here are three more angles: 100°, 20°, 70°.
      Explain why these three angles do not come from a triangle.

10 **Reasoning / Communication** What type of triangle is this? Explain.

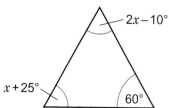

11 In the box are sets of three angles, written as expressions in x.
For each set, which of these is true?
These three angles can always/sometimes/never make a triangle.

x, $3x$, $5x$
x, x, x
$x + 60°$, $x - 60°$, $60° - 2x$
$x + 60°$, $x - 60°$, $x + 180°$
$60° - x$, $x + 90°$, $30°$

12 **Exam-style question**

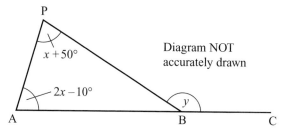

All angles are measured in degrees.
ABC is a straight line.
Angle APB = $x + 50$
Angle PAB = $2x - 10$
Angle PBC = y

a Show that $y = 3x + 40$
Give reasons for each stage of your working. **(3 marks)**

b Given that $y = 145$,
i work out the value of x
ii work out the size of the largest angle in triangle ABP. **(4 marks)**

Nov 2011, Q6, 1380/4H

Exam hint
'Show that' means you need to write down all of your working and explain your reasons as you go along. You can answer part **b** even if you cannot answer part **a**.

13 **Problem-solving / Communication** Triangle BDC is an isosceles triangle.
Triangle ACE is a right-angled triangle.
Show that triangle ABC is an equilateral triangle.

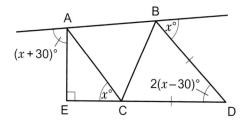

Q13 hint Start by finding the missing angles in triangle AEC and triangle CBD.

14 **Problem-solving** ABD is a triangle.
C is a point on DB.
Find the size of angle ABD.

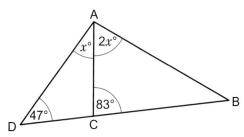

6 Problem-solving

Objective
· Use x for the unknown to help you solve problems.

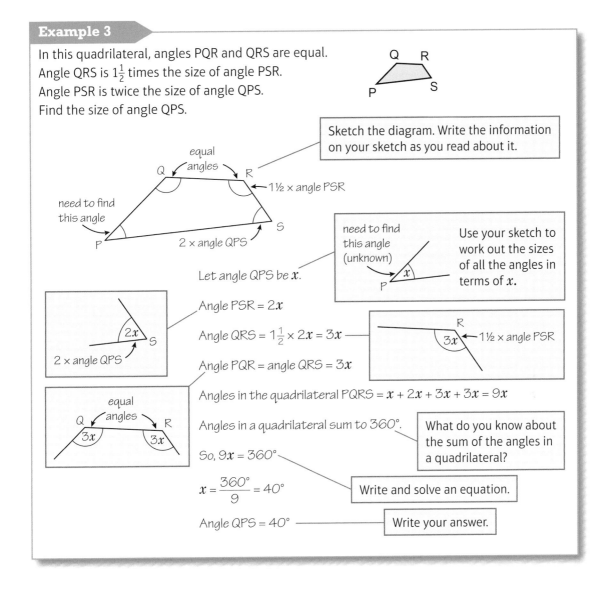

Example 3

In this quadrilateral, angles PQR and QRS are equal.
Angle QRS is $1\frac{1}{2}$ times the size of angle PSR.
Angle PSR is twice the size of angle QPS.
Find the size of angle QPS.

Sketch the diagram. Write the information on your sketch as you read about it.

equal angles
1½ × angle PSR
need to find this angle
2 × angle QPS

Let angle QPS be x.

need to find this angle (unknown)

Use your sketch to work out the sizes of all the angles in terms of x.

2 × angle QPS

Angle PSR = $2x$

Angle QRS = $1\frac{1}{2} × 2x = 3x$

1½ × angle PSR

Angle PQR = angle QRS = $3x$

equal angles

Angles in the quadrilateral PQRS = $x + 2x + 3x + 3x = 9x$

Angles in a quadrilateral sum to 360°.

What do you know about the sum of the angles in a quadrilateral?

So, $9x = 360°$

$x = \dfrac{360°}{9} = 40°$

Write and solve an equation.

Angle QPS = 40°

Write your answer.

---

1 In this triangle, angle BAC is 25% of the size of angle ABC. Angle ACB is 90° less than angle ABC.

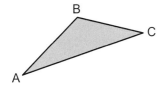

Q1 hint Sketch the diagram.
Label the 'unknown' that you are trying to find, x.
Use your sketch to write all the other angles in terms of x.
What do you know about the sum of the angles in a triangle? Write and solve an equation.

a   What is the size of angle ABC?
b   What kind of triangle is it?

2 Ross bought a new pair of trousers and a shirt. He spent £58 on the pair of trousers. This was £16 less than twice the cost of the shirt. How much did the shirt cost?

Q2 hint Write and solve an equation.

3   Louise spent £25 on a taxi fare. This was £10 more than 3 times the amount she could have spent on a train fare. How much was the train fare?

4   The perimeter of this triangle is 3 times the length of AB. Find the length of AB.

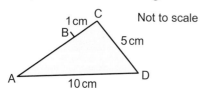

Not to scale

> **Q4 hint** Sketch the diagram. Label AB as x cm.

5   In a survey people were asked their favourite country for a holiday.
Here is a sketch of a pie chart that shows the information.

3 times as many people chose France as chose the UK.
Half as many people chose 'Other' as chose the UK.
9 times as many people chose Spain as chose 'Other'.
90 people were surveyed in total.
Draw an accurate pie chart.

Not to scale

> **Q5 hint** Let the number of people who chose the UK be x.

6   **Reflect**   Choose **A**, **B** or **C**.
Solving problems by using x for the unknown is
A always easy
B sometimes easy, sometimes hard
C always hard
Discuss with a classmate or your teacher what you find easy or hard.

6 Check up

Log how you did on your Student Progression Chart.

Angles between parallel lines

1   Work out the size of angle x in each diagram.

a

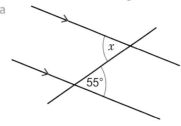

b

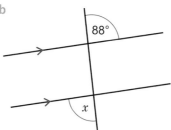

2   Work out the sizes of the angles marked with letters.
Give at least one reason for each answer.

a

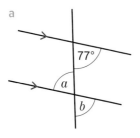

b

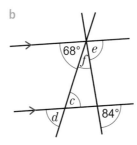

c

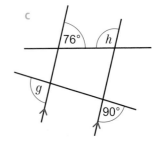

Active Learn   Homework, practice and support: Foundation 6 Check up

Triangles and quadrilaterals

3 Work out the size of angle x in each diagram.

a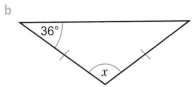

b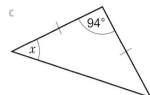

c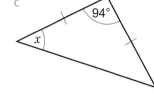

4 Find the sizes of angles a, b and c.
Choose at least one reason from the list to explain each answer.

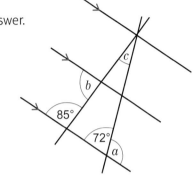

Reason 1	Angles are alternate.
Reason 2	Angles are corresponding.
Reason 3	The three angles in a triangle add up to 180°.
Reason 4	Angles on a straight line add up to 180°.
Reason 5	The four angles of a quadrilateral add up to 360°.
Reason 6	Co-interior angles.

5 The diagram shows a triangle.

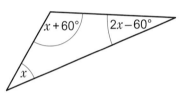

a Write an equation in terms of x.
b Solve your equation to find the value of x.
c Write down the sizes of the three angles in the triangle.

6 Work out the sizes of the angles marked with letters. State any angle facts that you use.

a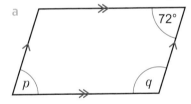

b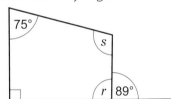

Interior and exterior angles

7 a How many sides does this regular polygon have?

b A regular polygon has 18 sides. What is the size of its exterior angle?

8 Work out the sizes of the angles marked with letters.

a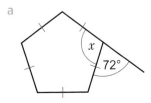

b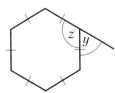

9 Find the sum of the interior angles of a polygon with 10 sides.

171

10 **Reasoning** Find the size of angle a.

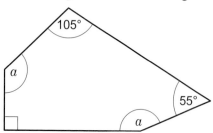

11 How sure are you of your answers? Were you mostly

Just guessing 😟   Feeling doubtful 😐   Confident 😊

What next? Use your results to decide whether to strengthen or extend your learning.

*Challenge

12 The diagram shows an irregular hexagon.

line of
symmetry

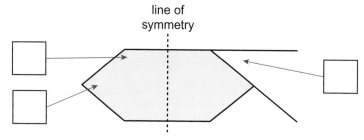

Here are some angle cards.

Make some copies of the diagram. Use the angle cards to complete the diagram in different ways.
How many ways can you find? You can use the same angle card more than once.

---

6 Strengthen

Angles between parallel lines

1 The diagrams show angles with straight lines.
Copy each diagram. Decide whether each pair of angles
are alternate, corresponding or vertically opposite.

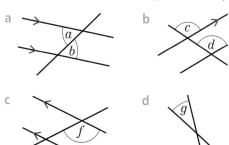

> **Q1 hint** Look for Z, X and F
> shapes and draw them on
> your diagrams.
>
> **Z N**
> alternate (Z shape)
>
> **✕**
> vertically opposite (X shape)
>
> **F ⤳ ⤳**
> corresponding (F shape)

2 Write down the sizes of the angles marked with letters. Part of each diagram is drawn in red to help you.
For each answer, choose the angle fact you have used from the box.

> alternate angles, corresponding angles, vertically opposite angles, angles on a straight line add up to 180°

a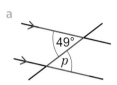

b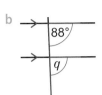

c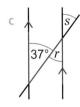

d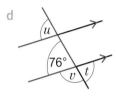

3 Find the sizes of the lettered angles in each diagram. Give at least one reason for each answer.

a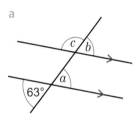

b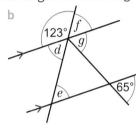

c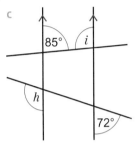

> **Q3 hint**
> Look at each diagonal separately. Copy the diagrams. Label the angles you find.

Triangles and quadrilaterals

1 Copy and complete.
  a The three angles of any triangle add up to ☐°.
  b The four angles of any quadrilateral add up to ☐°.

2 Work out the sizes of the angles marked with letters in these triangles and quadrilaterals.

a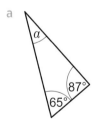

b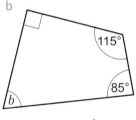

> **Q2a hint** Use the rules from Q1.
> Angle a = 180° − ☐ − ☐ = ☐°

c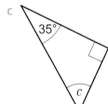

d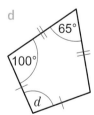

> **Q2d hint** Which other angle is the same as 100° in the kite? Use symmetry.

3 Which pair of angles are equal in each of these isosceles triangles?

a

b

c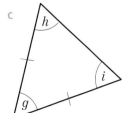

> **Q3 hint**
>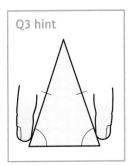

4 a What do the angles at Q and R add up to?
  b What is the size of the angle at Q?

> **Q4 hint** What do the three angles of a triangle add up to? Which two angles are equal?

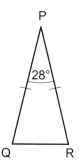

5 Work out the sizes of the angles marked with letters.

 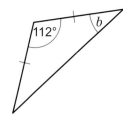 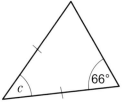

6 Find the sizes of all three angles in this triangle.
  Explain each stage of your working by choosing reasons from the box.

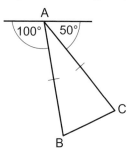

> **Q6 hint** Use the angles given to find angle BAC first.

> Reason 1   Angles of a triangle add up to 180°.
> Reason 2   Two angles in an isosceles triangle are equal.
> Reason 3   Angles on a straight line add up to 180°.

7 Here is a triangle.

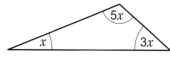

  a Write an expression for the sum of the three angles.
  b Solve the equation to find the value of x.
  c Write down the sizes of the three angles of the triangle.

> **Q7a hint** Sum of 3 angles = 180°

> **Q7c hint** When $x = \boxed{}°$
> $3x = \boxed{}°$
> $5x = \boxed{}°$

8 a Trace this equilateral triangle.
  b Rotate your tracing to compare the other angles with x.
  c Write an equation for the sum of the three angles.
  d Solve it to find the angles in an equilateral triangle.

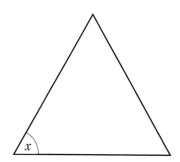

Interior and exterior angles

1 a Copy and complete.
The exterior angles of a polygon add up to ☐°.

b Use the information in part **a** to find the number of sides of this polygon.

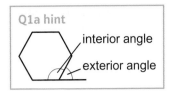

Q1a hint

interior angle

exterior angle

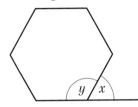

Q1b hint Use the formula

$$\text{exterior angle of regular polygon} = \frac{360°}{\text{number of sides}}$$

c What is the size of the exterior angle of a regular polygon that has 30 sides?

2 The diagram shows a regular hexagon.

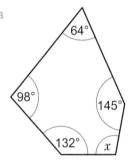

a Work out the size of the exterior angle, x.
b Work out the size of the interior angle, y.

Q2b hint The interior and exterior angles lie on a straight line. What do they add up to?

3 a

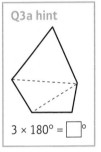

64°
98°
145°
132°
x

b

140°
150°
80°
95°
x 100°

Q3a hint

$3 \times 180° = $ ☐°

i Sketch each shape.
ii Divide each shape into triangles using diagonals from the same vertex.
iii Work out what the angles in each shape add up to.
iv Work out the size of angle x in each shape.

4 a Trace this regular octagon.
b Rotate your tracing to compare the other interior angles with x.
c Divide your octagon into triangles.
d Work out the angle sum for the octagon.
e Work out the size of angle x.

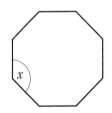

x

6 Extend

1 Work out the sizes of the angles in this triangle.

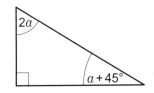

$2a$

$a + 45°$

2 **Reasoning / Communication**
Hannah says, 'Angle x is 51°.'
Is Hannah correct? Explain your answer.

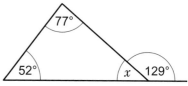

3 **Real** Shaojun wants to build a skate ramp. The diagram shows a cross-section of the ramp.

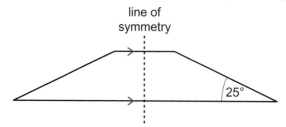

> **Q3 Communication hint**
> The **angle of elevation**
> is the angle between the
> horizontal base and the
> ramp.

The angle of elevation must be 25°.
Work out the sizes of the other interior angles.

4 **Problem-solving / Reasoning**
Find the size of angle x in this diagram.
Give reasons for each stage of your working.

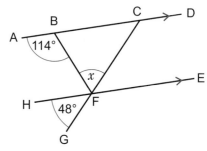

5 **Problem-solving** Find the size of angle ABJ.

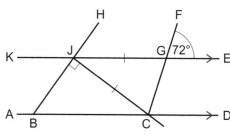

6 **Problem-solving** In triangle ABC, angle A is 80°
and angle B is 60% of angle A.
What is the size of angle C?

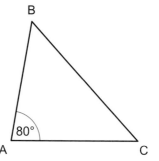

7 a What is the sum of the interior angles of a heptagon?
Explain how you worked out your answer.
   b The diagram shows a heptagon.
Work out the value of x.
   c Work out the sizes of all the angles
in the heptagon.
   d Show how to check that your answers
to part **c** are correct.

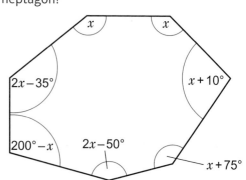

8 **Problem-solving**

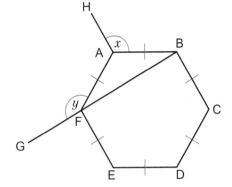

a What sort of triangle is triangle ABD?

b Write an expression in terms of x for angle ADB.

c Follow these steps to write an expression in terms of y for angle ADB.

i Write an expression in terms of x for angle CDB.
Use angles on a straight line.

ii Write an expression in terms of x for angle BCD.
Use angles in a triangle.

iii Write an equation for x in terms of y. Use angles on a straight line.

iv Substitute your expression for x into your answer to part b.

9 **Reasoning / Communication**

'Can a polygon have an angle sum of 500°?'

Explain how you know.

> **Q9 hint** 180° × ☐ = 500°

10 **Reasoning** ABCDEF is a regular hexagon.

Angle HAF = angle HAB

G, F and B lie on a straight line.

Work out the sizes of angles x and y.

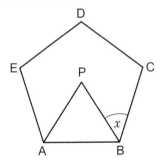

11 **Exam-style question**

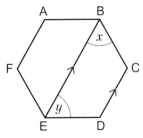

ABCDE is a regular pentagon.

ABP is an equilateral triangle.

Work out the size of angle x. **(4 marks)**

March 2013, Q8, 5MB2H/01

Exam hint

Write the sizes of the angles of the equilateral triangle and the pentagon on the diagram.

12 **Problem-solving** ABCDEF is a regular hexagon.

BE is parallel to CD.

Find x and y.

6 Knowledge check

- Two shapes are **congruent** when they are exactly the same shape *and* size. Two shapes are **similar** when they are the same shape. Similar shapes may be different sizes. .... *Mastery lesson 6.1*

- **Parallel lines** are shown with arrows. .... *Mastery lesson 6.2*

- **Alternate angles** are equal. .... *Mastery lesson 6.2*

- **Corresponding angles** are equal. .... *Mastery lesson 6.2*

- The angles in a triangle always add up to 180°. .... *Mastery lesson 6.3*

- An **exterior angle** of a triangle equals the sum of the two interior angles on the opposite side of the triangle. .... *Mastery lesson 6.3*

$$a = b + c$$

- A **regular polygon** has all equal sides and all equal interior angles. ... *Mastery lesson 6.4*

- An **irregular polygon** has unequal sides and unequal interior angles. .... *Mastery lesson 6.4*

- The exterior angles of any polygon always add up to 360°. .... *Mastery lesson 6.4*

- The exterior angle of a regular polygon $= \dfrac{360°}{\text{number of sides}}$. .... *Mastery lesson 6.4*

- Any pair of interior and exterior angles always add up to 180°. .... *Mastery lesson 6.4*

- For shapes to fit together (**tessellate**), all the angles where the shapes meet must add up to 360°. .... *Mastery lesson 6.5*

- The sum of the interior angles of any n-sided polygon is $(n − 2) \times 180°$ .... *Mastery lesson 6.5*

Reflect

'Notation' means symbols. Mathematics uses a lot of notation.

For example:

= means 'is equal to'   ° means degrees   ⌐ means a right angle

Look back at this unit. Write a list of all the maths notation used.

Why do you think this notation is important?

Could you have answered the questions in this unit without understanding the maths notation?

6 Unit test

1   Work out the sizes of the angles marked with letters.

   a 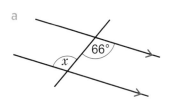   b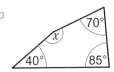

   (2 marks)

2   a   The exterior angle of a regular polygon is 12°. How many sides does the polygon have?
   b   A regular polygon has 36 sides. Work out the sizes of the exterior and interior angles.

   (3 marks)

3   Work out the size of angle x in each diagram.

   a 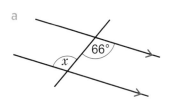   b 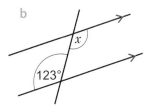   c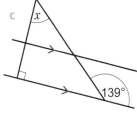

   (4 marks)

4   **Communication** The diagram shows
   two triangles with an extended line.
   The lines AB and DC are parallel.
   Show that triangle ABC is a right-angled
   triangle.
   Explain each stage of working.

   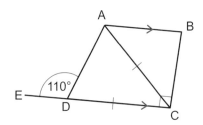

   (4 marks)

5   **Communication / Reasoning** Here is an irregular hexagon.

   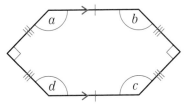

   a   Angle a = angle b = angle c = angle d. Explain why.
   b   Work out the size of this angle.

   (4 marks)

6   a   What is the sum of the interior angles of a pentagon?
       Explain how you worked out your answer.
   b   The diagram shows a pentagon. Work out the value of x.

   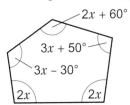

   c   Work out the sizes of all the angles in the pentagon.

   (4 marks)

7 **Reasoning** The diagram shows angles with parallel lines.
   a Work out the value of x.
   b Work out the sizes of the angles.

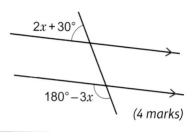

(4 marks)

8

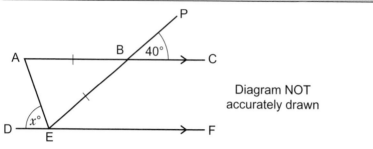

Diagram NOT
accurately drawn

ABC is parallel to DEF.
EBP is a straight line.
AB = EB
Angle PBC = 40°
Angle AED = $x°$
Work out the value of x.
Give a reason for each stage of your working.   **(5 marks)**

March 2012, Q10, 5MB2H/01

Exam hint
Every time you
write an angle on
the diagram, write
down a full reason
for this.

9 Find an expression for angle $D\hat{A}B$ in terms of x and y.
   $X\hat{C}D = y°$
   $C\hat{D}Y = 90°$

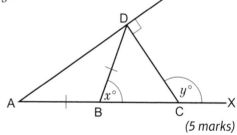

(5 marks)

Sample student answers

Which student's answer is better? Explain why.

   a Work out the size of angle x.

   b Give a reason for your answer.

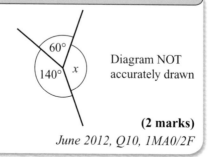

Diagram NOT
accurately drawn

(2 marks)
June 2012, Q10, 1MA0/2F

Exam hint
When giving reasons
for angles, remember
that you must state
the angle rule you
have used, in a
sentence, and *not* just
the calculations done.

Student A

a 160°

b Angles around a point should add up to 360°

Student B

a 160°

b Because 60° + 140° + 160° = 360°

7 AVERAGES AND RANGE

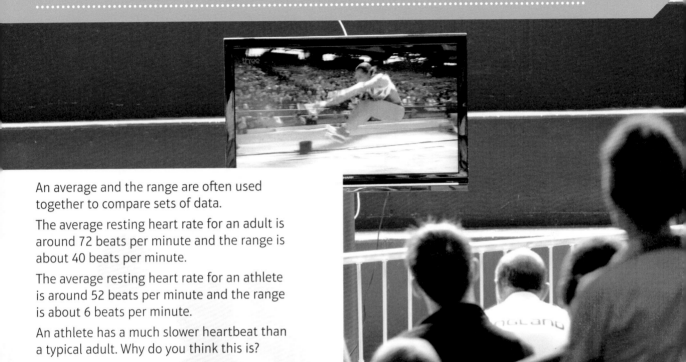

An average and the range are often used together to compare sets of data.

The average resting heart rate for an adult is around 72 beats per minute and the range is about 40 beats per minute.

The average resting heart rate for an athlete is around 52 beats per minute and the range is about 6 beats per minute.

An athlete has a much slower heartbeat than a typical adult. Why do you think this is?

7 Prior knowledge check

Numerical fluency

1   Work out
    a $412 \div 10$
    b $2.48 \div 10$
    c $29 \div 4$
    d $25.3 \div 11$

2   Work out
    a $2.7 + 1.8 + 16$
    b $14 + 7.8 + 1.2$
    c $0.8 + 3.1 + 11$
    d $4.5 + 0.8 + 18$

3   Round to 1 decimal place.
    a   0.76
    b   12.45
    c   8.762
    d   21.039

4   Round to 2 decimal places.
    a   2.568
    b   0.546
    c   0.049
    d   0.499

5   Work out
    a   10% of 120
    b   10% of 48
    c   5% of 90
    d   5% of 36

6   Write these numbers in order of size, putting the smallest first.
    6.3, 8, 5.9, 3, 6.2, 7

7   Find the value of each calculation correct to 2 decimal places.
    a $\dfrac{16 + 27 + 9 + 15 + 11 + 8}{6}$
    b $\dfrac{11 \times 4.6 + 16 \times 3.8}{27}$

Algebraic fluency

8   Work out the value of $\dfrac{n+1}{2}$ when $n = 49$.

9   Work out the value of $\dfrac{n+1}{2}$ when $n = 60$.

Fluency with data

10   Decide whether each of these is qualitative or quantitative data.
    a   eye colour
    b   number of siblings
    c   height
    d   shoe size

11   Isabel measured the height, in cm, of 9 pencils. Here are her results.
    12, 13, 10, 13, 15, 11, 12, 13, 14
    a   What is the mode?
    b   Work out the range.
    c   What is the median?

181

12 Harry measured the weight, in g, of 8 kittens. Here are his results.

130, 115, 135, 105, 140, 115, 120, 130

What is the median?

13 The chart shows the scores achieved in a test by students in Class 11X.

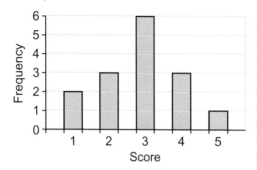

a What is the modal score?
b Work out the range.
c How many students took the test?

* Challenge

14 These five cards each show a number.

| 4.1 | 6.2 | 5 | 4.6 | x |

Find the value of x, if
a the sum of the numbers is 23.8
b the median value is 4.6 and the range is 2.4.

7.1 Mean and range

Objectives

- Calculate the mean from a list and from a frequency table.
- Compare sets of data using the mean and range.

Did you know?

The size of the average family in England and Wales is getting smaller. Over the last 50 years, the mean number of children per family has fallen from 2.4 to 1.9.

Fluency

How could you share these counters into 4 equal piles?

1 Work these out, rounding your answers to 2 decimal places where necessary.

a $3\overline{)21.6}$ b $\dfrac{2.7}{4}$ c $11.2 \div 3$

2 The table shows the number of cars of different colours in a car park.
a Which colour was the mode?
b How many cars were in the car park all together?

Colour	Black	Blue	Green	Red	White
Frequency	25	19	16	34	24

Key point 1

The mean of a set of values is the total of the set of values divided by the number of values.

Example 1

Work out the mean of 3, 6, 7, 7 and 8.

$3 + 6 + 7 + 7 + 8 = 31$ — Add the values first to find the total.

$\dfrac{31}{5} = 6.2$ — There are 5 values, so divide the total by 5.

The mean is 6.2

Active Learn   Homework, practice and support: Foundation 7.1

Questions in this unit are targeted at the steps indicated.

3 Here are the numbers of books a teacher marked each day over a 10-day period.
0, 31, 45, 30, 58, 60, 0, 52, 30, 46
Work out the mean number of books he marked each day.
Discussion Should the answer be rounded?

4 Jayme went on a four-day hike in Peru.
The table shows the distance she walked each day.
Work out the mean distance Jayme walked each day.

Q4 hint On a calculator, press '=' to find the total before you divide.

Day	Mon	Tue	Wed	Thu
Distance (km)	8.6	18.2	10.5	5.7

5 **Exam-style question**

The table shows the midday temperature on each day for ten days.

Day	1	2	3	4	5	6	7	8	9	10
Temperature (°C)	13	14	12	10	13	16	14	13	18	16

a Find the range of temperatures.
b Write down the mode.
c Work out the mean temperature. **(5 marks)**
June 2011, Q5, 5MB1F/01

Exam hint
Even if you are using a calculator, write down the calculation for each part to gain the marks for working.

6 Here are the times, in minutes, some patients waited at a doctors' surgery.
5, 7, 7, 6, 8, 5, 7, 6, 7, 8, 9, 6, 8, 7, 9, 7, 9, 6, 5, 8
a Work out the mean waiting time.
b Work out the range.

Q6 hint If your calculator has stats mode, find out how to use it to calculate the mean.

7 The waiting times in **Q6** are shown again in this table.

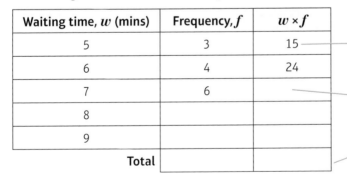

Waiting time, w (mins)	Frequency, f	$w \times f$
5	3	15
6	4	24
7	6	
8		
9		
Total		

Q7a hint 3 people waited 5 minutes = 3 × 5 = 15 minutes in total.

Q7a hint 7 × 6 = ☐.

Q7a hint Add the numbers from the two columns.

a Copy and complete the table.
b Find the total waiting time of all the patients.
c Find the total number of patients.
d Use your answers to parts **b** and **c** to work out the mean waiting time.

Q7c hint Total number of patients = total frequency

Q7d hint
Mean = $\dfrac{\text{total waiting time}}{\text{total number of patients}}$

Discussion What is the same and what is different about the methods used in **Q6** and **Q7**?
Which provides the most efficient way to find the answers?

8   100 children were asked how many portions of fruit and vegetables they eat each day.
    The results are shown in the table.

Number of portions, p	Frequency, f	$p \times f$
4	12	
5	33	
6	41	
7	14	
Total		

a   Work out the mean.   b   Work out the range.

> **Q8b hint** Look at the largest and smallest numbers of portions.

9   **Reasoning** The charts show the test scores achieved by students in 11W and 11Y.

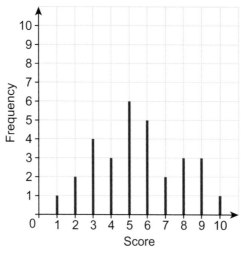

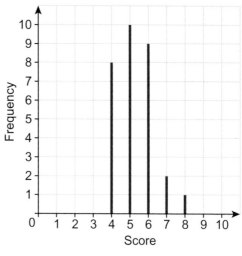

a   Copy and complete the frequency table for 11W.

b   Work out the mean and range for 11W. Give your answer correct to 1 decimal place.

c   Make a frequency table for 11Y.

d   Work out the mean and range for 11Y.

e   Choose words from the box to fill in the gaps.
    The mean score for 11W is .... to the mean score for 11Y.
    The range of the scores for .... is larger than the range of the scores for ....
    A larger range shows the data is more .... out.

Score, s	Frequency, f	$s \times f$
1	1	1
2	2	4
3	4	12
Total		

> different   spread   11W   close   11Y   size

> ### Key point 2
> To compare two sets of data, compare an average (mode, median or mean) for each set of data and compare the range of each set of data.

10 **Real / Reasoning** Tess and Jo play cricket. Here are the runs they scored in the last 8 matches.

Tess: 33   40   52   45   37   50   61   38

Jo: 62   70   0   52   7   85   22   96

a Find the mean score for each player.

b Find the range for each player.

c Which player is more consistent?

d Which player would you like to have on your team? Give a reason.

> **Q10c communication hint**
> A consistent player gets similar results every time.

Discussion Do you have enough information to compare two sets of data if you only know their mean values?

11   **Exam-style question**

Ed has 4 cards.

There is a number on each card.

| 12 | 6 | 15 | ? |

The mean of the 4 numbers on Ed's cards is 10.

Work out the number on the 4th card.   **(3 marks)**

June 2013, Q20, 1MA0/1F

> **Q11 strategy hint**
> Work out what the total of the 4 cards must be.

12   **Exam-style question**

The table gives information about the number of children in 25 families.

Number of children	Frequency
0	4
1	8
2	9
3	3
4	1

Work out the mean number of children per family.

Give your answer correct to 1 decimal place.   **(3 marks)**

> **Exam hint**
> Add an extra column to the table for number of children × frequency. Check that your answer is between 0 and 4.

7.2 Mode, median and range

Objectives

- Find the mode, median and range from a stem and leaf diagram.
- Identify outliers.
- Estimate the range from a grouped frequency table.

Did you know?

In 2014, the top nine Premier League clubs had a median income from TV rights and prize money of £90m with a range of £19m.

Fluency

Here are some readings taken in a science experiment.

2.6, 2.1, 2.7, 2.7, 2.9, 15, 3.1, 2.5

What is the mode?

One of the readings is incorrect. Which one do you think it is likely to be?

1   Here are the pulse rates of 20 runners at the end of a half marathon.

| 164 | 172 | 158 | 186 | 160 | 165 | 152 | 177 | 149 | 155 |
| 176 | 180 | 163 | 157 | 148 | 167 | 173 | 181 | 175 | 162 |

Draw a stem and leaf diagram for the data.

2   Find the median and range of these data sets.

a   5, 3, 7, 11, 9, 6, 10, 3, 6   b   21, 17, 19, 26, 29, 25, 22, 28

3   x and y are integers. List all the possible values of x and y.

a $5 \leqslant x < 10$ b $0 \leqslant y < 3$

> **Q3 communication hint**
> An integer is a whole number or zero.

4   This stem and leaf diagram shows the recorded mileage of cars, to the nearest 1000 miles, serviced at a garage in one week.

```
1 | 6  8
2 | 3  7  7
3 | 7  8  8  9
4 | 0  1  4  4  7  8      Key
5 | 0  0  3  4  6  6      1|6 means 16 000 miles
6 | 1  1  2  3  7
7 | 6  9
```

> **Q4b hint** What are the highest and lowest mileages?

a   How many cars are there?

b   Work out the range of the mileages.

Key point 3

The median is the middle value when the data is written in order.

Example 2

This stem and leaf diagram shows the times, in seconds, for a group of swimmers to swim 100 m. Find the median and the mode.

```
55 | 2  3  6
56 | 3  3  7  8
57 | 0  2  6  6  6  7    Key
58 | 4  4  5             55|2 means 55.2 seconds
59 | 3
```

> Count the number of values; 17.

> The median is the $\frac{n+1}{2}$ th value.
> There are 17 values, so $n = 17$.

$\frac{17+1}{2} = 9$

The median is the 9th value.

> In a stem and leaf diagram the data is in order. So count up to the 9th value.

The median is 57.2 seconds.

The mode is 57.6 seconds.

> Look for repeated values in the rows.
> 57| 0 2 **6 6 6** 7

5   **Reasoning** Why is the mode not a useful average for the data in **Q4**?

6   This stem and leaf diagram shows the lengths, in cm, of some pencils.

```
5 | 7  7  9
6 | 5  6  8  8  9
7 | 0  0  0  2  5  5  7  8   Key
8 | 4  4  5                  5|7 means 5.7 cm
```

Find   a   the mode   b   the median   c   the range.

Reflect Did you use the key to help you understand the data in the stem and leaf diagram? How does the key help you?

7 **Exam-style question**

Yan recorded the ages, in years, of a sample of people at a fairground.
He drew this stem and leaf diagram for his results.

```
1 | 5  5  7  7  7  7  9
2 | 0  3  7  8  8
3 | 4  6  7  7           Key
4 | 2  5  9              1|5 represents 15 years of age
5 | 0  5
```

a Write down the number of people in the sample. **(1 mark)**
b Write down the mode. **(1 mark)**
c Work out the range. **(2 marks)**

Nov 2012, Q18, 1MA0/1F

Exam hint
For parts **b** and **c**
you must use the
key to work out
the age.

8 Each act at a talent contest was given a score. This stem and leaf diagram shows the results.

```
5 | 4  5  7  7  9
6 | 0  2  2  3  7  8
7 | 3  4  5  7  9  9         Key
8 | 2  4  6  6  7  7  7  8   5|4 means 54 points
9 | 5  5  6
```

Find   a   the mode   b   the median   c   the range.

> **Q8b hint** There are 28 values.
> The position of the median
> is $\frac{28+1}{2} = 14.5$. So the
> median is halfway between
> the 14th and 15th values.

Key point 4

An outlier is an extreme data value that doesn't fit the overall pattern.

9 **Reasoning** a Identify the outlier in each set of experiment data.
    i   43, 41, 52, 55, 42, 23, 46, 50
    ii   128, 112, 133, 116, 195, 131, 120, 115
    iii   6.43, 6.72, 6.57, 2.76, 7.01, 6.88, 6.39, 6.98, 8.85, 7.11
   b   For each set of data work out
    i   the range
    ii   the range ignoring the outlier.

> **Q9a hint** The
> outliers may be
> large or small
> compared to the
> rest of the data.

Discussion Which range value do you think best describes the results of the experiment?

10 Each of these sets of data was found by experiment. The outliers are likely to be errors.
Work out the range of each data set, ignoring any outliers.
   a   50, 47, 56, 58, 59, 45, 86, 49, 53
   b   5.8, 6.2, 2.9, 5.7, 6.1, 6.3, 5.8, 5.9

11 **STEM** The diagram shows the results
of a physics experiment.
Find the range of the voltages,
ignoring any outliers.

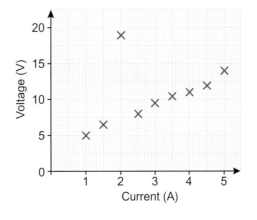

Key point 5

To estimate the range of grouped data, work out maximum possible value – minimum possible value.

12 A rugby team captain records the points scored by his team in a season.
Estimate the range of points scored.
Discussion Why can't you work out the exact value of the range from this table?

Points scored	Frequency
0–10	1
11–20	8
21–30	7
31–40	5
41–50	3

13 Jill measured the heights of some seedling for a project. The table shows her results.
a Write down a possible value of the tallest seedling that Jill measured.
b Write down a possible value for the shortest seedling that Jill measured.
c Estimate the range.
Discussion From the table, what is the maximum possible height?

Height, h (cm)	Frequency
$0 < h \leqslant 5$	31
$5 < h \leqslant 10$	40
$10 < h \leqslant 15$	55
$15 < h \leqslant 20$	38
$20 < h \leqslant 25$	16

Q13 hint A seedling can't actually have a height of 0 cm but it makes sense to use this to give an estimate.

14 **Real** Jim runs a spin class at a sports centre.
He recorded the ages of the people in one session.
Estimate the range of the ages.

Age, a (years)	Frequency
$15 \leqslant a < 20$	2
$20 \leqslant a < 25$	8
$25 \leqslant a < 30$	5
$30 \leqslant a < 35$	3
$35 \leqslant a < 40$	4

Q14 hint For your estimate use 40 years as the oldest possible.

7.3 Types of average

Objectives
- Recognise the advantages and disadvantages of each type of average.
- Find the modal class.
- Find the median from a frequency table.

Did you know?
The median time to run the London Marathon is between 4 hrs 15 mins and 4 hrs 40 mins.

Fluency

Faye counted the number of matches in 10 boxes:
46, 48, 49, 48, 47, 46, 48, 47, 46, 48
Find the mode.

1 Find the median of these test scores.
43, 29, 25, 56, 63, 72, 68, 82, 59

2 This table gives information about the test scores achieved by some Y11 students.

Score (%)	Number of students
0–19	6
20–39	24
40–59	22
60–79	8

How many students scored 40 or more?

3 Find the median of the first 20 odd numbers.

> **Q3 hint** List the first 20 odd numbers in order. Use $\dfrac{n+1}{2}$ to find the position of the median.

4 Find the mode of each data set.
  a 21, 29, 25, 23, 25, 28, 25, 30, 22
  b 108, 132, 110, 126, 132, 111, 129
  c Is the mode a good representative value for both data sets?

5 Find the mean of each data set.
  a 26, 28, 27, 153, 24   b 43, 52, 37, 44, 50
  c Is the mean a good representative value for both data sets?

6 Find the median of each data set.
  a 11, 12, 14, 17, 18, 18, 20, 21, 23
  b 1.7, 1.8, 1.8, 1.9, 5, 8, 14
  c Is the median a good representative value for both data sets?
Discussion Why do we need three types of average?

7 **Reasoning** Work out the most appropriate average for each data set.
  a 39, 43, 37, 40, 46, 38, 125, 41   b 1.7, 1.8, 1.9, 3, 6
  c 46, 48, 47, 48, 49, 48, 48, 50   d blue, yellow, blue, green, yellow, blue, red

8 a Find the median of 24, 26, 31, 32, 35
  b Find the median of 24, 26, 31, 32, 45
  c Find the median of 24, 26, 31, 32, 55
Discussion What do your answers to parts **a**, **b** and **c** tell you about the median?

9 Copy and complete the table by putting one boxed statement in each blank position.

> May not change if a data value changes.

> Every value makes a difference.

> Easy to find; not affected by extreme values, and can be used with non-numerical data.

> There may not be one.

> Not affected by extreme values.

> Affected by extreme values.

Average	Advantages	Disadvantages
Mean		
Median		
Mode		

> **Key point 6**
>
> The modal class is the class with the highest frequency.

10 Members of a karate club did as many press-ups as they
could in one minute. This table shows the results.
What is the modal class?

Press-ups	Frequency
20–29	4
30–39	5
40–49	8
50–59	7
60–69	5

11 This table shows the points scored by the top 25 decathletes in the London 2012 Olympics.

Points scored, p	Frequency
$7500 \leqslant p < 7750$	4
$7750 \leqslant p < 8000$	10
$8000 \leqslant p < 8250$	4
$8250 \leqslant p < 8500$	4
$8500 \leqslant p < 8750$	2
$8750 \leqslant p < 9000$	1

Write down the modal class.

Q11 hint Write the modal class as $\boxed{} \leqslant p < \boxed{}$.

> **Example 3**
>
> Tom rolled a dice 25 times. This table shows his scores.
>
> | Score | Frequency |
> |-------|-----------|
> | 1 | 4 |
> | 2 | 4 |
> | 3 | 6 |
> | 4 | 4 |
> | 5 | 2 |
> | 6 | 5 |
>
> Find the median score.
>
> $\dfrac{25 + 1}{2} = 13$ ——————— The median is the 13th score.
>
> | Score | Frequency |
> |-------|-----------|
> | 1 | 4 |
> | 2 | 4 |
> | 3 | 6 |
> | 4 | 4 |
> | 5 | 2 |
> | 6 | 5 |
>
> 4
> $4 + 4 = 8$ ——————— 8 of the scores are 2 or less.
> $4 + 4 + 6 = 14$ ——————— 14 of the scores are 3 or less.
> ——————— Find the 13th score in the table.
>
> The 13th score is 3.
>
> The median score is 3.

12 Jess spun a five-sided spinner 45 times. The table shows her scores.

Score	Frequency
1	11
2	8
3	9
4	7
5	10

Find the median score.

13 Lily recorded the number of people in each car that passed her house in one hour.
This table shows her results.

Number of people	Frequency
1	23
2	16
3	6
4	3
5	1

> **Q13a hint** Find the total number of cars first by adding the frequencies.

a Find the median number of people per car.

> **Q13c hint** Add an extra column to the table for number of people × frequency.

b What is the mode?

c Work out the mean number of people per car.

14 Visitors at a school fete were asked to guess the number of marbles in a jar.

Guess	Frequency
100–125	9
126–150	12
151–175	15
176–200	8
201–225	3

> **Q14 hint** Use the same method, adding frequencies. Write the class as ☐–☐.

Which class contains the median?

15 **Exam-style question**

This table gives information about the distances, in metres, Ted hit a golf ball.

Distance, d (m)	Frequency
$200 \leqslant d < 220$	4
$220 \leqslant d < 240$	10
$240 \leqslant d < 260$	9
$260 \leqslant d < 280$	8

a Write down the modal class interval. **(1 mark)**

b Find the class interval that contains the median. **(2 marks)**

c How many times did Ted hit the golf ball less than 240 m? **(2 marks)**

Exam hint

Write the class as
$☐ \leqslant d < ☐$

7.4 Estimating the mean

Objective

- Estimate the mean of grouped data.

Why learn this?

It is much easier to see patterns and trends in large amounts of data if the data is grouped.

Fluency

Find the value halfway between each of these pairs.

a 16, 20 **b** 20, 25 **c** 18, 20 **d** 4, 5

 1 This six-sided spinner is spun 50 times. The results are shown in the table.

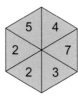

Score	Frequency
2	16
3	9
4	6
5	10
7	9

Work out the mean score.

 2 Work out the mean of these test scores. Give your answer correct to 1 decimal place.

8 11 5 14 5 5 9 12 14 13 10 18 16 20 17
10 11 9 4 15 18 20 17 12 6 16 5 19 18 14

 3 Write each of these in hours and minutes to the nearest minute.

a 3.74 hours   b 2.831 hours   c 4.658 hours

> **Q3 hint**
> 0.74 hours = 0.74 × 60 minutes

Example 4

The test scores given in **Q2** have been grouped in this table.

Score	Frequency
1–5	5
6–10	6
11–15	9
16–20	10

Add a column to calculate the midpoint of each class. Use this as an estimate of the scores, because you don't know the exact values in each class

Work out an estimate for the mean.

Score	Frequency, f	Midpoint of class, m	$m \times f$
1–5	5	3	15
6–10	6	8	48
11–15	9	13	117
16–20	10	18	180
Total	**30**	Total	**360**

Add a column, $m \times f$, to calculate an estimate of the total score for each class.

Estimate of mean = $\dfrac{360}{30}$ = 12 — Divide the total of the $m \times f$ column by the total frequency.

Discussion Why is this answer different to the one you worked out in **Q2**

ActiveLearn Homework, practice and support: Foundation 7.4

> **Key point 7**
>
> When the data is grouped, you can calculate an **estimate** for the mean.

4 **Real** In a survey, 30 small companies were asked how many employees they had. This table shows the results.

Number of employees	Frequency, f	Midpoint of class, m	$m \times f$
1–5	12		
6–10	7		
11–15	6		
16–20	5		
Total		Total	

Calculate an estimate for the mean number of employees per company.

5 **Real** This table gives information about cooking times, in minutes, of some frozen ready-meals.

Time (min)	Frequency, f
11–15	3
16–20	7
21–25	6
26–30	4

> **Q5 hint** Copy the table and add the extra columns needed as in **Q4**.

Work out an estimate of the mean cooking time to the nearest minute.

6 **Real** 28 members of a running club took part in a marathon. Information about their times is given in the table.

Finishing time, t (hrs)	Frequency, f
$2 \leq t < 3$	4
$3 \leq t < 4$	17
$4 \leq t < 5$	5
$5 \leq t < 6$	2

> **Q6 hint** The midpoint of $2 \leq t < 3$ is 2.5

Work out an estimate of the mean time. Give your answer in hours and minutes to the nearest minute.

7 **Real** Here are the temperatures, in °C, at 30 different locations in the UK one day in June.
16.1 17.2 18.3 18.7 18.8 19.1 19.2 20.0 20.4 21.2 21.7 21.8 22.3 22.5 22.5
22.8 22.8 23.0 23.1 23.4 23.4 23.7 24.2 24.2 24.3 24.9 25.3 25.3 25.8 25.9
Copy and complete the frequency table.

Temperature, t (°C)	Frequency, f		
$16 \leq t < 18$			
$18 \leq t < 20$			
$20 \leq t < 22$			
$22 \leq t < 24$			
$24 \leq t < 26$			

a Work out an estimate of the mean temperature.
b What is the modal class?
c Which class contains the median?
d Estimate the range of the temperatures.

8 **Exam-style question**

This table gives information about the lengths, in cm, of some eels.

Length, l (cm)	Frequency, f		
$10 \leqslant l < 20$	23		
$20 \leqslant l < 30$	29		
$30 \leqslant l < 40$	25		
$40 \leqslant l < 50$	17		

a Work out an estimate of the mean length of the eels.
  Give your answer to an appropriate level of accuracy. **(5 marks)**

b Explain why your answer is only an estimate. **(1 mark)**

Exam hint
Use the empty columns in the table to help you calculate the mean.

9 **Reflect** When you answered the questions in this lesson, did you make any mistakes? If so, check you understand where you went wrong.

7.5 Sampling

Objectives

• Understand the need for sampling.
• Understand how to avoid bias.

Why learn this?

At a sweet factory, some sweets are taken from every batch as a sample in order to check for quality.

Fluency

Explain how random numbers are used in the National Lottery.

Warm up

1 Which way of picking one student from a class is most fair?
  A Choosing the last name on the register
  B Choosing the tallest student
  C Putting all the names in a hat and choosing one without looking

Key point 8

In a survey, a **sample** is taken to represent the **population**. A sample that is too small can **bias** the results.

2 **Real / Reasoning** A TV company wants to find out what percentage of people living in the Midlands watch Eastenders.
  a What is meant by the population in this case?
  b Explain why a sample is needed.
  c Around 10 million people live in the Midlands. How many people should they choose for the sample?

     100   500   1000   100 000

  Discussion Would it make sense for the entire sample to be taken from one Midlands town?

Q2a communication hint
The **population** is the whole of the group that you are interested in.

Q2c hint The sample needs to be large enough to represent the population, but small enough to keep costs within a budget.

ActiveLearn Homework, practice and support: Foundation 7.5

3 **Real / Communication** Sean is writing an article for his local newspaper.
   He wants to comment on the popularity of a proposal to build a new supermarket.
   He carries out a survey in his office to test opinion.

   a What is meant by the population in this case?

   b Comment on the sample Sean is using.

4 **Real / Communication** Adam wants to find out if people in his town are
   satisfied with the new arrangements for refuse collection. He asks all of the people
   in one street for their opinions.

   a Explain why Adam's sample is not reliable.

   b How could Adam improve his survey?

5 **Real / Reasoning** Helen and John are conducting a survey to find out how far
   people drive in a typical week.
   Helen chooses a sample of people at a London shopping centre.
   John chooses a sample of people at a motorway service area on the M6.
   Would you expect Helen and John to get similar results?

   Discussion What does this say about the choice of location for a sample?

6 A teacher wants to know students' views on school dinners.
   She plans to give questionnaires to her Year 10 tutor group.

   a Will this give fair or biased results?

   b How can she improve her survey?

Key point 9

In a **random sample**, every member of the population has an equal chance of being included.

---

7 **Real** Tristan and Sue are both selecting a sample of five students
   from Class 11W.

   a Tristan selects his sample by putting all 11W students' names
      into a hat and picking out five names without looking.
      Will this give a random sample?
      Explain your answer.

   > **Q7a hint** Does every student have the same chance of being picked?

   b Sue chooses her sample by picking the first five names on
      the 11W register. Would this give a random sample?
      Explain your answer.

   > **Q7b hint** Think what would happen if you repeated this process.

8 a Here is a list of random numbers.
      103 883 041 033 381 839 122 109
      Write down 2 digits from each number.

   > **Q8a hint**
   >

   Discussion How could you use this method to pick 6 people
   at random from a numbered list?

   b Alix has a list of 40 people who bought theatre tickets.
      She wants to pick 5 at random to win a prize.
      Explain how she could use the random numbers in part **a** to pick 5 people from
      her list.

9 **Exam-style question**

A hotel has 200 rooms.
The manager wants to carry out a survey using a random sample
of 20 guests.
Describe two ways she could select a random sample. **(2 marks)**

10 **Reasoning / Real** In 1936, a Literary Digest poll used a sample of 2.4 million people to predict the result of that year's US presidential election.

They picked 10 million people from telephone directories, lists of magazine subscribers and members of clubs and associations and sent them all a mock ballot paper.

From the 2.4 million responses they received, they predicted Roosevelt would get 43% of the vote and Landon 57%.

In fact, Roosevelt got 62% of the vote and Landon 38%.

a   Who did the Literary Digest poll predict would win the election?

b   Who actually won?

Discussion Was the sample biased?

7 Problem-solving: Watching statistics

Objective • Analyse and compare the distributions of two data sets.

1 A television drama has just finished its second series. The production company wants to evaluate how many people are watching and enjoying the programme.
The tables below show data on the first and second series. Each series had 8 episodes and a Christmas Special (CS).
Write a brief summary of the data for the production company.
Here are a few things to think about.

- What statistics (mean, median, mode, range) could you find?
- Are there any outliers?
- What will you do about any outliers?
- Would it be useful to draw a graph? If so, what kind of graph?
- Can you comment on trends or patterns?
- What is the same/similar, or different about the two series?

> **Q1 strategy hint** Show all your working and write a summary of what you find.

2 Predict the success of a third series. Show all your workings and give reasons for your predictions.

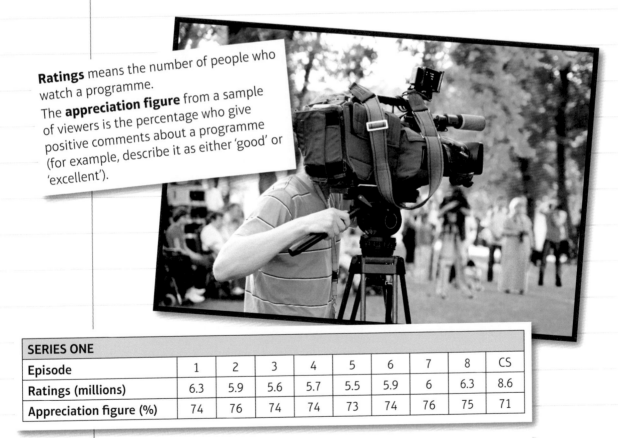

Ratings means the number of people who watch a programme.
The **appreciation figure** from a sample of viewers is the percentage who give positive comments about a programme (for example, describe it as either 'good' or 'excellent').

SERIES ONE									
Episode	1	2	3	4	5	6	7	8	CS
Ratings (millions)	6.3	5.9	5.6	5.7	5.5	5.9	6	6.3	8.6
Appreciation figure (%)	74	76	74	74	73	74	76	75	71

SERIES TWO									
Episode	1	2	3	4	5	6	7	8	CS
Ratings (millions)	7.2	7.1	6.9	6.9	6.7	7.0	7.1	7.5	10.1
Appreciation figure (%)	74	76	76	75	76	74	75	72	70

7 Check up

Averages and range

1   These are the numbers of people in the queue for each checkout at a supermarket.

    5   4   7   3   6   3   4   2   4   5

Work out the mean number of people in a queue.

2   Mark recorded the duration, in minutes, of each of his favourite television programmes.

    35   40   45   15   45   30   30   30   25   35   60   50   60   65

    a   Work out the mean duration.   b   Work out the range.

3   This table shows the number of goals scored in 25 Premier League football matches.

Goals scored	Number of matches
0	3
1	5
2	9
3	6
4	2

Work out the   a   mean   b   range   c   median.

4   Ria and Jade both play for a cricket team.

Here are the numbers of runs that Ria made in her last 5 matches.

    27   42   16   11   53

Jade scored a mean of 32.4 runs with a range of 28 in her last 5 matches.

    a   Who scored the most runs? Explain your answer.

    b   Who is the more consistent player? Explain your answer.

5   This stem and leaf diagram shows weights of parcels in a post office van.

```
0 | 5  7  7
1 | 2  3  5  5  6  9
2 | 0  1  6                Key
3 | 4  4  4  5  7          1|2 means 1.2 kg
4 | 2  3  6  8
```

Find   a   the mode   b   the range   c   the median.

6   Ed plays snooker. Here are the scores from his last 7 breaks.

    28   21   105   21   32   35   38

    a   Which would be the most appropriate average to use for Ed's scores?

    b   Explain why the other two averages do not represent the scores very well.

Averages and range for grouped data

7   The table shows the number of emails Sophie received per day in one month.

    a   Work out an estimate for the mean number of emails Sophie received per day.

    b   Work out an estimate for the range.

    c   Which is the modal class?

    d   Which class contains the median?

Number of emails	Frequency
1–10	3
11–20	5
21–30	7
31–40	16

ActiveLearn   Homework, practice and support: Foundation 7 Check up

Sampling

8 Dan wants to select three students at random from Class 8W.
   There are 30 children in the class.
   Describe how he can make the choice fairly.

9 **Communication** Jo wants to find out people's views on whether VAT should be charged
   for hairdressing. She asks a random sample of ten people at a shopping centre on a Monday
   morning for their views.
   Give two reasons why the results of her survey may not be reliable.

10 How sure are you of your answers? Were you mostly

   Just guessing   Feeling doubtful   Confident ☺

   What next? Use your results to decide whether to strengthen or extend your learning.

* Challenge

11 This table gives information about the ages of members of a gym.

Age	Frequency, f	Class mid-point, m	$f \times m$
18–24	32		672
25–		27	1107
–40	70		
41–59			2650
60–70	14		

   a Copy and complete the table to fill in the missing values.
   b Work out an estimate of the mean age.
   c Which class contains the median?
   d Which is the modal class? Explain why this may be misleading.

7 Strengthen

Averages and range

1 Lisa is training for a 200 m race. These are her latest times,
   in seconds, over that distance.
   24, 25, 24, 24, 23, 24, 25, 23
   a Copy and complete this sentence to predict the mean.
      I think Lisa's mean time will be about ....
   b How many times are there?
   c Work out the total of all her times.
   d Work out Lisa's mean time for the 200 m..

> **Q1a hint** The mean will be
> somewhere between the lowest
> value and the highest value.
> Your answer will be in seconds.

> **Q1d hint**
> $$\text{Mean} = \frac{\text{total of times}}{\text{number of times}}$$

2 Here are the top ten midday temperatures, in °C,
   in Bridlington during October 2014.
   16, 16, 16, 15, 14, 14, 14, 13, 12, 12
   Work out the mean of these temperatures.

> **Q2 hint** Predict the mean first.
> How many temperatures are there?
> Check your answer ☐°C with your
> prediction.

3 These are the amounts that Anya earned in the last 6 weeks.

£86   £94   £110   £80   £75   £88

Calculate Anya's mean weekly wage.

> **Q3 hint** Follow the method as in **Q2**. Round your answer to the nearest penny.

4 Here are the number of goals scored by Tom's hockey team in the first few games of the season.

2   0   3   1   5   2   0   4

Work out the mean number of goals scored.

> **Q4 hint** Make sure that you count the zero scores.

5 **Communication** Here are the numbers of runs scored by two cricketers in three matches.

Ellie: 35, 40, 36   Sam: 62, 0, 34

a Who had the highest score in a match?

b Work out the range of scores for Ellie and for Sam.

c Whose results are the most consistent?

d Who had the highest mean score?

e Who would you pick for your team? Explain why.

> **Q5c communication hint** Consistent means more or less the same every time. Results are more consistent when the range is small.

6 Kyle and George both play for a cricket team.

Here are the numbers of runs that Kyle made in the last five matches.

18   53   19   14   62

In the last five matches, George scored a mean of 31.6 runs with a range of 21.

a Who scored more runs? Explain your answer.

b Who is the more consistent player? Explain your answer.

> **Q6b hint** Compare the ranges.

7 The table shows the number of Valentines cards a group of boys claimed to have received.

Number of cards	Frequency	Number of cards × frequency
1	8	1 × 8 = ☐
2	5	2 × 5 = ☐
3	3	
4	2	
Total		

a i How many boys claimed to have received 1 card?

   ii How many cards did these boys claim to receive in total?

b i How many boys claimed to have received 2 cards?

   ii How many cards did these boys claim to receive in total?

c Copy and complete the table.

d How many boys, in total, were in the group?

e How many cards did all the boys claim to receive in total?

f Work out the mean number of cards the boys claimed to receive.

> **Q7d hint** The frequency column shows the total numbers of boys.

> **Q7f hint** Mean = total number of cards ÷ total number of boys.

8 This table shows the number of new tyres fitted to cars at a garage during one week.

Number of tyres	Frequency	Number of tyres × frequency
1	10	
2	16	
3	5	
4	8	
Total		

a How many cars in total had tyres fitted?

b How many tyres were fitted altogether?

c Work out the mean number of tyres fitted per car.

> **Q8 strategy hint** Read the question carefully and then look at the table. How many cars had 1 tyre fitted? What about 2 tyres?

9   A company has an office building with 32 rooms. This table shows the number of people working in each room.

Number of people	Frequency
1	7
2	10
3	8
4	6
5	1

> **Q9a hint** Copy the table. Draw an extra column for the number of people × frequency. The total of your new column gives the number of people in the building.

a   How many people work in the building in total?

b   What is the total frequency?

c   Work out the mean number of people per room.

> **Q9c hint** Range = largest number of people in a room – smallest number of people in a room.

d   Work out the range.

e   The data could be written in a list like this:

i   How many values are there in the list?

ii   Use $\dfrac{n+1}{2}$ to find the position of the middle one.

iii   What is the median value?

10   This table shows the results for two competitors in a javelin competition.

	1st throw	2nd throw	3rd throw	4th throw
Liz	28.45 m	29.20 m	27.38 m	28.35 m
Katie	30.24 m	29.76 m	29.84 m	20.47 m

a   Work out the mean distance thrown by Liz and by Katie.

b   Work out the median distance for Liz and for Katie.

> **Q10c hint** Is there any data value that stands out as being very high or very low?

c   Which value affected Katie's mean distance?

d   Why did it not affect Katie's median distance thrown?

e   Which average best represents their performances?

f   Who do you think is the better javelin thrower?

> **Q10d hint** Which type of average is not affected by an outlier?

11   This stem and leaf diagram shows weights of fish caught in a competition.

```
0 | 3  4  5
1 | 1  2  2  4
2 | 4  6  6  8  8  8  9     Key
3 | 2  3  5  5  9           0|3 means 0.3 kg
4 | 0  2
```

> **Q11a hint** Which number appears most in the diagram?

> **Q11c hint** Count the number of values. The position of the median is (☐ + 1) ÷ 2 = ☐.

Find   a   the mode   b   the range   c   the median.

Averages and range for grouped data

1 This table gives information about the number of puzzles completed by Y11 students in 30 minutes.

Puzzles completed	Frequency, f	Midpoint of class, m	$m \times f$
1–5	11	3	$3 \times 11 = 33$
6–10	15		
11–15	23		
16–20	16		
Total		**Total**	

a Copy and complete the table.

b How many students took part in total?

c Estimate the total number of puzzles solved.

d Explain why your answer to part **c** cannot be worked out exactly.

e Work out an estimate of the mean using $\dfrac{\text{estimate of total puzzles solved}}{\text{total number of students}}$.

f Which is the modal class?

g Work out $\dfrac{\text{total frequency} + 1}{2}$ to find the position of the median.

h Which class contains the median?

> **Q1a hint** midpoint
>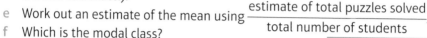
> 6 7 8 9 10

> **Q1d hint** Do you know how many students actually completed 3 puzzles in the given time? Perhaps all 11 students in the 1–5 class completed 5 puzzles.

2 Steve plays snooker. This table gives information about the number of points he scored on each visit to the table during a match.

Score	Frequency, f	Midpoint of class, m	$m \times f$
0–20	9	10	$10 \times 9 = 90$
21–40	6	30.5	
41–60	2		
61–80	2		
Total		**Total**	

> **Q2 hint** Look back at the method used for **Q12** if you need help answering this question.

a Copy and complete the table.

b How many times did Steve visit the table?

c Estimate the number of points that Steve scored in total.

d Explain why Steve's points total cannot be worked out exactly.

e Work out an estimate of Steve's mean score per visit.

f Estimate the range of Steve's scores.

3 This table shows the distances that students in Class 11Y travel to school.

Distance, d (miles)	Frequency, f	Midpoint of class, m	$m \times f$
$0 < d \leqslant 2$	9	1	$1 \times 9 = 9$
$2 < d \leqslant 4$	7	3	$3 \times 7 = 21$
$4 < d \leqslant 6$	8		
$6 < d \leqslant 8$	6		
Total		**Total**	

a Copy and complete the table.

b How many students are in Class 11Y?

c Estimate the combined distance travelled by all of the students in Class 11Y.

d Work out an estimate of the mean distance travelled.

Sampling

1 A printer produces books in batches of 1000.
A sample from each batch is checked for quality.

   a Explain why sampling is used.

   b How many books should be included in the sample?

      10   30   200   500

   c If the books from the batch are in a line, what would
be wrong with choosing all of the sample from one end?

   d Describe a better way to choose the sample.

> **Q1a hint** Imagine testing all 1000 books every time a batch is produced. What can you say about the cost and the time it would take?

> **Q1c hint** Remember, you want to have a good chance of finding a fault wherever it occurs in the batch.

2 Jenny wants to find out how many people do their weekly shopping online.
She asks people in her local town centre on a Saturday morning to complete her questionnaire.
Explain why her results may be biased.

> **Q2 hint** Do you think that the people in a town centre on a Saturday morning include a fair proportion of people who do their weekly shopping online?

7 Extend

1 Find the mean of the first 5 square numbers.

2 **Problem-solving** The mean of the numbers shown on these cards is 6.

| 7 | 6 | 9 | 4 | ? |

Work out the value on the fifth card.

3 Here are some fractions.

   $\frac{2}{3}$ $\frac{5}{12}$ $\frac{3}{8}$ $\frac{5}{6}$ $\frac{3}{4}$

   a Work out the mean.

   b Work out the range.

4 **Exam-style question**

Julie is x years old.
Kevin is $x + 3$ years old.
Omar is $2x$ years old.
Write an expression, in terms of x, for the mean of their ages. **(2 marks)**

Nov 2011, Q3, 5MB1H/01

> **Exam hint**
> Find the total by collecting like terms. Divide the total by the number of people.

5 Which average should you use for each of these? Choose mean, median or mode.
Explain your choices.

   a One value appears a lot more often than any other and it is not at one end of the data set.

   b There are no extreme values and every value should be taken into account.

   c The data is not numerical.

   d The data is fairly evenly spread but there is one extreme value.

6 Problem-solving This back-to-back stem and leaf diagram shows the heights, in cm, of Class 11W students.

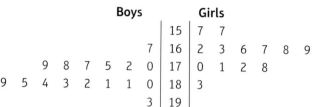

		Boys						Girls						
						15	7	7						
					7	16	2	3	6	7	8	9	9	
9	8	7	5	2	0	17	0	1	2	8				
9	5	4	3	2	1	1	0	18	3					
						3	19							

Q6 hint Compare an average and the range.

Key

For the boys: For the girls:

7|16 means a height of 167 cm   15|7 means a height of 157 cm

Compare the distribution of heights of the boys with the distribution of heights of the girls.

7 **Exam-style question**

A rugby team played six games.
Here are the number of points they scored in each game.

24   8   18   6   12   19

a Work out the median score for these six games.   **(2 marks)**

b Work out the mean score for these six games.   **(2 marks)**

The rugby team played one more game.

The mean score for all seven games is 16.

c Work out the number of points the team scored in the seventh game.

                                 (2 marks)

March 2012, Q15, 1380/2F

Exam hint
In part **c** first work out the total number of points scored in the 7 games.

8 This table gives information about the prices of cars at a car supermarket.

a Work out an estimate of the mean price.

b Which is the modal class?

c Which class contains the median?

d Work out an estimate of the range.

Price, p (£)	Frequency
$7000 < p \leqslant 8000$	652
$8000 < p \leqslant 9000$	583
$9000 < p \leqslant 12000$	976
$12000 < p \leqslant 15000$	354
$15000 < p \leqslant 20000$	185

Q8c hint When n is large, use $\dfrac{n}{2}$ to give the position of the median.

9 **Problem-solving** The histogram shows the distribution of times for an Austrian downhill ski race.

a Work out an estimate of the mean time.

b Work out an estimate of the range.

Q9a hint
Make a frequency table for these times.

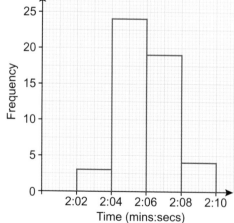

10 1200 people attend a concert. The organisers would like some feedback from members of the audience to help them plan their next event. They decide to sample 15% of the audience.

a How many people should they include in their sample?

b How could they select a representative sample?

Key point 10

The population may be divided into groups, e.g. men and women.
A **stratified sample** contains members of each group in proportion to the size of the group.

11 There are 30 women and 20 men in a dance class.
    a What fraction of the dancers are women?
    b A stratified sample of 15 dancers is selected.
       How many women should there be in the sample?

> **Q11a hint** Write the proportion as a fraction.

> **Q11b hint** Find the same fraction of 15.

12 There are 1200 students in a school.
    There are 200 students in Year 11.
    A stratified sample of 300 students is taken.
    How many Year 11 students should there be in the sample?

> **Q12 hint** Use the same method as in **Q11**.

7 Knowledge check

- To work out the mean of some values:
  - add the values together
  - divide your answer by the number of values. .... *Mastery lesson 7.1*

- To calculate the mean from a frequency table, use mean = $\dfrac{\sum f \times x}{\sum x}$

  where f is the frequency, x is the variable and $\sum$ means 'the sum of'.
  .... *Mastery lesson 7.1*

- To compare two sets of data, compare an average (mean, mode or median) and the range. .... *Mastery lesson 7.1*

- The median is the middle value when the data is written in order.
  The median is in the $\dfrac{n+1}{2}$ th position. .... *Mastery lesson 7.2*

- An outlier is an extreme value that doesn't fit the overall pattern. ...... *Mastery lesson 7.2*

- Estimate the range of grouped data using maximum possible value – minimum possible value. .... *Mastery lesson 7.2*

- The modal class is the class with the highest frequency. .... *Mastery lesson 7.3*

- When the data is grouped, you can calculate an **estimate** for the mean using the midpoints of the classes as estimates for data values.
  .... *Mastery lesson 7.4*

- In a survey, a **sample** is taken to represent the **population**.
  The sample must be chosen carefully to avoid **bias**. .... *Mastery lesson 7.5*

- When considering sample size, you need to balance the need for accuracy with the costs and time involved in taking a large sample. .... *Mastery lesson 7.5*

- In a **random sample**, every member of the population has an equal chance of being included. .... *Mastery lesson 7.5*

Look back at this unit.
Which lesson made you think the hardest? Write a sentence to explain why.
Begin your sentence with: Lesson _____ made me think the hardest because _____

Reflect

7 Unit test

1   Peter rings 20 people between 10 am and 12 pm on a Monday. He asks their views on traffic in the town. Explain why his sample is likely to be biased. *(2 marks)*

2   Work out the mean of 2, 5, 5, 6 and 7. *(1 mark)*

3   The mean of 6 numbers is 11.
    The mean of a different set of 9 numbers is 6.
    What is the mean of all 15 numbers? *(2 marks)*

4   Here are some fractions.
    $\frac{2}{3}$ $\frac{5}{6}$ $\frac{3}{8}$ $\frac{7}{12}$ $\frac{1}{4}$
    a   Work out the mean.
    b   Work out the range. *(2 marks)*

5   This stem and leaf diagram shows the amounts of money spent at a newsagents in one hour.

0	35	50	75	82	90
1	23	29			
2	18	45	64	71	
3	27	48	78	92	99
4	64	70	74		
5	37	45	54	69	
6	28	55			

    Key
    0|35 means £0.35

    a   Find the median amount spent. *(2 marks)*
    b   Work out the range. *(1 mark)*

6   **Problem-solving** This back-to-back stem and leaf diagram shows the lengths, in cm, of some grass snakes.

Male		Female
6 5 2 0	7	
9 8 7 7 5 3 3 1	8	6 8
5 5 4 3 3	9	0 2 2 3 5 6 8
	10	3 5 5 5 6 7
	11	0 1 1

    Key
    For the males: For the females:
    1|8 means a length of 81 cm   8|6 means a length of 86 cm
    Compare the distribution of lengths of the male grass snakes with the distribution of lengths of the female grass snakes. *(3 marks)*

7   This table shows the number of goals scored in each match of a knockout tournament.
    a   Work out the mean number of goals scored. *(2 marks)*
    b   Find the median. *(2 marks)*
    c   Find the modal score. *(1 mark)*
    d   Work out the range. *(1 mark)*

Goals scored	Frequency
0	1
1	3
2	7
3	11
4	8
5	6

8
Exam-style question

5 female giraffes have a mean weight of x kg.

7 male giraffes have a mean weight of y kg.

Write down an expression, in terms of x and y, for the mean weight of all 12 giraffes. **(2 marks)**

Nov 2012, Q10, 5MB1H/01

9 This table gives information about the number of visitors to a National Trust property in June.

Number of visitors, v	Frequency
$200 \leqslant v < 300$	4
$300 \leqslant v < 400$	11
$400 \leqslant v < 500$	12
$500 \leqslant v < 600$	3

a   Work out an estimate of the mean number of visitors.   *(2 marks)*

b   Work out an estimate of the range.   *(1 mark)*

c   Which is the modal class?   *(1 mark)*

d   Which class contains the median?   *(1 mark)*

10   **Problem-solving** This histogram shows the distribution of marks in a Y11 Maths mock exam.

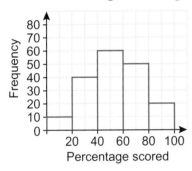

a   Work out an estimate of the mean score.   *(3 marks)*

b   Work out an estimate of the range.   *(1 mark)*

11   A school has 600 students in KS3 and KS4.

240 of the students are in KS4.

a   What fraction of the students are in KS4?

A sample of 50 students is selected.

b   How many KS4 students should be in the sample?   *(2 marks)*

12   Look back at the questions you answered in this test.

a   Which one are you most confident that you have answered correctly? What makes you feel confident?

b   Which one are you least confident that you have answered correctly? What makes you least confident?

Discussion Discuss the question you feel least confident about with a classmate. How does discussing it make you feel?

Reflect

Sample student answer

a   What will the student get 2 marks for?

b   Why won't the student get the 3rd mark?

Exam-style question

14 students did a history test.

Here are the results.

Girls	3	8	2	4	3	4	4	6
Boys	3	6	3	3	1	4		

Adele says, 'The range of the girls' marks is 1 more than the range of the boys' marks.'

Is Adele right? You must show your working.   **(3 marks)**

Nov 2012, Q7, 5MB1F/01

Student answer

Girls range is 8 – 2 = 6

Boys range is 6 – 1 = 5

| Master p.211 | Problem-solve p.228 | Check p.229 | Strengthen p.231 | Extend p.235 | Test p.240 |

8 PERIMETER, AREA AND VOLUME 1

On a plane, you can only take bottles smaller than 100 ml in your hand luggage.

How many of these small bottles can you fill from this 340 ml bottle of shampoo?

8 Prior knowledge check

Numerical fluency

1 Work out the missing number.

a $9 \times \boxed{} = 63$
b $3 \times \boxed{} \times 4 = 36$

2 Work out

a half of 16
b $\frac{1}{2} \times 12$
c $\frac{1}{2} \times 6 \times 3$
d $\frac{1}{2} \times (5 + 3)$
e $\frac{1}{2} \times (4 + 6) \times 5$

3 Round to 1 decimal place.

a 5.32
b 15.37
c 1.95
d 21.119

4 Round to 2 decimal places.

a 6.375
b 0.788
c 80.022
d 2.199

5 Round to 3 decimal places.

a 1.2377
b 78.0268
c 0.9321
d 60.6899

6 Round 13 655.97 to

a the nearest whole number
b the nearest 1000
c 1 decimal place
d 3 significant figures.

7 Calculate an estimate for

$$\frac{3.7 \times 18.2}{5.1}$$

8 Write down the values shown by the arrows on each scale.

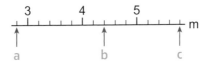

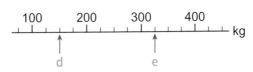

9   Anna is measuring the mass of a chemical for a science experiment. She says, 'The mass is between 10 g and 20 g, so a good estimate is 15 g.'

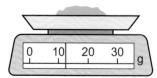

Is Anna correct? Explain your answer.

10 Measure the length of each line. Write each length in centimetres and in millimetres.

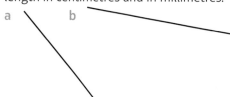

11 Which of these estimates are sensible? Give a better estimate where necessary.
    a   A cupcake has a mass of 500 g.
    b   The capacity of a milk bottle is 45 ml.
    c   A pencil is 18 cm long.
    d   The mass of a bag of sugar is 1 kg.

12 Copy and complete
    a   6 kg = ☐ g
    b   36 m = ☐ cm
    c   1.75 m = ☐ mm
    d   30 500 cm = ☐ km

13 Convert
    a   890 ml into litres   b   3.6 litres into ml.

Geometrical fluency

14 Sketch a kite. Draw a line to divide it into two congruent triangles.

15 Name these 3D solids.

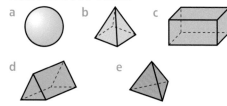

16 Find the area of each shape drawn on the centimetre grid. Give your answer in cm².

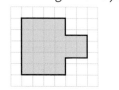

17 Estimate the area of each shape drawn on the centimetre grid.

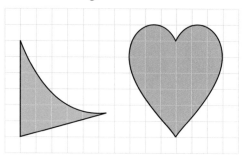

18 Work out the perimeter and area of each rectangle.

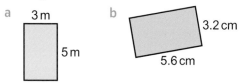

Give your answers to 1 decimal place.

19 a   A rectangle has length 7 cm and width 4 cm.
       What is its perimeter?
    b   A rectangle has area 28 cm² and length 7 cm.
       What is its width?

Algebraic fluency

20 Solve these equations.
    a $5h = 20$ b $\frac{1}{2}m = 10$

21 Substitute the values $a = 3$, $b = 4$ and $h = 6$ into
    a $a + b$ b $a \times b$
    c $\frac{1}{2}bh$ d $\frac{1}{2} \times (a + b) \times h$

22 Substitute the values $x = 6$, $y = 2$ and $z = 24$ into these formulae.
    Solve to find the value of the unknown.
    a $z = xt$ b $py = x$ c $\frac{1}{2} \times x \times r = z$

✱ Challenge

23 On centimetre squared paper, draw
    • two different rectangles with area 16 cm²
    • a triangle with area 16 cm².
    What other shapes can you draw with area 16 cm²?

    Q23 hint Try square, L-shape, T-shape, parallelogram, trapezium, rhombus.

8.1 Rectangles, parallelograms and triangles

Objectives

- Calculate the perimeter and area of rectangles, parallelograms and triangles.
- Estimate lengths, areas and costs.
- Calculate a missing length, given the area.

Did you know?

The word 'perimeter' comes from two Greek words – 'peri', which means 'around' and 'metron', which means measure.

Fluency

- What does perpendicular mean?
- Which of these show perpendicular lines?

1   Work out the **perimeter** of each shape.

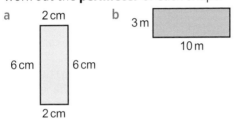 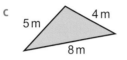 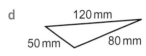

Q1 hint Make sure you write the correct units.

2   Work out the **area** and perimeter of each shape.

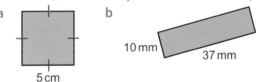

3   **Reasoning / Modelling**
   a   Which of these expressions is for
      i   the area of this rectangle
      ii the perimeter of this rectangle?
   b   Write a formula for the area of a rectangle. Start $A =$
   c   Write down a formula for the perimeter of a rectangle. Start $P =$

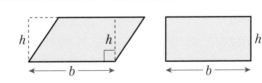

$$2l + 2w \qquad l \times w$$

Key point 1

The base of a parallelogram is b and its **perpendicular height** is h.

Cutting a triangle from one end of a parallelogram and putting it on the other end makes a rectangle.

Area of parallelogram = base length × perpendicular height

$$A = bh$$

Questions in this unit are targeted at the steps indicated.

4   Calculate the area of each parallelogram.

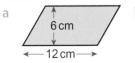 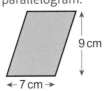 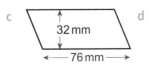

5   Calculate the perimeter and area of each shape.
    All lengths are in centimetres.

    a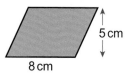

    b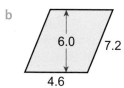

    Reflect How did you know which measurements to use in your area calculations?

6   **Reasoning**

    a   Work out the area of this parallelogram.

    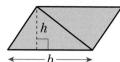

    b   Trace the parallelogram. Join opposite vertices with a
        straight line to make two triangles.

    c   What fraction of the parallelogram is each triangle?

    d   Write down the area of one of the triangles.

    > **Q6b communication hint**
    > Vertices are corners.

    > **Q6c hint** Are the triangles
    > the same shape and size?

Key point 2

The diagonal splits a parallelogram into two identical triangles.
Area of 2 triangles = $b \times h$
Area of a triangle = $\frac{1}{2} \times b \times h$

Area of a triangle = $\frac{1}{2}bh$

Example 1

Calculate the area of each triangle.

$b = 7, h = 4$ ──── Write down the values of b and h.

  Area $= \frac{1}{2}bh$

     $= \frac{1}{2} \times 7 \times 4$ ──── Substitute them into the formula for area of a triangle.

     $= 14\,cm^2$ ──── Write the units with your answer.

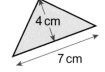

---

7   Calculate the area of each triangle.

    > **Q7 hint** Find the base first. The height
    > is perpendicular to the base.

    a   3 cm   9 cm

    b   5 cm   8 cm

    c   9.8 cm   3.5 cm

    d   21 mm   28 mm   35 mm

    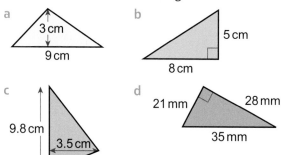

    > **Q7d hint** The base and height
    > measurements are
    > perpendicular to each other.

    Discussion Is the base measurement you use always along the bottom of the triangle?

8 Calculate the perimeter and area of each shape.
All lengths are in centimetres.

a   b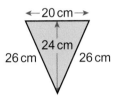

9 A triangle has base length 6.4 cm and height 5 cm.
Calculate its area.

> **Q9 strategy hint** Sketch a diagram of the triangle.

10 **Real / Problem-solving** Emma makes bunting from triangles of fabric like this.
Each has braid all around the three sides.

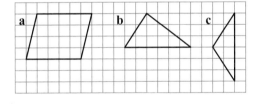

> **Q10 strategy hint** Work out the amount for one triangle first.

She makes 20 of these triangles.
Work out
a the area of fabric she needs   b the length of braid she needs.

11 **Problem-solving** Sketch and label three triangles that have area 18 cm^2.

12 ⌜ **Exam-style question** ⌝

These shapes are drawn on centimetre squared paper.
Calculate the area of each one.   **(6 marks)**

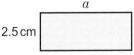

Exam hint
The question says 'Calculate', so you need to do calculations and show working out.

13 **Problem-solving / Modelling** Each shape has an area of 25 cm^2.
Calculate the lengths marked by letters.

a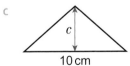
2.5 cm

b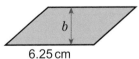

c

> **Q13 strategy hint** Write down the area formula for the shape. Substitute the values you know into the formula. Solve the equation to find the missing length.

14 **Finance / Modelling** A rectangular room measures 3.7 m by 2.9 m, and is 2.6 m high.
a Sketch four rectangles to represent the four walls. Label their lengths and widths.
b Calculate an estimate for the total area of the walls.
c A tin of paint costs £11.99 and covers 6 m^2. Estimate the cost of paint for the room.
Discussion What assumptions have you made?

8.2 **Trapezia and changing units**

Objectives

- Calculate the area and perimeter of trapezia.
- Find the height of a trapezium given its area.
- Convert between area measures.

Why learn this?

Scientists may need to convert between mm² and cm² to compare the number of bacteria in a sample.

Fluency

- Work out 3.6×100 $52 \div 100$ $0.4 \times 10\,000$ $56\,400 \div 10\,000$
- Convert   2.6 cm to mm   6.1 m to cm   54 cm to mm   127 cm to m

Warm up

1 Work out

   a $\frac{1}{2} \times 16$ **b** $\frac{1}{2}(3 + 9)$ **c** $\frac{1}{2} \times (2 + 4) \times 10$

2 Solve

   a $5h = 20$ **b** $6.4x = 19.2$

3 Work out the area of this shape.
Round your answer to 1 decimal place.

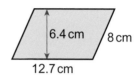

Key point 3

This **trapezium** has parallel sides a and b and perpendicular height h.
Two trapezia put together make a parallelogram, with base $(a + b)$ and perpendicular height h.
Area of 2 trapezia = base × perpendicular height = $(a + b) \times h$
Area of a trapezium = $\frac{1}{2}(a + b)h$

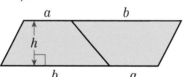

> **Communication hint**
> Trapezia is the plural of trapezium.

Example 2

Calculate the area of this trapezium. All lengths are in centimetres.

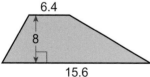

Area = $\frac{1}{2}(a + b)h$ | Write down the formula for the area of the trapezium.

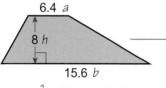

                    | Sketch the trapezium. Label a, b and h.

Area = $\frac{1}{2} \times (6.4 + 15.6) \times 8$

     = $\frac{1}{2} \times 22 \times 8$ | Substitute the values in the formula.

     = $88 \, \text{cm}^2$

4 Calculate the areas of these trapezia.
All lengths are in centimetres.
Round answers to 1 decimal place where necessary.

a 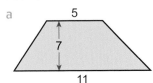   b 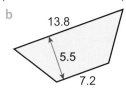   c 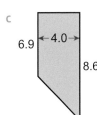   d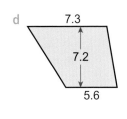

Discussion How can you remember how to find the area of a trapezium?

5 Calculate the area and perimeter of this isosceles trapezium.

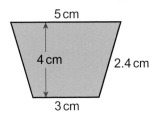

> **Q5 communication hint** An **isosceles trapezium** has one line of symmetry. Its two sloping sides are equal.

6 **Problem-solving / Modelling** This trapezium has area 40 cm².

a Substitute the values for A, a and b into the formula
$A = \frac{1}{2}(a + b)h$.

b Simplify your answer to part **a**.

c Solve your equation from part **b** to find h.
Give the units with your answer.

> **Q6 hint**
> Work out $\frac{1}{2}(6 + 10)$.

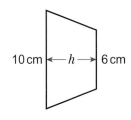

7 **Problem-solving / Modelling** This trapezium has area 35 mm².
Work out its height.

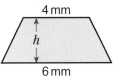

Key point 4

These two squares have the same area.
To convert from cm² to mm², multiply by 100.
To convert from mm² to cm², divide by 100.

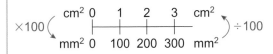

Area = 1 cm × 1 cm
= 1 cm²

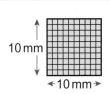

Area = 10 mm × 10 mm
= 100 mm²

---

8 Convert

a 2750 mm² to cm²

b 1.4 cm² to mm².

> **Q8 hint** ☐ × 100 = ☐

9 a Calculate the area of this square in cm².

b Convert your answer to mm².

2.1 cm

10 a Use these diagrams to help you work out the number of cm² in 1 m².

> Q10a hint Work out the area of each sqaure.

1 m

100 cm

1 m

100 cm

b Copy and complete the double number line.

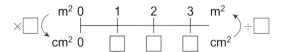

11 Convert
   a 2.2 m² to cm²
   b 5000 cm² to m²

12 a Choose four decimal measures between 7 m² and 8 m².
     Convert each measure to cm².
   b Choose four measures between 10 000 cm² and 12 000 cm².
     Convert each measure to m².

13 The diagram shows a yoga mat.
   a Work out the area of the mat in m².
   b Convert your answer to part **a** to cm².
   c Convert the length and the width of the mat to cm.
     Use these measurements to work out the area of the mat in cm².
   d **Reflect** Which do you think is the easier way to find the area in cm²?
     • Find the area in m², then convert to cm².
     • Convert the lengths to cm, then find the area.

2.1 m

0.5 m

14 **Problem-solving** Work out these areas in cm².

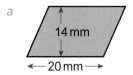

a   14 mm   ←—20 mm—→

b   1.5 m   2.7 m

c   3.2 cm   20 mm   8 cm

Discussion Which method did you use and why?

15
> **Exam-style question**
>
> Work out the area of this rectangle
>
> 5.5 cm
>
> 2.4 cm
>
> **a** in cm²
> **b** in mm². **(3 marks)**

Exam hint
Make sure you answer both parts of the question. Write down the calculations before you do them on a calculator.

16 **STEM / Problem-solving** A biologist counts
170 bacteria on slide A with area 340 mm².
On slide B she counts 120 bacteria in an area of 2 cm².
Which sample has more bacteria per square millimetre?
Show your working to explain.

> Q16 communication hint
> The number of 'bacteria per square millimetre' means the number of bacteria in 1 mm².

8.3 Area of compound shapes

Objectives

- Calculate the perimeter and area of shapes made from triangles and rectangles.
- Calculate areas in hectares, and convert between ha and m^2.

Did you know?

You can find the area of a trapezium by dividing it into two triangles and a rectangle.

Fluency

- How many metres are there in 1 km?
- Work out $24 \times 10\,000$ $800\,000 \div 100\,000$ $3.2 \times 10\,000$ $4760 \div 100\,000$

1   Work out the area and perimeter of these shapes. All lengths are in cm.

a

b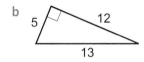

2   Convert these areas into the units given.
   a $5\,m^2$ to cm^2
   b $2.2\,m^2$ to cm^2
   c $20\,000\,cm^2$ to m^2
   d $7200\,cm^2$ to m^2

Key point 5

1 hectare (ha) is the area of a square 100 m by 100 m.
$1\,ha = 100\,m \times 100\,m = 10\,000\,m^2$.
Areas of land are measured in hectares.

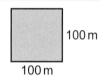

3   Convert these areas to the units given.
   a   8 ha to m^2
   b   3.5 ha to m^2
   c $40\,000\,m^2$ to ha
   d $225\,000\,m^2$ to ha

Q3 hint You could use this double number line.

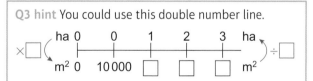

4   a   Use these diagrams to help you work out the number of m^2 in $1\,km^2$.

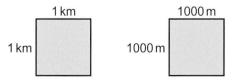

   b   Copy and complete the double number line.

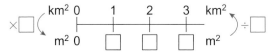

Warm up

5   The diagram shows 1 km² divided into 100 m squares.

   a   What is the area of each 100 m square?

   b   How many hectares are there in 1 km²?

   c   Copy and complete the double number line.

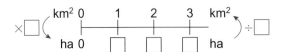

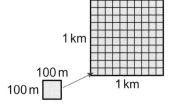

6   **Reasoning** Windermere and Coniston Water are two lakes in the Lake District.
    The area of Windermere is 1473 ha. The area of Coniston Water is 4.9 km².

   a   Which is larger, Windermere or Coniston Water?

   b   What is the difference in area?

Key point 6

A **compound shape** is made up of simple shapes.
To find the area of a compound shape, split it into simple shapes like rectangles and triangles.
Find the area of each shape and then add them all together.

Example 3

Calculate the perimeter and area of this compound shape.

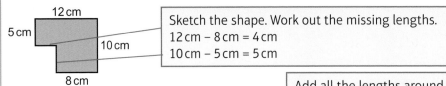

Sketch the shape. Work out the missing lengths.
12 cm – 8 cm = 4 cm
10 cm – 5 cm = 5 cm

Perimeter = 12 + 10 + 8 + 5 + 4 + 5 = 44 cm²

Add all the lengths around the shape to work out the perimeter.

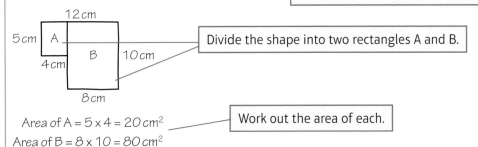

Divide the shape into two rectangles A and B.

Area of A = 5 × 4 = 20 cm²
Area of B = 8 × 10 = 80 cm²
    Total area = 100 cm²

Work out the area of each.

---

7   Here is the shape from Example 3, split
    into two different rectangles C and D. Work out

   a   the areas of C and D

   b   the total area

   c   the total perimeter.

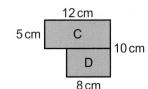

   Discussion Does it matter how you divide up the shape into rectangles to find the area?

8   Calculate the area and perimeter of these compound shapes.

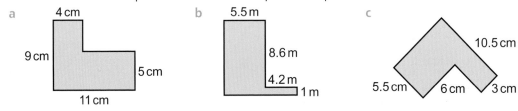

9   **Reflect** Here is a compound shape.   Here are two ways of working out its area.

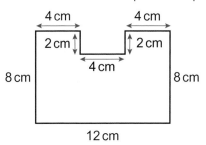

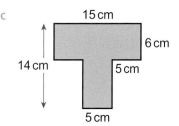

Method 1
Work out the area of
the whole rectangle.
Subtract area E.

Method 2
Work out the areas
of F, G and H.
Add them together.

a   Work out the area using both methods.

b   Which do you prefer? Explain why.

10   Calculate the area and perimeter of each shape.

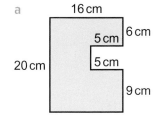

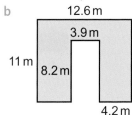

Q10 strategy hint Use the method from **Q9** that you liked better.

11   Work out the area of each shape.

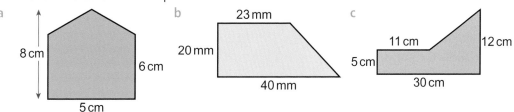

Q11 strategy hint Divide each shape into rectangles and triangles.

12   **Reflect**

a   Sketch this isosceles trapezium.

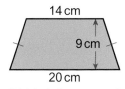

b   Divide it into two triangles and a rectangle.

c   Work out the area of each part, then add to find the total area.

d   How could you work out the area of a trapezium if you forget
the area formula?

Q12c hint Work
out the base of
the triangle first.

13   **Problem-solving** Tia makes this photo frame by cutting a rectangle 3 cm by 4 cm out of a
larger rectangle of card.

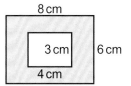

Q13 strategy hint
Work out the area of the larger
rectangle then subtract the
area of the cut out rectangle.

Work out the area of card in the frame.

14 **Exam-style question**

This wooden frame holds two photos, each 8 cm by 15 cm.
Work out the area of wood in this picture frame.

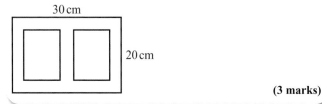

30 cm

20 cm

(3 marks)

Exam hint

Show clearly your calculations for the area of the two photos and how this leads to the area of the frame.

8.4 Surface area of 3D solids

Objectives

- Calculate the surface area of a cuboid.
- Calculate the surface area of a prism.

Did you know?

Distinctive packaging makes products stand out on the shelves.

Fluency

- Describe and name these 3D solids.
- How many faces do they have?
  What shapes are they? How many edges and vertices?

Warm up

1 Sketch the net of this cuboid.
Label the lengths on your net.

3 cm
6 cm
4 cm

2 Work out the area of these shapes.

a

2 cm
6 cm

b
2 cm
8 cm
←6 cm→

c
5 cm
3 cm
7 cm

Key point 7

A prism is a 3D solid that has the same cross-section all through its length.

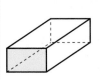 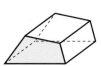

Key point 8

The **surface area** of a 3D solid is the total area of all its faces.
To find the surface area of a 3D solid, sketch the net and work out the areas of the faces.

Example 4

Work out the surface area of this cuboid.

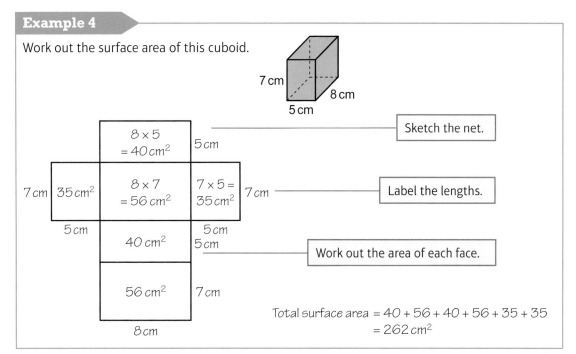

8 × 5
= 40 cm² 5 cm

7 cm 35 cm²   8 × 7
= 56 cm²   7 × 5 =
35 cm² 7 cm

5 cm   5 cm

40 cm² 5 cm
5 cm

56 cm² 7 cm

8 cm

Sketch the net.

Label the lengths.

Work out the area of each face.

Total surface area = 40 + 56 + 40 + 56 + 35 + 35
= 262 cm²

3   Work out the surface area of this cuboid.

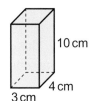

10 cm
4 cm
3 cm

4   **Reasoning**
   a   i   What is the area of the top of this cuboid?
       ii Which other face of the cuboid is identical to this one?
   b   i   What is the area of the front of this cuboid?
       ii Which other face of the cuboid is identical to this one?
   c   i   What is the area of the side of this cuboid?
       ii Which other face of the cuboid is identical to this one?

   TOP 2 cm
   FRONT SIDE 4 cm
   9 cm

   d   Copy and complete this calculation to work out the total surface area of the cuboid.
       2 × ☐ + 2 × ☐ + 2 × ☐ = ☐ cm²

   Discussion How can you calculate the surface area of a cuboid without drawing its net?

5   Calculate the surface area of each cuboid.

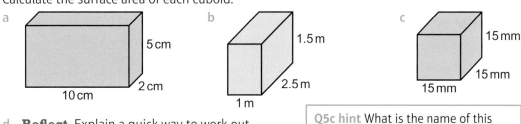

a   5 cm 2 cm 10 cm

b   1.5 m 2.5 m 1 m

c   15 mm 15 mm 15 mm

   d   **Reflect** Explain a quick way to work out
       the surface area of a cube.

   Q5c hint What is the name of this
   shape? How many faces are identical?

6   a   Sketch the net of this triangular prism.
       Label the lengths.
   b   Work out
       i   the area of each face
       ii the total surface area.

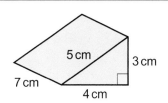

5 cm 3 cm
7 cm 4 cm

7   Calculate the surface area of each 3D solid.
    All the measurements are in centimetres.

    a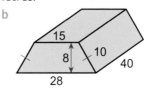

    b

8   **Real / Problem-solving** A central heating oil tank is in the shape of a cuboid with
    cross-section 120 cm tall and 100 cm wide. The tank is 150 cm long.
    Pete paints all the faces except the base.

    | Q8a hint Sketch the tank. |

    a   Work out the total area he paints, in square metres.
    b   One tin of paint covers 6 m². How many tins of paint does he need?

9   **Problem-solving** These soft play blocks are covered
    with plastic fabric.

    a   Work out how many square metres of fabric you will
        need to cover them both.

    The plastic fabric costs £12.99 per square metre.

    b   Estimate the cost of the fabric needed to
        cover them both.

    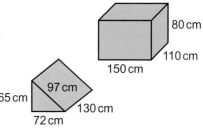

10 **Problem-solving / Reasoning** This cube has surface area 384 cm².
    What is the length of one side of the cube?

    | Q10 hint First work out |
    | the area of one face. |

11 **Exam-style question**

    Children's building blocks are 4 cm wooden cubes, painted
    on all faces.

    a Work out the area to be painted on one building block.

    b One litre of paint covers an area of 5 m².
       How many cubes can be painted with one litre
       of paint?   **(4 marks)**

    Exam hint
    In exam questions, the
    answer to the first part often
    helps you with the next part.
    Use your answer to part **a** to
    help you answer part **b**.

8.5 Volume of prisms

Objectives

- Calculate the volume of a cuboid.
- Calculate the volume of a prism.

Why learn this?

Scientists use the volume and mass of a block
of metal to calculate its density.

Fluency

- What shape is the cross-section of a
  cuboid   cube   triangular prism   pentagonal prism?
- Is this solid a prism?   • Is a cuboid a prism?

1   Work out the area of each shape.

  a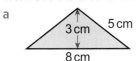
  3 cm   5 cm
  8 cm

  b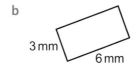
  3 mm
  6 mm

  c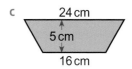
  24 cm
  5 cm
  16 cm

2   Work out

  a $8 \times 4 \times 2$ b $6 \times 5 \times 7$ c $9 \times 4 \times 2$

3   Convert each length to the units given.

  a   240 cm to m   b   52 cm to m   c   4.3 m to cm   d   0.7 m to cm

---

Key point 9

The volume of a 3D solid is the amount of space inside it.

---

4   This cuboid is made of centimetre cubes.
   Each cube has volume $1 \, cm^3$.
   Work out the volume of the cuboid.

  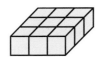

  Q4 hint Work out the number of centimetre cubes in the cuboid.

Key point 10

Volume is measured in cubic units:
millimetre cubed (mm^3), centimetre cubed (cm^3), metre cubed (m^3).

---

5   Here are some more cuboids made of $1 \, cm^3$ cubes.
   Work out their volumes.

  Q5 hint Count the cubes in the top layer. Multiply by the number of layers.

  a   b   c

Key point 11

Volume of a cuboid = length × width × height = lwh

l w h

---

6   Work out the volume of each cuboid. Give the correct units with your answer.

  a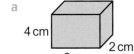
  4 cm   2 cm   6 cm

  b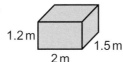
  1.2 m   1.5 m   2 m

  c
  50 mm   15 mm   10 mm

  d
  5 cm   5 cm   5 cm

  Q6b hint
  m × m × m = m^3

  Reflect Explain a quick way to work out the volume
  of a cube of side 4 cm.

7   This prism is made from centimetre cubes. Work out the volume of the prism.

  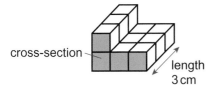
  cross-section   length 3 cm

  Q7 hint There are 3 'slices' with 4 cubes in each 'slice'.   1 cm   1 cm   1 cm

  Discussion What is the area of the cross-section of this prism?
  What does area of cross-section × length give you? Does this work for the cuboids in Q6?

Key point 12

Volume of a prism = area of cross-section × length

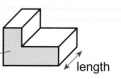

area of cross-section — length

Example 5

Work out the volume of this prism.

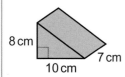

8 cm
10 cm
7 cm

Volume = area of cross-section × length — | Write down the formula. |

Area of △ = $\frac{1}{2}$ × 10 × 8 — | Work out the area of the cross-section. |

= 5 × 8

= 40 — | Substitute the area of the cross-section and the length into the formula. |

Volume = 40 × 7

= 280 cm³ — | Write the units. |

8 Calculate the volume of each prism.

a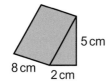
5 cm
8 cm   2 cm

b
area 18 cm²
4 cm

c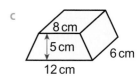
8 cm
5 cm
12 cm
6 cm

9 **Problem-solving / Reasoning** The cross-section of a prism is an equilateral triangle with area 6.4 cm².
The prism is 9.2 cm long.
Work out the volume of the prism.

| **Q9 strategy hint** Sketch the prism and label the measures you know. |

10 **Exam-style question**

Tim builds a cuboid shaped sandpit.
The sandpit is 40 cm tall, 1.6 m long and 2.3 m wide.

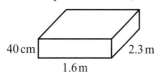
40 cm
2.3 m
1.6 m

Tim puts sand 35 cm deep in the sandpit.

a Work out how many cubic metres of sand he needs,
to 1 decimal place. **(3 marks)**

b Sand costs £54 per cubic metre, plus £20 delivery.
How much will the sand cost for the sandpit? **(2 marks)**

| **Q10 strategy hint** Mark the depth of the sand on the diagram. Before you start your working out, check what units you need to use for your answer. Do you need to change any units before you do the calculation? |

11 **Problem-solving** A cube has volume 343 cm³. How long is one side of the cube?

12 **Problem-solving** Sketch and label the dimensions of three cuboids with volume 24 cm³.

| **Q12 communication hint** The dimensions of a solid are the lengths of its sides. |

8.6 More volume and surface area

Objectives

- Solve problems involving surface area and volume.
- Convert between measures of volume.

Why learn this?

Calculating volume helps you work out the amount of water in a swimming pool.

Fluency

- Work out
  100 × 100 × 100
  7.2 × 1000
  6 × 1000
  3.71 × 1 000 000
  5 × 1 000 000
  35 000 000 ÷ 1 000 000
- How many ml are there in 1 litre?

Warm up

1 Match each object to the amount of liquid it can hold.

| teaspoon | drink can | bucket | juice carton |

| 330 ml | 5 ml | 5 litres | 1 litre |

2 Solve these equations.
   a $54 = 6h$ b $222 = 74h$

3 For this prism, work out
   a the surface area
   b the volume.
   Give your answers to 1 decimal place.

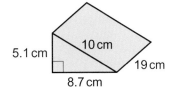

10 cm
5.1 cm
19 cm
8.7 cm

Key point 13

Volume is measured in mm³, cm³ or m³. These two cubes have the same volume.

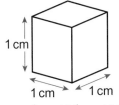

1 cm
1 cm
1 cm

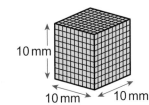

10 mm
10 mm
10 mm

Volume = 1 cm × 1 cm × 1 cm
       = 1 cm³

Volume = 10 mm × 10 mm × 10 mm
       = 1000 mm³

1 cm³ = 1000 mm³

4 Convert
   a 24 cm³ to mm³   b 30 000 mm³ to cm³
   c 5.7 cm³ to mm³   d 4810 mm³ to cm³.

Q4 hint
×1000
cm³ ⟷ mm³
÷1000

5 a Work out the volume of this cuboid in cm³,
     then convert it to mm³.
   b Convert the measurements of the cuboid to mm.
     Then work out the volume in mm³.

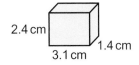

2.4 cm
3.1 cm
1.4 cm

   Discussion Which method for finding the volume in mm³ did you prefer? Why?

6 a Use these diagrams to help you work out the number of cm³ in 1 m³.

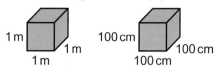

Q6a hint Work out the volume of each cube.

   b Copy and complete
      1 m³ = ☐ cm³

Reflect How can you use your answer to part **b** to help you convert 7 000 000 cm³ into metres?

7 Convert
   a 3.5 m³ into cm³   b 6.87 m³ into cm³
   c 8 200 000 cm³ into m³   d 3 159 000 cm³ into m³.

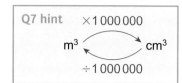

Q7 hint
$\times 1\,000\,000$
m³ ⟶ cm³
$\div 1\,000\,000$

Key point 14

1 cm³ = 1 ml   1000 cm³ = 1 litre

---

8 Convert
   a 0.5 litres into cm³   b 4900 cm³ into litres
   c 33 m³ into cm³   d 2 m³ into litres
   e 3.4 m³ into litres   f 4000 litres into m³
   g 2.4 litres into cm³   h Copy and complete
                                   1 m³ = ☐ litres

Q8d hint Convert 2 m³ to cm³ first.

Q8g hint Convert 2.4 litres to ml first.

9 **STEM** This beaker can hold 200 ml. What is the capacity of this beaker in cm³?

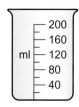

Q9 communication hint **Capacity** is the amount of liquid a 3D object can hold.

Discussion What do you think are the units for capacity?

10 This cuboid is a fuel tank. How many litres of fuel does it hold?

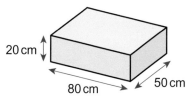

Q10 hint Work out the volume in cm³. Then convert to ml and then to litres.
$\times 1000$
l ⟶ ml
$\div 1000$

11 Work out the capacity of this vase
    a in ml   b in litres.
    Discussion Which are the most sensible units for the capacity of the vase?

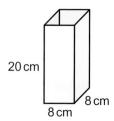

12 **Problem-solving** Here is the design for a swimming pool.

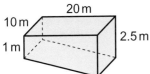

a How much water will the swimming pool hold?

All the sides of the pool will be tiled.

b How many square metres of tiles are needed?

Reflect Did you need to work out the area of two of the sides for part **b**?

> **Q12a hint** Is your answer in sensible units?

> **Q12b hint** Sketch all the surfaces to be tiled. Work out the area of each.

13 **Modelling** Tracey pours melted chocolate into moulds like this.

a What volume of chocolate does she need for each mould?

She melts this block of chocolate.

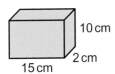

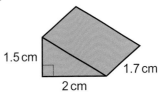

b How many moulds can she fill with the chocolate?

Discussion What assumptions have you made?

Is it sensible to round the number of moulds up or down?

Example 6

This cuboid has volume 2640 cm³.
Work out its height.

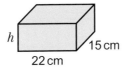

Volume = $l \times w \times h$ ──── Write down the formula.

$2640 = 22 \times 15 \times h$ ──── Write in the values you know.

$2640 = 330 \times h$

$h = \dfrac{2640}{330} = 8\,cm$ ──── Solve the equation to find h.

---

14 Work out the missing measurements in these cuboids.

a

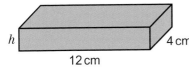

Volume = 96 cm³

b

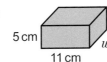

Volume = 440 cm³

15 **Exam-style question**

This cuboid has volume 600 cm³.

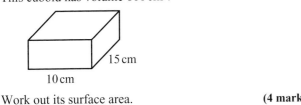

Work out its surface area.   **(4 marks)**

> **Q15 strategy hint**
> Copy the cuboid. Work out the missing measurement and write it on your diagram. Use your diagram to help you work out the surface area.

8 Problem-solving

Objective • Use a flow diagram to help you solve problems.

Example 7

A blue cube has side 8 cm. The side of the yellow cube is 50% longer than the side of the blue cube.
What is the surface area of the yellow cube?

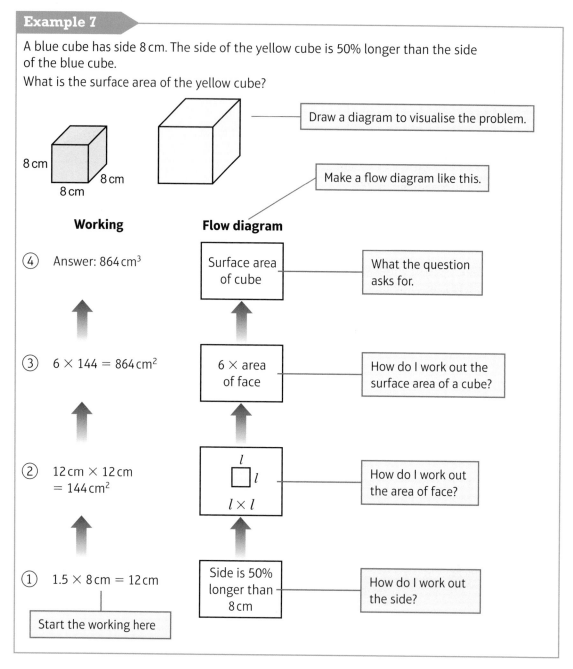

Draw a diagram to visualise the problem.

Make a flow diagram like this.

Working   **Flow diagram**

④ Answer: 864 cm³

Surface area of cube — What the question asks for.

③ 6 × 144 = 864 cm²

6 × area of face — How do I work out the surface area of a cube?

② 12 cm × 12 cm = 144 cm²

$l \times l$ — How do I work out the area of face?

① 1.5 × 8 cm = 12 cm

Side is 50% longer than 8 cm — How do I work out the side?

Start the working here

1 A regular hexagon and a regular octagon are joined together.

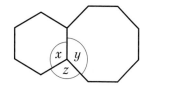

Work out the size of angle z.

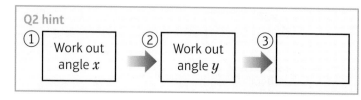

Q2 hint

① Work out angle x ② Work out angle y ③

2   The bar chart shows the amount people spend on clothes each month.
    What percentage of these people spend more than £20?

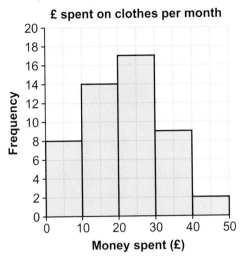

£ spent on clothes per month

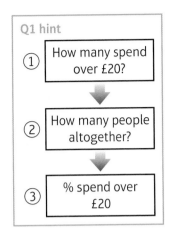

Q1 hint

① How many spend over £20?

② How many people altogether?

③ % spend over £20

3   The area of this triangle is 6 cm².
    Work out its perimeter.

Q3 hint You could draw a flow diagram to help you.

4   A water trough is in the shape of a prism with trapezium cross-section.
    4 litres of water run into the trough every minute.
    How long will it take to fill the trough?
    Give your answer to the nearest minute.

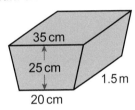

Q4 hint 1000 cm³ = 1 litre

5   **Reflect** Did the flow diagrams help you?
    Is this a strategy you would use again to solve problems?
    What other strategies did you use to solve these problems?

8 Check up

Log how you did on your
Student Progression Chart.

2D shapes

1   For each shape, work out
    i   the perimeter
    ii the area.
    Give your answers to the nearest whole number.

a
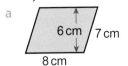
6 cm 7 cm
8 cm

b 3.1 cm
5.1 cm
9.8 cm
6.7 cm

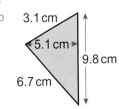

2 **Problem-solving** The area of this parallelogram is 27 cm².
Work out the length marked x.

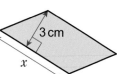

3 **Problem-solving** The area of this triangle is 15 cm².
Work out its height.

4 For this shape, work out
a the perimeter
b the area.

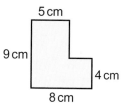

5 Work out the area of this arrow shape.

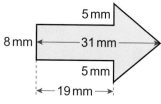

6 Work out the area of this trapezium.

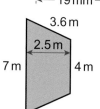

7 This wooden mirror frame was made by
cutting out a right-angled triangle from
the centre of a rectangle.
Work out the area of wood in the frame.

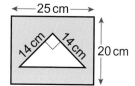

3D solids

8 For this cuboid, calculate
a the surface area
b the volume.

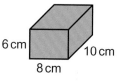

9 Work out the volume of this 3D solid.

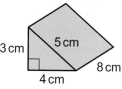

Measures

10 Convert these areas to the units given.
a 15 cm² = ☐ mm²   b 54 700 cm² = ☐ m²
c 1.2 m² = ☐ cm²   d 980 mm² = ☐ cm²

11 Convert these volumes to the units given.
a 24 m³ = ☐ litres   b 450 cm³ = ☐ ml   c 0.2 litres = ☐ cm³
d 8400 mm³ = ☐ cm³   e 549 000 cm³ = ☐ m³   f 1.6 m³ = ☐ cm³

12 How sure are you of your answers? Were you mostly

Just guessing 😞   Feeling doubtful 😐   Confident 😊

What next? Use your results to decide whether to strengthen or extend your learning.

* Challenge

13 Ria cuts a square of side 1 cm from each corner of a
10 cm by 8 cm piece of card.

Then she folds along the dashed lines to make an
open cuboid like this.

a Work out the length, width and volume of the cuboid.

b Predict the volume of the cuboid made by cutting
3 cm squares from a piece of card, 10 cm by 8 cm.
Check by calculating the volume.

c Can you make a box by cutting 4 cm squares from
each corner? Draw diagrams to explain.

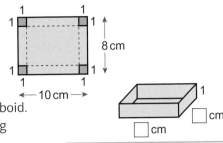

Q13b hint Sketch and label
a diagram to work out the
length, width and height.

---

8 Strengthen

2D shapes

1 Work out the perimeter of these shapes.

a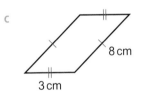
7 cm
5 cm
6 cm

b
☐ cm
3 cm   ☐ cm
5 cm

Q1 hint Start at one corner
and trace your finger along
the sides of the shape. Add
up the lengths as you go.

c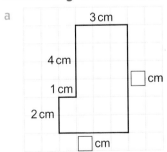
8 cm
3 cm

d
8 cm
3.5 cm

Q1c hint Matching dashes
show equal lengths.

2 **Reasoning** Work out the missing lengths marked with ☐ in each diagram.

a
3 cm
4 cm
1 cm
2 cm
☐ cm
☐ cm

b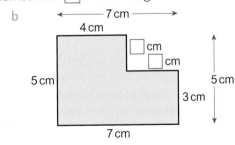
7 cm
4 cm
☐ cm
☐ cm
5 cm
3 cm
5 cm
7 cm

3 **Problem-solving** Work out the perimeter of each shape.

a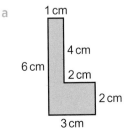
1 cm
4 cm
6 cm
2 cm
2 cm
3 cm

b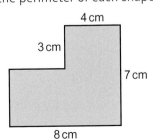
4 cm
3 cm
7 cm
8 cm

Q3b strategy hint Work out
the missing lengths first.

4 a Copy the diagram. Draw a vertical line to divide the shape into two rectangles.

    b Work out the area of each rectangle.

    c Work out the total area of the shape.

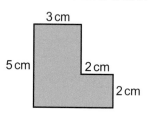

Q4 hint

Area of rectangle = length × width
Total area = area of A + area of B

5 **Problem-solving** Copy each diagram.
    i Work out the missing lengths.
    ii Work out the area of each rectangle.
    iii Work out the area of the whole shape.

a

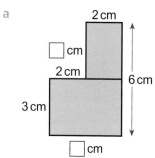

b

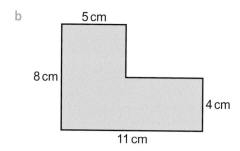

6 Copy the diagrams. In each one, mark two lines
    that are perpendicular to each other.

a   b 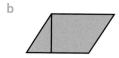   c

7 a In this triangle the base is labelled b and the height is labelled h.

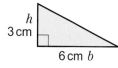

Q7 hint The base and the height
are perpendicular to each other.

      Use the formula $A = \frac{1}{2}bh$ to work out the area, A.

    b This is the same triangle turned on its side.
      Use the formula $A = \frac{1}{2}bh$ to work out the area, A.
      Check that you get the same answer as in part **a**.

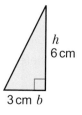

  Discussion When the base and height are perpendicular,
  does it matter which you label b and which you label h?

8   i Copy the diagram for each parallelogram.
    ii Find two perpendicular lengths. Label them b and h.
    iii Work out the area of the parallelogram using $A = bh$.

a 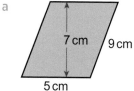   b 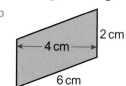   c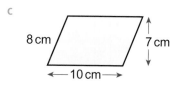

9   Copy the diagrams.
    i   Draw a line to divide the shape into a rectangle and a triangle.
    ii   Work out the area of the rectangle and the triangle.
    iii   Work out the total area of the shape.

a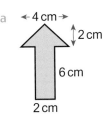
b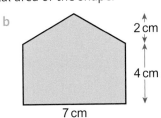

> **Q9 hint** Total area =
> area of rectangle + area of triangle

10   a   Work out the area of this blue rectangle.

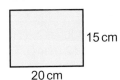

b   Pat cuts a white rectangle out of the centre of the blue rectangle.
Work out the area of the white rectangle.

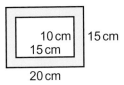

> **Q10c hint** Area remaining =
> area of blue rectangle – area of white rectangle

c   Work out the area of the blue rectangle that is left when the white one is cut out.

11   For each trapezium
    i   copy the diagram
    ii   label the two parallel sides a and b
    iii   label the height h
    iv   use the formula $A = \frac{1}{2}(a + b)h$ to work out the area.

a
b

12   a   This parallelogram has area 12 cm².
Copy and complete to find its height.

$A = bh$

$12 = \square \times h$

$h = \square$

> **Q16a hint** Write down the
> area formula. Substitute
> the values you know.

b   This triangle has area 20 cm².
Copy and complete to find its base.

> **Q16b hint** $\frac{1}{2} \times 8 = \square$

$A = \frac{1}{2}bh$

$\square = \frac{1}{2} \times b \times 8$

$\square = \frac{1}{2} \times 8 \times b$

$\square = \square \times b$

$b = \square$

3D solids

1 a Copy and complete these sketches of the front, side and top faces of this cuboid.

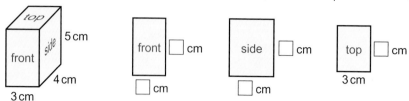

b Make labelled sketches of the back, other side, and bottom faces.

Q1d hint
Some are identical.

c Which faces are identical?
d Work out the areas of the faces.
e Work out the total surface area.

Q1e hint Add the areas of all the faces.

2 a Sketch all the faces of this triangular prism.

Q2a hint It has 3 rectangular faces and 2 triangular faces.

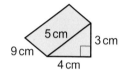

b Work out their areas.
c Work out the total surface area.

Q2b hint Are any faces identical?

3 Work out the volume of these prisms.
   i Work out the area of the front face.
   ii Multiply the area by the length of the prism.

Q3 hint Sketch the end face to help you work out its area.

a 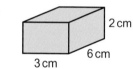 b 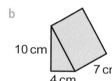 c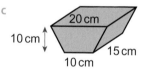

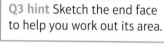

Measures

1 a Work out the area of each rectangle in cm².

b Work out the area of each rectangle in mm².
c Copy and complete this double number line for cm² and mm².

Q1c hint Use the areas of the rectangles from parts **a** and **b**. Follow the pattern.

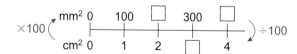

2 a Work out the area of these rectangles in cm² and m².

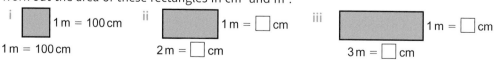

b Copy and complete this double number line for cm² and m².

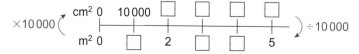

3 Copy and complete these conversions.

 a $5 \text{ cm}^2 = \square \text{ mm}^2$

 b $300 \text{ mm}^2 = \square \text{ cm}^2$

 c $6 \text{ m}^2 = \square \text{ cm}^2$

 d $80\,000 \text{ cm}^2 = \square \text{ m}^2$

4 Convert these volumes.

 a $3 \text{ cm}^3 = \square \text{ mm}^3$

 b $9000 \text{ mm}^3 = \square \text{ cm}^3$

 c $4.72 \text{ cm}^3 = \square \text{ mm}^3$

 d $2 \text{ m}^3 = \square \text{ cm}^3$

 e $1.2 \text{ m}^3 = \square \text{ cm}^3$

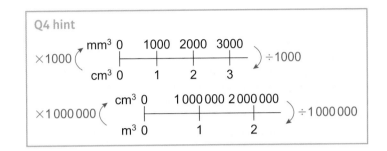

Q4 hint

5 Convert these volumes.

 a $9\,000\,000 \text{ cm}^3 = \square \text{ m}^3$

 b $3\,200\,000 \text{ cm}^3 = \square \text{ m}^3$

 c $5000 \text{ cm}^3 = \square \text{ ml} = \square \text{ litres}$

 d $875 \text{ cm}^3 = \square \text{ ml} = \square \text{ litres}$

 e $2.3 \text{ litres} = \square \text{ ml} = \square \text{ cm}^3$

 f $1.345 \text{ litres} = \square \text{ ml} = \square \text{ cm}^3$

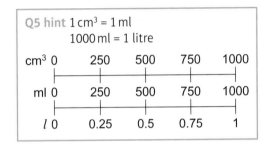

Q5 hint $1 \text{ cm}^3 = 1 \text{ ml}$
$1000 \text{ ml} = 1 \text{ litre}$

8 Extend

1 **Problem-solving** A water tank is in the shape of a cuboid.

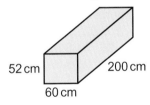

52 cm 200 cm 60 cm

All the faces except the base are to be painted.
Work out the total area to be painted, in square metres.
Give your answer to 1 decimal place.

2 **STEM / Problem-solving** A botanist counts
4 dandelion plants in a square of lawn measuring 50 cm by 50 cm.
The whole lawn has area 15 m².
Estimate the number of dandelion plants on the lawn.

Q2 strategy hint
Work out the area
of the square in m².
How many of these
would fit in the lawn?

3 **Problem-solving** Jake is tiling this wall in a swimming pool.
He has 170 m² of tiles. Is this enough?

25 m
3 m
8 m
5 m

4 | **Exam-style question**

The diagram shows the floor of a village hall.

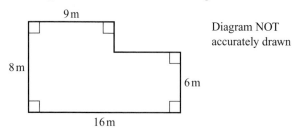

Diagram NOT
accurately drawn

The caretaker needs to polish the floor.

One tin of polish normally costs £19.
One tin of polish covers 12 m² of floor.

There is a discount of 30% off the cost of the polish.

The caretaker has £130.

Has the caretaker got enough money to buy the polish for the floor?
You must show all your working. **(5 marks)**

Nov 2013, Q24, 1MA0/1F

> **Exam hint**
> Show all the different steps in your calculation clearly. Underline the answer in each part.

5 **Modelling / Problem-solving**

a Write an expression for the perimeter of this shape.

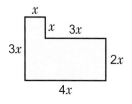

> **Q5a hint** Add up all the lengths of the sides.

b The perimeter of the shape is 28 cm.
Use your expression from part **a** to write an equation for the perimeter of the shape.
Solve the equation to find x.

c Work out the area of the shape.

> **Q5c hint** Sketch the shape and label all the lengths.

6 **Problem-solving** Work out the area of card used to make this box, in the shape of a square-based pyramid.

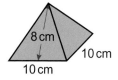

7 **Modelling / Problem-solving** The area of this shape is 35 cm².

a Write an equation for the area of the shape.

b Find the value of x.

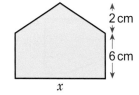

236

8   Work out the side length of each square. Round any decimal answers to 1 decimal place.

a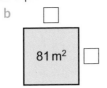

$\square \times \square = 4$

b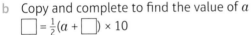

c

9   **Real / Reasoning** In a barn, each sheep and its lambs need 2.2 m² of floor space.

a   What area of floor space do you need for 40 sheep and their lambs?

b   Sketch and label a rectangle with this area.
Work out the length of fence needed to make
this rectangle.

> **Q9b hint** Use 2 numbers that multiply to give the area as the length and width.

c   Sketch and label another rectangle with this area.
Work out the length of fence needed.
Does this take more or less fence than your rectangle in part **b**?

d   What length of fence is needed to make a square with this area?

e   Which shape takes least fence to make?

10 The area of this trapezium is 50 cm².

a   Substitute $A = 50$ and the values
of b and h into the formula
$A = \frac{1}{2}(a + b)h$

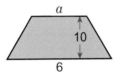

b   Copy and complete to find the value of a

$\square = \frac{1}{2}(a + \square) \times 10$

$\square = \frac{1}{2} \times 10 \times (a + \square)$

$\square = 5 \times (a + \square)$

Divide both sides by 5

$\square = a + \square$

$a = \square$ cm

> **Q10b hint** You can do multiplication in any order.

11 Write an expression for the area of each shape in mm².

a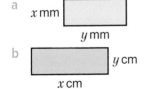

x mm
y mm

b y cm
x cm

> **Q11b hint** y cm $= y \times \square$ mm.

Write an expression for the volume of each shape in cm³.

c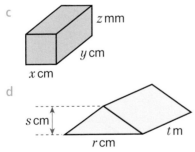

z mm
y cm
x cm

> **Q11c hint** z mm $= \square$ cm.

d s cm
t m
r cm

12 **Problem-solving** Estimate the volume of this cuboid in cm³.

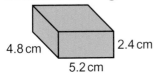

4.8 cm   2.4 cm
5.2 cm

> **Q12 hint** Round all the measurements to the nearest centimetre.

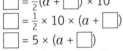

13 **STEM / Modelling** Rose marks out a square 20 cm by 20 cm
on an area of leaf litter.
Inside the square she collects leaf litter to a depth of 10 cm.
a Calculate an estimate for the volume of leaf litter she collects.
b Rose counts 43 woodlice in the leaf litter.
Estimate the number of woodlice in a 24 m² area of leaf litter
that is 40 cm deep.

> **Q13a hint** Model the
> leaf litter she collects
> as a cuboid.
> Model the 24 m² area as
> a prism, height 40 cm.

14 **Problem-solving** A box of teabags is a cuboid
20 cm by 15 cm by 5 cm.
24 boxes of teabags are put into a larger box.
The larger box is a cuboid 40 cm by 50 cm by 20 cm.
Work out the volume of empty space in the larger box.

> **Q14 hint** Work out the volume of
> one box of teabags, then 24 boxes.
> Subtract from the volume of the
> larger box.

15 **Problem-solving** How many of these boxes will
fit in the larger box?

> **Q15 hint** How many can you fit
> across the bottom of the box?

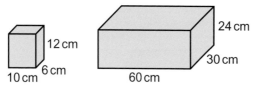

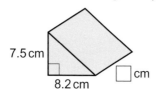

Reflect Is there more than one way to work out the answer?
Which way is best?

16 **Problem-solving** This prism has volume 381.1 cm³.

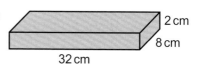

Work out its length, to 1 decimal place.

17 **STEM / Reasoning** This steel bar is melted down to
make ball bearings.
Each ball bearing has volume 0.75 cm³.
How many ball bearings can be made from the bar?

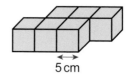

18 A rectangular field measures 300 m by 800 m.
Work out the area of the field in hectares.

> **Q18 hint** 1 hectare (ha) = 10 000 m²

19 Six identical cubes of side 5 cm are joined together like this.
For this solid, work out
a the volume
b the surface area.

20 **Reasoning** A cube of side 6 cm has a cuboid shaped
hole cut through it from front to back.
Work out the volume of the remaining shape.

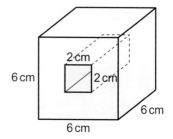

8 Knowledge check

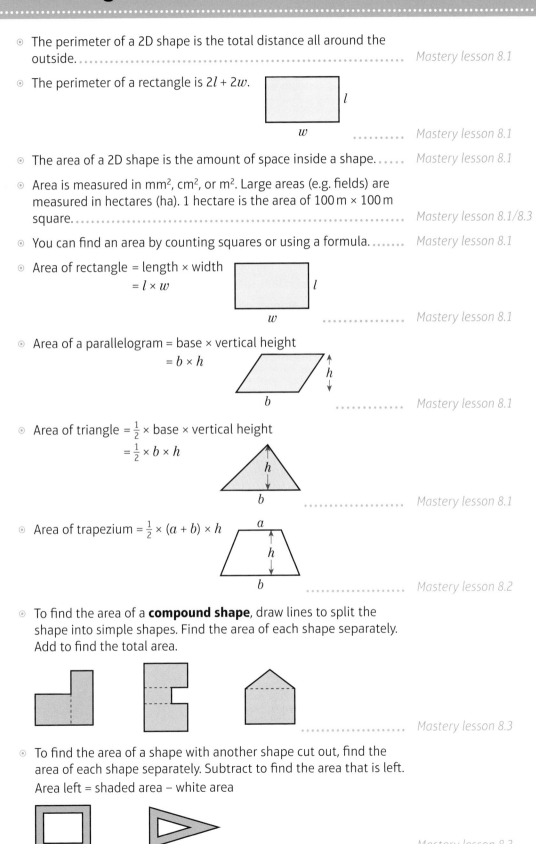

- The perimeter of a 2D shape is the total distance all around the outside. .... *Mastery lesson 8.1*

- The perimeter of a rectangle is $2l + 2w$. .... *Mastery lesson 8.1*

- The area of a 2D shape is the amount of space inside a shape. ...... *Mastery lesson 8.1*

- Area is measured in mm², cm², or m². Large areas (e.g. fields) are measured in hectares (ha). 1 hectare is the area of 100 m × 100 m square. .... *Mastery lesson 8.1/8.3*

- You can find an area by counting squares or using a formula. .... *Mastery lesson 8.1*

- Area of rectangle = length × width
  = $l \times w$ .... *Mastery lesson 8.1*

- Area of a parallelogram = base × vertical height
  = $b \times h$ .... *Mastery lesson 8.1*

- Area of triangle = $\frac{1}{2}$ × base × vertical height
  = $\frac{1}{2} \times b \times h$ .... *Mastery lesson 8.1*

- Area of trapezium = $\frac{1}{2} \times (a + b) \times h$ .... *Mastery lesson 8.2*

- To find the area of a **compound shape**, draw lines to split the shape into simple shapes. Find the area of each shape separately. Add to find the total area.

  .... *Mastery lesson 8.3*

- To find the area of a shape with another shape cut out, find the area of each shape separately. Subtract to find the area that is left.
  Area left = shaded area − white area

  .... *Mastery lesson 8.3*

⊙ To find the **surface area** of a 3D solid, work out the area of each
face. Add to find the total surface area. . . . . *Mastery lesson 8.4*

⊙ The **volume** of a 3D solid is the amount of space it takes up.
Volume is measured in mm³, cm³ or m³. . . . . *Mastery lesson 8.5*

⊙ Volume of a cuboid = length × width × height
  = $l \times w \times h$

. . . . *Mastery lesson 8.5*

⊙ A prism is a 3D object whose cross-section is the same all
through its length.

. . . . *Mastery lesson 8.5*

⊙ Volume of a prism = area of cross-section × length . . . . *Mastery lesson 8.5*

⊙ The **capacity** of a 3D object is the amount of liquid it can hold.
Capacity is measured in cm³, ml or litres. . . . . *Mastery lesson 8.6*

⊙ 1 cm³ = 1 ml, 1000 cm³ = 1 litre . . . . *Mastery lesson 8.6*

⊙ Converting areas:
  100 mm² = 1 cm²   10 000 cm² = 1 m²   10 000 m² = 1 hectare . . . . *Mastery lesson 8.3*

⊙ Converting volumes:
  1000 mm³ = 1 cm³   1 000 000 cm³ = 1 m³ . . . . *Mastery lesson 8.6*

How can you make sure you remember the formulae for areas, volumes and converting units
from this unit? Would it help to write each formula in a diagram of the shape, so you can visualise
them?

How are the parallelogram and triangle formulae the same or different from the area of a
rectangle formula?

Can you make a phrase using the letters in the formula to help you remember?

8 Unit test

Log how you did on your
Student Progression Chart.

1   Calculate the area of these shapes. All lengths are in centimetres.
    Give your answers to 1 decimal place.

a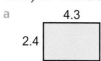

b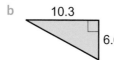

c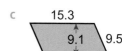

d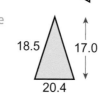

e

(10 marks)

ActiveLearn   Homework, practice and support: Foundation 8 Unit test

2 **Reasoning** The diagram shows a triangular picture frame made of wood.
All measurements are in cm.

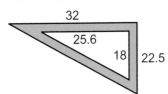

The smaller triangle has been cut out of the larger triangle.
Calculate the area of wood in the frame. *(5 marks)*

3 The perimeter of this triangle is 30 cm.
Work out

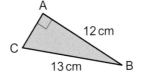

a the length of the third side *(2 marks)*
b the area. *(2 marks)*

4 Convert 3700 cm² to m². *(1 mark)*

5 The diagram shows a prism.
All measurements are in cm.

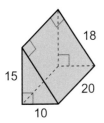

Calculate

a the volume *(2 marks)*
b the surface area. *(3 marks)*

6 A freezer has a volume of 1.5 m³. Write this volume in cm³. *(1 mark)*

7 **Reasoning** These shapes both have area 21 cm².

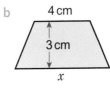

Work out the values of the letters. *(5 marks)*

8 a **Problem-solving** How many litres of water will this tank hold? *(2 marks)*

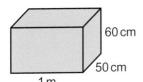

b The top and sides of the tank need painting.
Work out the surface area to be painted. *(3 marks)*
c 1 litre of paint covers 10 m².
Dan has $\frac{1}{2}$ litre of paint.
Is this enough to paint the tank with two coats of paint? *(3 marks)*

241

9 A park is a rectangle, 350 m by 420 m.

Work out the area of the park in

a square metres   b hectares.

(2 marks)

10 This shape is made from 1 cm cubes.

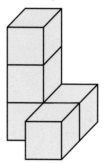

Work out its volume and surface area.

(2 marks)

Sample student answer

a How could you present the answer to part **a** in a much clearer way?

b Explain why the answer to part **b** is wrong.

a Ellen would like to make the shape of a house with a garage out of material. How much material does she need?

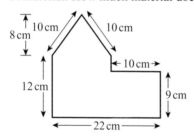

b Ellen would like to put a ribbon around the perimeter of her shape. How much ribbon will she need? **(4 marks)**

Student answer

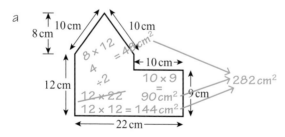

b 10 + 10 + 10 + 9 + 22 + 12 = 73 cm

Ellen will need 73 cm of ribbon.

9 GRAPHS

A space ship is at location (5, 3).

The four spacemen are at locations A (2, 5), B (5, 1), C (3, 6), D (0, 3).

Each spaceman can move once along either a horizontal or a vertical line.

Which spacemen can get back to the space ship?

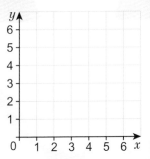

9 Prior knowledge check

Numerical fluency

1   a   Input these numbers. What is the output?

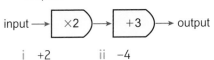

input → ×2 → +3 → output

     i   +2   ii   −4

   b   What is the function for each machine?

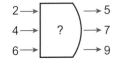

input function output

2 →
4 → ? → 5, 7, 9
6 →

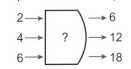

input function output

2 →
4 → ? → 6, 12, 18
6 →

Fluency with measures

2   What time is indicated by each arrow?

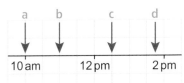

Graphical fluency

3   a   Write down the coordinates of points A, B, C, D and E.

   b   Write down the letter of the points with these coordinates.

     i   (2, −3)   ii   (5, 3)   iii   (−4, 5)

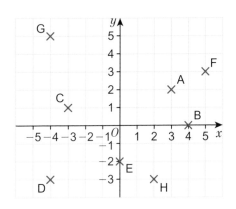

4   The table shows the cost of potatoes.

Number of kg	1	2	3	4	5
Cost in pence	80	160	240	320	400

   a   Draw a graph for this data. The x-axis should go from 0 to 6. The y-axis should go from 0 to 500.

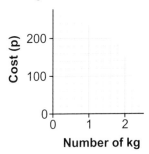

   b   Work out the cost of 3.5 kg of potatoes.

   c   How much do 6 kg of potatoes cost?

243

5 5 miles is the same distance as 8 km.

   a Copy and complete the table of values.

Number of miles	0	5	20	50	100
Number of km					

   b Draw a graph for this data.

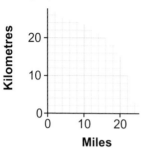

   c Use your graph to convert 70 miles to kilometres.

Algebraic fluency

6 Work out the value of each expression when $x = 3$

   a $-x$

   b $2x$

   c $3x + 1$

   d $4x - 2$

7 Solve these equations.

   a $3x = 6$

   b $5x = 30$

Geometrical fluency

8 **a** Draw a coordinate grid from −5 to +5 on both axes.

   b Plot the points A (2, 3), B (5, 3) and C (4, 1).

   c Mark the position of D on the graph to make ABCD a parallelogram.

   d Write down the coordinates of D.

* Challenge

9 Write down the coordinates of points to make geometrical shapes.

How many of these can you find?

triangle, square, rectangle, parallelogram, rhombus, trapezium, kite

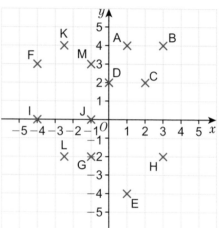

9.1 **Coordinates**

Objectives

- Find the midpoint of a line segment.
- Recognise, name and plot straight-line graphs parallel to the axes.
- Recognise, name and plot the graphs of $y = x$ and $y = -x$.

Why learn this?

Global positioning systems (GPS) use coordinates to pinpoint a position.

Fluency

Work out half of 8, 14 and 9.

1 $y = -x$. Work out the value of y when

   a $x = 2$ **b** $x = 5$ **c** $x = -1$ **d** $x = -4$

2 Add each pair of numbers and then halve the answer.

   a 6 and 4   **b** 3 and 9   **c** 2 and 5

   d −5 and 7   **e** −2 and 8   **f** −3 and 7

*Active*Learn Homework, practice and support: Foundation 9.1

Questions in this unit are targeted at the steps indicated.

3 **Reasoning** a David uses this rule to generate coordinates.
The x-coordinate is always 1, no matter what the y-coordinate is.
Which of these coordinate pairs satisfy David's rule?
(1, 5), (5, 1), (1, 1), (−1, 3), (1, 0), (1, 4), (3, 1), (1, 2)

b Draw a coordinate grid from −5 to +5 on both axes.
Plot the points from part **a** that satisfy David's rule.
Reflect What do you notice about the points you have plotted?

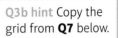

Q3b hint Copy the grid from **Q7** below.

c Charlie uses this rule to generate coordinates.
The x-coordinate is always 3, for any y-coordinate.
Charlie generates the coordinates (3, 0), (3, −2), (3, 4) and (3, 2).
Where do you expect these points to be on the grid?

d Plot the points on the same grid. Were you correct?

Key point 1

On the line with equation $y = 1$ the y-coordinate is always 1. The line is **parallel** to the x-axis.
On the line with equation $x = 3$ the x-coordinate is always 3. The line is parallel to the y-axis.

4 a Write down the integer coordinates of all the
points you can see on line A.

b What do you notice about the coordinates
on line A?

c Copy and complete.
i The equation of line A is $x = ...$
ii The equation of line B is $x = ...$
iii The equation of line C is $y = ...$

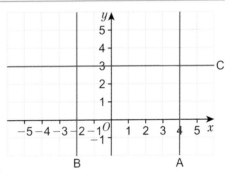

5 Write the equations of the lines labelled P, Q, R and S.

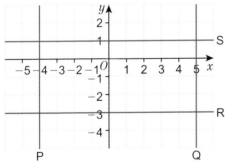

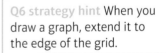

Q6 strategy hint When you
draw a graph, extend it to
the edge of the grid.

Q6d hint What is the
y-coordinate of every point
on the x-axis?

6 **Reasoning** Draw a coordinate grid from −5 to +5 on
both axes. Draw and label these graphs.
a $y = 5$ b $x = -3$ c $y = -4$
d What is the equation of the x-axis?

7 **Reasoning** a Write down the integer coordinates
of the points you can see on this line.

b What do you notice about the coordinates?

c Complete the missing coordinates
of these points on the line.
(6, □), (□, −8)

d The equation of the line is $y = ...$

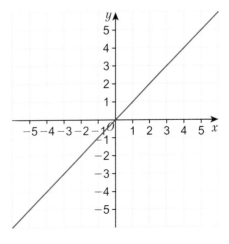

8 These coordinates are on the line $y = -x$.
   (3, −3) (2, −2) (0, 0) (−1, 1) (−5, 5)
   a Draw a coordinate grid from −5 to +5 on both axes.
   b Plot the points and join them with a straight line.
   c Write the coordinates of three more points on the line $y = -x$
   Discussion Did everyone write the same points for part **c**?

Key point 2

The **midpoint** of a line segment is the point exactly in the middle.

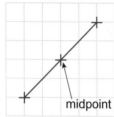
midpoint

9 Find the midpoints of each of these line segments.
   Make a table to help you.

> **Q9 communication hint**
> A **line segment** has a start and end point.

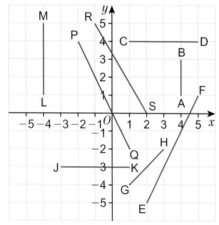

Line segment	Start point	End point	Midpoint
AB	(4, 1)	(4, 3)	
CD	(1, 4)	(5, 4)	
EF			
GH			
JK			
LM			
PQ			
RS			

Discussion How can you find the midpoint of a line segment using the start point and end point?

Example 1

Find the midpoint of a line segment with start point (3, 2) and end point (7, 9).

$\dfrac{(3 + 7)}{2} = 5$ ——— Add the x-coordinates and divide by 2.

$\dfrac{(2 + 9)}{2} = 5.5$ ——— Add the y-coordinates and divide by 2.

Midpoint = (5, 5.5)

10 **Exam-style question**

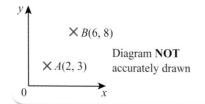

Diagram **NOT** accurately drawn

The point A has coordinates (2, 3).
The point B has coordinates (6, 8).
M is the midpoint of the line AB.
Find the coordinates of M.   **(2 marks)**
June 2014, Q1, 1MA0/2H

> **Exam hint**
> You can use the diagram or do the calculation as in Example I.

11 Work out the midpoints of the line segments with these start and end points.

   a (3, 5) and (7, 9)   b (2, 7) and (5, 10)

   c (−3, 4) and (1, 6)   d (−2, −5) and (0, 3)

> **Q11c strategy hint** $\dfrac{(-3 + 1)}{2} = \square$

9.2 Linear graphs

Objectives

- Generate and plot coordinates from a rule.
- Plot straight-line graphs from tables of values.
- Draw graphs to represent relationships.

Why learn this?

Scientists plot results on graphs to look for relationships between quantities.

Fluency

Give three inputs and functions for this function machine.

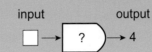

1 Work out

   a −2 + 3   b 4 × 5   c −3 × 6

2 Work out the value of each expression when

   i $x = 2$ ii $x = -1$

   a $3x + 2$ b $-2x + 1$ c $\frac{1}{4}x + 3$

3 Write the number each arrow is pointing to.

4 This function machine generates coordinates.

input   function   output   coordinates

$x \rightarrow$ $\rightarrow y \rightarrow (x, y)$

$-3 \rightarrow$ $\rightarrow \square \rightarrow (-3, \square)$

$0 \rightarrow$ $+5$ $\rightarrow \square \rightarrow (0, \square)$

$1 \rightarrow$ $\rightarrow \square \rightarrow (1, \square)$

   a Work out the missing coordinates.

   b Draw a coordinate grid from −3 to +6 on both axes.

   c Plot the coordinates from part **a** on the grid.

   d Join the points and extend the line to the edges of the grid.

   e Write the coordinates of two more points that lie on the line.

   f Write a rule using algebra for this function machine.

   g Label your graph with this rule.

> **Q4f hint** Your rule should start $y = ...$ What do you do to the x-value to get the y value?

> **Q4g communication hint** Labelling a graph means writing the equation of the graph next to the line.

5 Repeat **Q4** for this two-step function machine.

input   functions   output   coordinates

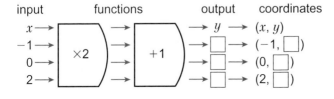

Example 2

a Complete this table of values for the equation $y = 3x + 2$

x	−2	−1	0	1
$y = 3x + 2$	−4	−1	2	5

When $x = -2$, $y = 3 \times -2 + 2 = -4$
When $x = -1$, $y = 3 \times -1 + 2 = -1$
and so on…

b Draw the graph of $y = 3x + 2$

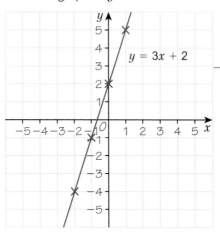

Plot the coordinate pairs from the table with crosses (−2, −4), (−1, −1), (0, 2), (1, 5). Join them with a straight line and extend it to the edge of the grid.

6 a Copy and complete the tables of values for these straight-line graphs.

Q6 strategy hint If your points are not on a straight line, check your working.

i

x	−3	−2	−1	0	1	2	3
$y = x + 1$			0	1			

ii

x	−3	−2	−1	0	1	2	3
$y = 2x - 3$			−5	−3			

Q6b hint

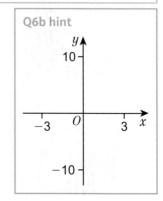

b Draw a coordinate grid with −3 to +3 on the x-axis and −10 to +10 on the y-axis.
Draw and label the graphs of $y = x + 1$ and $y = 2x - 3$, using your tables of values from part **a**.

7 Draw and label these straight-line graphs for $x = -3$ to +3.
Copy the coordinate grid from **Q6**. Draw all four graphs on the same grid.

a $y = 3x - 2$ b $y = 2x + 4$
c $y = 4x - 6$ d $y = 0.5x + 1$

Q7a strategy hint Make a table of values like the ones in **Q6**.

8 a Copy and complete these tables of values for straight-line graphs.

i

x	−3	−2	−1	0	1	2	3
$y = -x + 1$	4			1			

ii

x	−3	−2	−1	0	1	2	3
$y = -2x - 3$	3			−3			−9

iii

x	−3	−2	−1	0	1	2	3
$y = -4x + 2$	14			2			−10

b Draw a coordinate grid with −3 to +3 on the x-axis and −10 to +10 on the y-axis.
Draw the graphs of $y = -x + 1$, $y = -2x - 3$ and $y = -4x + 2$ using your tables of values from part **a**.

9 Draw and label the graphs of these straight lines on the same grid, for values of x from -4 to $+4$.

   a $y = -3x + 2$ b $y = -2x - 1$

   c $y = -4x + 6$ d $y = \frac{1}{4}x + 1$

   e **Reflect** Would you expect these graphs to slope upwards or downwards from left to right?

      i $y = 5x - 2$ ii $y = -4x + 3$

> **Q9 strategy hint** Make a table of values for each graph, with x-values from -4 to $+4$. Find the lowest and highest values of y in your tables. Draw the y-axis between these values.

10 **Exam-style question**

   a Complete the table of values for $y = 2x + 2$

x	-2	-1	0	1	2	3	4
y	-2				6		

   b On the grid, draw the graph of $y = 2x + 2$ **(4 marks)**

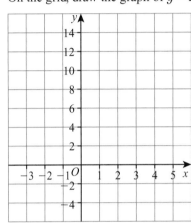

March 2013, Q18, 5MB2F/01

> **Exam hint**
> Plot your points accurately and use a ruler. Extend the line to the edges.

11 The diagram shows the graph of $y = 1 - \frac{1}{2}x$.

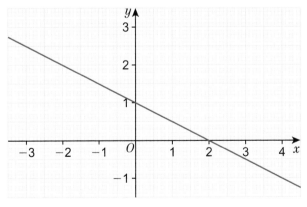

   a Use the graph to find an estimate for the value of y when

      i $x = 1.6$ ii $x = 2.2$ iii $x = -0.8$

   b Use the graph to find an estimate for the value of x when

      i $y = 1.8$ ii $y = 0.6$ iii $y = -0.4$

Discussion Why are your readings estimates for the values of x and y?

   c Use the graph to find approximate solutions to these equations.

      i $1 - \frac{1}{2}x = -1$ ii $1 - \frac{1}{2}x = 2.5$ iii $1 - \frac{1}{2}x = 1.2$

> **Q11ai hint** Find the point on the line where $x = 1.6$ and read across to the y-axis to find the value of y.

> **Q11ci hint**
> $-1 = 1 - \frac{1}{2}x$ $y = 1 - \frac{1}{2}x$
> Read the x-value when $y = -1$

9.3 Gradient

Objectives
- Find the gradient of a line.
- Identify and interpret the gradient from an equation.
- Understand that parallel lines have the same gradient.

Did you know?
Steep hills have warning signs showing the gradient.

Fluency

What types of lines will never intersect? Why?

1 Work out
   a $10 \div 2$ b $-6 \div 3$ c $8 \div 6$ d $2 \div 4$

2 Which line is steepest?

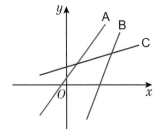

3 Look at the graph.
   a How many squares does line A go up for every
      i   1 square across
      ii 2 squares across
      iii 3 squares across?
   b How many squares does line B go down for every
      i   1 square across
      ii 2 squares across
      iii 3 squares across?
   Discussion What do you notice?

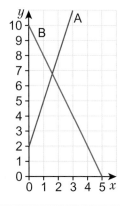

Key point 3

The steepness of a graph is called the **gradient**.
To find the gradient work out how many units the graph goes up for each unit it goes across.

4 Here are the graphs of $y = 2x - 1$ and $y = 2x + 3$.
   a Fill in the missing word:
      The lines $y = 2x + 3$ and $y = 2x - 1$ are _____
   b What are the gradients of the lines?

   > **Q4b hint** Choose two points on the line. Find the gradient of the line segment between them.

   c Fill in the missing word.
      Parallel lines have the same _____

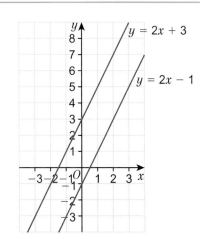

Active Learn Homework, practice and support: Foundation 9.3

Key point 4

positive gradient   negative gradient

5   Work out the gradient of each line segment by calculating $\dfrac{\text{total distance up}}{\text{total distance across}}$.

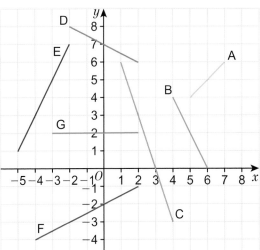

Q5D hint A gradient can be a fraction.

Discussion Is it easier to work out the gradient from a whole line or a line segment?

6   On squared paper draw line segments with these gradients.

a   1   b   3   c   −2

d $\frac{1}{2}$ e $\frac{-2}{3}$

7   **Exam-style question**

Find the gradient of the straight line drawn on this grid.

(2 marks)

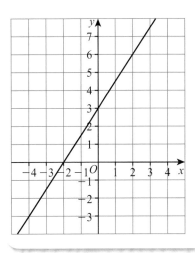

March 2013, Q22c, 1MA0/2F

Exam hint
Is the gradient
positive or negative?
Write your answer:
Gradient = ☐

8   **Real** These sketches show ramps for wheelchair access.
Work out the gradient of each ramp.

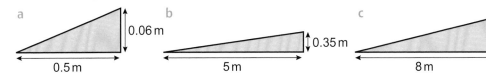

a   0.06 m   0.5 m
b   0.35 m   5 m
c   0.5 m   8 m

9 a Copy and complete these tables of values.

i

x	–3	–2	–1	0	1	2	3
$y = x + 1$	–2			1			

ii

x	–3	–2	–1	0	1	2	3
$y = -2x + 3$	9			3			

b Draw the graphs of   i $y = x + 1$ ii $y = -2x + 3$

c Work out the gradient of each line.

d **Reflect** What do you notice about

   i the gradient

   ii how the y-value changes when x increases by 1

   iii the coefficient of x in the equation?

> **Q9d communication hint**
> The **coefficient** is the number in front of the x.

10 **Reasoning** Choose from these equations

  a the line with the steepest gradient

  b a pair of parallel lines

  c a line that slopes down from left to right.

> $y = 3x + 2$ $y = 2x + 8$ $y = 3x$
> $y = 5x + 2$ $y = -3x + 4$ $y = x + 5$
> $y = 4$ $y = \frac{1}{2}x + 1$

Example 3

Find the gradient of the line joining the points A (–2, 1) and B (4, 3).

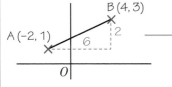

Sketch a diagram. Draw in lines across and up. Work out the distances across and up.

$$\text{Gradient} = \frac{\text{total distance up}}{\text{total distance across}} = \frac{2}{6} = \frac{1}{3}$$

---

11 Find the gradient of the line joining the points C (1, 2) and D (7, 5).

9.4 $y = mx + c$

Objectives

- Understand what m and c represent in $y = mx + c$.
- Find the equations of straight-line graphs.
- Sketch graphs given the values of m and c.

Why learn this?

Game designers use equations of straight lines to program characters' movements.

Fluency

Which of these equations give parallel lines? $y = x + 5$ $y = 2x + 5$ $y = 2x - 3$

1 Work out the value of y when $x = 0$.

  a $y + x = 6$ b $2y + x = 10$ c $5y - 3x = 20$

2 Work out the value of x when $y = 0$.

  a $y + x = 6$ b $y + 2x = 10$ c $5y - 4x = 20$

3 On squared paper, draw line segments with these gradients.

  a 4   b –2

Warm up

Active Learn Homework, practice and support: Foundation 9.4

4   The diagram shows five straight-line graphs and their equations.

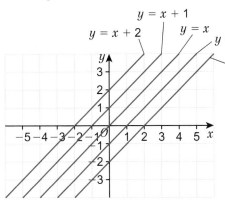

Q4 communication hint The y-intercept is the point where the line crosses the y-axis.

a   What do you notice about the equation of the line and the y-intercept?
b   Where would you expect $y = x + 4$ to cross the y-axis?
c   What is the gradient of each line?
Discussion How can you tell the gradient and y-intercept of a line from its equation?

Key point 5

A **linear equation** produces a straight line graph. The equation of a straight line is $y = mx + c$, where m is the gradient and c is the y-intercept.

Example 4

Write the equation of each line.

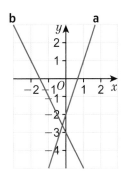

a $y = mx + c$

   [Work out the gradient, m.]

   $m = 3$

   [The line crosses the y-axis at −2.]

   $c = -2$

   Equation of line is $y = 3x - 2$

   [Substitute $m = 3$ and $c = -2$ into $y = mx + c$]

b $y = mx + c$

   $m = -2$ [This line slopes down so its gradient is negative.]

   $c = -3$

   Equation of line is $y = -2x - 3$

5   Work out the equations of lines A, B, C, D and E.

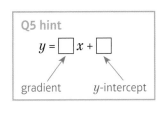

Q5 hint

$y = \boxed{}\, x + \boxed{}$

gradient y-intercept

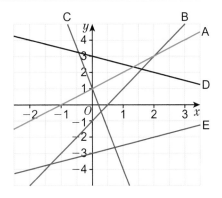

6   Write down the equation of a straight line that is parallel to $y = 5x + 6$.
    Discussion Did everyone write the same equation?

7   Which of these lines pass through
    a   (0, 0)
    b   (0, −3)?

$$y = 3x \qquad y = 3x − 3 \qquad y = 3x + 3$$
$$y = −3x \qquad y = −3x + 3 \qquad y = x − 3$$

8   In each of these you are given one point that lies on a line and the gradient of the line.
    Work out the equation of the line.
    a   (2, 1) and $m = 3$
    b   (4, 6) and $m = 2$
    c   (−1, 2) and $m = 4$
    d   (−2, 10) and $m = \frac{1}{2}$
    e   (5, −12) and $m = 3$

> **Q8a hint** $x = 2$, $y = 1$ and $m = 3$.
> Use $y = mx + c$ to work out c.
> $1 = 3 \times 2 + c$, so $c = −5$
> The equation is $y = \boxed{}\, x + \boxed{}$

9   In each of these you are given two points that lie on a line.
    Work out the equation of the line.
    a   (2, 3) and (4, 7)
    b   (−6, 0) and (2, 4)
    c   (3, 0) and (4, 2)
    d   (−2, 4) and (3, −1)

> **Q9 hint** First work out the gradient of the line (m). Then use the method from **Q8**.

10 Draw these graphs from their equations.
    a $y = 2x − 7$
    b $y = −3x + 5$
    c $y = \frac{1}{2}x − 4$
    d $y = −x + 6$

> **Q10 strategy hint** Draw a coordinate grid from −8 to 8 on both axes. Plot a point at the y-intercept. From the y-intercept, draw a line with the given gradient. Extend your line across the grid and label it with its equation.

11 Draw a coordinate grid from 0 to 8 on both axes.
    Plot these graphs.
    a $x + y = 8$ b $x + y = 6$ c $x + y = 4$
    Reflect What do you notice?
    Discussion How do you find the gradient from the equation?

12 Draw a coordinate grid from −8 to +8 on both axes.
    Plot these graphs.
    a $2x + y = 8$ b $3x + 2y = 6$
    c $4x + 2y = 4$ d $2x + 4y = 6$

> **Q12a hint** When $x = 0$: $0 + y = 8$, so $y = 8$
> When $y = 0$: $2x + 0 = 8$, so $x = 4$
> The graph goes through (0, 8) and (4, 0).

9.5 **Real-life graphs**

Objective

• Draw and interpret graphs from real data.

Why learn this?

Modelling a relationship allows you to make an informed choice.

Fluency

Write the value each arrow is pointing to.

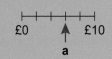

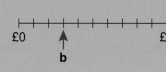

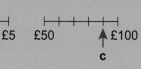

1   Work out

   a   5 kg at 70p per kg   b   4 hours at £30 per hour

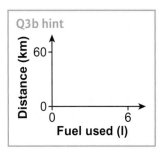

Q1a hint 70p per kg means 70p for each kg.

2   Draw the graph of $y = \frac{1}{5}x - 4$

3   **Real / Problem-solving** A car uses 1 litre of fuel for every 10 kilometres it travels.

Fuel used (litres)	0	1	2	3	4	5
Distance travelled (km)	0	10				50

   a   Copy and complete the table.

   b   Draw a graph of fuel used against distance.

   c   How much fuel is used on a 60 kilometre journey?

   d   At the start of a 100 kilometre journey there are 27 litres of fuel in the tank. How many litres will be left at the end of the journey?

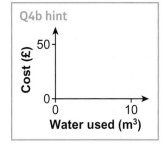

Q3b hint

4   **Real** A water company charges a flat rate of £30 per month plus £2 per cubic metre of water used.

   a   Copy and complete the table.

Water used (m³)	0	1	2	3	4	5	6	7	8	9	10
Cost (£)	30										50

   b   Draw a graph of water used against cost.

   c   Lianne's water bill is £41 this month. Use your graph to find the number of cubic metres of water she has used.

   d   How many litres of water has Lianne used?

Q4d hint 1 m³ = 1000 litres

Q4b hint

5   **Reasoning** The graph shows the cost of buying different quantities of tomatoes.

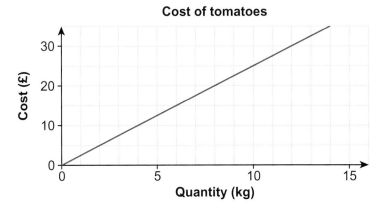

Cost of tomatoes

   a   What is the cost of

      i   10 kg of tomatoes

      ii 1 kg of tomatoes?

   b   What is the gradient of the graph?

Discussion What does the gradient of the graph represent?

6 **Real / Reasoning** The graphs show the costs of grapes at different supermarkets.

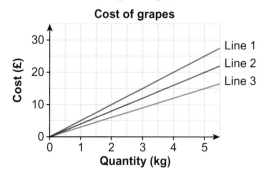

Supermarket A: Green grapes £4 per kg
Supermarket B: Organic grapes £5 per kg
Supermarket C: Black grapes £3 per kg

a Match each line on the graph with the correct supermarket.
b Copy and complete this statement.
   The steeper the slope the …. the unit price.

7 **Reasoning / Real** An electrician charges a £60 callout fee and £30 per hour.
a Copy and complete the table of values.

Hours worked	0	1	2	3	4	5
Total cost (£)	60					210

b Draw a graph of hours worked against total cost.
c Work out the equation of the line.
d What does the gradient represent?
e What does the y-intercept represent?

Q7b hint Put hours worked on the x-axis and total cost on the y-axis.

8 **Real** The graph shows the costs of hiring two different plumbers.

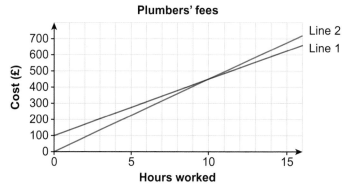

Plumber A charges £100 callout fee and £35 per hour.
Plumber B has no callout fee but charges £45 per hour.

a Match each line on the graph with the correct plumber.
b Alice thinks her plumbing work will take about 12 hours.
   Explain which plumber she should use.

9 **Real / Modelling** Sam has a landline telephone.
He pays £16 per month plus 12p per minute for calls. Sam uses the phone for about 80 minutes each month.
He sees this offer: £20.50 per month plus 6p per minute for calls.

a Should he change to the rate in the offer?
b **Reflect** How could you see at a glance when each payment option was better value?

10 **Communication / Real** The graph shows two different ways to pay for golf at a golf club.

  a Explain in words what each option means.

  b John expects to play about 45 games of golf in a year. Which option should he choose?

11 **Modelling** A company works out the monthly pay for sales staff using the equation $y = \dfrac{x}{10} + 600$, where y is the monthly wage (£) and x is the value of the sales made in the month.

  a Draw the graph for the equation.

  b What is the monthly pay when no sales are made?

  c Pete's sales average £12 000 per month. He is offered a different job where the monthly pay is £2000. Should he take it?

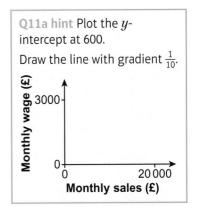

Q11a hint Plot the y-intercept at 600.
Draw the line with gradient $\frac{1}{10}$.

12 **Exam-style question**

The graph shows the cost of getting work done by three different companies.

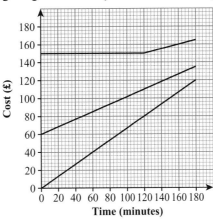

Here are the charges of the three companies.

Company A: Callout charge of £60 and every hour costs £25.

Company B: No callout charge but every hour costs £40.

Company C: Callout charge of £150 which includes the first 2 hours and then every hour costs £15.

  a Label each line on the graph with the letter of the company it represents.

Ned uses company A for 2 hours of work.

  b Find the total cost.

Fiona needs 8 hours of work done.

  c Explain which company would be the cheapest for her to use. You *must* give reasons for your answer.

(4 marks)

9.6 Distance–time graphs

Objectives

- Use distance–time graphs to solve problems.
- Draw distance–time graphs.
- Interpret rate of change on graphs.

Why learn this?

Scientists use graphs to identify types of movement.

Fluency

A car is travelling at 40 km per hour. How far will it travel in 3 hours?
How long will it take to do a 200 km journey?

Warm up

1 Work out

   a 60 ÷ 2   b 60 ÷ 3   c 60 ÷ 5   d 60 ÷ 6

2 Copy and complete

   a 6 × ☐ = 60   b 5 × ☐ = 60   c 30 × ☐ = 60   d 2 × ☐ = 60

Key point 6

A **distance–time graph** represents a journey. The vertical axis represents the distance from the starting point. The horizontal axis represents the time taken.

$$\textbf{Average speed} = \frac{\text{distance travelled}}{\text{time taken}}$$

Example 5

Jenny walks 500 metres in 5 minutes, then arrives at the bus stop.
She waits 5 minutes for the bus.
She travels 3000 metres on the bus and gets off 16 minutes after she left home.

a Draw a distance–time graph for her journey.

b Work out the average speed in km/h of Jenny's walk.

a

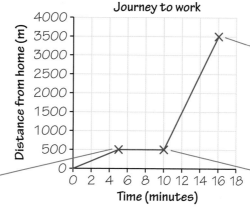

After 16 minutes she has travelled 500 + 3000 m. Plot the point (16, 3500) and join your points with straight lines.

After 10 minutes she has still only travelled 500 m. Plot the point (10, 500).

After 5 minutes she has travelled 500 m. Plot the point (5, 500).

b Jenny walks 500 m in 5 minutes.

×12 (5 min → 60 min, 500 m → 6000 m) ×12

6000 m = 6 km

Average walking speed = 6 km/h

3   **Reasoning** Satbir leaves the house at 9 am. She cycles 5 km
    to town. It takes her 15 minutes.
    She goes to one shop for 30 minutes and then cycles 2 km
    further to her friend's house.
    The journey to her friend's house takes her 10 minutes.
    a   Draw a distance–time graph for Satbir's journey.
    b   Work out the average speed in km per hour for each part
        of her journey.
    c   Compare the average speeds to decide in which
        part of the journey she was cycling uphill.

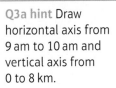

Q3a hint Draw
horizontal axis from
9 am to 10 am and
vertical axis from
0 to 8 km.

Q3b hint How many
15 minutes are there
in 60 minutes?

4   This distance–time graph shows Isaac's journey on his bicycle.

    a   How far did Isaac ride his bike on the first part of the journey?
    b   At what time did he stop to rest?
    c   How long did the first part of his journey take?
    d   What was his average speed on the first part of the journey?
    e   How many minutes did Isaac rest for?
    f   How long did the last part of his journey take?
    g   How far did he ride on the last part of the journey?
    h   What was his average speed for the last part of the journey?
    Discussion What does the gradient of the line represent? How can you tell when Isaac
    was travelling fastest?

Key point 7

On a distance–time graph the gradient represents the speed of the journey.

5   This graph shows a fire engine's journey to and
    from a fire.
    a   Work out the average speed of the fire engine
        in km/h on the way to the fire.
    b   For how long was the fire engine at the fire?
    c   Work out the average speed of the fire engine
        in km/h on the way back to the fire station.
    d   How far did the fire engine travel altogether?
    Discussion On a distance–time graph, what
    does a horizontal line represent?
    What about a negative gradient?

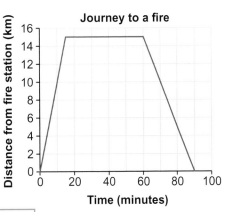

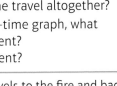

Q5d hint The fire engine travels to the fire and back again.

6 Mike walks to the doctor's surgery. He stops to talk to a friend on the way home.

The graph shows his journey.

Mike's journey

a How far is the doctor's surgery from Mike's home?

b How long does Mike spend in the surgery?

c How long does he spend talking to his friend?

d On which part of the journey is Mike walking fastest?

7 Bob leaves home at 10 am and cycles 5 km to the sports centre. It takes him $\frac{1}{4}$ of an hour.
He spends $2\frac{1}{2}$ hours at the sports centre.
He starts to cycle home on the same route and after 4 km stops at a friend's house.
He arrives at 1 pm and stays $1\frac{1}{2}$ hours.
He gets home at 2.35 pm.

a Draw a distance–time graph to show Bob's journey.

b On which part of the journey was he cycling the fastest?

8 The graph shows two bus journeys.

One bus leaves Southampton at 9 am and travels to Winchester.

The other bus leaves Winchester at 9 am and travels to Southampton.

a Which line shows the Winchester to Southampton bus?

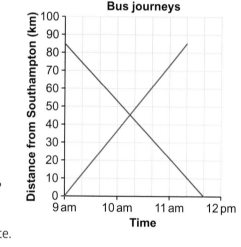

Bus journeys

> **Q8a hint** The vertical axis shows the distance from Southampton.

b What time does each bus get to its destination?

c Which bus journey is faster?

9 Sasha and Kelly go hill walking along the same route.

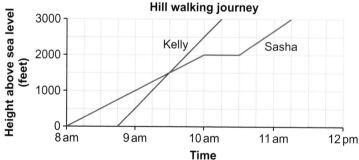

Hill walking journey

The graph shows their journeys.

a How high do they climb?

b How long does it take Sasha to reach the top?

c How long does it take Kelly to reach the top?

d At what time does Kelly overtake Sasha?

> **Q9d hint** At what time are they both at the same height?

10 On Friday, Molly cycles from her house to the gym. The travel graph shows her journey.

Molly stays at the gym for 50 minutes.
She then cycles back to her house at 24 km/h.

a Copy and complete the travel graph.

b What time does Molly leave home?

c How far is the gym from Molly's house?

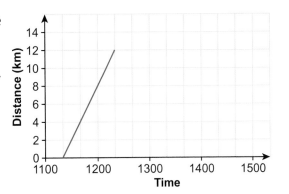

11 (Exam-style question)

Debbie drove from Junction 12 to Junction 13 on a motorway.

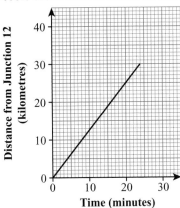

Exam hint
You could plot Ian's journey on the graph to compare the speeds. Or, you could work out Debbie's speed from the graph.

The travel graph shows Debbie's journey.
Ian also drove from Junction 12 to Junction 13 on the same motorway.
He drove at an average speed of 66 km/hour.
Who had the faster average speed, Debbie or Ian?
You must explain your answer.   **(4 marks)**

June 2013, Q11, 1MA0/1H

Key point 8

A **rate of change graph** shows how a quantity changes over time.
On a **velocity–time graph** the gradient represents the acceleration.

12 Match each velocity–time graph to one of the sentences.

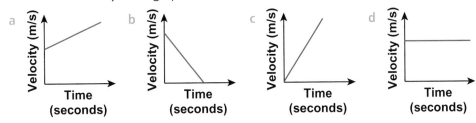

i Starts from rest and velocity steadily increases.

ii Travels at a steady velocity.

iii Starts from 5 m/s and steadily increases in speed.

iv Starts from 8 m/s and velocity steadily decreases.

Q12 communication hint
Velocity means speed in a particular direction.

13 The graph shows the movement of a particle over time.

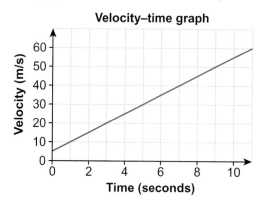

Velocity–time graph

a What was the starting speed of the particle?
b What was the acceleration of the particle?
c What is the equation of the line?

> **Q13a hint** Starting speed is the speed when time = 0.

> **Q13b hint** Find the gradient of the line.

> **Q13c hint** The equation of a line is $y = mx + c$. Look back at Example 4 Lesson 9.4.

9.7 **More real-life graphs**

Objectives

- Draw and interpret a range of graphs.
- Understand when predictions are reliable.

Why learn this?

Finding the equation of a line of best fit on a scatter graph helps you predict values more accurately.

Fluency

When was the person travelling faster?

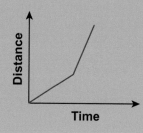

1 What type of correlation do these show?

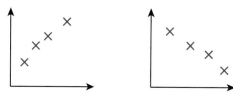 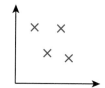

2 Find the equation of this line.

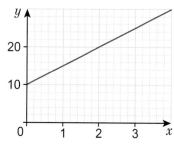

3 **Reasoning** The graph shows the depth of water for three containers, A, B and C, filling at a constant rate.

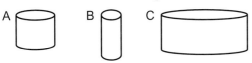

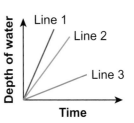

a   Which container will fill fastest?
b   Match each container to a line on the graph.

Q3 hint What does the gradient represent?

Q3 communication hint
Constant rate means the same amount flows in every second.

Discussion How did you match the lines to the containers?

4 **Reasoning** This graph shows the depth of water for two containers, A and B, with sloping sides filling at the same constant rate.

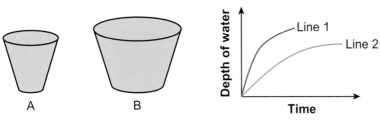

Q4 hint The depth in a container with sloping sides makes a curved graph. Which container will take longer to fill?

Match each container to a line on the graph.

5 **Problem-solving** Some containers are filled with water at a constant rate.
a   Sketch a graph of depth of water against time for each of these containers.

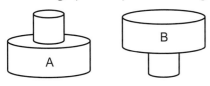

b   Match each container to a line on the graph.

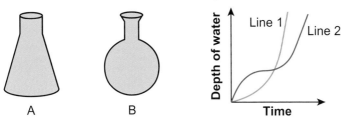

6 **STEM / Communication** The graph shows pressure against temperature for a gas at a constant volume.
a   Estimate the temperature when the pressure is 84 cm of mercury.
b   Estimate the pressure when the temperature is −80 °C.
c   Billy says, 'The rate of increase of pressure with temperature is always the same.' What evidence from the graph supports this?
d   Work out the gradient of the line.
Discussion What does the gradient tell you about the pressure?

Pressure–temperature graph

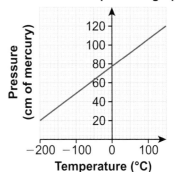

7 Exam-style question

Water flows out of a cylindrical tank at a constant rate.

The graph shows how the depth of water in the tank varies with time.

a Work out the gradient of the straight line.

b Write down a practical interpretation of the value you worked out in part **a**.

(3 marks)

Nov 2010, Q8, 5MB1H/01

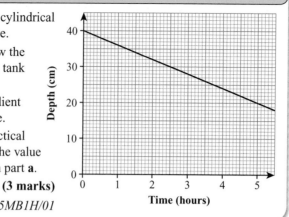

Exam hint
Part **b** is asking you to use your calculated value from part **a** to explain what is happening to the water in the tank.

8 **Reasoning / Modelling** The scatter graph shows the hours spent on study and test scores for some students.

a Describe the correlation between hours of study and test scores.

b Work out the equation of the line of best fit.

c Use your equation to predict the test score for someone who studies for 4.5 hours.

Q9c hint Using the equation gives a more accurate value than reading from the graph.

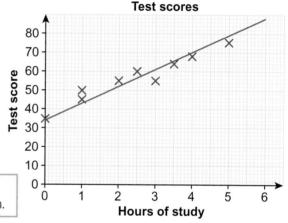

d Explain why you cannot estimate the hours of study for someone with a test score of 20.

9 **Modelling** The table shows the results for 9 students in their maths and science exams.

Maths	32	45	52	58	64	69	73	80	86
Science	27	38	46	49	57	60	65	71	75

a Draw a scatter graph for the data.

b Draw a line of best fit.

c Find the equation of your line of best fit.

d A student scored 60 in the maths exam but was absent for the science exam. Predict what they might have scored for science.

e Another student scored 15 in the maths exam. Predict their science score.

Discussion How reliable are these predictions?

10 **Modelling** The table shows the average household size in the UK from 1961 to 2011.

Year	1961	1971	1981	1991	2001	2011
Average household size	3.1	2.9	2.7	2.5	2.4	2.3

a Use this information to predict the average household size in the year 2021.

Q10a strategy hint Plot a graph of the data. Put year on the x-axis and average household size on the y-axis. Draw a line of best fit through your plotted points and extend the line.

b How reliable are your results?

Q10b hint Explain any assumptions you have made.

9 Problem-solving: Dinosaur trackways

Objective • Draw and interpret graphs of linear functions.

Sets of fossilised dinosaur footprints are called dinosaur trackways. Scientists can deduce how a dinosaur was moving by comparing the size of each footprint in a trackway with the gaps between steps.

1 Draw a set of axes where the horizontal axis represents hind foot length in cm (from 0 to 100) and the vertical axis represents stride length in cm (from 0 to 450).

If the stride length is less than 8 times the size of the hind foot length, the dinosaur was walking.

2 Add the graph of $y = 8x$ to your axes.
Any data points which fall below this line and above the x-axis represent a walking dinosaur.
Label this area 'WALKING'.

> **Q2 hint** You can check your answer by imagining where the graphs of $y = 7x$, $y = 6x$ and $y = 5x$ would fall on your axes. These should all go through the 'walking' section.

If the stride length is more than 11.6 times the size of the hind foot length, the dinosaur was running.

3 Add the graph of $y = 11.6x$ to your axes.
Any data points which fall to the left of this line and to the right of the y-axis represent a running dinosaur.
Label this area 'RUNNING'.
Label the remaining area in the middle 'TROTTING'.

> **Q3 hint** Draw a table of values or calculate some points to help you draw this graph.

The table at the bottom of the page contains measurements from four dinosaur trackways.

4 By plotting these points on your graph, determine whether these dinosaurs were walking, trotting or running.

5 A particular stegosaurus had a hind foot which was 35 cm long. How large could its stride length get before it stopped walking and started to trot?

6 One scientist suggests that the measurements of the stride lengths may be inaccurate by as much as 10%. Would this change any of your answers to **Q4**? Explain your decision.

Dinosaur	Hind foot length (cm)	Stride length (cm)
Allosaurus	44	280
Tyrannosaurus rex	80	420
Velociraptor	20	250
Microceratus	7	43

9 **Check up**

Log how you did on your
Student Progression Chart.

Algebraic straight-line graphs

1 a Copy and complete this table of values for
the equation $y = 2x - 1$.

x	−2	−1	0	1	2
$y = 2x - 1$					

 b Draw a coordinate grid from −5 to +5 on
both axes. Plot the line $y = 2x - 1$.

2 Find the midpoint of the line segment PQ where P is (2, 5) and Q is (6, 3).

3 Write the equation of lines A, B, C and D.

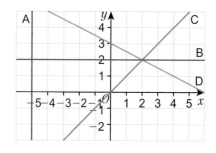

4 $y = 3x + 4$ $\qquad y = x + 3$ $\qquad y = 4x + 3$ $\qquad y = 3x - 7$

 a Which of these lines are parallel?

 b Which of these lines have the same y-intercept?

 c Which of these lines is the steepest?

Distance–time graphs and scatter graphs

5 The graph shows Rhoda's walk.

 a For how long did Rhoda stop to talk
to a friend?

 b Between what times was Rhoda
walking fastest?

 c What was her average speed in km/h
on the first part of the journey?

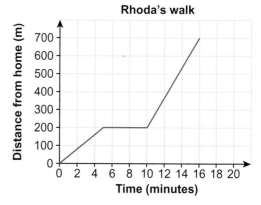

6 The scatter graph shows the marks for
10 students in their Spanish and
French exams.

 a Find the equation of the line
of best fit.

 b Use your equation to predict the
French mark of a student
who gets 48 in Spanish.

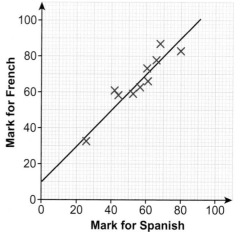

*Active*Learn Homework, practice and support: Foundation 9 Check up

Real-life graphs

7 **Reasoning** The graph shows the cost of hiring two handymen, Fred and Joe. Both men have a fixed charge for callout and then a rate per hour.

a How much does Joe charge for callout?

b What is Fred's hourly rate?

c Estimate the difference in their total charges for work that takes 7 hours.

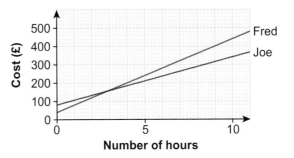

8 How sure are you of your answers? Were you mostly

Just guessing 😞   Feeling doubtful 😐   Confident 😊

What next? Use your results to decide whether to strengthen or extend your learning.

* Challenge

9 Use a coordinate grid to design a puzzle map and a set of clues that allows a treasure hunter to get from point A to point B locating at least 5 items of treasure on the way and avoiding at least 5 hazards.

Here are two examples of a possible first clue.

- Walk 4 units north.
- Walk 4 units along the line $x = -6$.

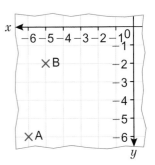

---

9 Strengthen

Algebraic straight-line graphs

1 a Copy and complete this table of values for the function $y = 3x$.

x	0	1	2	3	4
$y = 3x$	0	3			

> **Q1a hint** $y = 3x$ so multiply each x-value by 3 to get the y-value.

b Write down the five pairs of coordinates from your table.

c Draw a grid with x-axis from 0 to 5 and y-axis from 0 to 15. Plot the coordinates on the grid.

d Join the points with a straight line. Extend it to the edge of the grid.

e Label the line $y = 3x$.

2 a Copy and complete this table of values for the function $y = 4x - 3$.

x	0	1	2	3	4
$y = 4x - 3$	-3	1			

> **Q2a hint** Input the x-coordinates to work out the y-coordinates.

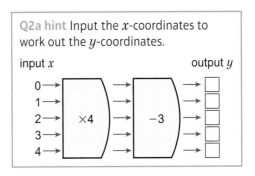

b Draw a grid with x-axis from 0 to 5 and y-axis from -5 to +20. Plot the coordinates on the grid.

c Join the points with a straight line.

d Label the line $y = 4x - 3$.

e What are the coordinates of the point where the graph crosses the y-axis?

> **Q2c hint** Extend the lines to the edges.

3 a Draw a grid with x and y-axes from −5 to +5. Plot four points with x-coordinate 3.
  b Join the points with a straight line. Where does the line cross the x-axis?
  c The equation of the line is $x = 3$. Which of these points lie on the line?
    (3, 4) (0, 3) (−3, 4) (3, 2) (5, 3)
  d Holly says (4, 3) lies on the line. Explain why Holly is wrong.

4 Draw a grid with x and y-axes from −5 to +5.
  Draw these lines.
  a $x = 1$ b $y = 2$ c $x = -4$

  > **Q4b hint** Plot some points with y-coordinate 2. Draw a line through them.

5 The steepness of a graph is called the gradient.
  a A line that goes upwards from left to right has a .... gradient.
  b A line that goes downwards from left to right has a .... gradient.

  > **Q5 hint** Positive gradients slope uphill left to right. /
  > Negative gradients slope downhill left to right. \

6 a Are the gradients of these lines positive or negative?
  b Work out the gradients of lines A, B, C and D by counting how many squares the line goes up or down for every 1 across.
  c What do you notice about lines B and D?

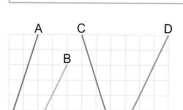

7 These lines have fraction gradients. Decide if each gradient is positive or negative. Work out the gradient of each line.

  > **Q7 hint**
  > $$\text{Gradient} = \frac{\text{squares up}}{\text{squares across}} = \frac{1}{\square}$$

  a   b   c

8 These lines are all parallel.
  a Work out their gradients. What do you notice?
  b Copy and complete this table.

Line	y-intercept
$y = 2x + 3$	
$y = 2x + 1$	
$y = 2x$	
$y = 2x - 2$	
$y = 2x - 4$	

  > **Q8b hint** The y-intercept is the value where the line crosses the y-axis.

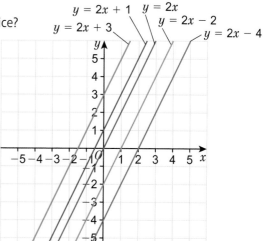

  c Where do you think the line $y = 2x - 5$ will cross the y-axis?

9 a Work out the gradient of these lines.
  b Find the y-intercept.
  c Write the equation of the line.

  > **Q9c hint** $y = \square x + \square$
  > gradient y-intercept

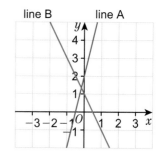

10 Match each equation with a line on the graph.

$y = 4x - 3$ $y = 3x + 2$

$y = -2x + 4$ $y = \frac{1}{2}x - 1$

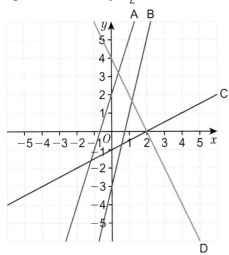

11 a Write down the gradient of each line.

    i $y = 5x + 4$ ii $y = 3x - 5$ iii $y = -\frac{1}{2}x + 1$

    iv $y = 3x - 3$ v $y = \frac{1}{2}x + 2$

   b Which lines are parallel?

> **Q11b hint** Parallel lines have the same gradient but any y-intercept.

12 Write equations of lines parallel to

   a $y = 4x + 3$ b $y = -2x + 3$ c $y = \frac{1}{2}x - 2$

> **Q12 hint** Copy the gradient. Choose any y-intercept.

13 a Draw and label the line segment AB with start point A (1, 3) and end point B (5, 1).

   b Label the midpoint of the line M.

   c Work out the coordinates of M.

   d Check your answer matches the midpoint M on your line.

> **Q13c hint**
>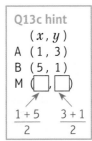

14 Work out the midpoints of these line segments.

   a From C (2, 9) to D (8, 3)   b From E (0, 4) to F (4, 6)

   c From G (−2, 7) to H (4, −3)   d From J (3, 8) to K (2, 5)

> **Q14 hint** Follow the steps in **Q13**. Coordinates can include fractions.

Distance–time graphs and scatter graphs

1 The graph shows Ali's car journey to his friend's house and home again.

   a How can you tell from the graph when Ali is at his friend's house?

   b How long does it take Ali to get to his friend's house?

   c How long does Ali spend at his friend's house?

   d How far from Ali's home does his friend live?

   e How long does it take Ali to get home from his friend's house?

   f Which part of the journey was fastest?

   g Ali left home at 7:30 pm. What time did he get home again?

   h What was his average speed in miles per hour for the first part of his journey?

Ali's journey

> **Q1h hint**
>

2   The graph shows sales of ice cream and temperature on each of 10 days.

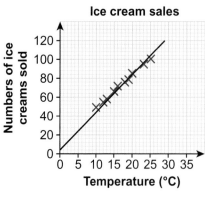

Q2a hint

$$y = \boxed{}x + \boxed{}$$

gradient   y-intercept

a   For the line of best fit
   i   Write down the y-intercept   ii   Work out the gradient   iii   Write the equation.
b   Use your equation to estimate the number of ice creams sold
   when the temperature was 27 °C.

Q2b hint $x = 27°$

Real-life graphs

1   **Reasoning** The graph shows costs to have documents translated. The translators charge a fixed fee, and then a rate per word.

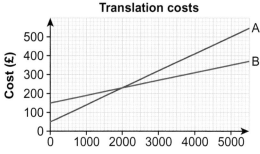

a   Copy and complete this table.
b   Who is cheaper for fewer than 2000 words?
c   Who is cheaper for more than 2000 words?

	Translator A	Translator B
Fixed fee		
Cost for 1000 words		
Rate per word		
Cost for 1500 words		
Cost for 2500 words		

2   Match each container to a statement, then match each container to a graph.

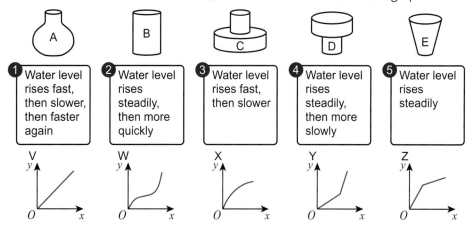

9 Extend

1   **STEM** The diagram shows the distance travelled by a particle (A) over a period of time.

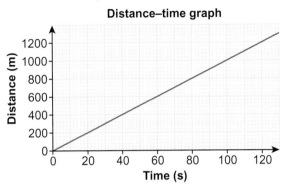

Distance–time graph

> **Q1 hint** The graph is distance (m) against time (s), so its gradient is speed (m/s).

a   Work out the speed of the particle in m/s.

Another particle (B) travelled at a speed of 15 m/s.

b   Which particle moved faster?

2   Here is some information about the gradients of three hills.

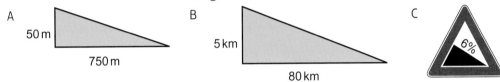

A   50 m   750 m
B   5 km   80 km
C   6%

Which hill is the steepest? You must show working to explain your answer.

3   **Reasoning / STEM** In a science experiment, the total length of a spring is measured when different masses are attached.

The graph shows the results.

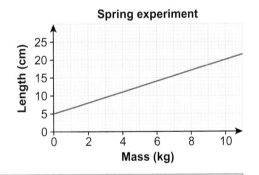

Spring experiment

a   How long is the spring when no mass is attached?

b   What mass gives a spring length of 17 cm?

c   What is the equation of the line?

d   Copy and complete this formula connecting mass (m) and length (l).

$l = \Box m + \Box$

> **Q3d hint** Use the equation of the line to help you.

e   A mass of 12 kg is attached. Predict the length of the spring.

f   How reliable is your answer to part **e**?

g   Write down a practical interpretation of the gradient of the line.

4   **Reasoning** a   Copy and complete the table for the perimeter of this rectangle for different values of x.

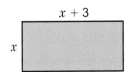

$x + 3$

x

Width in cm (x)	1	2	3	4	5	6
Perimeter in cm (y)	10					

b   Draw a graph for these values of x and y.

c   Work out the equation of the line.

d   What width gives a perimeter of 20 cm?

> **Q4d hint** The width is x.

5 **Real / Modelling** The table shows the total area of ice (in km²) on Mount Kilimanjaro between 1912 and 2011.

Q5 strategy hint
Make a scatter graph to model the data.

Year since 1900	12	62	75	84	93	100	111
Area of ice (km²)	11.4	7.32	6.05	4.82	3.8	2.9	1.76

If this trend continues by what year will all the ice have melted?

Q5 Communication hint The **trend** is the general direction in which something is developing. A line of best fit is used to show the trend.

6 **Problem-solving** The lines $x = 1$, $x = 4$, $y = 3$ and $y = -2$ drawn on a centimetre-squared grid enclose a rectangle. What is the area of the rectangle?

Q6 strategy hint Start by drawing the lines on a coordinate grid.

7 **Problem-solving** The lines $y = -x$, $x = -4$ and $y = 0$ drawn on a centimetre-squared grid enclose a triangle. What is the area of the triangle?

8 a Plot the points (2, 1), (2, 3), (0, 3) and (−2, −1) on a coordinate grid.
   b Join them to make a quadrilateral. What is the name of the quadrilateral?
   c Draw a line of symmetry on the quadrilateral.
   d What is the equation of the line of symmetry?

9 Repeat **Q8** for the points (3, 2), (5, 2), (6, 4) and (2, 4).

10 The velocity of a particle in metres/second starting from rest is given by the formula $v = at$, where v = velocity in m/s, a = acceleration in m/s² and t = time in seconds.
   a Find the value of v when $a = 4$ and
      i $t = 0$ ii $t = 5$
   b Plot these points on a velocity–time graph and join them.
   c A different particle starts at the same time and moves at a constant speed of 12 m/s. Show this on the graph.
   d After how many seconds is the speed of both particles the same?

Q10b hint
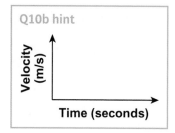

11 A straight line passes through (5, 3) and has a gradient of −2. What is the equation of the line?

12 a Make y the subject of the formula $x + y = 7$.
   b What is the gradient and y-intercept of the line $x + y = 7$?

Q12b hint Rearrange the formula into the form $y = mx + c$

13 The graph shows the movement of a particle over time.

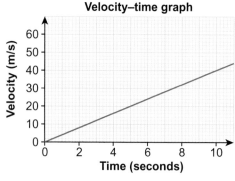

Velocity–time graph

Q13c hint The enclosed area gives the distance travelled. Find the area of the triangle shown.

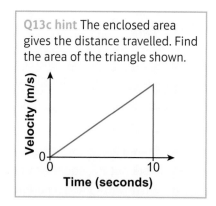

   a What is the equation of the line?
   b What is the speed of the particle after 10 seconds?
   c How far did the particle travel in the first 10 seconds?

14 **Reasoning** The graph shows the population of a town from 1951 to 2011.

a In what year was the population lowest?

b Gary says, 'From 2001 to 2011 the population increased by more than 14%.'

   Is Gary correct?

   You must show your working.

> **Q15b hint** Increase the 2001 population by 14% to check Gary's statement.

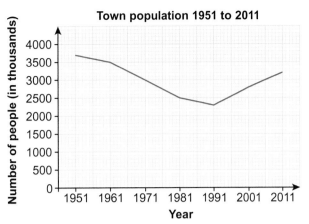

Town population 1951 to 2011

15 **Exam-style question**

Simon went for a cycle ride.
He left home at 2 pm.
The travel graph represents part of Simon's cycle ride.
At 3 pm Simon stopped for a rest.

a How many minutes did he rest? **(1 mark)**

b How far was Simon from home at 5 pm? **(1 mark)**

At 5 pm Simon stopped for 30 minutes.
Then he cycled home at a steady speed.
It took him 1 hour 30 minutes to get home.

c Complete the travel graph. **(2 marks)**

March 2013, Q21, 1MA0/2F

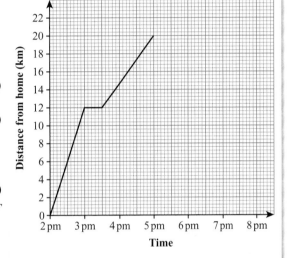

Exam hint

He should end 0 km from home.

16 **Exam-style question**

On the grid, draw the graph of $y = \frac{1}{2}x + 5$ for values of x from -2 to 4. **(3 marks)**

June 2013, Q29, 1MA0/1F

> **Q16 strategy hint** Use the y-intercept and gradient to plot the graph.

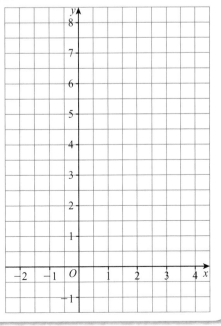

9 Knowledge check

⊙ The equation $y = 1$ means the y-coordinate is always 1.
The line is parallel to the x-axis.
The equation $x = 3$ means the x-coordinate is always 3.
The line is parallel to the y-axis. .... *Mastery lesson 9.1*

⊙ The **midpoint** of a line segment is the point exactly in the middle.
To find the coordinates of the midpoint, add the x-coordinates of the
end points and divide by 2, and add the y-coordinates of the endpoints
and divide by 2. .... *Mastery lesson 9.1*

⊙ The steepness of a graph is called the **gradient**.

You can find the gradient using: $\dfrac{\text{units up or down}}{\text{units across}}$

Gradients can be positive (/) or (\) negative. .... *Mastery lesson 9.3*

⊙ To find the gradient work out
how many units the graph
goes up or down for each
unit it goes across.

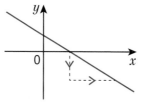

.... *Mastery lesson 9.3*

⊙ **Parallel lines** have the same gradient. .... *Mastery lesson 9.3*

⊙ A linear equation produces a straight-line graph. The equation of a
straight line is

$$y = mx + c$$

gradient y-intercept .... *Mastery lesson 9.4*

⊙ You can use the **y-intercept** and gradient to write the equation of a line
of best fit and other real-life graphs, and use this equation to
make predictions. .... *Mastery lesson 9.7*

⊙ A distance–time graph represents a journey. The vertical axis represents
the distance from the starting point. The horizontal axis represents the
time taken. A horizontal line on a distance-time graph represents an
object at rest. .... *Mastery lesson 9.6*

⊙ Average speed $= \dfrac{\text{distance travelled}}{\text{time taken}}$ .... *Mastery lesson 9.6*

⊙ On a distance–time graph, the gradient represents the speed of
the journey. .... *Mastery lesson 9.6*

⊙ A rate of change graph shows how a quantity changes over time. .... *Mastery lesson 9.7*

⊙ On a velocity–time graph the gradient represents the acceleration. .... *Mastery lesson 9.6*

In this unit, which was easier, 'plotting and drawing graphs' or 'reading and interpreting graphs'?
Copy and complete this sentence to explain why:
I find _____ graphs easier, because _____

Reflect

9 Unit test

1   a   What is the equation of line A?
    b   Draw a coordinate grid from
        −5 to +5 on both axes. On this grid,
        draw and label the line with
        equation $y = 1$.
    c   On your grid from part **b**, draw and
        label the line with equation $y = x$.

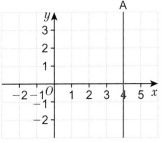

(3 marks)

2   A has coordinates (0, 1). B has
    coordinates (4, 5).
    Work out the midpoint of the line AB.

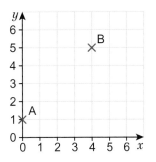

(2 marks)

3   a   Copy and complete the table of values for $y = 4x − 1$.
    b   Draw a grid from −1 to 3 on the x-axis and from
        −6 to 18 on the y-axis.
        Plot the graph of $y = 4x − 1$.

x	−1	0	1	2	3
y		−1			11

(4 marks)

4   Use the graph of $y = 4x − 1$ you drew in **Q3**. Estimate the value of
    a y when $x = 2.4$ b x when $y = 1.6$ *(4 marks)*

5   **Problem-solving** From the list of equations,
    write down the equations of
    a   the steepest line
    b   a pair of parallel lines
    c   a line that slopes downwards from left to right
    d   a line with a negative gradient
    e   a line that crosses the y-axis at 2.   *(5 marks)*

$$y = 2x + 3 \qquad y = 5x + 3$$
$$y = −4x + 6 \qquad y = 3x + 1$$
$$y = 2x − 5 \qquad y = 0.5x + 2$$

6   Find the equation of each line.

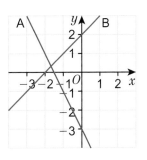

(4 marks)

7   Find the equation of the straight line passing through the points A (−1, −3)
    and B (3, 5).   *(3 marks)*

8   Draw a coordinate grid with x and y-axes from −6 to +6. Plot the graph of $2y + x = 6$.
                                                           (3 marks)

9 ⬭ **Exam-style question**

The graph shows Paul's journey from his home to town.

a At what time did Paul leave his home?

b The distance from Paul's home to town is 1.5 km. What was Paul's speed on the journey to town?

c For how long did Paul stay in town?

d Describe fully Paul's return journey home.

e For which part of the return journey was Paul travelling the fastest? You must give a reason for your answer. **(8 marks)**

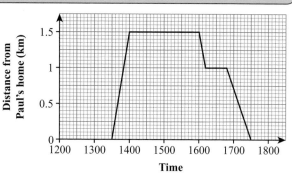

10 STEM The graph shows the relationship between the speed of sound and the height above sea level.

a What is the speed of sound at 5000 m above sea level?

b Copy and complete the statement.

As height above sea level increases, the speed of sound _____.

c Use the graph to estimate the rate of change of speed with height above sea level.

(4 marks)

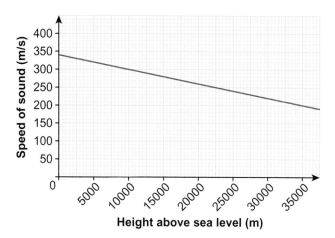

Sample student answer

This student did not get full marks.

a What did they forget to do?

b Can you find a different way to answer the question?

⬭ **Exam-style question**

On the grid, draw the graph of $y = 2x - 1$ for values of x from -2 to $+3$. **(3 marks)**

June 2013, Q18, 5MB/2F

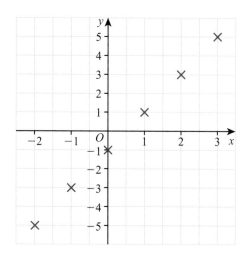

Student answer

x $-2, -1, 0, 1, 2, 3$

y $-5, -3, -1, 1, 3, 5$

10 TRANSFORMATIONS

Tangrams are puzzles consisting of seven shapes. The objective of the puzzle is to form a specific shape using all seven pieces.

Can you make each black shape by reflecting, rotating and translating the seven coloured shapes?

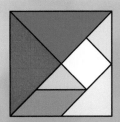

10 Prior knowledge check

Numerical fluency

1 Work out
   a 6×0.5 b $4 \times 1\frac{1}{2}$
   c $7 \times \frac{1}{2}$ d 8×1.5

2 Work out
   a   80% of 30 cm   b   120% of 40 cm

Geometrical fluency

3 Copy and complete each diagram so that the final pattern is symmetrical about the given line of symmetry.

  a 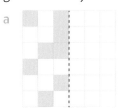   b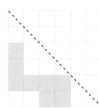

4 Copy each diagram. Draw the reflection of the shape in the red mirror line.

  a 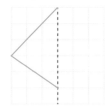   b

5 State if each reflection is correct or incorrect.

  a   b
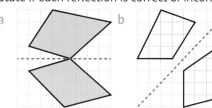

6 Copy each diagram. Move the shape as described.

  a   b

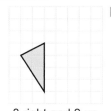 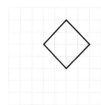

   3 right and 2 up   3 left and 3 down

7 Shape A has been rotated onto shape B. State the size and direction of turn. The first one has been started for you.

  a   b   c

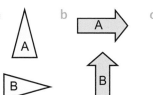

  a $\frac{1}{4}$ turn _____

8   Copy each diagram.
    Rotate the shape as described.

a
    Rotate 90° clockwise

b
    Rotate 180°

c
    Rotate 90° clockwise

d
    Rotate 90° anticlockwise

Graphical fluency

9   Draw a coordinate grid with axes labelled from −8 to 8.
    On the grid draw and label these lines.
    a $x = -2$ b $y = 7$
    c $y = x$ d $y = -x$

10 Plot and label these points on your grid from **Q9**.
    a   A(0, 2)   b   B(−6, 4)
    c   C(5, −2)   d   D(−3, −1)

★ Challenge

11 Copy the diagram.
    a   Draw a reflection of the shape in the red mirror lines to make a shape with four lines of symmetry.
    b   Create your own design using a different shape.

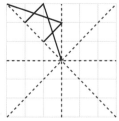

10.1 Translation

Objectives

- Translate a shape on a coordinate grid.
- Use a column vector to describe a translation.

Why learn this?

Robots and machines use translation to move components from place to place.

Fluency

What are the four types of transformation?
Which arrow points right and which points left?

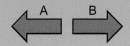

1   Find the missing letters.
    a   A is translated 5 right, 2 up to ☐.
    b   A is translated 3 right, 2 down to ☐.
    c   B is translated 2 left, 4 down to ☐.
    d   C is translated 3 left, 2 up to ☐.

2   Describe how to move between each pair of points.
    a   (1, 2) and (3, 6)
    b   (0, 5) and (2, 3)
    c   (−2, 3) and (5, 8)

    Discussion Do you have to draw a grid? How else can you work out the answers?

Q2a strategy hint
Draw a grid.

Active Learn   Homework, practice and support: Foundation 10.1

Warm up

3 Describe how to move from

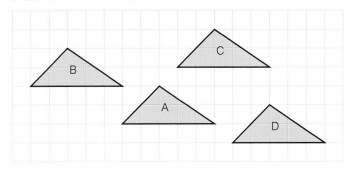

Q3a hint ☐ right, ☐ up

a shape A to shape C   b shape A to shape B   c shape A to shape D.

Questions in this unit are targeted at the steps indicated.

Key point 1

You can use a **column vector** to describe a translation.
The top number describes the movement to the left or right, and the bottom number describes the movement up or down. For example:

$\begin{pmatrix} 3 \\ 2 \end{pmatrix}$ means 3 right, 2 up $\begin{pmatrix} -4 \\ -5 \end{pmatrix}$ means 4 left, 5 down.

Example 1

Translate shape P by the column vector $\begin{pmatrix} 7 \\ -1 \end{pmatrix}$.

$\begin{pmatrix} 7 \\ -1 \end{pmatrix}$ means 7 right, 1 down.

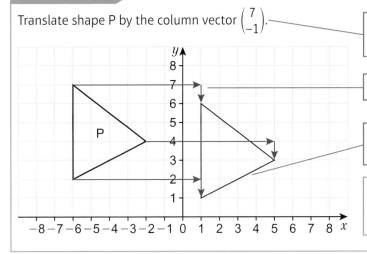

Translate each vertex separately.

Join up the new vertices to make the new shape.

Communication hint A **vertex** is a corner. The plural of vertex is **vertices**.

4 Copy the diagram.

a Translate shape A by $\begin{pmatrix} 2 \\ 3 \end{pmatrix}$.

b Translate shape B by $\begin{pmatrix} 4 \\ -3 \end{pmatrix}$.

c Translate shape C by $\begin{pmatrix} -1 \\ 2 \end{pmatrix}$.

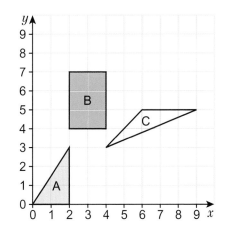

Discussion Here is Austin's answer to **Q4a**.
What did he do wrong?

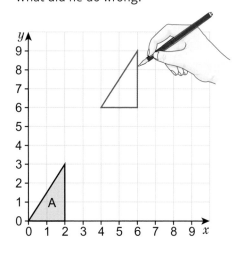

5   Copy the grid.

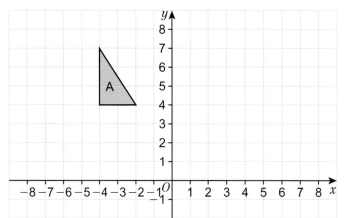

**Q5 communication
hint** A **transformation**
transforms a shape to a
different position.

Q5 communication hint
The **image** is the shape
after a transformation.

Translate shape A by each of these column vectors.

a $\begin{pmatrix} 2 \\ 1 \end{pmatrix}$ Label the image B.   b $\begin{pmatrix} -3 \\ -2 \end{pmatrix}$ Label the image C.   c $\begin{pmatrix} 0 \\ -5 \end{pmatrix}$ Label the image D.

d $\begin{pmatrix} 7 \\ -3 \end{pmatrix}$ Label the image E.   e $\begin{pmatrix} 5 \\ 0 \end{pmatrix}$ Label the image F.

6   Write the column vector that translates shape A to shape B.

a 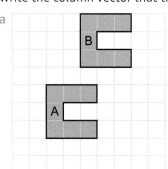   b 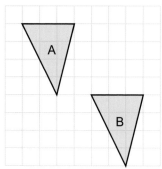   c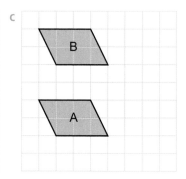

7 Write the column vector
that maps shape

a   A onto B   b   A onto C

c   A onto D   d   A onto E

e   B onto D   f   C onto D

g   D onto C.

> **Q7 communication hint**
> The original shape **maps**
> onto the image.

Discussion What do you
notice about the column
vectors that translate C onto D
and D onto C?

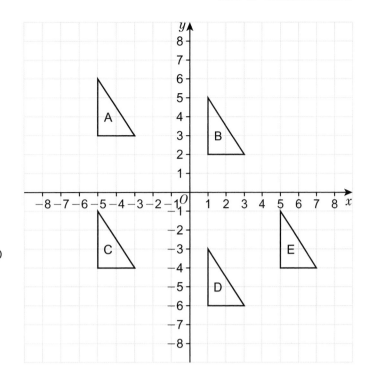

8

Exam-style question

Describe fully the single transformation that maps triangle P onto triangle Q.

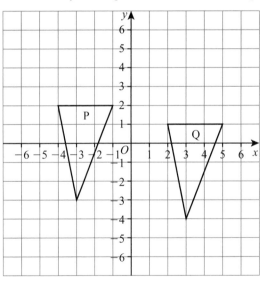

(2 marks)

March 2013, Q10, 1MA0/2H

Exam hint

The first
mark is for
the type of
transformation;
the second
mark is for the
vector.
'Describe fully'
means write
the type of
transformation
(translation)
and the vector.

9 Copy the grid and shape A from **Q7**.

a   Translate shape A by the column vector $\begin{pmatrix} 1 \\ -8 \end{pmatrix}$. Label the image P.

b   Translate shape P by the column vector $\begin{pmatrix} 6 \\ 1 \end{pmatrix}$. Label the image Q.

c   Describe fully the transformation that maps A onto Q.

Discussion When a shape is translated twice how can you find the final vector without
drawing a diagram?

10 **Reflect** Sian draws this diagram to help her remember how to use a column vector.

$\binom{3}{4}\overset{\rightarrow}{\underset{\uparrow}{}}$ $\binom{-2}{-5}\overset{\leftarrow}{\underset{\downarrow}{}}$

a Do you think it is a good diagram? Explain.

b What other diagram would you draw?

> **Q10a hint** Write:
> Yes, this is a good diagram because _____
> No, this is not a good diagram because _____

10.2 Reflection

Objectives

- Draw a reflection of a shape in a mirror line.
- Draw reflections on a coordinate grid.
- Describe reflections on a coordinate grid.

Why learn this?

Reflection is used by scientists and engineers to measure distance. Radar, telescopes and X-ray machines all use reflection to construct maps and pictures.

Fluency

What does perpendicular mean?

1 Which reflection is correct? Explain why.

A 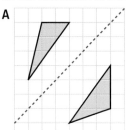 B 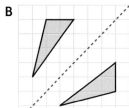 C 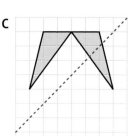 D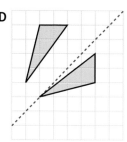

2 Copy each diagram. Draw the reflection of the shape in the mirror line.

a b c

3 Copy the diagram.
   a Reflect shape A in the x-axis.
     Label the image B.
   b Reflect shape A in the y-axis.
     Label the image C.
   c Reflect shape B in the y-axis.
     Label the image D.

   Discussion David says, 'If I reflect a point in the x-axis, its x-coordinate stays the same.
   If I reflect a point in the y-axis, its y-coordinate stays the same.' Is he correct?

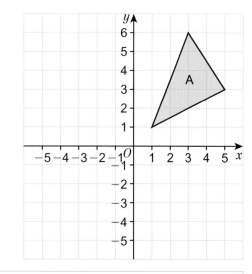

*Active*Learn Homework, practice and support: Foundation 10.2

4 Copy the diagram from **Q3**.

   a Draw the line $x = 1$.

   b Reflect shape A in the line $x = 1$. Label the image X.

   c Draw the line $y = -1$.

   d Reflect shape A in the line $y = -1$. Label the image Y.

> **Q4d hint** You may need to extend the y-axis to draw the image.

5 **Reasoning**

Copy the diagram.

   a Reflect shape P in the line $y = 2$. Label the image Q.

> **Q5a strategy hint**
> Draw the line $y = 2$ first.

   b Reflect shape Q in the line $x = -1$. Label the image R.

   c Shona starts with shape P and reflects it first in $x = -1$ and then in $y = 2$. Does she get the same final image?

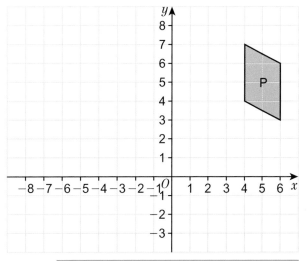

6 Copy the diagram from **Q5**.
Reflect shape P in the line $y = 4$.

> **Q6 hint**
> The new vertices are on the opposite side of the mirror line. For example:

7 Copy the diagram.

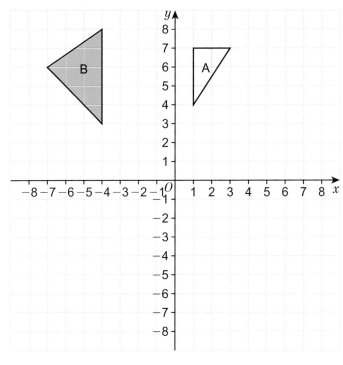

> **Q7 strategy hint** Count the perpendicular distance of each vertex from the mirror line, then count the same again the other side of the line.
>
>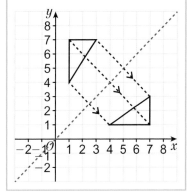

   a Reflect shape A in the line $y = x$. Label the image A'.

   b Reflect shape B in the line $y = -x$. Label the image B'.

8 ┌─── **Exam-style question** ───────────────────────

Reflect shape P in the line $y = x$.

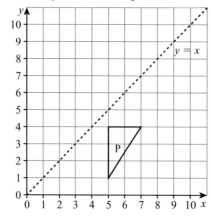

(2 marks)
Nov 2012, Q2a, 1MA0/2H

Exam hint
Turn the page so that the given mirror line is vertical. Reflect each vertex. Join them with a ruler.

9 **Reflect** Write down three steps for reflecting a shape.

╔═══ **Key point 2** ═══╗

To describe a reflection on a coordinate grid you need to give the equation of the **mirror line**.

╔═══ **Example 2** ═══╗

Describe fully the transformation that maps shape A onto shape B.

┌──────────────────────────────────┐
│ Find the mirror line halfway   │
│ between the vertices of the   │
│ image (B) and the original (A). │
└──────────────────────────────────┘

┌──────────────────────────────────┐
│ Write down the type of   │
│ transformation (reflection) and │
│ the equation of the mirror line. │
└──────────────────────────────────┘

Reflection in the line $y = 2$.

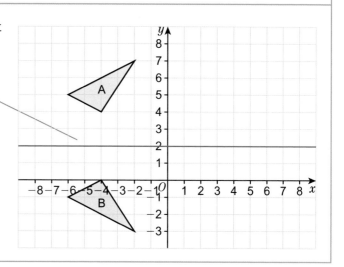

10 **Reasoning** Describe fully the transformation that maps shape
  a   A onto B
  b   C onto F
  c   D onto F
  d   E onto F.

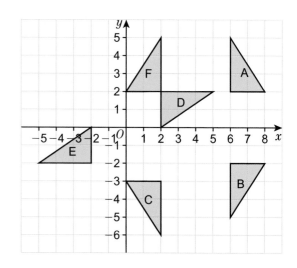

11 **Reasoning** Describe fully the transformation that maps shape A onto shape B.

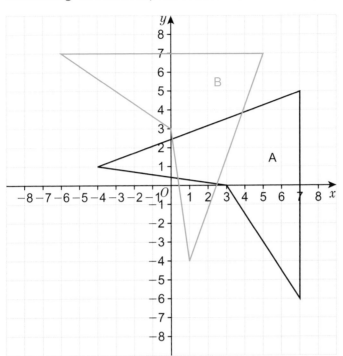

12 **Reflect** What two pieces of information are needed to describe a transformation that is a reflection?

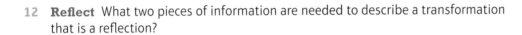

10.3 **Rotation**

Objectives

- Rotate a shape on a coordinate grid.
- Describe a rotation.

Why learn this?

Computer game programmers and film animators all use rotation to make 3D animations more realistic.

Fluency

How many degrees are there in
- a full turn
- a half turn
- three quarters of a turn?

1 For each turn, write the number of degrees and whether the direction is clockwise or anticlockwise.

a
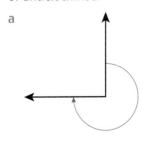

b

c
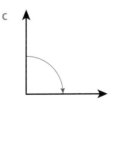

Warm up

Key point 3

You rotate a shape by turning it around a point called the **centre of rotation**.

Example 3

Rotate the shape 90° anticlockwise about the point (1, 2).

(1, 2) is the centre of rotation.

Mark the point (1, 2) with a cross.

Trace the shape.

Rotate the tracing paper 90° anticlockwise about (1, 2).

Lift up the tracing paper and draw the image on the grid.

2 Copy each diagram. Use the centre of rotation to draw the image of each shape after the rotation given.

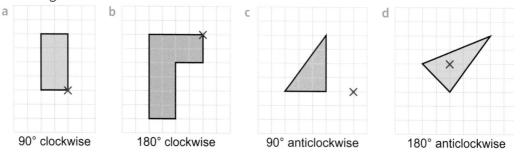

a   90° clockwise   b   180° clockwise   c   90° anticlockwise   d   180° anticlockwise

3 Copy the grid and shape A.
Rotate shape A 90° clockwise
about each centre of rotation.
a (0, 0)   b (2, 0)
c (0, −5)   d (−3, −3)
Discussion How does changing
the centre of rotation affect
the image?

4 Copy the diagram and shapes B
and C from **Q3**.
a Rotate shape B 90°
anticlockwise about (−2, 0).
Label the image B′.
b Rotate shape C 180°
clockwise about (0, 1).
Label the image C′.
Discussion In part **c** what
happens if you rotate shape C
180° anticlockwise about (0, 1)?

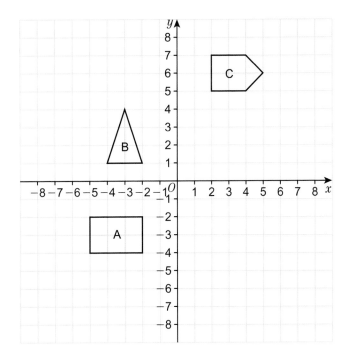

5 **Reasoning** Shape P can be
rotated onto shape Q.
a What is the direction and
what is the angle of rotation?

Q5a strategy hint Trace shape P.
Rotate the tracing paper until the
shape is the same way round as
image Q.

b Where is the centre of rotation?

Q5b strategy hint Try holding
the tracing paper at different
centres of rotation.

c Describe fully the
transformation that maps
shape P onto shape Q.

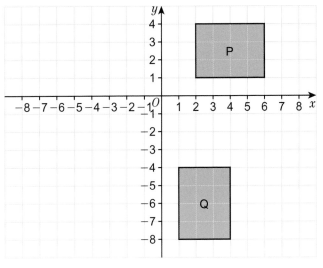

Q5c hint 'Describe fully' means you need
to state (1) the type of transformation
(rotation), (2) the angle and direction of
rotation, and (3) the centre of rotation.

6 **Reasoning** Describe fully the transformation
that maps shape
a A onto B
b A onto C
c A onto D
d B onto C.

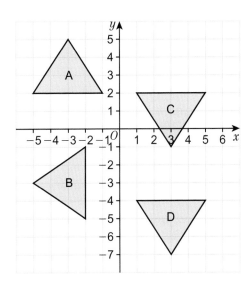

7   **Exam-style question**

Describe fully the single transformation that maps triangle A
onto triangle B.

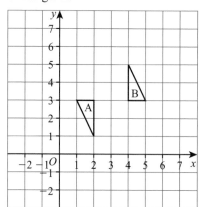

(3 marks)

June 2012, Q9, 1MA0/1H

Exam hint

For 3 marks you must
include (1) the *one* type of
transformation, (2) the angle
and direction and (3) the
coordinates of the centre
of rotation.

10.4 Enlargement

Objectives

*   Enlarge a shape by a scale factor.
*   Enlarge a shape using a centre of
    enlargement.

Why learn this?

Architects, graphic designers and engineers all
use enlargement to produce scale drawings of
buildings, logos and machines.

Fluency

What are the missing numbers?
Height of B = __ × height of A
Base of B = __ × base of A

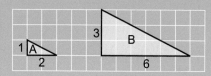

1   Work out the area of triangle A and the area of triangle B in the fluency question.

2   Work out
   a $4 \times 1\frac{1}{2}$ b $6 \times \frac{1}{2}$ c $8 \times 2\frac{1}{2}$ d 6×1.5 e 4×0.5 f 10×2.5

Key point 4

To enlarge a shape you multiply all the side lengths by the same number.
The number that the side lengths are multiplied by is called the **scale factor**.

3   Copy the diagrams. Enlarge each shape by the scale factor.

a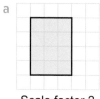

Scale factor 2

b

Scale factor 3

c

Scale factor $1\frac{1}{2}$

d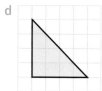

Scale factor 0.5

Discussion Does enlargement always make a
shape bigger?
What happens when the scale factor is greater
than 1, less than 1, equal to 1?

Q3d hint Triangle height is 4.
Enlargement height is $4 \times 0.5 =$ __

*Active*Learn   Homework, practice and support: Foundation 10.4

Key point 5

When you enlarge a shape using a **centre of enlargement**, you multiply the distance from the centre to each vertex by the scale factor.

Example 4

Enlarge shape A by scale factor 2, using centre of enlargement (1, 3). Label the image B.

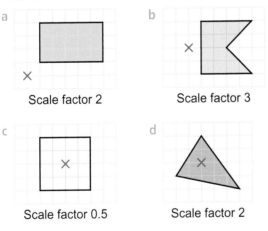

Count the squares from the centre of enlargement to each vertex. Multiply all the distances from the centre by the scale factor.

The distance to the top vertex changes from 4 up, 2 right to 8 up, 4 right.

Mark the centre of enlargement.

The distance to the bottom vertex changes from 2 right to 4 right.

Check that the lengths of the image are twice as long as the original.

4   Copy each diagram. Enlarge each shape by the scale factor from the centre of enlargement.

a

Scale factor 2

b

Scale factor 3

c

Scale factor 0.5

d

Scale factor 2

Q4 hint To check your answer, draw lines from the centre of enlargement through the vertices on the original shape and across the grid. These lines should go through the vertices of the image.

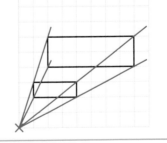

5   Copy the diagram.
    Enlarge shape A by scale factor 2 from the given centre of enlargement.
    a   (1, 2)   b   (0, 3)

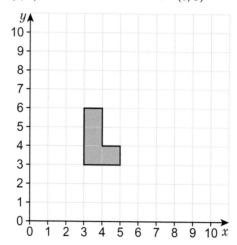

    Discussion How does changing the centre of enlargement affect the image?

6   **Reflect** You can check an enlargement by:
    • drawing lines from the centre (see **Q4** hint)
    • checking all the lengths have been multiplied by the scale factor (see Example 4).
    Which method do you prefer? Why?

7   Copy the diagram.
    Enlarge each shape by the scale factor from the centre of enlargement.

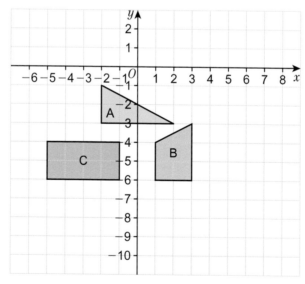

    a   Shape A   Scale factor 2, centre (−4, 2)
    b   Shape B   Scale factor 3, centre (4, −4)
    c   Shape C   Scale factor $1\frac{1}{2}$, centre (−5, −4)

8   **Real** Raj enlarges this shape on a photocopier.
    a   He enlarges it by 120%.
        Work out the length and width of this enlargement.
    b   He enlarges the original shape by 80%.
        Work out the length and width of this enlargement.

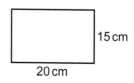

10.5 **Describing enlargements**

Objectives

- Identify the scale factor of an enlargement.
- Find the centre of enlargement.
- Describe an enlargement.

Did you know?

A scale model is an enlargement of the real object by a fractional scale factor.

Fluency

Which shape is not an enlargement of A?

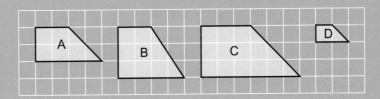

1   Simplify

   a $\frac{10}{5}$
   b $\frac{3}{6}$
   c $\frac{12}{8}$
   d $\frac{18}{6}$
   e $\frac{2}{8}$

Example 5

Write the scale factor of this enlargement.

The bottom of the image is 12 squares long.

The bottom of the object is 4 squares long.

Scale factor $= \frac{12}{4} = 3$

Write $\dfrac{\text{length on image}}{\text{length on object}}$ and simplify.

Choose the same side on the image and object and count the number of squares.

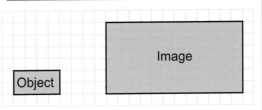

2   Shape B is an enlargement of shape A.
    Write the scale factor of enlargement for each pair of shapes.

   a   b

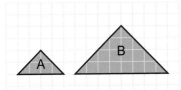   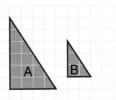

3   **Reasoning** Triangle ABC is enlarged to triangle DEF.

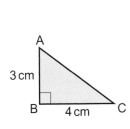   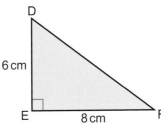

   a   What is the scale factor of the enlargement?
   b   AC = 5 cm. Work out the length of DF.

4   Shape B is an enlargement of shape A.

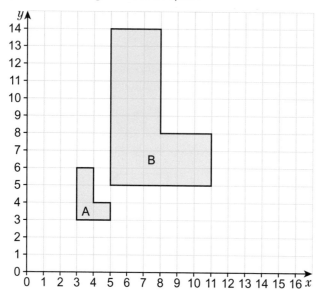

Q4b strategy hint
Draw lines through each vertex on the image and the equivalent vertex on the original. All the lines should meet at the **centre of enlargement**.

Q4c hint 'Describe fully' means write down the scale factor *and* the coordinates of the centre of enlargement.

a   What is the scale factor of the enlargement?
b   What is the centre of enlargement?
c   Describe fully the enlargement that maps shape A onto shape B.

5   **Reasoning** Describe fully the transformation that maps shape A onto shape B.

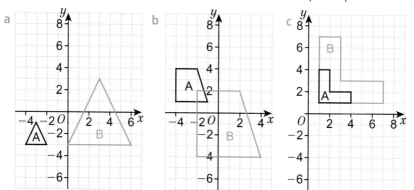

Q5 hint Write:
enlargement,
scale factor ☐,
centre (☐,☐)

6   **Exam-style question**

Describe fully the single transformation that maps shape P onto shape Q.

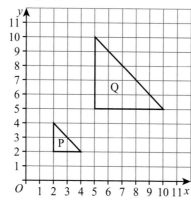

(3 marks)

Nov 2012, Q6, 1MA0/1H

7 **Reasoning** For each diagram
   i describe fully the transformation that maps shape A onto shape B
   ii describe fully the transformation that maps shape B onto shape A.

a   b

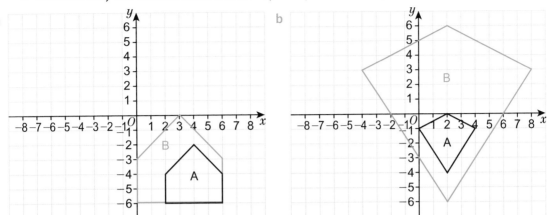

Discussion A shape has been enlarged by scale factor 4. What scale factor returns the shape to its original size?

10.6 **Combining transformations**

Objectives

- Transform shapes using more than one transformation.
- Describe combined transformations of shapes on a grid.

Why learn this?

Artists, architects and designers all use a variety of transformations to create inspirational designs.

Fluency

Match each type of transformation to the information needed to describe it.

Translation	Equation of a line
Reflection	Column vector
Rotation	Scale factor and centre
Enlargement	Angle, direction and centre

1 Which type of transformation has been used to map shape A onto shape B?

a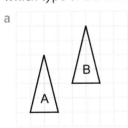

b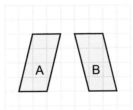

c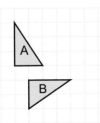

d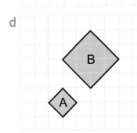

Warm up

2 **Reflect** Adam writes:

Transformations all transform shapes.

Translations <u>s</u>lide shapes left, right, up, down.

a Do you think this is a good way to remember translations? Explain.

b How can you remember the other transformations?

3 What information is missing from each description?

a Rotate shape A 90°, centre of rotation (1, 2).　　b Enlarge shape B by scale factor 3.

c Translate shape C 4 across.　　d Reflect shape D in the mirror line.

Example 6

Triangle A is reflected in the line $y = 1$ to give triangle B.

Triangle B is reflected in the line $x = -1$ to give triangle C.

Describe fully the single transformation that maps triangle A onto triangle C.

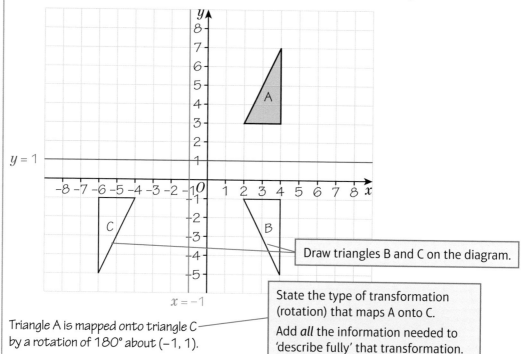

Draw triangles B and C on the diagram.

State the type of transformation (rotation) that maps A onto C.

Triangle A is mapped onto triangle C by a rotation of 180° about (−1, 1).

Add *all* the information needed to 'describe fully' that transformation.

---

4 Copy the grid and shape A.

a Reflect shape A in the y-axis.
Label the image B.

b Reflect shape B in the line $y = x$.
Label the image C.

c Describe fully the single transformation that maps shape A onto shape C.

5 Copy the grid and shape D from **Q4**.

a Reflect shape D in the line $x = 2$.
Label the image E.

b Reflect shape E in the line $x = -1$.
Label the image F.

c Describe fully the single transformation that maps shape D onto shape F.

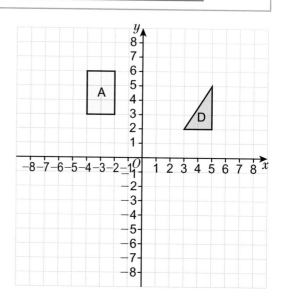

6   Copy the diagram.

   a   Rotate shape A 180° about the origin.
       Label the image B.

   b   Reflect shape B in the x-axis.
       Label the image C.

   c   Describe fully the single transformation
       that maps shape A onto shape C.

   > **Q6a communication hint**
   > The **origin** is the point (0, 0).

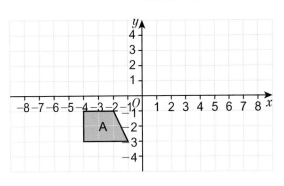

7   Copy the diagram.

   a   Enlarge shape A by scale factor 2,
       centre (7, 8). Label the image B.

   b   Enlarge shape B by scale factor 0.5,
       centre (5, 0). Label the image C.

   c   Describe fully the single transformation
       that maps shape A onto shape C.

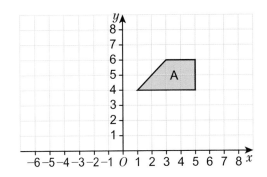

8   **Exam-style question**

   Shape P is reflected in the line $x = -1$ to give shape Q. Shape
   Q is reflected in the line $y = 0$ to give shape R. Describe fully
   the single transformation that maps shape P onto shape R.

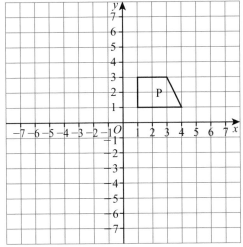

   (3 marks)

   March 2013, Q18, 1MA0/1H

Exam hint
Draw the reflections on
the grid. $y = 0$ is the
x-axis. Remember to give
only *one* transformation.

9   **Problem-solving** Which single transformation
    is the same as

   a   reflecting in two parallel lines

   b   reflecting in two perpendicular lines

   c   two translations?

> **Q9a hint** Look back at **Q5**.

> **Q9b hint** Look back at Example 6.

> **Q9c strategy hint** Draw a shape
> and translate it twice.

10 Problem-solving: Exploding shapes

Objectives
- Be able to transform shapes using the four different transformations.
- Be able to fully describe transformations.

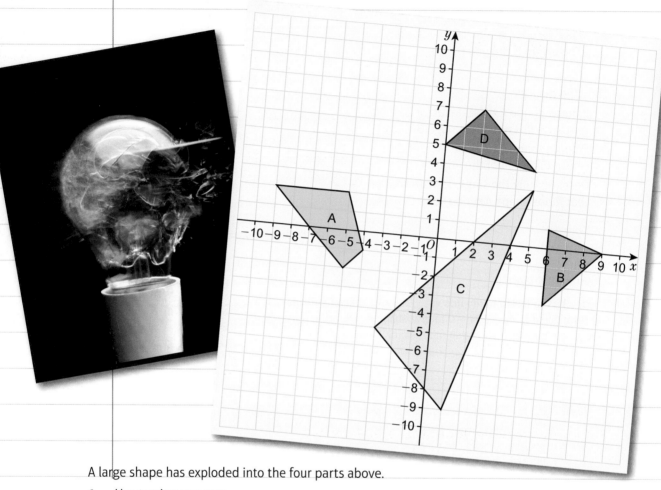

A large shape has exploded into the four parts above.

1   Use tracing paper to:
   a   reflect shape A in the line $x = -2$
   b   rotate shape B 90° anticlockwise, centre (3, −1)
   c   enlarge shape C, scale factor $\frac{1}{2}$, centre (−1, 7)
   d   translate shape D by the column vector $\begin{pmatrix} -4 \\ -6 \end{pmatrix}$.

2   The transformations in **Q1** have recreated the original shape before it exploded. Lift your tracing paper off the diagram to see it. What is the name of this shape?

3   Draw your own shape (or one of your own initials) onto coordinate axes. Explode the shape into four parts in a similar way to the example. Describe the transformations which would re-create your original shape. Try to include all four different types of transformation.

   Reflect What did you have to think about when exploding your shape? Make a list. Compare your list with others in your class.

> **Q3 hint** Drawing your shape or initial using straight lines will make this activity easier to complete.

> **Q3 hint** Include all the information for each transformation.
> For example, for a rotation you need to include the angle, direction and centre of rotation.

10 Check up

Translations and reflections

1   Copy the grid and shape A.
    a   Reflect shape A in the x-axis.
        Label the image B.
    b   Reflect shape A in the
        line $x = 2$.
        Label the image C.
    c   Reflect shape A in the
        line $y = x$.
        Label the image D.
    d   Translate shape A by the
        column vector $\begin{pmatrix} -5 \\ -1 \end{pmatrix}$.

        Label the image E.

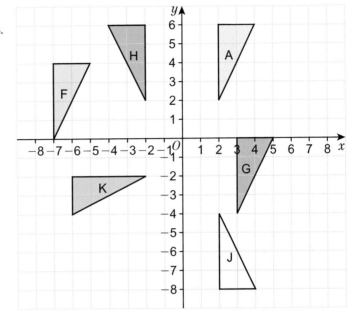

2   Look at the grid and shapes in **Q1**. Write down the column vector that translates
    a   A to F   b   A to G   c   F to G.

3   Look at the grid and shapes in **Q1**. Describe fully the reflection that maps
    a   A onto H   b   A onto J   c   A onto K.

Enlargements and rotations

4   Copy the diagram. Enlarge the shape by scale factor 3.

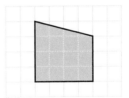

5   Copy the grid and shape A.
    a   Rotate shape A 90°
        clockwise about (0, 0).
    b   Rotate shape A 90°
        anticlockwise about (−2, −2).

6   **Reasoning** Look at the grid
    and shapes in **Q5**. Describe
    fully the transformation
    that maps
    a   A onto B
    b   A onto C.

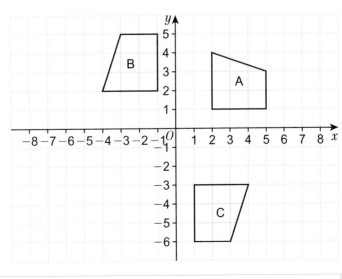

7   Copy the diagram.
   a   Enlarge shape A by scale factor 2, centre of enlargement (1, 1).
   b   Enlarge shape A by scale factor $\frac{1}{2}$, centre of enlargement (3, 3).

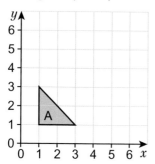

8   **Reasoning** Describe fully the transformation that maps
   a   A onto B   b   B onto C.

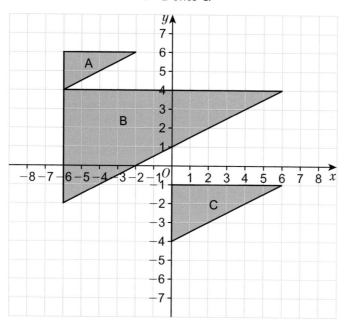

9   How sure are you of your answers? Were you mostly

   Just guessing 😞   Feeling doubtful 😐   Confident 🙂

   What next? Use your results to decide whether to strengthen or extend your learning.

✱ Challenge

10 Design your own combined transformations GCSE exam question.
   a   Draw a coordinate grid with axes labelled from −8 to 8.
   b   Draw a shape and label it A.
   c   Write the first transformation and carry it out on the grid.
       Label the image B.
   d   Write the second transformation and carry it out on the grid.
       Label the image C.
   e   If there is a single transformation that maps shape A onto shape C, write it down.
   f   Repeat the steps using different types of transformations.

10 Strengthen

Translations and reflections

1 Copy and complete these descriptions of how the yellow shapes translate to the red shapes.

a 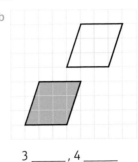   b

3 _____ , 1 _____   3 _____ , 4 _____

2 Write each of these as a column vector.

a   3 right, 2 up   b   2 left, 1 down
c   4 right, 3 down   d   6 left, 5 up

3 Copy the diagram.

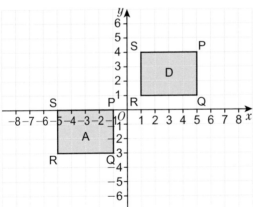

a   Translate shape A by the column vector $\begin{pmatrix} 6 \\ -4 \end{pmatrix}$.
    Label the new shape B.

b   Translate shape A by the column vector $\begin{pmatrix} -2 \\ 5 \end{pmatrix}$.
    Label the image C.

c   Copy and complete this table.

d   Write the column vector that translates shape A to shape D.

Discussion Do you need to find the coordinates of all the vertices (corners) of shapes A and D to work out the column vector in part **d**?

Coordinates of vertex	P	Q	R	S
Shape A	(–1, 0)			
Shape D	(5, 4)			

4   Copy each diagram.
     Draw the reflection of points A, B and C in the red mirror line. Label the new points A', B' and C'.

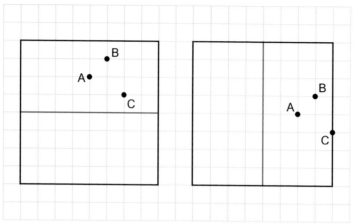

5   Copy the diagram.

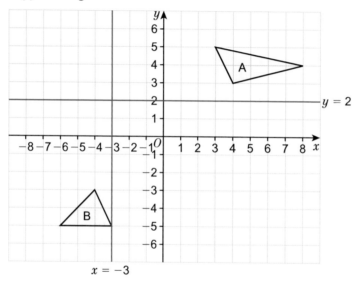

$x = -3$

a   Reflect shape A in the line $y = 2$. Label the image A'.
b   Reflect shape B in the line $x = -3$. Label the image B'.

Q5a hint $y = 2$ is the
mirror line. Reflect points
A, B, C in this line. Join
them to draw the image.

6   Copy the diagram.

Q6 communication hint
A **transformation** moves
a shape so that it is in a
different position.

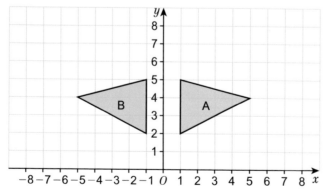

a   Draw in the mirror line for the reflection that maps A onto B.
     Write the name of the line.
b   Copy and complete the description of the transformation that maps A onto B:
     Reflection in the _____

7   Copy the diagram. Reflect points A, B and C in the line $y = x$.
    Label the new points A', B' and C'.

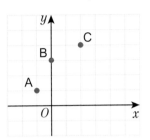

Q7 hint For a diagonal line of symmetry
hold your ruler perpendicular to the mirror
line and count the diagonal squares.

Enlargements and rotations

1   Copy each diagram.
    Rotate each shape by the turn given.

a   b   c

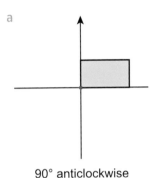

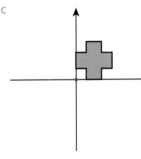

90° anticlockwise   180° clockwise   270° anticlockwise

Q1a hint Trace the shape and the arrow.
Hold the tracing paper on the dot with your pencil.
Turn the traced shape.
Stop when the arrow has turned through the correct angle.

Q1 communication hint
The point where you hold
your tracing paper is called
the **centre of rotation**.

2   Copy the diagram.
    a   Trace shape A.
    b   Hold the tracing paper at (0, 0) with
        your pencil.
        Rotate shape A 90° clockwise.
    c   Lift up the tracing paper and copy the
        rotated shape onto the coordinate grid.
        Label the image B.
    d   Repeat steps **b** and **c** holding the
        tracing paper at (0, 6).
        Label the image C.

    Discussion What changes when you
    hold the tracing paper in a different place?

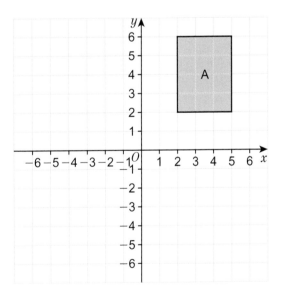

3   Use your diagram from **Q2**.
    a   Rotate shape A 180° about (0, 0).
        Label the image D.
    b   Rotate shape A 90° anticlockwise about (−1, 2).
        Label the image E.

Q3 hint Follow the steps in **Q2**.

301

4 a Trace shape A.

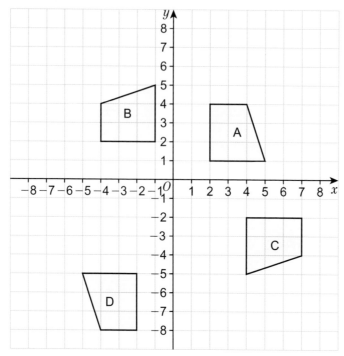

Q4 communication hint A **transformation** moves a shape so it is in a different position.

Q4b strategy hint Hold the tracing paper at different points.

Q4c hint ____° clockwise or ____° anticlockwise?

b Turn the tracing paper and rotate shape A onto shape B. What is the centre of rotation that maps A onto B?

c Copy and complete the description of the transformation that maps A onto B:

Rotation ____° _____ around (__ , __)

d Describe fully the transformation that maps
   i A onto C   ii A onto D.

Q4d hint To 'describe fully' give (1) the type of transformation (rotation), (2) the angle and direction of turn, and (3) the centre of rotation.

5 Shape A is drawn on a centimetre square grid.

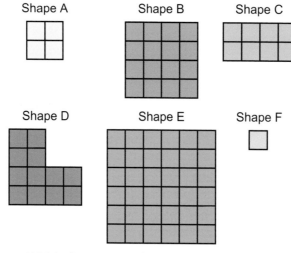

Shape A   Shape B   Shape C

Shape D   Shape E   Shape F

Q5 communication hint The word **enlargement** is used even when the new shape is smaller than the original shape.

Q5a hint Is B the same shape (square) but a different size to A? Is C the same shape but a different size to A?

Q5b hint What have the lengths on shape A been multiplied by?

a Which shapes are enlargements of shape A?
b State the scale factor of any enlargements.
c Rectangle C is enlarged by scale factor 4.
   i What is the new length of the rectangle?
   ii What is the new width of the rectangle?

Q5c hint Multiply the original length and width by the scale factor.

length × 4

width × 4

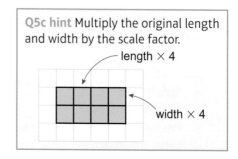

6 Copy each diagram.
Enlarge the shape by the scale factor.

a
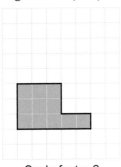
Scale factor 2

b
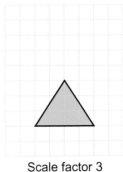
Scale factor 3

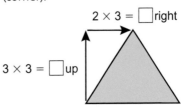

Q6b hint Count the squares up and across between each vertex (corner).

$2 \times 3 = \square$ right

$3 \times 3 = \square$ up

c
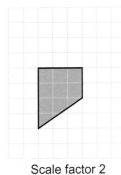
Scale factor 2

d
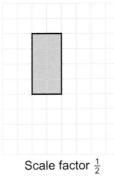
Scale factor $\frac{1}{2}$

Q6d hint Find $\frac{1}{2}$ of the original length and width.

7 Copy the diagram.

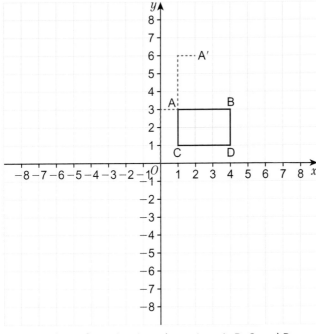

Q7a hint Draw up and across from (0, 0) to each vertex (corner).

Q7b hint Draw each line twice to enlarge by scale factor 2.

Q7c hint
Join the points A', B', C' and D'.

a Draw lines from (0, 0) to the points A, B, C and D.
b Extend each line by scale factor 2. Label the new points A', B', C' and D'.
c Enlarge rectangle ABCD by scale factor 2 from (0, 0).

8 Use the same grid as you used for **Q7**.
a Draw lines from (7, 0) to the points A, B, C and D.
b Enlarge rectangle ABCD by scale factor 2 from (7, 0).
Discussion How does the centre of enlargement affect the image?

9 **Reasoning** Shape B is an enlargement of shape A.

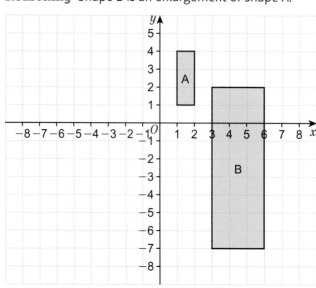

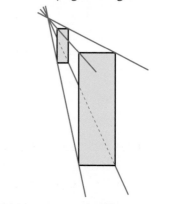

Q9b hint Position your ruler so that it joins corresponding vertices (corners) on shapes A and B. Repeat for all four vertices (corners). What point does your ruler always go through?

a What is the scale factor of the enlargement?

b Where is the centre of enlargement?

c Describe fully the transformation that maps shape A onto shape B.

Q9c hint To 'describe fully' give (1) the type of transformation (enlargement), (2) the scale factor, and (3) the centre of enlargement.

10 Extend

1 **Reasoning** The diagram shows triangles ABC and DEF.

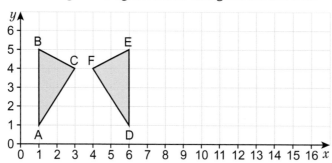

a Describe fully the transformation that maps triangle ABC onto triangle DEF.

Triangle DEF is reflected in the line $x = 8$ to become triangle GHI.

b Copy and complete this table showing the coordinates of the vertices of the triangles.

Triangle ABC	A(1, 1)	B(1, 5)	C(3, 4)
Triangle DEF	D(☐, ☐)	E(☐, ☐)	F(☐, ☐)
Triangle GHI	G(☐, ☐)	H(☐, ☐)	I(☐, ☐)

Triangle GHI is reflected in the line $x = 12$ to become triangle JKL.

c Without drawing triangle JKL, work out the coordinates of the vertices of triangle JKL. Explain how you worked out your answer.

2 **Problem-solving / Reasoning** Rectangles A, B and C are drawn on a centimetre square grid. Rectangles B and C are both enlargements of A.

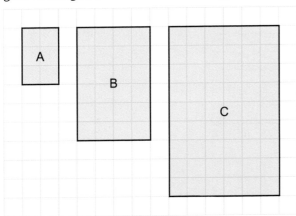

a What is the scale factor of enlargement from A to B?
b What is the scale factor of enlargement from A to C?
c What is the scale factor of enlargement from B to C?
d Copy and complete this table.

Rectangle	A	B	C
Perimeter (cm)			

e Olivia drew this shape on a centimetre square grid.

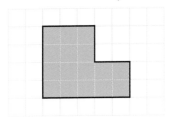

She enlarged the shape by scale factor 4.
Without drawing the enlargement, work out the perimeter of the image.

3 **Reasoning** Shape A has a perimeter of 24 cm. Shape A is enlarged by scale factor n to give shape B. Write an expression for the perimeter of shape B.

4 **Real** Meredith enlarges a photograph.

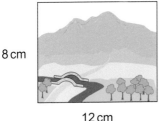

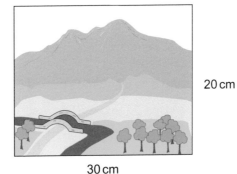

8 cm

20 cm

12 cm

30 cm

a What is the scale factor of this enlargement?
b She enlarges the original photograph by 150%.
Work out the length and width of the enlarged photograph.

Q4a hint Give the scale factor as a mixed number or a decimal.

5 (Exam-style question)

Describe fully the single transformation that maps triangle A onto triangle B.

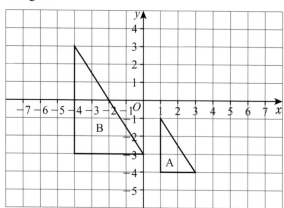

(3 marks)

March 2013, Q7, 5MB3H/01

Exam hint
For 3 marks
you must state
(I) the type of
transformation, (2)
the scale factor
from A to B and
(3) the centre of
enlargement.

6 **Reasoning** a What is the mathematical name of this shape?

Q6a hint Is it regular or irregular?

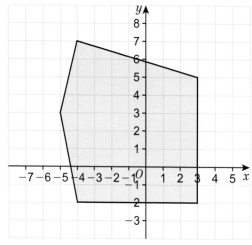

b Find its area.

c Copy the shape. Reflect it in the line $y = -x$.

d Does the reflection have the same area as the original shape? Explain.

Q6b strategy hint Trace the shape.
Draw a vertical line to split the shape
into a triangle and a trapezium.

7 **Problem-solving**

a A translation is described by the column vector $\begin{pmatrix} x \\ y \end{pmatrix}$.

Write the vector that would map the image
back to the original shape.

b A ship sails from A to B, to C and then back to A.

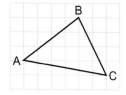

Q7a hint: The vector $\begin{pmatrix} 4 \\ -2 \end{pmatrix}$
describes the translation from
A to B. What vector describes
the translation from B to A?

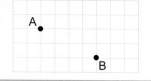

The vector from A to B is $\begin{pmatrix} 4 \\ 3 \end{pmatrix}$. The vector from B to C is $\begin{pmatrix} 2 \\ -4 \end{pmatrix}$.

What is the vector from C to A?

8 **Problem-solving** Describe fully at least three different transformations that map shape A onto shape B.

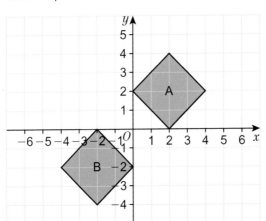

9 **Problem-solving** A regular hexagon ABCDEF has centre O.
Describe fully the transformation that maps triangle ABO onto triangle BCO.

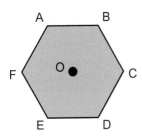

10 **Problem-solving** A shape is translated twice using the column vector $\begin{pmatrix} a \\ b \end{pmatrix}$ followed by $\begin{pmatrix} c \\ d \end{pmatrix}$. Write a vector using the letters a, b, c and d to describe the complete translation.

> **Q10 hint** Try with small numbers for a, b, c and d.

11 **Reasoning** a Rectangle A has length 4 cm and width 3 cm.
Rectangle B is an enlargement of rectangle A and one of its sides has length 24 cm.
What are the possible scale factors of the enlargement?

b Draw a rectangle with length 3 cm and width 2 cm. Label the rectangle A.
  i What is the area of rectangle A?
  ii Enlarge rectangle A by scale factor 2. Label the image B.
  iii What is the area of rectangle B?

Discussion To find the area of an enlarged shape, do you multiply the area of the original shape by the scale factor?

12 **Problem-solving** This shape has an area of 4 cm^2.

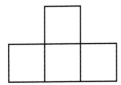

The shape is enlarged by scale factor 3.
Predict the area of the enlargement.

Draw the enlargement to check your answer.

13 (**Exam-style question**

Triangle ABC is drawn on a centimetre grid. A is the point (2, 2).
B is the point (6, 2). C is the point (5, 5). Triangle PQR is an enlargement
of triangle ABC with scale factor $\frac{1}{2}$ and centre (0, 0).
Work out the area of triangle PQR.

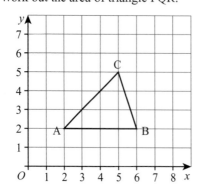

(3 marks)
June 2012, Q18, 1MA0/1H

Exam hint
Work out the
area of triangle
ABC first.

10 Knowledge check

⊙ You can use a **column vector** to describe a transformation. The top
number describes the movement to the left or right, and the bottom
number describes the movement up or down. For example:

$\binom{3}{2}$ means 3 right, 2 up $\qquad \binom{-4}{-5}$ means 4 left, 5 down. .... *Mastery lesson 10.1*

⊙ To describe fully a reflection on a coordinate grid you need to give
the equation of the **mirror line**. .... *Mastery lesson 10.2*

⊙ You rotate a shape by turning it around a point called the **centre of
rotation**. .... *Mastery lesson 10.3*

⊙ To describe fully a rotation, you need to give the angle, the direction
and the centre of rotation. .... *Mastery lesson 10.3*

⊙ To enlarge a shape you multiply all the side lengths by the same
number. The number that the side lengths are multiplied by is called
the **scale factor**. .... *Mastery lesson 10.4*

⊙ When you enlarge a shape using a centre of enlargement, you
multiply the distance from the centre to each vertex by the
scale factor. .... *Mastery lesson 10.4*

⊙ To describe fully an enlargement, you need to give the scale factor
and the centre of enlargement. .... *Mastery lesson 10.5*

Write down a word that describes how you feel

a before a maths test   b during a maths test   c after a maths test.

Hint Here are some possible words: OK, worried, excited, happy, focused, panicked, calm.

Beside each word, draw a face, or , to show if it is a good or a bad feeling.

Discuss with a classmate what you could do to change feelings to feelings.

Reflect

10 Unit test

1 Shape B is an enlargement of shape A.
   Write the scale factor of the enlargement.   *(1 mark)*

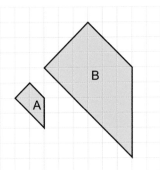

2 Copy the diagram.

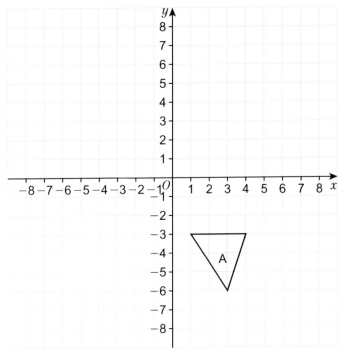

a Translate triangle A by vector $\begin{pmatrix} -5 \\ 6 \end{pmatrix}$.
   Label the image B.
b Reflect triangle A in the line $y = 1$.
   Label the image C.
c Rotate triangle A 90° anticlockwise around (1, −1).
   Label the image D.   *(6 marks)*

3   **Reasoning** Look at the diagram.

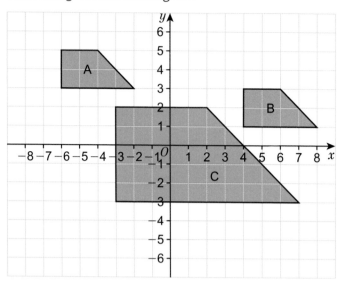

a   Describe fully the transformation that maps shape A onto shape B.

b   Describe fully the transformation that maps shape A onto shape C.   *(5 marks)*

4   Copy the diagram. Enlarge shape A by scale factor 2, centre of enlargement (6, 5).   *(3 marks)*

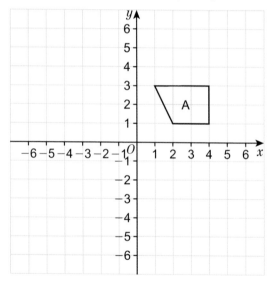

5   **Reasoning** Describe fully the transformation that maps triangle A onto tringle B.   *(2 marks)*

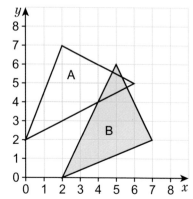

6   Copy the diagram.

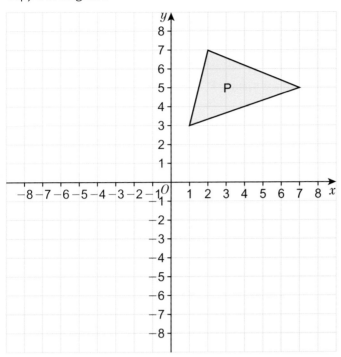

a   Reflect triangle P in the line $y = x$. Label the image Q.
b   Reflect triangle Q in the line $y = -x$. Label the image R.
c   Describe fully the single transformation that maps triangle P onto triangle R.   *(7 marks)*

7   **Problem-solving** Triangle A is enlarged by scale factor 2 from centre (0, 0).
Work out the area of the enlarged triangle.   *(3 marks)*

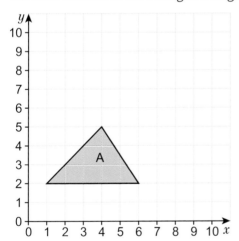

Sample student answers

a   Which student has the correct answer? Explain why.

b   What could the student draw on the diagram to help ensure they get the correct answer?

Exam-style question

Describe fully the single transformation that maps triangle **P** onto triangle **Q**.

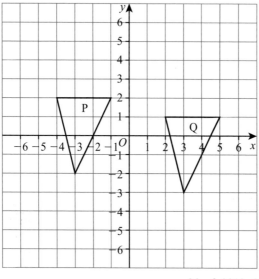

(2 marks)

March 2013, Q10, 1MA0/2H

Student A

Translation $\begin{pmatrix} 3 \\ -1 \end{pmatrix}$

Student B

Translation $\begin{pmatrix} 6 \\ -1 \end{pmatrix}$

11 RATIO AND PROPORTION

The engine performance of cars can be compared by looking at the power-to-weight ratio.

Cars with a high power-to-weight ratio accelerate well.

The times taken for cars to accelerate from 0 to 100 km/h are given in the table.

Car make and model	Time (0–100 km/h)
Ferrari 458 Italia	3.4 s
Lamborghini Aventador	2.9 s
Porsche 918 Spyder	2.6 s

Which car accelerated the fastest?

11 Prior knowledge check

Numerical fluency

1 Copy and complete
   a $10 \times \boxed{} = 60$
   b $\boxed{} \times 9 = 900$

2 Find the highest common factor (HCF) of each pair of numbers.
   a 25 and 35
   b 24 and 40

3 Work out
   a 24 kg ÷ 3
   b 72 mm ÷ 8
   c £2.80 ÷ 4
   d 300 g ÷ 5

4 Work out
   a 17.5 ÷ 5
   b 5.4 × 6

5 Work out
   a 40 ÷ 5 × 3
   b 36 ÷ 4 × 8

6 What fraction of each diagram is shaded?
   a
   b

7 Copy and complete
   a $\frac{3}{5} = \frac{\boxed{}}{20}$
   b $\frac{24}{32} = \frac{3}{\boxed{}}$

Fluency with measures

8 How many
   a g in 1 kg
   b ml in 1 litre?

Geometrical fluency

9 What is the scale factor of this enlargement?

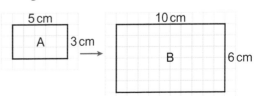

Graphical fluency

10 Plot a line graph for the values in the table.

x	1	2	3	4	5
y	3	6	9	12	15

  a What is the gradient of the line?

  b What is the equation of the line?

＊Challenge

11 Design a set of 20 loop cards. The first two have been done for you.

  The last card must end with $\frac{3}{4}$ or 0.75. Explain why.

11.1 Writing ratios

Objectives

- Use ratio notation.
- Write a ratio in its simplest form.
- Solve simple problems using ratios.

Why learn this?

Outdoor pursuits need to have the correct supervision ratios.

Fluency

- Work out 28 ÷ 4, 45 ÷ 5
- Chloe has 3 red beads and twice as many blue beads. How many blue beads does she have?

Warm up

1 The bar chart shows the number of teenagers doing each activity at an outdoor pursuits centre on one day.

  a How many teenagers are taking part in activities at the centre altogether?

  b Which activity is the most popular?

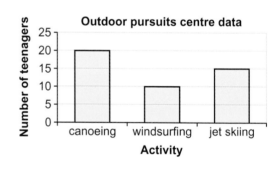

Key point 1

A **ratio** is a way to compare two or more quantities.

Questions in this unit are targeted at the steps indicated.

2 Write down each ratio of red tins to yellow tins.

  a
  b

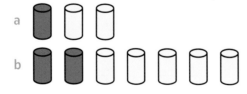

Q2a hint There is 1 red tin and 2 yellow tins. Write the ratio 'red : yellow' using the numbers.

3 Draw tins to show these ratios of red to yellow.

  a 4 : 3   b 3 : 4

  Discussion Is the ratio 4 : 3 the same as the ratio 3 : 4?

Active Learn Homework, practice and support: Foundation 11.1

4   A necklace has 30 beads.
    There is 1 purple bead for
    every 4 blue beads.

    How many beads are
    a   purple   b   blue?

    Q4 hint How many sets
    of '1 purple, 4 blue' are
    there?

Key point 2

You **simplify** a ratio by making the numbers as small as possible.
Divide the numbers in the ratio by their **highest common factor (HCF)**.

5   Write each ratio in its simplest form.
    a   4 : 12   b   16 : 8
    c   27 : 9   d   7 : 42
    e   15 : 20   f   63 : 28
    g   18 : 48   h   36 : 45

    Q5a hint
    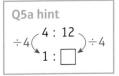

    Q5e hint

6   **Problem-solving / Reasoning** The bar charts show the numbers of gold medals
    and other medals won at a competition by each group from a gym club.

    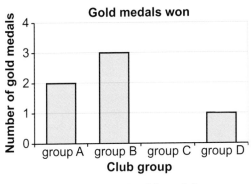

    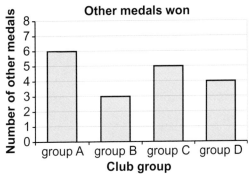

    a   What is the ratio of gold medals won to other medals won?
        Write your answer in its simplest form.
    b   The club's coach says that they won twice as many other
        medals as gold medals.

        Is the coach correct? Explain your answer.

    Q6a hint What do
    you need to work out
    before you can find
    the ratio?

7   **Real** An after-school club at a primary school is attended by
    32 children. It is run by 4 adults. The guidelines say that the
    adult-to-child ratio should be 1 : 8.

    Does the club have enough adults?

    Q7 hint Show your
    working.

8   Which of these ratios are equivalent?
    A 36 : 16   **B** 135 : 60
    C 28 : 16   **D** 126 : 56
    E 49 : 28

    Q8 communication hint
    Ratios are **equivalent** if they
    have the same simplest form.

9   Write each ratio in its simplest form.
    a   20 : 25 : 15   b   36 : 24 : 30
    c   56 : 42 : 35   d   16 : 40 : 56

    Q9a hint The highest common factor of
    20, 25 and 15 is 5, so divide all the parts by 5.

10 **Communication** Show that these ratios are equivalent.

10 : 15 : 25   12 : 18 : 30   18 : 27 : 45

11 **STEM** A recipe for shortbread uses 125 g of butter, 55 g of sugar and 180 g of flour.
Write the ratio of butter : sugar : flour in its simplest form.

12 **Exam-style question**

There are 80 marbles in a bag. Of these, 25 are china and the rest are glass. Write the ratio of china marbles to glass marbles in its simplest form. **(2 marks)**

Q12 strategy hint
Start by writing
C : G and the
information you
know.

11.2 Using ratios 1

Objective

• Solve simple problems using ratios.

Why learn this?

You need to mix paints in the same ratio to get the same colour each time.

Fluency

• 1 m = ☐ cm
• 1 m = ☐ mm
• 1 cm = ☐ mm
• 2500 mm = ☐ m

Warm up

1 Copy and complete

a 2 × ☐ = 100   b 3 × ☐ = 75   c 2 × ☐ = 150   d 6 × ☐ = 300

2 Work out

a 1.5 × 10   b 2.45 × 100   c 3.71 × 10   d 9.37 × 100

3 Find the HCF of each pair of numbers.

a 24 and 30   b 35 and 49   c 18 and 45   d 64 and 80

Example 1

To make orange paint Maria mixes yellow paint with red paint in the ratio 3 : 1.
She uses 4 tins of red paint. How many tins of yellow paint does she use?
Write down the ratio. Use Y for yellow and R for red.

Y : R

×4 (3 : 1) ×4
    12 : 4

Multiply each part by the same number to get an equivalent ratio.

Maria uses 12 tins of yellow paint.

4 **STEM** The ratio of oyster sauce : fish sauce in a stir fry recipe is 2 : 1.
Clare uses 5 tablespoons of fish sauce.
How many tablespoons of oyster sauce does she use?

5   **Real** Thomas makes a model of the Eiffel Tower
     using a ratio of 1 : 1280.

     The height of his model is 250 mm.

     What is the height of the Eiffel Tower in metres?

> Q5 hint 1 : 1280 means that 1 mm in the model
> represents 1280 mm in real life.

Q5 hint

Model : Real

1 : 1280

☐ : ☐

Key point 3

Ratios in their **simplest form** only have whole numbers.

Example 2

Write 1.5 : 8 as a whole number ratio in its simplest form.

×10 (1.5 : 8) ×10 ——— 1.5 has 1 decimal place so multiply both sides
     15 : 80   of the ratio by 10 to get a whole number.
÷5 (3 : 16) ÷5

The HCF is 5 so divide both sides by 5.

---

6   Write each ratio as a whole number ratio in its simplest form.

     a   0.4 : 6   b   3.5 : 4.2   c   45 : 13.5   d   25.6 : 46.4

     Discussion What should you multiply by if a number in a ratio has 2 decimal places?

7   Write each ratio as a whole number ratio in its simplest form.

     a   0.25 : 3.1   b   1.4 : 0.28   c   1.62 : 1.8   d   4.8 : 11.2

8   **STEM / Problem-solving** Old computer monitors had a width : height ratio of 4 : 3.
     New computer monitors have a width : height ratio of 16 : 9.

     Is each of these monitors old or new?

     a   b
     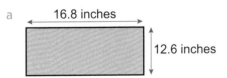   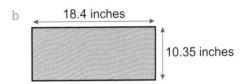

9   **STEM** Sterling silver is made from silver and copper in the ratio 92.5 : 7.5.

     Write this ratio in its simplest form.

10   **STEM** The ratio of chick peas : broad beans in a hummus
     recipe is 2 : 9.

     Jack uses 150 g of chick peas.
     How many grams of broad beans should he use?

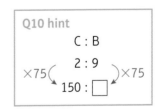

Q10 hint

         C : B

×75 (2 : 9   ) ×75
       150 : ☐

11   **STEM / Reasoning** The ratio of okra : sweet potato in a korma recipe is 15 : 8.

     Raj uses 160 g of sweet potato. He has 350 g of okra.

     Will he use all of the okra?

12 **Finance** Anna splits her monthly net pay into rent money and money left in the ratio 7 : 13. Her rent is £420. How much money does she have left each month?

13 **Exam-style question**

5 schools sent some students to a conference.

One of the schools sent both boys and girls.
This school sent 16 boys. The ratio of the number of boys it sent to the number of girls it sent was 1 : 2

The other 4 schools sent only girls.
Each of the 5 schools sent the same number of students.

Work out the total number of students sent to the conference by these 5 schools. **(4 marks)**

March 2013, Q12, 1MA0/1H

Q13 hint Start by working out how many girls the school sent.

Exam hint
Deal with the information one paragraph at a time.

11.3 **Ratios and measures**

Objectives

- Use ratios to convert between units.
- Write and use ratios for shapes and their enlargements.

Why learn this?
You can use ratios to work out your money for holidays abroad.

Fluency

- 1 kg = ☐ g, 1 m = ☐ cm, 1 litre = ☐ ml, 1 cm = ☐ mm, 1 day = ☐ hours
- 9 = ☐², 25 = ☐², 8 = ☐³

1 Faaiz has 7 red marbles and 2 blue marbles.
   a How many marbles does he have altogether?
   b What fraction are   i red   ii blue?

2 Work out the area of this shape.

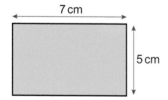

3 Work out the volume of this cube.

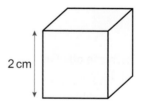

4 Write these ratios in their simplest form.
   a 15 minutes : 1 hour
   b 400 g : 1 kg
   c 2 hours : 30 seconds

Q4 hint Both parts of the ratio need to be in the same units before you simplify.

Q4a and c hint There are 60 seconds in 1 minute and 60 minutes in 1 hour.

5 A mug contains 250 ml of coffee and a jug contains 1 litre of coffee.
   Write down the ratio of the amount of coffee in the mug to the amount of coffee in the jug.
   Give your answer in its simplest form.

Key point 4

You can use **ratios** to convert between **units**.

Example 3

Convert 8 m to cm.

1 m is 100 cm.

m : cm

×8 (1 : 100) ×8
     8 : 800

The ratio of m : cm is 1 : 100.

So 8 m is 800 cm.

6   Complete these conversions.
    a   2.5 km = ☐ m   b   34 000 g = ☐ kg
    c   3.8 litres = ☐ ml   d   95 mm = ☐ cm

Q6a hint
km : m = ☐ : ☐

7   1 mile ≈ 1.6 km. Convert
    a   5 miles to km   b   16 km to miles.
    c   **Real** The length of the London Marathon is 26.219 miles. How far is this in kilometres?

8   A recipe needs 2 lb of flour.
    a   1 lb = 16 oz (ounces). Convert 2 lb to ounces.
    b   1 oz ≈ 25 g. Convert your answer to part **a** into grams.

9   (**Exam-style question**)

Linda is buying wool to knit a baby's blanket.
She needs 1800 yards of wool.

Linda chooses some balls of wool.
There are 245 metres of wool in each ball of wool.

She knows that
•   1 yard is 36 inches
•   1 inch is 2.54 centimetres

How many balls of wool does Linda need to buy?
You must show all your working.   **(4 marks)**

Nov 2013, Q21, 1MA0/2F

Exam hint
Present your working clearly and remember to show the units next to each number. Explain what you are doing at each step so it is easy for an examiner to follow.

10 **Real / Finance** On the day Lucy buys some
    US dollars for her holiday, £1 buys US $1.68.
    How many dollars does Lucy get for £500?

Q10 hint
£ : $
×500 (1 : 1.68) ×500
       500 : ☐

11 **Real / Finance** Choose an amount in £ from the cloud.
    Choose an exchange rate from the table.

£200   £350
£475   £690

Currency	£1
Dollars	1.68
Euros	1.23
Rupees	99.12
Yen	172.2
Swiss francs	1.50

Work out how much money
you will get. Repeat for two
more amounts and currencies.

Q11 communication hint
The exchange rate is the amount of money in another currency that your currency will buy or sell for.

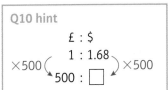

12 **Real / Problem-solving** Use the table in **Q11**.
   a   How many £ does Sue get for 300 Swiss francs?
   b   Which is worth more, 49 560 rupees or 85 700 yen?

13 **Problem-solving** The diagram shows two rectangles.

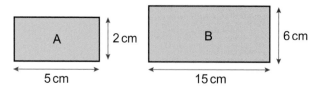

Write these ratios in their simplest form.
   a   The width of rectangle A to rectangle B.
   b   The length of rectangle A to rectangle B.
   c   The area of rectangle A to rectangle B.

   Q13c hint 3^2 = ?

Discussion What do you notice about the results?

14 The diagram shows two cubes.
   Write these ratios in their simplest form.
   a   Height of A to height of B.
   b   Area of a face of A to area of a face of B.
   c   Volume of A to volume of B.

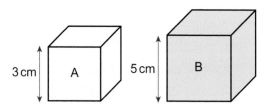

Discussion What do you notice?

15 A bag contains 11 red counters and 3 white counters.
   a   What is the ratio of red counters : white counters?
   b   What fraction of the counters are red?
   c   What fraction of the counters are white?

Discussion What do you notice about your answers?

16 In a box of mints and toffees, $\frac{3}{4}$ of the sweets are mints and $\frac{1}{4}$ are toffees.
   a   What is the ratio of mints : toffees?
   b   There are 8 toffees in the box. How many mints are there?
   c   How many sweets are in the box altogether?

   Q16 hint

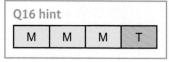

Reflect Is there more than one way to work out the answer to part **c**?

17 **Reasoning** In a band, $\frac{3}{5}$ of the members are male and the rest are female.
   a   What is the ratio of males : females in the band?
   b   Maya thinks there are 18 people in the band.
       Explain why she must be wrong.

11.4 **Using ratios 2**

Objectives

- Divide a quantity into 2 parts in a given ratio.
- Divide a quantity into 3 parts in a given ratio.
- Solve word problems using ratios.

Why learn this?

The ratio of gold to other metals in a piece of jewellery is used to work out the value of the item.

Fluency

Write down the ratio of green : blue : yellow counters on the dish.

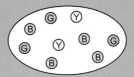

1. Work out
   - a $35 \div 7$
   - b $10 \div 4$
   - c $4.5 \div 3$
   - d $2.8 \div 4$

2. Round these figures.
   - a 12.468 to 2 decimal places
   - b 4.973 to 2 decimal places
   - c 3.1584 to 3 decimal places
   - d 0.012 439 to 3 decimal places

3. Write these ratios in their simplest form.
   - a $25 : 10$
   - b $100 : 350$
   - c $6000 : 2000$
   - d $4.5 : 6$

Example 4

Share £25 in the ratio 3 : 2.

$3 + 2 = 5$ parts

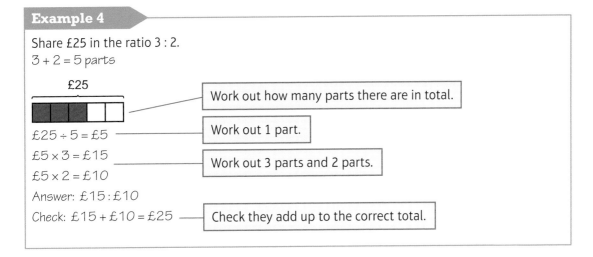

£25

£25 ÷ 5 = £5 ——— Work out 1 part.

£5 × 3 = £15
£5 × 2 = £10 ——— Work out 3 parts and 2 parts.

Answer: £15 : £10

Check: £15 + £10 = £25 ——— Check they add up to the correct total.

——— Work out how many parts there are in total.

---

4. Share these amounts in the ratios given.
   - a £18 in the ratio 2 : 1
   - b £42 in the ratio 1 : 6
   - c £27 in the ratio 4 : 5
   - d 35 kg in the ratio 2 : 3
   - e 60 m in the ratio 5 : 7
   - f 7.5 litres in the ratio 2 : 3

   > **Q4 hint** How many parts are there?

5. **STEM / Real** Before 2012, 10p coins were made from copper and nickel in the ratio 3 : 1.

   The coins had a mass of 6.5 g. What was the mass of
   - a copper
   - b nickel?

---

6 Purple paint is mixed from red paint and blue paint in the ratio 5 : 3.

A painter needs 20 litres of purple paint.

a How many litres of red paint should he use?

b How many litres of blue paint should he use?

Discussion How did you check your answers?

7 A fruit drink is made from orange, pineapple and apple juice in the ratio 1 : 2 : 4.

Rita wants to make 35 litres of fruit drink. How much of each type of juice does she need?

8 **Exam-style question**

Talil is going to make some concrete mix.
He needs to mix cement, sand and gravel in the ratio 1 : 3 : 5 by weight.
Talil wants to make 180 kg of concrete mix. Talil has
- 15 kg of cement
- 85 kg of sand
- 100 kg of gravel

Does Talil have enough cement, sand and gravel to make the concrete mix?

(4 marks)

Nov 2012, Q29, 1MAO/1F

> **Q8 strategy hint** You could work out how many kg of each he needs to make 180 kg.

9 Share these amounts in the ratios given.

a £72 in the ratio 2 : 3 : 4

b 100 g in the ratio 2 : 3 : 5

c 360 ml in the ratio 3 : 4 : 5

d £25 in the ratio 1 : 3 : 4

> **Q9 hint** How many parts are there?

Discussion How should you round if the ratio is money? What about litres? Why?

10 Share these amounts in the ratios given. Round your answers sensibly.

a £80 in the ratio 2 : 5

b 70 litres in the ratio 2 : 7

11 **Reasoning** Bob and Phil buy a greyhound for £4500.

Bob pays £3000 and Phil pays £1500.
The greyhound wins an £18 000 prize.

a Write the amounts they each pay as a ratio.

b Write the ratio in its simplest form.

c Divide the prize money in this ratio.
How much does each of them get?

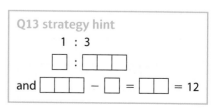

Q11 hint

Bob : Phil

Discussion Is this a fair way to share the prize money?

12 **Problem-solving** Andrea and Penny buy a statue for £350.
Andrea pays £140 and Penny pays £210.
They sell the statue 3 years later for £475.
How should they share the money fairly?

13 **Problem-solving / Reasoning** Two numbers are in the ratio 1 : 3 and their difference is 12.
What are the numbers?

> **Q13 strategy hint**
> 1 : 3
> ☐ : ☐☐☐
> and ☐☐☐ − ☐ = ☐☐ = 12

14 Rob and Simon share £50 in the ratio 7 : 3.
   a How much do they each receive?   b How much more does Rob get than Simon?
   Discussion How could you check your answer?

15 ┌───┐
    │ **Exam-style question**   │
    │   │
    │ Pat and Julie share some money in the ratio 2 : 5   │
    │ Julie gets £45 more than Pat.   │
    │ How much money did Pat get?   **(3 marks)**   │
    │   *June 2012, Q22, 1MA0/2F*   │
    └───┘

   ┌─────────────────────────┐
   │ Q15 strategy hint   │
   │ Draw a diagram.   │
   └─────────────────────────┘

Reflect How can drawing a diagram help you to answer ratio questions?

11.5 Comparing using ratios

Objectives

- Use ratios involving decimals.
- Compare ratios.
- Solve ratio and proportion problems.

Why learn this?

You can use this to calculate and compare the performance of different cars.

Fluency

A bracelet has gold and silver links in the ratio 2 : 7. Which type of link is used most?

1 What are the missing numbers?
   a $\dfrac{3}{5} = \dfrac{\square}{10}$ b 1 whole $= \dfrac{\square}{5}$ c $\dfrac{1}{6} + \dfrac{\square}{6} = \dfrac{6}{6}$ d $\dfrac{2}{7} + \dfrac{1}{7} + \dfrac{\square}{7} = \dfrac{7}{7}$

2 Work out
   a $2.5 + 3.5$ b $1 - \dfrac{2}{9}$ c $\dfrac{3}{4} \times 100$ d $\dfrac{2}{5} \times 20$

3 Simplify these ratios.
   a $2 : 2.5 : 1.5$ b $5.5 : 4.5 : 3$ c 3 litres : 500 ml   d 70 cm : 1 m

┌──┐
│ **Key point 5**   │
│   │
│ A **proportion** compares a part with the whole.   │
└──┘

4 In a canoeing lesson $\frac{2}{7}$ of the group are girls and
   the rest are boys.
   What is the ratio of girls to boys in the group?

   ┌─────────────────────────────────────┐
   │ Q4 hint   │
   │ ┌──┬──┬──┬──┬──┬──┬──┐   │
   │ │ G│ G│ B│ B│ B│ B│ B│   │
   │ └──┴──┴──┴──┴──┴──┴──┘   │
   └─────────────────────────────────────┘

5 Copy and complete the table for different groups of teenagers having canoeing lessons.

Fraction of group that are girls	Ratio of girls : boys
$\frac{5}{9}$	
	3 : 2
$\frac{7}{10}$	
	4 : 7

Warm up

6   **STEM** In a gluten-free pizza base, $\frac{7}{12}$ of the flour is rice flour,
    $\frac{3}{12}$ is potato flour and the rest is tapioca flour.
    Work out the ratio of rice flour : potato flour : tapioca flour.

7   **STEM / Real** A 5p coin used to be made of
    copper and nickel in the ratio 3 : 1.
    What fraction of the coin was
    a   copper   b   nickel?

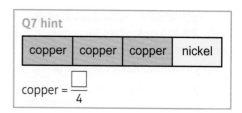

Q7 hint

copper	copper	copper	nickel

copper = $\dfrac{\square}{4}$

8   Clare and Fiona share a cash prize in the ratio 4 : 3.
    What fraction of the prize should   a   Clare get   b   Fiona get?
    Discussion How can you check your answer?

9   **STEM / Real** A 20p coin is made of copper and nickel in the ratio 21 : 4.
    What fraction of the coin is   a   copper   b   nickel?

10 **Reasoning / Communication** The ratio of office staff to
    manufacturing staff at a company is 2 : 5.
    Jo says that $\frac{2}{5}$ of the people are office staff.
    Is she correct? Explain your answer.

    Q10 strategy hint
    Draw a diagram.

11 **STEM / Reasoning** Red brass is made from a mix of copper, zinc, lead and tin
    in the ratio 17 : 1 : 1 : 1.
    a   What fraction of red brass is
        i   copper   ii   zinc?
    b   In 100 g of red brass, what are the masses of copper, zinc, lead and tin?
    c   Lily has plenty of copper and zinc, but only 80 g of lead and 60 g of tin.
        What is the maximum amount of red brass she can make?
    d   **Reflect** Look back at your working for part **c**.
        Have you clearly shown how you got your answer?

12 **STEM / Reasoning** A topping for shortbread uses milk, chocolate, butter and syrup in the
    ratio 4 : 3.5 : 1.5 : 1.
    a   What fraction of the topping is chocolate?
    b   In 500 g of the topping what are the masses of milk, chocolate,
        butter and syrup?
    c   You have plenty of milk and chocolate, but only 200 g of butter
        and 100 g of syrup.
        What is the maximum amount of topping you can make?

    Q12 hint Write the
    ratio with whole
    numbers first.

Key point 6

You can compare ratios by writing them as **unit ratios**. In a unit ratio, one of the numbers is 1.

13 Copy and complete to write these as unit ratios.

    a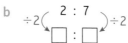
        $\div 3 \left(\begin{array}{c} 5 : 3 \\ \square : 1 \end{array} \right) \div 3$

    b $\div 2 \left(\begin{array}{c} 2 : 7 \\ \square : \square \end{array} \right) \div 2$

14 Write each of these in the form $m : 1$.

Give each answer to a maximum of 2 decimal places.

a 2 : 5   b 7 : 4   c 16 : 9

d 5 : 36   e 9 : 42   f 11 : 56

Q14 hint Write each of these as a unit ratio, with the second value 1.

Example 5

Molly makes a blackcurrant drink by mixing 30 ml of blackcurrant with 450 ml of water.

Hope makes a blackcurrant drink by mixing 40 ml of blackcurrant with 540 ml of water.

Whose drink is the stronger? Explain your answer.

Molly   Hope

blackcurrant : water   blackcurrant : water

$\div 30 \left(\dfrac{30 : 450}{1 : 15}\right) \div 30$ $\div 40 \left(\dfrac{40 : 540}{1 : 13.5}\right) \div 40$ ─── Simplify to a unit ratio.

Hope's drink is the stronger because it uses less water for every millilitre of blackcurrant.

Compare the quantity of water per ml of blackcurrant.

---

15 **Reasoning / Communication** Anna makes orange squash by mixing 50 ml of squash with 850 ml of water.

Jeevan makes orange squash by mixing 60 ml of squash with 1110 ml of water.

Whose squash is the stronger? Explain your answer.

16 **Reasoning / Communication** Josh makes purple paint by mixing 2 litres of red paint and 500 ml of blue paint.

Dexter makes purple paint by mixing 1.5 litres of red paint and 400 ml of blue paint.

Whose paint is the darker purple? Explain your answer.

Q16 hint Darker purple paint has a higher concentration of blue paint.

17 **Reasoning** Raj makes concrete using aggregate to cement in the ratio 1930 : 265.

Sunil makes concrete using aggregate to cement in the ratio 935 : 175.

Whose concrete has the higher proportion of cement?

18 **Reasoning** Billy makes mortar by mixing 360 kg of cement with 90 kg of sand.

Sam makes mortar by mixing 340 kg of cement with 85 kg of sand.

Whose mortar has the higher proportion of cement?

19 **Exam-style question**

STEM Electrum is $\frac{2}{5}$ gold and $\frac{3}{5}$ silver.

Jessica says, 'The amount of silver is one and a half times the amount of gold.'

Is she correct? Explain your answer.   **(3 marks)**

Exam hint

Explain your workings and answer by writing your reasons alongside your calculations rather than listing them at the end.

11.6 Using proportion

Objectives

- Use the unitary method to solve proportion problems.
- Solve proportion problems in words.
- Work out which product is better value for money.

Why learn this?

Working out which product is the better buy helps you get the best value for money.

Fluency

In each pair, which number is bigger?
3.28 or 3.7? 2.9 or 2.45? 1.06 or 1.009?

Warm up

1 Work out
a 840 ÷ 7
b 450 × 9
c 2.7 × 5
d 4.8 ÷ 3
e 10 × ☐ = 20
f 10 ÷ ☐ = 5
g 5 ÷ ☐ = 1
h 7.2 ÷ ☐ = 1

2 Write each of these as a unit ratio.
a 4 : 18
b 11 : 5
c 14 : 35
d 18 : 8

3 What is the HCF of each pair of numbers?
a 500 and 350
b 600 and 800

4 Which is better value if you are shopping?
a Paying 1.8p per gram or 2.1p per gram?
b Getting 3.25 ml for 1p or 2.85 ml for 1p?

Example 6

A recipe for 6 people uses 900 g of mince. How much mince is needed for

a 12 people

P : M

×2 (6 : 900 g) ×2
12 : 1800 g

b 3 people

P : M

÷2 (6 : 900 g) ÷2
3 : 450 g

c 9 people?

6 people + 3 people = 9 people
900 + 450 = 1350 g

Discussion How would you work out the amounts for 18 people and 15 people?

5 A recipe for 4 people uses 6 eggs. How many eggs are needed for
a 8 people
b 2 people
c 6 people
d 10 people?

Key point 7

In the **unitary method** you find the value of one item before finding the value of more.

6 5 tickets to a theme park cost £125.
How much will 18 tickets cost?
Discussion Is there more than one way to work out the answer?
Which is the best method? Why?

Q6 hint Find the cost of 1 ticket first.

Tickets : Cost

÷☐ (5 : £125) ÷☐
1 : ☐

7   20 litres of fuel cost £25.38.

What do 35 litres cost?

> **Q7 hint** Find the cost of 1 litre first.
>
> L : £
>
> $÷\square$ 20 : 25.38 $÷\square$
>
> 1 : $\square$

8   Maria gets £31.55 for working for 5 hours.

How much will she get for working for

a   9 hours   b   26 hours   c   30 hours?

Reflect Did you do part **c** the same way as parts **a** and **b**?

Can you show a quicker way?

9   **Reasoning** A gym coach orders a tracksuit for each of the 25 gymnasts in the club.

The total cost is £1247.50.

2 more gymnasts join the club so he orders an extra 2 tracksuits.

What is the total value of the order now?

Key point 8

You can use the unitary method to work out which product gives better value for money.

---

10   **Reasoning** Washing powder comes in two sizes.

a   Copy and complete to find the unit ratios.

i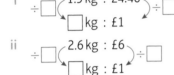
  $÷\square$ 1.3 kg : £4.40 $÷\square$
  $\square$ kg : £1

ii $÷\square$ 2.6 kg : £6 $÷\square$
  $\square$ kg : £1

> **Q10b hint** Which gives more grams for £1?

1.3 kg for £4.40   2.6 kg for £6

b   Which size is the better buy?

11   **Reasoning** Jenna can buy her favourite blend of coffee in two sizes: 500 g for £14.90 or 300 g for £8.85.

Which is better value for money?

Discussion How did you work it out?

12   **Reasoning** David can buy blackcurrant squash in two sizes: 2 litres for £4.99 or 600 ml for £1.98.

Which is better value for money?

> **Q12 hint** Convert both to the same units.

13   **Reasoning** Mary gets paid £31.50 for 5 hours of work. Alice gets paid £50 for 8 hours of work.

They both work 36 hours a week. Who earns more money?

14   **Exam-style question**

Brenda works in an office.
She finds out the prices of folders from two companies, Office Deals and Paper World.

OFFICE DEALS	PAPER WORLD
Packs of 20 folders £10.80	Pack of 15 folders £8.40

Brenda needs to buy exactly 60 folders. She wants to buy the folders as cheaply as possible.

Which company should Brenda buy the folders from?

You must explain your answer.

(4 marks)

Nov 2013, Q16, 5MB3F/01

Exam hint

Highlight or underline key information in a question. What information not given in the boxes should you highlight here?

11.7 Proportion and graphs

Objectives

- Recognise and use direct proportion on a graph.
- Understand the link between the unit ratio and the gradient.

Why learn this?

Knowing simple relationships allows us to make predictions.

Fluency

- What is the equation of a straight line through the origin (0, 0) with gradient 5?
- What is the gradient of the line $y = 7x$?

Warm up

1   The graph changes between gallons and pints.
    Use the graph to work out these conversions.
    a   5 gallons = ☐ pints
    b   9 gallons = ☐ pints
    c   120 pints = ☐ gallons

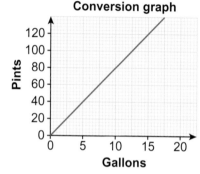

Conversion graph

2   Plot a line graph for the values in the table.
    a   What is the gradient of the line?
    b   What is the equation of the line?

x	1	2	3	4	5
y	4	8	12	16	20

Key point 9

When two values are in **direct proportion**, if one value is zero so is the other.
When one value doubles, so does the other.

3   Look at the graph in **Q1**.
    Are gallons and pints in direct proportion?
    Explain your answer.

> **Q3 hint** Does doubling one value double the other? When one value is zero, what is the other?

Key point 10

When two quantities are in direct proportion, plotting them as a graph gives a straight line through the origin. The origin is the point (0, 0) on a graph.

4   **Modelling** The table shows some temperatures in both Celsius and Fahrenheit.

Celsius	5°	10°	20°	25°
Fahrenheit	41°	50°	68°	77°

> **Q4a hint** Plot Celsius on the horizontal axis and Fahrenheit on the vertical axis. Use sensible scales.

    a   Plot a line graph for these values.
    b   Are Celsius and Fahrenheit in direct proportion?
        Explain your answer.
    c   Ice melts at 0 °C. What is this temperature in Fahrenheit?

  *Active*Learn   Homework, practice and support: Foundation 11.7

5   **Modelling** The table shows the price of pears at a market.

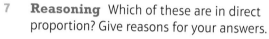

Mass of pears (kg)	0.5	1	5	10
Price of pears (£)	1	2	10	20

   a   Plot a line graph for these values.

   b   Is the price of the pears in direct proportion to the mass? Explain your answer.

6   **Modelling** This line graph shows the
price of grapes by mass.

   a   Are price and mass in direct
proportion?

   b   Work out the gradient of the line.

   c   How much does 1 kg of grapes cost?

   Discussion How does your answer to **b**
help you answer **c**?

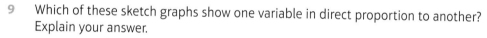

Price of grapes

7   **Reasoning** Which of these are in direct
proportion? Give reasons for your answers.

   a   Metres and centimetres

   b   Age and favourite music type

   c   Number of teens in swimming pool and time of day

   d   Euros and pounds

8   **Problem-solving** A plumber charges a callout fee of £50 plus
£35 per hour he works.

Is his total charge (C) in direct proportion to the number of hours (h)
he works?

> **Q8 hint** You might
> wish to draw a graph.

9   Which of these sketch graphs show one variable in direct proportion to another?
Explain your answer.

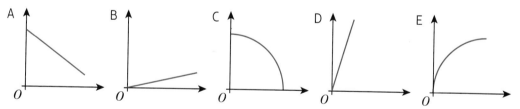

10   **Modelling** The table shows the amount paid for different numbers of litres of fuel
on one day.

Number of litres (n)	5	10	20	30	50
Cost (C)	£6.60	£13.20	£26.40	£39.60	£66

   a   Draw the graph.

   b   Work out the equation of the line.

   c   Are C and n in direct proportion?

   d   Write a formula linking the number of litres (n) and the cost (C).

   e   Use the formula to work out the cost of 82 litres of fuel on that day.

> **Q10d hint**
>
> $C = \boxed{} \times n$

Reflect Look at the equation of the line and the formula for C and n.
What is the same? What is different?

11 **Modelling** Look back at the graph in **Q1**.

   a Write the ratio gallons : pints in its simplest form.

   b Work out the gradient of the graph.

   c Write a formula that links gallons (G) and pints (P).

   Discussion What do you notice about the gradient, the unit ratio and the formula?

12 **Modelling** The ratio of miles to km is 5 : 8.

   a Copy and complete this table of values for miles and kilometres.

Miles	0	5	10
Kilometres			

> **Q12a hint** Are miles and km in direct proportion?

   b Draw a conversion graph for miles to kilometres.

   c Write 5 : 8 as a unit ratio.

   d Write a formula linking kilometres (y) and miles (x).

> **Q12b hint** Put miles on the x-axis and km on the y-axis.

13 **Exam-style question**

   STEM / Problem-solving A formula to change between temperature in kelvins (K) and in degrees Celsius (C) is

   $K = C + 273.15$.

   Are kelvins and degrees Celsius in direct proportion?
   Explain your answer.   **(2 marks)**

> **Exam hint**
> You won't get any marks for just 'Yes' or 'No'. You must write 'Yes' or 'No' because
> ....

11.8 Proportion problems

Objectives

- Recognise different types of proportion.
- Solve word problems involving direct and inverse proportion.

Why learn this?

A manager needs to understand proportion to allocate the right number of people to a job.

Fluency

Will 5 men take more or less time to build a wall than 3 men?

1 Work out

   a £20 ÷ 8   b £36 ÷ 5   c £2.45 × 3   d £5.90 × 3

2 How many hours and minutes in

   a 315 minutes   b 200 minutes   c $2\frac{3}{4}$ hours   d $4\frac{1}{3}$ hours?

> **Key point 11**
> When two values are in **inverse proportion**, one increases at the same rate as the other decreases. For example, as one doubles (×2) the other halves (÷2).

> **Example 7**
> It takes 2 painters 7 days to paint a house. How many days does it take 1 painter to paint an identical house?
>
> 2 people take 7 days, so 1 person takes 2 × 7 = 14 days.
>
> It takes 1 person twice (×2) as long as 2 people.

3   3 people dig a ditch in 7 hours. How long will it take

   a   1 person   b   5 people?

> **Q3a hint** Will it take 1 man more or less time than 3 men?

4   2 workers take 5 hours to tile a room. How long will it take

   a   4 workers   b   3 workers?

   Give your answers in hours and minutes.

> **Q3 hint**
> 60 minutes = 1 hour

5   4 concert tickets cost £50.

   How much will 9 concert tickets cost?

> **Q5 hint** Do you expect them to cost more or less?

6   **Real / Reasoning** A forestry supervisor needs to transplant 420 tree seedlings.

   He wants all the work done in 7 hours or less.

   He knows that 1 person can transplant 8 seedlings in an hour.

   How many people does he need for this job?

> **Q6 strategy hint**
> How many could 1 person transplant in 7 hours?

7   A band with 4 people can play a song in 4 minutes 30 seconds.

   How long will it take a band with 8 people to play the same song?

8   It takes 5 hill walkers 1 hour 20 minutes to walk up a hill.

   How long will it take 8 hill walkers?

   Discussion Compare your answer with other people. Is this a proportion problem?

9   **Real / Reasoning** A farmer has enough food for 250 chickens for 20 days.

   He buys 50 more chickens.

   How many days will the food last for?

> **Q9 strategy hint**
> How many chickens does he have to feed?

10   The weight of 30 identical coins is 150 g.

   What is the weight of 800 of these coins? Give your answer in kg.

11   **Real / Reasoning** A contractor needs to decorate 20 rooms.

   He knows that 2 people can decorate a room in 3 hours.

   He needs all the work to be done in 8 hours.

   How many people does he need for the job?

12   **Reasoning** Frank is paid £43.61 for 7 hours of work.

   How many hours does he work to earn £112.14?

13   **Reasoning** It takes 5 shredders 4 hours to shred 625 sheets of paper.

   Rick gets 3 more shredders.

   How long will it take to shred 625 sheets of paper now?

14   **Reflect** a Write and solve your own proportion problem.

   b   Give your question to a classmate. Mark their answer. Did they show their working clearly?

15   **Exam-style question**

   It takes 5 machines 2 days to complete a harvest.

   How long would it take 4 machines?   **(2 marks)**

> **Q15 strategy hint** Start by deciding if you expect it to take more or less time. This will tell you whether it is direct or inverse proportion.

11 Problem-solving

Objective · Use bar models to help you solve problems.

Example 8

Miguel and Jean share a tin of biscuits in the ratio 3 : 5. Jean has 6 more biscuits than Miguel. How many biscuits were in the tin?

all the biscuits

> Draw a rectangular bar to represent all the biscuits in the tin.

all the biscuits

Miguel   Jean

> Split the bar into the ratio 3 : 5 for Miguel and Jean.

Miguel

Jean

> Compare the bars for Miguel and Jean.
> Label difference between the bars as 6 biscuits.

6 biscuits

1 section = 6 ÷ 2 = 3 biscuits

> 2 sections represent 6 biscuits.
> Work out 1 section.

All the biscuits = 8 × 3 = 24 biscuits

> 8 sections represent all the biscuits.

Check: Miguel has: 3 × 3 = 9 biscuits

> Check your answer works.

Jean has: 3 × 5 = 15 biscuits

Jean has 15 − 9 = 6 more biscuits

Total number of biscuits = 9 + 15 = 24 biscuits ✓

1 In an orchestra, 10% of the musicians play brass instruments, 15% play woodwind and 5% play percussion. The other 28 musicians play stringed instruments.

How many musicians are in the orchestra altogether?

> **Q1 hint** Draw a bar to represent all the musicians in the orchestra. Split the bar into 10% sections. One section = ☐ musicians. Total number of musicians = ☐

2 The sizes of the angles in a triangle are in the ratio 1 : 3 : 5.

Sketch the triangle and mark the angles in degrees.

> **Q2 hint** Draw a bar to represent the sum of all the angles in a triangle. Split the bar into sections in the ratio 1 : 3 : 5. One section = ☐°

3 Hilary and Ruth share a flat with a monthly rent of £885. Hilary's bedroom is twice the size of Ruth's, so she pays twice as much rent as Ruth.

How much do they each pay?

> **Q3 hint**
>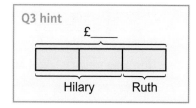

332

4   Amjit buys 3 bunches of tulips and 1 bunch of roses.
    Bunches of roses cost twice as much as bunches of tulips.
    Amjit hands over a £20 note and receives £5 change.
    How much is a bunch of roses?

Q4 hint

tulips   tulips   tulips

5   Toby and Isy share some money in the ratio 5 : 7.
    Isy gets £60 more than Toby.
    How much money does Toby get?

Q5 hint Draw a bar to show
the ratio 5 : 7.
Label the sections.
How many sections
represent £60?
One section = £☐

6   A farmer has sheep, cows and pigs in the ratio 12 : 3 : 2.
    The farmer has 150 more sheep than pigs.
    How many cows has she got?

7   8 boys and 12 girls go to swimming lessons.
    In one lesson, the mean number of lengths swum
    by the boys is 4.5, and the mean number of lengths
    swum by the girls is 2.
    Work out the mean number of lengths swum by
    all the children in that lesson.

Q7 hint Draw a bar to represent all
the children.
Split the bar into sections that show
the numbers of boys and girls.
Work out how many lengths the
boys swam in total and how many
lengths the girls swam in total.
Work out how many lengths the
children swam altogether.

8   **Reflect** How did drawing bar models help you?
    Is this a strategy you would use again to
    solve problems?

11 Check up

Log how you did on your
Student Progression Chart.

Simple proportion and best buys

1   A recipe for 4 people uses 150 g of sugar. How much sugar is needed for
    a   8 people   b   2 people   c   6 people?

2   **Reasoning** Helen can buy 300 ml of shampoo for £3.50 or 75 ml of shampoo for £1.
    Which is better value for money?

Ratio and proportion

3   A necklace has 24 beads. There is 1 red bead for every 3 blue beads.
    How many beads are
    a   red   b   blue?

4   Write each ratio in its simplest form.
    a   24 : 32   b   27 : 18 : 54   c   2.5 : 3.5

5   Copy and complete these conversions.

   a   4.2 km = ☐ m

   b   0.05 litres = ☐ ml

   c   £200 = ☐ euros (£1 = 1.23 euros)

   d   25 km ≈ ☐ miles (8 km ≈ 5 miles)

6   Iona makes a scale model of a sailing boat using the ratio 1 : 25.
   Her model is 60 cm long.
   How long is the real sailing boat? Give your answer in metres.

7   Hazel mixes pink paint using red paint and white paint in the ratio 1 : 4.
   How much white paint should she use with 5 tins of red paint?

8   Naadim and Bal share £60 in the ratio 2 : 3.

   a   How much do they each receive?

   b   What fraction of the total amount does Naadim receive?

   c   Show how you checked your answer is correct.

9   Write each of these as a unit ratio. Give your answers to a maximum of 2 decimal places.

   a   8 : 5   b   7 : 19

10   The ratio of primary school children to secondary school children in a club is 2 : 3.
   There are 12 primary school children in the club.

   a   How many secondary school children are there?

   b   What is the total number of children in the club?

11   **Reasoning** Luke makes lemon squash using 40 ml of squash and 380 ml of water.
   Joseph makes lemon squash using 50 ml of squash and 475 ml of water.
   Who has made the stronger squash? Explain your answer.

Proportion, graphs and inverse proportion

12   **Reasoning** The table shows the number of
   Barbadian dollars you get for different numbers of
   US dollars.

US dollars	5	10	50
Barbadian dollars	10	20	100

   a   Draw a conversion graph with US dollars on the
      x-axis and Barbadian dollars on the y-axis.

   b   Are US dollars and Barbadian dollars in direct proportion?

   c   Write a formula linking US dollars and Barbadian dollars.

13   It takes 4 people 3 days to lay the cables for high-speed fibre broadband.
   How long would it take

   a   1 person   b   6 people?

14   **Reasoning** It takes 3 machines 4 minutes to print 600 pages.
   How long will it take 8 machines to print 600 pages?

15   How sure are you of your answers? Were you mostly

   Just guessing 😟   Feeling doubtful 😐   Confident 🙂

   What next? Use your results to decide whether to strengthen or extend your learning.

* Challenge

16 a How many different ways can you divide this rectangle
into two parts in the ratio 1 : 3?

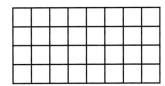

b How many of your answers show a symmetrical pattern?

11 Strengthen

Simple proportion and best buys

1 A bus ticket costs £3. Work out the cost of

   a 2 tickets   b 5 tickets.

2 It costs £32 for 4 teenagers
to go to the cinema.
How much does it cost for

   a 8 teenagers

   b 2 teenagers

   c 10 teenagers?

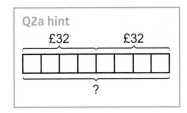

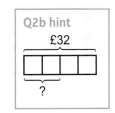

3 Harry can wash 6 cars in 1 hour.
How many cars can he wash in $3\frac{1}{2}$ hours?

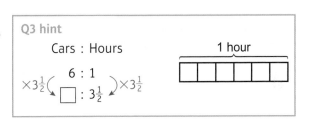

4 **Reasoning** Which is better value for money:
200 g of lotion for £2.50 or
300 g of lotion for £3.60?

Ratio and proportion

1 Andie makes a bracelet with
green beads and yellow beads.

She uses 1 green bead for every 2 yellow beads.

The bracelet has 12 beads altogether.
How many beads are

   a green   b yellow?

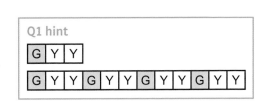

2   Write each ratio in its simplest form.

    a   2 : 6   b   6 : 9   c   10 : 15

> **Q2b hint** You cannot divide by 2. Try dividing by 3.

    d   24 : 36 : 60   e   2.5 : 6   f   2.4 : 1.6

> **Q2e hint** Choose a number to multiply by to give whole numbers first.

Q2a hint

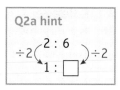

3   Complete these conversions.

    a   3.8 m = ☐ cm   b   2.5 litres = ☐ ml

    c   £600 = ☐ euros (£1 = 1.23 euros)   d   24 pints = ☐ gallons (1 gallon = 8 pints)

Q3a hint

Q3c hint

4   Jim mixes red paint and blue paint in the ratio 1 : 5.
    He uses 3 tins of red paint.
    How much blue paint should he use?

Q4 hint 1 red for every 5 blue.

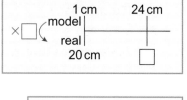

5   Jenny makes a scale model of a train carriage using
    the ratio 1 : 20.
    Her model is 24 cm long.

    a   How long is the real train carriage in cm?

    b   Write the real length in metres.

Q5a hint

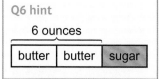

6   The ratio of butter to sugar in a recipe is 2 : 1.
    Ian uses 6 ounces of butter.
    How many ounces of sugar should he use?

Q6 hint

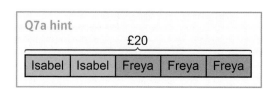

7   Isabel and Freya share £20 in the ratio 2 : 3.

    a   How much do they each receive?

    b   Show how you have checked your answer.

> **Q7b hint** Add your answers for Isabel and
> Freya's shares together.

Q7a hint

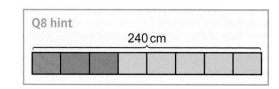

8   A piece of cloth is cut into three pieces in
    the ratio 1 : 2 : 5.
    The piece of cloth is 240 cm long.

    a   How long is each piece?

    b   Show how you have checked your answer.

Q8 hint

9 A shop sells jam doughnuts and toffee doughnuts.

The ratio of jam doughnuts to toffee doughnuts sold is 2 : 3.

The shop sold 18 jam doughnuts.

a How many toffee doughnuts were sold?

b How many doughnuts were sold altogether?

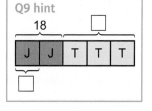

Q9 hint

10 The ratio of quad bikes to go-karts at an activity centre is 3 : 7.

a What fraction of the vehicles are go-karts?

b What fraction of the vehicles are quad bikes?

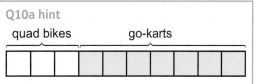

Q10a hint

quad bikes   go-karts

11 $\frac{3}{4}$ of the bikes at a cycle shop are road bikes. The rest are mountain bikes.

a What is the ratio of road bikes : mountain bikes?

b There are 27 road bikes. How many mountain bikes are there?

c How many bikes are there in the cycle shop altogether?

Q11a strategy hint
Draw a picture to show the fractions.

12 Write each of these as a unit ratio. Give your answers to a maximum of 2 decimal places.

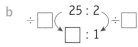

a $\div 4 \overset{\frown}{\underset{\smile}{\left(\frac{4 : 26}{1 : \square} \right)}} \div \square$

b $\div \square \overset{\frown}{\underset{\smile}{\left(\frac{25 : 2}{\square : 1} \right)}} \div \square$

c 13 : 5

d 7 : 22

Q12 strategy hint
Divide both numbers in the ratio by the smaller number.

13 **Reasoning** Amar makes lemon squash using 20 ml of squash and 180 ml of water.

Ben makes lemon squash using 50 ml of squash and 500 ml of water.

Who has made the stronger squash? Explain your answer.

Q13 strategy hint
Write them both as unit ratios.

Proportion, graphs and inverse proportion

1 The sketch graph shows an object moving at a constant speed.

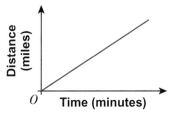

a Is this graph a straight line?

b Does the graph go though (0, 0)?

c Is the distance travelled in direct proportion to the time taken?

Q1c hint Did you answer 'Yes' to parts **a** and **b**?

2 **Reasoning** An electrician charges a callout fee of £60 and £30 per hour she works.

Hours worked	0	1	2	3
Total charge (£)	60	90	120	150

a Draw a graph for the table of her charges.

b Is her total charge in direct proportion to the hours she works?

Q2b hint Answer **Q1a** and **Q1b** for the graph you have drawn.

337

3   A shop sells material for bridesmaid
dresses at £4 per metre.

Is the price of material proportional
to the length bought?

Answer **Q1a** and **Q1b** for the
graph you have drawn.

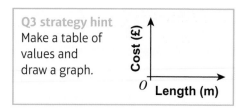

Q3 strategy hint
Make a table of
values and
draw a graph.

4   The ratio of kilograms to pounds is 1 : 2.2.

Write a formula that shows the relationship
between kilograms (k) and pounds (p).

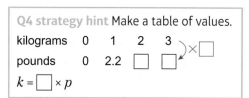

Q4 strategy hint Make a table of values.

| kilograms | 0 | 1 | 2 | 3 |
| pounds | 0 | 2.2 | | |

$k = \boxed{} \times p$

5   It takes 3 people 4 hours to lay some pipes.
Copy and complete the table.

Number of people	Longer or shorter?	Hours
3	—	4
1		
2		
6		

Q5 strategy hint If
longer, then multiply.
If shorter, then divide.

6   **Reasoning** It takes 4 washing machines 12 hours to wash the
hotel laundry.

How long will it take 6 machines?

Q6 hint Draw a table
like the one in Q5.

11 Extend

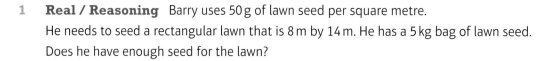

1   **Real / Reasoning** Barry uses 50 g of lawn seed per square metre.

He needs to seed a rectangular lawn that is 8 m by 14 m. He has a 5 kg bag of lawn seed.

Does he have enough seed for the lawn?

2   **Real / Problem-solving** Freda is planning a trip for her gymnasts.

The recommended adult-to-child ratios are

• Age 4 to 8 – a minimum of 1 adult for every 6 children

• Age 9 to 12 – a minimum of 1 adult for every 8 children

• Age 13 to 18 – a minimum of 1 adult for every 10 children

Here are the ages of the gymnasts.

Age	Number of gymnasts
5 or 6	18
7 or 8	12
10 or 11	13
12	3
14 and over	23

Work out the minimum number of adults that need to go on the trip.

3

Ketchup is sold in three different sizes of bottle.

Small bottle   Medium bottle   Large bottle

A small bottle contains 342 g of ketchup and costs 88p
A medium bottle contains 570 g of ketchup and costs £1.95
A large bottle contains 1500 g of ketchup and costs £3.99

Which bottle is the best value for money?
You must show your working.   **(4 marks)**

June 2013, Q9, 5MB3H/01

> **Q3 strategy hint**
> Write your final answer clearly '__ bottle is the best value for money'.

4 **Reasoning** The lengths of the sides of a triangle are in the ratio 3 : 4 : 5.
The perimeter of the triangle is 18 cm.
What is the length of the longest side?

5

Here is a list of ingredients for making cherry scones.

Makes 8 cherry scones

200 grams flour

60 grams margarine

40 grams sugar

60 grams cherries

160 ml milk

Chen wants to make 20 cherry scones.
a Work out how much milk he will need.

Sophie has 80 grams of sugar and 300 grams of flour.
She has plenty of the other ingredients.
b What is the greatest number of cherry scones she can make?
You must show all your working.

(5 marks)

Nov 2013, Q5, 5MB2H/01

> **Exam hint**
> Most of the marks are for your method so write down all the steps of the working.

6 **Reasoning** 3 painters take 12 hours to paint a house.
a How long would it take 4 painters to paint the house?
b The charge for the work is £18 per painter per hour. What is the total cost?

7 **Problem-solving** The sizes of the angles in a triangle are in the ratio 2 : 3 : 4.
What size is each angle?

8 **Real / Reasoning** Potting compost is made using loam, peat and sand in the ratio 7 : 3 : 2.
Molly uses $2\frac{1}{2}$ litres of sand.
a How much loam and peat should she use?
b What is the total volume of the compost?

9 **Reasoning** Sunil and Karl shared some money in the ratio 3 : 5.
Sunil gave $\frac{1}{6}$ of his money to Lucas.
Lucas got £12.50.
How much money was shared originally by Sunil and Karl?

> **Q9 strategy hint**
> Start with £12.50. How much did Sunil have?

10 **Modelling / Problem-solving** 3 boys did some gardening. They earned £129.60.
They shared the money in the ratio of the number of hours they each worked.
Ollie worked for $9\frac{1}{2}$ hours, Sam worked for $8\frac{1}{4}$ hours and Peter worked for $6\frac{1}{4}$ hours.
How much money did each boy receive?

11 **Modelling / Problem-solving** Adam is tiling a room. He can fit 80 tiles in 1 hour.
He takes a 15-minute morning break and $\frac{3}{4}$ hour for lunch.
He has to fit 540 tiles. He starts work at 8:30 am. At what time will he finish work?

12 **Problem-solving** A box contains 480 counters.
There are twice as many red counters as yellow counters.
There are three times as many blue counters as red counters.
The rest of the counters are green.
The box contains 50 yellow counters.
How many green counters are in the box?

> **Q12 strategy hint**
> Start with the
> 50 yellow counters.
> What can you work
> out next?

13 **Problem-solving** The sizes of the angles in a quadrilateral are in the ratio 2 : 4 : 5 : 7.
What size is the largest angle?

14 **Reasoning / Problem-solving** Sasha wants to buy a new camera.
She can either buy it from a local dealer or order it from the USA online.

Local dealer	£399
Imported from USA	$529 plus taxes and duties of 20%

Both prices include postage
The exchange rate is £1 = $1.68. She wants to pay the cheaper total cost.
Which option should she choose?

15 **Problem-solving** The chart shows the proportions
of sand and cement needed to mix mortar for brickwork.
A builder uses 78 kg of cement to make his mortar.
How much sand should he use?

Key:
☐ sand
☐ cement

(Pie chart showing 80° and 280°)

> **Q15 hint** Write down the ratio of sand : cement.

16 **Problem-solving** 50 people go on a trip to the cinema.
The ratio of adults to teenagers on the trip is 2 : 3.
30 per cent of the adults are senior citizens.
The table shows the ticket prices.
What is the total cost for all 50 people to visit the cinema?

Adult	£8.45
Teenager	£6.75
Senior or child	£6.35

11 Knowledge check

⊙ A **ratio** is a way to compare two or more quantities. .... *Mastery lesson 11.1*

⊙ You **simplify** a ratio by making the numbers as small as possible.
Divide the numbers in the ratio by their **highest common factor**
(**HCF**). .... *Mastery lesson 11.1*

⊙ Ratios in their **simplest form** only have whole numbers. .... *Mastery lesson 11.2*

⊙ You can use **ratios** to convert between units. .... *Mastery lesson 11.3*

⊙ A **proportion** compares a part with the whole. .... *Mastery lesson 11.5*

⊙ You can compare ratios by writing them as **unit ratios**.
In a unit ratio, one of the numbers is 1. .... *Mastery lesson 11.5*

⊙ In the **unitary method** you find the value of one item before
finding the value of more. .... *Mastery lesson 11.6*

⊙ You can use the unitary method to work out which product gives
better value for money. .... *Mastery lesson 11.6*

⊙ When two values are in **direct proportion**, if one value is zero so is
the other. When one value doubles, so does the other. .... *Mastery lesson 11.7*

⊙ When two quantities are in direct proportion, plotting them as a
graph gives a straight line through the origin. .... *Mastery lesson 11.7*

⊙ When two values are in **inverse proportion**, one increases at the
same rate as the other decreases. For example, as one doubles (×2)
the other halves (÷2). .... *Mastery lesson 11.8*

For each statement, A, B and C, choose a score:

1 – strongly disagree; 2 – disagree; 3 – agree; 4 – strongly agree

A I always try hard in mathematics

B Doing mathematics never makes me worried

C I am good at mathematics

For any statement you scored less than 3, write down two things you could do so that you
agree more strongly in the future

11 Unit test

Log how you did on your
Student Progression Chart.

1 At the 2014 Commonwealth Games, Australia won 42 silver medals and
India won 30 silver metals.

Write the ratio of the number of silver medals won by Australia to the
number of silver medals won by India. Give your ratio in its simplest form. *(2 marks)*

2 1 kg of cheddar cheese costs £6.40. How much does 100 g cost? *(2 marks)*

3 There are 45 lemons and limes in a box.
The ratio of the number of lemons to the number of limes is 1 : 4.
Work out the number of limes in the box. *(3 marks)*

4 Greg changed £475 into euros. The exchange rate was £1 = 1.2 euros.
How many euros did he get? *(2 marks)*

5 **Reasoning** At Mini Mart, 4 tubs of ice cream cost £9.40.

At Dave's Deli, 3 tubs of the same ice cream cost £7.50.

At which shop is ice cream better value for money? *(3 marks)*

6 A school with 900 students has 150 computers. Write the ratio of the number of
students to the number of computers in the form $m : 1$. *(1 mark)*

7 **Reasoning** There are 250 marbles in a tin.
40 marbles are metal, 60 are stoneware and the rest are glass or china.
There are twice as many glass marbles as china marbles.
How many are   a china   b glass? *(4 marks)*

8 **Reasoning** Sam, Jack and Ali share £45 in the ratio 2 : 3 : 4.

a What fraction does Sam get?

b How much does Ali get? *(4 marks)*

9 **Reasoning** James, Isaac and Lucas share £30 in the ratio of their ages.

James is 10 years old, Isaac is 8 years old and Lucas is 7 years old.

Isaac gives a third of his share to his Dad.

How much money does Isaac have now? *(4 marks)*

10 **Reasoning** Here are the ingredients needed to make
a beef pie for 6 people.

a How much beef is needed to make
a beef pie for 15 people? *(3 marks)*

b If you have 5 eggs and plenty of the
other ingredients, can you make
a beef pie for 8 people? *(3 marks)*

Beef pie (serves 6 people)
180 g flour
320 g beef
80 g butter
4 eggs
160 ml milk

11 **Reasoning** A company makes fruit drinks.

A machine fills crates of 12 bottles with drink. In 1 hour the machine can fill 18 000 bottles.

How many seconds does the machine take to fill a crate of 12 bottles? *(4 marks)*

12 **Reasoning** The graph shows a car moving at a constant speed.

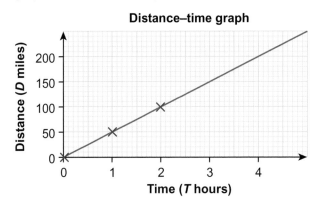

Distance–time graph

a Does the graph show distance and time in direct proportion?
Explain your answer. *(1 mark)*

b Write down the equation of the line. *(1 mark)*

13 **Reasoning** 3 machines can print a batch of leaflets in 2 hours.

How long would it take 4 machines to print the same batch of leaflets? *(3 marks)*

14 ⟨ **Exam-style question** ⟩

Each day a company posts some small letters and some large letters.

The company posts all the letters by first class post.

The tables show information about the cost of sending a small letter by first class post and the cost of sending a large letter by first class post.

Small letter

Weight	First class post
0–100 g	60p

Large letter

Weight	First class post
0–100 g	£1.00
101–250 g	£1.50
251–500 g	£1.70
501–750 g	£2.50

One day the company wants to post 200 letters.

The ratio of the number of small letters to the number of large letters is 3 : 2

70% of the large letters weigh 0–100 g.

The rest of the large letters weigh 101–250 g.

Work out the total cost of posting the 200 letters by first class post. **(5 marks)**

Nov 2013, Q11, 1MA0/1H

Sample student answers

Which student gives the best answer and why?

⟨ **Exam-style question** ⟩

Potatoes cost £9 for a 12.5 kg bag at a farm shop.
The same type of potatoes cost £1.83 for a 2.5 kg bag at a supermarket.

Where are the potatoes the better value, at the farm shop or at the supermarket? You must show your working. **(4 marks)**

June 2012, Q23, 1MA0/2F

Exam hint

Remember that there are marks for showing your method and for communicating your answer clearly, as well as for getting the answer right.

Student A Farm shop: £9 for 12.5 kg

Supermarket: ×5 ⟨ £1.83 for 2.5 kg ⟩ ×5
                        £9.15 for 12.5 kg

The potatoes are better value at the farm shop.

Student B 12.5 ÷ 2.5 = 5

£9 ÷ 5 = £1.80

Student C Farm shop: 12.5 ÷ 9 = 1.388… kg for £1

Supermarket: 2.5 ÷ 1.83 = 1.366 kg… for £1

The potatoes are better value at the supermarket.

12 RIGHT-ANGLED TRIANGLES

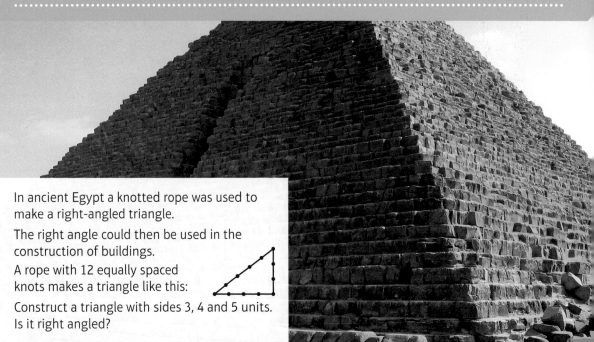

In ancient Egypt a knotted rope was used to make a right-angled triangle.

The right angle could then be used in the construction of buildings.

A rope with 12 equally spaced knots makes a triangle like this:

Construct a triangle with sides 3, 4 and 5 units. Is it right angled?

12 Prior knowledge check

Numerical fluency

1 Round these numbers to 1 decimal place.
   a 1.95
   b 18.52

2 Round these numbers to 3 significant figures.
   a 12.327 86
   b 0.364 999 999

3 Work out
   a 5^2
   b 11^2
   c $5^2 + 6^2$
   d $(5 + 6)^2$
   e $\sqrt{36}$
   f $\sqrt{13^2 - 12^2}$

4 Work out
   a $\sqrt{8}$
   b $\sqrt{1 + 2^2}$
   c $\sqrt{6^2 - 5^2}$
   d $\sqrt{3^2 + 7^2}$
      i give your answers in surd form
      ii as a decimal.

5 Work out
   a $1.9^2 + 2.3^2$
   b $\sqrt{15.21}$
   c $\sqrt{(12^2 + 16^2)}$

Geometrical fluency

6 Which of these are right-angled triangles?

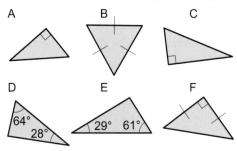

7 For triangle ABC write

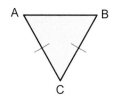

   a the equal sides
   b the equal angles
   c the side opposite angle BAC
   d the type of triangle.

Fluency with measures

8 Measure the length of each line AB to the nearest millimetre.

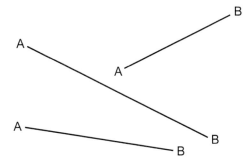

9 Use a protractor to draw these angles accurately.
   a 30°   b 47°

Algebraic fluency

10 $c^2 = 3^2 + 4^2$
    Work out the value of
    a c^2 b c

11 $x = 6$ and $y = 9$.
    Work out $\frac{x}{y}$.
    Give your answer as a fraction in its simplest form.

12 Solve these equations.
    a $0.3 = \frac{x}{10}$ b $0.2 = \frac{x}{12}$ c $0.5 = \frac{2}{x}$

✱ Challenge

13 One of these sets of measurements cannot be used to construct a triangle.

> **Q13 hint** Try constructing each triangle accurately.

   A 10 cm, 5 cm, 12 cm
   B 15 m, 19 m, 30 m
   C 21 mm, 16 mm, 40 mm
   a Which set is it?
   b Write a rule that could be used to test whether a set of three measurements could be used to construct a triangle.

12.1 Pythagoras' theorem 1

Objectives

- Understand Pythagoras' theorem.
- Calculate the length of the hypotenuse in a right-angled triangle.
- Solve problems using Pythagoras' theorem.

Did you know?

Pythagoras was a mathematician who lived in Greece around 2700 years ago. He is still remembered today for his work on geometry.

Fluency

a Work out 6^2
b Work out
   • $\sqrt{49}$ • the square root of 81
c What is the area of this square?

 12 cm

1 $x = 7$ and $y = 24$. Work out
   a $x^2 + y^2$ b $\sqrt{x^2 + y^2}$

2 Find the positive solution of each equation.
   a $c^2 = 65$ b $c^2 = 11^2 + 15^2$
   Give your answers correct to 3 significant figures.

Warm up

Key point 1

In a right-angled triangle the longest side is called the **hypotenuse**.

Questions in this unit are targeted at the steps indicated.

3   a   Draw each triangle on centimetre squared paper.

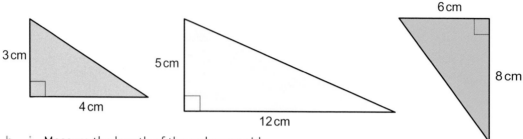

    b   i   Measure the length of the unknown side.
        ii   What is the length of the hypotenuse?
    Discussion Will the largest angle in a right-angled triangle always be the right angle?
    c   **Reflect** What do you notice about the relationship between the right angle and the hypotenuse in right-angled triangles?

4   Write the length of the hypotenuse for each of these triangles.

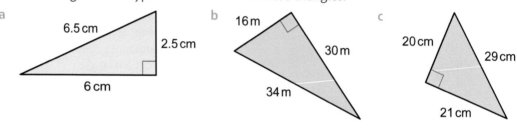

5   A square is drawn on each side of a right-angled triangle.
    a   Check that the measurements match your answer to **Q3a**.
    b   Find the area of each square.
    c   Copy and complete:
        $\_\_^2 = 3^2 + 4^2$

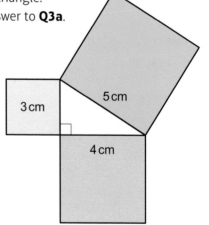

6   **Reasoning** Copy and complete this statement.
    The square on the hypotenuse has the _____ area as the sum of the areas of the _____ on the other two sides.

> Q6 hint Look back at **Q5**.

Key point 2

Pythagoras' theorem shows the relationship between the lengths of the three sides of a right-angled triangle.
$c^2 = a^2 + b^2$

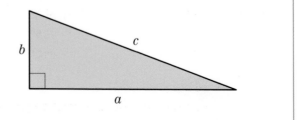

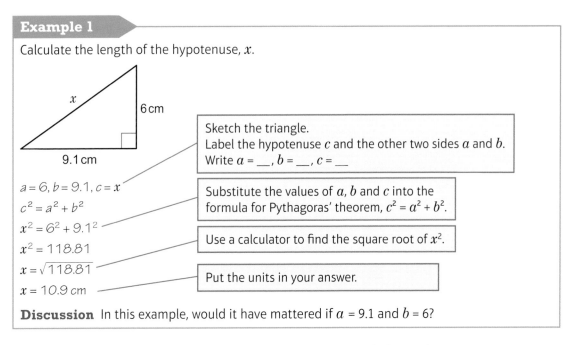

Example 1

Calculate the length of the hypotenuse, x.

x 6 cm

9.1 cm

Sketch the triangle.
Label the hypotenuse c and the other two sides a and b.
Write $a = $ __ , $b = $ __ , $c = $ __

$a = 6, b = 9.1, c = x$

Substitute the values of a, b and c into the formula for Pythagoras' theorem, $c^2 = a^2 + b^2$.

$c^2 = a^2 + b^2$

$x^2 = 6^2 + 9.1^2$

$x^2 = 118.81$

Use a calculator to find the square root of x^2.

$x = \sqrt{118.81}$

Put the units in your answer.

$x = 10.9$ cm

Discussion In this example, would it have mattered if $a = 9.1$ and $b = 6$?

7 Calculate the length of the hypotenuse, x, in each right-angled triangle.

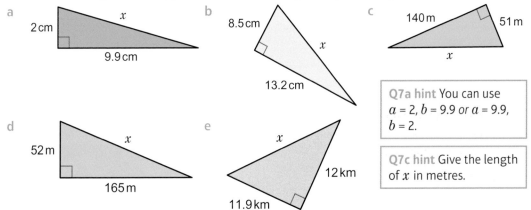

a   2 cm x 9.9 cm

b   8.5 cm x 13.2 cm

c   140 m   51 m x

d   52 m x 165 m

e x 12 km   11.9 km

Q7a hint You can use $a = 2, b = 9.9$ or $a = 9.9$, $b = 2$.

Q7c hint Give the length of x in metres.

8 Calculate the length of PR in each right-angled triangle. Give your answers correct to 2 decimal places.

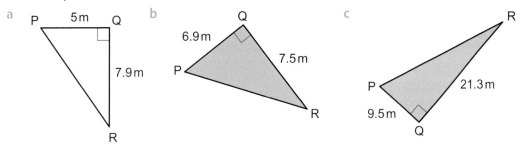

a   P   5 m   Q   7.9 m   R

b   Q   6.9 m   7.5 m   P   R

c   R   21.3 m   P   9.5 m   Q

Discussion Why does this question ask you to round to 2 decimal places rather than 3 significant figures? What if the lengths were given in cm? What would be a sensible rounding then?

9 **Real / Problem-solving** A school is adding a wheelchair ramp to the front entrance. Safety regulations require a wheelchair ramp to rise 1 m over a distance of 12 m.
How long must the ramp be? Give your answer to an appropriate degree of accuracy.

Q9 strategy hint
Sketch a diagram.

10 **Exam-style question**

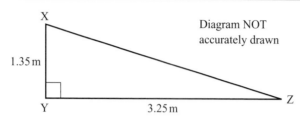

Diagram NOT
accurately drawn

XYZ is a right-angled triangle.
Calculate the length of XZ.
Give your answer correct to 3 significant figures. **(3 marks)**

June 2013, Q27 1MA0/2F

Exam hint
Do not round before
taking the square root.
Use all the figures on
your calculator display.

11 **Exam-style question**

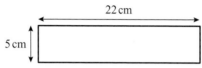

The opening of a letterbox is a rectangle 22 cm wide and 5 cm high.
An envelope is 25 cm long and 22.4 cm wide. Will the envelope
go through the letterbox without bending?
Show your working. **(4 marks)**

Q11 strategy hint
Copy the diagram.
Draw in a diagonal and
mark in a right angle.

12.2 Pythagoras' theorem 2

Objectives

- Calculate the length of a line segment AB.
- Calculate the length of a shorter side in a
  right-angled triangle.

Why learn this?

Builders use Pythagoras' theorem to check the
layout of a building whose corners have to be
right angles.

Fluency

- What does ≠ mean?
- Here is a calculator display. What does it mean?

$3\sqrt{2}$

1 Here is a right-angled triangle.
  a Sketch the triangle. Label the hypotenuse c.
    Label the other two sides a and b.
  b Calculate the length of AC.
    Give your answer correct to 3 significant figures.

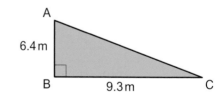

2 Solve these equations.
  a $13^2 = 5^2 + b^2$ b $5^2 = 3^2 + b^2$
  c $13^2 = a^2 + 12^2$ d $10^2 = a^2 + 8^2$

Q2a hint $13^2 - 5^2 = 5^2 + b^2 - 5^2$
$13^2 - 5^2 = b^2$

Key point 3

A **surd** is a root whose value cannot be worked out exactly.

ActiveLearn Homework, practice and support: Foundation 12.2

3   Are these numbers surds?

Q3 hint $\sqrt{4}$ is not a surd as $\sqrt{4} = 2$.

   a $\sqrt{16}$ b $\sqrt{9}$ c $\sqrt{3}$
   d $\sqrt{2}$ e $\sqrt{25}$ f $\sqrt{7}$

4   Simplify these expressions. Give each answer as a surd.

Q4 hint Work out the calculation inside the square root first.

   a $\sqrt{1^2 + 2^2}$ b $\sqrt{2^2 + 3^2}$ c $\sqrt{5^2 - 2^2}$

5   **Problem-solving** The points A(1, 2) and B(5, 4) are shown on a centimetre square grid.
   Work out the length in centimetres of AB. Give your answer correct to 3 significant figures.

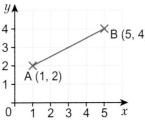

Q5 hint Start by sketching a right-angled triangle with AB as the hypotenuse.

6   **Modelling / Problem-solving** A ship sails from point X to point Y.
   Calculate XY, the distance it sails. The points are shown on a kilometre square grid.

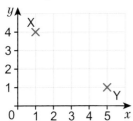

Q6 hint Start by writing the coordinates of X and Y.
Sketch the right-angled triangle.

Q7a hint Sketch and label a diagram.

7   The points P and Q are plotted on a centimetre square grid.
   For each set of points, calculate the length of PQ.
   a   P(3, 7) and Q(6, 3)   b   P(−5, 2) and Q(7, 7)   c   P(−7, 5) and Q(8, −3)

Key point 4

You can use Pythagoras' theorem to work out the length of a shorter side in a right-angled triangle.

Example 2

Work out the length of the unknown side in this right-angled triangle.
Give your answer to 1 decimal place.

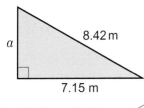

Sketch the triangle.
Label the hypotenuse c and the other two sides a and b.
Write $b =$ __, $c =$ __

$b = 7.15, c = 8.42$

$c^2 = a^2 + b^2$

Substitute the values of a, b and c into Pythagoras' theorem, $c^2 = a^2 + b^2$

$8.42^2 = a^2 + 7.15^2$

$8.42^2 - 7.15^2 = a^2 + 7.15^2 - 7.15^2$ ─── Solve the equation.

$a^2 = 8.42^2 - 7.15^2$

$a^2 = 19.7739$

$a = \sqrt{19.7739}$ ─── Use a calculator to find the square root.

$a = 4.4467...$

$a = 4.4 \text{ m (to 1 d.p.)}$ ─── Give your answer correct to 1 d.p. and include the units.

8

Exam-style question

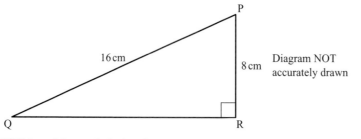

Diagram NOT accurately drawn

PQR is a right-angled triangle.
PQ = 16 cm
PR = 8 cm
Calculate the length of QR.
Give your answer correct to 2 decimal places. **(3 marks)**

March 2012, Q23 1380/2F

Exam hint
Show your unrounded answer before rounding to 2 dp.

9 Work out the length of the unknown side in each right-angled triangle. Give your answers to an appropriate degree of accuracy.

a
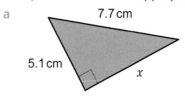

b
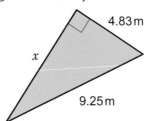

Q9 hint State the degree of accuracy after each answer e.g. 1 d.p.

Discussion How do the values in the question help you decide on a suitable degree of accuracy for the answers?

10 **Problem-solving** Work out the perimeter of triangle ABC.

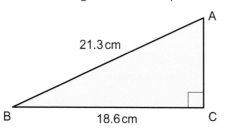

Q10 hint Work out the length of AC first.

11 Find the length of the unknown side in each right-angled triangle. Give your answers in surd form.

Q11 hint Your answer should look like this:
$x = \sqrt{\square}$ cm

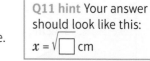

a
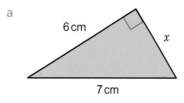

b
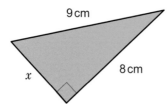

c
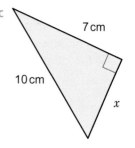

Q11 communication hint Giving an answer in 'surd form' means that you *do not* work out the square root.

Key point 5

A triangle with sides a, b and c, where c is the longest side, is right-angled only *if* $c^2 = a^2 + b^2$.

12 **Problem-solving** Here are the lengths of sides of triangles. Which of these triangles are right-angled triangles? The first one has been done for you.

a 5 cm, 6 cm, 8 cm

$8^2 = 64$

$5^2 + 6^2 = 61$

$8^2 \neq 5^2 + 6^2$

The triangle is not right-angled.

b 6 km, 8 km, 10 km

c 4 m, 5 m, 6 m

> **Q12a hint** If the triangle is right-angled, the longest side will be the hypotenuse.

12.3 Trigonometry: the sine ratio 1

Objectives

- Understand and recall the sine ratio in right-angled triangles.
- Use the sine ratio to calculate the length of a side in a right-angled triangle.
- Use the sine ratio to solve problems.

Why learn this?

Computer game developers use trigonometry to control character movements around the screen.

Fluency

Simplify these fractions.

a $\frac{15}{30}$ b $\frac{12}{36}$ c $\frac{14}{21}$

1 Convert each fraction to a decimal. Give your answer correct to 1 decimal place where appropriate.

a $\frac{9}{18}$ b $\frac{5}{12}$ c $\frac{21}{32}$

2 Solve these equations.

a $0.67 = \frac{x}{10}$ b $0.413 = \frac{x}{3}$ c $0.325 = \frac{x}{20}$

Key point 6

The side opposite the right angle is called the **hypotenuse**.
The side opposite the angle θ is called the **opposite**.
The side next to the angle θ is called the **adjacent**.

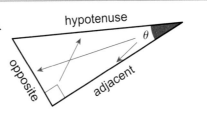

3 Copy these right-angled triangles. For each triangle, label the **hyp**otenuse '**hyp**' and the side **opp**osite the angle θ '**opp**'. The first one has been started for you.

> **Q3 communication hint** θ is the Greek letter 'theta'.

a

opp

b
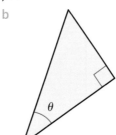

c
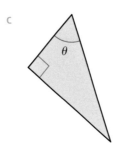

4 **Reasoning**

a Draw these triangles accurately using a ruler and protractor.

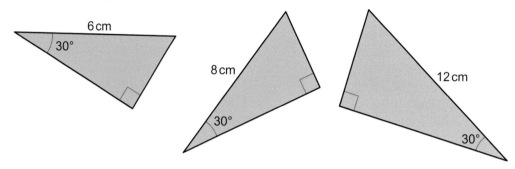

> **Q4a hint**
> Draw the hypotenuse first. Then draw a line at 30° to the hypotenuse.
> Complete the right-angled triangle

b Label the hypotenuse (hyp) and opposite side (opp).

c Measure the opposite side.

d i Write the fraction $\dfrac{\text{opposite}}{\text{hypotenuse}}$ for each triangle and convert it to a decimal.

   ii What do you notice?

   iii What do you think the fraction will be for any right-angled triangle with an angle of 30°?

   iv Test your prediction by drawing some more triangles.

Discussion Why are the fractions for the same angle the same? What does this tell you about the ratio of the sides?

Key point 7

In a right-angled triangle the sine of an angle is the ratio of the **opp**osite side to the **hyp**otenuse.
The sine of angle θ is written as **sin θ**.

$\sin \theta = \dfrac{\text{opp}}{\text{hyp}}$

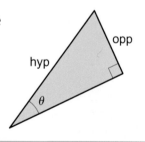

---

5 Use your calculator to find, correct to 3 decimal places

   a sin 50°   b sin 38°   c sin 46.8°

Reasoning Use your answers to **Q4** to write the value of sin 30°.
Check using a calculator.

> **Q5a hint** On your calculator enter
>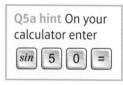

6 Write sin θ as a fraction for each triangle.

   a   b   c

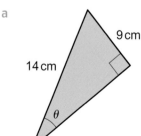
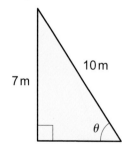
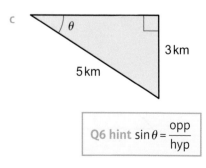

> **Q6 hint** $\sin \theta = \dfrac{\text{opp}}{\text{hyp}}$

Example 3

Use the sine ratio to work out the value of x.

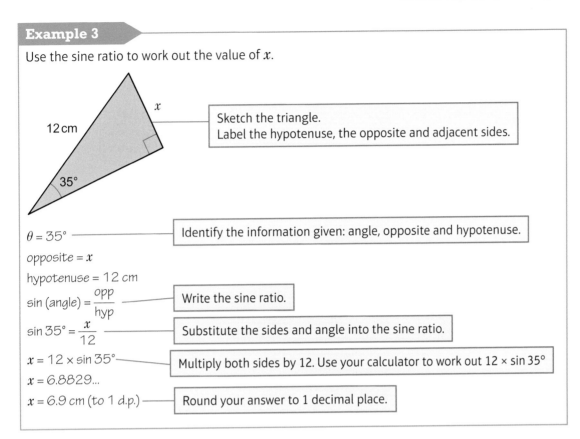

Sketch the triangle.
Label the hypotenuse, the opposite and adjacent sides.

$\theta = 35°$ —— Identify the information given: angle, opposite and hypotenuse.

opposite = x

hypotenuse = 12 cm

$\sin (\text{angle}) = \dfrac{\text{opp}}{\text{hyp}}$ —— Write the sine ratio.

$\sin 35° = \dfrac{x}{12}$ —— Substitute the sides and angle into the sine ratio.

$x = 12 \times \sin 35°$ —— Multiply both sides by 12. Use your calculator to work out $12 \times \sin 35°$

$x = 6.8829...$

$x = 6.9$ cm (to 1 d.p.) —— Round your answer to 1 decimal place.

7 Find the value of x in each triangle. Give your answers correct to 1 decimal place.

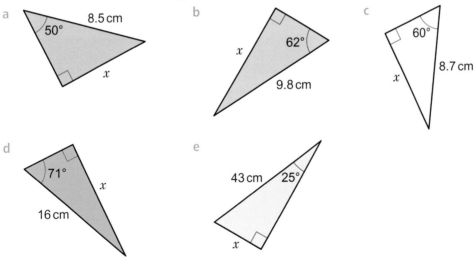

8 **Modelling / Real**
A ladder of length 3 m leans against a vertical wall.
The ladder makes an angle of 68° with the horizontal ground.
How far is the top of the ladder from the ground?

3 m

68°

Q8 hint State the degree of accuracy after your answer.

353

9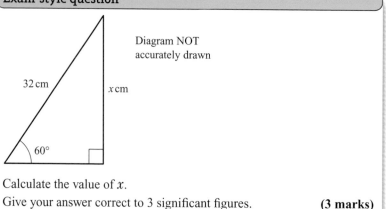

Exam-style question

Diagram NOT
accurately drawn

32 cm

x cm

60°

Calculate the value of x.

Give your answer correct to 3 significant figures. **(3 marks)**

Nov 2012, Q17, 1MA0/2H

Exam hint
You get I mark
for writing an
equation using
the correct
ratio, I mark for
rearranging and
I mark for the
correct value
of x.

10 **Reasoning / Modelling** The top of a vertical flagpole is secured to the ground by a rope.
The rope makes an angle of 60° with the ground.
Will a rope of length 8 m reach the top of a 7.5 m flagpole?

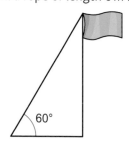

60°

12.4 Trigonometry: the sine ratio 2

Objectives

- Use the sine ratio to calculate an angle in a
  right-angled triangle.
- Use the sine ratio to solve problems.

Why learn this?

Civil engineers use trigonometry to calculate
angles.

Fluency

Use the diagram to give sin 30° as a decimal.

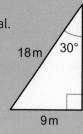

18 m

30°

9 m

1 Use the diagram to write sin 22.6° as a fraction.

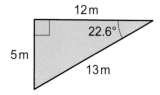

12 m

22.6°

5 m

13 m

*Active*Learn Homework, practice and support: Foundation 12.4

2 Use your calculator to find, correct to 3 decimal places where necessary
  a $\sin 20°$ b $\sin 45°$ c $\sin 32.4°$
  d $\sin 90°$ e $\sin 0°$ f $\sin 7.2°$

3 Find angle θ. Each one is a multiple of 10°.
  a $\sin \theta = 0.642\,787\,6097$
  b $\sin \theta = 0.939\,692\,6208$
  c $\sin \theta = 0.173\,648\,1777$

> **Q3 hint** Use the 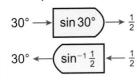 key on your calculator. You only need to try angles such as 10°, 20°, 30°, …

4 Find angle θ. Each one is a multiple of 5°.
  a $\sin \theta = 0.906\,307\,787$
  b $\sin \theta = 0.422\,618\,2617$
  c $\sin \theta = 0.258\,819\,0451$

Key point 8

When you know the value of $\sin \theta$, you can use **$\sin^{-1}$** on a calculator to find θ.

$30° \rightarrow \boxed{\sin 30°} \rightarrow \frac{1}{2}$

$30° \leftarrow \left(\sin^{-1} \frac{1}{2}\right) \leftarrow \frac{1}{2}$

---

5 Use $\sin^{-1}$ on your calculator to check your answers to **Q3** and **Q4**.
  The first one has been started for you.
  a $\sin \theta = 0.6427876097$
     $\theta = \sin^{-1}(0.6427876097)$
     $\theta =$

> **Q5a hint**
>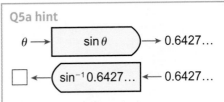
> $\theta \rightarrow \boxed{\sin \theta} \rightarrow 0.6427…$
> $\boxed{} \leftarrow \left(\sin^{-1} 0.6427…\right) \leftarrow 0.6427…$

6 Use $\sin^{-1}$ on your calculator to find the value of θ correct to 0.1°.
  a $\sin \theta = 0.345$
  b $\sin \theta = 0.8241$
  c $\sin \theta = 0.8672$

> **Q6 hint** Correct to 0.1° means give your answer to 1 decimal place.

7 Use your calculator to find the value of θ correct to 0.1°.
  a $\sin \theta = \frac{11}{12}$
  b $\sin \theta = \frac{5}{8}$
  c $\sin \theta = \frac{3.7}{5.9}$

> **Q7a hint**
> Enter $\sin^{-1}\left(\frac{11}{12}\right)$

8 Copy and complete these diagrams. The first one has been done for you.

a
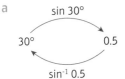
  $\sin 30°$
  $30° \quad\quad 0.5$
  $\sin^{-1} 0.5$

b
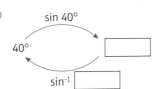
  $\sin 40°$
  $40° \quad\quad \boxed{}$
  $\sin^{-1} \boxed{}$

c
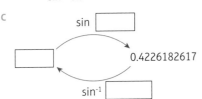
  $\sin \boxed{}$
  $\boxed{} \quad\quad 0.4226182617$
  $\sin^{-1} \boxed{}$

Example 4

Calculate the size of angle x.

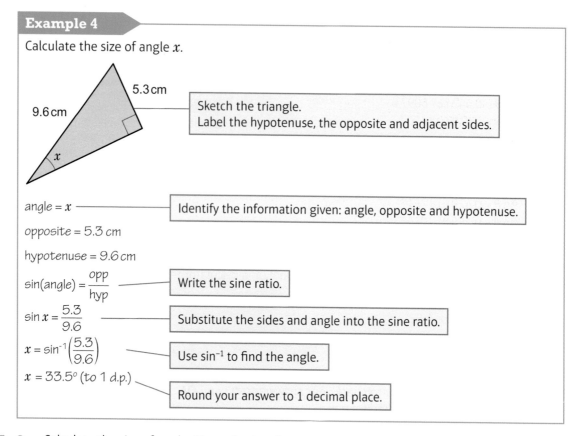

5.3 cm

9.6 cm

x

Sketch the triangle.
Label the hypotenuse, the opposite and adjacent sides.

angle = x

Identify the information given: angle, opposite and hypotenuse.

opposite = 5.3 cm

hypotenuse = 9.6 cm

$\sin(\text{angle}) = \dfrac{\text{opp}}{\text{hyp}}$

Write the sine ratio.

$\sin x = \dfrac{5.3}{9.6}$

Substitute the sides and angle into the sine ratio.

$x = \sin^{-1}\left(\dfrac{5.3}{9.6}\right)$

Use $\sin^{-1}$ to find the angle.

$x = 33.5°$ (to 1 d.p.)

Round your answer to 1 decimal place.

9 Calculate the size of angle θ in each triangle.

a

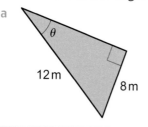

θ

12 m

8 m

b

25 km

31 km

θ

c

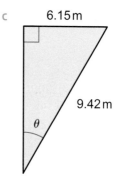

6.15 m

9.42 m

θ

Q9 hint State the degree of accuracy after your answer. Angles are normally given to the nearest 0.1°.

10 Real / Problem-solving
The track for the Lisbon tram rises 1 m for every 7.47 m travelled. What angle does the track make with the horizontal?

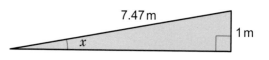

7.47 m

x

1 m

11 **Exam-style question**

LMN is a right-angled triangle.

MN = 8.5 cm

LN = 10 cm

Calculate the size of the angle marked x.

Give your answer correct to 1 decimal place.

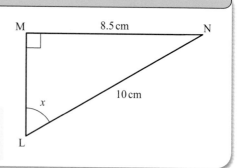

M   8.5 cm   N

10 cm

x

L

Exam hint
Write the calculations down to get marks for working.
Make sure your calculator is in degree mode.

12.5 Trigonometry: the cosine ratio

Objectives

- Understand and recall the cosine ratio in right-angled triangles.
- Use the cosine ratio to calculate the length of a side in a right-angled triangle.
- Use the cosine ratio to calculate an angle in a right-angled triangle.
- Use the cosine ratio to solve problems.

Did you know?

Astronomers use trigonometry to calculate distances in space.

Fluency

Name the hypotenuse and adjacent side in each triangle.

Warm up

1 Solve these equations.

a $0.8 = \dfrac{a}{12}$ b $0.5 = \dfrac{5}{a}$ c $0.763 = \dfrac{a}{10}$ d $0.25 = \dfrac{10}{a}$

2 Write the fraction $\dfrac{\text{adj}}{\text{hyp}}$ for each triangle.

a

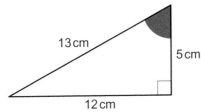

b

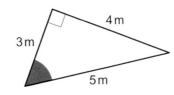

 3 **Reasoning**

a Draw these triangles accurately using a ruler and protractor.

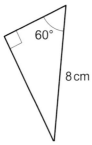 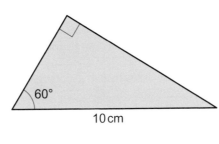 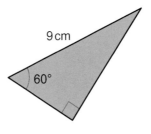

b Label the hypotenuse (hyp) and adjacent side (adj).

c Measure the adjacent side.

d i Write the fraction $\dfrac{\text{adjacent}}{\text{hypotenuse}}$ for each triangle and convert it to a decimal.

ii What do you notice?

Discussion Why are the fractions for the same angle the same? What does this tell you about the ratio of the sides?

Key point 9

In a right-angled triangle the **cosine** of an angle is the ratio of the **adj**acent side to the **hyp**otenuse.

The cosine of angle θ is written as **cos θ**.

$$\cos \theta = \frac{adj}{hyp}$$

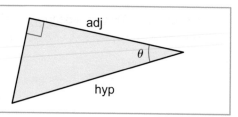

4 Use your calculator to find, correct to 3 decimal places where necessary

    a   cos 37°   b   cos 82°   c   cos 5.8°

    d   cos 0°   e   cos 90°   f   cos 45°

Q4a hint On your calculator enter cos 3 7 = key

5 Write cos θ as a fraction for each triangle.

    a

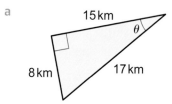

    b

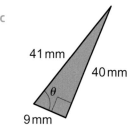

    c

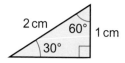

6   **Reasoning** Use this triangle to show why cos 60° = sin 30°.

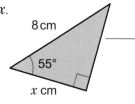

Example 5

Use the cosine ratio to work out the value of x.

    Sketch the triangle. Label the hypotenuse, the opposite and adjacent sides.

angle = 55° ⟶ Identify the information given: angle, adjacent and hypotenuse.

adjacent = x

hypotenuse = 8 cm

$\cos (angle) = \dfrac{adj}{hyp}$ ⟶ Write the cosine ratio.

$\cos 55° = \dfrac{x}{8}$ ⟶ Substitute the sides and angle into the cosine ratio.

$x = 8 \times \cos 55°$ ⟶ Multiply both sides by 8. Use your calculator to work out 8 × cos 55°.

$x = 4.5886.....$

$x = 4.6$ cm (to 1 d.p.) ⟶ Round your answer to 1 decimal place.

7 Find the value of x in each triangle. Give your answers correct to 1 decimal place.

    a

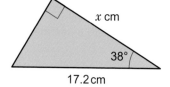

    b   c

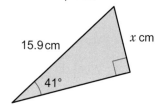

Q7 hint Your final answer will be x = __ without units as the unknown side is labelled x cm not simply x.

8 **Modelling / Problem-solving**
A ship sails 50 km on a bearing of 058°.
How far north does the ship travel?

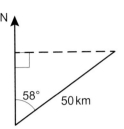

> **Key point 10**
>
> When you know the value of cos θ, you can use **cos⁻¹** on a calculator to find θ.

9 Use cos⁻¹ on your calculator to find the value of θ
correct to 0.1°.
a cos θ = 0.362 b cos θ = 0.5729 c cos θ = 0.6735

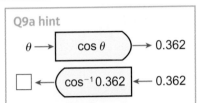

Q9a hint

10 Use your calculator to find the value of θ correct to 0.1°.
a $\cos \theta = \frac{5}{8}$ b $\cos \theta = \frac{24}{25}$ c $\cos \theta = \frac{9.8}{15.6}$

Q10a hint
Enter $\cos^{-1}\left(\frac{5}{8}\right)$

11 Copy and complete these diagrams.
The first one has been done for you.

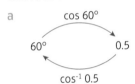 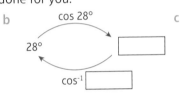 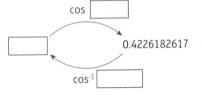

12 Calculate the size of angle θ in each of these triangles.

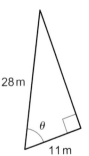 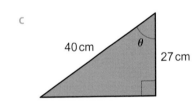

Q12 hint First sketch the triangle. Label the hyp, opp and adj.
Write the cosine ratio and substitute the values you are given into it.
Use cos⁻¹ to find the angle.

13 **Exam-style question**

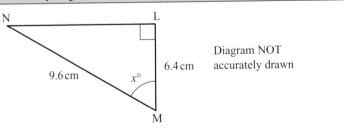

Diagram NOT
accurately drawn

LMN is a right-angled triangle.
MN = 9.6 cm   LM = 6.4 cm
Calculate the size of the angle marked x°.
Give your answer correct to 1 decimal place.   **(3 marks)**
June 2012, Q16, 1MA0/2H

Exam hint
Use all the
figures on
your calculator
display. Round
your final answer.

12.6 **Trigonometry: the tangent ratio**

Objectives

- Understand and recall the tangent ratio in right-angled triangles.
- Use the tangent ratio to calculate the length of a side in a right-angled triangle.
- Use the tangent ratio to calculate an angle in a right-angled triangle.
- Solve problems using an angle of elevation or angle of depression.

Did you know?

The tangent ratio can be used to calculate the height of a building only using measurements at ground level.

Fluency

Name the opposite and adjacent sides in these triangles.

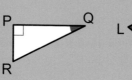
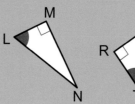

1 Write the fraction $\dfrac{\text{opp}}{\text{adj}}$ for each triangle.

a

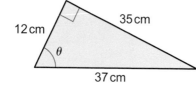

b

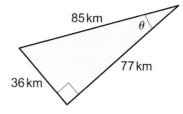

2 **Reasoning**

a Draw these triangles accurately using a ruler and protractor.

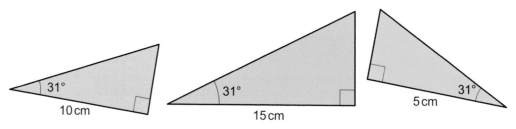

b Label the opposite side (opp) and adjacent side (adj).

c Measure the opposite side.

d i Write the fraction $\dfrac{\text{opposite}}{\text{adjacent}}$ for each triangle and convert it to a decimal (correct to 1 decimal place).

ii What do you notice?

Discussion Why are the fractions for the same angle the same? What does this tell you about the ratio of the sides?

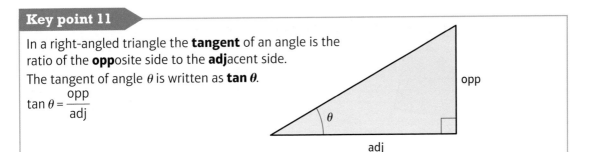

Key point 11

In a right-angled triangle the **tangent** of an angle is the ratio of the **opp**osite side to the **adj**acent side.
The tangent of angle θ is written as **tan θ**.

$$\tan \theta = \frac{\text{opp}}{\text{adj}}$$

opp

adj

θ

3   Use your calculator to find, correct to 3 decimal places where necessary

Q3a hint On your calculator

enter [tan] [2] [5] [=]

    a   tan 25°   b   tan 50°   c   tan 75°
    d   tan 0°   e   tan 89°   f   tan 45°

Reasoning Use your answers to **Q2** to write the value of tan 31° and tan 59°.
Check using a calculator.

4   Write tan θ as a fraction for each triangle.

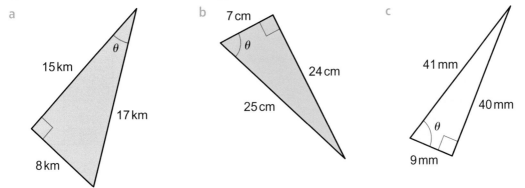

a

15 km

8 km

17 km

θ

b   7 cm

24 cm

25 cm

θ

c

41 mm

40 mm

9 mm

θ

Example 6

Use the tangent ratio to work out the value of x.

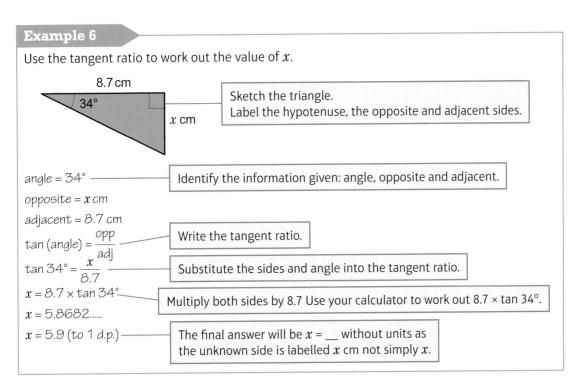

8.7 cm

34°

x cm

Sketch the triangle.
Label the hypotenuse, the opposite and adjacent sides.

angle = 34°

opposite = x cm

adjacent = 8.7 cm

Identify the information given: angle, opposite and adjacent.

$$\tan (\text{angle}) = \frac{\text{opp}}{\text{adj}}$$

Write the tangent ratio.

$$\tan 34° = \frac{x}{8.7}$$

Substitute the sides and angle into the tangent ratio.

$x = 8.7 \times \tan 34°$

Multiply both sides by 8.7 Use your calculator to work out 8.7 × tan 34°.

$x = 5.8682.....$

$x = 5.9$ (to 1 d.p.)

The final answer will be $x =$ __ without units as the unknown side is labelled x cm not simply x.

5   Find the value of x in each triangle. Give your answers correct to 1 decimal place.

a
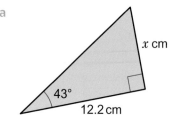

b
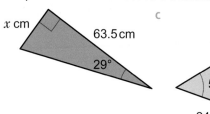

c
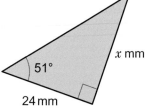

6   **Real / Problem-solving**

A flagpole is secured to the ground by a rope.

The end of the rope is 4 m from the base of the flagpole.

The rope makes an angle of 50° with the ground.

At what height does the rope attach to the flagpole?

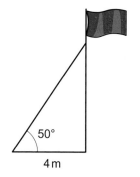

Key point 12

The angle of **elevation** is the angle measured upwards from the horizontal.

The angle of **depression** is the angle measured downwards from the horizontal.

The angle of elevation of the top of the building from point A on the ground is 64°.

The angle of depression of point B from the top of the building is 72°.

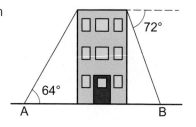

7   Which angle is an angle of elevation, which is an angle of depression?

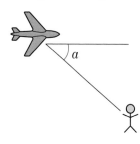
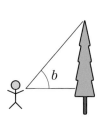

8   **Problem-solving / Real**

The angle of elevation of a cliff top, C, from a small boat, B, out at sea is 36°.

The boat is 100 m from the bottom of the cliffs.

a   Copy the diagram and label it with this information.

b   How high are the cliffs? Give your answer correct to the nearest metre.

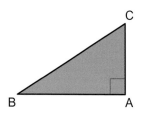

> **Q8 hint** Assume that the surface of the sea is horizontal and that the cliffs are vertical.

9 **Problem-solving / Real** From the top of Lantern Tower in York Minster, Georgia can see a boat on the river. The angle of depression from Georgia to the boat is 10°. The boat is 400 m away.

How high is the tower?

Give your answer correct to the nearest metre.

> **Q9 hint** Sketch and label a right-angled triangle to represent the information.

Key point 13

When you know the value of tan θ, you can use **tan⁻¹** on a calculator to find θ.

---

10 Use tan⁻¹ on your calculator to find the value of θ correct to 0.1°.

a $\tan \theta = 0.853$ b $\tan \theta = 1.725$

c $\tan \theta = \frac{7}{8}$ d $\tan \theta = \frac{30}{17}$.

> **Q10c hint**
> Enter tan⁻¹ $\left(\frac{7}{8}\right)$

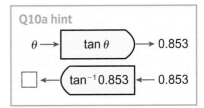

11 Copy and complete these diagrams. The first one has been done for you.

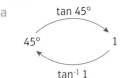

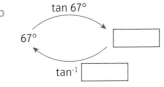

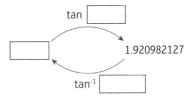

12 Calculate the size of angle θ in each of these triangles.

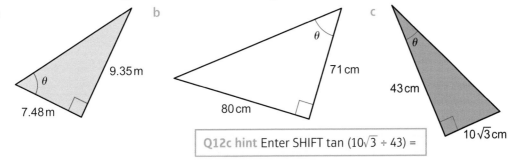

> **Q12c hint** Enter SHIFT tan (10√3 ÷ 43) =

13 **Exam-style question**

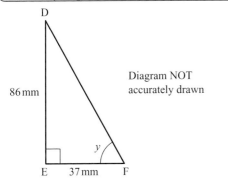

Diagram NOT accurately drawn

DEF is a right-angled triangle.

DE = 86 mm   EF = 37 mm

Calculate the size of the angle marked y.

Give your answer correct to 1 decimal place. **(3 marks)**

Nov 2013, Q11, 5MB3H/01

> **Exam hint**
> Show all stages of your working including showing sin⁻¹y or cos⁻¹y or tan⁻¹y after your first equation.

12.7 Finding lengths and angles using trigonometry

Fluency

Match sine, cosine and tangent to one of these ratios.

$$\frac{\text{opposite}}{\text{adjacent}} \qquad \frac{\text{adjacent}}{\text{hypotenuse}} \qquad \frac{\text{opposite}}{\text{hypotenuse}}$$

Warm up

1 Write $\sin \theta$, $\cos \theta$ and $\tan \theta$ as fractions for this triangle.

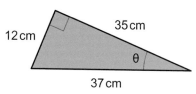

2 Solve these equations. Give each answer correct to 3 significant figures.

    a $0.827 = \dfrac{9}{x}$ **b** $0.765 = \dfrac{17.3}{x}$ **c** $0.259 = \dfrac{78.4}{x}$

3 Work out

    a $\sqrt{3^2 + 1^2}$ **b** $\sqrt{3^2 - 1^2}$ **c** $\sqrt{2^2 + 5^2}$

    Give each answer as a surd.

Key point 14

You need to know these ratios and be able to choose the one you need to solve a problem.

$$\sin (\text{angle}) = \frac{\text{opp}}{\text{hyp}} \qquad \cos (\text{angle}) = \frac{\text{adj}}{\text{hyp}} \qquad \tan (\text{angle}) = \frac{\text{opp}}{\text{adj}}$$

Example 7

Calculate the value of x. Give your answer correct to 3 significant figures.

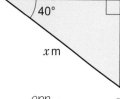

Identify the information given.

angle = 40°   opposite = 8.63 m   hypotenuse = x m

$\sin (\text{angle}) = \dfrac{\text{opp}}{\text{hyp}}$

Decide on the ratio (sin, cos, tan) you need to use. You are given 'opp' and 'hyp' so use the sine ratio.

$\sin 40° = \dfrac{8.63}{x}$

Substitute the sides and angle into the sine ratio.

$x \times \sin 40° = \dfrac{8.63}{x} \times x$

$x \times \sin 40° = 8.63$

Multiply both sides by x.

$x = \dfrac{8.63}{\sin 40°} = 13.4258\ldots = 13.4$ (to 3 s.f.)

Divide both sides by sin 40°. Then round to 3 significant figures.

4 Calculate the value of x in each triangle. Give your answers correct to 3 significant figures.

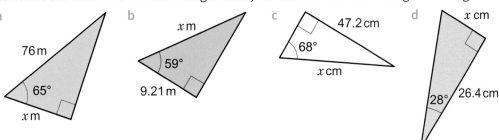

a   76 m   65° x m

b x m   59°   9.21 m

c   47.2 cm   68° x cm

d x cm   28°   26.4 cm

5 Calculate the size of angle x in each of these triangles.

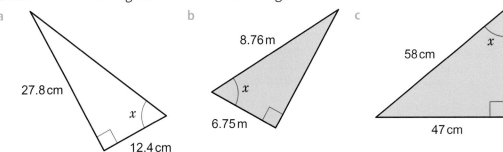

a   27.8 cm x 12.4 cm

b   8.76 m x 6.75 m

c   58 cm x 47 cm

6 **Reflect** The mnemonic SOHCAHTOA can be used to remember the sine, cosine and tangent ratios. You could also use:
'Some Old Horse Caught Another Horse Taking Oats Away';
'Some Of Her Children Are Having Trouble Over Algebra'.
Which do you prefer? Do you have one of your own?

Q6 communication hint
A mnemonic is a special word or phrase that you use to help remember something.

7 **Exam-style question**

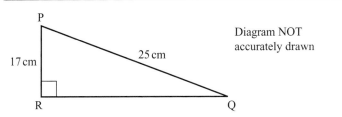

P   17 cm   25 cm   R   Q

Diagram NOT accurately drawn

PQR is a right-angled triangle.
PR = 17 cm
PQ = 25 cm
Work out the size of angle RPQ.
Give your answer correct to 1 decimal place.   **(3 marks)**

March 2013, Q13b, 1MA0/2H

Exam hint
You need to correctly identify angle RPQ. The middle letter of RPQ should help you.

8 **Real / Problem-solving** An aeroplane at a height of 300 m approaches a runway.
The aeroplane is at a horizontal distance of 1700 m from the end of the runway. Find the angle of elevation of the aeroplane.

Q8 hint Sketch and label a right-angled triangle to show this information.

9 **Real / Problem-solving** A ship is spotted from the top of some cliffs.
The cliffs are 380 m high and the angle of depression to the ship is 6°.
How far is the ship from the base of the cliffs?
Give your answer, in km, correct to 3 significant figures.

Q9 hint Work out the answer in metres then convert it to kilometres.

Key point 15

You need to know these values.

$\sin 0° = 0$ $\qquad \sin 30° = \dfrac{1}{2}$ $\qquad \sin 45° = \dfrac{1}{\sqrt{2}}$ $\qquad \sin 60° = \dfrac{\sqrt{3}}{2}$ $\qquad \sin 90° = 1$

$\cos 0° = 1$ $\qquad \cos 30° = \dfrac{\sqrt{3}}{2}$ $\qquad \cos 45° = \dfrac{1}{\sqrt{2}}$ $\qquad \cos 60° = \dfrac{1}{2}$ $\qquad \cos 90° = 0$

$\tan 0° = 0$ $\qquad \tan 30° = \dfrac{1}{\sqrt{3}}$ $\qquad \tan 45° = 1$ $\qquad \tan 60° = \sqrt{3}$

10 Find the missing angles.

a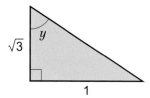

b

> **Q10 hint** Write the cos, sin or tan ratio. Compare it to the values in the key point.

c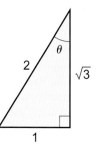

d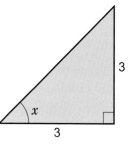

11 Find the lengths of the sides marked with letters.

a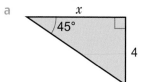

b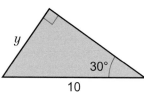

c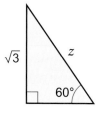

12 **Reasoning** What type of triangle is ABC?

> **Q12 hint** Find the angles.

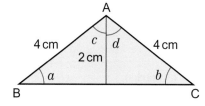

13 **Modelling** A bird is on the ground, 8 m from a tree. The tree is 8 m tall. What is the angle of elevation from the bird to the top of the tree?

> **Q13 hint** Draw a diagram.

Discussion What assumptions have you made?

12 Problem-solving

Objective   • Use a formula to help you solve problems.

Example 8

The hypotenuse of a right-angled triangle is 8.5 cm. Its height is twice its base.
What is its height?

height =
2 × base
= 2b

8.5 cm

b

> Sketch the shape. Include all the information on your sketch as you read the question.

> Underline all the words in the question that describe the shape. Ignore any numbers.
> Write down any formula you know that connects the words you have underlined.

Pythagoras' theorem: $c^2 = a^2 + b^2$

$8.5^2 = (2b)^2 + b^2$

> Substitute the information on your shape into the formula.

$72.25 = 4b^2 + b^2$

> Simplify.

$72.25 = 5b^2$

$\dfrac{72.25}{5} = \dfrac{5b^2}{5}$

> Divide both sides of the equation by 5.

$14.45 = b^2$

> Square root both sides of the equation.

$\sqrt{14.45} = b$

$b = 3.80$ cm (2 d.p.)

> State the units (cm) and accuracy (2 d.p.).

Height = $2b = 2 \times 3.80 = 7.60$ cm (2 d.p.)

> Check the question. It asks for the height.

1 The height of a parallelogram is 3 times its base.
   The area of the parallelogram is 108 mm².
   What is its height?

Q1 hint Follow the method in the Example.

2 A right-angled triangle has a height of 3 cm.
   Its base is 80% of the length of its hypotenuse.
   Work out its base.

Q2 hint Write 80% as a decimal.

3 A rectangle's height is 2.9 cm.
   This is 1.6 cm shorter than its diagonal.
   Work out the area of the rectangle.

Q3 hint Sketch the shape. Include all the information on your sketch. Write down the formula for the area of a rectangle.

4 A duathlon includes a run, a cycle, then a run.
   In training, one athlete starts at 9 am.
   She records:
   Run 1: 10 km at 12.5 km/hour
   Cycle: 50 km at 31.25 km/hour
   Run 2: 9 km at 15 km/hour
   What time did the athlete finish?

Q4 hint Use a table (rather than a sketch) to represent the information.

	Distance (km)	Speed (km/h)
Run 1		
Cycle		

What formula connects the information in the question?

5   96 cm² of card is used to make
    a box in the shape of a cube.
    What is its volume?

> **Q5 hint** 96 cm² is the surface area of the box.
> What do you need to know to find the volume?

6   An isosceles triangle has a base of 12 cm, and two
    other sides, each of length 9 cm.
    a   What are the angles in the isosceles triangle?
    b   **Reflect** Compare your working with others in
        your class. Did they use the same formulae as you?
        Does it matter what formulae you use to solve
        a problem?

> **Q6 hint** Sketch the triangle and
> draw the perpendicular height.
> Use trigonometry to find one angle.
> You will need more than one
> formula to find all the angles.

7   An isosceles trapezium has an area of 70 cm² and
    parallel sides with lengths 6.2 cm and 13.8 cm.
    What is its perimeter?

> **Q7 hint** Area of a trapezium = $\frac{1}{2}(a + b)h$
> Divide the trapezium into a rectangle and
> two right-angled triangles to help you:

12   Check up

Log how you did on your
Student Progression Chart.

Pythagoras' theorem

1   Calculate the value of x in each of these right-angled triangles.

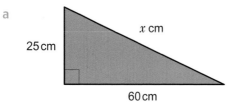

a

25 cm
x cm
60 cm

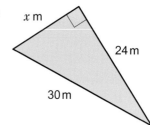

b

x m
24 m
30 m

2   The points A(−3, 2) and B(4, −6) are plotted on a centimetre square grid.
    Work out the length of AB.

3   **Reasoning** Geoff thinks that this triangle is right-angled.
    Is Geoff correct? Explain your answer.

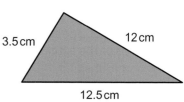

3.5 cm
12 cm
12.5 cm

Finding lengths using trigonometry

4   a   Use the sine ratio to
        find x.

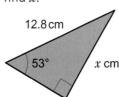

12.8 cm
53°
x cm

    b   Use the tangent ratio to
        find x.

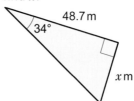

48.7 m
34°
x m

    c   Use the cosine ratio to
        find x.

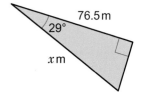

76.5 m
29°
x m

5   **Reasoning** A ladder of length 5 m leans against a vertical wall.
    The ladder makes an angle of 50° with the ground.
    How far is the top of the ladder from the ground?

5 m
x
50°

*Active*Learn   Homework, practice and support: Foundation 12 Check up

6 The angle of elevation of the top of a building from a point P is 75°.
The point P is on level ground 30 m from the base of the building.
Calculate the height of the building.

Finding angles using trigonometry

7 **a** Use the sine ratio to find θ.

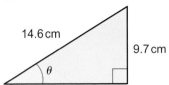

14.6 cm
9.7 cm
θ

b Use the cosine ratio to find θ.

24.9 cm
θ
32.6 cm

8 Write the exact value of
   a tan 45°
   b sin 30°
   c cos 30°
   d sin 0°

9 How sure are you of your answers? Were you mostly

Just guessing   Feeling doubtful   Confident

What next? Use your results to decide whether to strengthen or extend your learning.

✱ Challenge

10 Draw a square of side 5 cm.
Draw the diagonal.
Calculate tan θ.
Explain why tan θ = 1 for any square.

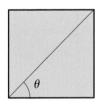

θ

---

12 Strengthen

Pythagoras' theorem

1 Write the length of the hypotenuse for each of these triangles.

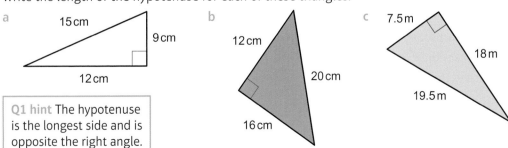

a
15 cm
9 cm
12 cm

> **Q1 hint** The hypotenuse is the longest side and is opposite the right angle.

b
12 cm
20 cm
16 cm

c
7.5 m
18 m
19.5 m

2   A square is drawn on each side of a right-angled triangle.
    Copy and complete these statements.

    a   The coloured square is on the .... of the
        right-angled triangle.

    > **Q2a hint** What is the side opposite
    > the right angle called?

    b   The square on the 5 cm side has
        area 5 cm × .... = ... cm²
    c   The square on the 6 cm side has
        area .... × .... = ...
    d   The coloured square has
        area .... + ... = ...

    > **Q2d hint** Add the areas of the smaller squares
    > to get the area of the coloured square.

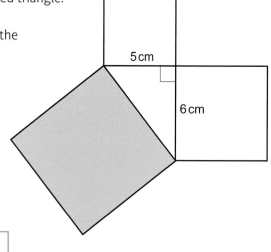

3   Calculate the area of the coloured
    square drawn on the hypotenuse
    of this right-angled triangle.

    > **Q3 hint** Use the same
    > method as in **Q2**.

4   The hypotenuse of this right-angled triangle is labelled c.
    Copy and complete these steps to find the value of c.

    $c^2 = 8^2 + \square^2$

    $c^2 = \square$

    $c = \sqrt{\square}$

    $c = \square$ km

    > **Q4 hint** You can use the
    > $\boxed{\sqrt{}}$ key on your calculator
    > to find the value of c.

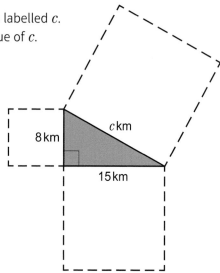

5 Calculate the value of c in each of these right-angled triangles.

Q5 hint Use the same method as in **Q4**.

a

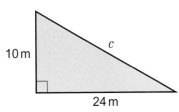

b

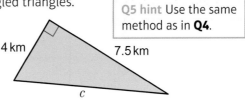

6 Right-angled triangle ABC is drawn on a centimetre square grid.

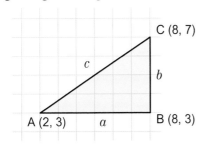

Q6 strategy hint Work out the lengths of AB (a) and BC (b) first.
AB = a = 8 – ☐
BC = b = 7 – ☐
Use Pythagoras' theorem to find the length of AC (c).
$c^2 = a^2 + b^2$

Work out the length of AC.

7 The points A(5, –3) and B(–1, 8) are plotted on a centimetre square grid.
Work out the length of AB.

Q7 hint First draw a sketch. Label the ends of the line and draw a right-angled triangle.

8 Work out the length of the unknown side in each right-angled triangle.
Give your answers correct to 3 significant figures.

a

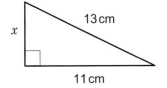

b

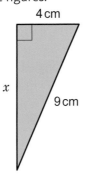

Q8 strategy hint Sketch the triangle.
Label the hypotenuse c and the other two sides a and b.
Substitute the values of a, b and c into Pythagoras' theorem, $c^2 = a^2 + b^2$.
Solve the equation.

9 **Communication** Show that this triangle is right-angled.

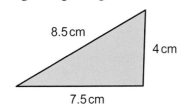

Q9 hint The longest side has length 8.5 cm. If the triangle is right-angled, this will be the hypotenuse.
Work out 8.5^2.
Work out $7.5^2 + 4^2$.
If your answers are the same, you can say that the triangle is right-angled.
If your answers are not the same, you can say that the triangle is not right-angled.

Finding lengths using trigonometry

1 Sketch each triangle and label
  a 'opp' on the side opposite to angle θ
  b 'adj' on the adjacent side
  c 'hyp' on the hypotenuse.

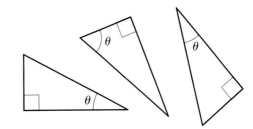

2 **a** Write the tangent ratio for each triangle.

i

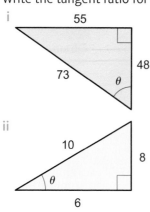

ii

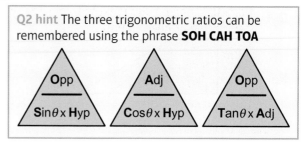

b Write the sine ratio for each triangle.

c Write the cosine ratio for each triangle.

3 Calculate the value of x in each triangle. Give your answers correct to 3 significant figures.

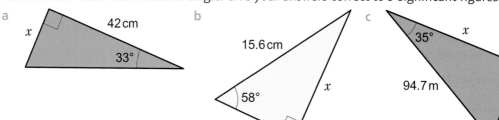

4 **Real / Problem-solving** Fergus wants to find the height of a bridge above the ground. The angle of elevation of the top of the bridge is 68° from a point at ground level 40 m from the bridge.

a Sketch and label a right-angled triangle to show this information.

b Work out which trigonometric ratio Fergus needs to use.

c Calculate the height of the bridge from the ground.

5 Follow these steps to work out the length of AC.

a Choose the trigonometric ratio you are going to use.

b Write the ratio.

c Rearrange to make AC the subject.

d Use your calculator to work out the missing side.

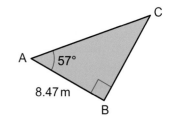

Finding angles using trigonometry

1 Use the inverse functions on your calculator to work out θ correct to 0.1°.

   a $\sin \theta = \frac{1}{5}$ **b** $\sin \theta = 0.7$ **c** $\cos \theta = \frac{3}{7}$

   d $\cos \theta = 0.2$ **e** $\tan \theta = \frac{2}{3}$ **f** $\tan \theta = 0.5$

2 Calculate the size of angle θ in each of these triangles.

   a

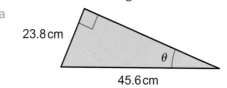

   b

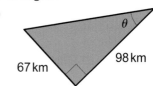

3 Follow these steps to work out the size of angle θ.
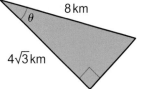
  a Copy the triangle and label the sides that you have been given or need to find.
  b Choose the trigonometric ratio you are going to use.
  c Write the ratio.
  d Rearrange to make θ the subject.
  e Use your calculator to work out the missing angle.

4 Use your calculator to find these values as fractions or surds. Copy and complete the table

Angle	0°	30°	45°	60°	90°
sin					
cos					
tan					

12 Extend

1 **Real / Modelling** A ladder is leaning against a wall.
The foot of the ladder is 2 m from the wall.
The top of the ladder is 6 m above the ground.
  a How long is the ladder, in metres, correct to 2 decimal places?
  b Write any assumptions that you have made to answer part **a**.

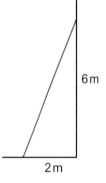

2 **Real / Problem-solving** An aeroplane leaves an airport and flies 50 km due North.
It then flies due East a further 120 km to reach its destination.
Calculate the direct distance from the airport to the destination.
Give your answer correct to 3 significant figures.

3 **Problem-solving** Calculate the area of this triangle.

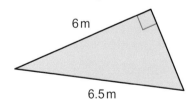

4 **Problem-solving** Calculate the length of AC.
Give your answer to 1 decimal place.

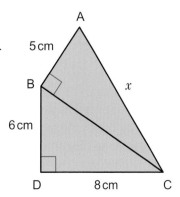

373

5   A square has area 2 cm².
    a   What is the length of one side?
    b   Work out $\sqrt{2} \times \sqrt{2}$

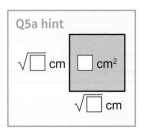

Q5a hint

$\sqrt{\square}$ cm $\square$ cm²

$\sqrt{\square}$ cm

6   Simplify
    a $\sqrt{5} \times \sqrt{5}$ b $(\sqrt{3})^2$
    c $2\sqrt{3} \times 2\sqrt{3} = 2 \times 2 \times \sqrt{3} \times \sqrt{3} = \square$ d $3\sqrt{2} + 4\sqrt{2}$
    e $5\sqrt{3} + \sqrt{3}$ f $4\sqrt{5} - \sqrt{5}$

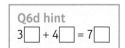

Q6d hint

$3\square + 4\square = 7\square$

7   A triangle has side lengths $3\sqrt{2}$ cm, $3\sqrt{2}$ cm and 6 cm.
    Is it a right-angled triangle?

8   **Communication**
    a   Show that triangle ABD is right-angled.
    b   Work out the length of DC.
        Give your answer in surd form.
    c   Work out the length of AC.
        Give your answer in surd form.
    d   Show that triangle ABC is right-angled.

Q8c hint
$\sqrt{5} + \sqrt{5} = 2\sqrt{5}$
$2\sqrt{5} + \sqrt{5} = 3\sqrt{5}$

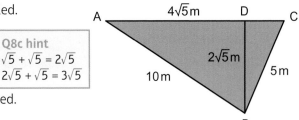

9   Find the exact values of x and y.

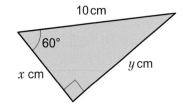

10 Find the exact values of x and y.

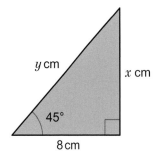

11 Copy and complete these statements.
    a   As angle x increases from 0° to 90°, sin x ......
    b   As angle x increases from 0° to 90°, cos x .....

Q11a hint Write down the values of
sin x when $x = 0, ..., 90°$.
Write a statement about how the
value of sin x changes.

12 **Problem-solving** The angle of elevation from point A to the top of a tower is 38°.
    A flagpole stands on top of the tower.
    The angle of elevation to the top of the flagpole is 45°.
    The distance from point A to the base of the tower is 30 m.
    What is the height of the flagpole?

Q12 hint Sketch and
label a diagram to show
this information.

13 **Communication**
   a   Show that angle ABC is a right angle.
   b   Calculate the size of angle BCA.

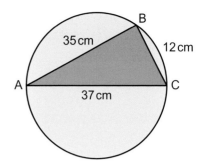

14 The diagram shows a triangular prism.
   a   Calculate the size of angle θ.
   b   Calculate the surface area of the prism.

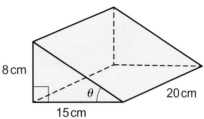

15 This octagon is made from isosceles triangles. Calculate its area.

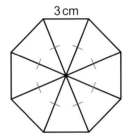

Q15 hint Work out the height of one triangle, and then its area.

16 Find the exact value of θ in this triangle.

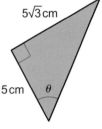

17 **Exam-style question**

The diagram shows a ladder leaning against a vertical wall.

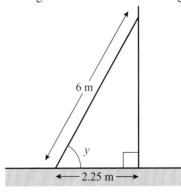

Diagram NOT accurately drawn

Exam hint
After you have shown all your calculation, write:
Yes, the ladder is safe to use
or
No, the ladder is not safe to use.

The ladder stands on horizontal ground.
The length of the ladder is 6 m.
The bottom of the ladder is 2.25 m from the bottom of the wall.
A ladder is safe to use when the angle marked y is less than 75°.
Is the ladder safe to use?
You must show all your working.   **(3 marks)**

June 2013, Q20 1MA0/2H

12 Knowledge check

◉ In a right-angled triangle the **hypotenuse** is the longest side and is opposite the right angle. .... *Mastery lesson 12.1*

◉ Pythagoras' theorem shows the relationship between the lengths of the three sides of a right-angled triangle.

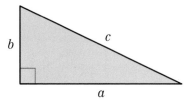

$c^2 = a^2 + b^2$ .... *Mastery lesson 12.1*

◉ A triangle with sides a, b and c, where c is the longest side, is right-angled only if $c^2 = a^2 + b^2$. .... *Mastery lesson 12.2*

◉ In a right-angled triangle, the side opposite the angle θ is called the **opposite**. The side next to the angle θ is called the **adjacent**.

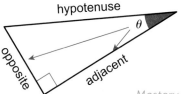

.... *Mastery lesson 12.3*

◉ The **sine** of an angle is the ratio of the opposite side to the hypotenuse. The sine of angle θ is written as $\sin \theta$.

$$\sin \theta = \frac{\text{opp}}{\text{hyp}}$$ .... *Mastery lesson 12.3*

◉ The **cosine** of an angle is the ratio of the adjacent side to the hypotenuse. The cosine of angle θ is written as $\cos \theta$.

$$\cos \theta = \frac{\text{adj}}{\text{hyp}}$$ .... *Mastery lesson 12.5*

◉ The **tangent** of an angle is the ratio of the opposite side to the adjacent side. The tangent of angle θ is written as $\tan \theta$.

$$\tan \theta = \frac{\text{opp}}{\text{adj}}$$ .... *Mastery lesson 12.6*

◉ You can use $\sin^{-1}$, $\cos^{-1}$ or $\tan^{-1}$ to find the size of an angle. .... *Mastery lesson 12.4, 12.5, 12.6*

◉ The **angle of elevation** is the angle measured upwards from the horizontal. The **angle of depression** is the angle measured downwards from the horizontal.

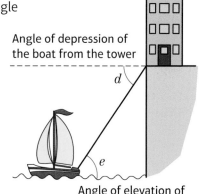

Angle of depression of the boat from the tower

d

e

Angle of elevation of the tower from the boat

.... *Mastery lesson 12.6*

○ The sine, cosine and tangent of some angles may be written exactly.

	30°	45°	60°	90°
sin	$\frac{1}{2}$	$\frac{1}{\sqrt{2}}$	$\frac{\sqrt{3}}{2}$	1
cos	$\frac{\sqrt{3}}{2}$	$\frac{1}{\sqrt{2}}$	$\frac{1}{2}$	0
tan	$\frac{1}{\sqrt{3}}$	1	$\sqrt{3}$	

Mastery lesson 12.7

Look back at this unit.
Which lesson did you like most? Write a sentence to explain why.
Which lesson did you like least? Write a sentence to explain why.
Begin your sentence with: I liked lesson … most/least because …

Reflect

12 Unit test

Log how you did on your
Student Progression Chart.

1 The diagram shows a right-angled triangle.
  a Calculate the value of x.
    Give your answer correct to 3 significant figures.
  b Calculate the area of the triangle.
    Give your answer correct to 3 significant figures.

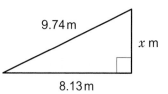

9.74 m

x m

8.13 m

(4 marks)

2 **Reasoning** Karen thinks that triangle LMN is a right-angled triangle.

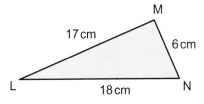

M

17 cm

6 cm

L

18 cm

N

Is Karen correct? Explain your answer. *(3 marks)*

3 Use the tangent ratio to find
  a angle x
  b length y.

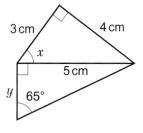

3 cm

4 cm

x

5 cm

y 65°

(3 marks)

4 Calculate the value of x in each triangle.
  a
  b *(6 marks)*

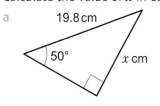

19.8 cm

50°

x cm

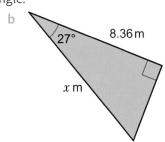

27°

8.36 m

x m

5 The diagram shows the positions A, B and C of Alice,
Bob and Charlie at the side of a canal.
Charlie is directly opposite Bob.
Alice and Bob are 8 m apart on the same side of the canal.
The direction of Charlie from Alice makes an angle of 41°
with the side of the canal.
Calculate the width of the canal.
Give your answer correct to 3 significant figures. (3 marks)

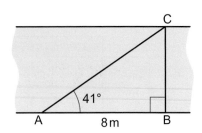

6 Calculate the size of angle θ in this triangle. (4 marks)

7 **Reasoning** The cliffs at Bempton are 400 feet high.
The angle of depression of a fishing boat, B, is 15° from the top of the cliffs.
Calculate the horizontal distance of the boat from the base of the cliffs.
Give your answer correct to the nearest 10 feet. (4 marks)

8 The diagram shows triangle ABC.
a Use trigonometry to work out the exact value of
  i angle ABD   ii angle DAC.
b Use Pythagoras' theorem to work out the length of
  i BD   ii AC.
c What kind of triangle is ABC? Explain your answer. (5 marks)

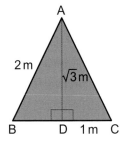

Sample student answers

a Which student has the correct answer?
b How has the diagram and the labelling helped the student?

Exam-style question

A ladder is 6 m long.
The ladder is placed on horizontal ground, resting against a vertical wall.

The instructions for using the ladder say that the bottom of the ladder must *not* be
closer than 1.5 m from the bottom of the wall.

How far up the wall can the ladder reach?
Give your answer correct to 1 decimal place. **(3 marks)**

Mock Paper 2010, Q26, 1MA0/2F

Student A

$c^2 = a^2 + b^2$

$6^2 = a^2 + 1.5^2$

$6^2 - 1.5^2 = a^2$

$33.75 = a^2$

$a = 5.8$ m

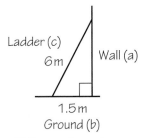

The ladder can reach to 5.8 m up the wall.

Student B

$c^2 = a^2 + b^2$

$6^2 + 1.5^2 = 38.25$

$\sqrt{38.25} = 6.2$ m

13 PROBABILITY

What is the chance of picking your favourite doughnut if there are 12 different flavours in a box?

13 Prior knowledge check

Numerical fluency

1 Simplify
   a $\frac{2}{4}$ b $\frac{15}{20}$ c $\frac{64}{100}$ d $\frac{18}{30}$

2 Convert
   a 0.75 to a fraction
   b 23% to a decimal
   c $\frac{4}{5}$ to a percentage
   d $\frac{34}{50}$ to a decimal
   e 52% to a fraction
   f 0.31 to a percentage.
   Simplify your answers where necessary.

3 Which is bigger, $\frac{3}{5}$ or $\frac{16}{25}$?

4 Order from smallest to largest.
   $\frac{1}{3}$ $\frac{2}{5}$ $\frac{7}{20}$ $\frac{3}{7}$

5 Work out
   a 45% of 120
   b 18 as a percentage of 25
   c 60% of 70
   d 24 as a percentage of 40.

6 Work out
   a 45 × 0.1
   b $\frac{3}{4} \times \frac{2}{3}$
   c 0.2 × 0.4
   d 0.3 × 150

Fluency with probability

7 How would you describe the probability of each event?
   Choose from impossible, unlikely, even chance, likely or certain.
   a Getting a head when you toss a coin.
   b Getting a 7 when you roll a standard fair dice.
   c Rain in March.
   d The next vehicle to pass the school gates being a purple car.

8 Copy the probability scale. Mark each event from **Q7** on your scale.

   | Impossible | Even chance | Certain |

9 Write the probability that this spinner lands on
   a red
   b green
   c yellow.

✳ Challenge

10 Design a spinner with 5 sectors where the probability of spinning 'red' is 0.4 and the probability of spinning 'yellow' is 0.2.

13.1 Calculating probability

Objectives

- Calculate simple probabilities from equally likely events.
- Understand mutually exclusive and exhaustive outcomes.

Did you know?

Car insurance companies use probability to assess how likely a driver is to have an accident.

Fluency

What is the probability of this spinner landing on green as a
a a fraction **b** a decimal **c** a percentage?

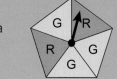

Warm up

1 Work out

    a $\frac{3}{13} + \frac{10}{13}$ b $\frac{1}{6} + \frac{1}{3}$ c $1 - \frac{1}{4}$ d $1 - 0.7$

Key point 1

$$\text{probability} = \frac{\text{number of successful outcomes}}{\text{total number of possible outcomes}}$$

> Questions in this unit are targeted at the steps indicated.

2 **Reasoning / Communication**
This regular 5-sided spinner is spun.
    a Are the events 'red' and 'blue' equally likely?
    b Which is more likely, red or green?
Explain your answers.

> **Q2 hint** 'Equally likely' means having the same chance of happening.

3 **Modelling** For this regular 7-sided spinner.
    a Write the probability of the spinner landing on blue.
    b Write the probability of the spinner landing on either blue or green.
    c Write the probability of the spinner landing on purple.
    d Which colour is twice as likely as red?

4 The letters from the word PROBABILITY are written on cards and placed in a bag.
Ella picks one card at random from the bag. Work out
    a P(R)
    b P(B)
    c P(O or A)

> **Q4 communication hint** P(R) means probability of picking R.

Discussion Ella says that the probability of selecting a card with a vowel on it is $\frac{1}{2}$ because there are two types of letters, vowels and consonants. Is Ella correct?

5 A bag contains 4 purple balls, 3 yellow balls and 5 green balls. Steven picks one ball from the bag at random.
What is the probability of Steven
    a picking a green ball
    b picking a yellow ball
    c picking a ball that is *not* yellow?

> **Q5c hint** What is the probability of purple or green?

Discussion What do you notice about your answers to parts **b** and **c**?

If the probability of an event happening is P, then the probability of it not happening is 1 – P.

6   A bag contains coloured counters. The probability of picking a blue counter is $\frac{4}{7}$.
    What is the probability of picking a counter that is *not* blue?

7   **Exam-style question**

Exam hint
Write your answers
as fractions.
You don't need to
simplify them.

Pippa has a bag of 30 coloured sweets. 17 of the sweets are red, 7 are
green and 6 are yellow. Pippa takes a sweet from the bag at random.

Write the probability that Pippa
a takes a red sweet   **(1 mark)**
b does *not* take a green sweet   **(1 mark)**
c takes a purple sweet.   **(1 mark)**

8   **Reasoning** Amber has a hat containing red, yellow and blue cards.
    She selects a card at random.
    a   P(red) = $\frac{11}{18}$. P(blue) = $\frac{5}{18}$. What is P(yellow)?
    b   Amber adds blue cards to the hat so that the probability of picking a yellow card is $\frac{1}{10}$.
        How many blue cards does she add?

Events are **mutually exclusive** when they cannot happen at the same time.

9   A fair six-sided dice is rolled. Are the following pairs
    of events mutually exclusive?
    a   Rolling an even number and rolling an odd number.
    b   Rolling an even number and rolling a prime number.
    c   Rolling a factor of six and rolling an odd number.

Q9a hint Can you
roll a number
that is both odd
and even at the
same time?

Events are **exhaustive** if they include all possible outcomes.

10 A fair eight-sided dice numbered 1 to 8 is rolled.
    Are these pairs of events exhaustive?
    a   Rolling an even number and rolling a factor of 8.
    b   Rolling a number less than 5 and rolling a number greater than 5.
    c   Rolling an odd number and rolling an even number.
    d   Work out
        i   P(even number)
        ii   P(odd number)
        iii P(even or odd number)

The probabilities of an **exhaustive** set of **mutually exclusive** events sum to 1.

11 The probability that Stratworth Town win a football match is $\frac{1}{3}$.
    The probability that they draw the match is $\frac{1}{6}$.
    Work out the probability that they lose the match.

> **Key point 6**
>
> Predicted number of outcomes = probability × number of trials.

12 Oscar's spinner has scores from 1 to 4. The table shows the probabilities of scores 1, 2 and 3.

Outcome	1	2	3	4
Probability	0.1	0.5	0.1	

a Work out P(4).

b Write P(even number) as
   i a percentage
   ii a fraction.

c **Reflect** Does it matter if a probability is given as a fraction, a decimal or a percentage?

d Oscar spins the spinner 100 times. Predict the number of times it will land on 2.

13 **Problem-solving / Real** A safe is opened by entering a two-digit number into a keypad numbered from 0 to 9.

a How many possible combinations are there?

b What is the probability of getting the combination right first time?

c It is known that the number begins with 1 or 2. What is the probability of getting the combination right?

Discussion How many possible combinations are there if it is known that the first digit is even and the second digit is odd?

> **Q13a hint** How many two-digit numbers are there between 00 and 99?

13.2 Two events

Objectives

- Use two-way tables to record the outcomes from two events.
- Work out probabilities from sample space diagrams.

Why learn this?

Calculating probabilities can help you work out if games are fair.

Fluency

- What are the possible outcomes of flipping a coin?
- Simplify $\frac{6}{16}$

1 A fair coin is flipped. Write the probability of getting a head.

2 A fair, six-sided dice is rolled.
   a List all the possible outcomes. How many outcomes are there?
   b Are the outcomes equally likely?
   c Work out
      i P(5 or 6)
      ii P(odd number)
      iii P(prime number).

*Active*Learn Homework, practice and support: Foundation 13.2

Key point 7

A sample space diagram shows all the possible outcomes. You can use it to find a theoretical probability.

Example 1

A fair dice is rolled and a fair coin is flipped.

a Draw a sample space diagram.

b Write the probability of getting a tail on the coin and an odd number on the dice.

c Write the probability of getting a head on the coin or an even number on the dice.

a

	1	2	3	4	5	6
Head	H, 1	H, 2	H, 3	H, 4	H, 5	H, 6
Tail	T, 1	T, 2	T, 3	T, 4	T, 5	T, 6

Draw a two-way table to show the possible outcomes.

b $P(T, odd) = \frac{3}{12} = \frac{1}{4}$

There are 12 equally likely outcomes. Three are tail and an odd number: 'T, 1', 'T, 3' and 'T, 5'

c $\frac{9}{12} = \frac{3}{4}$

6 outcomes are 'H, –' and 3 outcomes are 'T, even number'. 3 + 6 = 9

3 Two fair coins are flipped.

a Draw a two-way table to show the possible outcomes.

b How many possible outcomes are there?

c Write the probability of getting two heads.

d Write the probability of getting one head and one tail.

e Write the probability of getting at least one head.

Discussion How can you use P(no heads) to work out P(at least one head)?

Q3a hint

	Head	Tail
Head		
Tail		

4 **Reasoning / Communication** Two spinners are spun.

a Copy and complete the two-way table showing the possible outcomes.

	Red	Green	Blue	Yellow
Red	R, R	R, G		
Green	G, R			
Blue				

Work out

b P(R, R)

c P(at least one green)

d P(Y, Y)

e Mischa says, 'The probability of getting at least one blue is $\frac{7}{24}$ because there are 24 letters in the table and 7 of them are B.'
Explain why Mischa is wrong.

Q4e hint Show any working then write 'Mischa is wrong because … '.

5 Two 3-sided spinners are spun.

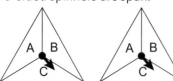

Q5 hint One possible outcome is (A, B).

Q5 communication hint A theoretical probability is calculated using equally likely outcomes.

a Draw a sample space diagram to show all the possible outcomes.
b Find the probability that both spinners land on the same letter.
c Predict the number of times both spinners will land on the same letter in 30 spins.

6 **Exam-style question**

Ishah spins a fair 5-sided spinner. She then throws a fair coin.

a List all the possible outcomes she could get.
The first one has been done for you. (1, head) **(2 marks)**

b Ishah spins the spinner once and throws the coin once.
Work out the probability that she will get a 1 and a head. **(1 mark)**

March 2009, Q2, 5381F/5B

Exam strategy hint
Check that the number on the spinner comes first in your list of outcomes.

7 **Real** Emily is deciding what to wear. She has a choice of three different skirts and three different tops.
a Draw a sample space diagram showing Emily's options.
b How many combinations has she to choose from?
c Emily chooses a skirt and top at random.
Write the probability that she wears
  i only red
  ii a green top and a blue skirt
  iii some blue.

Discussion Emily also has two possible scarves to wear with her outfit. How many different combinations does she have now?

Skirts

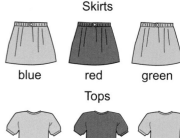

blue   red   green

Tops

blue   red   green

Q7 discussion hint Emily could wear either scarf with all of the possible combinations of skirt and top.

8 A fair dice is numbered 2, 4, 6, 8, 10 and 12. Another fair dice is numbered 1, 3, 5, 7, 9 and 11. The dice are rolled together, and the scores added.
a Copy and complete the sample space diagram showing the possible total scores.
b Find the probability of scoring
  i a total of 7
  ii more than 12
  iii an even number
  iv a multiple of 3.
c Which is more likely, scoring a multiple of 3 or a total of 7?

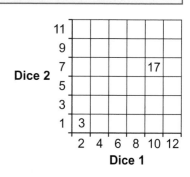

9   **Problem-solving** A bag contains two 10p coins and two 20p coins.
    Another bag contains one 10p coin and three 20p coins.
    Olivia takes one coin from each bag.
    What is the probability of her getting 40p?

> **Q9 hint** Draw a sample space diagram showing the possible outcomes.

10 **Problem-solving / Communication** A stall at a fete is running a game called 'Spin it
    to win it'. Contestants spin two numbered wheels and multiply the two numbers they score
    together. If the overall score is odd, they win a prize.

> **Q10 hint** Draw a sample space diagram showing the possible total scores. Write a sentence or two explaining what your sample space diagram shows and if Mr Dixon is correct.

    Mr Dixon claims that winning and losing are equally likely. Is he correct?

11 **Modelling** Two ordinary fair dice are rolled and the difference in the scores recorded.
    Work out the probability of the difference being more than 1.

13.3 Experimental probability

Objectives

- Find and interpret probabilities based on experimental data.
- Make predictions from experimental data.

Why learn this?

Modelling probabilities by experimenting can help scientists predict the outcomes of random events.

Fluency

- What is   **a** $\frac{12}{25}$ as a percentage.   **b** 34% as a decimal.
- Which is bigger, $\frac{2}{3}$ or $\frac{3}{4}$?
- What is the theoretical probability of rolling 3 on a fair dice?

1   Work out
    a $\frac{3}{4} \times 80$ b $\frac{2}{5} \times 65$ c $\frac{3}{8} \times 40$

2   Bella recorded the type and colour of vehicles
    passing the school gate. Here are her results.
    a   How many cars passed the gate?
    b   How many vehicles passed the gate?
    c   What proportion of the vans were silver?

	Red	Silver	White	Blue
Car	5	10	3	2
Van	2	8	15	0
Bus	17	2	1	5

> **Key point 8**
>
> You can estimate the probability of an event from the results of an experiment or survey.
>
> estimated probability = $\dfrac{\text{frequency of event}}{\text{total frequency}}$
>
> This estimated probability is also called the **experimental probability**.

3   Eric rolled a six-sided dice 20 times.

a   Copy and complete the frequency table to record his outcomes.

6, 3, 4, 6, 1, 2, 2, 4, 6, 4, 3, 2, 3, 1, 3, 5, 6, 3, 2, 5

Outcome	1	2	3	4	5	6
Frequency						

Discussion Eric rolls the dice 100 times. How many times would you expect it to land on 6?

b   **Reflect** Think carefully about what 'frequency' tells you. Write a definition in your own words.

Key point 9

The **relative frequency** of an event is also an estimate of the probability.

$$\text{relative frequency} = \frac{\text{number of 'successful' trials}}{\text{total number of trials}}$$

---

4   Stephan drops a drawing pin 60 times and records whether it lands 'point up' or 'point down'. He records his results in a table.

	Point up	Point down
Frequency	35	25

The relative frequency of 'point up' is $\frac{35}{60}$. Write the relative frequency of 'point down'.

Discussion Stephan states that the probability of a drawing pin landing 'point up' is $\frac{7}{12}$. Is he correct?

5   Jack experiments with a spinner. He records his results in a table.

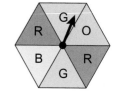

	Red	Blue	Green	Orange
Frequency	7	3	6	2

a   How many trials did Jack do?

b   What is the relative frequency of him spinning red?

6   Anjenna and Belinda are both experimenting with a spin wheel. Their results are recorded in the table.

	Green	Yellow
Anjenna	72	28
Belinda	120	40

a   How many times did Anjenna spin the wheel?

b   Write the estimated probability of green from Anjenna's results.

c   How many times did Belinda spin the wheel?

d   Write the estimated probability of green from Belinda's results.

e   Altogether Anjenna and Belinda spin the wheel 260 times. What is the estimated probability of green for their combined results?

f   Which is the best estimate for P(Green)? Explain.

7   **Communication** Esme is interested in the saying, 'Toast usually lands butter side down.' She decides to test its truth by carrying out the experiment 20 times.

The toast lands butter side down 12 times.

Esme concludes that the statement is true.

Do you agree? How could she get a more accurate measure of probability?

> **Q7 hint** Show your working. Write a sentence or two comparing your working to Esme's conclusion.

8   Arthur repeatedly rolls a standard four-sided dice and notes the number that is rolled.

	1	2	3	4
Frequency	14	18	12	16

   a   What is the experimental probability of Arthur rolling a 2?
   b   What is the theoretical probability of Arthur rolling a 2 on a fair four-sided dice?
   Discussion Is the dice biased? How could Arthur improve his experiment?

9   **Modelling / Communication** Two dice are rolled 120 times.
    The difference in the scores is recorded as being 0 or not 0.

	0	Not 0
Frequency	42	78

> **Q9b hint** Draw a sample space diagram to show the possible outcomes.

   a   What is the experimental probability of a difference of 0?
   b   What is the theoretical probability of a difference of 0?
   c   How many 0 differences would you predict from 120 rolls of the dice?
   d   Do you think the dice are biased? Explain your answer.

10   **Problem-solving** A charity game at a fete is won when a ball is rolled down a board and lands in an even-numbered pot.
    Mr Barker claims that the probability of winning on any go is $\frac{1}{3}$. He charges 20p per go and awards a prize of 40p to winning rolls.

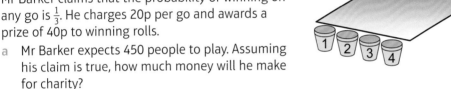

   a   Mr Barker expects 450 people to play. Assuming his claim is true, how much money will he make for charity?
   b   Class 10B think that Mr Barker is wrong. They decide to conduct an experiment. The table shows the results.

	1	2	3	4
Frequency	51	22	60	27

   Estimate the probability of winning from this experiment.
   Discussion Compare this estimate with Mr Barker's claim. Is Mr Barker correct?

11   **Exam-style question**

   a Tony throws a biased dice 100 times.
      The table shows his results.
      He throws the dice once more.
      Find an estimate for the probability
      that he will get a 6.   **(1 mark)**

Score	Frequency
1	12
2	13
3	17
4	10
5	18
6	30

   b Emma has a biased coin.
      The probability that the biased coin will land on a head is 0.7.
      Emma is going to throw the coin 250 times.
      Work out an estimate for the number of times the coin will land on a head.   **(2 marks)**

   Nov 2004, Q18, 5504/04

> **Exam hint**
> Show your working. If you calculate your answer incorrectly, you may still get marks for method.

12 The table shows some information about year 11 students.

	Staying at 6th form	Going to college	Total
Male	116	59	175
Female	88	65	153
Total	204	124	328

a Work out the probability that a student picked at random is
  i   a male
  ii a female going to college
b A male student is picked at random. What is the probability that the student is staying at 6th form?

> **Q12a hint**
>
> $\dfrac{\text{number of males}}{\text{total number of students}}$

> **Q12b hint**
>
> $\dfrac{\text{number of males staying}}{\text{total number of males}}$

13 Look back at the table of vehicles passing in **Q2**.
a Estimate the probability that the next vehicle to pass the gate is
  i   a red car
  ii a blue bus.
b Which is more likely to pass the school gates, a silver van or a blue bus?
c Which type of vehicle is most likely to be red?
d Which type of vehicle is most likely to be silver?
e What is the most likely van colour?

13.4 Venn diagrams

Objectives

- Use Venn diagrams to work out probabilities.
- Understand the language of sets and Venn diagrams.

Did you know?

John Venn developed Venn diagrams in the late 19th century while lecturing at the University of Cambridge.

Fluency

- What are the first five   **a** prime numbers   **b** multiples of three?
- What is an integer?

1 Twenty cards numbered from 1 to 20 are placed in a hat. One card is drawn out at random.
a What is the probability that it is
  i   a factor of both 12 and 18
  ii   a prime number
  iii an even number
  iv   a square number?
b The first card is replaced and a second card drawn. What is the probability that both cards drawn are odd?

Key point 10

Curly brackets { } show a set of values.

ϵ means 'is an element of'.

$5 \in \{$odd numbers$\}$ means '5 is in the set of odd numbers'.

An element is a 'member' of a set. Elements are often numbers, but could be letters, items of clothing or even body parts.

ξ means the universal set – all the elements being considered.

2   Set A = {odd numbers under 10}
    Set B = {square numbers under 10}
    a   List the numbers in each set. A = {1, 3, …}, B = {1, 4, …}
    b   Write 'true' or 'false' for each statement:
        i   6 ∈ A   ii   9 ∈ B   iii   11 ∈ A
    c   Which numbers are in both sets?
    d   Which numbers are in set A only?
    e   Which numbers are in either A or B (or both)?
    f   Which numbers under 10 are not in A or in B?

3   The Venn diagram shows 2 sets,
    X and Y, and ξ, the set of all numbers
    being considered.
    a   Copy and complete these sets.
        X = {2, 4, 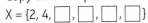}
        Y = {3, ☐, ☐, ☐}
        ξ = {1, 2, 3, …}

> **Q3a hint**
> ξ includes all
> the numbers
> in X and Y too.

    b   Match each set to its description.

| X | {integers 1 to 12} | Y | {multiples of 2 up to 12} | ξ | {multiples of 3 up to 12} |

4   Copy and complete the Venn diagram for these sets.
    ξ = {integers from 1 to 15}
    A = {even numbers from 1 to 15}
    B = {multiples of 3 from 1 to 15}

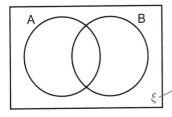

> Write the numbers that are in
> ξ but not in A or B outside.

Key point 11

A ∩ B means A intersection B. This is all the elements that are in A *and* in B.

A ∪ B means A union B. This is all the elements that are in A *or* B *or* both.

A' means the elements *not* in A.

---

5   For the Venn diagram in **Q3**, copy and complete these sets.
    a   X ∩ Y = {6, ☐}   b   X ∪ Y = {2, 3, …}   c   X' = {1, …}
    d   Y' = {1, 2, …}   e   X' ∩ Y = {…}

> **Q5e hint**
>

6   For the Venn diagram you drew in **Q4**, write these sets.
    a   A ∩ B   b   A ∪ B   c   A'
    d   B'   e   A' ∩ B   f   A ∩ B'

Key point 12

You can calculate probabilities from Venn diagrams.

Example 2

The Venn diagram show the numbers of students studying French and Spanish.

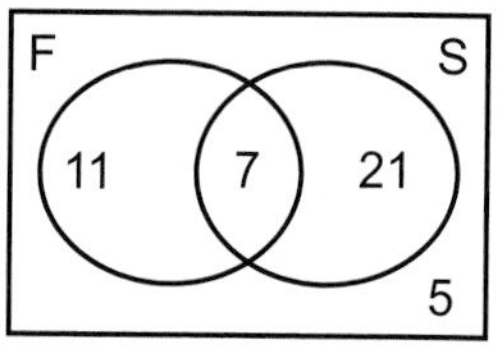

Work out

a the total number of students represented in the diagram

b P(F)

c P(F ∪ S)

d P(F′)

F   S

11 7 21

5

Add all the numbers in each part of the diagram.

a 11 + 7 + 21 + 5 = 44

b P(F) = $\frac{18}{44}$

$\dfrac{\text{number of students in F}}{\text{total number of students}}$

c P(F ∪ S) = $\frac{39}{44}$

$\dfrac{\text{number of students in F ∪ S}}{\text{total number of students}}$

d P(F′) = $\frac{26}{44}$

21 + 5 students are not in F

---

7 **Real / Communication** Jessica asks 25 people in her school if they have a mobile phone or tablet.

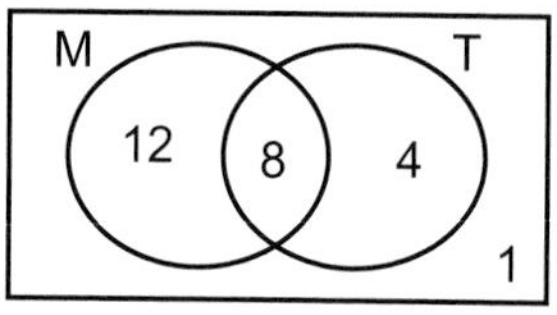

The Venn diagram shows her results.

M   T

12 8 4

1

a How many people have a tablet?

b How many people have either a mobile phone or a tablet?

c What is the probability that a person chosen at random has a tablet and a mobile phone?

d What is the probability that a person chosen at random just has a tablet?

e Jessica's friend Anneliese says, 'The Venn diagram shows that 12 people have mobile phones.' Explain why Anneliese is not correct.

8 **Real** Adrianna is doing a survey of music tastes. She asks people if they like rock (R) or pop (P). Of the 30 people she surveyed, 17 said they like both rock and pop, 8 said they like neither and 3 said they like only pop music.

a Draw a Venn diagram to show Adrianna's findings.

b What is the probability that a person chosen at random likes only pop?

c What is the probability that a person chosen at random likes rock?

Discussion What is the probability that a person chosen at random likes classical music?

9 **Exam-style question**

X is the set of students who read science fiction books. Y is the set of students who read romantic fiction.

The Venn diagram shows the number of students in each set.

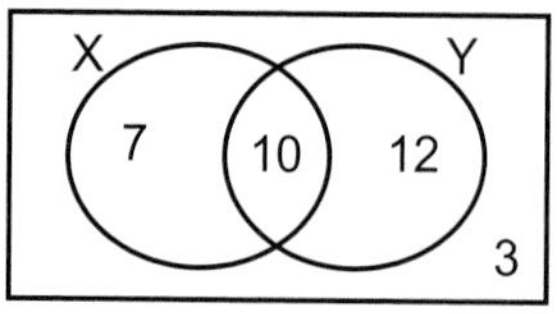

X   Y

7 10 12

3

a Work out P(X ∩ Y). **(2 marks)**

b Work out P(X′ ∪ Y′). **(2 marks)**

Q9 strategy hint
First work out the total number of students

13.5 **Tree diagrams**

Objectives

- Use frequency trees and tree diagrams.
- Work out probabilities using tree diagrams.
- Understand independent events.

Why learn this?

If we know how likely sequences of events are it can help us plan.

Fluency

Work out   **a** $\frac{1}{3} \times \frac{1}{3}$ **b** $\frac{1}{2} \times \frac{1}{2}$ **c** $\frac{2}{9} + \frac{4}{9}$

Warm up

1   Traffic lights can show red, amber or green. On her journey to school, Anna passes through two sets of traffic lights. Write all the possible combinations of lights that Anna could encounter.

2   A bag contains 10 coloured sweets. There are 7 red sweets and 3 green sweets.
   1 sweet is picked from the bag at random. Write
   a   the probability of getting a red sweet
   b   the probability of *not* getting a red sweet.

3   A fair 6-sided spinner has 2 red and 4 yellow sections. Work out
   a   P(red)
   b   P(yellow)
   c   P(red or yellow)

Key point 13

Two events are **independent** when the results of one do not affect the results of the other.

4   In a box of chocolates, there are 2 hard centres and 5 soft centres.
   a   What is the probability of picking a soft centre?
   David picks a chocolate at random. It is a soft centre and he eats it.
   b   How many chocolates are left in the box?
   c   How many soft centres are left in the box?
   He picks another chocolate at random.
   d   What is the probability that he picks a soft centre this time?
   Discussion   Are the two events 'picking a soft centre first time' and 'picking a soft centre second time' independent?

5   Which of these pairs of events are independent?
   A   Rolling a 5 on a dice and then rolling another 5.
   B   Getting full marks in a physics test and then getting full marks in a biology test.
   C   Getting a head on a coin on the first flip and getting a head on a coin on the second flip.
   D   Picking a jelly sweet from a bag, eating it and then picking a second jelly sweet from a bag.

Key point 14

A frequency tree shows the numbers of options for different choices.

6   Amarta has a box of toys containing 4 soft ones and
    5 hard ones. She chooses a toy at random, puts it
    back and then chooses another toy at random.
    The frequency tree shows the possible outcomes.

Q6 hint S stands for
soft toy and H for
hard toy.

    a   Copy this frequency tree.

    b   4 is on the first branch as there are 4 soft toys.
        How many hard toys are there?
        Use your answer to complete the branches for
        the 1st choice box on the frequency tree.

    c   Amarta puts her first toy back in the box.
        How many hard toys and soft toys are in
        the box when she chooses another one?
        Complete the branches for the 2nd choice by filling
        in the rest of the numbers.

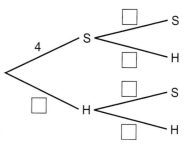

    Discussion How many ways are there of picking a soft toy?

Key point 15

When there are 3 ways of making the first choice
and 3 ways of making the second, there are 3 × 3 = 9
ways of choosing two objects.

Sample space diagram

---

7   Jennie has a bag of balls containing 7 yellow balls and 4 red balls.
    Jennie takes one ball from the bag, replaces it and then takes a second ball from the bag.

    a   Copy and complete this frequency tree.

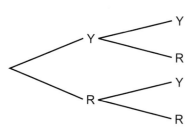

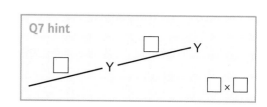

Q7 hint

    b   How many ways are there of picking two yellow balls?

8   Philomena has 3 red tops and 6 green tops. She picks a top at random, puts it back and picks
    another top.
    Draw a frequency tree to show this.

Example 3

There are 3 blue and 4 green counters in a bag.

Steven picks one counter at random, notes its colour and replaces it.

He then chooses a second counter and notes its colour.

a Draw a tree diagram to show this.

b Work out the probability that Steven picks 2 blue counters.

c Work out the probability that Steven picks 1 blue counter and 1 green counter.

a

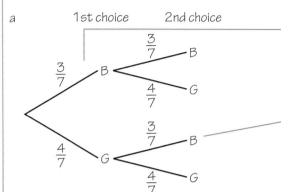

A tree diagram is the same as a frequency tree except the branches are labelled with the probabilities, not the frequencies.
$P(B) = \frac{3}{7}$ and $P(G) = \frac{4}{7}$

The two events are independent so the probabilities for the second choice are the same as for the first choice.

b $\frac{3}{7} \times \frac{3}{7} = \frac{9}{49}$

When events are independent, multiply the probabilities together along the branches of the tree to work out the probability of the final outcome.

c BG: $\frac{3}{7} \times \frac{4}{7} = \frac{12}{49}$

GB: $\frac{4}{7} \times \frac{3}{7} = \frac{12}{49}$

Total: $\frac{12}{49} + \frac{12}{49} = \frac{24}{49}$

Look at the events 'BG' and 'GB'. Work out each probability and add them together.

9 The tree diagram shows the probabilities that Mrs Johnson has to stop at two sets of traffic lights.

a What is the probability she stops at the first set?

b What is the probability that she stops at both sets?

Q9b hint Follow the branches 'stop' and 'stop'. Multiply the probabilities.

c What is the probability that she doesn't stop at either set?

d What is the probability that she gets stopped at just one set? Explain your answer.

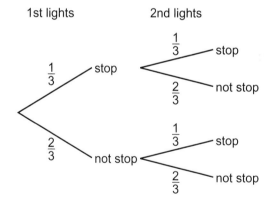

10 Melanie travels to school by bus. On each day, the probability that the bus is late is $\frac{1}{4}$.

a Write the probability that the bus is not late.

b Copy and complete the tree diagram.

c Work out the probability that the bus is not late two days running.

d Work out the probability that the bus is late on the first day and not late on the second.

Discussion How can you check the probabilities on your diagram are correct?

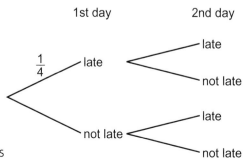

11 Frances has a bag of marbles. The bag contains 8 green marbles and 5 purple marbles. She picks a marble at random, notes its colour and replaces it in the bag. She then picks a second marble.

   a Copy and complete the tree diagram.

   b What is the probability that Frances picks one marble of each colour?

   Discussion How can you work out the probability of picking at least one green?

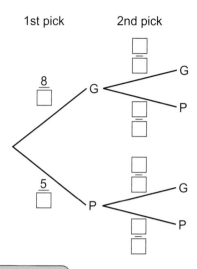

12 **Exam-style question**

   Archie rolls a fair black dice once and a fair red dice once.

   a Copy and complete the probability tree diagram to show the outcomes. Label clearly the branches of the probability tree diagram.

   b Calculate the probability that Archie gets a six on both the red dice and the black dice.

   c Calculate the probability that Archie gets at least one six.

   Exam hint
   Complete the tree diagram carefully – you will need it to answer the rest of the question.

13.6 More tree diagrams

Objectives

- Understand when events are not independent.
- Solve probability problems involving events that are not independent.

Why learn this?

Your first choice can change the options for your second choice.

Fluency

- Work out **a** $\frac{2}{3} \times \frac{1}{2}$ **b** $\frac{2}{7} \times \frac{5}{6}$

1 A bag contains 8 coloured sweets. There are 5 red sweets and 3 green sweets.
   Fiona picks a red sweet, changes her mind and puts it back in the bag. She then chooses another sweet at random.

   a Copy and complete the tree diagram.

   b Work out the probability that she picks 2 green sweets.

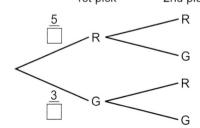

2 A bag contains 6 orange and 4 pink sweets.
   a Debbie picks a sweet at random. What is P(pink)?
   Debbie's sweet is pink. She eats it.
   b How many sweets are left in the bag?
   c How many pink sweets are left in the bag?
   d Now what is
      i P(pink)   ii P(orange)?

> **Key point 16**
>
> When the outcome of one event changes the possible outcomes of the next event, the two events are not independent.

3 Are these pairs of events dependent or independent?
   a Picking a blue sock at random from a drawer, putting it on, then picking another sock.
   b Oversleeping and being late to school.
   c Rolling an even number on a fair dice and rolling a second even number.
   d Rolling a 6, then rolling another 6.
   e **Reflect** In your own words, write a definition for the terms independent and dependent.

4 Yasmin has a box of chocolates containing 3 soft centres and 5 hard centres. The frequency tree shows what happens when she chooses a chocolate at random, eats it and then chooses another chocolate.
   Copy and complete her frequency tree.

> Q4 hint The total number of chocolates on the second branch pairs will go down by 1.

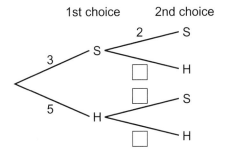

5 Jaime has a bag of balls containing 7 yellow balls and 3 green balls. She takes one ball from the bag, puts it to one side and then takes a second ball from the bag.

   a Copy and complete the frequency tree to show this.

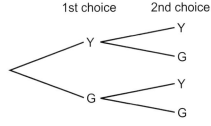

   b How many ways are there of choosing two yellow?

Example 4

Alexander has seven coins in his pocket. There are three £1 coins and four €1 coins.
He picks a coin from his pocket at random, keeps it out and then picks another coin at random.
a Draw a tree diagram to show this.
b Work out the probability that he picks two £1 coins.
c Work out the probability that he picks one of each type of coin.

a

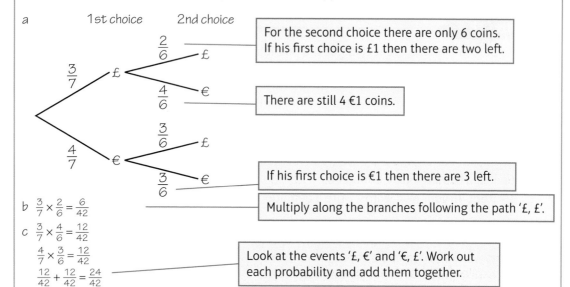

1st choice   2nd choice

For the second choice there are only 6 coins.
If his first choice is £1 then there are two left.

There are still 4 €1 coins.

If his first choice is €1 then there are 3 left.

b $\frac{3}{7} \times \frac{2}{6} = \frac{6}{42}$

Multiply along the branches following the path '£, £'.

c $\frac{3}{7} \times \frac{4}{6} = \frac{12}{42}$

$\frac{4}{7} \times \frac{3}{6} = \frac{12}{42}$

$\frac{12}{42} + \frac{12}{42} = \frac{24}{42}$

Look at the events '£, €' and '€, £'. Work out
each probability and add them together.

6   Eleri has a bag of sweets. There are 7 toffees and 3 jellies.
    She picks a sweet at random and eats it. She then picks a
    second sweet at random and eats it.
    Use the tree diagram to find the probability that Eleri eats
    a   2 toffees
    b   1 of each type of sweet.

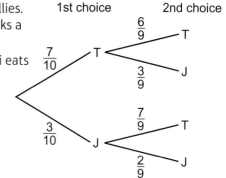

7   **Modelling** June has these numbered cards. She picks 2 cards at random without replacing
    them. She records whether the number on each card is odd or even.

    [1] [1] [3] [4] [5] [6] [6]

    a   Copy and complete the tree diagram.
    b   Find the probability that she picks
        i   2 even numbered cards
        ii 1 even numbered card and
            1 odd numbered card.

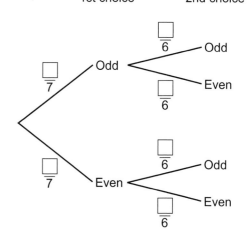

8   **Real / Reasoning** Mr Brown passses through two sets of traffic lights on his drive to work.
The probability that he stops at the first set of lights is 0.4.
If he stops at the first set, the probability he stops at the second set is 0.6.
If he does not stop at the first set, the probability he stops at the second set is 0.25.

a   Copy and complete the tree diagram.

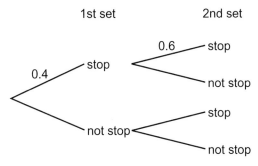

b   Find the probability that Mr Brown stops at just one set of traffic lights

9
┌──┐
Exam-style question

There are 10 socks in a drawer.
7 of the socks are brown.
3 of the socks are grey.
Bevan takes at random two socks from the drawer at the same time.

a   Complete the probability tree diagram.

```
            1st sock            2nd sock

                     _____  Brown
         7    Brown ⟨
        ──               _____  Grey
        10
                     _____
     ⟨
                     _____  Brown
        ____  Grey  ⟨
                     _____  Grey
                                              (2 marks)
```

b   Work out the probability that Bevan takes two socks of the
same colour.   (3 marks)

Nov 2010, Q12, 5MB1H/01
└──┘

Exam hint
Make sure
you label each
calculation
clearly, for
example
Brown, Brown
= or B, B =

397

13 Problem-solving: Ciphers, language and probability

Objectives

- Estimate probabilities from experimental data.
- Compare experimental data with given probabilities.

1 Some letters in the English language occur more often than others. The table at the bottom shows the approximate relative frequency of letters in English words.
   a A letter is picked at random from a page of English text.
      i What is the probability that it is a D? Write your answer as a decimal.
      ii What is the probability that it is a vowel?
   b Four English letters are missing from the table. What are they?
   Discussion Why do you think these letters are missing from the table?

2 The text below is a substitution cipher. This means that each of the original capital letters of the message has been substituted with a lower case letter.
   a Calculate the relative frequency of q in the cipher. Then use the table to decide which capital letter has been substituted by q.
   b Crack the substitution cipher.

   > **Q2b hint** Start by looking for letters with high relative frequencies. You need to use your knowledge of the English language.

3 The relative frequencies of the letters in the message do not match exactly with those in the table. Why might it be easier to use the table to crack a longer code?

4 Are the letters used in English words independent? Why might a cryptographer also look at the probabilities of pairs or sets of letters?

   > **Q4 hint** think about how the letter q occurs in English words.

Letter	Relative Frequency (%)	Letter	Relative Frequency (%)
E	13	C	3
T	9	U	3
A	8	M	2
O	8	W	2
I	7	F	2
N	7	G	2
S	6	Y	2
H	6	P	2
R	6	B	2
D	4	V	1
L	4	K	1

Facts

- A cipher is a message written in secret code.
- Cryptographers (code-breakers) can use relative frequency to crack substitution ciphers.

```
q xqoo aiic cdi hgfcgt pibt cdi tgbh ktqhli.
di dbe fbed bph q dbyi cdi hgfzaipce.
qj vgz eii bpvgpi ioei, tzp!
hgp'c cbwi b ctbqp gt kgbc.
vgzt tibo aqeeqgp: cioo cdia pgcdqpl.
```

Log how you did on your
Student Progression Chart.

Calculating probabilities

1   The probability of getting a 4 on a spinner is $\frac{1}{4}$. Write the probability of not getting a 4.

2   The letters from the word STATISTICS are written on cards and placed in a bag. One card is chosen at random.
Work out
a   P(S)   b   P(a vowel)   c   P(not S or T)

3   The table shows the probabilities of certain outcomes on a spinner with 3 colours.

Colour	Red	Yellow	Blue
Probability	0.3	0.2	

Work out P(blue).

Experimental probability

4   Pierre and Guillermo do an experiment with a spinner.
The table shows their results.

	Yellow	Blue	Green	Red
Pierre	32	25	22	31
Guillermo	18	15	17	20

a   Estimate the experimental probability of spinning blue using Pierre's results.
b   Estimate the experimental probability of spinning blue using Guillermo's results.
c   Which experimental results are more accurate?

5   Luca flips a coin 200 times. The coin comes down heads 80 times.
a   Write the experimental probability of getting a head.
b   Write the theoretical probability of getting a head.
c   How many heads would you predict in 200 flips of a fair coin?
d   Is Luca's coin biased? Explain.

Probability diagrams

6   Two fair coins are flipped.
a   Draw a sample space diagram.
b   What is the probability that both coins land on tails?
c   What is the probability that one coin lands heads and the other lands tails?

7   Annabel and Elizabeth each have a set of cards. Florence picks one card from each set and adds the numbers.

Annabel

| 1 | 2 | 4 | 5 |

Elizabeth

| 2 | 3 | 4 | 6 |

a   Draw a sample space diagram showing the possible total scores.
b   How many possible outcomes are there?
c   What is the probability that Florence picks cards with
  i   a total of 7   ii   a total greater than 5?

8   Zainab asks students in her class whether they like football or tennis. The Venn diagram shows the results.

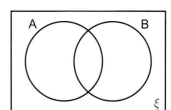

a   How many people are in Zainab's class?
b   How many people like football?
c   What is the probability that a person chosen at random likes football but not tennis?
d   Work out
     i   P(F ∪ T)   ii   P(F ∩ T)

9   a   Copy and complete the Venn diagram for these sets.
       ξ = {integers < 10}
       A = {odd numbers < 10}
       B = {factors of 8}
    b   Write these sets.
        i   A ∩ B   ii   A ∪ B   iii   A′

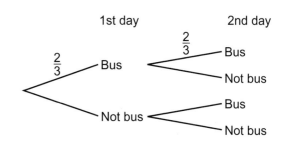

10 The probability that Julienne gets the bus to school each day is $\frac{2}{3}$.
    a   Copy and complete the tree diagram.
    b   What is the probability that Julienne gets the bus on both days?
    c   What is the probability that she gets the bus on just one day?

Dependent events

11 **Reasoning** A box contains 7 necklaces and 3 bracelets. Philippa chooses an item from the box at random, puts it on and then chooses another item.
    a   Copy and complete the tree diagram.
    b   What is the probability that she is wearing two necklaces?
    c   What is the probability that she is wearing one necklace and one bracelet?

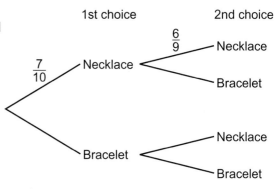

12 How sure are you of your answers? Were you mostly

    Just guessing 😕   Feeling doubtful 😐   Confident 🙂

    What next? Use your results to decide whether to strengthen or extend your learning.

✳ Challenge

13 A fairground game involves spinning two wheels. You win if your total score is less than 7.
    a   Is the game fair? Explain.
    b   The fairground game operator suggests changing the rules so that you win if the difference in the scores is less than 3. Is the game fair now? Explain.

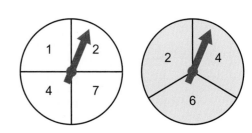

13 Strengthen

Calculating probabilities

1 The letters from the word MATHEMATICS are placed in a hat.

   a One letter is selected at random.
How many possible outcomes are there?

> **Q1a hint** How many letters are there in the word?

   b What is the probability of

     i selecting an M

     ii selecting a C

     iii selecting a vowel

     iv *not* selecting an A?

> **Q1b i hint** $\dfrac{\text{Number of Ms}}{\text{Total number of letters}}$

> **Q1b iv hint** How many letters are *not* 'A'?

2 A bag contains 7 blue balls, 5 red balls and 1 green ball. What is

   a P(blue)?

   b P(blue or red)?

   c P(not red)?

> **Q2 hint** P(blue) means probability of blue.

3 Cards numbered 1–5 are put in a bag.
One card is picked at random.

Work out

   a P(odd)

   b P(even)

   c P(square number)

   d P(not square number)

   | 1 | 2 | 3 | 4 | 5 |

Discussion Are all the numbers 1–5 odd or even? Can a number be both odd and even?
What is P(even) + P(odd)?

4 The probability of getting a six on a fair dice is $\frac{1}{6}$.
Write the probability of not getting a six.

> **Q4 hint**
>
>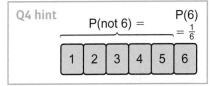
>
> P(not 6) = [1 2 3 4 5]   P(6) = $\frac{1}{6}$ [6]

5 The table shows the probabilities of trains arriving early, late or on time.

Arrival	Early	Late	On time
Probability	0.1		0.6

Work out the probability of a train arriving late.

> **Q5 hint** Early, late or on time are the only possible outcomes.

Experimental probability

1 **Communication** Maddie rolls a dice and records the number of times she gets a 2. Freya does the same with a different dice.

	Maddie	Freya
Number of rolls	60	90
Number of 2s	12	30

   a Write the theoretical probability of rolling a 2.

   b Write the experimental probability of

     i Maddie getting a 2

     ii Freya getting a 2?

> **Q1a hint** What is the probability of rolling a 2 on a fair dice?

   c Compare the experimental probabilities with the theoretical probability. Do you think either dice is biased? Explain your answer.

2   In an experiment, Hermione spins these two spinners 120 times.
    Her results are shown in the table.

	1	2	3	4
Green	12	14	13	10
Blue	13	9	12	6
Red	11	9	6	5

   a   How many times did she get green and
       2 together?

   b   Estimate the experimental probability of
          i   green and 2
          ii red and 3
          iii green and odd.

> **Q2b i hint** $\dfrac{\text{number of 'green and 2' results}}{\text{total number of spins}}$

3   **Reasoning** Seb flips a coin 10 times and gets 7 heads.
    a   Explain why he thinks the coin might be biased.
    He flips the same coin 200 times and gets 102 heads.
    b   Explain why he now thinks the coin is fair.
    c   Which is his most accurate estimate of the
        experimental probability of getting a head? Explain.

> **Q3c hint** More _____ → more accurate estimate.

Probability diagrams

1   Rachel spins these two spinners. One possible outcome is (2, 3).

   a   Complete the sample space diagram to show all the possible outcomes.

	2	4	6
3	(2, 3)		
6			
9			

   b   How many possible outcomes are there?
   c   Work out the probability of
          i   getting (4, 6)
          ii both numbers being at least 4
          iii getting two even numbers.

> **Q1d hint** Draw a new sample space diagram to show the totals. How many are more than 8?

   d   Rachel now adds the scores together. What is the probability of getting a total greater
       than 8?

2   Denise rolls two fair, six-sided dice and records the product of the two numbers.
    a   Copy and complete the sample space diagram.
    b   How many possible outcomes are there?
    c   What is the probability of getting a
           i   product of 1
           ii product greater than 20
           iii product that is a square number?

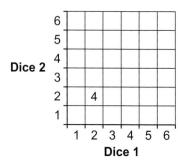

> **Q2 communication hint**
> Product means multiply.

3   At a sports club, 15 members play hockey, 17 play cricket and 8 play both. 10 members play neither.

   a   Copy and complete the Venn diagram.

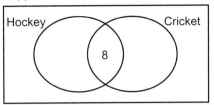

> **Q3a hint** Subtract the members who play both from the total for hockey and from the total for cricket. How many members are left over?

   b   Work out the total number of members?

> **Q3b hint** Add up all the numbers in every section.

   c   What is the probability that a member chosen at random

      i   plays hockey

      ii   plays both hockey and cricket

> **Q3c hint** $\dfrac{\text{number who play hockey}}{\text{total number of members}}$

     iii   plays hockey but not cricket.

     iv   plays neither hockey nor cricket?

4   Francis asks 25 students in his class if they play music or sports outside school. The Venn diagram shows the results.

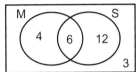

> **Q4a hint**

> **Q4b hint**

> **Q4c hint**

Work out the probability that a person picked at random

   a   plays music or sport or both.

   b   plays music and sport.

   c   doesn't play music.

5   Look at this Venn diagram.

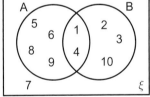

> **Q5c hint** In A ∪ B means in A or B or both.

> **Q5d hint** In A ∩ B means in A and B.

> **Q5e hint** ξ includes A and B.

List the numbers

   a   in A   **b**   in B   **c**   in A ∪ B   **d**   in A ∩ B   **e**   in ξ

6   Menna picks a ball from a bag containing 4 green balls and 6 red balls.
She replaces the ball and then selects another.
The tree diagram shows the probabilities.

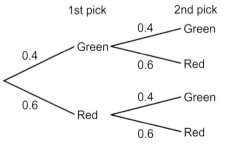

> **Q6a hint** Move along the branches for green, then green. Are 2 greens more or less likely than 1 green? Do you add or multiply?

> **Q6b ii hint** This means (red, green) or (green, red). Is the probability of these two outcomes greater than the probability of just one of them? Do you add or multiply?

   a   Work out the probability of picking 2 green balls.

   b   Work out the probability of picking

      i   red then green

      ii   1 red ball and 1 green ball *in any order*.

7   Daphne plays online chess against a friend. The probability of Daphne winning each game is $\frac{3}{5}$. The friends never draw a game.

a   What is the probability that Daphne loses a game?

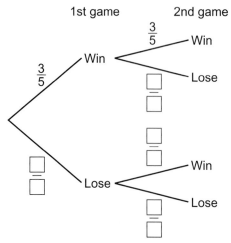

1st game   2nd game

b   Copy and complete the tree diagram showing the possible outcomes from two games.
c   What is the probability that Daphne wins both games?
d   What is the probability that Daphne loses both games?
e   What is the probability that Daphne wins one game and loses one game?

Dependent events

1   **Reasoning** Geeta's pack of sweets contains 7 jelly sweets and 9 boiled sweets. She picks one sweet at random, eats it and then picks a second sweet.

a   If her first sweet is jelly, the probability that her second sweet is jelly is $\frac{6}{15}$. Explain why.
b   If her first sweet is boiled, the probability that her second sweet is boiled is $\frac{8}{15}$. Explain why.
c   Copy and complete the tree diagram.

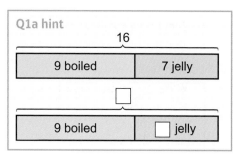

Q1a hint

1st sweet   2nd sweet

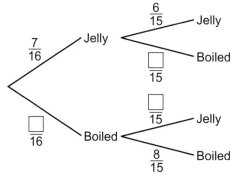

d   Work out the probability that Geeta eats
i   2 jelly sweets   ii   1 jelly sweet and 1 boiled sweet.

13 **Extend**

1 The probability of rain tomorrow is 26%.
What is the probability that it will not rain?

2 **Problem-solving** Lynn has a bag containing 7 blue counters and 8 red counters.
   a Lynn picks a counter. Write the probability that the counter is red.
   b Lynn adds blue counters to the bag so that the probability of picking a red counter is now $\frac{1}{3}$.
   How many blue counters did she add?

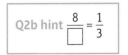
Q2b hint $\frac{8}{\square} = \frac{1}{3}$

3 **Modelling** A dice is rolled twice.
   a Write the probability of getting a 6 on one roll.
   b Copy and complete the tree diagram.
   c Write a calculation for the probability of getting a 6 on both rolls.
   d The dice is rolled a third time. Write a calculation for the probability of getting a 6 on all three rolls.

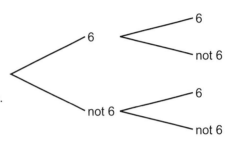

1st roll   2nd roll

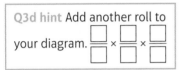
Q3d hint Add another roll to your diagram. $\frac{\square}{\square} \times \frac{\square}{\square} \times \frac{\square}{\square}$

   e The dice is rolled n times. Write an expression for the probability of getting a 6 on all n rolls.

4 **Problem-solving** Amit has a bag of red and yellow balls in the ratio 2:3. He picks one ball at random.
   a Write the probability that he picks
      i a red ball   ii a yellow ball.
   b Amit replaces the ball, then picks another. He does this 20 times. Predict how many times he will pick a red ball.

5 **Modelling / Communication** The table shows the probabilities of picking coloured counters from a box.

	Blue	Green	Red	Yellow
Probability	0.1	0.15	0.5	0.25

   a Explain why there must be more than 10 counters in the box.
   b What is the smallest number of counters that could be in the box?

Q5a hint Calculate the expected results for 10 counters.

6 **Real / Reasoning** The probability of a football team winning a match is 0.5 The probability that they lose the match is 30% and the probability that they draw is 0.2. They play two matches.
   a Copy and complete the tree diagram.
   b Work out the probability that they
      i win both matches
      ii win the first match and lose the second
      iii draw one match and win the other.

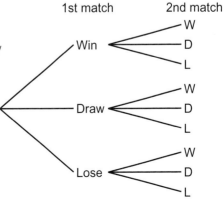

1st match   2nd match

7 **Real** Sylvie conducts a survey of her classmates favourite chocolates. She records her results in a Venn diagram.

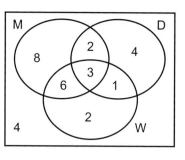

  a   How many students are in Sylvie's class?
  b   How many students liked milk chocolate?
  c   How many students liked white and dark chocolate?
  d   Work out the probability that a student picked at random
      i   likes milk chocolate
      ii likes all three types of chocolate
      iii likes milk and dark but not white chocolate
      iv likes all chocolate except milk chocolate
      v   doesn't like any of these types of chocolate.

8 **Modelling / Communication** Layla rolls two dice. She adds the scores. Which is more likely – a total score greater than 7 or less than 7? Explain your answer.

9 A bus company records the numbers of times its buses are early, late or on time one month.

Early	Late	On time
176	527	964

Calculate the experimental probability that a bus selected at random is on time. Give your answer as a percentage to the nearest whole number.

10 **Modelling** On her way to school, Miss Stevens passes through three sets of traffic lights. The probability that she gets stopped by any of the sets of lights is 0.4.

  a   Copy and complete the tree diagram.
  b   Work out the probability that Miss Stevens is stopped by
      i   all three sets of lights
      ii none of the sets of lights
      iii just one set of lights.
  c   Using your answers to parts **b i** to **iii**, work out the probability that Miss Stevens is stopped by two sets of lights.

11 **Real** Brendan is often late for school. The probability that he oversleeps is $\frac{2}{5}$. If he oversleeps he is late to school $\frac{1}{2}$ of the time. If he gets up on time he is late $\frac{1}{4}$ of the time.

  a   Copy and complete the tree diagram.
  b   What is the probability that he oversleeps and is late for school?
  c   What is the probability that he gets up on time and is late for school?
  d   What is the probability that Brendan is late for school?

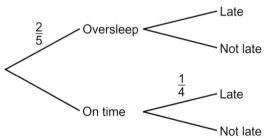

12 **Modelling** Patrick has x cards. Jim has 3 times as many cards as Patrick.

  a   Write an expression, in terms of x, for the number of cards Jim has.
  b   They place the cards into a bag. Jed chooses one card at random. What is the probability that he chooses one of Jim's cards?

13 (Exam-style question)

The probability that Rebecca will win any game of snooker is p.
She plays two games of snooker.

a Complete, in terms of p, the probability tree diagram. **(2 marks)**

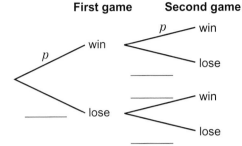

First game   **Second game**

p — win
win
lose

win
lose
lose

b Write down an expression, in terms of p, for the probability that
Rebecca will win both games. **(1 mark)**

c Write down an expression, in terms of p, for the probability that
Rebecca will win exactly one of the games. **(2 marks)**

June 2012, Q12, 5MB1H/01

Exam hint
For part **a**
what must all
outcomes add
up to?
To help you
answer part **b**,
write down all
the possible
outcomes, for
example (win,
win), (win,
lose),
...

13 Knowledge check

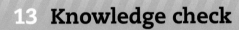

⊙ The probability of an event happening is a number between
  0 and 1. .... *Mastery lesson 13.1*

⊙ If an event is certain, the probability is 1, P = 1.
  If an event is impossible, the probability is 0, P = 0. .... *Mastery lesson 13.1*

⊙ Events are **mutually exclusive** when they cannot happen at the
  same time. .... *Mastery lesson 13.1*

⊙ Events are **exhaustive** if they include all possible outcomes.
  The probabilities of an exhaustive set of mutually exclusive
  events sum to 1. .... *Mastery lesson 13.1*

⊙ Equally likely outcomes have the same probability of happening. .... *Mastery lesson 13.1*

⊙ For equally likely outcomes, the probability that an event will

  happen is P = $\dfrac{\text{number of successful outcomes}}{\text{total number of possible outcomes}}$ .... *Mastery lesson 13.1*

⊙ If the probability of an event happening is P, the probability of it not
  happening is 1 – P. .... *Mastery lesson 13.1*

⊙ A sample space diagram shows all the possible outcomes for one or
  more events. You can use it to find a theoretical probability. .... *Mastery lesson 13.2*

⊙ You can estimate the probability of an event from the results of an
  experiment or survey.

  relative frequency = $\dfrac{\text{number of successful trials}}{\text{total number of trials}}$

  estimated probability = $\dfrac{\text{frequency of event}}{\text{total frequency}}$. .... *Mastery lesson 13.3*

⊙ Estimated probability is also called **experimental probability**. ...... *Mastery lesson 13.3*

⊙ A larger number of trials gives a more accurate estimate
  of probability. .... *Mastery lesson 13.3*

⊙ Predicted number of outcomes = probability × number of trials ...... *Mastery lesson 13.3*

⊙ The elements of two (or more) sets can be shown together in a Venn diagram. Curly brackets { } show a set of values.
ϵ means is an element of.
ξ means the universal set – all the elements being considered.
A ∩ B means A intersection B. This is all the elements that are in A *and* in B.
A ∪ B means A union B. This is all the elements that are in A *or* B or both. A' means the elements *not* in A. .... *Mastery lesson 13.4*

⊙ A set is a list of things that share certain characteristics. .... *Mastery lesson 13.1*

⊙ You can calculate probabilities from a Venn diagram using
$$\frac{\text{number in the set}}{\text{total number in the Venn diagram}}.$$ .... *Mastery lesson 13.4*

⊙ Two events are **independent** when the results of one do not affect the results of the other. .... *Mastery lesson 13.5*

⊙ When the outcome of one event changes the possible outcomes of the next event, the two events are not independent. .... *Mastery lesson 13.5*

⊙ When there are 3 ways of making the first choice and 3 ways of making the next, there are 3 × 3 = 9 ways of choosing 2 objects. ...... *Mastery lesson 13.5*

⊙ A frequency tree shows the number of options for different choices. .... *Mastery lesson 13.5*

⊙ A probability tree diagram shows all possible outcomes of an experiment. .... *Mastery lessons 13.5 and 6*

Choose A B or C to complete each statement about probability.

In this unit, I did…	A well	B OK	C not very well
I think probability is…	A easy	B OK	C hard
When I think about doing probability, I feel…	A confident	B OK	C unsure

Did you answer mostly As and Bs? Are you surprised by how you feel about probability? Why?
Did you answer mostly Cs? Find the three questions in this unit that you found the hardest.
Ask someone to explain them to you. Then complete the statements above again.

13 **Unit test**

Log how you did on your Student Progression Chart.

1 The letters of the word MATHEMATICS are written onto cards and placed in a bag.
What is the probability that a card chosen at random

   a is T   b is a vowel   c is *not* M or S?   *(3 marks)*

   d The card is returned to the bag. A second card is chosen.
      What is the probability of picking two Ts?   *(2 marks)*

2 Richard spins these two spinners and records the total score.

   a Draw a sample space diagram.   *(2 marks)*

   b What is the probability that Richard spins
     i an even total   *(1 mark)*
     ii a total greater than 7   *(1 mark)*
     iii a total that is a prime number?   *(1 mark)*

*Active*Learn Homework, practice and support: Foundation 13 Unit test

3 20 students are surveyed and asked whether they like pop or rock music. 12 students say they like both pop and rock, 3 students say they like just pop and 4 students say they like neither.

    a How many students said they liked just rock? *(1 mark)*

    b Draw a Venn diagram. *(2 marks)*

    c A student is chosen at random. Work out the probability that they

        i like pop music *(1 mark)*

       ii like rock music *(1 mark)*

      iii like neither pop nor rock. *(1 mark)*

4 Evie surveys the vehicles passing the school gates one morning. The table shows her results.

	Blue	Silver	Red
Car	5	13	7
Van	8	2	5

    a How many vehicles did she record? *(1 mark)*

    b Estimate the probability of a red van. *(1 mark)*

    c Estimate the probability that the next vehicle is

        i a car   ii silver *(2 marks)*

5 A bag contains green, blue and purple balls.

The probability of picking a blue ball is $\frac{1}{4}$. The probability of picking a green ball is $\frac{1}{3}$.

    a Write the probability of picking a purple ball. *(1 mark)*

    b There are 6 blue balls. How many balls are there in total? *(1 mark)*

    c Sam adds 12 balls to the bags so that

       P(purple) = $\frac{1}{2}$ and P(blue) = $\frac{1}{4}$

       How many balls of each colour does he add? *(5 marks)*

6 This table shows the destination of students leaving Clopton High School.

	Sixth form	College	Employment	Apprenticeships
Male	85	26	15	14
Female	92	80	12	15

    a A student is chosen at random. What is the probability that

        i they went to sixth form *(1 mark)*

       ii they *did not* go to college? *(1 mark)*

    b A female student is chosen at random. What is the probability that she had an apprenticeship? *(2 marks)*

7 **Reasoning** Dr Goode can either walk or drive to work. The probability that he drives is $\frac{3}{5}$. If he drives, the probability that he is late is $\frac{1}{6}$. If he walks, the probability that he is late is $\frac{1}{3}$.

    a Copy and complete the tree diagram. *(3 marks)*

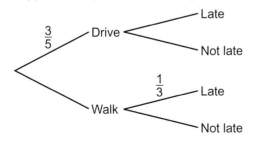

    b What is the probability that Dr Goode drives and is late? *(2 marks)*

    c What is the probability that Dr Goode walks and is on time? *(2 marks)*

8 Anu and Beth are experimenting with a spinner. Their results are shown in the table.

	Red	Blue	Green
Anu	17	6	7
Beth	36	11	13

    a Work out the experimental probability of getting red for each. *(1 mark)*

    b Whose results are likely to be more reliable?
       Explain your answer. *(2 marks)*

    c The spinner has 10 sections. How many are likely to be red? *(2 marks)*

9 Augustus is a keen archer. Each time he fires an arrow at the target, the probability that he hits the bullseye is 0.8. He fires two arrows.

    a Copy and complete the tree diagram showing the probabilities. *(3 marks)*

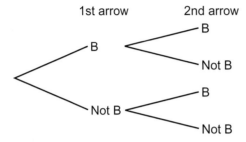

    b What is the probability that he hits two bullseyes? *(2 marks)*

    c What is the probability that he hits just one bullseye? *(2 marks)*

    d He fires a third arrow. What is the probability that he hits the bullseye with all three arrows? *(3 marks)*

10 A bag contains Red, Yellow and Black counters in the ratio 2 : 3 : 5
    A counter is picked at random. Work out

    a P(R)

    b P(Y or B)

    c P(not black) *(3 marks)*

Sample student answer

The student will only get 1 mark for the method.

a What mistake did he make when adding up the decimals?

b How could he set out his calculation better to ensure that this doesn't happen?

> **Exam-style question**
>
> Issac's pencil case contains pens which are blue, red, green or orange.
>
> The table shows each of the probabilities that a pen taken at random from the pencil case will be blue or red or orange.
>
> | Colour | Blue | Red | Green | Orange |
> |--------|------|-----|-------|--------|
> | Probability | 0.4 | 0.21 | | 0.16 |
>
> Issac takes at random a pen from the bag.
> Work out the probability that he will take a
> green pen. **(2 marks)**

Student answer

0.4 + 0.21 + 0.16 = 0.41   1 − 0.41 = 0.59   *Probability of getting a green is 0.59*

Master
p.413

Problem-solve
p.427

Check
p.428

Strengthen
p.430

Extend
p.434

Test
p.437

14 MULTIPLICATIVE REASONING

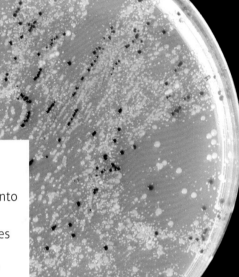

Multiplicative reasoning involves solving problems using multiplication and division.

In bacterial growth, a bacterium cell divides into two daughter cells, which then divide again.

The number of bacteria in one sample doubles every day. In the morning of day 3 there are 1500 bacteria in the sample. How many were there in the morning of day 1?

Discussion What assumptions have you made in arriving at your answer?

14 Prior knowledge check

Numerical fluency

1 Write each percentage as a decimal.
   a 20%   b 17.5%
   c 8.25%   d 145%

2 Work out
   a $\frac{1}{3}$ of 45   b $\frac{2}{5}$ of 55

3 Write each ratio in its simplest form.
   a 8:20   b 10:35
   c 18:63   d 22:99

4 Write each ratio as a unit ratio.
   a 3:12   b 4:6   c 12:5   d 4:17

Fluency with measures

5 Is each of these a unit of length, area, volume or mass?
   a centimetre   b mm²   c gram
   d inch   e litre   f m³

6 Copy and complete
   a 5400 g = ☐ kg
   b 6235 mm = ☐ m
   c 500 ml = ☐ litres
   d 750 s = ☐ minutes ☐ s
   e 340 minutes = ☐ hours ☐ minutes

7 Write these times in hours as decimals.
   a 2 hours 15 minutes
   b 3 hours 30 minutes
   c 1 hour 45 minutes

8 Write these ratios in their simplest form.
   a 500 ml : 3 litres
   b 450 g : 5 kg
   c 400 s : 4 minutes
   d 8 hours : 20 minutes

9   What is the reading on each scale?

a

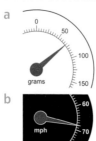

b

c

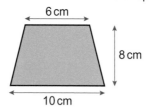

10   Find the area of this shape.

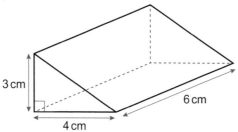

11   Work out the volume of this triangular prism.

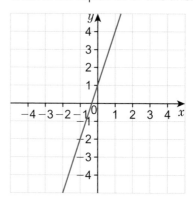

Algebraic fluency

12   Work out these when $a = 3$ and $b = -2$.

a $2a - 4$ b a^2

c $3a + 5b$ d $2ab$

13   Solve these equations.

a $12 = 2x$

b $5 = \dfrac{x}{4}$

c $3 = \dfrac{6}{x}$

14   Rearrange to make x the subject.

a $y = 4x$

b $y = \dfrac{x}{2}$

c $y = \dfrac{3}{x}$

Graphical fluency

15   a   What is the gradient of this line?

b   What is the equation of this line?

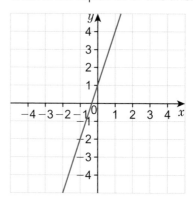

16   Which of these graphs show direct proportion?

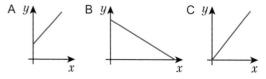

∗ Challenge

17   Currency exchange rates change every day. The bar chart shows the average monthly exchange rate from British pounds to euros for the first 8 months of 2014.

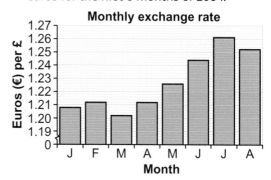

The average exchange rate for each month can be shown as a conversion line graph, with € on the vertical axis and £ on the horizontal axis. How does a change in the rate affect the gradient of the graph? List the conversion graphs for these months in order, from flattest to steepest.

> **Q17 hint** The average exchange rate for January was £1 = €1.209.
> What is the gradient of the conversion graph for January?
> Is the gradient of the conversion graph for February flatter or steeper?

14.1 **Percentages**

Objectives

- Calculate percentage profit or loss.
- Express a given number as a percentage of another in more complex situations.
- Find the original amount given the final amount after a percentage increase or decrease.

Did you know?

Companies describe their success or failure as an increase or decrease in percentage profit levels.

Fluency

Write these as decimals.

75%   80%   125%   1.5%

1   Express
   a   8 as a percentage of 20   b   6 as a percentage of 24   c   9 as a percentage of 45

2   VAT is charged at a rate of 20%. Work out
   a   £350 + VAT   b   £275 + VAT   c   £630 + VAT

3   A shop has a sale with 15% off the normal prices.
    Work out the sale price of an item with a normal price of
   a   £50   b   £85   c   £62

4   What percentage of the original amount will there be after
   a   an increase of 15%   b   a decrease of 25%
   c   an increase of 3.2%   d   a decrease of 1.75%?

Warm up

Key point 1

The original amount is always 100%. If the amount is *increased* the new amount will be *more* than 100%. If the amount is *decreased* the new amount will be *less* than 100%.

Example 1

A shop offers a 20% discount in its sale. The sale price of a jumper is £32.
What was the original price?

Sale price = 100% – 20% = 80% = 0.8

Original price → ×0.8 → £32 ──── Draw a function machine and use inverse operations to find the original price.

£40 ← ÷0.8 ← £32

Questions in this unit are targeted at the steps indicated.

5   **Finance** The price of a TV is £624 after 20% VAT is added.
    What was the price before VAT was added?

    Q5 hint

    £☐ → ×1.2 → £624

6   **Finance** A shop offers a 30% discount in its sale.
    A wallet has a sale price of £14. What was the original price?

7   **Finance** A house has 7% stamp duty (a tax) added to its price.
    The price including the tax is £2 621 500.
    What was the price of the house before the tax was added?

8 **Exam-style question**

The normal price of a television is reduced by 30% in a sale.
The sale price of the television is £350.
Work out the normal price of the television. (3 marks)

June 2013, Q16, 1MA0/1H

Exam hint
Check that your answer is more than £350.

9 **Finance / Problem-solving** Paul buys a phone that is reduced by 25%.
He pays £292.50.

a What was its original price before the reduction?

b Matt pays £360 for a phone that is reduced by 20%.
Who saves more money?

10 **Finance / Reasoning** Katie earns £45 000 a year.
The first £10 000 of her earnings is free of tax.
She pays 20% income tax on the next £31 865.
She then pays 40% income tax on the rest.
Katie's employer deducts the tax monthly. How much tax does Katie pay each month?

Q10 hint Taxed pay = £45 000 − £10 000 = ☐
Total tax for the year is
20% of £31 865 + 40% of (£45 000 − ☐ − ☐)
To find the tax for a month, divide by ☐.

Key point 2

You can calculate a **percentage change** using the formula

$$\text{percentage change} = \frac{\text{actual change}}{\text{original amount}} \times 100$$

11 **Finance** Zidan invests £3200.
When his investment matures he receives £3334.40.
Copy and complete the calculation to find the percentage increase in his investment.

Actual change = £3334.40 − £3200 = £☐

$$\text{Percentage change} = \frac{\text{actual change}}{\text{original amount}} \times 100 = \frac{\square}{3200} \times 100 = \square\%$$

Q11 communication hint
An investment 'matures' when the investment period (e.g. 5 years) ends.

12 **Finance** Clara invests £2500.
When her investment matures she receives £2420.
Calculate the percentage decrease in her investment.

Q12 hint Percentage increase and decrease, profit and loss can all be calculated using the formula for a percentage change.

13 **Finance** Sam invests £1200.
When her investment matures she receives £1281.60.
Work out the percentage increase in her investment.

14 **Finance** The table shows the prices a shopkeeper pays for some items (cost price) and the prices he sells them for (selling price).

Item	Cost price	Selling price	Actual profit	Percentage profit
ring	£5	£8		
bracelet	£12	£18		
necklace	£20	£30		
watch	£18	£25		

Copy and complete the table to work out the percentage profit on each item.

Discussion Is the item with the greatest actual profit the item with the greatest percentage profit?

15   **Finance** Kellie bought a car for £15 000. A year later she sold it for £13 900.
     Work out her percentage loss. Give your answer to 1 decimal place (1 d.p.).

16   **Real** In 2000, the UK Census recorded a population of 58 789 194.
     In 2013, it recorded a population of 64 105 700.
     What was the percentage increase in the population?
     Give your answer to 3 significant figures (3 s.f.).

17   **Finance / Problem-solving** Satvir buys 30 T-shirts for £4 each.
     She sells $\frac{1}{2}$ of them for £6 each, $\frac{1}{3}$ of them for £5 each and the rest for £4.50 each.
     What is her percentage profit?

18   **Problem-solving / Reasoning** The table shows information about visitor numbers to
     a cinema in 2012 and in 2013.

Year	Total number of visitors	Ratio of children to adults	Price of child ticket	Price of adult ticket
2012	13 275	2 : 1	£5.60	£8.60
2013	10 410	3 : 2	£7	£10

   a   Work out the percentage change in the total number of visitors from 2012 to 2013.
       Give your answer to 1 decimal place (1 d.p.).
   b   Does your answer to part **a** show a percentage increase or decrease?
   c   How many children visited the cinema in 2012?
   d   Work out the percentage change in the amount of
       money taken in ticket sales from 2012 to 2013.
       Give your answer to 1 d.p.

   > **Q18d strategy hint** What
   > information do you need? How can
   > you use the figures in the table?

19   **Reflect** Emma says, 'I find it difficult to remember the percentage change formula.
     Is it actual change divided by original amount or actual change divided by final amount?'
     Matt says, 'I know that if I invest £10 and get £15 in return, I've made £5. This is 50% of £10.
     So there has been a 50% change. That helps me to remember the formula.'
     Matt sketches a diagram.

£10	£5

   50%

     Try both of Emma's suggestions for the formula using Matt's numbers.
     Which formula is correct?
     What do you think of Matt's strategy? Do you have another way of remembering the formula?
     If so, what is it?

14.2 Growth and decay

Objectives

- Find an amount after repeated percentage
  changes.
- Solve growth and decay problems.

Why learn this?

Percentage changes often happen over a
period of time. For example, the value of a car
depreciates over time.

Fluency

Work out $3 \times 3 \times 3 \times 3 = 3^{\square}$ $2 \times 2 \times 2 \times 2 \times 2 = 2^{\square}$

1 Copy and complete
   a ☐% of 120 = 0.6 × 120
   b ☐% of 30 = 0.015 × 30
   c ☐% of 250 = 1.30 × 250

2 Work out the multiplier for
   a an increase of 40%
   b a decrease of 12%
   c an increase of 6.4%
   d a decrease of 1.5%

> **Q2a hint** 100% + 40% = 140%
> 140% = ☐ as a decimal number

> **Q2b hint** 100% − 12% = ☐%
> ☐% = ☐ as a decimal number

3 **Real** Sunir bought a car for £6000.
   It lost 30% of its value in the first year.
   It lost 10% of its value in the second year.
   a What is the multiplier to find the value of the car at the end of the first year?
   b What was the value of the car at the end of the first year?
   c What is the multiplier to find the value of the car at the end of the second year?
   d What was the value of the car at the end of the second year?
   e What is the single decimal number that the original value of the car can be multiplied by to find its value at the end of the 2 years?

> **Q3a hint** Decrease by 30%.

> **Q3c hint** Decrease by ☐%.

> **Q3e hint**
> original value ⟶ × ☐ ⟶ value at the end of 2 years

Discussion What do you notice about your answers to parts **a**, **c** and **d**?

4 **Finance** Munir has a job with an annual salary of £25 000.
   At the end of the first year he is given a salary increase of 2.5%.
   At the end of the second year he is given an increase of 3%.
   a Write the single number, as a decimal, that Munir's original salary can be multiplied by to find his salary at the end of the 2 years.
   b Work out Munir's salary at the end of 2 years.

> **Q4 communication hint**
> **Annual** means 'yearly'.
> A **salary** is a fixed regular payment, often paid monthly.

5 **Reasoning** Mia says, 'An increase of 12% followed by an increase of 20% is the same as an increase of 32%.'
   Is Mia correct? Explain your answer.

6 **Modelling** Stevie's manager says she can have either a 2% pay rise this year and then a 1.5% pay rise next year, or a 2.5% pay rise this year. Which is the better offer? Explain.

> **Q6 strategy hint** Decide on a salary figure to work with.

Key point 3

Banks and building societies pay **compound interest**. At the end of the first year, interest is paid on the money in the account. The interest is added to the amount in the account.
At the end of the second year, interest is paid on the original amount in the account *and* on the interest earned in the first year, and so on.

Example 2

£2000 is invested for 2 years at 5% per annum compound interest.
Work out the total interest earned over the 2 years.

Communication hint
Per annum means year.

100% + 5% = 105% so the multiplier is 1.05 —— Work out the multiplier for an increase of 5%.

After 1 year: £2000 × 1.05

After 2 years: £2000 × 1.05 × 1.05 —— Multiply the original amount by 1.05² to find the amount in the account after 2 years.

= £2000 × 1.05²

= £2205 —— Use a calculator.

£2205 – £2000 = £205 —— Subtract the original amount to find the interest.

The total interest earned over the 2 years is £205.

---

7   **Finance / Problem-solving** £2500 is invested for
     2 years at 4% per annum compound interest.
     Work out the total interest earned over the 2 years.

Q8 hint In compound interest
questions you could be asked to
work out either the *total interest*
or the *total amount in the account*
at the end of a period of time.

8   **Finance** £3000 is invested for 2 years at
     4.5% per annum compound interest.
     Work out the total amount in the account after 2 years.

9   **Finance** £500 is invested at 4% compound interest. Copy and complete the table.

Year	Amount at start of year	Amount plus interest	Total amount at end of year
1	£500	500 × 1.04	£520
2	£520	520 × 1.04 = 500 × 1.04²	£540.80
3	£540.80	540.80 × 1.04 =	
4			
5			

Q9 hint Do not
round **during** your
working. Only round
your final answer.

10   Work out the multiplier, as a single decimal number, that represents
     a   an annual increase of 15% for 3 years
     b   an annual decrease of 20% for 4 years
     c   an increase of 5% followed by an increase of 2%
     d   a decrease of 25% followed by a decrease of 12%
     e   an increase of 8% followed by a decrease of 5%

11   **STEM** 2000 bacteria are put into a Petri dish.
     The number of bacteria doubles every hour.
     How many bacteria will be in the dish after 8 hours?

Q11 hint
• 1 hour = 2000 × 2
• 2 hours = 2000 × 2$^\square$

12   **STEM** The level of activity of a radioactive source
     decreases by 5% per hour.
     The activity is 1400 counts per second.
     What will it be 10 hours later?

Q12 hint Counts per second after
• 1 hour = 1400 × 0.95
• 2 hours = 1400 × 0.95$^\square$

13   **STEM** The level of activity of a sample containing a
     radioactive isotope is 120 000 counts per minute.
     The half-life is 2 days. What will the count rate be after 10 days?

Q13 communication hint
The **half-life** is the time
taken for the count rate to
fall to half its starting value.

Q13 hint What is the multiplier for a half-life?

14 **Real** In 2014 a fast-food chain has 160 outlets in the UK. The number of outlets increases at a rate of 8% each year. How many outlets will it have in 2020?

Discussion What is an appropriate degree of accuracy for this question?

15 **Exam-style question**

Bill buys a new machine.

The value of the machine depreciates by 20% each year.

a Bill says, 'After 5 years the machine will have no value.'

Bill is wrong.

Explain why. **(1 mark)**

Bill wants to work out the value of the machine after 2 years.

b By what single decimal number should Bill multiply the value of the machine when new? **(2 marks)**

Nov 2005, Q13, 5525/05

> **Exam hint**
> Part **b** is worth 2 marks, so you must state your answer and show how you worked it out.

14.3 Compound measures

Objective

- Solve problems involving compound measures.

Why learn this?

Compound measures are used when we want to see how a quantity changes in relation to another quantity. For example, we use grams per cm^3 when calculating density.

Fluency

Copy and complete.

Payment of £80 for an 8-hour day   Rate is £☐ per hour

Travelling 240 km on 12 litres of petrol   Rate is ☐ km per ☐

1 $a = \dfrac{b}{c}$

Find

a a when $b = 15$ and $c = 3$ b b when $a = 12$ and $c = -2$ c c when $a = -4$ and $b = 0.5$

2 a Karl pays £192 for 8 football tickets. What is the cost per ticket?

b 24 sweets weigh 300 g. What is the weight per sweet?

3 A cuboid has dimensions 2 m by 3 m by 4 m. Work out its volume.

4 Work out the area of the trapezium.

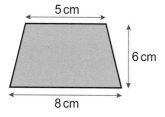

5 Work out the volume of the prism.

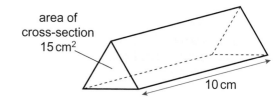

area of cross-section 15 cm²

10 cm

6 Convert

a 350 mm³ to cm³   b 54 000 cm³ to m³

7 Rearrange to make x the subject

a $y = 3x$ b $y = \dfrac{x}{4}$ c $y = \dfrac{c}{x}$

ActiveLearn Homework, practice and support: Foundation 14.3

8 **Finance /Reasoning** George works a basic 35-hour week. He is paid 'time and a half' for each hour he works on a Saturday and 'double time' for each hour he works on a Sunday.

His basic hourly rate of pay is £7.50.

How much is George paid when he works a basic 35-hour week, plus 4 hours on Saturday and 3 hours on Sunday?

Q8 hint
Rate of pay for Saturday = 1.5 × ☐ = ☐
Rate of pay for Sunday = 2 × ☐ = ☐

9 **Real** Water is leaking from a water butt at the rate of 3 litres per hour.

a Work out how much water leaks from the water butt in

  i 20 minutes

  ii 50 minutes.

Initially there are 120 litres of water in the water butt.

b Work out how long it takes for all the water to leak from the water butt.

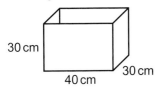

Q9a i hint   Minutes   Litres

÷☐ ⟨ 60   3 ⟩ ÷☐
       20   ☐

Think of the rate as 3 litres per 60 minutes, because the question asks for minutes.

10 (**Exam-style question**)

The diagram shows a water tank in the shape of a cuboid.

30 cm

40 cm   30 cm

The measurements of the cuboid are 30 cm by 40 cm by 30 cm.

a Work out the volume of the water tank. **(1 mark)**

Water is poured into the tank at a rate of 5 litres per minute.

1 litre = 1000 cm³

b Work out the time it takes to fill the water tank completely. Give your answer in minutes. **(2 marks)**

Exam hint
In part **b** first find out how many litres the tank will hold.

11 **Real** A car travels 300 km and uses 20 litres of petrol.

a Work out the average rate of petrol usage. State the units with your answer.

b Estimate the amount of petrol used when the car has travelled 65 km.

Key point 4

Density is a compound measure. It is the mass of substance contained in a certain volume.

$$\text{density} = \frac{\text{mass}}{\text{volume}} \quad \text{or} \quad D = \frac{M}{V}$$

Density is usually measured in g/cm³. To calculate density in g/cm³, you need to know the mass in grams (g) and the volume in cubic centimetres (cm³). It can also be measured in kg/m³.

12 **Real** A sample of brass has a mass of 2 kg and a volume of 240 cm³.
What is its density in g/cm³?

Q12 hint Density is in g/cm³ so convert the mass to grams and use the formula in Key point 4.

13 A cubic metre of concrete weighs 2400 kg.
What is the density of the concrete in g/cm³?

Q13 hint 1 m³ = ☐ cm³

Example 3

The diagram shows a block of wood in the shape of a cuboid.

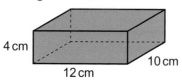

4 cm

10 cm

12 cm

The density of the wood is 0.6 g/cm³.
Work out the mass of the block of wood.

Volume of block = $l \times w \times h$

$= 12 \times 10 \times 4 = 480 \, cm^3$ ——— Work out the volume of the block in cm³.

Density $= \dfrac{mass}{volume}$

$0.6 = \dfrac{mass}{480}$ ——— Substitute values into the formula.

$0.6 \times 480 = \dfrac{mass \times \cancel{480}}{\cancel{480}}$ ——— Multiply both sides by 480.

Mass = 288 g

14 STEM Iron has a density of 7.87 g/cm³.
The mass of a piece of iron is 5.4 kg.
Work out its volume.

> **Q14 hint** Rearrange the formula
> $D = \dfrac{M}{V}$ to get $V = \dfrac{\square}{\square}$

15 STEM / Reasoning 1 cm³ of gold has a mass of 19.32 g.
1 cm³ of platinum has a mass of 21.45 g.
  a Write down the density of each metal.
  b Which metal is more dense? Explain your answer.

16 STEM An iron bar has volume 650 cm³ and density 7.87 g/cm³.
Work out the mass of the iron. Give your answer in
  a grams   b kilograms.

17 **Exam-style question**

The diagram shows a prism.

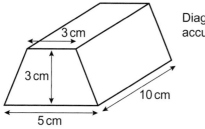

3 cm

Diagram NOT
accurately drawn

3 cm

10 cm

5 cm

The length is 10 cm.
The cross-section is a trapezium.
The lengths of the parallel sides of the trapezium are 3 cm and 5 cm.
The distance between the parallel sides of the trapezium is 3 cm.
  a Work out the volume of the prism.   **(3 marks)**
The prism is made out of gold. Gold has a density of 19.3 grams per cm³.
  b Work out the mass of the prism.
    Give your answer in kilograms.   **(2 marks)**

> **Exam hint**
> Make sure you
> state the correct
> units with your
> answer.

> **Key point 5**
>
> **Pressure** is a compound measure. It is the force applied over an area.
> To calculate pressure, you need to know the force in newtons (N) and the area in square metres (m²).
>
> $\text{pressure} = \dfrac{\text{force}}{\text{area}}$ or $P = \dfrac{F}{A}$
>
> Pressure is usually measured in N/m².

18 STEM

  a A force of 45 N is applied to an area of 2.6 m².
Work out the pressure in N/m².

  b A force is applied to an area of 4.5 m². It produces
a pressure of 20 N/m². Work out the force in N.

> **Q18b hint** Substitute into
> the formula $P = \dfrac{F}{A}$
> Rearrange to find F.

19 STEM Copy and complete the table.

Force (N)	Area (m²)	Pressure (N/m²)
60	2.5	
	4.8	15
100		12

14.4 Distance, speed and time

Objectives

- Convert between metric speed measures.
- Calculate average speed, distance and time.
- Use formulae to calculate speed and acceleration.

Did you know?

The speed of a common snail is 1 millimetre per second, which is 0.001 metres per second.

Fluency

What is the speed, in kilometres per hour, for a car that travels
- 20 kilometres in 30 minutes?
- 15 kilometres in 20 minutes?

1 Copy and complete these conversions.

  a 180 cm = ☐ m   b 4.7 km = ☐ m   c 28 000 cm = ☐ km   d 54 600 m = ☐ km

  Discussion When converting from a smaller unit to a larger unit, do you divide or multiply?

2 $d = st$

  a Find d when $s = 10$ and $t = 0.3$

  b Find s when $d = 400$ and $t = 80$

  c Find t when $d = 150$ and $s = 50$

3 Write these times in hours (as decimals).

  a 30 minutes   b 20 minutes   c 75 minutes

> **Q3 hint**
> $\dfrac{\square}{60} = \square$

4 Write these times in hours and minutes.

  a 0.2 hours   b 1.75 hours   c 3.4 hours

Active Learn Homework, practice and support: Foundation 14.4

Warm up

> ### Key point 6
>
> You can calculate speed using the formula
>
> $$\text{speed} = \frac{\text{distance}}{\text{time}} \quad \text{or} \quad S = \frac{D}{T}$$
>
> Speed is often measured in metres per second (m/s), kilometres per hour (km/h) or miles per hour (mph).

> ### Example 4
>
>
>
> A car travels 90 kilometres in 2 hours 15 minutes. What is the average speed in km/h?
>
> Use $\text{speed} = \dfrac{\text{distance}}{\text{time}}$
>
> Answer in km/h
>
> Distance = 90 km
>
> Time = 2 hours 15 minutes = 2.25 hours ⟶ Convert to a decimal to make the calculation easier.
>
> Average speed = $\dfrac{90}{2.25}$ = 40 km/h ⟶ Substitute values into the formula.

5 Copy and complete the table.

Distance (km)	Speed (km/h)	Time
280		3 h 30 min
	48	2 h 45 min
350	60	☐ h ☐ min

Q5 hint To find distance in row 2, rearrange the formula $S = \dfrac{D}{T}$ to make D the subject.

How do you need to rearrange the formula to find time (T) in row 3?

6 **Modelling** Work out the average speed for these journeys.
   a A car travels 75 miles in $2\frac{1}{2}$ hours.
   b A man cycles 45 km in 3 hours 45 minutes.
   c A train travels 200 miles in 2 hours 6 minutes.

Q6 hint Time must be a decimal.
60 mins = 1 hour
6 mins = $\frac{1}{10}$ hour
6 mins = 0.1 hour

7 **Modelling** Work out the distance travelled for these journeys.
   a An aeroplane travels for 3 hours 20 minutes at an average speed of 600 mph.
   b A tram travels for 50 minutes at an average speed of 21 km/h.
   c A man walks for 35 minutes at an average speed of 4.8 km/h.

8 **Modelling** Work out the time taken for these journeys.
   a A spider climbs a 3.2 m wall at an average speed of 8 cm/s.
   b A swallow flies 0.5 km at an average speed of 11 m/s.
   c A snail slithers 20 cm at an average speed of 1 mm/s.

Q8a hint The speed is in cm/s so convert the distance to cm.

9 Convert these speeds from metres per second (m/s) to metres per hour (m/h).
   a 1 m/s
   b 12 m/s
   c 8 m/s

Q9a hint Would more or fewer metres be travelled in 1 hour than in 1 second? 1 m/s means travelling:

×☐ (1 m in 1 second) ×☐
×☐ (☐ m in 60 seconds) ×☐
(☐ m in 60 minutes)

10 Karl travels the first 35 kilometres of his journey in 45 minutes. He then travels the last 65 km in $1\frac{1}{2}$ hours. What is his average speed for the whole journey in km/h?

Q10 hint
Average speed = $\dfrac{\text{total distance}}{\text{total time}}$

11 Paul lives 504 metres from the tram station. It takes him 6 minutes to walk to the station. What is his average speed in m/s?

Q11 hint The question asks for the speed in m/s. What do you need to convert time to?

12 (Exam-style question)

On Monday Ravi drives for 4 hours.
His average speed is 30 mph.
  a How far does Ravi drive on Monday? **(2 marks)**
On Tuesday Ravi drives 200 km.
5 miles ≈ 8 kilometres
  b On which day did Ravi drive further? **(3 marks)**
Nov 2013, Q18, 1MA0/1F

Exam hint
In part **b** you will need to explain your conclusion and include the correct units in your explanation.

13 Convert these speeds from metres per second (m/s) to kilometres per hour (km/h).
  a 5 m/s   b 18 m/s   c 30 m/s

14 Convert these speeds from kilometres per hour (km/h) to metres per second (m/s).
  a 54 km/h
  b 72 km/h
  c 9 km/h

Q14a hint

$\times\,\Box$ 54 km in 1 hour
$\rightarrow\;\Box$ m in 1 hour $\Big\}\div\Box$
$\div\,\Box\;\rightarrow\;\Box$ m in 1 minute $\Big\}\div\Box$
$\div\,\Box\;\rightarrow\;\Box$ m in 1 second

15 **Real** A commercial jet aircraft has a cruising speed of 250 m/s.
What is this speed in km/h?

16 **Problem-solving / Reasoning** A Formula 1 racing car has a top speed of 350 km/h.
A peregrine falcon (the fastest bird) has a top speed of 108 m/s.
Which is faster? Explain your answer.

Key point 7

You can use the **kinematics formulae** for calculations with moving objects.
• $v = u + at$ • $s = ut + \frac{1}{2}at^2$ • $v^2 = u^2 + 2as$
a is a constant **acceleration**, u is the initial **velocity**, v is the final velocity, t is the time taken and s is the displacement from the position when $t = 0$.

Communication hint Velocity is speed in a given direction. It is often measured in m/s.
The **initial velocity** is the speed in a given direction at the start of the motion. It may be zero.
Acceleration is the rate of change of velocity. It is often measured in m/s².

17 Use the formula $v = u + at$ to work out v when
  a $u = 3$, $a = 2$ and $t = 4$ b $u = 4$, $a = 0.5$ and $t = 10$

18 Use the formula $v = u + at$ to work out u when
  a $v = 5$, $a = 1$ and $t = 3$ b $v = 4$, $a = 0.5$ and $t = 6$

Q18a hint First substitute into the formula, then solve.
$5 = u + 1 \times 3$
$5 = u + \Box$

19 Use the formula $v = u + at$ to work out a when
  a $v = 7$, $u = 1$ and $t = 2$ b $v = 4$, $u = 0.2$ and $t = 5$

20 Copy and complete the table using the formula $s = ut + \frac{1}{2}at^2$

s (m)	u (m/s)	a (m/s²)	t (s)
10		2	1
8	2		1
15		3	2
12	4		2

Q20 hint For row 1, substitute into the formula, then solve.
$10 = u \times 1 + \frac{1}{2} \times 2 \times 1^2$
$10 = u + \Box$

21 Copy and complete the table using the formula $v^2 = u^2 + 2as$

v (m/s)		6	9	5	7
u (m/s)	8		4		3
a (m/s²)	2	1		3	
s (s)	9	5	7	4	6

> **Q21 hint**
> For column 1:
> $v^2 = 8^2 + 2 \times 2 \times 9$
> $v^2 = \boxed{}$
> $v = \sqrt{\boxed{}}$

22 a **STEM** A car starts from rest and accelerates at 5 m/s² for 200 m. Work out the final velocity in m/s.

b **Reflect** Did the strategy hint in part **a** help you to answer the question? What other strategies do you find helpful when solving problems with the formula? List at least two.

> **Q22a strategy hint**
> The car starts from rest so initial velocity (u) = 0.
> You are given a (= 5) and s (= 200), and need to find v. t is not given so use $v^2 = u^2 + 2as$

14.5 Direct and inverse proportion

Objectives

- Use ratio and proportion in measures and conversions.
- Use inverse proportion.

Did you know?

The force of gravity and the distance from Earth are in inverse square proportion. The further you move away from Earth, the weaker the pull of gravity.

Fluency

Which of these graphs shows direct proportion?

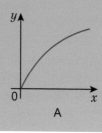
A

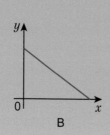
B

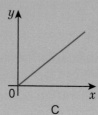
C

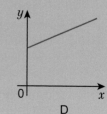
D

1 Write each ratio as a unit ratio.
  a $3:12$
  b $6:2$
  c $11:2$
  d $\frac{3}{5}:1$

2 The ratio of kg to pounds is $1:2.2$.
  a Copy and complete this table of values for kg and pounds.

kg	0	5	10
pounds			

  b Draw a conversion graph for kg to pounds.
  c Are kg and pounds in direct proportion? Explain.
  d Write a formula linking kg (x) and pounds (y).

> **Q2b hint** Put kg on the x-axis and pounds on the y-axis.

3 **STEM** The table shows the pressure for different forces on an area of 4 m².

Force (N)	4	10	16	20
Pressure (N/m²)	1	2.5	4	5

  a Write the ratio force : pressure for each force.
  b Write each ratio from part **a** in its simplest form. What do you notice?
  c Write a formula linking the values of F and P in the table.
  Discussion How can the unit ratio help you write the formula?

> **Q3c hint** F : P
>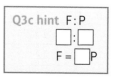

ActiveLearn Homework, practice and support: Foundation 14.5

Key point 8

When two variables are in direct proportion, pairs of values are in the same ratio.

4 For each ratio, write a formula connecting the variables.

   a $x:y$
       $3:1$

   b $s:t$
       $1:6$

   c $p:q$
       $1:3.5$

5 Write each ratio as a unit ratio.
   Write a formula linking each pair of variables.

   a $F:m$
       $5:2$

   b $x:y$
       $4:5$

   c $r:s$
       $4:7$

6 **STEM / Reasoning** The table shows the distance (d) travelled by a car over a period of time (t).

Distance, d (miles)	8	16	24	32	40
Time, t (minutes)	10	20	30	40	50

> Q6a hint $d:t$

   a Write the ratios for the pairs of values in the table.
   b Is d in direct proportion to t? Explain your answer.
   c Write a formula that shows the relationship between distance (d) and time (t).
   d Use your formula to work out the distance travelled after 25 minutes.

Key point 9

When
- y varies as x
- y varies directly as x
- y is in direct proportion to x

you can write $y \propto x$
- $y \propto x$ means 'y is proportional to x'. When $y \propto x$, then $y = kx$, where k is the constant of proportionality.

> **Literacy hint**
> In the formula $y = kx$, k is the constant of proportionality.
> Its value is constant (stays the same) when x and y vary.

Example 5

The number of dollars (D) is directly proportional to the number of pounds (P).

a One day, $1 = £0.64$. Write a formula linking D and P.
b Use your formula to convert £15 to dollars.

a $D \propto P$ — Write the proportional relationship using the $\propto$ symbol. Then write $D = kP$.

   $D = kP$

   When $D = 1, P = 0.64$ — Substitute the values given for D and P into $D = kP$.

   $1 = k \times 0.64$

   $k = \dfrac{1}{0.64} = 1.5625$ — Solve the equation to find k.

   $D = 1.5625\,P$ — Rewrite the equation using the value of k.

b $D = 1.5625 \times 15$ — Use the formula to convert £15 to dollars.

   $= 23.4375$

   $= \$23.44$

7   a   Write each statement using the $\propto$ symbol.
    b   Write each statement using k, the constant of proportionality.
        i F is proportional to m
        ii V is proportional to x

8   x is proportional to t.
    When $x = 6$, $t = 4$.
    a   Write the statements of proportionality for x and t.
    b   Use the given values of x and t to find k.
    c   Find x when $t = 10$.

9   r is directly proportional to s. When $r = 5$, $s = 4$.
    a   Write a formula for r in terms of s.   b   Find r when $s = 7$.

    > **Q9a hint** Start $r =$

10 **Finance** The number of pounds (P) is directly proportional to the number of euros (E).
    a   One day €1 = £0.80.
        Write a formula connecting P and E.
    b   Use your formula to convert €80 to pounds.

Key point 10

When two variables X and Y are in inverse proportion,

$$X \propto \frac{1}{Y} \qquad X = \frac{k}{Y} \qquad Y = \frac{k}{X}$$
$$XY = k \text{ (constant)}$$

---

11 **Reasoning** A and B are in inverse proportion. Copy and complete the table.

A	10	20	15		
B	15			60	12

> **Q11 hint**
> AB = constant

12 Does each equation represent direct proportion,
    inverse proportion or neither?
    a $y = 2x$
    b $y = \dfrac{4}{x}$
    c $x + y = 7$
    d $xy = 5$
    e $\dfrac{y}{x} = 8$

    > **Q12c hint** Rearrange the
    > equation to $y = \square$.

13 a   Write each statement using the $\propto$ symbol.
    b   Write the statement using k, the constant of proportionality.
        i y is inversely proportional to x
        ii P is inversely proportional to A

14 y is inversely proportional to x.
    When y is 6, x is 10.
    a   Write the statements of proportionality for y and x.
    b   Use the given values of x and y to find k.
    c   Find y when $x = 15$.

15 **Exam-style question**

    s is inversely proportional to t.
    $s = 20$ when $t = 0.4$.
    Calculate the value of s when $t = 0.5$.   **(3 marks)**

    Exam hint
    The question has 3 marks.
    One of these will be for the
    correct answer. The others will
    be method marks. Make sure
    you show your working clearly.

16 **STEM** In an electrical circuit, the current, I, is inversely proportional to the resistance, R.
    When the resistance in the circuit is 12 ohms, the current is 8 amperes.
    Find the current in amperes when the resistance is 6.4 ohms.

14 Problem-solving: Second-hand cars

Objective   • Solve problems involving percentage decrease/increase and compound percentages.

Nick wants to know how to price the cars on his second-hand car lot. To help him make a decision, he has been collecting data for the same make of car over 7 years. Nick's data is shown in Table 1.

1   Plot the data shown in Table 1 on a scatter graph.

2   Nick says that the price decreases by approximately £4500 each year. Explain why Nick is wrong.

3   Another second-hand car dealer, Mary, says that the price decreases by 10% each year.
    a   Copy and complete the table to show the prices after a 10% decrease each year.

Age (years)	Price (£)	
0 (new)	26 065	
1	23 459	(90% of 26 065)
2	21 113	(90% of 23 459)
3		(90% of 21 113)

    b   Plot the points from your table on the axes from **Q1**.
    c   Does Mary's model fit the data better than Nick's model?

4   Can you find a graph that matches the data even more closely?

> **Q4 hint** You could start by plotting the price of a car which decreases by 20% each year. Is this a better or worse fit for the data than 10%? Choose a new percentage decrease to try.

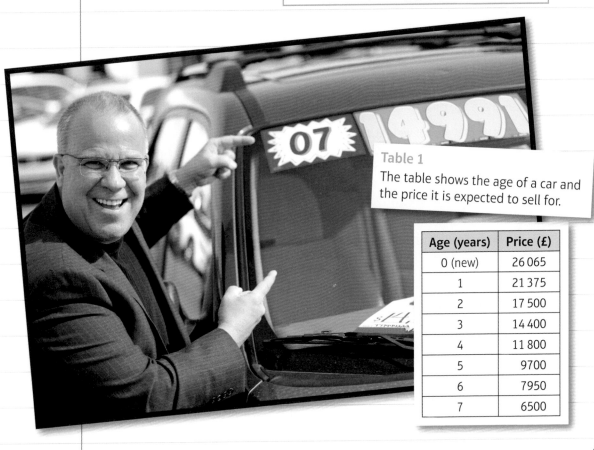

Table 1
The table shows the age of a car and the price it is expected to sell for.

Age (years)	Price (£)
0 (new)	26 065
1	21 375
2	17 500
3	14 400
4	11 800
5	9700
6	7950
7	6500

14 Check up

Log how you did on your Student Progression Chart.

Percentages

1 Louise buys a pack of 24 bottles of a fruit drink for £12.49. She sells all of them for 95p each. Work out her percentage profit. Give your answer to 3 significant figures (3 s.f.).

2 A dishwasher originally cost £300. It is reduced to £250 in a sale. What is the percentage decrease in price?

3 £2500 is invested for 2 years at 3.5% compound interest. Work out the total amount in the account after 2 years.

4 The number of bees in a beehive decreases by 2% each year. There are 6500 bees at the beginning of 2014. How many will there be at the end of 2020?

Compound measures

5 **Reasoning** Sarah works a 30-hour week, Monday to Friday. Her hourly pay is £10.20.
She is paid 'time and a quarter' for each hour she works on a Saturday, and 'time and a half' for each hour she works on a Sunday.
Sarah works her 30-hour week, plus 3 hours on Saturday and 4 hours on Sunday.
How much is she paid?

6 These two metal blocks each have a volume of 0.5 m³.

 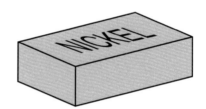

The density of the copper block is 8900 kg per m³.
The density of the nickel block is 8800 kg per m³.
Calculate the difference between the masses of the blocks.

7 A bottle of water of weight 19.6 N rests on a table.
The area of the base of the bottle in contact with the table is 7 cm².
What pressure in N/cm² does the bottle of water exert on the table?

Distance, speed and time

8 a Steffan drives 35 km in 45 minutes. What is his average speed?
  b Michelle walks 10 km at an average speed of 3.2 km per hour. How long does it take her?

9 A car starts from rest and accelerates uniformly for 8 seconds for a distance of 200 m. Work out the acceleration of the car.

> **Kinematics formulae**
> $v = u + at$
> $s = ut + \frac{1}{2}at^2$
> $v^2 = u^2 + 2as$

10 The fastest recorded speed of Usain Bolt is 13.4 m/s.
   The fastest speed of a great white shark is 40 km/h.
   Which is the faster speed? Explain your answer.

Diverse and inverse proportion

11 The table shows the amounts received when British pounds are changed to euros.
   a Copy and complete the table.

Pounds (£)	Euros (€)
150	189
400	504
320	
	352.80

   b Is the number of pounds in direct proportion to the number of euros?
     Explain your answer.
   c Write a formula connecting pounds to euros.

12 f is directly proportional to g, so $f = kg$ for some value of k.
   a When $f = 8$, $g = 2$. Find the value of k.
   b Work out f when $g = 1.5$.

13 y is inversely proportional to x. When $x = 5$, $y = 9$.
   a Write a formula for y in terms of x.
   b Find y when $x = 15$.

14 How sure are you of your answers? Were you mostly

Just guessing   Feeling doubtful   Confident

What next? Use your results to decide whether to strengthen or extend your learning.

*Challenge

15 £650 is invested at 3.4% compound interest.
   a Copy and complete the table.

Year	Amount at start of year	Multiplier in index form	Total amount at end of year
1	£650	1.034	
2		$1.034^{\square}$	
3		$1.034^{\square}$	
4			
5			

   b What is the multiplier for year 10?
   c What is the multiplier for year n?
   d Write a formula for the amount in the account at the end of year n, when £P is
     invested at r%.

14 Strengthen

Percentages

1 Copy and complete the working to find the percentage profit made on each item.

original price = ☐

actual profit = ☐

$$\text{percentage profit} = \frac{\text{actual profit}}{\text{original price}} \times 100 = \frac{\square}{\square} \times 100 = \square\%$$

a Bought for £12, sold for £15
b Bought for £25, sold for £32.50
c Bought for £140, sold for £260

Check your answers.

2 Work out the percentage loss made on each item.
a Bought for £16, sold for £12
b Bought for £450, sold for £360
c Bought for £60, sold for £42

Check your answers.

> **Q2 hint** Follow the same method as in **Q1**.
> Replace 'actual profit' with 'actual loss'.

3 Steve bought a box of 12 pineapples for £5.
He sold all of them for £2.50 each.
Work out his percentage profit.

> **Q3 hint** First work out how
> much he sold them for:
> 2.50 × ☐ = ☐

4 A shop has a sale with 10% off the normal prices.
Copy and complete the working to find the normal price of a coat with a sale price of £45.

$$100\% - 10\% = 90\%$$

$$
\begin{array}{c}
\div\square \\
\times\square
\end{array}
\left(
\begin{array}{c}
90\% = £45 \\
1\% = £0.50 \\
100\% = \square
\end{array}
\right)
\begin{array}{c}
\div 90 \\
\times\square
\end{array}
$$

5 The price of a house increases by 20% to £360 000.
Copy and complete the working to find the price of the house before the increase.

$$100\% + 20\% = 120\%$$

$$
\begin{array}{c}
\div\square \\
\times\square
\end{array}
\left(
\begin{array}{c}
120\% = £360\,000 \\
1\% = £3000 \\
100\% = \square
\end{array}
\right)
\begin{array}{c}
\div 120 \\
\times\square
\end{array}
$$

6 Work out the original price of each of these items.
a Discount 15%, sale price £51
b Discount 25%, sale price £70
c Discount 40%, sale price £96

> **Q6a hint**
>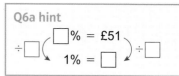

7 Work out the original value of each of these before the increase.
a Increase of 15% to 460 g
b Increase of 30% to 390 km
c Increase of 45% to 580 litres

8 **Finance** Saima invests £700 for 2 years at 3.8% per annum compound interest.
Copy and complete the table to work out the total amount in the account
after 2 years:

Year	Amount at start of year	Interest at end of year	Total at end of year
1	£700	0.038 × 700 =	
2		0.038 ×	

> **Q8 hint** Write your final answer to money calculations rounded to the nearest penny. Only round your final answer.

9 **Finance** £750 is invested for 2 years at 4.3% per annum.
Work out
a the *total amount* in the account after 2 years
b the *total interest* earned.

> **Q9 hint** The total amount in the account is the original amount invested plus the total interest.

> **Q9b hint**
> Interest earned = final amount − original amount.

10 Write the multiplier for each of these percentage *increases* and *decreases* as a decimal number.

a 15%
b 8%
c 2.6%
d 21%
e 7%
f 4.5%

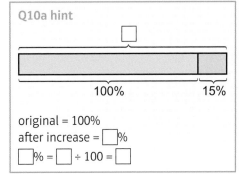

> **Q10a hint**
>
> 100%   15%
>
> original = 100%
> after increase = ☐%
> ☐% = ☐ ÷ 100 = ☐

11 **Modelling** A population of rabbits increases by 20% each month.
At the beginning of January the population is 15 rabbits.
How many rabbits would you expect there to be in the population at the end of December?

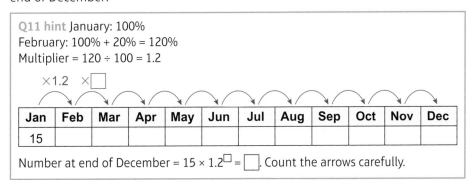

> **Q11 hint** January: 100%
> February: 100% + 20% = 120%
> Multiplier = 120 ÷ 100 = 1.2
>
> ×1.2 ×☐
>
> | Jan | Feb | Mar | Apr | May | Jun | Jul | Aug | Sep | Oct | Nov | Dec |
> |-----|-----|-----|-----|-----|-----|-----|-----|-----|-----|-----|-----|
> | 15 | | | | | | | | | | | |
>
> Number at end of December = 15 × 1.2$^{☐}$ = ☐. Count the arrows carefully.

12 **Real / Modelling** The winter Arctic ice area is decreasing at a rate of 4.2% every 10 years.
In 1979 the area was 16 million square metres.
What is the winter area expected to be in 2019?

> **Q12 hint** The 1979 area was 100%.
> The 1989 area was 100 − 4.2 = ☐%
> Multiplier is ☐

13 **Reflect** Some of the hints in this section suggest using bars and arrows to help you answer the question. Did you find them helpful? Explain why.

Compound measures

1 **Finance** Jamal works a basic 35-hour week. His hourly rate of pay is £6.60.

He is paid 'time and a quarter' for each hour he works on a Saturday and 'time and a half' for each hour he works on a Sunday.

How much is Jamal paid for a week when he works a basic 35-hour week, plus 4 hours on Saturday and 3 hours on Sunday?

> **Q1 hint** Basic pay = £6.60
> Pay at 'time and a quarter' = £6.60 × 1.25 = ☐
> Pay at 'time and a half' = £6.60 × 1.5 = ☐
> Total pay = 35 × £6.60 + 4 × ☐ + 3 × ☐ = ☐

2 **STEM / Modelling** Copy and complete this table of mass, volume and density.

Metal	Mass (g)	Volume (cm³)	Density (g/cm³)
aluminium		10	2.70
copper	448		8.96
zinc	427.8	60	

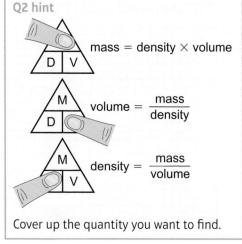

> **Q2 hint**
>
> mass = density × volume
>
> volume = $\dfrac{\text{mass}}{\text{density}}$
>
> density = $\dfrac{\text{mass}}{\text{volume}}$
>
> Cover up the quantity you want to find.

3 **STEM / Modelling** Copy and complete this table of force, area and pressure.

Force (N)	Area (cm²)	Pressure (N/cm²)
	20	11
60	15	
45		9

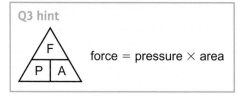

> **Q3 hint**
>
> force = pressure × area

Distance, speed and time

1 **Reasoning** Is the answer to each of these conversions going to be higher or lower than the number given?

a 30 km/h in m/h   b 4000 m/h in m/min   c 80 m/min in m/s

> **Q1a hint** If you travel 30 km in 1 hour, do you travel further or less far than 30 m?

2 **Modelling** Copy and complete this table of distance, time and speed.

Distance (miles)	Time (hours)	Speed (mph)
	4	45
	2.5	58
150	3	
120	1.25	
45		30
154		56

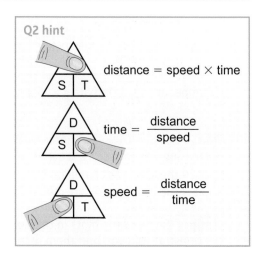

> **Q2 hint**
>
> distance = speed × time
>
> time = $\dfrac{\text{distance}}{\text{speed}}$
>
> speed = $\dfrac{\text{distance}}{\text{time}}$

3 Copy and complete this table to convert from km/h to m/s.

km/h	m/h	m/min	m/s
9			
			5
12			
			8

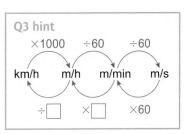

Q3 hint

×1000   ÷60   ÷60

km/h → m/h → m/min → m/s

÷☐   ×☐   ×60

4 Choose the correct formula from the panel to work out each of these.
  a Find a when $s = 30$, $t = 3$ and $u = 5$.
  b Find u when $s = 12$, $v = 8$ and $a = 2$.
  c Find v when $u = 4$, $a = 3$ and $t = 7$.

$$v = u + at$$
$$s = ut + \tfrac{1}{2}at^2$$
$$v^2 = u^2 + 2as$$

Q4a hint You are not given a value for v, or asked to find it.
Look for the formula that does *not* include v.

5 **Modelling** Copy and complete the table.

s (cm)	t (s)	u (cm/s)	v (cm/s)	a (cm/s²)
30	3	5		
12			8	2
	7	4		3
15	2			1
8		3		2
9		2	8	
	1		10	4
	2	2	12	

Q5 hint Use the formulae from **Q4**. The grey boxes show you which variable is *not* used.

Direct and inverse proportion

1 A and B are in direct proportion.
Work out the values of W, X, Y and Z.

A	B
4	7
8	W
X	42
16	Y
Z	35

Q1 hint

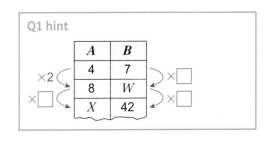

A	B
4	7
8	W
X	42

×2 ↻   ×☐ ↻
×☐ ↻   ×☐ ↻

2 A and B are in inverse proportion.
Work out the values of W, X, Y and Z.

A	B
6	8
8	W
12	X
Y	16
Z	12

Q2 hint If A and B are in inverse proportion then $A \times B$ = a constant

A	B	$A \times B$
6	8	48
8	W	$8W$
12	X	$12X$

$8W = 48$, so $W = 48 \div 8 = \boxed{}$
$12X = 48$, so $X = 48 \div \boxed{} = \boxed{}$

3 Match the equivalent statements.

y is proportional to x $y = kx$ $y \propto \dfrac{1}{x}$

$y = \dfrac{k}{x}$ $y \propto x$ y is inversely proportional to x

4 a Write y is directly proportional to x using algebra.

Q4a hint Use $\propto$

b Write a formula showing that y is directly proportional to x.
Use the form $y = \square x$.

c When $x = 7$, $y = 21$. Substitute these values into your formula from part **b**.
Solve to find k.

d Use your answer from part **b** to rewrite your formula as $y = \square x$.

e Use your formula to find y when $x = 4$.

14 Extend

1 **Real** A train travels at 50 km/h for $1\frac{1}{2}$ hours and then at 60 km/h for 45 minutes.
What is the train's average speed for the whole journey?

Q1 hint
Average speed = $\dfrac{\text{total distance}}{\text{total time}}$

2 **Real / STEM** An object starts from rest and travels for a distance of 15 m with an acceleration of 5 m/s².
Work out its final velocity.

Q2 hint Use a kinematics formula.
u = initial velocity = 0 as the object starts from rest.

3 **Exam-style question**

Viv wants to invest £2000 for 2 years in the same bank.

The International Bank
Compound Interest
4% for the first year
1% for each extra year

The Friendly Bank
Compound Interest
5% for the first year
0.5% for each extra year

At the end of 2 years, Viv wants to have as much money as possible.
Which bank should she invest her £2000 in? **(4 marks)**
June 2013, Q14, 1MA0/1H

Exam hint
You must make clear which working is for which bank and make sure you state which is better and why.

4 **Real** The value of a new car is £15 000.
The value of the car depreciates by 20% per year.
Work out the value of the car after 3 years.

5 **STEM** A 25 g block of aluminium has volume 9260 mm³.
Calculate its density in g/cm³.

Q6 hint What units do you need for volume?

6   **STEM / Problem-solving** The diagram shows a piece of plastic cut into the shape of a trapezium.

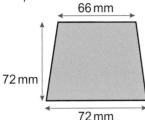

Work out the force needed to exert a pressure of 15 N/cm² on the trapezium.

7   **Problem-solving** A car is travelling at 15 m/s. The speed limit is 64 km/h.
    Is the car travelling below the speed limit? Show your working.

8   A cheetah covers 100 metres in 7.19 seconds. What is its average speed
    a   in m/s   b   in km/h?
    Give your answers to 3 significant figures (3 s.f.).

9   m is directly proportional to r. When $m = 42$, $r = 10$.
    a   Write a formula for m in terms of r.
    b   Find m when $r = 12$.
    c   Find r when $m = 21$.

    > **Q9c hint** Substitute $m = 21$ into your formula and solve.

10 **STEM** The extension E of a string (in mm) is directly proportional to the mass m on the string in grams.
    A mass of 250 g gives an extension of 12 mm.
    a   Write a formula for E in terms of m.
    b   Find the extension of the string for a mass of 600 g.
    c   Find the mass that extends the string by 36 mm.

11 y is inversely proportional to x.
    When $x = 2$, $y = 36$.
    Work out
    a   the value of y when $x = 12$ b   the value of x when $y = 9$

12 **STEM** The volume V (m³) of a gas is inversely proportional to the pressure P (N/m²).
    When $P = 600$ N/m², $V = 5$ m³.
    a   Write a formula for V in terms of P.
    b   Find the volume when the pressure is 300 N/m².
    c   Calculate the pressure on 7.5 m³ of the gas.

13 **Exam-style question**

    A shop is having a sale. Each day, prices are reduced by 20% of the price on the previous day.
    Before the start of the sale, the price of a fridge is £550.
    On the first day of the sale, the price is reduced by 20%.
    a   Work out the price of the fridge on
      i   the first day of the sale
      ii the third day of the sale.   **(5 marks)**
    On the first day of the sale, the price of a table is £200.
    b   Work out the price of the table before the start of the sale.   **(2 marks)**

    Exam hint
    Before you start part **b** think whether your answer will be more or less than £200.

14 Knowledge check

⊙ The original amount is always 100%. If the amount is *increased* the new amount will be *more* than 100%. If the amount is *decreased* the new amount will be *less* than 100%. .... *Mastery lesson 14.1*

⊙ You can calculate a **percentage change** using the formula

$$\text{percentage change} = \frac{\text{actual change}}{\text{original amount}} \times 100$$ .... *Mastery lesson 14.1*

⊙ Percentage increase and decrease, profit and loss can be calculated using the formula for percentage change. .... *Mastery lesson 14.1*

⊙ Banks and building societies pay **compound interest**. At the end of the first year, interest is paid on the money in the account. The interest is added to the amount in the account. At the end of the second year, interest is paid on the original amount in the account *and* on the interest earned in the first year, and so on. .... *Mastery lesson 14.2*

⊙ **Density** is a compound measure. It is the mass of substance contained in a certain volume. It is usually measured in grams per cubic centimetre (g/cm³). .... *Mastery lesson 14.3*

⊙ To calculate density, you need mass in g and volume in cm³:

$$\text{density} = \frac{\text{mass}}{\text{volume}} \quad \text{or} \quad D = \frac{M}{V}$$ .... *Mastery lesson 14.3*

⊙ **Pressure** is a compound measure. It is the force applied over an area. It is usually measured in newtons (N) per square metre (N/m²). .... *Mastery lesson 14.3*

⊙ To calculate pressure, you need force in N and area in m²:

$$\text{pressure} = \frac{\text{force}}{\text{area}} \quad \text{or} \quad P = \frac{F}{A}$$ .... *Mastery lesson 14.3*

⊙ You can calculate speed using the formula:

$$\text{speed} = \frac{\text{distance}}{\text{time}} \quad \text{or} \quad S = \frac{D}{T}$$ .... *Mastery lesson 14.4*

⊙ Average speed $= \dfrac{\text{total distance}}{\text{total time}}$

⊙ You can use the **kinematics formulae** for calculations with moving objects.
- $v = u + at$
- $s = ut + \frac{1}{2}at^2$
- $v^2 = u^2 + 2as$

> **Hint** These are given on the exam paper.

where a is a constant acceleration, u is the initial velocity, v is the final velocity, t is the time taken and s is the displacement from the position when $t = 0$. .... *Mastery lesson 14.4*

⊙ $y \propto x$ means 'y is proportional to x'. When $y \propto x$, then $y = kx$, where k is the constant of proportionality. .... *Mastery lesson 14.5*

⊙ $X \propto \dfrac{1}{Y}$ means X and Y are in inverse proportion. This means that $XY = k$ (constant). .... *Mastery lesson 14.5*

14 Unit test

1 Michael's hourly rate of pay increases from £15.40 to £18.00.
What is the percentage increase in his rate of pay?
Give your answer correct to 1 decimal place (1 d.p.). *(3 marks)*

2 A metal cube of side 20 cm exerts a force of 6 N on a table.
Work out the pressure in N/m². *(4 marks)*

3 (Exam-style question)

Peter goes for a walk.
He walks 15 miles in 6 hours.
a Work out Peter's average speed.
Give your answer in miles per hour. **(2 marks)**
Sunita says that Peter walked more than 20 km.
5 miles = 8 km.
b Is Sunita right?
Show all your working. **(2 marks)**

March 2013, Q7, 1MA0/2H

4 Dom cycles 44.5 km at an average speed of 15.5 km/h.
He then cycles 500 m at an average speed of 25 km/h.
What is his average speed for the whole journey? *(5 marks)*

5 (Exam-style question)

Derek buys a house for £150 000. He sells it for £154 500.
a Work out Derek's percentage profit. **(3 marks)**
Derek invests £154 500 for 2 years at 4% per year
compound interest.
b Work out the value of the investment at the end of 2 years. **(3 marks)**

March 2013, Q16, 1MA0/2H

Exam hint
In part **a** first find
the profit before
working out the
percentage profit.

6 Jamie got a pay rise of 5%. His new pay was £1785 per month.
Work out his pay per month before the pay rise. *(2 marks)*

7 (Exam-style question)

Katie invests £200 in a savings account for 2 years.
The account pays compound interest at an annual rate of
3.3% for the first year
1.5% for the second year
a Work out the total amount of money in Katie's account at
the end of 2 years. **(3 marks)**
Katie travels to work by train. The cost of her weekly train ticket
increases by 12.5% to £225.
Her weekly pay increases by 5% to £535.50.
*b Compare the increase in the cost of her weekly train ticket
with the increase in her weekly pay. **(3 marks)**

June 2014, Q18, 1MA0/2H

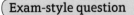

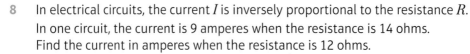

8   In electrical circuits, the current I is inversely proportional to the resistance R.
     In one circuit, the current is 9 amperes when the resistance is 14 ohms.
     Find the current in amperes when the resistance is 12 ohms.   *(5 marks)*

9   The wavelength of a wave is inversely proportional to its frequency.
     a   A radio wave of wavelength 1000 metres has a frequency of 300 kHz.
         A different radio wave has a frequency of 600 kHz.
         What is its wavelength?   *(2 marks)*
     b   Calculate the frequency of a radio wave with wavelength 842 metres. *(3 marks)*

This unit is call multiplicative reasoning.
'Multiplicative' means involving multiplication or division'.
Reasoning is being able to explain why you have used multiplication or division in this unit.
a   List three ways you have used multiplication or division in this unit.
b   Why is it good to reason in mathematics?

Sample student answers

Which student gave the correct answer? Explain what the other student did wrong.

> ### Exam-style question
>
> Mr and Mrs Adams sold their house for £168 000.
> They made a profit of 12% on the price they paid for the house.
>
> Calculate how much they paid for the house.   **(3 marks)**
>
> *Nov 2012, Q13, 5MB3H/01*

Student A

12% of 168 000 = 0.12 × 168 000
                = 20 160

   168 000
−   20 160
   147 840

They paid £147 840 for the house.

Student B

100 + 12 = 112%

price   → × 1.12 → 168 000
150 000 ← ÷1.12 ← 168 000

They paid £150 000 for the house.

15 CONSTRUCTIONS, LOCI AND BEARINGS

Abed has a sat-nav system in his car.
Draw a sketch of these instructions.
Instruction 1: Go south for 4 miles.
Instruction 2: Go east for 3 miles.
Instruction 3: Go north for 2 miles.
Instruction 4: Go west for 5 miles.
Where is Abed now?

15 Prior knowledge check

Numerical fluency

1 Write these ratios in the form $1:m$.
   a 16:32   b 9:72

2 Write these ratios in their simplest form.
   a 1 m:50 cm   b 20 mm:5 cm

Fluency with measures

3 a Copy and complete this diagram showing the eight points of the compass.

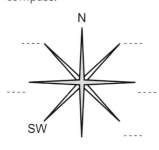

   b What angle does the compass move through if it turns anticlockwise from north to south-west?

4 Measure the length of each line.
   a _____
   b _____

5 Draw a line exactly 8.7 cm long.

6 Draw these angles accurately.
   a 54°   b 163°

7 Sketch and label the net of this cuboid.

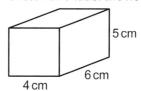

Geometrical fluency

8 Describe each turn giving clockwise or anticlockwise and the fraction of a turn.

a   b

9 Which of these shapes are congruent?

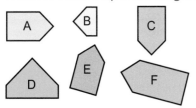

10 Use compasses to draw a circle with:
   a a diameter of 6 cm
   b a radius of 4.2 cm.

11 Choose pairs of angles that work for angles a and b.

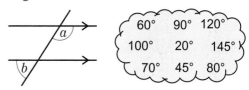

15.1 3D solids

Objectives

- Recognise 3D shapes and their properties.
- Describe 3D shapes using the correct mathematical words.
- Understand the 2D shapes that make up 3D objects.

Did you know?

Architects use mathematical shapes and symmetry to make buildings more interesting.

Fluency

What are the names of these two-dimensional shapes?

1 Draw these shapes accurately on squared paper.
   a A square of side length 3 cm.
   b A rectangle 4 cm by 6.5 cm.

Key point 1

The flat surfaces of 3D shapes are called **faces**, the lines where two faces meet are called **edges** and the corners at which the edges meet are called **vertices** (the singular of vertices is **vertex**).

Questions in this unit are targeted at the steps indicated.

2 Here are two ordinary dice.
   a How many faces does each dice have?
   b How many edges does each dice have?

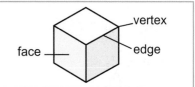

3 **Exam-style question**

Here is a cuboid.

Copy and complete each of the following sentences by inserting the correct number.
   a A cuboid has _____ faces.
   b A cuboid has _____ edges.
   c A cuboid has _____ vertices. **(3 marks)**

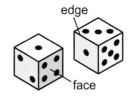

Nov 2013, Q7, 1MA0/1F (part a only)

Exam hint
Use the diagram to check that your answer makes sense.

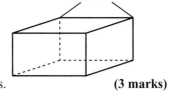

4   Here is a cuboid.

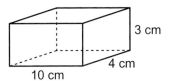

a   What is the name of the shape of its faces?
b   Write down the dimensions of each of the faces.

Q4b communication hint
A **dimension** is the size of something in a particular direction. Length, width, height and diameter are all dimensions of shapes or solid objects.

5   Here are some three-dimensional solids.

a   b 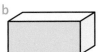   c   d   e

cylinder
sphere
cube
cuboid
cone

Match the correct name to each solid.

Key point 2

Pyramids have a base that can be any shape and sloping triangular sides that meet at a point.
In a **right prism**, the sides are perpendicular (at right angles) to the base.

6   **Reasoning** Here are some more three-dimensional solids, this time pyramids and prisms.

tetrahedron   square-based   triangular   hexagonal
                pyramid   prism   prism

a   What shape is each face of the tetrahedron?
b   What shape are the side faces of the triangular prism?
c   Copy and complete this table.

Q6a hint When the shape is a triangle be specific about which type.

	Number of faces	Number of edges	Number of vertices
Cube	6	12	8
Tetrahedron	4		
Square-based pyramid			
Triangular prism			
Hexagonal prism			

d   Write a rule connecting the number of faces (F), vertices (V) and edges (E).

Q6d hint Add together the number of faces and vertices and compare this with the number of edges.

7   **Reasoning** Here is another prism.
Cathy says, 'It has twice as many edges as it has faces.'
Is she correct?
Explain your answer.

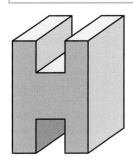

15.2 **Plans and elevations**

Objectives

- Identify and sketch planes of symmetry of 3D shapes.
- Understand and draw plans and elevations of 3D shapes.
- Sketch 3D shapes based on their plans and elevations.

Did you know?

If you apply for planning permission to build a new house or an extension, you need to send plan and elevation drawings of the new building to the council for approval.

Fluency

What 2D shapes are the faces of these 3D objects?

1   Group these into 2D and 3D shapes.

> cube   triangle   sphere   cuboid   rectangle   square   pentagon   tetrahedron   circle

2   How many lines of symmetry does each of these shapes have?

Example 1

A triangular prism has ends that are equilateral triangles with side lengths 4 cm. It is 12 cm long. Sketch the prism.

To sketch a 3D shape, it is best to draw the biggest or most complicated part of the shape first.
In this case, start by drawing an equilateral triangle, which will be one end of the prism.

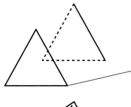

Pick one corner of your triangle, and move it up and to the right.
Now draw a second, identical triangle from this point.
The gap between the triangles will show the length of the prism.
For longer prisms, move the second triangle further away.

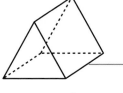

Now join the corners of the two triangles with straight lines.

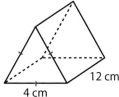

12 cm

4 cm

Remember to put the lengths on your diagram.

Key point 3

A **plane** is a flat (2D) surface. A solid shape has a **plane of symmetry** when a plane cuts the shape in half so that the part of the shape on one side of the plane is an identical reflection of the part on the other side of the plane. The planes of symmetry for this cuboid are shown in blue.

  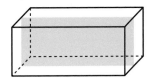

3   Here are two right prisms and two pyramids.

a 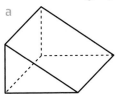   b 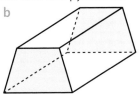   c 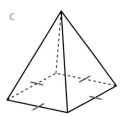   d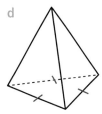

Copy the diagrams and draw in all the planes of symmetry for each shape.

4   This diagram represents a shed (not drawn to scale). The shed is in the shape of a prism.

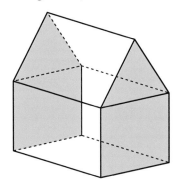

> **Q4 hint** Use a ruler to find the midpoint of the lengths. Join these points to make a plane.

The cross-section of the shed is an equilateral triangle on top of a square.
The shed has two planes of symmetry. Copy the diagram and draw in the two planes of symmetry.

5   **Reasoning** Twins George and Gemma were given this cake on their 8th birthday.
How many ways can the cake be cut to give each twin exactly half of the cake?

> **Q5 hint** Sketch the cake to show each different cut.

Discussion Is the top loop of the 8 exactly the same size as the bottom loop? Does it make a difference to your answer if it is?

Reflect Look at your answers to questions 3 to 5.
Work out the number of lines of symmetry for the base of each prism or pyramid. Now compare this number to your answers. Is there a connection?

Key point 4

The **plan** is the view from above an object.
The **front elevation** is the view of the front of an object.
The **side elevation** is the view of the side of an object.

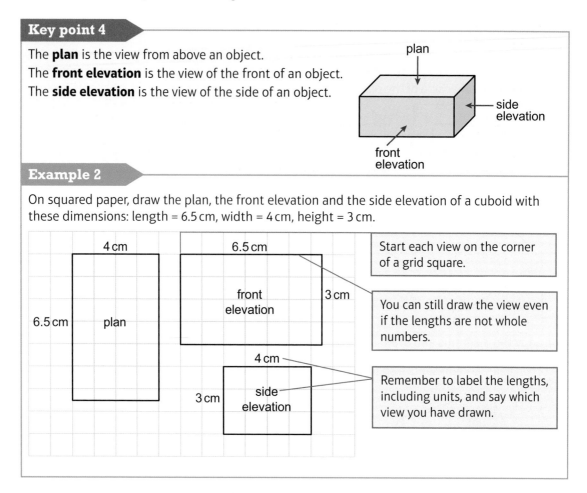

Example 2

On squared paper, draw the plan, the front elevation and the side elevation of a cuboid with these dimensions: length = 6.5 cm, width = 4 cm, height = 3 cm.

Start each view on the corner of a grid square.

You can still draw the view even if the lengths are not whole numbers.

Remember to label the lengths, including units, and say which view you have drawn.

6   Draw the plan view, the front elevation and side elevation of these cuboids on squared paper.

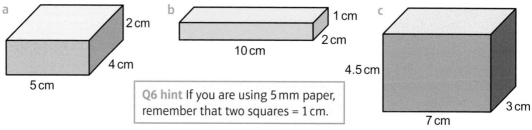

Q6 hint If you are using 5 mm paper, remember that two squares = 1 cm.

7   Here are the plan views of some solids. What solid could they be?

Discussion Is there more than one possible answer for each image?

8   On squared paper, draw the plan, front elevation (shown by the arrow) and side elevation of this prism.

9 Here are the plan, front elevation and side elevation views of a prism, drawn on cm squared paper.

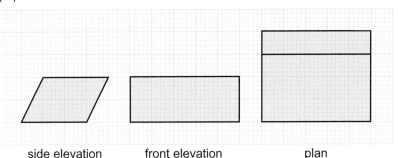

side elevation   front elevation   plan

> **Q9 hint** Use the plan to draw the base of your shape.

Sketch the prism. Label its lengths.

10 Here are the plan, front elevation and side elevation views of a prism.

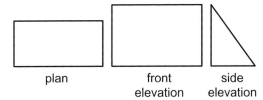

plan   front   side
            elevation   elevation

Sketch the prism.

11 Here is a box in the shape of a cuboid.
Draw an accurate diagram of:
a   the front elevation (shown by the arrow)
b   the plan view.

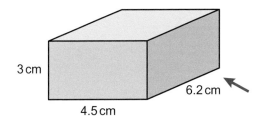

3 cm
4.5 cm
6.2 cm

12 **Exam-style question**

Here are the front elevation, side elevation and plan of a 3D shape.

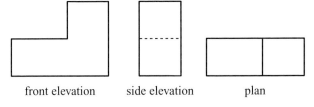

front elevation   side elevation   plan

Draw a sketch of the 3D shape.   **(2 marks)**

June 2011, Q13, 5384F/12F

> **Exam hint**
> Remember to use a HB pencil and a ruler to ensure your shape is drawn neatly.

13 This diagram shows a greenhouse in the shape of a prism. The cross-section of the greenhouse is a trapezium.
Draw a sketch of the plan of the greenhouse.
Show the dimensions of the plan on your sketch.

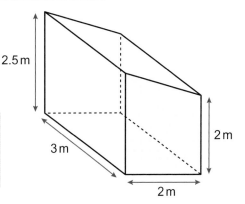

2.5 m
3 m
2 m
2 m

> **Q13 hint** Imagine you are flying over the greenhouse directly above it. When you look down, what shape do you see? The shape of the top of the greenhouse or the shape of the bottom?

15.3 **Accurate drawings 1**

Objectives

- Make accurate drawings of triangles using a ruler, protractor and compasses.
- Identify SSS, ASA, SAS and RHS triangles as unique from a given description.
- Identify congruent triangles.

Did you know?

Sailors use compasses to draw triangles on navigation charts when plotting a course.

Fluency

What does 'congruent' mean?

Warm up

1 Draw these lines accurately using a ruler.
  a 7.5 cm
  b 47 mm

2 Use compasses to draw:
  a a circle of radius 6 cm
  b half a circle of radius 4.5 cm.

3 Draw these angles accurately using a ruler and protractor.
  a 75°
  b 48°
  c 112°

Key point 5

You can draw an accurate diagram of a triangle with a ruler and protractor if you know three measurements. For example, if you know the length of two sides of the triangle and the size of one angle, or the length of one side of the triangle and the size of two angles.

Example 3

Make an accurate drawing of this triangle.

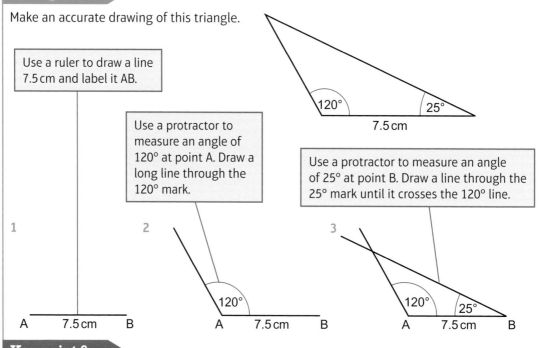

Use a ruler to draw a line 7.5 cm and label it AB.

Use a protractor to measure an angle of 120° at point A. Draw a long line through the 120° mark.

Use a protractor to measure an angle of 25° at point B. Draw a line through the 25° mark until it crosses the 120° line.

Key point 6

The kind of triangle drawn in the worked example is sometimes called an **ASA** triangle, because you are asked to make an accurate drawing given an **A**ngle, a **S**ide length and another **A**ngle.

4 Make an accurate drawing of triangle PQR using a ruler and protractor.

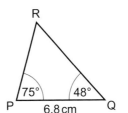

Q4 strategy hint
Do not erase your construction lines.

> **Key point 7**
>
> In an **SAS** triangle you are given two **S**ide lengths and the **A**ngle in between.

5 Make an accurate drawing of triangle XYZ using a ruler and protractor.

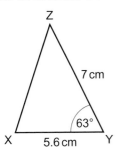

Q5 hint Start by drawing and labelling the line XY, and then measure the angle XYZ from point Y using a protractor. Next, accurately draw the line YZ from Y through the measured point and label the end of this line Z. Finally, join X and Z with a straight line to complete your triangle.

6 **Exam-style question**

The diagram shows a sketch of triangle ABC.

Diagram NOT accurately drawn

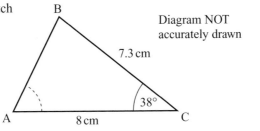

a Make an accurate drawing of triangle ABC. **(2 marks)**
b Measure the size of angle A on your diagram. **(1 mark)**

June 2004, Q1, 5504/04

7 **Reasoning** a Which of these five triangles are ASA triangles and which are SAS triangles?

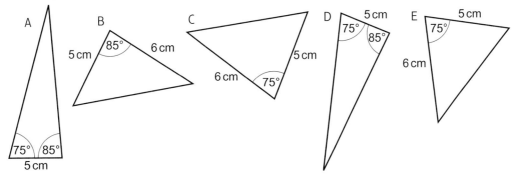

b Make accurate drawings of all the triangles using a ruler and protractor.
c Which pairs of triangles are congruent?
Discussion Are these statements always, sometimes or never true?
• Two ASA triangles are congruent if the given side length and two angle sizes are identical in each triangle.
• Two SAS triangles are congruent if the given angle size and two side lengths are identical in each triangle.

> **Key point 8**
>
> In an **SSS** triangle, you are given all three **S**ide lengths but none of the angles.

8   Make an accurate drawing of triangle TUV using a ruler and pair of compasses.

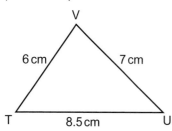

> **Q8 hint** Start by drawing and labelling the line TU. Then use your compasses to draw an arc, radius 6 cm with centre T and another arc, radius 7 cm, with centre U. Label the point where the two arcs intersect as V. Finally, join T to V and U to V to complete your triangle.

> **Q8 strategy hint** Do not erase your arcs.

9   Use compasses and a ruler to draw an accurate diagram of this sketch.

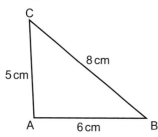

10 Make an accurate drawing of an equilateral triangle with side lengths 5 cm.

   Discussion What angle have you constructed here? Is this a useful method to remember?

> **Q10 hint** Draw one line 5 cm long and then use compasses set at length 5 cm to draw an arc from each end of this line.

> **Key point 9**
>
> Triangles with a right angle can be referred to as **RHS** triangles if you are given the **R**ight angle, the **H**ypotenuse length and another **S**ide length. The **hypotenuse** is the longest side of a right-angled triangle.

11 a   Make an accurate drawing of triangle VWX.

   b   How long is side VW?

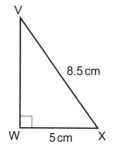

> **Q11a hint** Use a protractor to draw the right angle.

12 **Reasoning** Here is a sketch of an RHS triangle and an SSS triangle.

   Discussion Are these statements always, sometimes or never true?

   • Two RHS triangles are congruent if one given side length and hypotenuse length are identical in each triangle.

   • Two SSS triangles are congruent if the given three side lengths are identical in each triangle.

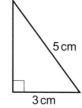

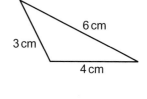

15.4 Scale drawings and maps

Objectives

- Draw diagrams to scale.
- Correctly interpret scales in real-life contexts.
- Use scales on maps and diagrams to work out lengths and distances.
- Know when to use exact measurements and estimations on scale drawings and maps.
- Draw lengths and distances correctly on given scale drawings.

Did you know?

Scale diagrams are a useful planning tool. For example, you would use one when designing a new kitchen or arranging furniture in a house.

Fluency

Triangle B is an enlargement of triangle A. What is the scale factor of enlargement?

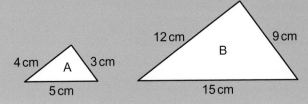

Warm up

1   Write each ratio as 1:m.
    a   7:14   b   15:375   c   25:5

2   Convert:
    a   1.25 cm to mm   b   125 000 mm to m.

3   Copy and complete these equivalent ratios.
    a   1:20   3:☐   b   1:50   ☐:200

4   On a scale drawing, 1 cm represents 2 m.
    a   What line length on the drawing represents:
        i   4 m   ii   9 m?
    b   On the same drawing, what actual length do these represent?
        i   5 cm   ii   6.8 cm?

    Q4 hint ×☐ ⟲ 1 cm : 2 m ⟳ ×☐
             ☐ : 4 m

5   Here is a sketch of two rooms.

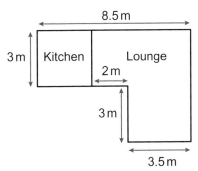

On cm squared paper, draw a scale diagram of the rooms using the scale 2 cm represents 1 m.

Key point 10

A scale is a ratio that shows the relationship between a length on a map or drawing and the actual length. Scale 1:25 000 means 1 cm on the map represents 25 000 cm in real life.

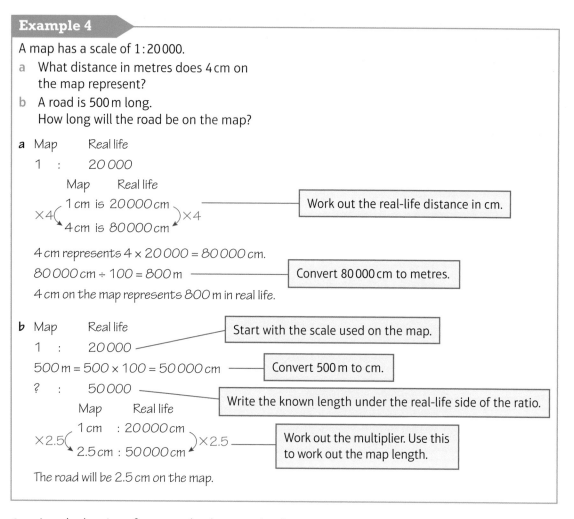

Example 4

A map has a scale of 1 : 20 000.
a   What distance in metres does 4 cm on
    the map represent?
b   A road is 500 m long.
    How long will the road be on the map?

a   Map   Real life
    1   :   20 000

       Map   Real life
    ×4(1 cm is 20 000 cm)×4 — ⟶ Work out the real-life distance in cm.
       4 cm is 80 000 cm

    4 cm represents 4 × 20 000 = 80 000 cm.
    80 000 cm ÷ 100 = 800 m ——— Convert 80 000 cm to metres.
    4 cm on the map represents 800 m in real life.

b   Map   Real life ——— Start with the scale used on the map.
    1   :   20 000
    500 m = 500 × 100 = 50 000 cm ——— Convert 500 m to cm.
    ?   :   50 000 ——— Write the known length under the real-life side of the ratio.
       Map   Real life
    ×2.5(1 cm   : 20 000 cm)×2.5 ——— Work out the multiplier. Use this
        2.5 cm : 50 000 cm   to work out the map length.

    The road will be 2.5 cm on the map.

6   A scale drawing of a room plan has a scale of 1 : 25.
    a   The length of the room in the plan is 24 cm. How long is the actual room?
    b   The width of the actual room is 4 m. What is the width of the room on the plan?

7   This diagram shows a scale plan of Jackie's bedroom, drawn
    on cm squared paper. Her bed is 2 m by 1.5 m.
    a   Copy and complete this statement:
        1 cm on the diagram represents ☐ m.
    b   Write the scale used as a ratio 1 : x.
    c   Use the scale to work out:
        i   the dimensions of the bedroom
        ii the dimensions of the wardrobe
        iii the area of carpet which covers the whole floor.
    d   Copy the diagram and draw these items in the bedroom:
        i   a bedside table 40 cm × 50 cm
        ii a radiator 10 cm × 1.5 m.

8   On a scale diagram, 5 cm represents a real-life length of 1.25 m.
    a   Write the two lengths as a ratio scale diagram : real-life
    b   Write as a ratio in the form 1 : m.

Q8 hint Both
measurements should
be in the same units.

9 **Real** A plane makes cargo deliveries in Ireland. The map of the area has a scale of 100 km : 5 cm.

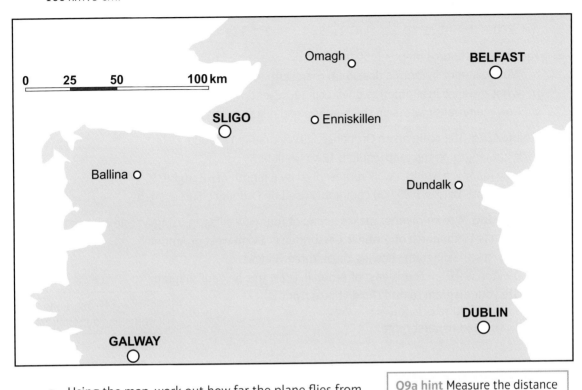

a Using the map, work out how far the plane flies from
  i Dublin to Belfast
  ii Belfast to Galway
  iii Galway back to Dublin.

> **Q9a hint** Measure the distance on the map in centimetres. Then use the map scale to work out the real distance.

b Which airport is 180 km from Dublin and 150 km from Belfast?

Discussion Are these distances exact?

10 An archaeologist makes a scale drawing of her dig site using a scale of 1 : 10.
   She digs a trench in the site 1 m wide by 3.5 m long.
   a What are the dimensions of the trench on the scale drawing?
   b An object on the drawing is 83 mm long. How long is this object in real life?

11 **Reasoning** Helen buys a new flat. These are the dimensions of the rooms.
   Living room 3.5 m × 2.5 m   Hall 1.5 m × 2 m   Bedroom 3 m × 3 m
   Kitchen 2 m × 2 m   Bathroom 3 m × 1.5 m   All doors are 1 m wide.
   A scale diagram is drawn on cm squared paper.
   a This scale diagram is incorrect.
     Explain why.
   b Draw a correct version of the diagram
     on squared paper.

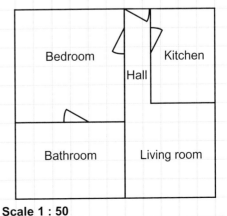

Scale 1 : 50

12 **Reasoning / Real** Simon is designing a new kitchen and wants to draw a scale diagram of it on a sheet of A4 paper. The dimensions of the kitchen are 5 m by 8.5 m.
Which of these scales should Simon use to draw the scale diagram? Explain why.
A 1:200   B 1:40   C 1:25   D 1:125   E 1:5

13 The scale factor on a map is 1:50 000.
    a What distance in metres does 5 cm represent?
    b What distance in kilometres does 7 cm represent?
    c How many centimetres on the map is a real distance of 12.5 km?

> **Q13 hint** 1 km = 1000 m and 1 m = 100 cm.

14 **Reasoning** The scale on an Ordnance Survey map is 1:25 000.
    a What length on the map would a 12 km walk be?
    b How many km would a walk represented by a length of 32 cm be?
    Discussion Why is 1:25 000 commonly used on Ordnance Survey maps?

15 **Reasoning** A town planner makes a map of four new villages, using a scale of 1 cm to 2 km. Bradfield is 15 km north of Amphill. Carsbridge is 24 km west of Amphill.
    a Draw a scale diagram showing these three villages.
    b Dreyton is 9.6 km south-east of Amphill. Draw this on your diagram.
    c Use your diagram to find the distance from
      i Bradfield to Dreyton
      ii Dreyton to Carsbridge.

> **Q15c hint** Measure the lengths on your diagram.

15.5 Accurate drawings 2

Objectives

- Accurately draw angles and 2D shapes using a ruler, protractor and compasses.
- Construct a polygon inside a circle.
- Recognise nets and make accurate drawings of nets of common 3D objects.

Did you know?

Food manufacturers look at the nets for food cartons very carefully in order to work out the most efficient way to cut the cardboard.

Fluency

Work out the missing numbers.
a 55 + 55 + ☐ = 180   **b** 150 + 75 + ☐ = 360   **c** 122 + 76 + 89 + ☐ = 360

1 What 3D solid will each of these nets make?
    a   b

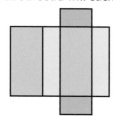

2 **Reasoning** This is one net of a cube.
How many different nets can you find for a cube?

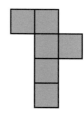

    Active Learn   Homework, practice and support: Foundation 15.5

3 In triangle FGH, FG = 10.3 cm, GH = 9.4 cm and angle GFH = 62°.
   a Draw a sketch of triangle FGH showing all three given measurements.
   b Now make an accurate drawing of triangle FGH.
   c Measure the length of FH.

4 Using compasses, make an accurate drawing of triangle JKL, where JK = 11.3 cm, KL = 9.7 cm and JL = 4.8 cm.
   Discussion What problems do you find? Why?

> **Q4 strategy hint**
> Draw a sketch first.

5 Here is a quadrilateral, ABCD.

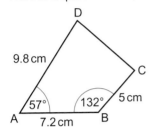

   a Make an accurate drawing of this quadrilateral.
   b Measure the length of CD.
   c Measure the size of angle BCD.

6 Follow these instructions to construct a regular hexagon inside a circle.
   a Draw a circle with radius 7 cm. Label the centre O.
   b Draw a vertical line from O up to a point on the circumference. Label this point A.

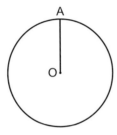

   c To work out the angle you need to measure in the middle, divide 360° by the number of sides on a hexagon. Call this angle x.
   d Starting at line OA, measure angle x at the centre of the circle. Draw a line from O to the circumference through your measured angle. Label the new point on the circumference B and the angle $x°$.

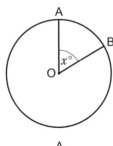

   e Join points A and B together with a straight line. Line AB is the first side of your hexagon.
   f Repeat parts **d** and **e** until you have drawn the whole hexagon.

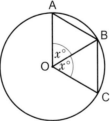

   Reflect How accurately did you measure the angles? How can you tell?

> **Q6 communication hint**
> **Construct** means draw accurately using a ruler and pair of compasses.

> **Q6 hint** Remember to use a ruler to draw all of the straight lines.

> **Q6d hint** Make sure you measure every angle carefully. Turn your book around to move the start line, so it is easier to measure each angle.

7 JKLM is a cyclic quadrilateral.
   OJ = OK = OL = OM = 6 cm. JOK = 72° and KOL = 48°.
   Line KOM is a diameter of the circle.
   a Using a ruler, compasses and a protractor, draw polygon JKLM.
   b Using your diagram, measure angle JOM and angle KLM.
   Discussion Are the angles you measured in part **b** accurate? How can you tell? Is there a way to check?

> **Q7 communication hint**
> A **cyclic quadrilateral** is a four-sided polygon whose vertices are all on the circumference of a circle.

8   On squared paper, draw an accurate net for this square-based pyramid.

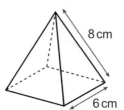

8 cm

6 cm

> **Q8 hint** Draw the square first then use compasses to draw the triangles.

9   **Problem-solving / Real** A manufacturer makes boxes of chocolates. The box looks like this.

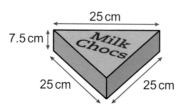

25 cm

7.5 cm

25 cm   25 cm

Milk Chocs

Use a ruler and compasses to draw a scaled diagram of an accurate net for the box.
Use a scale of 1 cm to represent 5 cm.

15.6 **Constructions**

Objectives

- Draw accurately using rulers and compasses.
- Bisect angles and lines using rulers and compasses.

Did you know?

Architects use a variety of constructions in order to accurately draw plans for new buildings.

Fluency

Which pair of lines are perpendicular to each other?   **a**   **b**   **c**

Warm up

1   Use a ruler to draw lines with these measurements.
    a   5.5 cm   b   47 mm

2   What size are the angles in an equilateral triangle?

Key point 11

Constructions are accurate drawings made using a ruler and pair of compasses. **Bisect a line** means to cut a line exactly in half. A **perpendicular bisector** cuts a line in half at right angles.

Example 5

Draw a line that is 6 cm long. Then construct its perpendicular bisector.

1

6 cm

2

3

4

| 1 | Use a ruler to draw a 6 cm line. |

| 2 | Open your compasses greater than half the length of the line. Place the point on one end of the line and draw an arc, roughly half a circle, above and below the line. |

| 3 | Keeping the compasses the same distance, move them to the other end of the line and draw another arc. |

| 4 | With a straight line, join the points where the two arcs intersect. This vertical line divides the horizontal line exactly in half. Do not rub out the arcs. These are your **construction lines**. |

3 a Use a ruler and compasses to construct the perpendicular bisector of a line 7 cm long.

  b Check, by measuring, that your line cuts the original line exactly in half.

Key point 12

You can also use a ruler and compasses to **construct** a perpendicular from a point to a line.

Example 6

Draw a line 8 cm long and a point P that is 3 cm above the line. Then construct a perpendicular line to your original line that passes through point P.

1
P
•

2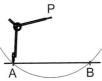

3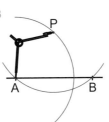

4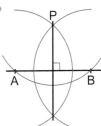

| 1 | Use a ruler to draw an 8 cm line and then mark point P, 3 cm above this line. |

| 2 | Open your compasses to a radius larger than the distance from P to the line (the larger the distance, the more accurate your diagram). Then, with your compasses on point P, draw an arc that cuts the line twice. Label the two intersection points A and B. |

| 3 | Put the compasses on Point A and draw an arc above and below the horizontal line. Repeat, with compasses the same distance, but this time with the compass point at B. |

| 4 | Now draw a vertical line through the intersection points of these two arcs. This line will go through P and will be perpendicular to the horizontal line. |

4   Draw a line 9 cm long and a point Q, 4 cm above
     this line. Using a ruler and compasses, construct
     a perpendicular line to your original line which
     passes through point Q.

> **Q4 hint** Don't forget to label the
> two intersection points on your
> original line as A and B

Reflect Did you use the worked example to help you answer this question? Did it help you
work carefully through each step of the construction?

5   a   On the diagram you drew in **Q4**, label the point where the perpendicular bisects the
          original line as point X.
     b   Add in point C somewhere between points A and X.
     c   Add in point D somewhere between point A and the end of the line.
     d   Measure the distance from Q to each of the other labelled points.
     Discussion What is the shortest distance from a point to a line.

6   **Reasoning** An engineer needs to draw an accurate diagram showing a new road leading
     from a Sports Centre to a main road.
     The new road needs to be the shortest distance possible.
     Here is part of the diagram.

                         Main road
     ───

                              •
                         Sports Centre

     Scale: 1 cm represents 500 m.

     a   Trace the diagram and complete it using a ruler and compasses.
     b   How long will the new road be in km?

7   a   Copy or trace this diagram.

     ───────────────────────────────────────
                         P
                         •

> **Q7 hint**
> See Example 5.

     b   Put your compasses on point P and draw arcs so that P is the midpoint of the line
          segment between the arcs.
     c   Construct the perpendicular bisector for this new line segment.

Key point 13

An **angle bisector** cuts an angle exactly in half.

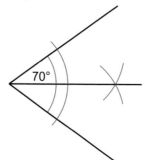

Example 7

Draw an angle of 80°. Construct the angle bisector.

1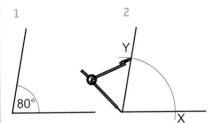

2

3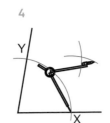

4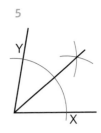

5

| 1 | Draw the angle using a protractor. |

| 2 | Open your compasses and place the compass point at the vertex of the angle. Draw an arc that cuts both arms of the angle. Label these intersection points X and Y. |

| 3 | Keep the compasses set at the same length. Put the compass point on Y and draw an arc in the middle of the angle. |

| 4 | Repeat step 3, but this time with the compass point on X. |

| 5 | Join the intersection of these two arcs with the vertex of the angle. This straight line is the angle bisector. Each part of the angle on either side of this line will be the same size. |

8   a   Draw an angle of 70° using a protractor. Construct the angle bisector.
    b   Check, by measuring with a protractor, that your
      angle bisector cuts the angle exactly in half.

9   Copy the diagrams.
   Then construct the bisector of angle Q in
   each triangle.

10   **Exam-style question**

Trace this angle and then use a ruler and compasses to **construct** the
bisector of the angle.
You must show all your construction lines.   **(2 marks)**

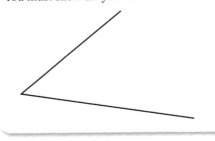

Exam hint
Use the
equipment
they tell you
to use. Using
a protractor
for this
question gains
no marks.

11   Draw an 85° angle. Then construct its angle bisector.
    Discussion Does a well-drawn construction bisect an angle accurately?

12   **Reasoning** How could you use your knowledge of constructing perpendicular bisectors and
    angle bisectors to construct a 90° and a 45° angle?

13 a Draw a 60° angle by constructing an equilateral triangle.
   b Bisect your 60° angle to construct a 30° angle.
   Discussion How would you draw a 120° angle?

Q13a hint Remember that all sides are the same length in equilateral triangles. You drew SSS triangles in 15.3.

15.7 Loci and regions

Objectives

- Draw loci for the path of points that follow a given rule.
- Identify regions bounded by loci to solve practical problems.

Did you know?

Air-sea rescue services use loci to help them identify which areas to search for missing survivors.

Fluency

The scale on a map is 1 cm represents 5 km.
How many kilometres does a distance of 4 cm on the map represent?
How many centimetres on the map is a real distance of 7.5 km?

1 Draw a circle with a radius of 6 cm.

2 Draw a line 76 mm long. Construct its perpendicular bisector.

3 What is the locus of points a fixed distance of 5 cm from a point P?

Q3 hint Draw a point P, and then draw different points which are all 5 cm from P. When you join these points up, what shape do you make?

4 **Reasoning / Real** A road is planned to be **equidistant** from two towns A and B.
   a Draw towns A and B 8 cm apart.
   b Construct the path of the road.

Communication hint Equidistant means the same distance.

5 a Draw a line 8 cm long.
   b A point moves so that it is always exactly 2 cm away from the line segment. Construct the locus.

6 a Draw an 80° angle, ABC. Make each arm of the angle at least 6 cm long.
   b What shape is the locus of points equidistant from AB and BC? Make a sketch of the locus.
   c Now draw the locus accurately on your diagram using a ruler and compasses.
   Discussion What construction have you made?

Q6b hint Think about students standing the same distance from two walls in the corner of a classroom. What shaped line would they make?

7 **Real** This diagram shows a plan of a garden.
Treasure is buried at the intersection point of the loci that are:
- equidistant from lines PQ and QR
- equidistant from lines SR and RQ.
Copy the diagram and draw in the loci.
Then mark the location of the treasure with an X.

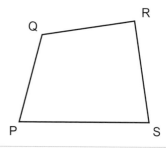

ActiveLearn Homework, practice and support: Foundation 15.7

Key point 14

A **locus** is a set of all points that obey a given rule. This produces a path followed by the points. The plural of locus is **loci**.

- The locus of all points a given distance (d) from a fixed point (P) is a circle, with centre P and radius d.

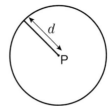

- A locus that is equidistant from two given points, X and Y, is the perpendicular bisector of the line between the two points.

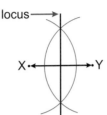

- The locus of all points 1 cm from a line, AB, is two parallel lines 1 cm from the line and a semi-circle at each end radius 1 cm.

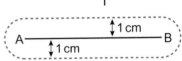

- A locus that is equidistant from two intersecting lines, OX and OY, is the angle bisector of the two lines.

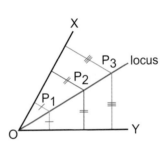

8   **Reasoning** This diagram shows a wheel. The point C is at the centre of the wheel and the point P lies on its edge.

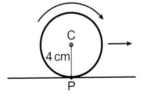

> **Q8a hint** Think about the distance from the centre of the wheel to the surface as the wheel moves.

a   The wheel rolls along a flat surface. Sketch the locus of points that C moves as the wheel rolls along the surface.

b   Sketch the locus of points that P moves as the wheel rolls along the surface.

> **Q8b hint** Mark a 2p coin at one point, P, on its edge. Observe the position of P as the coin rolls. Think about the distance of P from the surface.

9   A square piece of card is placed on a horizontal surface.

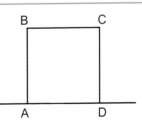

a   The card is first rotated 90° clockwise about vertex D.
   Copy or trace the diagram in colour 1. Draw the new position of the square in colour 1.
   Draw the locus of the vertex A in colour 2.

b   The card is then rotated 90° clockwise about vertex C and then 90° clockwise about vertex B.
   Draw the new positions of the card in colour 1 and the remainder of the locus of vertex A in colour 2.

> **Key point 15**
>
> A **region** is an area bounded by loci.

10 **Reasoning** Draw points A and B, 7.5 cm apart. Now draw a circle centre A radius 4 cm and a circle centre B radius 5 cm. Shade in the region where the circles intersect.

Decide which of these statements are true.

A The shaded region represents the area which is more than 4 cm from point A and more than 5 cm from point B.

B The shaded region represents the area which is less than 4 cm from point A and less than 5 cm from point B.

C The shaded region represents the area which is less than 5 cm from point A and less than 4 cm from point B.

11 (**Exam-style question**

The diagram shows an accurate scale drawing of two towns, Exe and Wye.

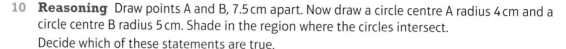

Scale: 1 cm to 2 km

A new cinema is to be built. The cinema will be
• less than 6 km from Exe and
• less than 9 km from Wye.
Copy the diagram and shade the region where the cinema
can be built.   **(3 marks)**

Exam hint
Draw your circles to scale. Show all your working.

12 **Problem-solving / Real** A man reports seeing a swimmer in distress from his boat at point B. A helicopter sets out to search for the swimmer in the region of sea within 500 metres from point B.

Scale: 1:10 000

B.

sea | land

Copy the diagram and shade in the area where the helicopter will search for the swimmer.

13 A new mobile phone tower is to be built equidistant from two towns that are 6 km apart.
The tower has a range of 5 km.
Draw a scale diagram showing all possible locations of the new tower so that both towns have full mobile phone reception.
Use a scale of 1 cm : 1 km.

Q13 strategy hint First draw in the position of the two towns 6 km apart. (Use the given scale.)

Discussion What kind of construction have you made in this answer?

14 **Exam-style question**

The diagram represents a triangular garden ABC.

The scale of the diagram is 1 cm represents 1 m.

A tree is to be planted in the garden so that it is
- nearer to AB than to AC
- within 5 m of point A.

Copy the diagram and shade the region where the
tree may be planted.

(3 marks)

Exam hint

Make sure you read
the question carefully
to ensure you draw
the correct loci
and then shade the
correct region on
your diagram.

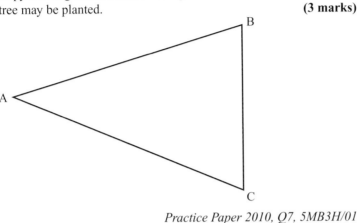

Practice Paper 2010, Q7, 5MB3H/01

15.8 **Bearings**

Objectives

- Find and use three-figure bearings.
- Use angles at parallel lines to work out bearings.
- Solve problems involving bearings and scale diagrams.

Did you know?

Air-sea rescue services rely on
accurate bearings in order to locate
areas to search for survivors.

Fluency

Work out: a 360 − 45 = ☐   b 240 − 180 + 3 = ☐   c 180 − 68 + 43 = ☐

1   Work out angles a, b and c.

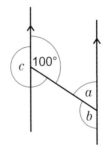

Key point 16

A **bearing** is an angle measured in degrees clockwise from north.
A bearing is always written using three digits.
This bearing is 025°.

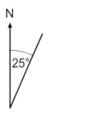

2 Write the bearing of B from A in these diagrams.

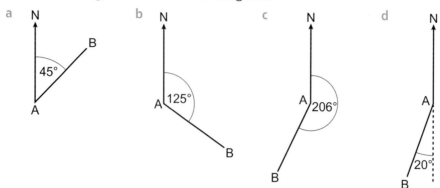

a 45°

b 125°

c 206°

d 20°

> **Q3 hint** Always measure the angle of a bearing at the 'from' point.

3 The bearing of P from Q is 080°.
Work out the bearing of Q from P.

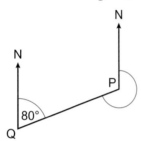

80°

4 **Exam-style question**

Work out the bearings of X from Y.

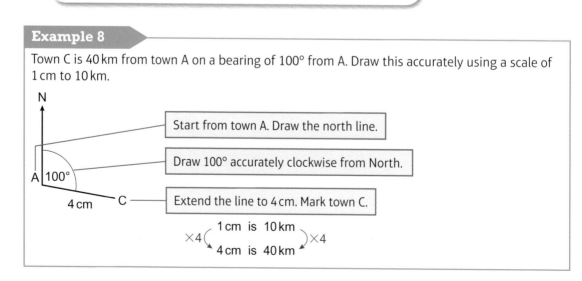

120°   130°   200°

Example 8

Town C is 40 km from town A on a bearing of 100° from A. Draw this accurately using a scale of 1 cm to 10 km.

N

Start from town A. Draw the north line.

Draw 100° accurately clockwise from North.

A 100°

4 cm   C

Extend the line to 4 cm. Mark town C.

×4 (1 cm is 10 km
      4 cm is 40 km) ×4

5 **Exam-style question**

Work out the bearing of

a B from P

b P from A. **(4 marks)**

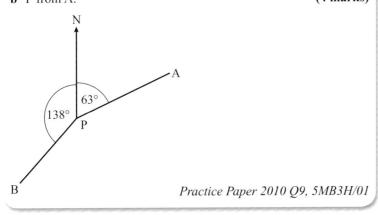

Practice Paper 2010 Q9, 5MB3H/01

Exam hint

Draw a north line at point A. Can you use angles at parallel lines to help you work out the bearing?

Exam hint

Always measure your bearings clockwise, not anticlockwise.

6 **Real** Draw these bearings accurately. Use the scale 2 cm = 100 km.

a Venice Marco Polo airport is 420 km away from Rome Ciampino airport on a bearing of 356°.

b Naples airport is 175 km away from Rome Ciampino airport on a bearing of 128°.

7 Town P is on a bearing of 280° from a lookout post (L) 5 km away.
Town Q is on a bearing of 180° from L 3 km away.

a Draw an accurate diagram showing the towns and the lookout post. Use the scale 1 cm = 1 km.

> Q7a strategy hint Make a sketch first. Start with the Lookout point L and draw in the north line. Measure the bearing and correct distance to each town, using the scale.

Use your diagram to find

b the distance between towns P and Q

> Q7b hint Draw in the line between the two towns.

c the bearing of Q from P

> Q7c hint Remember to draw in a North line at P and at Q.

d the bearing of P from Q

Reflect The strategy hint suggested making a sketch first. Did the sketch help you draw an accurate map? Write a sentence explaining how.

> Q7d hint Remember to measure the bearing angle at the 'from' point.

8 **Real / Problem-solving** A yacht is sailing in a race. The bearing of the yacht from a lighthouse is 162°. The bearing of the yacht from the coastguard station is 225°. The lighthouse is 81 km due west of the coastguard station.

> Q8a hint Draw the lighthouse and coastguard station first, then draw in the bearings to the yacht. The yacht's position will be where your two bearing lines intersect. Remember to use the scale carefully.

a Draw an accurate diagram of the lighthouse, the yacht and the coastguard station. Use a scale of 1 cm : 10 km.

b Find the bearing of the coastguard station from the yacht.

c Is the yacht closer to the coastguard station or the lighthouse? By how many km?

9 **Problem–solving** Town A is on a bearing of 026° from town C. The distance between town A and C is 9.2 miles and the distance between towns B and C is 14.6 miles.

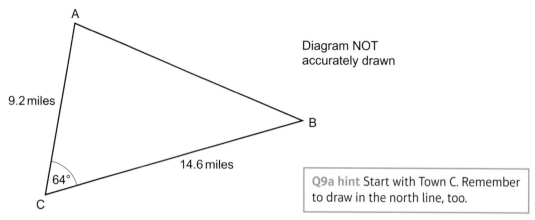

Diagram NOT accurately drawn

9.2 miles

A

64°

C

14.6 miles

B

> Q9a hint Start with Town C. Remember to draw in the north line, too.

a Draw an accurate diagram of the position of towns A, B and C.

Use your diagram to find:

b the bearing of town A from B.

c the distance of town A from B.

10 HMS Ocean (O) and HMS Albion (A) are on a disaster-relief mission.

The diagram shows the position of the ships and the harbour (H) where they need to deliver their supplies.

a What is the bearing of HMS Ocean from the harbour?

b What is the bearing of HMS Albion from HMS Ocean?

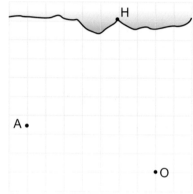

H

A •

• O

11 **Real / Problem-solving** This map shows the positions of helicopter search and rescue stations X and Y.

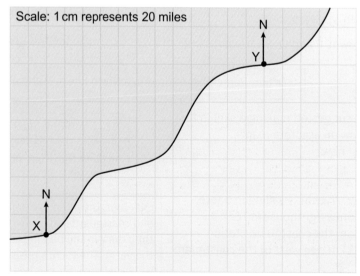

Scale: 1 cm represents 20 miles

N

Y

N

X

Boat B is in distress. An emergency signal goes out to both rescue stations.

Boat B is on a bearing of 093° from station X and a bearing of 155° from station Y.

Both stations respond to emergencies within a radius of 140 miles.

a Make an accurate scale drawing of the map shown above.

b Mark the exact position of B on your diagram using the bearings.

c Show the region covered by station X and the region covered by station Y on your diagram.

d Which station should respond to the call?

15 Problem-solving

Objective • Use arrow diagrams to help you solve problems.

Example 9

This is a scale drawing of a football pitch. In real life, the perimeter of the pitch is 320 m.
What is the scale of the drawing?

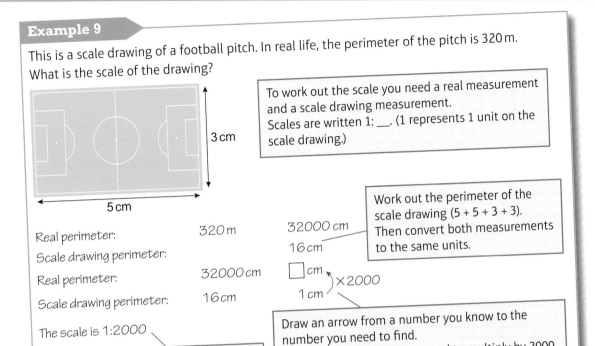

To work out the scale you need a real measurement and a scale drawing measurement.
Scales are written 1: ___. (1 represents 1 unit on the scale drawing.)

Work out the perimeter of the scale drawing (5 + 5 + 3 + 3).
Then convert both measurements to the same units.

Real perimeter:	320 m
Scale drawing perimeter:	16 cm
Real perimeter:	32000 cm
Scale drawing perimeter:	16 cm

The scale is 1:2000

Write the scale.

Draw an arrow from a number you know to the number you need to find.
32 000 ÷ 16 = 2000, so you need to multiply by 2000.

1   This is a 5 km cycle path through a park. What is the scale of the map?

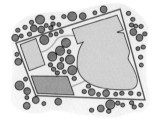

Q1 hint Measure the length of the red line. You will need to convert km to cm. First convert km to m, and then m to cm.

2   Car A travels 498 miles and uses 6 gallons of petrol. Car B travels 402 miles and uses 5 gallons of petrol. Which car is more cost effective to run? Show your working.

Q2 hint Draw an arrow diagram like this for each:
miles: 498   ?
gallons: 6   1   ) × □
Which car travels furthest on 1 gallon of petrol?

3   A box of chicken feed can be bought in two sizes. There is an offer on one of the sizes.

Which size is the best buy if you want to buy:
a   1 box   b   2 boxes?
Show your working.

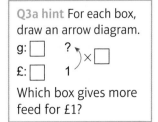

Q3a hint For each box, draw an arrow diagram.
g: □   ?
£: □   1   ) × □
Which box gives more feed for £1?

4   Here is the High Street exchange rate for £1.
     US Dollar $ 1.622
     Euro €   1.183
     How many Euros can you buy with 120 US Dollars?

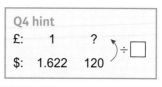

Q4 hint

£:	1	?
$:	1.622	120

÷ □

5   India's 'Golden Triangle' is a tourist route between
     Delhi, Agra and Jaipur. Delhi is at the top of the
     triangle, Agra bottom right and Jaipur bottom left.
     The distances between the cities are:

Q5 hint Decide on a sensible scale.
Use your scale to draw arrow diagrams
for each distance.
Convert km to cm and write your
scale 1: ___ beside your scale drawing.

Delhi		
180 km	Agra	
240 km	220 km	Jaipur

Construct a scale drawing of India's Golden Triangle.

6   A couple have action figures of themselves made. The action figures are in a ratio of 1 : 8.
     The male action figure is 24 cm tall. In real life, the female is 20 cm shorter than the male.
     How tall is the female action figure?

7   **Reflect** Lou says he often uses arrow diagrams when a
     problem includes a scale. Renee says she often uses arrow
     diagrams when a problem includes two different types of
     units. Are Lou and Renee's strategies sensible?

Q7 hint Look back at the questions
in this problem-solving section. Do
Lou and Renee's strategies work?

15 Check up

Log how you did on your
Student Progression Chart.

3D solids

1   Here is a triangular prism.

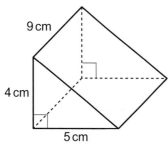

9 cm

4 cm

5 cm

a   How many vertices does this triangular prism have?
b   How many edges does it have?   c   How many faces does it have?

2   Here are the plan, front elevation and side elevation views of a solid shape drawn on 1 cm
     squared paper.

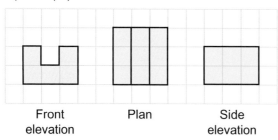

Front   Plan   Side
elevation   elevation

Sketch the shape. Label its lengths.

3   Sketch the prism in **Q1**. Draw in all the planes of symmetry.

Constructions

4   Draw a line 10 cm long. Use a ruler and compasses to construct its perpendicular bisector.

5   Draw a 110° angle using a ruler and protractor. Use a ruler and compasses to bisect the angle.

6   Make an accurate drawing of the .
    triangle PQR using a ruler and protractor.

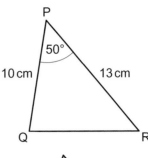

7   Construct this triangle using a ruler and compasses.

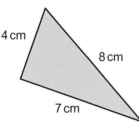

Loci and regions

8   a   Draw the locus of points 4 cm away from a fixed point.
    b   Shade the region less than 4 cm from the point.

9   Draw two points that are exactly 10 cm apart. Construct the locus of points that are equidistant from these two points.

10 Trace this diagram.
    a   Construct the locus of points which are equidistant from lines AB and BC.
    b   Shade in the region that is closer to line AB than AC.

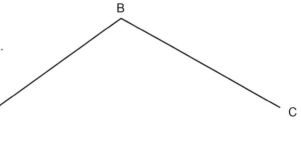

Scale drawings and bearings

11 This sketch shows a plan for a new youth club.
    a   Jim is making an accurate scale drawing using a scale of 2 cm to 1 m. How long will the basketball court be on the scale drawing?
    b   The café is 5 cm wide on the drawing. How wide is it in real life?

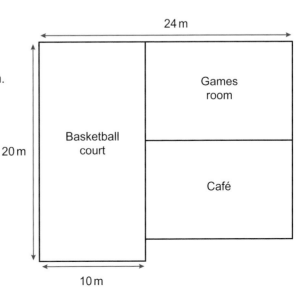

Reflect

12 A map has scale 1:50 000.
a What is the actual length of a road 6 cm long on the map?
b A footpath is 8 km long. How long is the footpath on the map?

13 a Draw point P in the middle of a new page.
b Draw point Q on a bearing of 245° from P.
c What is the bearing of P from Q?

14 Work out
a the bearing of P from Q
b the bearing of Q from P
c the bearing of R from Q.

15 How sure are you of your answers? Were you mostly

Just guessing 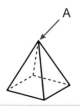   Feeling doubtful   Confident

What next? Use your results to decide whether to strengthen or extend your learning.

✱ Challenge

16 Matthew has some toy bricks. Each brick is a cube of side 1 cm.

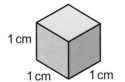

Diagram NOT
accurately drawn

He uses some of the bricks to make this solid shape.
a He rearranges the blocks to make a cuboid 1 cm deep.
   What are the dimensions of Matthew's cuboid?
Discussion How many different possible solutions
are there?
Cardel rearranges the same blocks to make a different
cuboid. None of the dimensions are 1 cm.
b What are the dimensions of Cardel's cuboid?
c On squared paper, draw the plan, side elevation and
   front elevation views of the original 3D object.
   (The front is shown by an arrow in the diagram above.)

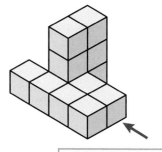

> **Q16a hint** 1 × 2 × 3 is
> the same as 2 × 3 × 1.

---

15 Strengthen

3D solids

1 Match each letter to one of the words in the cloud.

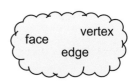

face   vertex   edge

*Active*Learn   Homework, practice and support: Foundation 15 Strengthen

2 Here are the front elevation, side elevation and plan views of a 3D solid.

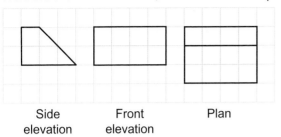

Side elevation   Front elevation   Plan

> **Q2a hint** Use the lengths from the 2D views to work out the dimensions of the 3D shape. In the sketch, make all vertical lines parallel. All horizontal lines going in the same direction should be parallel, too.

a Draw a sketch of the solid.

b Copy and complete these sentences.
The plan view shows the _____ of the object.
This has the same overall shape as the _____.

Constructions

1 Draw a line 10 cm long. Follow these instructions to construct the perpendicular bisector of your line.

> **Q1 strategy hint** Remember this diagram.

a Draw the line. Open your comasses to more than half the length of the line.

b Draw the first arc.

c Draw the second arc.

10 cm

> **Q1 hint** Check your answer by measuring to make sure the angle is 90° and that the line has been cut into two equal pieces.

2 Draw a line 12.5 cm long and construct its perpendicular bisector.

3 Draw a 50° angle. Follow these instructions to construct the bisector of the angle.

a Draw the angle. **b** Draw an arc.   **c** Draw the first arc between the two sides of the angle.

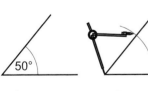

50°

d Draw the second arc.   **e** Draw the angle bisector.

4 Draw a 120° angle. Construct the bisector of the angle.

5   Follow these instructions to construct accurately a triangle with sides 5 cm, 6 cm and 7 cm.

a Use a ruler to draw accurately the longest side.   **b** Open your compasses to exactly 6 cm and draw an arc from one end of the line.   **c** Open your compasses to exactly 5 cm and draw an arc from the other end.   **d** Use the point where the arcs cross to create the triangle.

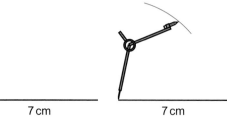
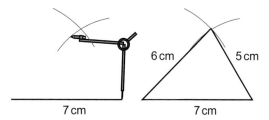

7 cm   7 cm   7 cm   7 cm

> **Q5 hint** Make the longest side the base. Turn the book round so the picture is the right way up.

6   Here is a sketch of a right-angled triangle. Construct an accurate diagram of the triangle.

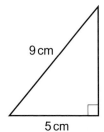

9 cm

5 cm

> **Q6 hint** Draw the 5 cm base, then the right angle. Use compasses for the 9 cm side.
>

7   Make an accurate construction of this triangle.

> **Q7 hint** Draw the angle first.

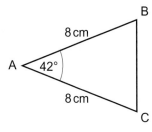

B

8 cm

A   42°

8 cm

C

Loci and regions

1   Copy this diagram. Then draw the locus of all points the same distance from line FG and line GH.

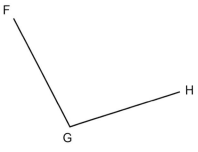

F

H

G

> **Q1 hint** Draw your diagram nice and large. Use counters or pencil marks to find points the same distance from each line. What shape do these points make?

2   Mark a point in the middle of your page and label it K. Draw an accurate diagram of the region that shows all points more than 5 cm but less than 8 cm from K. Shade in this region.

> **Q2 hint** First work out how to draw the locus of all points exactly 5 cm from K. Where will all the points more than 5 cm from K be?

Scale drawings and bearings

1   This sketch shows a plan for a new community centre.

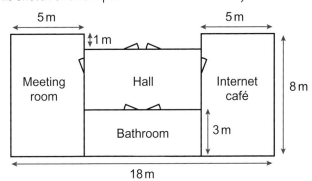

Elsie is making an accurate scale drawing of the community centre.
She uses a scale of 1 cm represents 2 m.

a   How long will the internet café be on
   Elsie's scale drawing?

b   What are the dimensions of the hall in
   real life?

c   Some furniture has been donated to the
   internet café. Elsie has been given eight
   tables that each measure 1.6 m × 2 m.
   Can she fit all eight tables in the café?

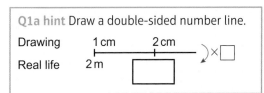

Discussion You could draw the internet café and the tables to the right scale to work this
problem out. Is there another way to work it out?

2   a   The length of a hike on a map is 20 cm.
      The map scale is 1 : 50 000.
      How many kilometres long is the
      actual hike?

   b   The actual distance of another hike
      is 4.5 km.
      How long would this be on the map?

   c   How many cm long would this distance
      be on a map if the
      scale with 1 : 100 000?

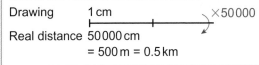

3   Write the bearing of B from A in each of these diagrams.

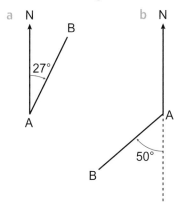

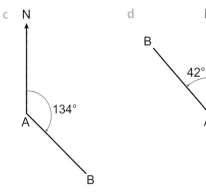

> **Q3 hint** Always measure bearings clockwise from north and write them as three-figure numbers.
> Bearing angles that are less than 100° need to use zeroes to fill in the missing digits.

4   Trace these points.   ◆Z

   a   What is the bearing of Y from Z?

> **Q4a hint** Join Y to Z. Draw a north line at point Z.

   b   Draw an accurate bearing of 210° from Z

   c   Draw an accurate bearing of 115° from Y.   Y◆

   d   Label the part where these two lines intersect, M.
       What is the bearing of point M from X?

                                                                  ◆X

5   The bearing of B from A = 040°.

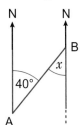

> **Q5 hint** Use alternate angles to find angle x.
> Now add this angle to 180° to find the bearing.

> **Q5 strategy hint** Remember to measure
> the bearing clockwise from the north line.

What is the bearing of A from B?

15 Extend

1   **Problem-solving** Here is a scale diagram of a barn.

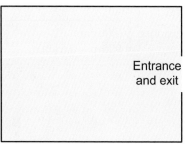

Scale: 1 : 250

A farmer wants to divide the barn into rows of rectangular bays to keep horses in.
Each horse bay should be 3 m long and 2 m wide.
There must be at least 2 m between rows of bays.
The farmer wants to have at least 9 bays in the barn.
Is this possible? Explain your answer.

2

Clare plans to hire a car and drive from New York to Connecticut.
The map below shows the roads between New York and Connecticut.
Using her ruler, Clare marks the map with the approximate route she
will drive.

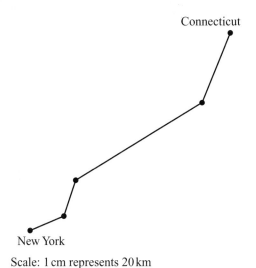

Scale: 1 cm represents 20 km

Clare hires the car in New York.
The car does 50 km per gallon of fuel. A full tank holds 10 gallons of fuel.
When she gets the car the fuel gauge shows the tank is half full.
Does the car have enough fuel in the tank for the journey
from New York to Connecticut? **(5 marks)**

May 2011, Q6, FSM02/01

Exam hint

Write what you
are doing with
each calculation.
Eg Total distance
= ... cm = ...km.
This distance will
use ... gallons.

3 **Real / Reasoning** This is a sketch of a disabled ramp 2.5 metres wide.

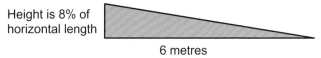

Height is 8% of
horizontal length

6 metres

a Draw an accurate scale diagram of the plan, front elevation and side elevation views of
the ramp.

b Sketch the 3D ramp.

4 Use a ruler and compasses to make an accurate
drawing of this isosceles trapezium.

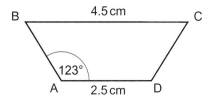

5 Draw an accurate net for this tetrahedron.

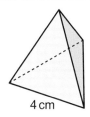

Q5 hint Use a ruler and compasses
to construct equilateral triangles
with side length 4 cm.

6 Follow these instructions to find the locus of points which are equidistant from a line **and** a point not on the line.
a Draw any line and a point P, which is not on the line.
b Draw a circle with a pair of compasses. The centre must be P and the radius a distance less than the distance between P and the line.
c Draw a line parallel to the first line and between the first line and P. Make sure the distance between this line and the first line is the same as the radius of your circle.
d Mark the points of intersection between your parallel line and the circle. These new points are on the locus you are trying to find.
e Rub out the circle and parallel line. Draw a different parallel line and circle to find another two points on the locus.
f Repeat the process until you have found enough points to sketch the locus.
g This locus is a very important curve. Can you find out the name of this curve?

7 **Reasoning / Problem-solving**
The towns of Finchfield and Greenborough each have a television transmitter. The transmitter at Greenborough (G) has a range of 35 miles and the one at Finchfield (F) has a range of 28 miles.
a What do the circles represent?
b Describe, in context, the shaded region bounded by these loci.
c How far away from Greenborough is Finchfield?

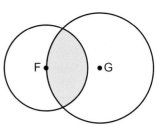

8 **Exam-style question**

Here is a map.   Burford   Scale: 1 cm represents 10 km
The map shows two
towns, Burford and
Hightown.

A company is going to
build a warehouse.

The warehouse will be
less than 30 km from
Burford and less than
50 km from Hightown.

On a copy of the map,   •
shade the region where   Hightown
the company can build the warehouse.   **(3 marks)**

Nov 2012, Q10, 1MA0/1H

Q8 hint First show the set of points 30 km from Burford and the set of points 50 km from Hightown.

Exam hint
Draw accurately with compasses and a sharp pencil. Then shade the required region.

9 **Problem-solving** The diagram shows a sketch of the position of towns D, E and F.

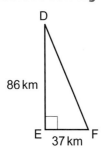

Q9 hint Draw the North line where you measure the bearing from.

Town D is 86 km North of town E. Town F is 37 km due East of town E.
Construct an accurate scale diagram showing positions D, E and F using protractor and ruler.
a What is the bearing of town D from town F?
b What is the distance in km from town D to town F?

10 **Real** This diagram shows the position
of ships M and P, and lighthouse L.

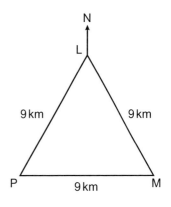

a Construct a scale diagram of the position of L, M and P
using a ruler and pair of compasses. Use a scale of 1 cm
to represent 1 km.

b Use your diagram to find the bearing of ship M and
ship P from the lighthouse.

---

15 Knowledge check

- ⊚ The flat surfaces of 3D shapes are called **faces**, the lines where two
faces meet are called **edges** and the corners at which the edges
meet are called **vertices** (the singluar of vertices is **vertex**). .... *Mastery lesson 15.1*
- ⊚ **Pyramids** have a base that can be any shape and sloping triangular
sides that meet at a point. In a **right prism**, the sides are
perpendicular (at right angles) to the base. .... *Mastery lesson 15.1*
- ⊚ A **plane** is a flat (2D) surface. A solid shape has a **plane of symmetry**
when a plane cuts the shape in half so that the part of the shape on
one side of the plane is an identical reflection of the part on the
other side of the plane. .... *Mastery lesson 15.2*
- ⊚ The **plan** is the view from above an object. The **front elevation** is
the view of the front of an object. The **side elevation** is the view of
the side of an object. .... *Mastery lesson 15.2*
- ⊚ You can draw an accurate diagram of a triangle with a ruler and
protractor if you know three details about the triangle. For example,
if you know the length of two sides and the size of one angle, or the
length of one side and the size of two angles. .... *Mastery lesson 15.3*
- ⊚ An **ASA** triangle has a given **A**ngle, a **S**ide length and another **A**ngle.
An **SAS** triangle is one where you are given two **S**ide lengths and the
Angle in between. In an **SSS** triangle, you are given all three **S**ide
lengths but none of the angles. .... *Mastery lesson 15.3*
- ⊚ Triangles with a right angle can be referred to as **RHS** triangles if you
are given the **R**ight angle, the **H**ypotenuse length and another **S**ide
length. The **hypotenuse** is the longest side of a right-angled triangle. *Mastery lesson 15.3*
- ⊚ To find the **scale factor** of a drawing make a ratio of length on the
drawing to length in real life. Simplify this so that the drawing side
of the ratio is 1. .... *Mastery lesson 15.4*
- ⊚ A **net** is a 2D shape that folds up to make a hollow 3D shape. .... *Mastery lesson 15.5*
- ⊚ **Constructions** are accurate drawings made using a ruler and pair of
compasses. **Bisect a line** means to cut a line exactly in half. A
perpendicular bisector cuts a line in half at right angles. You can
also use a ruler and compasses to **construct** a perpendicular from a
point to a line. An **angle bisector** cuts an angle exactly in half. .... *Mastery lesson 15.6*
- ⊚ A **locus** is a set of all points that obey a given rule. This produces a
path followed by the points. The plural of locus is **loci**. .... *Mastery lesson 15.7*
- ⊚ The locus of all points a given distance (d) from a fixed point (P) will
be a circle, with point P as its centre and d as its radius. .... *Mastery lesson 15.7*

⊙ A locus that is equidistant from two given points is the perpendicular bisector between the two points. .... *Mastery lesson 15.7*

⊙ A locus of all points a given distance away from a line is two parallel lines a fixed distance from the given line and a semi-circle at each end of the line with a radius equalling the fixed distance. .... *Mastery lesson 15.7*

⊙ A locus that is equidistant from two intersecting lines is the angle bisector of the two lines. .... *Mastery lesson 15.7*

⊙ A **region** is an area bounded (or surrounded) by limiting lines. These lines may be circles or straight lines and can be described as distances from particular points. .... *Mastery lesson 15.7*

⊙ A **bearing** is an angle measured in degrees clockwise from north. A bearing is always written using three digits. .... *Mastery lesson 15.8*

In this unit you have done a lot of drawing. Write down at least three things to remember when doing drawings in mathematics.
Compare your list with a classmate. What else can you add to your list?

15 Unit test

Log how you did on your Student Progression Chart.

1 Here is a sketch of a shed in the Roberts' family garden.
   a Trace the sketch.
     Draw in any planes of symmetry. *(1 mark)*
   b How many vertices does the shed have? *(1 mark)*
   c On cm squared paper, draw a side view and plan for the shed using a scale of 1 cm to represent 1 m. *(2 marks)*

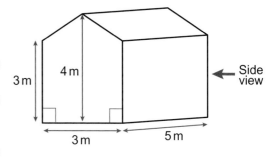

2 Here is a plan of a drama club, using a scale of 1 cm to represent 2 metres.
   Work out the real dimensions of the stage. *(2 marks)*

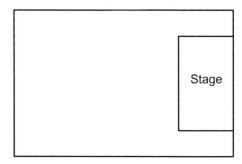

3 Draw a line segment 8 cm long. Construct the locus of points which are 2 cm away from the line segment. *(2 marks)*

4 a Draw two points, G and H, 7 cm apart. *(1 mark)*
   b Construct the locus of all points equidistant from G and H on your diagram. *(1 mark)*

5 a Construct an accurate drawing of this triangle. *(2 marks)*
   b How long is side EG? *(1 mark)*

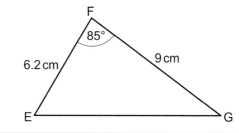

6   Construct the triangle STU with sides ST = 5.3 cm, TU = 6.7 cm and SU = 8.4 cm.   *(2 marks)*

7   Point P is 6 cm along an 11 cm line.

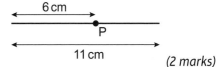

a   Copy the diagram.

b   Construct the perpendicular through the point P.

(2 marks)

8   Draw a 10 cm line and a point P in any position which does not lie on the line.
    Construct the perpendicular from P to the line.   *(2 marks)*

9   **Exam-style question**

PQRS is a square.

Shade the set of points inside
the square which are **both**
more than 3 centimetres from
the point *P* **and** less than
2 centimetres from the line *PS*.

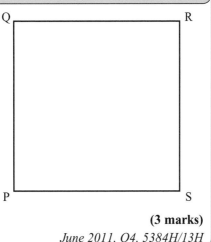

(3 marks)

June 2011, Q4, 5384H/13H

10   Write the bearing of

a   A from B

b   B from A.

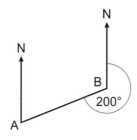

(3 marks)

11   **Reasoning / Communication** Aircraft B and C are flying into airport A.
     Both planes are flying directly towards the airport.
     B is 216 km from the airport on a bearing of 242° from A.
     C is 300 km from the airport on a bearing of 345° from B.

a   Draw a diagram showing A, B and C. Use a scale of 1 cm = 20 km.   *(2 marks)*

b   Aircraft X has recently taken off from the airport. X is 90 km from A on a bearing
    of 038° from B. Draw in the position of aircraft X on your diagram.   *(1 mark)*

c   X is on a bearing of 297° from A. Why is this not a good course for aircraft X to take?
    Suggest an alternative bearing and explain your choice.   *(2 marks)*

12   The diagram shows the flight of a plane from A to B to C.
     Copy and complete the distances and directions:

From A, fly ___ km _____

From B, fly___ km _____

From C, fly back to A on a bearing of ___°   *(3 marks)*

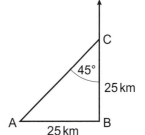

13 **Problem-solving / Real** A plane flies 100 km due West from an airfield.
Then it flies on a bearing of 030° until it is North of the airfield.
a Sketch the plane's flight path. *(2 marks)*
b In what direction does the plane need to fly to return to the airfield? *(1 mark)*
c How far is it back to the airfield? *(2 marks)*

14 **Real** A ship starts at A and sails 50 km South then 30 km East.
Then it sails back to its starting point.

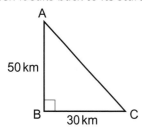

a Copy the diagram. Draw a North line at C. *(1 mark)*
b Use $\tan \theta = \dfrac{\text{opp}}{\text{adj}}$ to calculate angle ACB. *(2 marks)*
c Work out the bearing the ship sails on to A from C. *(1 mark)*
d Use Pythagoras' theorem to calculate the distance AC. *(2 marks)*

Sample student answer

Describe four things that are wrong with this answer.

Exam-style question

Here is a scale drawing of a rectangular garden *ABCD*.

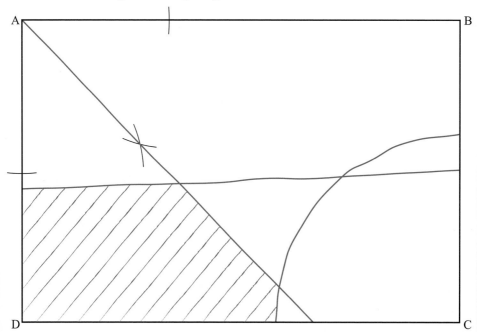

Scale: 1 cm represents 1 metre

Jane wants to plant a tree in the garden at least 5 m from point *C*, nearer to *AB* than to
AD and less than 3 m from *DC*.
On the diagram, shade the region where Jane can plant the tree. **(4 marks)**

February 2013, Q21, 1MA0/1F

16 QUADRATIC EQUATIONS AND GRAPHS

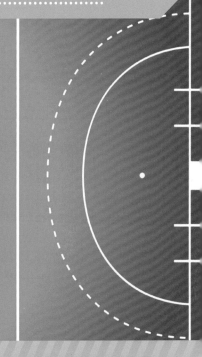

The width of this hockey pitch is x m.

The length is $(x + 36)$ m.

Work out an expression for the area by multiplying the expressions. Multiply each term in the brackets by the width.

16 Prior knowledge check

Numerical fluency

1 Square

   a   3   b   –7

2 Work out

   a $\sqrt{49}$ b $\sqrt{16}$

3 Work out

   a $-6 + 10$ b $-5 - 3$ c $2 + (-3)$

   d $-9 + (-4)$ e 3×-4 f -7×-3

   g $\dfrac{16}{-8}$ h $\dfrac{-9}{-3}$

4 a   i   Write all the positive and negative factor pairs of 9.

        ii   Use your answers to part **i** to write a pair of numbers with sum –6 and product 9.

   b   i   Write all the positive and negative factor pairs of –4.

        ii   Use your answers to part **i** to write a pair of numbers with sum 3 and product –4.

Algebraic fluency

5 Simplify

   a $3x + 5x$ b $9x - 10x$

   c $3x^2 + 7x + 2x - 2x^2$ d $3 + 8x + 4x^2 - x$

   e $7 \times x$ f $x \times -5$

   g $x \times x$ h $\dfrac{16x}{x}$ i $\dfrac{-18x}{2x}$

6 When $a = 4$ and $b = -2$, work out

   a a^2 b $a - b$

   c $b^2 + 2a$ d $a^2 + a + 1$

7 Expand

   a $2(x + 3)$ b $5(a - 2)$

   c $x(x + 4)$ d $d(d - 9)$

8 Factorise

   a $10t + 5$ b $12h^2 + h$

   c $14y - 7$ d $cd - d$

9 Solve

   a $x + 1 = 3$ b $x - 2 = 0$

   c $3x = 21$ d $2x + 3 = 15$

Geometrical fluency

10 Work out the area and perimeter of this shape.

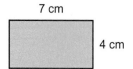
7 cm
4 cm

11 How many lines of symmetry do these shapes have?

a   equilateral triangle
b   regular hexagon

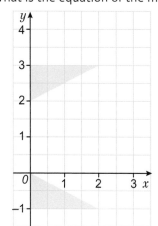

12 What is the equation of the mirror line?

Graphical fluency

13 Copy and complete the table and plot the graph of $y = x + 1$. Label your graph.

x	−2	−1	0	1	2
y		0			

14 a   What is the y-intercept of a function?
   b   Write the y-intercepts of the functions
       i $y = x + 3$
       ii $y = x - 5$
       iii $y = 7x$
       iv $y = 4$

* Challenge

15 Multiply the linked terms.

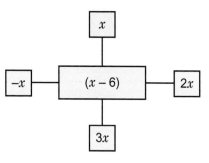

16.1 Expanding double brackets

Objectives

- Multiply double brackets.
- Recognise quadratic expressions.
- Square single brackets.

Why learn this?

Expanding two brackets is a skill needed for graphing and analysing quadratic functions.

Fluency

What is the area of each shape?

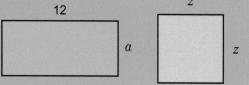
12
a
z
z

1   Simplify
    a $-9 \times a$
    b $3b \times -4$
    c $z \times -3z$
    d $8a + 10a$
    e $-3y + 11y$
    f $7t - 9t - t$
    g $m^2 + 2m + 7m + 4$
    h $y^2 - 9y - 11y - 3$

2   Expand
    a $2(a + 2)$
    b $3(x + 4)$
    c $m(m + 8)$
    d $b(2b - 5)$

Questions in this unit are targeted at the steps indicated.

3   Write an expression for the area of each rectangle.

a 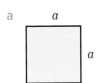   b 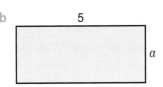   c   d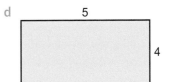

e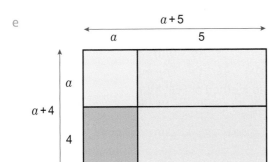

Q3e hint $(a + 5)(\square + \square)$

Discussion Look at rectangles **a** to **d** and then **e**. What do you notice?

4   **Reasoning**

i   Write an expression for the area of each small rectangle.

ii Write an expression for the area of the large rectangle. Collect like terms and simplify.

Q4i hint Look at **Q3a** to **d** above.

Q4ii hint Look at **Q3e** above.

a 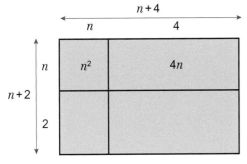   b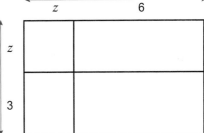

c 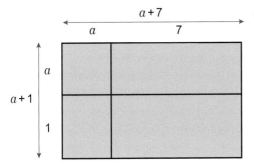   d

| Key point 1 |
To expand or multiply double brackets, multiply each term in one bracket by each term in the other bracket.

481

Example 1

Expand and simplify $(x + 2)(x + 6)$.

Grid method

$(x + 2)(x + 6)$

×	x	+2
x	x^2	+2x
+6	+6x	+12

$= x^2 + 2x + 6x + 12$

$= x^2 + 8x + 12$

FOIL: Firsts, Outers, Inners, Lasts

$(x + 2)(x + 6) = x^2 + 6x + 2x + 12$

$= x^2 + 8x + 12$

5 Expand and simplify

a $(x + 1)(x + 2)$ b $(t + 3)(t + 4)$ c $(q + 6)(q + 9)$

d $(z + 12)(z + 1)$ e $(m + 8)(m + 11)$ f $(y + 7)(y + 10)$

Reflect Which method did you use: the grid method or FOIL? Why?

6 Expand and simplify

a $(z + 1)(z - 2)$ b $(m - 5)(m + 6)$ c $(a + 4)(a - 9)$

d $(n - 10)(n + 7)$ e $(x - 2)(x - 3)$ f $(y - 6)(y - 1)$

> **Q6a hint**
> $1 \times -2 = \boxed{}$
> $1 \times z = \boxed{}$

7 What is the missing term?

$(x + 3)(x + \boxed{}) = x^2 + 8x + 15$

8 **Reasoning** Isabella says that the answer to $(x + 3)(x - 7)$ is the same as $(x - 7)(x + 3)$.
Is she correct? Give reasons for your answer.

9 **Reasoning** Rex expands and simplifies $(a - 4)(a - 7)$.
He says that the answer is $a^2 + 11a - 28$.
Is Rex correct? Give reasons for your answer.

10 **Exam-style question**

   a Expand $3(2 + t)$ **(1 mark)**

   b Expand $3x(2x + 5)$ **(2 marks)**

   c Expand and simplify $(m + 3)(m + 10)$ **(2 marks)**

                     June 2013, Q4, 1MA0/1H

> **Exam hint**
> In parts **a** and **b** make sure you multiply *each* term in the bracket by the term outside the bracket.

Key point 2

Expanding double brackets often gives a quadratic expression.
A quadratic expression always has a squared term
(with a power of 2). It cannot have a power higher than 2.
It may have a term with a power of 1 that is the same letter
as the squared term.
It may also have a constant (number) term.

$x^2 + 8x + 10$

squared term   term with power 1   constant term

11 **Reasoning** Which of these are quadratic expressions? Give reasons for your answers.

a $x^2 + 3x - 2$ b $x^3 + x^2 - 3$ c $y^2 + 9y$

d $5 - 2x$ e $16 + 2z - z^2$ f m^2

Key point 3

To square a single bracket, multiply it by itself, then expand and simplify.
$(x + 1)^2 = (x + 1)(x + 1)$

12 Expand and simplify

  a $(x + 2)^2$ b $(a + 5)^2$ c $(y - 9)^2$ d $(m - 4)^2$

> **Q12c hint**
> $-9 \times -9 = \square$

13 Square these expressions. Simplify your answers.

  a $x + 6$ b $n + 12$ c $q - 4$ d $t - 10$

> **Q13a hint** The square of $x + 6$ is $(x + 6)^2$

14 Match the cards that have equivalent expressions.

A: $(x + 1)^2$	B: $(x + 7)^2$	C: $(x - 5)^2$	D: $(x - 1)^2$
E: $x^2 - 10x + 25$	F: $x^2 - 2x + 1$	G: $49 + 14x + x^2$	H: $x^2 + 2x + 1$

15 **Exam-style question**

The length of a side of a square paving slab is $x + 4$.

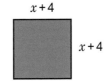

$x + 4$

$x + 4$

> **Exam hint**
> Expand and simplify all expressions. Show your working clearly.

  a Work out an expression for the area of the paving slab. Simplify your answer. **(2 marks)**

  b Ten paving slabs are used to make a patio. Write an expression for the total area of the patio. Simplify your answer. **(1 mark)**

> **Q15b strategy hint**
> Draw a picture.

16 **Problem-solving** The length of a rectangle is $n + 6$ and its width is $n + 3$.
Show that the area of the rectangle is $n^2 + 9n + 18$.

> **Q16 strategy hint**
> 'Show that …' means 'Show your working.'

17 **Reasoning** Luke expands and simplifies $(x + 4)(x - 4)$.
He says that the answer is $x^2 - 16$ and this is a quadratic expression.
Is he correct? What do you notice about the two terms in the answer?

16.2 Plotting quadratic graphs

Objectives

- Plot graphs of quadratic functions.
- Recognise a quadratic function.
- Use quadratic graphs to solve problems.

Why learn this?

A quadratic function will produce a curve that is called a parabola. The reflectors in headlights are 3D parabolas.

Fluency

Work out a 2^2 b 4^2 c 1^2 d $x \times x$

1 What is the equation of the mirror line?

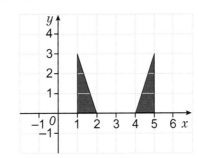

Warm up

2 $x = -2$, $m = 4$ and $t = 5$. Evaluate

    a x^2 **b** $x^2 + x$ **c** $t^2 + 2t$ **d** $m^2 + m - t$

3 **a** Copy and complete the table of values for the function $y = x + 2$

x	−3	−2	−1	0	1	2	3
y							

> **Q3 communication hint** A function describes the relationship between variables. For each input value there is an output value.

    b Plot a graph of the function $y = x + 2$

    c Draw the line $y = 1$ on your graph. What are the coordinates of the point where it crosses $y = x + 2$?

4 **a** Copy and complete the table of values for $y = x^2$.

x	−4	−3	−2	−1	0	1	2	3	4
y									

> **Q4a hint** Work out each y value by substituting each x value into the equation, e.g. when $x = -3$, then $y = x^2 = (\square)^2 = \square$

    b Plot the graph of $y = x^2$. Join the points with a smooth curve. Label your graph.

5 **a** Copy and complete the table of values for $y = x^2 - 3$

x	−4	−3	−2	−1	0	1	2
x^2	16				0		
−3	−3				−3		
y	13				−3		

> **Q5a hint** When quadratic functions have more than one step, include a row for each step in the table.

    b Plot the graph of $y = x^2 - 3$. Join the points with a smooth curve. Label your graph.

    Discussion What do you think the graph of $y = x^2 + 5$ would look like?

6 **Reasoning** **a** Copy and complete the table of values for $y = x^2 + 2x - 4$

    b Plot the function $y = x^2 + 2x - 4$

    c What do you notice about the y values in the table?

x	−4	−3	−2	−1	0	1	2
x^2		9					
$+2x$		−6					+4
−4		−4		−4			
y		−1			−1		

Key point 4

The turning point of the curve is where it turns in the opposite direction.

Example 2

For the graph of $y = x^2 - 6x + 5$ write down

a the equation of the line of symmetry

b the turning point

c the coordinates of the y-intercept.

> The y-intercept is the point where the graph crosses the y-axis.

> Sketch in the **line of symmetry**. Write its equation.

> Write down the coordinates of the point where the curve turns.

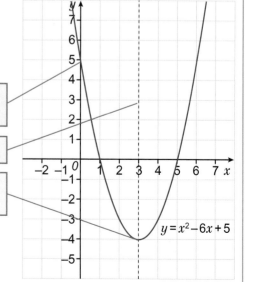

$y = x^2 - 6x + 5$

a line of symmetry $x = 3$

b turning point $(3, -4)$

c y-intercept $(0, 5)$

7   For each graph, write down
    i the equation of the line of symmetry   ii the turning point   iii the y-intercept.

a

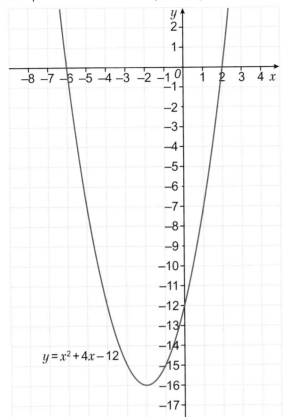

$y = x^2 + 4x - 12$

b

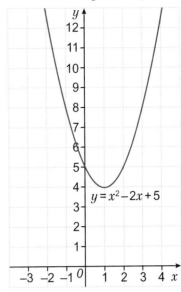

$y = x^2 - 2x + 5$

8   Copy and complete the table and plot the function $y = -x^2$

x	−3	−2	−1	0	1	2	3
y	−9				−1		

Discussion What is the same about the graphs of $y = x^2$ and $y = -x^2$. What is different?

Key point 5

A quadratic function has symmetrical U-shaped curve called a **parabola**.
A quadratic function with a $-x^2$ term has a symmetrical ∩-shaped curve.
The curve always has a minimum or maximum turning point.

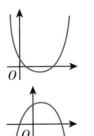

9   **Reasoning** Which of these are graphs of quadratic functions?

a

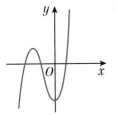

b

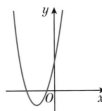

c

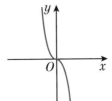

d

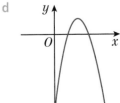

10 **Reasoning / Problem-solving**
The graph shows the height of a baseball against time.

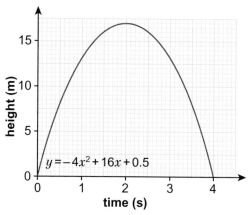

a i What is the maximum height that the baseball reaches?
ii At how many seconds does it reach its maximum height?
b At what times is the ball 12.5 m above the ground?
c When does the ball hit the ground?

$y = -4x^2 + 16x + 0.5$

11 **Real / Reasoning** The graph $y = x^2$ can be used to work out the area of a type of tile, where the y-axis represents area in cm^2, and the x-axis represents side length in cm.
a Use your graph of $y = x^2$ from **Q4** to work out area of a tile with side length
i 1 cm ii 4 cm iii 2.5 cm.
b What shape is the tile? Why?

12 **Problem-solving**
A toy rocket is fired from the ground. The table shows the height, h, of the rocket and the time, t, that it takes to travel.

Time, t (seconds)	0	1	2	3	4
Height, h (metres)	0	38	72	100	125

a Plot a graph of the height of the rocket against time.
b From your graph, estimate how long the rocket takes to reach 20 metres.
c From your graph, estimate the height of the rocket at 2.5 seconds.

> **Q12a hint** Put t on the horizontal axis and h on the vertical axis.
> Draw a smooth curve through the points.

16.3 Using quadratic graphs

Objectives

- Solve quadratic equations $ax^2 + bx + c = 0$ using a graph.
- Solve quadratic equations $ax^2 + bx + c = k$ using a graph.

Did you know?

Later in this unit you will learn to solve quadratic equations using algebra.

Fluency

- What is the origin on a graph?
- What is the line $y = 0$ known as? Why?

1 a Copy and complete the table of values for $y = x^2 - x - 2$.
b Plot the graph of $y = x^2 - x - 2$.

x	−3	−2	−1	0	1	2
y						

Warm up

Key point 6

To solve the equation $ax^2 + bx + c = 0$, read the x-coordinates where the graph crosses the x-axis. The values of x that satisfy the equation are called **roots**.

2 Look at your graph from **Q1**.

a Write down the coordinates of the points where the graph crosses the x-axis.

b Write down the roots of $x^2 - x - 2 = 0$

> **Q2b hint** Look at the x-coordinates when $y = 0$

3 Use the graphs to solve the equations

a $x^2 - x - 6 = 0$ b $x^2 + 3x - 10 = 0$ c $x^2 - 4x + 4 = 0$

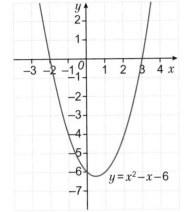

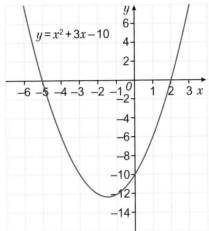

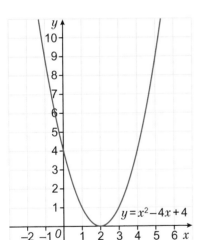

> **Q3c hint** $x^2 - 4x + 4$ only has one solution.

4 a On graph paper, draw the graph of $y = x^2 - 2x - 4$ from $x = -2$ to $x = 4$.

b From your graph, estimate the solutions to $x^2 - 2x - 4 = 0$.

> **Q4a hint** Draw a table of values.

5 **Exam-style question**

a Copy and complete the table of values for $y = x^2 - 5x + 3$

x	−1	0	1	2	3	4	5
y		3	−1		−3		

(2 marks)

b Draw the graph of $y = x^2 - 5x + 3$ for values of x from $x = -1$ to $x = 5$

(2 marks)

c Find estimates of the solutions of the equation $x^2 - 5x + 3 = 0$

(2 marks)

Nov 2013, Q9, 5MB3H/01

Exam hint

Make sure the bottom of the graph is a smooth curve and not a straight line. 'Find estimates of the solutions' means read the values of the solutions from the graph as accurately as possible.

Example 3

Solve the equation $x^2 - 2x - 2 = 6$

This means $y = 6$

Plot the function $y = x^2 - 2x - 2$
Draw the line $y = 6$ on the graph.

The solutions to the equation are $x = -2$ and $x = 4$

Write down the x-**coordinates** where the line and curve cross.

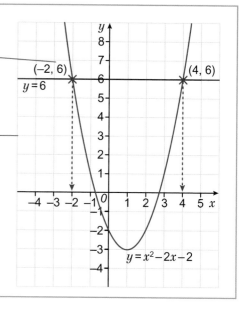

6   Use these graphs to solve the equation $x^2 + 3 = 4$

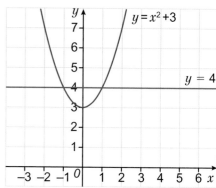

7   Use these graphs to solve the equation
$x^2 - 2x - 8 = -5$

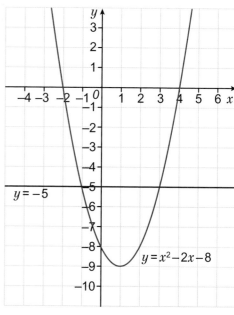

8   a   On graph paper draw the graph of $y = x^2 - 2x - 5$ for values of x between -2 and 4.
    b   Use the graph to estimate the solutions to $x^2 - 2x - 5 = 0$.
    c   Use the graph to estimate the solutions to $x^2 - 2x - 5 = 3$.

Q8c hint
Draw the line $y = \square$

9 Use this graph to solve the equations
   a $-x^2 + 3x = 0$
   b $-x^2 + 3x = -2$

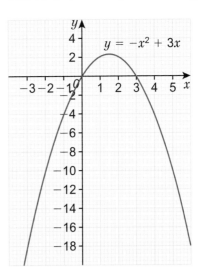

$y = -x^2 + 3x$

10 **Exam-style question**

a Copy and complete the table of values for $y = x^2 - 2x$

x	-2	-1	0	1	2	3	4
y		3	0			3	

   (2 marks)

b Draw the graph of $y = x^2 - 2x$ for values of x from $x = -2$ to 4. **(2 marks)**

c Solve $x^2 - 2x - 2 = 1$ **(2 marks)**

June 2013, Q15, 1MA0/2H

Exam hint
Plot the points with crosses. The graph should be symmetrical and the points joined with a smooth curve.

Q10c strategy hint Rearrange $x^2 - 2x - 2 = 1$ so that one side of the equation is the same as $x^2 - 2x$, then use your graph.

11 **Reasoning** a Copy and complete the table of values and plot the function $y = x^2 - 2x - 1$

x	-2	-1	0	1	2	3	4
y							

b Use your graph to solve the equation $x^2 - 2x - 1 = 2$

c Why are there no solutions to the equation $x^2 - 2x - 1 = -3$?

12 Write down the y-intercept of the function in **Q8**.

16.4 Factorising quadratic expressions

Objective
- Factorise quadratic expressions.

Why learn this?
Factorising an expression helps you solve the equation algebraically.

Fluency

What are the factor pairs of
a -1 b -2 c -3?

1   Expand
   a $(x + 6)(x + 1)$ b $(x - 5)(x + 8)$ c $(x - 2)(x - 1)$ d $(x + 3)(x - 3)$

2   Write the negative factor pairs for
   a   2   b   4   c   6

3   Write down a pair of numbers
   a   whose product is 8 and whose sum is 6
   b   whose product is 12 and whose sum is 7
   c   whose product is –6 and whose sum is 1
   d   whose product is –10 and whose sum is –9.

> **Q3 communication hint**
> 'Product' means multiply.
> The product of 2 and 3 is 6.

4   Copy and complete
   a $(\square + 2)(x + 4) = x^2 + 6x + 8$
   b $(x + 1)(x + \square) = x^2 + 8x + 7$
   c $(x - \square)(x + 3) = x^2 - 2x - 15$
   d $(\square - 7)(x - \square) = x^2 - 11x + 28$

> **Q4 hint** Use the grid method and write in the terms you are given to help you work out what is missing.
>
> | | | +2 |
> |-----|-------|-----|
> | x | x^2 | |
> | +4 | | 8 |

Example 4

Factorise $x^2 + 5x + 6$

> Work out all the factor pairs of 6, the number term.

> Write a pair of brackets with x in each one. This gives the x^2 term when multiplied.

$$x^2 + 5x + 6$$

$$(x \quad)(x \quad)$$

$1 \times 6 \qquad 2 \times 3$

> Work out which factor pair will **add** to give 5, the number in the x term.

> Then write one of the numbers in each of the brackets with x

$1 + 6 = 7 \qquad 2 + 3 = 5$

$(x + 2)(x + 3)$

> The expression is now factorised. Expand the brackets to check it is correct.

Check: $(x + 2)(x + 3) = x^2 + 5x + 6$

5   Factorise
   a $x^2 + 7x + 12$ b $x^2 + 6x + 8$ c $x^2 + 4x + 3$ d $x^2 + 7x + 10$

6   Factorise
   a $x^2 + x - 6$
   b $x^2 + 3x - 10$
   c $x^2 - 2x - 3$
   d $x^2 - x - 20$
   e $x^2 - 5x + 6$
   f $x^2 - 6x + 8$

> **Q6a hint** Which factor pair of –6 adds to give 1?
>
>
>
> $-\square \times \square = -6 \quad -\square + \square = 1$

> **Q6c hint** Which factor pair of –3 adds to give –2?
>
>
>
> $-\square \times \square = -3 \quad -\square + \square = -2$

7   **Exam-style question**

   Factorise $x^2 + 3x - 10$ **(2 marks)**

   June 2012, Q16b, 1MA0/1H

> **Exam hint**
> Expand your answer to check it is correct.

8   **Real / Problem-solving** An expression for the area of a square mat in cm² is $y^2 + 18y + 81$.
   a   What is the length of a side of the mat?   b   Work out the area when $y = 3$

9   Factorise
   a $x^2 + 11x + 30$ b $x^2 + 5x - 14$ c $x^2 - 3x - 4$

10 **Problem-solving** An expression for the area of a rectangle is $x^2 + 10x - 24$. Work out expressions for the length and height of the rectangle.

h

l

Q10 hint Factorising the expression will give the terms that multiply to give the expression.

11 Expand and simplify
   a $(x + 2)(x - 2)$ b $(x + 3)(x - 3)$ c $(x + 4)(x - 4)$ d $(x + 10)(x - 10)$
   Discussion What do you notice about the x term in these quadratics?

Key point 7

The **difference of two squares** is a quadratic expression with two squared terms, and one term is subtracted from the other. For example $x^2 - 25$

x^2 5^2

12 **Reasoning** Write three different expressions that are the difference of two squares using the terms in the box.

| x^2 | x | $- 4$ | $+10$ | -30 | -36 | $+9$ | $-y^2$ |

13 **Problem-solving** Josie cuts a small square of card from a larger square.
   a Write an expression for the remaining area of the large square.
   b What type of expression is your answer to **a**?

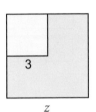

3

z

Q13a hint Work out the area of the large square and then subtract the area of the small square.

14 **Exam-style question**

   a Expand and simplify $(p + 9)(p - 4)$ **(2 marks)**
   b Factorise $x^2 - 49$ **(1 mark)**

   June 2012, Q14a and c, 1MA0/2H

Q14 strategy hint In part **a** write down all four terms in the expansion before simplifying.

16.5 Solving quadratic equations algebraically

Objectives

• Solve quadratic functions algebraically.

Why learn this?

You don't always need a graph to solve quadratic equations as they can be solved algebraically.

Fluency

• What are the square roots of **a** 9   **b** 49   **c** 100?
• What is   **a** 0×7 **b** 12×0 **c** 0×0?

1 Factorise
   a $x^2 + 7x + 10$ b $x^2 + x - 12$ c $x^2 - 9$

2 Solve
   a $d - 8 = 10$ b $x + 1 = 0$ c $p - 4 = 0$

Warm up

> **Key point 8**
>
> Solutions to quadratic equations can be found algebraically as well as from a graph.

3   Copy and complete to solve $x^2 - 9 = 0$
$$x^2 - 9 + \boxed{} = 0 + \boxed{}$$
$$x^2 = \boxed{}$$
$$x = \boxed{} \text{ or } x = -\boxed{}$$

> **Q3 hint** What number must you add to the left-hand side of the equation to get $x^2 = \boxed{}$?

> **Q3 hint** Square root both sides to find x.

4   Solve
   a $x^2 - 25 = 0$　　b $y^2 - 1 = 0$

5   Solve
   a $x^2 - 6 = 30$　　b $x^2 + 3 = 52$　　c $x^2 - 20 = 61$

> **Q5 hint** Rearrange so you only have x^2 on the left-hand side.

> **Example 5**
>
> Solve $x^2 + 2x - 15 = 0$
>
> $x^2 + 2x - 15 = 0$　　← Factorise the quadratic expression.
>
> $(x + 5)(x - 3) = 0$
>
> $x + 5 = 0 \text{ or } x - 3 = 0$　　← As the two expressions multiply to make 0, at least one of them must equal zero.
>
> $x = -5 \text{ or } x = 3$　　← Solve both equations.

6   Solve by factorising
   a $x^2 + 8x - 20 = 0$　　b $x^2 + 2x - 15 = 0$　　c $x^2 - 6x + 5 = 0$

7   Solve
   a $x^2 + 5x + 4 = 0$　　b $x^2 + 9x + 18 = 0$　　c $x^2 + x - 6 = 0$
   d $x^2 - 3x - 4 = 0$　　e $x^2 - 6x + 5 = 0$　　f $x^2 - 10x + 16 = 0$

8   **Reasoning** Match each equation to the correct solution.

A: $x = -5$ and $x = -7$	B: $x = -7$	C: $x = 1$ and $x = -1$	D: $x = -6$ and $x = 2$
E: $x^2 + 12x + 35 = 0$	F: $x^2 - 1 = 0$	G: $x^2 + 14x + 49 = 0$	H: $x^2 + 4x - 12 = 0$

9   > **Exam-style question**
>
> a   i   Factorise $x^2 - 12x + 27$
>   ii   Solve the equation $x^2 - 12x + 27 = 0$　　**(3 marks)**
> b   Factorise $y^2 - 100$　　**(1 mark)**
>
> *March 2013, Q20, 1MA0/1H*

> **Q9 strategy hint** In part **a** use your factorised answer from **i** to answer **ii**.

10   Solve by factorising
   a $x^2 - 36 = 0$　　b $z^2 - 81 = 0$
   c $p^2 - 64 = 0$　　d $d^2 - 121 = 0$

> **Q10a hint** $x^2 - 36 = x^2 - \boxed{}^2$
> $= (x + \boxed{})(x - \boxed{})$

Discussion How else can you solve $x^2 - 36 = 0$? Which method do you prefer?

11   **Problem-solving** Write a quadratic equation with solution $x = 3$ and $x = 4$

12   a   Plot the graph of $y = x^2 - 4$ for values of x from -3 to $+3$.
   b   Use your graph to solve the equation $x^2 - 4 = 0$.
   c   Solve $x^2 - 4 = 0$ by factorising. Show that the solutions match the solutions from your graph.

16 Problem-solving

Objective • Use graphs to help you solve problems.

Example 6

Earthquake survivors receive boxes of food, fitted with parachutes and dropped by planes. This table models the distance, in metres, a box falls each two seconds after the parachute opens.

Time (seconds)	0	2	4	6	8	10
Distance (m)	0	39	96	171	264	375

If the parachute opens at 215 m, after how many seconds will it hit the ground?

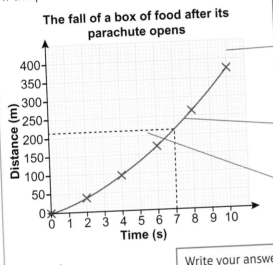

The fall of a box of food after its parachute opens

Draw the axes for your graph. The first row in the table is usually on the horizontal axis. Decide on a scale for each axis by looking at the smallest and largest number in each row. Give your graph a title.

Carefully plot the points in the table on your axes, then join the points with a curve.

Draw a line from 215 m on the distance axis to your graph.
Draw a line down to the time axis and read the time in seconds.

Write your answer.
Check your answer against the table. Where would 215 m appear in the table? Would its likely time be 7 seconds? Yes.

7 seconds

---

1 The table shows the approximate pressure a diver experiences at different depths in the sea.

Depth (metres)	10	15	35	50
Pressure (bar)	2	2.5	4.5	6

Q1 communication hint
A bar is a metric unit of pressure.

Approximately how much pressure does a diver experience at a depth of 45 m?

2 The shape of a bridge is modelled by the equation $y = -2x^2 + 4x + 3$, where x and y are measured in metres. How high is the highest point of the bridge?

Q2 hint When given an equation, decide what shape you think it will make. Sketch the shape.

Now draw and complete a table:

x	0	1	2
y			

Plot the graph.
Where is the highest point on the graph? Give your answer in metres.

3 A tourist records distances and fares for taxi rides around a city:

Distance (km)	1.5	2	3.5
Cost (£)	6.30	7.50	11.10

How much is a 5 km taxi fare?

4   Sally is considering renting a room in a house. For each advert, she notes the rent per month, and its distance from the city centre.

Rent	£550	£350	£470	£220	£300	£365
Distance (km)	0.5	2	1.2	3	2.5	2.1

Sally wants to pay rent of approximately £400 per month. Estimate how far she can expect to live from the city centre.

> **Q4 hint** Plot the points on axes. Do the points lie in a straight line or a curve? If yes, join the points. If no, then are you drawing a scatter graph? Can you draw a line of best fit?

5   The table models the path of a volleyball after it is hit.

Time (t) in seconds	0	0.25	0.5	0.75	1	1.25
Height (h) in metres	1	1.75	2.25	2.5	2.5	2.25

The net is at 2.4 m. At approximately what times does the ball clear the net?

6   **Reflect** What clue could you look for in a question that tells you a graph may be a good problem-solving strategy? What types of graphs might you draw?

16 Check up

Log how you did on your Student Progression Chart.

Quadratic graphs

1   Which of these are graphs of quadratic functions?

a 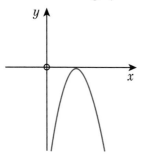   b 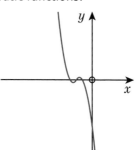   c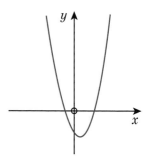

2   a   Copy and complete the table of values and plot graph of $y = x^2 - 6x$

x	−1	0	1	2	3	4	5	6	7
y									

b   Write the equation of the line of symmetry for your graph.
c   Write the coordinates of the turning point.
d   Write down the coordinates of the y-intercept.
e   Use your graph to write the solutions of $x^2 - 6x = 0$.

3   Here are the graphs of $y = x^2 + 5x + 3$ and $y = 7$.
a   Use the graphs to estimate the solutions to $x^2 + 5x + 3 = 0$
b   Use the graphs to estimate the solutions to $x^2 + 5x + 3 = 7$

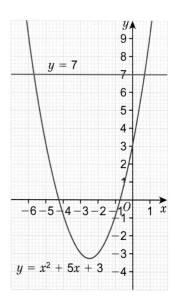

4   **Reasoning** The graph shows the height, h metres, of a stone thrown up in the air, at time t seconds.

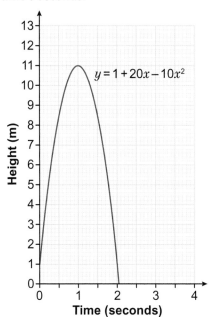

Use the graph.
a   Estimate the height of the stone at 0.5 seconds.
b   At how many seconds does the stone reach its maximum height?
c   At what times is the stone at a height of 1 m?
d   Estimate the number of seconds that the stone takes to hit the ground.

Quadratic equations

5   Expand and simplify
a $(t + 6)(t + 4)$ b $(f - 5)(f + 7)$ c $(a + 1)(a - 1)$ d $(n - 7)^2$

6   Factorise $x^2 - 49$.

7   Solve
a $x^2 - 9 = 0$ b $2x^2 - 72 = 0$

8   Factorise
a $x^2 + 15x + 56$ b $x^2 + 2x - 99$

9   Solve by factorising
a $x^2 + 13x + 30 = 0$ b $x^2 - 7x + 6 = 0$

10   How sure are you of your answers? Were you mostly

Just guessing   Feeling doubtful   Confident

What next? Use your results to decide whether to strengthen or extend your learning.

* Challenge

11   Write ten different quadratic expressions using combinations of the given solutions.
Solutions: $x = 3$ $x = -3$ $x = 2$ $x = -2$

16 Strengthen

Quadratic graphs

1 a Copy and complete the table of values for $y = x^2 + 1$

x	-3	-2	-1	0	1	2	3
x^2	9	4		0			
+1	+1	+1	+1	+1	+1	+1	+1
y	10		2		2		10

> **Q1a hint** Look at the first column:
> $x = -3$ so $x^2 = 9$
> Add 1 to get y: $y = 9 + 1 = 10$
> Do the same for the other columns.

b Plot the graph of the function.
Label your graph.

> **Q1b hint** Draw a set of axes that includes the lowest and highest values from all of your x- and y-coordinates. Plot the points and draw a smooth curve through them.
>
>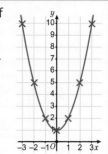

2 a Copy and complete the table of values for $y = x^2 + 2x + 3$

x	-4	-3	-2	-1	0	1	2
x^2	16		4	1			
+2x	-8	-6				2	
+3	+3			+3			
y	11	6			3		

b Plot the graph of the function. Label your graph.

3 a Copy and complete the table of values for $y = x^2 + x$

x	-3	-2	-1	0	1	2
x^2	9		1	0		
+x	-3	-2		0		2
y	6	2			2	

> **Q3b hint** This table of values doesn't show the lowest coordinate (turning point) on the curve. Draw the turning point by estimating where it will lie between the two lowest coordinates in your table.

b Plot the graph of the function.
Label your graph.

4 Name the labelled points on the graph of $y = ax^2 + bx + c$
Choose from:

y-intercept	line of symmetry	turning point	roots

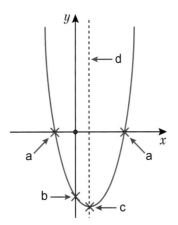

> **Q4 hint**
> a This is where the curve crosses the x-axis.
> b This is where the curve crosses the y-axis.
> c This is the minimum (or maximum) point of the curve.
> d This line divides the curve into two identical halves. It has the equation $x = $ 'a number'.

5   Use the graph to find
     a   the y-intercept
     b   the coordinates of the turning point
     c   the equation of the line of symmetry
     d   The solutions to $x^2 - 6x + 8 = 0$.

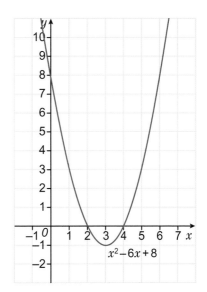

$x^2 - 6x + 8$

6   a   Copy and complete the table of values for $y = x^2 + 5x + 5$.

x	−5	−4	−3	−2	−1	0
x^2						
$+5x$						
$+5$						
y						

> **Q6ci hint** The line $y = 1$ is where all the y-coordinates are equal to 1.

> **Q6cii hint** The solutions to $x^2 + 5x + 5 = 1$ are the x-coordinates at the points where the curve $y = x^2 + 5x + 5$ and the line $y = 1$ cross.

     b   Plot the graph of the function. Label your graph.
     c   i   Draw the line $y = 1$ on your graph.
          ii   Find the solutions to $x^2 + 5x + 5 = 1$
     d   Estimate the roots of the function.

7   a   Copy and complete the table of values for $y = x^2 - 2x$.

x	−1	0	1	2	3	4	5
x^2	1						
$-2x$	2				−6		−10
y	3		−1				

     b   Plot the graph of the function. Label your graph.
     c   Use your graph to write the equation of the line of symmetry and the coordinates of the turning point of the function.
     d   Use your graph to write the solutions to $x^2 - 2x = 0$.

8   Which of these are graphs of quadratic functions?

a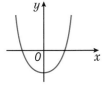
b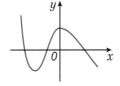

> **Q8 hint** Quadratic graphs are ∪ or ∩ shaped and symmetrical.

c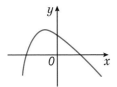
d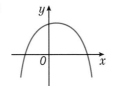

Quadratic equations

1 Emma uses this method to expand double brackets.
Draw in a smiley face to help you expand the brackets:
Copy and complete her calculation.

Eyebrow 1: $a \times a = \square$

Eyebrow 2: $2 \times 3 = \square$

Nose: $2 \times a = \square$

Smile: $a \times 3 = \square$

eyebrows

$(a + 2)$ $(a + 3)$

nose

smile

Eyebrow 1 + eyebrow 2 + nose + smile $= \square + \square + \square + \square = \square + \square + \square$

2 Expand and simplify

a $(a + 2)(a + 3)$ b $(t + 1)(t + 4)$

c $(x + 1)(x + 7)$ d $(y + 5)(y + 10)$

3 Expand and simplify

a $(z + 1)(z - 2)$ b $(f - 7)(f + 9)$ c $(x + 1)(x - 1)$

> **Q3a hint** Use the smiley face method.

4 Expand and simplify

a $(m + 4)^2$ b $(x + 8)^2$ c $(g - 10)^2$

> **Q4a hint** Rewrite as two brackets: $(m + 4)(m + 4)$

5 Solve

a $x^2 - 4 = 0$

b $a^2 - 9 = 0$

c $t^2 - 49 = 0$

> **Q5a hint** Rearrange to work out x.
> $x^2 - 4 + \square = 0 + \square$
> $x^2 = \square$
> $x = \sqrt{\square}$
> $x = \square$ or $x = -\square$

6 Match the expressions to their factorisations.

| $x^2 - 9$ | $(x - 2)(x + 2)$ | $(x + 4)(x - 4)$ | $x^2 - 49$ |

| $(x + 7)(x - 7)$ | $x^2 - 16$ | $(x - 3)(x + 3)$ | $x^2 - 4$ |

> **Q6 hint** Expand the brackets to check.

7 Factorise

a $x^2 - 16$ b $p^2 - 1$

c $y^2 - 81$ d $k^2 - 100$

8 Factorise

a $x^2 + 7x + 12$

b $x^2 + 8x + 12$

c $x^2 + 7x + 6$

> **Q8a hint** Look for the pair of factors of 12 that has a sum of 7.
>
>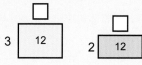

9 Factorise and solve

a $x^2 + 6x + 8 = 0$ b $x^2 + 6x + 5 = 0$

c $x^2 + 3x - 18 = 0$ d $x^2 - 7x + 12 = 0$

16 Extend

1 a Copy and complete the table and plot the function $y = -x^2 + 3x + 4$

x	-2	-1	0	1	2	3	4	5
$-x^2$	-4							
$+3x$	-6							
$+4$	+4							
y	-6							

b Use your graph.

 i Write down the y-intercept of the function.
 ii Write down the equation of the line of symmetry.
 iii Estimate the turning point of the function.
 iv Estimate the solutions to $-x^2 + 3x + 4 = 0$.

2

> **Exam-style question**
>
> **a** Factorise $6 + 9x$ **(1 mark)**
> **b** Factorise $y^2 - 16$ **(1 mark)**
>
> *June 2013, Q19a and b, 1MA0/2H*

3 **Real / Problem-solving** A hanging cable can be modelled by the equation $y = x^2 - 5x + 10$, where x is the horizontal distance in metres and y is the height in metres. The initial height of the cable is equal to its final height.

a Copy and complete the table of values and plot the function $y = x^2 - 5x + 10$

x	0	1	2	3	4	5
y						

b **i** What is the maximum height of the cable?
 ii Estimate the minimum height of the cable.
 iii Work out the horizontal distance that the cable spans.

4 **Real / Problem-solving** A length of 40 m of fencing is used to enclose a corner of a rectangular field. The area that the fencing can enclose can be modelled by the function $A = 20w - w^2$, where A is the area enclosed and w is the width of the enclosed area in metres. The intial height of the cable is equal to its final height.

a Copy and complete the table of values and plot the function $A = 20w - w^2$.

w	0	4	8	10	12	16	20
$20w$		80			240		
$-w^2$	-0	-16		-100		-256	
A		64					0

> **Q4a hint** Put w on the x-axis and A on the y-axis.

b Use your graph to find the maximum area that can be enclosed.
c Use your graph to find the width of the enclosure when the area is at its maximum.

> **Q4b hint** What is the highest y coordinate on the curve?

d Use your answers to parts **b** and **c** to work out the length of the enclosure when the area is at its maximum.

> **Q4d hint** width × ☐ = area

5 Copy and complete to solve $3x^2 - 12 = 0$

$3x^2 - 12 + \boxed{} = 0 + \boxed{}$
$3x^2 = \boxed{}$
$x^2 = \boxed{}$
$x = \boxed{}$ or $x = -\boxed{}$

6 Solve

a $4x^2 - 16 = 0$　　　b $2y^2 - 98 = 0$

c $4x^2 - 36 = 0$　　　d $6x^2 - 96 = 0$

7 A skateboard ramp can be modelled by the equation $y = x^2 - 2x + 1$, where x is the horizontal distance in metres and y is the height in metres. The intial height of the skateboard ramp is equal to its final height.

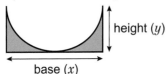

height (y)

base (x)

a Copy and complete the table of values and plot the function $y = x^2 - 2x + 1$

x	0	0.5	1	1.5	2
y					

b From the graph:

　i What is the maximum height of the ramp?

　ii How long is the base of the ramp?

　iii What is the minimum height of the ramp from the ground?

8 **Real / Problem-solving** A ball is thrown the air. After s, seconds, its height, h, in metres above ground, is given by the formula $h = -(4s + 1)(s - 2)$

a What is the height of the ball after 1 second?

b What is the height of the ball after 2 seconds?

c Use your answer to part **b** to say what has happened to the ball after 2 seconds.

> **Q9a hint** After 1 second means $s = 1$. Substitute into the equation to get h.

9 **Real / Problem-solving** The diagram shows a photo in a frame. The photo and frame are both square.

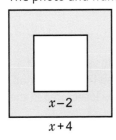

$x-2$

$x+4$

a Work out an expression for the area of the white photo.

b Using your answer to part **a**, work out and simplify an expression for the area of the blue frame surrounding the photo.

10 **Problem-solving** The area of the square and the area of the triangle are equal.

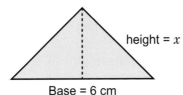

height = x

Base = x cm　　　　Base = 6 cm

a Write and simplify an equation in x to represent the areas of the shapes.

b Solve your equation.

c Work out the areas of the shapes.

> **Q10b hint** To solve, rearrange to = 0 and then factorise. One answer will be a sensible one, one will not!

11 Here is the graph of $y = 2x^2 - 3x - 7$.
    Use the graph to estimate the solutions to
    a $2x^2 - 3x - 7 = 0$
    b $2x^2 - 3x - 7 = -4$

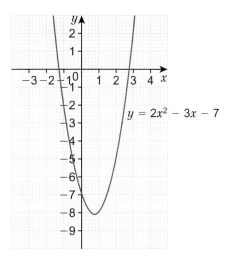

$y = 2x^2 - 3x - 7$

16 Knowledge check

⊙ To expand or multiply double brackets, multiply each term in one
  bracket by each term in the other bracket. .... *Mastery lesson 16.1*

⊙ To square a single bracket, multiply it by itself, then expand and
  simplify. $(x + 1)^2 = (x + 1)(x + 1)$ .... *Mastery lesson 16.1*

⊙ A quadratic expression always has a squared term (with a power of 2).
  It cannot have a power higher than 2. It may also have a term with a
  power of 1 that is the same letter as the squared term. It may also
  have a constant (number) term. $ax^2 + bx + c$ has a squared term, ax^2,
  a term with power 1, bx, and a constant term, c .... *Mastery lesson 16.1*

⊙ A quadratic function has symmetrical ∪-shaped curve called a **parabola**.
  A quadratic function with a $-x^2$ term has a symmetrical ∩-shaped
  curve. .... *Mastery lesson 16.2*

⊙ The curve always has a minimum or maximum turning point. .... *Mastery lesson 16.2*

⊙ To solve the equation $ax^2 + bx + c = 0$ using a graph, read the x-coordinates
  where the graph crosses the x-axis. These are called **roots**. .... *Mastery lesson 16.3*

⊙ To solve the equation $ax^2 + bx + c =$ 'a number' using a graph, read
  the x-coordinates where the graph crosses the line $y =$ 'a number'. .... *Mastery lesson 16.3*

⊙ To factorise a quadratic equation, $ax^2 + bx + c = 0$, you need to find
  two numbers whose product is c and whose sum is b. .... *Mastery lesson 16.4*

⊙ The **difference of two squares** is a quadratic expression with two
  squared terms, and one term is subtracted from the other.
  For example $x^2 - 25$ .... *Mastery lesson 16.4*

⊙ Solutions to quadratic equations can be found algebraically by
  factorising as well as from a graph. .... *Mastery lesson 16.5*

For this unit, copy and complete these sentences.
I showed I am good at ____.
I found ___ hard.
I got better at ____ by ___.
I was surprised by _____.
I was happy that ____.
I still need help with ____.

Reflect

16 Unit test

1 Which of these are quadratic functions?

a 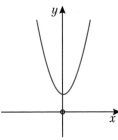 *(1 mark)*

b 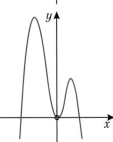 *(1 mark)*

c 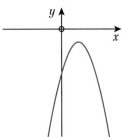 *(1 mark)*

d $(a + 1)(a + 3)$ *(1 mark)*

e $7(y - 2)$ *(1 mark)*

f $x(9x^2 + 3x)$ *(1 mark)*

2 Here is the graph of $y = x^2 - 5x$ *(4 marks)*

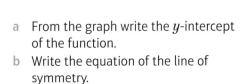

a From the graph write the y-intercept
of the function. *(1 mark)*

b Write the equation of the line of
symmetry. *(1 mark)*

c Estimate the coordinates of the
turning point of the function. *(1 mark)*

d Use the graph to solve the equation
$x^2 - 5x = 0$. *(2 marks)*

3 a Draw and complete the table of values and plot the function $y = x^2 + 7x + 6$ *(4 marks)*

x	−7	−6	−5	−4	−3	−2	−1	0
y								

b Use your graph to estimate the solutions to $x^2 + 7x + 6 = 3$ *(2 marks)*

4 Expand and simplify
   a $(y + 10)(y + 1)$ *(1 mark)*
   b $(z + 7)(z - 8)$ *(1 mark)*
   c $(x + 1)(x - 1)$ *(1 mark)*
   d $(n - 9)^2$ *(1 mark)*

5 Solve $a^2 - 25 = 0$ *(2 marks)*

6 Factorise $x^2 - 4x + 3$ *(2 marks)*

7 Solve $x^2 + 6x - 40 = 0$ *(2 marks)*

8 **Reasoning** An expression for the area of a rectangular tile is $x^2 - 10x + 24 \, \text{cm}^2$.
Write expressions for the length and width of the tile. *(3 marks)*

9 **Reasoning** The graph shows the height, h metres, of a practice missile fired from a ship, at time, t seconds.
From the graph:
   a Estimate the height of the missile at 10 seconds.
   b i What is the maximum height the missile reaches?
      ii At how many seconds is this height reached?

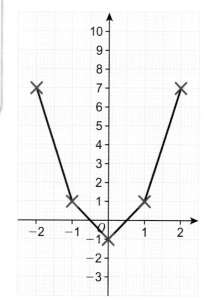

   c Estimate at what times the missile is at a height of 600 m.
   d How long does it take for the missile to fall from its maximum height into the sea? *(5 marks)*

Sample student answer

Suggest three ways to improve the graph in the student's answer.

> **Exam-style question**
>
> a Complete the table of values for $y = 2x^2 - 1$ **(2 marks)**
>
> | x | -2 | -1 | 0 | 1 | 2 |
> |---|---|---|---|---|---|
> | y | 7 | | | 1 | |
>
> b On the grid below, draw the graph of $y = 2x^2 - 1$
> for values of x from $x = -2$ to $x = 2$ **(2 marks)**
>
> *June 2012, Q11a and b, 5MB3H/01*

Student answer

x	-2	-1	0	1	2
y	7	1	-1	1	7

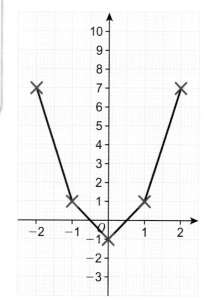

17 PERIMETER, AREA AND VOLUME 2

The crater of Mount Bromo in Indonesia is roughly in the shape of a circle. It has a diameter of about 800 metres.

What is the radius of the crater?

17 Prior knowledge check

Numerical fluency

1 Work out
   a 8^2 b 3^3 c $\sqrt{49}$

2 Round each number to the given accuracy.
   a 3.46 to 1 significant figure (1 s.f.)
   b 2.5347 to 3 s.f.
   c 1 538 926 to 1 s.f.

Fluency with measures

3 Copy and complete
   a 1 ml = ☐ cm³
   b 1000 ml = ☐ cm³
   c 1 litre = ☐ cm³

Algebraic fluency

4 Substitute into each formula to work out the unknown value.
   a $C = 3d$ Find C when $d = 5$.
   b $A = 3r^2$ Find A when $r = 2$.
   c $A = lw$ Find w when $l = 5$ and $A = 30$.
   d $A = \frac{1}{2}bh$ Find h when $A = 12$ and $b = 8$.

Geometrical fluency

5 Name each shape.
   a   b

   c   d

6 Sketch a circle. Draw and label
   a   the centre   b   the radius
   c   the diameter.

7 Work out the area of each shape.
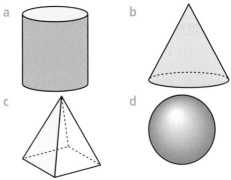
   a   b

8 Work out
   a the volume
   b the surface area of this cuboid.

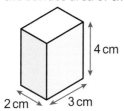

4 cm

2 cm   3 cm

9 Work out the volume of this prism.

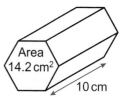

Area 14.2 cm²

10 cm

✱ Challenge

10 A sheet of A1 paper is 594 mm by 841 mm.
   a What is the largest cube that can be made from a sheet of A1 paper?
   b What assumptions have you made?

17.1 Circumference of a circle 1

Objectives

- Calculate the circumference of a circle.
- Solve problems involving the circumference of a circle.

Why learn this?

You need to know the circumference of a bike wheel to work out the distance travelled for each revolution of the wheel.

Fluency

- Round 3.7816 to 1 d.p. and to 2 d.p.
- Round 3487 to 3 s.f.

1 Make x the subject of $\frac{x}{5} = y$.

2 Work out $\frac{58.6}{14}$. Round your answer to 2 decimal places (2 d.p.).

Key point 1

The **circumference** is the **perimeter** of a circle.

> Questions in this unit are targeted at the steps indicated.

3 a The radius of the London Eye is 60 metres. Work out its diameter.
   b A Frisbee has a diameter of 10 inches. Work out its radius.
   c The diameter of a circle is d. Write a formula to give the radius r.
   d The radius of a circle is r. Write a formula to give the diameter d.
   e The length of a tennis court is 78 feet. 360 tennis balls fit touching each other along the length.
     What is the diameter of a tennis ball in inches?

   > Q3e hint
   > 1 foot = 12 inches

4 **Real / Reasoning** The table shows the diameter and circumference of some everyday objects.

Object	Diameter, d (cm)	Circumference, C (cm)
clock	25	78.6
CD	12	37.7
dartboard	45	141.4
bicycle wheel	65	204

   a Work out the value of $\frac{C}{d}$ for each object. Round your answers to 2 d.p.
   b **Reflect** What do you notice about the values of $\frac{C}{d}$?
   c Copy and complete the formula $C = \boxed{} \times d$.
   d The diameter of the centre circle on a football pitch is 18.3 metres.
     Use your formula to estimate the circumference of the centre circle correct to 1 d.p.

Warm up

> **Key point 2**
>
> The Greek letter π (pronounced pi) is the ratio of the circumference of a circle to the diameter. Its decimal value never ends, but starts as 3.141 592 653 589 7…
> The formula for the circumference of a circle is $C = \pi d$.
> If you know the radius you can use $C = 2\pi r$.

5   Work out the circumference of each circle. Use the π button on your calculator. Round your answers to 1 d.p.

a 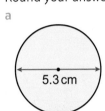   b 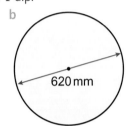   c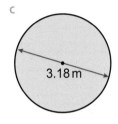

5.3 cm   620 mm   3.18 m

6   Work out the circumference of each circle. Use the π button on your calculator. Round your answers to 1 d.p.

a 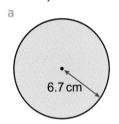   b 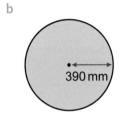   c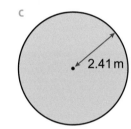

6.7 cm   390 mm   2.41 m

7   **Exam-style question**

Here is a circle.

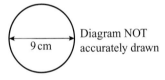

9 cm   Diagram NOT accurately drawn

The diameter of the circle is 9 cm.

Work out the circumference of this circle.

Give your answer correct to 3 significant figures.   **(2 marks)**

June 2014, Q4, 1MA0/2H

Exam hint
Write down all the numbers on your calculator display before you round.

8   **Modelling / Problem-solving** Kathy makes cakes for special events.
She uses a cake tin with a 20 cm diameter.
Kathy puts ribbon around each cake.
She allows an extra 2 cm of ribbon for joining.

a   How much ribbon does she need for each cake?

b   She buys the ribbon in rolls of 25 metres.
How many cakes can she decorate from one roll?

Q8a hint
Sketch a diagram.

Q8b hint
25 m = ☐ cm

9   **Modelling / Problem-solving** Peter is putting edging round a circular flower bed with diameter 460 cm.
The edging comes in 245 cm lengths. How many lengths does he need?

10   The radius of a racing bike wheel is 340 mm.
How many revolutions will the wheel make in a 500 m sprint race?

Q10 strategy hint Start by finding the distance travelled when the wheel makes one revolution.

17.2 Circumference of a circle 2

Objectives

- Calculate the circumference and radius of a circle.
- Work out percentage error intervals.

Did you know?

International Pi day is 14 March, because in the USA that date is written as 3.14 (e.g. 3.14.2016).

Fluency

- Round 2.5 m to the nearest metre.
- Round 3.49 m to the nearest metre.

Warm up

1 Solve these equations. Give your answers to 2 decimal places (2 d.p.) where necessary.

   a $\dfrac{d}{2} = 12.5$ b $3.142d = 70$ c $2r = 45$ d $\dfrac{C}{3.142} = 5$

2 $2 \leqslant n < 9$.
List all the integer values of n.

3 a Rearrange $A = lw$ to make l the subject of the formula.
   b Work out the value of l correct to 2 significant figures (2 s.f.) when $A = 30.5$ and $w = 4.2$.

4 a What is the smallest value that can be rounded up to
     i 8 cm to the nearest cm
     ii 20 km to the nearest km
     iii 38 mm to the nearest mm?
   b Write a value that rounds down to
     i 8 cm to the nearest cm
     ii 20 km to the nearest km
     iii 38 mm to the nearest mm.

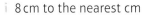

Q4a i hint

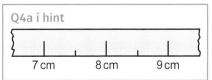

7 cm   8 cm   9 cm

Discussion What is the biggest value that rounds down to 8 cm?

Key point 3

Measurements given to the nearest whole unit may be inaccurate by up to one half of a unit below and one half of a unit above. For example, the range of possible values for a length given as 3 cm to the nearest cm is 2.5 cm ≤ length < 3.5 cm.

5 The length of a pen is 15 cm to the nearest cm. Write down the minimum possible length of the pen.

6 Bob's height is 1.63 m correct to the nearest cm.
   a What is his minimum possible height in metres?
   b What is the minimum possible height that would round to 1.64 m?
   c Write an inequality to show Bob's possible heights.

Q6c hint
$\square \leqslant h < \square$

7 There are 25 300 people at a concert, rounded to the nearest 100. Write an inequality to show the possible numbers of people.

Q7 hint Rounded to the nearest 100, so $\frac{1}{2}$ of 100 above and below.

8 Write an inequality to show the possible values for
   a 4.36 m (rounded to 2 d.p.)
   b 720 (rounded to the nearest 10)
   c 15.7 (rounded to 3 s.f.)
   d 450 (rounded to 2 s.f.)

9

Exam hint
Show all of your workings and explain what your workings show at each stage of your calculation.

10 a Write down the first 6 digits of π. Use your calculator.
   b Round your answer from part **a** to 3 d.p.
   c Use your value from part **b** to work out the circumference of this circle.

Discussion Do you get the same answer if you use the π button on your calculator to work out the circumference of the same circle?

5 cm

11 Real Work out the circumference of each of these circular objects. Round your answers to an appropriate degree of accuracy.
   a A pond with a diameter of 4.8 m.
   b A battery with a diameter of 6 mm.
   c The world's largest pancake with a radius of 750.5 cm.

Q11a hint 4.8 m is given to 1 d.p. so your answer should be to 1 d.p.

12 Modelling A crop circle at Hackpen Hill has a radius of 89.5 feet.
   a Write the circumference in terms of π.
   b Estimate the circumference of the crop circle.
   c Calculate the circumference to the nearest foot.

Q12a hint
$\pi d = \boxed{} \pi$

Q12b hint
Round the radius and the value of π to 1 s.f.

13 Mary says the circumference of this circle is 14π cm. Paul says the circumference of this circle is 7π cm.
   Reasoning Who is correct? Explain your answer.

7 cm

Example 1

The circumference of a circle is 60.8 cm. Work out the radius of the circle.

$C = 2\pi r$

$60.8 = 2\pi r$ — Substitute the values that you know.

$\dfrac{60.8}{2\pi} = r$ — Rearrange to make r the subject.

$r = 9.676\,620\,54$ — Enter $\dfrac{60.8}{2\pi}$ as a fraction on your calculator.

radius = 9.7 cm (to 1 d.p.) — Write the answer to the same degree of accuracy as the measurement given. Remember to include the units.

---

14 Work out, to 1 d.p., the diameter of a circle with circumference
   a 350 cm   **b** 2.8 m.

Q14 hint Start by substituting into $C = \pi d$.

15 Work out, to 3 s.f., the radius of a circle with circumference
   a 670 m   **b** 3 km.

16 Reasoning The circumference of a circle is 28 cm correct to the nearest cm.
   a Write the range of possible values for the circumference.
   b What are the possible values for the radius, correct to 1 d.p.?

17.3 Area of a circle

Objectives

- Work out the area of a circle.
- Work out the radius or diameter of a circle.
- Solve problems involving the area of a circle.
- Give answers in terms of π.

Why learn this?

You need to know the area of a trampoline to work out the amount of material needed.

Fluency

- Work out 5^2 40^2 $\sqrt{64}$ $\sqrt{900}$
- Estimate these by rounding each value to 1 s.f.
  3.14×4^2 2.3×5.9

1 Substitute into each formula to work out the unknown quantity.
  a $p = mh^2$ when $m = 10$ and $h = 2$
  b $t = as^2$ when $a = 3$ and $s = 6$
  c $k = 2bv^2$ when $b = 4$ and $v = 5$

2 Work out r when
  a $r^2 = 10.4$ b $r^2 = \dfrac{16}{9}$ c $r^2 = \dfrac{10}{3}$

3 Solve these equations.
  a $3y = 24$ b $\pi y = 52$

Key point 4

The formula for the area A of a circle with radius r is $A = \pi r^2$.

Example 2

A circle has a radius of 6.4 cm.
Work out the area of the circle. Give your answer correct to 3 significant figures (3 s.f.).

$A = \pi r^2$

$= \pi \times 6.4^2$ —— Write the substitution. Input it into your calculator.

$= 128.679\,635\,1$ —— Write down all the figures on the calculator display.

$= 129\,cm^2$ (to 3 s.f.) —— Round the answer to the required accuracy. Remember the units.

4 Work out the area of each circle. Round your answers to an appropriate degree of accuracy.

a   b   c   d

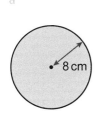
8 cm

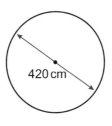
6.5 m

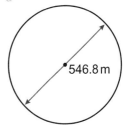
420 cm   546.8 m

Discussion What mistake have you made if your answer to part **b** is 417.0 m²?

Q4c hint Work out the radius first.

Warm up

5 a Estimate the area of each of these circular objects.
      i   The head of a nail with a radius of 3 mm.
      ii A trampoline with a diameter of 14 feet.
      iii The Caldera Crater at Yellowstone Lake with a diameter of 5 km.

   b Calculate each area in part **a** leaving your answers in terms of π.
   c Calculate each area in part **a** correct to 2 s.f.

> **Q5a i hint**
> $\pi \times 3^2 = 9\pi \approx 9 \times \Box$

6 **Reasoning** Rita says the area of the circle is $25\pi\ cm^2$.
   Jack says the area of the circle is $100\pi\ cm^2$.
   Who is correct? Explain your answer.

10 cm

7 **Real** A circular space has a radius of 4.5 m.
   a Work out the area of the space.
   b To make a lawn Grace sows 50 g of grass seed per m^2.
      How much grass seed does Grace need to cover this space?

8 **STEM / Modelling** A scientist wants to estimate the number of slugs
   in a field with an area of 4050 m^2.
   She uses a piece of string to mark out a circle with a radius of 1.5 m.
   a Work out the area of the circle.
   b She finds 35 slugs inside the circle. Estimate the number of slugs in the field.
   Discussion How reliable is your estimate?

> **Example 3**
>
> The area of a circle is 98.5 cm^2. Work out the radius of the circle.
> Give your answer correct to 1 d.p.
>
> $A = \pi r^2$
> $98.5 = \pi r^2$ —— Show the substitution.
> $\dfrac{98.5}{\pi} = r^2$ —— Make r^2 the subject.
> $\sqrt{\dfrac{98.5}{\pi}} = r$ —— Square root both sides.
> $r = 5.599\ 421\ 737$ —— Write down all the figures on the calculator display.
> $r = 5.6\ cm$ (to 1 d.p.) —— Round the answer to the required accuracy. Remember the units.

9 For each circle work out
   i   the radius
   ii the diameter.

a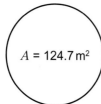

$A = 124.7\ m^2$

b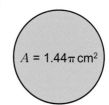

$A = 13.8\ cm^2$

c

$A = 1.44\pi\ cm^2$

10 **Modelling** A sprinkler covers a circular area of 380 m^2.
   How far from the sprinkler head does the water reach?
   Give your answer to an appropriate degree of accuracy.

> **Q10 hint** Sketch a
> diagram.

11 **Modelling** A circular rug has an area of 1.13 m².
What is the diameter of the rug in cm?
Discussion When did you convert the units?

> **Q11 hint** Start by working out the radius of the rug.

12 **Problem-solving** There is a circular pond of radius 2.5 m in a rectangular garden. The garden is 10 m by 15 m.
Max plans to grass the area of the garden not covered by the pond.
Work out the area of garden to be grassed.

> **Q12 hint** Work out the areas of the rectangle and the circle and then subtract.

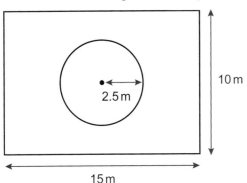

13 **Problem-solving** This circular mirror has a metal surround.
Work out the area of the metal correct to 3 s.f.

> **Q13 strategy hint** Start by working out the radius of the large circle.
> Required area = area of large circle – area of small circle

14 **Exam-style question**

Mr Weaver's garden is in the shape of a rectangle.
In the garden there is a patio in the shape of a rectangle and two ponds in the shape of circles with diameter 3.8 m.
The rest of the garden is grass.

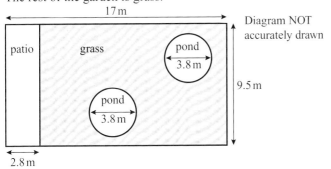

Diagram NOT accurately drawn

Mr Weaver is going to spread fertiliser over all the grass.
One box of fertiliser will cover 25 m² of grass.
How many boxes of fertiliser does Mr Weaver need?
You must show your working.   **(5 marks)**
June 2012, Q27, 1MA0/2F

Exam hint
Show all the stages of your working clearly. Most of the marks are for the methods you use.

15 **Reflect** You have used three formulae for circles:

$C = 2\pi r$ $A = \pi r^2$ $C = \pi d$

Write yourself some tips to help you remember them.

17.4 Semicircles and sectors

Objectives

- Understand and use maths language for circles and perimeters.
- Work out areas of semicircles and quarter circles.
- Solve problems involving sectors of circles.

Did you know?

On weather charts a warm front is shown with red lines and red semicircles.

Fluency

- What is the angle at the centre of
  a circle   a semicircle   a quarter circle?

1 Write each fraction in its simplest form.

  a $\dfrac{180}{360}$ b $\dfrac{90}{360}$ c $\dfrac{15}{60}$

2 Work out

  a $\frac{1}{4}$ of 360   b $10 \times \boxed{} = 360$ c $\dfrac{360}{5}$

3 Find the area and circumference of this circle.

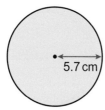

5.7 cm

Key point 5

A **chord** is a line that touches the circumference at each end.
An **arc** is a part of the circumference of a circle.
A **segment** is a part of a circle between an arc and a chord.
A **sector** is a slice of a circle between an arc and two radii.
A **tangent** is a line outside a circle that touches the circle at only one point.

4 Copy these diagrams. Write chord, arc, segment, sector and tangent in the correct places.

  a    b 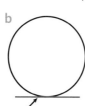   c 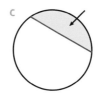   d   e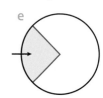

5 Work out the area of each semicircle.
Give your answers
  i in terms of π ii correct to 1 decimal place (1 d.p.).

  a 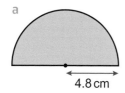   b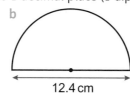

   4.8 cm   12.4 cm

> **Q5 communication hint**
> A **semicircle** is half a circle.

> **Q5 strategy hint** Start by finding the area of the whole circle, then halve the answer.

6 Work out the area of each quarter circle.
Give your answers   i   in terms of π   ii   correct to 3 significant figures (3 s.f.).

a

3.2 cm

b

19 mm

7 The diagram shows a semicircle.
a Work out what the circumference
would be if it were a whole circle.
b Halve your answer to part **a** to find
the length of the arc.
c Work out the perimeter of the semicircle
by adding your answer to part **b** to the diameter.

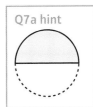

6 cm

Q7a hint

8 Work out the perimeter of this semicircle in terms of π.

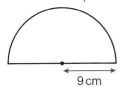

9 cm

Q8 hint Find the arc
length of half a circle
in terms of π. Add the
diameter to it.

9 Work out the perimeter of this quarter circle. Give your answer to 1 d.p.

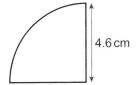

4.6 cm

Q9 hint Find the arc
length of a quarter of a
circle and add the radius
to it twice.

10 **Modelling** The minute hand on a clock is 7 cm long.
Work out, to the nearest whole cm, the distance the point end moves in

a 1 hour
b 4 hours
c 30 minutes
d 15 minutes
e 105 minutes.

Q10a hint The
radius is 7 cm.

Q10c hint 30 minutes = ☐ of an hour

11 **Reasoning** The diagram shows a pizza with a
radius of 12 cm and a cut slice.
a Work out the circumference of the pizza
in terms of π.
b Sam cuts slices of pizza with an angle size of 60°.
How many slices can he cut?
c What fraction of the pizza is each slice?

12 cm   60°   12 cm

Q11b hint $\frac{60}{360} = \frac{\square}{\square}$

12 **Reasoning** Repeat **Q11** parts **b** and **c** for these angle sizes.
a 30°
b 36°
c 120°

Key point 6

For a sector of a circle with an angle of $x°$ and radius r:

Arc length $= \dfrac{x}{360} \times 2\pi r$

Area of sector $= \dfrac{x}{360} \times \pi r^2$

r

$x°$

Example 4

For this sector of a circle, work out
a the arc length
b the perimeter
c the area.
Give your answers correct to 3 s.f.

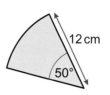

12 cm

50°

a Arc length $= \dfrac{x}{360} \times 2\pi r$

$\quad = \dfrac{50}{360} \times 2 \times \pi \times 12$ ⟶ Substitute angle size of 50 and radius of 12.

$\quad = 10.471\,975\,51$ ⟶ Write down all the figures on your calculator display.

$\quad = 10.5\,\text{cm}$ (to 3 s.f.) ⟶ Give the answer correct to 3 s.f. and include the units.

b Perimeter $= 10.471\,975\,51 + 12 + 12$ ⟶ Arc length + radius + radius
Use the unrounded value for the arc length.

$\quad = 34.471\,975\,51$

$\quad = 34.5\,\text{cm}$ (to 3 s.f.) ⟶ Give the answer correct to 3 s.f. and include the units.

c Area $= \dfrac{x}{360} \times \pi r^2$

$\quad = \dfrac{50}{360} \times \pi \times 12^2$

$\quad = 62.831\,853\,07$

$\quad = 62.8\,\text{cm}^2$ (to 3 s.f.)

---

13 Work out the arc length of each sector.
Give your answers to an appropriate degree of accuracy.

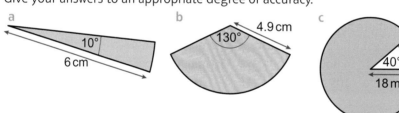

a

10°

6 cm

b

4.9 cm

130°

c

40°

18 mm

Q13c hint Work out the angle of the sector.

---

14 Work out the area of each sector. Give your answers correct to 3 s.f.

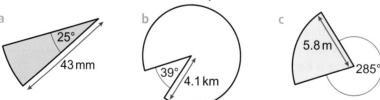

a

25°

43 mm

b

39°

4.1 km

c

5.8 m

285°

---

15 Work out the perimeter of each sector. Give your answers correct to 3 s.f.

a

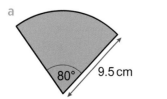

80°

9.5 cm

b

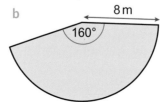

8 m

160°

16 **Exam-style question**

OAB is a sector of a circle, centre O.

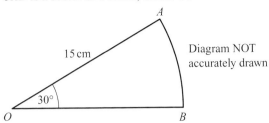

15 cm

30°

Diagram NOT
accurately drawn

The radius of the circle is 15 cm.
The angle of the sector is 30°.
Calculate the area of sector OAB.
Give your answer correct to 3 significant figures. **(2 marks)**

March 2013, Q19, 1MA0/2H

Exam hint
Consider what
fraction of the
circle the given
sector is.

17.5 Composite 2D shapes and cylinders

Objectives

- Solve problems involving areas and
  perimeters of 2D shapes.
- Work out the volume and surface area of
  cylinders.

Why learn this?

The volume of the cylinders in a car is directly
related to the power of the car.

Fluency

- How do you work out the volume of a prism?
- What units is volume measured in?

1 Sketch a net of this cylinder. What shapes are its faces?

2 Find the area of each shape. Leave your answers in terms of π where appropriate.

a

3 cm

8 cm

b

6 cm

c

4 cm

4 cm

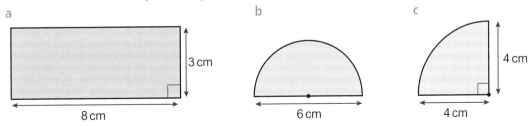

3 Find the perimeter of each shape in **Q2**.
Leave your answers in terms of π where appropriate.

Warm up

> **Example 5**
>
> For this shape work out, correct to 1 decimal place,
> a the perimeter
> b the area.
>
>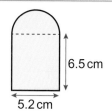
> 6.5 cm
> 5.2 cm
>
> a Perimeter of shape = 3 sides of rectangle + arc of semicircle
>
> $$= 6.5 + 6.5 + 5.2 + 0.5 \times \pi \times 5.2$$
> $$= 26.368\,140\,9$$
> $$= 26.4\,cm\ (to\ 1\ d.p.)$$
>
> | Write down all the figures on your calculator display. |
>
> | Round the answer to the required accuracy. Remember the units. |
>
> b Area of shape = area of rectangle + area of semicircle
>
> $$= 5.2 \times 6.5 + 0.5 \times \pi \times 2.6^2$$
> $$= 44.418\,583\,17$$
> $$= 44.4\,cm^2\ (to\ 1\ d.p.)$$

4 Work out the perimeter and area of each of these shapes. Give your answers to an appropriate degree of accuracy.

a
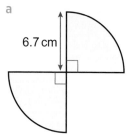
6.7 cm

b
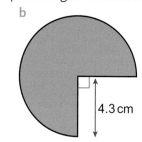
4.3 cm

> **Q4 strategy hint** Work out each shape separately and then combine the answers. Show all the working.

5 Modelling The diagram shows a school sports field. The field is a rectangle with semicircular ends. The rectangle is 80 m by 60 m.

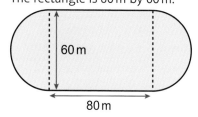
60 m
80 m

  a What is the diameter of the semicircle at each end?
  b Work out the area of the field.
  c Work out the perimeter of the field.
  d How many times must Bob run round the field to run at least 1 km?

6 Modelling The diagram shows the logo for a new company letterhead.

The logo is symmetrical. Each section is a quarter circle. Calculate the area that will be coloured green.

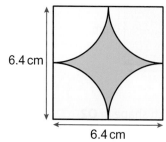
6.4 cm
6.4 cm

7 a The diagram shows a cylinder. What shape is its cross-section?

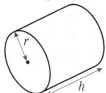

b Write a formula for the area of its cross-section.
c Write a formula for the volume of a cylinder.
$$V = \boxed{} \times h$$

> **Key point 7**
> The formula for the volume of a cylinder is $V = \pi r^2 h$.

8 Work out the volume of each cylinder. Give your answers to 3 significant figures (3 s.f.).

a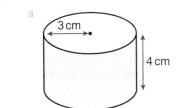

b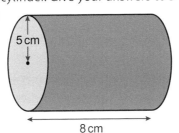

c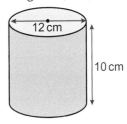

9 **Modelling** A cylindrical tank has a radius of 1.5 m and a height of 3 m.
a What is the volume of the tank?
b The tank is full of water. How many times can a bucket that holds 20 litres be filled from the tank?

> Q9b hint
> $1\,m^3 = 1000$ litres

10 **Reasoning** Which of these two cylinders has the greater volume?

A 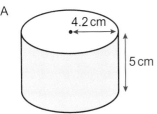   B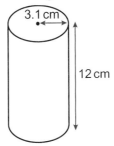

11

> **Exam-style question**
>
> A water butt is in the shape of a cylinder with a diameter of 510 mm and a height of 660 mm.
>
> How many litres of water does the water butt contain when it is completely full?
>
> 1 litre = 1000 cm³   **(4 marks)**

Exam hint
You will need to know the formula for the volume of a cylinder and understand how to use it.

12 **STEM / Modelling** A cylindrical oil drum has a radius of 50 cm and a height of 160 cm.
a Work out the volume of the oil drum.
The drum is completely filled with oil.
Oil has a density of 4.3 g/cm³.
b Work out the mass of the oil in the tank in kilograms.

> Q12b hint
> $$\text{Density} = \frac{\text{mass}}{\text{volume}}$$

Key point 8

The **surface area** of a prism is the total area of all its faces.

Example 6

Work out the surface area of this cylinder.

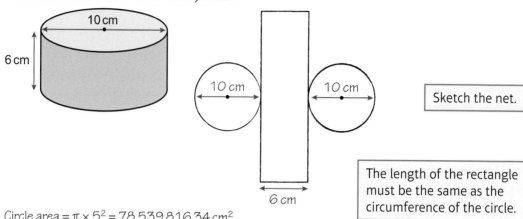

Sketch the net.

The length of the rectangle must be the same as the circumference of the circle.

Circle area = $\pi \times 5^2$ = 78.539 816 34 cm²

Circumference of circle = $\pi \times 10$ = 31.415 926 54 cm

Rectangle area = 6 × 31.41592654 = 188.495 559 2 cm

6 × circumference of circle

Total surface area of cylinder = 2 × circle area + rectangle area

  = 2 × 78.539 816 34 + 188.495 559 2

  = 345.575 191 9

  = 345.6 cm² (to 1 d.p.)

Use all the digits in the calculation.

Round the final answer to a suitable level of accuracy.

 13 Work out the total surface area of each cylinder A and B in **Q10**.

14 **Modelling** A poly tunnel is used for growing vegetables. Its shape can be approximated to a half-cylinder. Its cross-section has a diameter of 10 feet and its length is 60 feet.

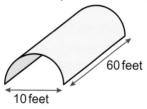

60 feet

10 feet

Work out in terms of π

a its volume   b its surface area.

Q14b hint The base of the half-cylinder is *not* part of the surface area of the poly tunnel.

 15 Work out the area of the label needed to cover the curved surface of this tin can.

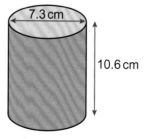

7.3 cm

10.6 cm

16 **Reflect** Write some tips to help you remember the formulae for the volume and surface area of a cylinder.

Q16 hint Diagrams may help.

17.6 Pyramids and cones

Objectives

- Work out the volume of a pyramid.
- Work out the surface area of a pyramid.
- Work out the volume of a cone.
- Work out the surface area of a cone.

Did you know?

The Great Pyramid at Giza is estimated to have 2 300 000 stone blocks that weigh between 2 and 30 tonnes each.

Fluency

How many faces does a square-based pyramid have? What shapes are they?

Warm up

1 Work out the area of each shape.

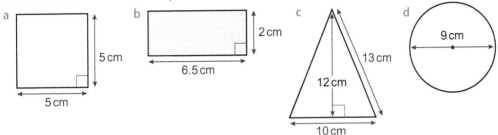

a — 5 cm, 5 cm

b — 2 cm, 6.5 cm

c — 13 cm, 12 cm, 10 cm

d — 9 cm

2 Write the answers in terms of π.

a $\frac{1}{3}$ of 18π

b $5\pi \times 9$

c $\frac{1}{3} \times 4^2 \times \pi \times 15$

Key point 9

The volume of a pyramid $= \frac{1}{3} \times$ area of base $\times$ vertical height

Example 7

A pyramid has a square base of side 7 cm and a vertical height of 12 cm.
Work out the volume of the pyramid.

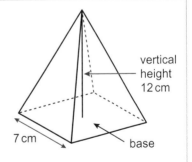

vertical height 12 cm

7 cm

base

Volume of pyramid $= \frac{1}{3} \times$ area of base $\times$ vertical height

$= \frac{1}{3} \times 7 \times 7 \times 12$

$= 196 \, \text{cm}^3$

Base is a 7 cm square, so area $= 7 \times 7 \, \text{cm}^2$. Height is 12 cm.

3 Work out the volume of each pyramid.

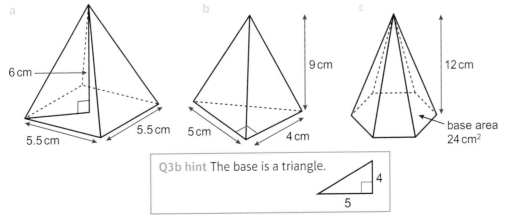

a — 6 cm, 5.5 cm, 5.5 cm

b — 9 cm, 5 cm, 4 cm

c — 12 cm, base area 24 cm²

Q3b hint The base is a triangle. 4, 5

4   **Real** The Great Pyramid at Giza has a square base 475 yards by 475 yards.
Its height is 286 yards.

Work out the volume of the Great Pyramid. Give your answer in cubic yards.

Example 8

Work out the surface area of this pyramid.

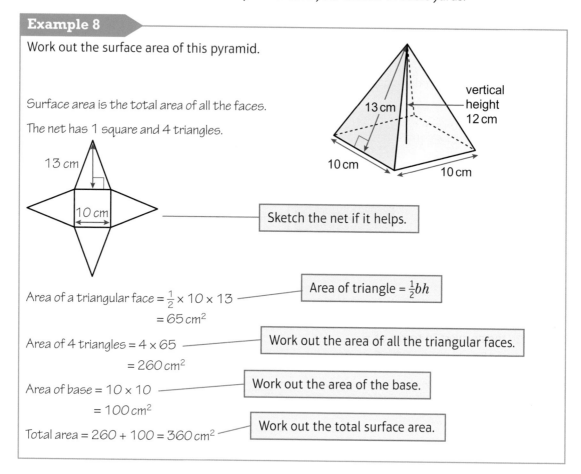

Surface area is the total area of all the faces.

The net has 1 square and 4 triangles.

Sketch the net if it helps.

Area of a triangular face = $\frac{1}{2}$ × 10 × 13 —— Area of triangle = $\frac{1}{2}bh$

= 65 cm²

Area of 4 triangles = 4 × 65 —— Work out the area of all the triangular faces.

= 260 cm²

Area of base = 10 × 10 —— Work out the area of the base.

= 100 cm²

Total area = 260 + 100 = 360 cm² —— Work out the total surface area.

---

5   Work out the surface area of each pyramid.

a   b

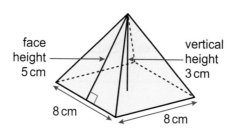

6   Work out the volume of each pyramid in **Q5**.

7   By rounding the dimensions to 1 significant
figure (1 s.f.) work out an estimate of
   a   the volume of the pyramid
   b   the surface area of the pyramid.

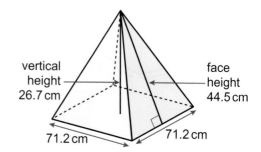

> **Key point 10**
>
> The volume of a cone = $\frac{1}{3}$ × area of base × vertical height

8 a What shape is the base of a cone?

b What is the formula for the area of a circle of radius r?

c **Reasoning** What is the formula for the volume, V, of a cone with base radius r and height h?

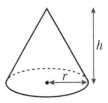

9 Work out the volume of each cone.

a 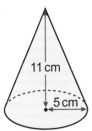 b 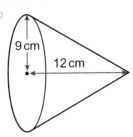 c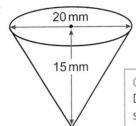

> **Q9c hint**
> Diameter = 20 mm, so radius = ☐ mm

10 **Real** A paper cone cup has a diameter of 8 cm and a height of 10 cm.

a What is the radius of the base of the cone?

b What is the volume of the cone? Give your answer correct to 3 s.f.

c How many of these cups can be filled from a 5 litre container?

> **Q10c hint**
> 1 cm³ = 1 ml
> 1 litre = 1000 ml

> **Key point 11**
>
> The area of the curved surface of a cone = π × base radius × slant height = π rl

> **Communication hint**
> The **slant height** of a cone is the length of the sloping side.

> **Example 9**
>
> Work out the total surface area of this cone in terms of π.

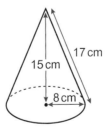

Area of curved surface = π × base radius × slant height

= π × r × l

= π × 8 × 17 — Use the slant height, not the vertical height.

= 136π

Area of base = π × 8²

= 64π

Total area = 136π + 64π — Total surface area = area of curved surface + area of base

= 200π cm²

Discussion How could you give the answer to 1 decimal place (1 d.p.)?

11 Work out the total surface area of each cone
   i   in terms of π ii   as a number correct to 3 s.f.

a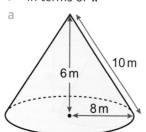

b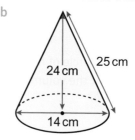

c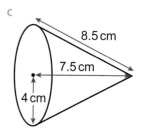

Discussion Do you need all the values on the diagrams?

12 Work out the volume of each cone in **Q11**
   i   in terms of π ii   as a number correct to 3 s.f.

13 **Exam-style question**

The diagram shows the conical roof of a building.

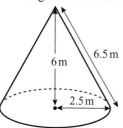

A litre of paint covers $15\,\text{m}^2$.
How many litres of paint does Uzma need to paint the curved
surface of the roof?   **(4 marks)**

Exam hint
Always read the
question carefully.
Here you will need to
find the area of the
curved part of the
cone, not the total
surface area.

**Q13 communication
hint**
Conical means cone
shaped.

17.7 Spheres and composite solids

Objectives

- Work out the volume of a sphere.
- Work out the surface area of a sphere.
- Work out the volume and surface area of
  composite solids.

Did you know?

The planet Jupiter has a polar diameter of
$133\,709\,\text{km}$. It is two and a half times as
massive as all the other planets in the Solar
System combined.

Fluency

- What are the formulae for the volumes of a cuboid, pyramid, cylinder and cone?
- What are the formulae for the area of a circle and the surface area of a cone?

1 Work out
   a $\frac{1}{3}$ of 24   b $\frac{4}{3}$ of 24   c 2^3 d 4×6^2

2 Work out
   a $\frac{4}{3} \times 6$ b $\frac{4}{3} \times 2^3$

3 Use a calculator to work out
   a $\frac{4}{3} \times \pi$ b $\pi \times 5.2^2$

  Homework, practice and support: Foundation 17.7

4 Work out the length x.

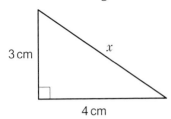

Key point 12

A **sphere** is a solid where all points on the surface are the same distance from the centre.

The volume of a sphere = $\frac{4}{3}\pi r^3$

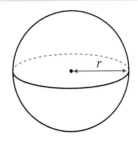

Example 10

The radius of a spherical ball is 11 cm.
Work out the volume of the ball to the nearest cm³.

Communication hint
Spherical means in the shape of a sphere.

Volume of ball = $\frac{4}{3}\pi r^3$

$= \frac{4}{3} \times \pi \times 11^3$ — Put this in your calculator.

$= 5575.279\,763$ — Give the answer in full first.

$= 5575\,\text{cm}^3$ — Round to the nearest cm³. Include the units.

 5 A spherical marble has a radius of 1.2 cm.
Work out the volume of the marble
a in terms of π
b to the nearest cm³.

 6 **Real** A spherical snooker ball has a diameter of 52.5 mm.
Work out the volume of the snooker ball to the nearest mm³.

Q6 hint Work out the radius first.

7 **Reasoning** A hemisphere has a radius of 6 m.
Moira says the volume is 144π m³.
Jill says the volume is 36π m³.
Who is correct?
Discussion What mistake has the other person made?

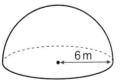

Key point 13

The surface area of a sphere = $4\pi r^2$

 8 Work out the surface area of the marble in **Q5**.
Give your answer to the nearest cm³.

9 **Modelling** The planet Venus can be modelled as a sphere with a diameter of 12 104 km.
Work out an estimate of the surface area of the planet Venus.

Q9 hint Round all values to 1 significant figure (1 s.f.)

Example 11

Work out the total surface area of a hemisphere with radius 15 cm.
Give your answer in terms of π.

Surface area of sphere = $4 \times \pi \times 15^2$

$\qquad = 900\pi$

Area of curved surface of hemisphere = $\dfrac{900}{2}\pi$ ———— | Halve the answer as a hemisphere is half a sphere. |

$\qquad = 450\pi$

Area of circle = $\pi \times 15^2$ ———— | The base of the hemisphere is a circle with radius 15 cm. |

$\qquad = 225\pi$

Total surface area = $450\pi + 225\pi$ ———— | The total area is the curved surface area + the circle area. |

$\qquad = 675\pi \, cm^2$

10 Work out the total surface area of the hemisphere in **Q7**. Give your answer
   a   in terms of π b   to 2 decimal places (2 d.p.)

11 Work out the volume of
   this solid.

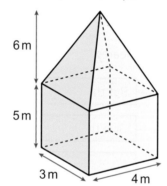

6 m

5 m

3 m

4 m

| Q11 hint Work out the volumes of the pyramid and the cuboid separately. |

12 Work out the volume of each solid. Give your answers in terms of π.

a

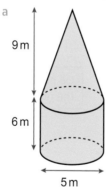

9 m

6 m

5 m

b

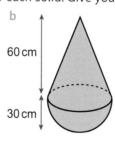

60 cm

30 cm

c

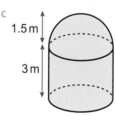

1.5 m

3 m

13 a   Use Pythagoras' theorem to work out the slant height of the cone.
      Copy and complete the working.

   $3^2 + 4^2 = \square$

   Slant height, $l = \sqrt{\square} = \square$

   b   Work out the area of the curved surface
      in terms of π.

   c   Work out the area of the base in terms of π.

   d   Work out the total surface area of the cone
      in terms of π.

   e   Work out the total surface area of the cone as a decimal correct to 1 d.p.

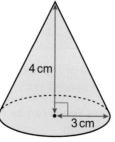

4 cm

3 cm

| Q13b hint Area of curved surface $= \pi \times r \times l$ |

| Q13c hint Area of base $= \pi r^2$ |

14 Work out the total surface area of each solid
   i in terms of π
   ii as a decimal correct to 1 d.p.

First use Pythagoras' theorem to work out the slant height. Then find the area of the curved surface and add it to the area of the base.

a
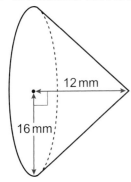
12 mm
16 mm

b
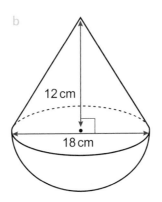
12 cm
18 cm

Q14b hint The circle where the cone and the hemisphere join is not part of the surface area.

15 Work out the total surface area of each solid
   i in terms of π
   ii as a number correct to 3 s.f.

Q15c hint Work out the slant height first.

a
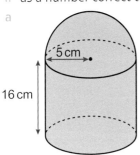
5 cm
16 cm

b
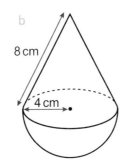
8 cm
4 cm

c
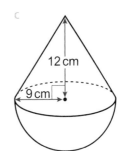
12 cm
9 cm

16 **Exam-style question**

The diagram shows a storage tank.

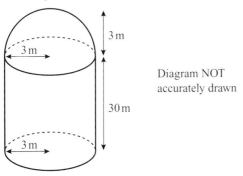
3 m
3 m
30 m
3 m

Diagram NOT accurately drawn

The storage tank consists of a hemisphere on top of a cylinder.
The height of the cylinder is 30 metres.
The radius of the cylinder is 3 metres.
The radius of the hemisphere is 3 metres.
Calculate the total volume of the storage tank.
Give your answer correct to 3 significant figures. **(3 marks)**

Nov 2008, Q24a, 5544H/15H

Exam hint
Do not round the answers to any calculations until you get your final answer.

17 **Reflect** What did you find easiest in this lesson? What was most difficult?
What made them easy/difficult?

17 Problem-solving: Designing a pencil case

Objectives
- Be able to find surface areas and volumes of prisms.
- Develop and improve solutions to more complicated problems.

A stationery company has asked you to design a new leather pencil case in the shape of a prism. The cross-section of the prism can be any shape, but the pencil case must be at least 18 cm long and open/close with a zip. The company would like the pencil case to have a lot of space inside.

Your design must not cost more than £4.50 to make.

- The leather costs 0.8p for each cm².
- It costs 7p to sew up each straight edge and 11p to sew up each curved edge.
- A 20 cm zip costs 15p and a 30 cm zip costs 25p. Zips can be cut down to size.

Draw a diagram of your design and show that it does not cost more than £4.50 to make.
Try to make the volume of your pencil case as large as possible.

> **Hint** You should start by working out the surface area of your design.

> **Hint** Experiment with different shapes, or try putting the zip in different places.

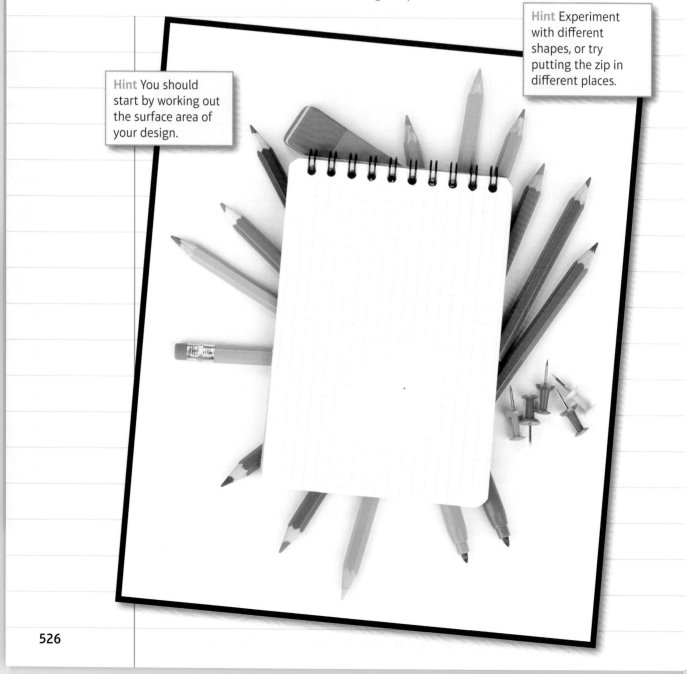

17 Check up

Accuracy

1 a Write an inequality to show the possible values for 11 cm, given to the nearest cm.
   b Write an inequality to show the possible values of 5.32 m to 2 decimal places (2 d.p.).

2 The length of a marker pen is 14 cm correct to the nearest cm.
   The length of the pencil case is 142 mm correct to the nearest mm.
   Explain why the marker pen may not fit lengthways in the pencil case.

Circles and sectors

3 Work out the circumference of a circle with a radius of 6 cm
   a in terms of π b as a number correct to 1 d.p.

4 The circumference of a circle is 44 cm.
   Work out the diameter of the circle. Give your answer to a sensible degree of accuracy.

5 Work out the area of a circle with a radius of 5 cm.
   Give your answer correct to 3 significant figures (3 s.f.).

6 A circle has an area of 49π m². Work out its radius.

7 The diagram shows a semicircle with a radius of 16 cm.
   Work out correct to 3 s.f.
   a its perimeter   b its area.

16 cm

8 Work out the shaded area correct to the nearest mm².

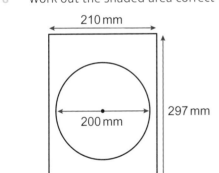

210 mm

297 mm

200 mm

9 The diagram shows a sector with a radius of 7 cm and an angle of 45°.

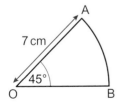

A

7 cm

45°

O   B

Work out to a suitable accuracy
a the area of the sector
b the length of the arc AB
c the perimeter of the sector.

Volumes and surface areas

10 Work out the volume of each of these solids. Give your answers
   i   in terms of π   ii   correct to 3 s.f.

a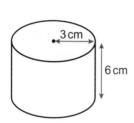

b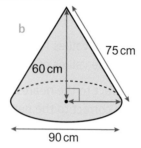

c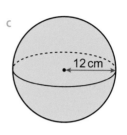

11 Work out the total surface area of each solid in **Q10**. Give your answers
   i   in terms of π   ii   correct to 3 s.f.

12 Work out the volume and surface area of this pyramid.
    Give your answers correct to the nearest cm³ or cm².

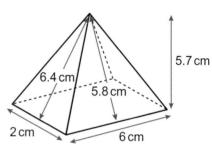

13 How sure are you of your answers? Were you mostly

   Just guessing 😟   Feeling doubtful 😐   Confident 🙂

   What next? Use your results to decide whether to strengthen or extend your learning.

* Challenge

14 **Real** A football pitch is 110 metres long and 70 metres wide.
    A marathon race is 42 km.
    How many times would you need to run round the football pitch to run a marathon?

17 Strengthen

Accuracy

1 a What is the smallest possible value for 4 cm, given to the nearest cm?
   b Write inequalities to show the possible values for
      i   4 cm, given to the nearest cm
      ii 2.5 cm, given to 1 decimal place (1 d.p.)
      iii 5.46 cm, given to 2 d.p.

> **Q1b hint** Use number lines to help.
> a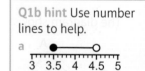

Circles and sectors

1 Work out the circumference
   of a circle with diameter 12 cm
   a in terms of π
   b correct to 1 d.p.

> **Q1 hint** Choose the formula with the diameter d in it.
> $C = \pi d$
> $C = \boxed{} \pi$

2 Work out the circumference of a circle
   with a radius of 4 cm
   a in terms of π
   b as a number correct to 1 d.p.

> **Q2a hint** Choose the formula with the radius r in it.
> $C = 2\pi r$

3   The circumference of a circle is 66 cm.
   a   Write down the formula connecting the circumference C
       and the diameter d.
   b   Work out the diameter of the circle.
       Give your answer to the nearest cm.

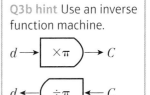

Q3b hint Use an inverse function machine.

4   A circle has a diameter of 20 cm.

   a   Work out the radius.
   b   Copy and complete: Area = $\pi \times \square^2$
   c   Work out the area of the circle
       correct to 1 d.p.

Q4b hint $A = \pi r^2$

5   A circle has area 25π m². Work out
   a   its radius
   b   its diameter.

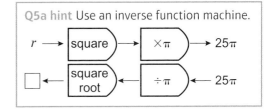

Q5a hint Use an inverse function machine.

6   What fraction of each circle is shaded?
   a   b   c   d

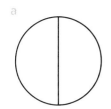
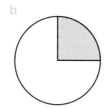
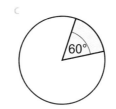
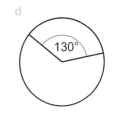

Q6c hint
$$\frac{\square}{360} = \frac{\square}{\square}$$

7   The diagram shows a semicircle with a radius of 10 cm.

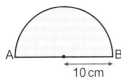

   a   Work out the area and circumference of a whole circle with radius 10 cm.
   b   Work out in terms of π
       i   the area of the semicircle
       ii   the arc length AB
       iii the diameter AB
       iv   the perimeter of the semicircle.

Q7b i and ii hint What fraction of the circle is it?

Q7b iv hint Perimeter = arc length + diameter

8   Work out, to a suitable degree of
    accuracy, the area
   a   of the rectangle
   b   of the circle
   c   of the shaded part.

Q8c hint Shaded area =
area of rectangle – area of circle

9 a Work out the circumference of a circle with radius 3 cm.

b Work out the length of the arc AB.

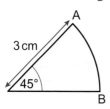

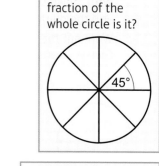

Q9b hint What fraction of the whole circle is it?

10 Work out in terms of π

a the area of a whole circle with radius 12 cm

b the area of the sector

c the circumference of a whole circle with radius 12 cm

d the arc length of the sector

e the perimeter of the sector.

Q10a hint $A = \square^2 \times \pi$

Volumes and surface areas

1 For this cylinder

a sketch the cross-section and label the radius

b work out the area of the cross-section

c work out the volume to the nearest whole number.

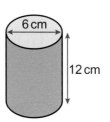

Q1 hint Radius = half of diameter
Area of cross-section = $\pi \times \square^2 = \square$
Volume = area of cross-section × height

2 a Sketch the faces of this cylinder.

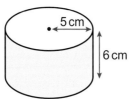

Q2b ii hint The rectangular face wraps right around the circle.

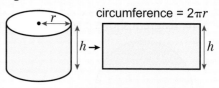

circumference = $2\pi r$

b Work out

i the area of the circle

ii the length of the rectangle

iii the area of the rectangle

iv the total surface area.

Q2b iv hint

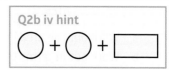

3 For each solid write down

i the vertical height

ii the slant height.

Q3 hint Slant height = sloping height
Vertical height is the distance from the vertex straight down to the base.

a

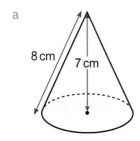

b

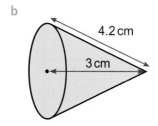

c

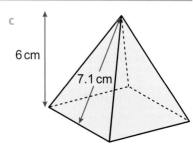

4 For this pyramid
   a work out the area of the base
   b write down the vertical height
   c work out the volume.

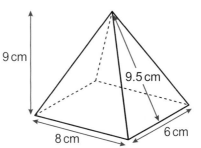

9 cm

9.5 cm

8 cm

6 cm

> **Q4c hint** Volume
> = $\frac{1}{3}$ × area of base
> × vertical height

5 For this cone

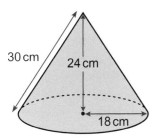

30 cm

24 cm

18 cm

   a work out the area of the base in terms of π
   b decide which value is the vertical height
   c work out the volume correct to 3 s.f.

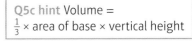

> **Q5c hint** Volume =
> $\frac{1}{3}$ × area of base × vertical height

6 For this cone

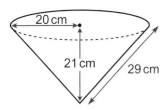

20 cm

21 cm

29 cm

   a work out the area of the base in terms of π
   b decide which value is the slant height
   c work out the curved surface area in
      terms of π
   d work out the total surface area as a number.

> **Q6c hint** Curved surface area = $\pi r l$

> **Q6d hint** Total surface area =
> curved surface area + base area

7 a What solid is this a net for?

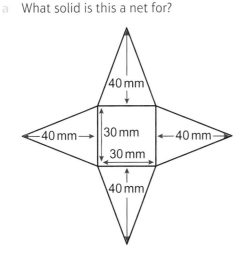

40 mm

40 mm

30 mm

30 mm

40 mm

40 mm

   b Work out the total surface area of the solid.

> **Q7b hint** Total surface area =
> △+△+△+△+☐

8 a Work out the volume of this sphere correct to 1 d.p.

b Work out the surface area of the sphere correct to 1 d.p.

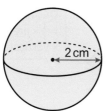

2 cm

> **Q8 hint**
> Volume is in cm³   Area is in cm²
> $\frac{4}{3}\pi r^3$ $4\pi r^2$

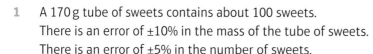

17 Extend

1 A 170 g tube of sweets contains about 100 sweets.
There is an error of ±10% in the mass of the tube of sweets.
There is an error of ±5% in the number of sweets.
a Work out the maximum and minimum possible masses of the tube of sweets.
b Work out the maximum and minimum numbers of sweets.

2 **Real** Mike's pond is a circle of diameter 5.7 m.
Pond edging is sold in 1.2 m strips.
Each strip costs £6.98.
Work out the total cost of the edging for the pond.

> **Q2 strategy hint** Start by working out the length of edging needed to go around the pond.

3 **Reasoning** A circle has a diameter of $(x + 9)$ cm and an area of 36π cm².
Work out the value of x.

> **Q3 hint** Start by rearranging πr^2 to give r in terms of A.

4 **Reasoning** The line joining $(-2, 5)$ and $(6, 5)$ is the diameter of a circle.
Work out the area of the circle in square units correct to 1 decimal place (1 d.p.).

> **Q4 hint** Work out the length of the line first.

5 **Exam-style question**

The diagram shows a circle drawn inside a square.

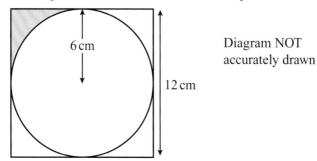

6 cm

12 cm

Diagram NOT accurately drawn

The circle has a radius of 6 cm.
The square has a side of length 12 cm.
Work out the shaded area.
Give your answer in terms of π. **(3 marks)**

November 2012, Q12, 1MA0/1H

> **Q5 strategy hint**
> Start by working out the area of the circle and the area of the square. What fraction of the area between the square and the circle is shaded?

6

The diagram shows a semicircle drawn inside a rectangle.

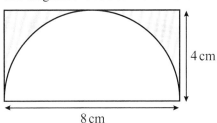

Diagram NOT
accurately drawn

4 cm

8 cm

The semicircle has a diameter of 8 cm.
The rectangle is 8 cm by 4 cm.
Work out the area of the shaded region.
Give your answer correct to 3 significant figures. **(4 marks)**

March 2013, Q23, 5MB3F/01

7 **Reasoning** The diagram shows a vase in the shape of a cylinder.

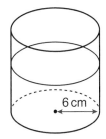

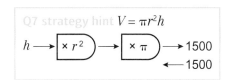

The vase has a radius of 6 cm. There is 1500 cm³ of water in the vase.
Work out the depth of the water in the vase. Give your answer correct to 1 d.p.

8 **Problem-solving** A water tank is in the shape of a cylinder with radius 35 cm and depth 140 cm.
It is filled at the rate of 0.4 litres per second.
Does it take longer than 1 hour to fill the tank? Show your working.

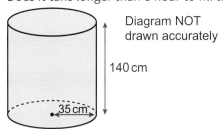

Diagram NOT
drawn accurately

140 cm

35 cm

9 The diagram shows a solid made from
two cones joined together.
Work out in terms of π
a   the total surface area of the solid
b   the volume of the shape.

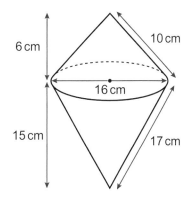

6 cm

10 cm

16 cm

15 cm

17 cm

10 Work out the total surface area of the cone correct to 3 significant figures (3 s.f.)

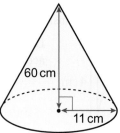

Q10 strategy hint Work out the slant height first.

11 The diagram shows a cone and a cylinder.

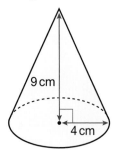
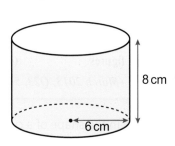

Q11 strategy hint First work out each volume in terms of π.

Find the ratio of the volume of the cone to the volume of the cylinder in its simplest form.

12 **STEM** The diagram shows a solid sphere made of ebony.
The radius of the sphere is 2.5 cm.
The mass of the sphere is 63 g.
An object will float in water if its density is less than 1 g/cm³.
Will this ebony sphere float on water?

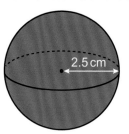

Q12 hint Density = $\dfrac{\text{mass}}{\text{volume}}$

13 **Real** Brenna has some containers in the shape of hemispheres with diameter 32 cm.
She is going to fill the containers completely with compost.
She has 90 litres of compost. Work out how many containers Brenna can completely fill with compost.
1 litre = 1000 cm³.

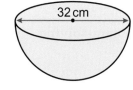

14 The diagram shows a triangle inside a quarter of a circle with radius 3 cm.
Work out the area of the shaded segment.

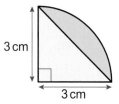

15 **Problem-solving** The diagram shows a pyramid and a cube.

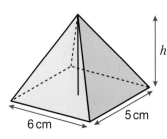
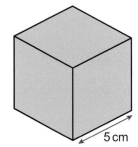

The ratio of the volume of the pyramid to the volume of the cube is 2 : 1.
Work out the height of the pyramid.

17 Knowledge check

○ Measurements given to the nearest whole unit may be inaccurate by up to one half of a unit below and one half of a unit above. For example, the range of possible values for a length given as 3 cm to the nearest cm is 2.5 cm ≤ length < 3.5 cm. .... *Mastery lesson 17.2*

○ The **circumference** is the **perimeter** of a circle. .... *Mastery lesson 17.1*

○ The Greek letter π (pronounced pi) is the ratio of the circumference to the diameter of a circle. Its decimal value never ends but starts as 3.1415926535897… .... *Mastery lesson 17.1*

○ To find the circumference of a circle (C) when given the radius (r) or the diameter (d), use $C = \pi d$ or $C = 2\pi r$. .... *Mastery lesson 17.1*

○ To find the diameter of a circle when given the circumference, use $d = \dfrac{C}{\pi}$. .... *Mastery lesson 17.2*

○ To find the area (A) of a circle when given the radius, use $A = \pi r^2$. .... *Mastery lesson 17.3*

○ To find the radius of a circle when given the area, use $r = \sqrt{\dfrac{A}{\pi}}$. .... *Mastery lesson 17.3*

○ A **chord** is a line through a circle that touches the circumference at each end.

.... *Mastery lesson 17.4*

○ An **arc** is a part of the circumference of a circle.

.... *Mastery lesson 17.4*

○ A **segment** is a part of a circle between an arc and a chord.

.... *Mastery lesson 17.4*

○ A **tangent** is a line outside a circle that touches the circle at only one point.

.... *Mastery lesson 17.4*

○ A **sector** is a slice of a circle between an arc and two radii. .... *Mastery lesson 17.4*

○ For a sector of a circle with an angle of $x°$ and radius r:
Arc length = $\dfrac{x}{360} \times 2\pi r$
Perimeter of sector = arc length + $2r$
Area of sector = $\dfrac{x}{360} \times \pi r^2$ .... *Mastery lesson 17.4*

⊙ The formula for the volume of a **cylinder** is $V = \pi r^2 h$.

.... *Mastery lesson 17.5*

⊙ The surface area of a prism is the total area of all its faces. .... *Mastery lesson 17.5*

⊙ Total surface area of a cylinder = $2\pi rh + 2\pi r^2$ .... *Mastery lesson 17.5*

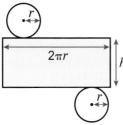

⊙ The volume of a pyramid = $\frac{1}{3} \times$ area of base × vertical height .... *Mastery lesson 17.6*

⊙ The volume of a cone = $\frac{1}{3} \times$ area of base × vertical height .... *Mastery lesson 17.6*

⊙ The area of the curved surface of a cone = $\pi \times$ base radius × slant height = πrl .... *Mastery lesson 17.6*

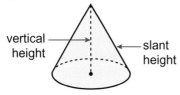

vertical height — slant height

⊙ Total surface area of a cone = $\pi rl + \pi r^2$ .... *Mastery lesson 17.6*

⊙ The volume of a **sphere** = $\frac{4}{3}\pi r^3$ .... *Mastery lesson 17.7*

⊙ The surface area of a sphere = $4\pi r^2$ .... *Mastery lesson 17.7*

⊙ A **hemisphere** is half of a sphere. .... *Mastery lesson 17.7*

⊙ Total surface area of a hemisphere
  = curved surface area + area of circular base
  = $2\pi r^2 + \pi r^2$ .... *Mastery lesson 17.7*

Look back at this unit.
Which lesson made you think the hardest?
Write a sentence to explain why.

Hint Begin your sentence with:
Lesson ____ made me think
the hardest because ____

---

17 Unit test

Log how you did on your
Student Progression Chart.

1 **a** A field is 80 m long, to the nearest 10 m.
Write an inequality to show the possible lengths.   *(2 marks)*

  b The area of a circle is 12.4 cm² to 1 d.p.
Write an inequality to show the possible values for the area.   *(2 marks)*

2 The circumference of a circular garden is 88 m to the nearest metre.
Fred wants to put a path from one point on the circumference to another
through the centre of the garden.

  a How long will the path be? Give your answer to the nearest metre.

  b Write an inequality to show the possible values.   *(3 marks)*

*Active*Learn   Homework, practice and support: Foundation 17 Unit test

3   Work out the area of a circle with a radius of 9 cm.
    Give your answer correct to 1 decimal place (1 d.p.).   *(2 marks)*

4   A circle has an area of 121π m². Work out its radius.   *(2 marks)*

5   The diagram shows a design for a window.
    The window has a rectangular bottom and
    a semicircular top.

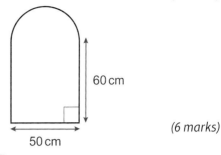

    a   Work out the area of the glass in the window.
    b   A frame of lead beading goes around the glass.
        How much lead beading is needed?
    Give your answers to 3 significant figures (3 s.f.).   *(6 marks)*

6   OAB is the sector of a circle with centre O.
    Work out in terms of π
    a   the perimeter of sector OAB
    b   the area of sector OAB.

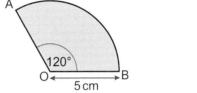

                                                                             (4 marks)

7   The diagram shows a container used to store rainwater.
    The container is in the shape of a cylinder of radius 50 cm.
    The height of the rainwater in the container is 70 cm.
    A gardener uses 80 litres of the rainwater.
    1 litre = 1000 cm³
    Work out the new volume of the rainwater in the container.
    Give your answer correct to 3 s.f.   *(3 marks)*

8   (**Exam-style question**)

    The diagram shows a solid hemisphere of radius 5 cm.

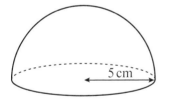

    Diagram NOT
    accurately drawn

    Find the *total* surface area of the solid hemisphere.
    Give your answer in terms of π.   **(3 marks)**
                                            March 2013, Q23, 1MA0/1H

9   The diagram shows a square-based pyramid.
    Work out the volume of the pyramid.

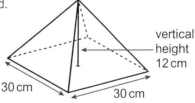

    vertical
    height
    12 cm

    30 cm   30 cm   *(2 marks)*

10 The diagram shows a cone.
    Work out
    a   its volume
    b   its total surface area.

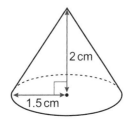

    2 cm

    1.5 cm   *(6 marks)*

Sample student answers

Which student gives the better answer and why?

Exam-style question

Inderpal is making two mirrors.

Mirror A is in the shape of a circle.
This mirror has a diameter of 60 cm.

Mirror B is in the shape of an isosceles triangle.
This mirror has base 48 cm and height 32 cm.

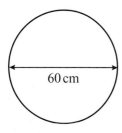

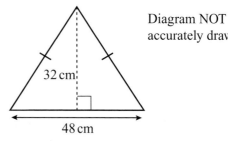

Diagram NOT
accurately drawn

Mirror A

Mirror B

Inderpal buys metal strips to put around the edge of each mirror.
The metal strip is sold in lengths of one metre.
Each one metre length of metal strip costs £5.68.
Work out the total amount Inderpal pays.
You must show all your working.

(7 marks)

June 2014, Q29b, 1MA0/2F

Student A

Mirror A: $\pi \times 60 = 188.5$ cm

Mirror B: side of triangle $= \sqrt{24^2 + 32^2} = 40$ cm

Length around triangle $= 48 + 40 + 40 = 128$ cm

Total length $= 316.5$ cm so need to buy 4 metres

Cost $4 \times 5.68 = £22.72$

Student B

Mirror A: $\pi \times 30^2 = 2827.4$

Mirror B: $\frac{1}{2} \times 48 \times 32 = 768$

Total $= 3595$ cm so need 36 metres

$36 \times 5.68 = £204.48$

18 FRACTIONS, INDICES AND STANDARD FORM

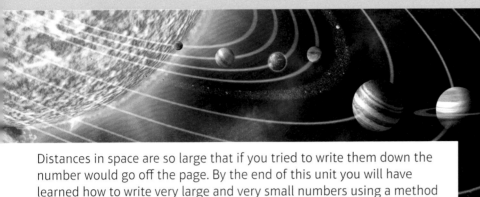

Distances in space are so large that if you tried to write them down the number would go off the page. By the end of this unit you will have learned how to write very large and very small numbers using a method that makes them easy to write.

The orbit of a planet (such as the Earth) around a star (such as the Sun) is never exactly circular.

The closest we come to the Sun is 147 million kilometres and the furthest is 152 million kilometres. What is our average distance from the Sun?

18 Prior knowledge check

Numerical fluency

1   Work out
    a -2×3 b $-2 - 3$
    c $5 - -3$ d $-7 + -4$

2   Write these decimals as fractions.
    a   0.1   b   0.01   c   0.001

3   Work out
    a $0.9 \times 10\,000$ b 1000×0.705

4   Work out
    a $36 \div 10$ b $0.81 \div 1000$

5   Simplify
    a $\frac{3}{15}$ b $\frac{10}{25}$ c $\frac{8}{20}$

6   Write $\frac{19}{5}$ as a mixed number.

7   Write $4\frac{2}{3}$ as an improper fraction.

8   Work out
    a $\frac{1}{2} \times \frac{3}{5}$ b $\frac{3}{7} \times \frac{2}{9}$

9   Work out
    a $\frac{1}{4} \div \frac{2}{3}$ b $\frac{4}{5} \div \frac{1}{10}$

10 Work out
    a $\dfrac{6}{0.3}$ b $\dfrac{1}{0.04}$

11 Write these products using powers.
    a $4 \times 4 \times 4$ b $9 \times 9 \times 9 \times 9 \times 9 \times 9$

12 Work out
    a 2^4 b 5^3 c 8^2

13 Write as a single power
    a $3^2 \times 3^3$ b $5^2 \times 5^2$
    c $\dfrac{6^7}{6^2}$ d $8^{11} \div 8^3$

✱ Challenge

14 Use the digits 1, 2, 3 and 4 once in each of the calculations below to make the largest possible solutions.

a   b

c

539

18.1 Multiplying and dividing fractions

Objective

- Multiply and divide mixed numbers and fractions.

Did you know?

Before decimalisation in the UK, measurements were given in whole inches and fractions of inches.

Fluency

- Which is the odd one out: $3\frac{1}{5}$, $\frac{4}{5}$ or $\frac{16}{5}$?
- Simplify $\frac{4}{10}$, $\frac{28}{35}$, $\frac{36}{60}$

Warm up

1 Work out the reciprocal of
   a 11   b $\frac{3}{7}$ c $3\frac{1}{2}$ d $2\frac{4}{5}$

2 Work out
   a $3 \times 2\frac{1}{2}$ b $2\frac{2}{3} \times 6$ c $\frac{3}{7} \times \frac{4}{5}$ d $\frac{3}{10} \times \frac{5}{6}$

3 Work out
   a $\frac{1}{4} \div \frac{3}{5}$ b $\frac{3}{8} \div \frac{9}{16}$ c $10 \div \frac{3}{8}$ d $5 \div 1\frac{3}{7}$

4 Work out the reciprocal of
   a 0.1   b 0.5
   c 0.8   d 2.5

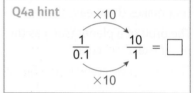

Q4a hint

$$\frac{1}{0.1} \xrightarrow{\times 10} \frac{10}{1} = \square$$

Key point 1

To multiply or divide mixed numbers change the mixed numbers to improper fractions first.

> Questions in this unit are targeted at the steps indicated.

5 Work out
   a $2\frac{3}{5} \times \frac{3}{7} = \frac{}{5} \times \frac{3}{7} =$ b $3\frac{1}{5} \times \frac{2}{3}$ c $1\frac{3}{10} \times \frac{1}{2}$
   d $\frac{5}{8} \times 2\frac{1}{3}$ e $\frac{2}{3} \times 1\frac{2}{3}$ f $4\frac{2}{3} \times \frac{1}{5}$

6 Work out
   a $2\frac{1}{7} \times \frac{1}{10}$ b $\frac{10}{11} \times 1\frac{2}{5}$ c $1\frac{1}{3} \times \frac{1}{2}$
   d $1\frac{1}{8} \times \frac{4}{9}$ e $\frac{4}{5} \times 3\frac{2}{3}$ f $\frac{2}{7} \times 2\frac{5}{11}$

   Q6a hint Where possible, simplify before multiplying.
   $$\frac{15}{7} \times \frac{1}{10} = \frac{\overset{3}{15} \times 1}{\underset{2}{10} \times 7}$$

7 Work out
   a $1\frac{3}{5} \times 2\frac{3}{5} = \frac{}{5} \times \frac{}{5} =$ b $2\frac{2}{3} \times 1\frac{2}{3}$ c $1\frac{1}{7} \times 2\frac{2}{3}$
   d $1\frac{1}{9} \times 1\frac{3}{5}$ e $2\frac{2}{5} \times 7\frac{1}{2}$ f $3\frac{1}{7} \times 1\frac{1}{5}$

Example 1

Work out $3\frac{1}{5} \div \frac{7}{10}$

$3\frac{1}{5} \div \frac{7}{10} = \frac{16}{5} \div \frac{7}{10}$

$\quad = \frac{16}{5} \times \frac{10}{7}$ ——— To divide by a fraction, multiply by its reciprocal.

$\quad = \frac{16 \times \overset{2}{10}}{\underset{1}{5} \times 7}$ ——— Simplify if possible.

$\quad = \frac{32}{7}$

$\quad = 4\frac{4}{7}$ ——— Write the answer as a mixed number.

8 Work out

a $1\frac{3}{5} \div \frac{1}{2}$ b $1\frac{1}{8} \div \frac{2}{3}$ c $2\frac{3}{5} \div \frac{7}{10}$

d $1\frac{4}{5} \div \frac{1}{10}$ e $\frac{4}{5} \div 1\frac{4}{10}$ f $\frac{3}{4} \div 1\frac{1}{8}$

> **Q8 hint** Change mixed numbers to improper fractions first.

9 Work out

a $1\frac{1}{3} \div 2\frac{3}{5}$ b $2\frac{1}{2} \div 2\frac{1}{4}$ c $1\frac{2}{3} \div 3\frac{3}{4}$

d $1\frac{2}{5} \div 1\frac{3}{10}$ e $3\frac{2}{3} \div 2\frac{1}{5}$ f $2\frac{1}{4} \div 5\frac{1}{2}$

10 Use a calculator to work out

a $4\frac{3}{5} \times \frac{6}{7}$

b $18\frac{1}{2} \times 14\frac{4}{9}$

c $19 \div 2\frac{3}{5}$

d $18\frac{4}{5} \div 12\frac{1}{3}$

> **Q10a hint** Make sure you know how to enter mixed numbers in your calculator.

11 **Problem-solving** a One of these cards doesn't have a pair. Which one is it?

$3\frac{1}{3} \times 4\frac{1}{5}$ $4\frac{2}{3} \div 5\frac{2}{5}$ $5\frac{2}{3} \times 20$ $16\frac{4}{5} \div 7$ 14

$42\frac{37}{40}$ $20\frac{1}{5} \times 2\frac{1}{8}$ $113\frac{1}{3}$ $\frac{70}{81}$ $2\frac{2}{5}$ $24\frac{4}{9}$

b Make up a calculation that gives the answer that doesn't have a calculation card.

12 **Real / Problem-solving** A roll of ribbon is cut into pieces measuring $1\frac{1}{4}$ inches. How many whole pieces can be cut from a $32\frac{1}{2}$ inch roll?

13 **Real / Problem-solving** A tile measures $1\frac{1}{2}$ inches by $2\frac{1}{4}$ inches. What is the area of the tile?

> **Q13 hint** The area will be in square inches.

14 Each of these calculations is missing a single digit. Work out the missing digits.

> **Q14 hint** Use the inverse operation.

a $\frac{2}{5} \times 1\frac{\square}{3} = \frac{8}{15}$ b $\frac{\square}{1}\frac{1}{5} \times 2\frac{1}{2} = 8$

c $\frac{2}{\square} \div 1\frac{1}{6} = \frac{4}{7}$ d $1\frac{1}{3} \div 2\frac{\square}{5} = \frac{5}{9}$

15 **Problem-solving / Reasoning** Use each of the numbers 1, 2, 3, 4, 5 and 6 once in the following calculation to give the largest solution possible.

> **Q15 hint** For example, $1\frac{2}{3} \times 4\frac{5}{6}$

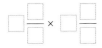

16 **Exam-style question**

A necklace is made of beads which are $1\frac{1}{4}$ inches long.
Each bead costs 30p and the thread costs £2.30.
The necklace is $21\frac{1}{4}$ inches long.
What is the cost of making the necklace?   **(3 marks)**

Exam hint
Break the question into steps. Show your working and clearly state your final answer.

17 **Problem-solving** There are $1\frac{3}{4}$ pints in a litre and 8 pints in a gallon. How many litres are there in a gallon?

18.2 **The laws of indices**

Objective

- To know and use the laws of indices.

Why learn this?

When dealing with very large numbers it is useful to be able to express them using indices. For example, $1\,000\,000\,000 = 10^9$ and $1024 = 2^{10}$.

Fluency

Work out
a 3^2 **b** 9^2 **c** 10^4 **d** 1^5

Warm up

1   Work out
   a $1 \div 0.2$ **b** $1 \div 0.05$ **c** $1 \div 0.4$ **d** $1 \div 2.5$

2   Write as a single power
   a $2^2 \times 2^4$ **b** $7^3 \div 7^2$ **c** $\dfrac{11^3 \times 11^2}{11^4}$ **d** $5^9 \div (5^2 \times 5^3)$

3   Work out
   a $\left(\dfrac{1}{6}\right)^2$ **b** $\left(\dfrac{1}{2}\right)^4$ **c** $\left(\dfrac{2}{5}\right)^3$ **d** $\left(\dfrac{3}{10}\right)^4$

4   **a**   Find the reciprocal of
     i   0.5   **ii**   0.2   **iii**   0.01
     iv   0.4   **v**   −3   **vi**   −2.5
   b   Multiply each number in part **a** by its reciprocal. What do you notice?
   Discussion Why is it not possible to find the reciprocal of zero?

> **Q4a hint** The reciprocal of a number is 1 divided by that number.

5   Copy and complete
   a $(3^4)^5 = 3^\square \times 3^\square \times 3^\square \times 3^\square \times 3^\square = 3^\square$
   b $(4^6)^4 = 4^\square \times 4^\square \times 4^\square \times 4^\square = 4^\square$
   Discussion Can you see a relationship between the powers and the answer?

Key point 2

To raise the power of a number to another power, multiply the indices.

Example 2

Work out $(7^6)^5$.

$(7^6)^5 = 7^{6 \times 5} = 7^{30}$ ———   Multiply the indices together.

---

6   Write as single power
   a $(5^4)^2$ **b** $(6^2)^3$ **c** $(7^8)^2$ **d** $(11^5)^5$

7   The answer to each question is one of the following.
     3^2 3^4 3^5 3^6 3^8 3^{12}
   Choose the correct answer for each expression.
   a $(3^2)^3$ **b** $(3^4 \times 3^2)^2$ **c** $(3^8 \div 3^6)^3$ **d** $(3^2)^2 \times 3$
   e $(3^{12} \div 3^{10}) \times 3^3$ **f** $(3^4 \div 3^2)^2$ **g** $(3 \times 3)^2 \times (3^2)^4$ **h** $\dfrac{3^5 \times 3^7}{(3^2)^3}$

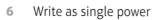

8 **Reasoning** Copy and complete the pattern.

$2^3 = 8$

$2^2 = \Box$

$2^1 = \Box$

$2^0 = 1$

$2^{-1} = \frac{1}{2^1} = \Box$

$2^{-2} = \frac{1}{2^2} = \Box$

$2^{-3} = \frac{1}{2^3} = \Box$

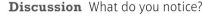

9 Use your calculator to work out

   a 4^0 b 100^0 c 57^0

   Discussion What do you notice?

> **Q9 hint** Use the $\boxed{x^y}$ key.

Key point 3

Any number (or term) raised to the power 0 is equal to 1.

Key point 4

Any number (or term) raised to the power −1 is the reciprocal of the number.
For example, $3^{-1} = \frac{1}{3}$.

10 Write as a fraction

   a 4^{-1} b 7^{-1} c 6^{-1} d x^{-1}

Key point 5

3^{-2} is the same as $(3^{-1})^2$.
To find a negative power, find the reciprocal and then raise the value to the positive power.

Example 3

Write down the value of

a 3^{-2}

$3^{-2} = \left(\frac{1}{3}\right)^2$ ──── The reciprocal of 3 is $\frac{1}{3}$.

$= \frac{(1 \times 1)}{(3 \times 3)}$ ──── Square the fraction.

$= \frac{1}{9}$

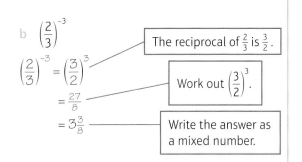

b $\left(\frac{2}{3}\right)^{-3}$

$\left(\frac{2}{3}\right)^{-3} = \left(\frac{3}{2}\right)^3$ ──── The reciprocal of $\frac{2}{3}$ is $\frac{3}{2}$.

$= \frac{27}{8}$ ──── Work out $\left(\frac{3}{2}\right)^3$.

$= 3\frac{3}{8}$ ──── Write the answer as a mixed number.

11 Work out the value of

   a 2^{-3} b 5^{-2} c 10^{-3}

   d $\left(\frac{1}{2}\right)^{-3}$ e $\left(\frac{1}{5}\right)^{-2}$ f $\left(\frac{2}{9}\right)^{-2}$

> **Q11 hint** Leave your answers as fractions where appropriate.

12 Write as a single power

   a $2^5 \times 2^{-2}$ b $(2^3)^{-1}$ c $8^{-3} \times 8^3$

   d $\frac{6^4 \times 6^{-2}}{6^2}$ e $(11^8 \div 11^{-5}) \times 11$ f $(3^{-2})^2 \times 3^{-4}$

> **Q12a hint** The laws of indices for negative numbers are the same as for positive numbers.

13 Write as a single power

   a $a^{-2} \times a^5$ b $b^{-2} \times b^{-1}$ c $\frac{m^{-3}}{m^2}$

   d $\frac{n^2}{n^{-3}}$ e $\frac{x^3 \times x^{-2}}{x^4}$ f $\frac{c^{-3} \times c^5}{c^2}$

> **Q13b hint** Write answers with negative powers as $\frac{1}{x^{\Box}}$.

14 Write these numbers from smallest to largest.

10^3 100^2 10^{-3} 100^{-2} 1^4 1^{-5}

15 Work out

a $6^{-1} \times 12$ b 300×10^{-1}

c 1800×10^{-4} d 18×23^0

16 Use your calculator to work out these calculations. Where necessary, give your answers to 3 significant figures.

a $0.5^{-2} \times 10^{-3}$ b $18^7 \div 20^6$

c $\dfrac{3^{-8} \times 5^7}{8^{-2}}$ d $2.5^7 \times 3.5^2 + 8.4^{-1}$

17 **Exam-style question**

   a Find the value of 3^3. **(1 mark)**

   b Write down the reciprocal of 7. **(1 mark)**

   c Work out the value of $\left(\dfrac{2}{3}\right)^{-2}$. **(2 marks)**

> **Exam hint**
> Part **c** is worth 2 marks so you should have at least 2 stages of working shown.

18 **Exam-style question**

   Write these numbers in order of size.
   Start with the smallest number.

   5^{-4} 0.5 5^{-1} 0.5^2 **(2 marks)**

> **Exam hint**
> Work out the value of each expression.

19 **Exam-style question**

   a Work out $4.5^{-2} + 9.8$.
   Write down all the figures on your calculator display. **(2 marks)**

   b Write your answer to part **a** to 3 significant figures. **(1 mark)**

> **Exam hint**
> Check you enter the number on your calculator correctly.

18.3 Writing large numbers in standard form

Objectives

- Write large numbers in standard form.
- Convert numbers from standard form into ordinary numbers.

Why learn this?

There are trillions of atoms in a litre of water. Using standard form enables us to read and write large numbers easily.

Fluency

Work out the value of 10^2, 10^3, 10^4 and 10^5.

1 Work out

   a 5×100 b 3.5×1000 c 8.75×10 d $9.03 \times 100\,000$

2 Write in figures

   a 1 million   b 1 billion

> **Q2b hint** 1 billion = 1000 million

3 a Work out the value of 10^3.

   b Use your answer to part **a** to work out 13.4×10^3.

Key point 6

Standard form is a way of writing very large or very small numbers.
A number in standard form looks like this.

$$8.4 \times 10^5$$

This part is a number between 1 and 10

This part is a power of 10

4   Write each number as an ordinary number. The first one has been started for you.
   a $3 \times 10^2 = 3 \times 100 = \Box$
   b 5×10^6
   c 7×10^4
   d 9×10^{11}

   Discussion Do you need two steps to write a number in standard form as an ordinary number?

Example 4

Write 4000 in standard form.

$4000 = 4 \times 1000$ ——— Write the number as a number between 1 and 10 multiplied by a power of 10.

$= 4 \times 10^3$ ——— Write the power of 10 using indices.

5   Write each number in standard form.
   a   500
   b   300 000
   c   9 000 000 000
   d   7 million

6   Write each number as an ordinary number.
   a $4.7 \times 10^4 = 4.7 \times 10 000 = \Box$
   b 9.21×10^6
   c 8.3×10^{11}
   d 9.23×10^5
   e 6.3×10^{11}
   f 9.05×10^6
   g 6.702×10^9
   h 4.07×10^8

   Q6f hint The 0 between the 9 and the 5 is essential: $9.05 \neq 9.5$

Example 5

Write 45 600 in standard form.

$45 600 = 4.56 \times 10^4$ ——— 4.56 lies between 1 and 10. Multiply by the power of 10 needed to give the original number.

$4\,5\,6\,0\,0$

7   Match the numbers on the left to those on the right.
   a   79 000
   b   791
   c   7910
   d   790 000 000
   e   7 910 000 000
   f   7 900 000
   g   79 100 000
   h   79 000 000 000

   A 7.91×10^3
   B 7.91×10^2
   C 7.9×10^{10}
   D 7.9×10^4
   E 7.91×10^7
   F 7.9×10^6
   G 7.91×10^9
   H 7.9×10^8

8   Write each number using standard form.
   a   450 000
   b   32 000
   c   150 000
   d   7 250 000
   e   6 291 000
   f   1 500 000
   g   703 000 000
   h   76 billion

9   **Real** Fiona uses a calculator to work out $752\,000 \times 59\,000$.
    Her calculator gives the answer

$$4.4368 \times 10^{10}$$

Write this number as an ordinary number.

10 **Reasoning** These numbers are not written correctly in standard form.
    Write them in standard form.

    a 33.4×10^3 b 27×10^5 c 42×10^2 | Q10a hint $33.4 \times 10^3 = 3.34 \times 10 \times 10^3$ |

    d 0.4×10^3 e 0.87×10^6 f 50.5×10^4

11 **Real / STEM** The table gives the average distances (to the nearest million km) between
    the Sun and each of the planets in our solar system.

Planet	Average distance from the Sun (km)
Mercury	58 000 000
Venus	108 000 000
Earth	150 000 000
Mars	228 000 000
Jupiter	778 000 000
Saturn	1 433 000 000
Uranus	2 871 000 000
Neptune	4 503 000 000

Write each distance in standard form.

12 A trillion is 10^{12}. Write 1 trillion

    a   as an ordinary number   b   in standard form.

13 **STEM** The table shows the meaning of prefixes for large numbers. Write the following
    measurements in standard form.

Prefix	Letter	Number
tera-	T	1 000 000 000 000
giga-	G	1 000 000 000
mega-	M	1 000 000
kilo-	k	1000

    a   8 terabytes in bytes   b   4.6 Mm in metres
    c   17.7 Gl in litres   d   0.95 kilograms in grams.

14 **Reasoning** Write $<$ or $>$ in the boxes.

    a 3.2×10^4 ☐ 5.4×10^8 b 7.05×10^7 ☐ 2.1×10^4 c 8.9×10^2 ☐ 9.1×10^2

    d 5.02×10^7 ☐ 5.2×10^7 e 6.35×10^3 ☐ 7.35×10^2 f 4.02×10^5 ☐ 2.04×10^5

15 **Exam-style question**

    Write these numbers in order of size. Start with the largest number.
    3.2×10^5 320 3.2×10^8 3.2×10^4 3.2×10^9 **(2 marks)**

    Exam hint
    First convert all to
    standard form or all to
    ordinary numbers.

16 **Exam-style question**

    a   Write $705\,000\,000$ in standard form.   **(1 mark)**
    b   Write 3.45×10^7 as an ordinary number.   **(1 mark)**

    Exam hint
    Only include zeros that
    are placeholders.

18.4 Writing small numbers in standard form

Objectives

- Write small numbers in standard form.
- Convert numbers from standard form with negative powers into ordinary numbers.

Did you know?

Biologists work with microscopic bacteria that are millionths of a metre long. Standard form enables them to write their size down in an easily understood form.

Warm up

Fluency

Work out

a $3000 \div 10$ **b** $920 \div 100$ **c** $891 \div 1000$ **d** $32.45 \div 1000$

1 Write these numbers as decimals.

   a $\frac{1}{10\,000}$ b $\frac{1}{10^6}$ c 10^{-2} d 10^{-10}

2 Write these numbers in order, starting with the smallest.

   10^{-2} 0.1 $\frac{1}{1000}$ 10^{-8}

3 Write each number as an ordinary number. The first one has been started for you.

   a $3 \times 10^{-5} = 3 \times 0.00001 = \square$ b 8×10^{-2} c 4×10^{-1} d 7×10^{-11}

Example 6

Write 0.00005 in standard form.

$0.00005 = 5 \times 0.00001$

$\qquad\quad = 5 \times 10^{-5}$

> Write the number as a number between 1 and 9 multiplied by a power of 10.

4 Write each number in standard form.

   a 0.03 b 0.005 c 0.0001 d 0.0000003

5 Write each number as an ordinary number.

   a $4.5 \times 10^{-5} = 4.5 \times 0.00001 = \square$ b 3.8×10^{-6} c 8.34×10^{-9} d 1.401×10^{-1}

Key point 7

To write a small number in standard form:
- Place the decimal point after the first non-zero digit.
- How many places has this moved the digit? This is the negative power of 10.

Example 7

Write 0.00352 in standard form.

$0.00352 = 3.52 \times 10^{-3}$

> 3.52 lies between 1 and 10.
> Multiply by the power of 10 needed to give the original number.
> 0.00352

6 Write each number in standard form.

   a 0.052 b 0.00071 c 0.000569 d 0.00241

   e 0.000014 f 0.00109 g 0.0000304 h 0.6102

7 Match these numbers to their equivalent written in standard form.
   a 0.003 03   A 3.3×10^{-4}
   b 0.000 33   B 3.3×10^{-9}
   c 0.000 003 03   C 3.3×10^{-1}
   d 0.000 000 003 3   D 3.03×10^{-3}
   e 0.303   E 3.03×10^{3}
   f 0.33   F 3.03×10^{-6}
   g 3030   G 3.3×10^{4}
   h 33 000   H 3.03×10^{-1}

8 **Reasoning** A question in an exam reads
   Write 0.000 907 in standard form.
   Here are some of the incorrect answers given.
   Lucy: 0.907×10^{-3}
   Ali: 9.7×10^{-4}
   Sam: 9.07×10^{4}
   a For each answer explain what the student did wrong.
   b What is the correct answer?

9 **STEM** The table shows the meaning of prefixes for small numbers. Write the following measurements in standard form.

Prefix	Letter	Number
deci-	d	0.1
centi-	c	0.01
milli-	m	0.001
micro-	μ	0.000 001
nano-	n	0.000 000 001
pico-	p	0.000 000 000 001

   a 7 picograms in grams   b 1.4 μs in seconds   c 593 dm in metres
   d 10.5 nV in volts   e 0.38 milliamps in amps f 9.9 centilitres in litres.

10 **STEM** The table gives the radius of one atom of different elements.
   a Write the radii as numbers in standard form.

Element	Radius (m)
Hydrogen	0.000 000 000 025
Lithium	0.000 000 000 145
Sodium	0.000 000 000 18
Phosphorus	0.000 000 000 1
Nitrogen	0.000 000 000 065
Chromium	0.000 000 000 14
Tin	0.000 000 000 145

   b Write the atoms in order of size, starting with the smallest.

11 **Exam-style question**
   a Write 0.000 000 705 in standard form.   **(1 mark)**
   b Write 3.2×10^{-4} as an ordinary number.   **(1 mark)**

18.5 Calculating with standard form

Objectives

- To multiply and divide numbers in standard form.
- To add and subtract numbers in standard form.

Did you know?

A googol is 10^{100}. A googolplex is 10^{googol}. A googolplexian is $10^{\text{googolplex}}$.

Fluency

Work out

a $\dfrac{8-4}{5}$　　**b** $3 \times 4 - 2$　　**c** $\dfrac{7}{5+6}$

1　Rewrite each number in standard form.
　　a 32×10^4　　　　**b** 180×10^7　　　**c** 0.9×10^4　　　**d** 59.6×10^{-6}

2　Write as a single power of 10
　　a $10^4 \times 10^5$　　　**b** $10^3 \times 10^{-2}$　　**c** $10^8 \div 10^4$　　　**d** $\dfrac{10^{-3}}{10^{-4}}$

3　Work out these calculations.
　　Give your answers in standard form.
　　a　$3 \times 2 \times 10^5 = \boxed{} \times 10^5$　　b　$5 \times 1.5 \times 10^{-7}$
　　c　$12 \times 5 \times 10^3$　　　　　　　　d　$9 \times 2 \times 10^{-6}$
　　e　$8 \times 10^{-3} \div 2 = \boxed{} \times 10^{-3}$　　f　$6 \times 10^3 \div 3$
　　g　$4 \times 10^8 \div 8$　　　　　　　　　h　$1 \times 10^{-11} \div 5$

> **Q3c hint** Check that your answer is in standard form.

4　**Problem-solving / Real** A grain of rice weighs 2×10^{-5} kg.
　　How many grains of rice are there in 1 kg?

Key point 8

To multiply and divide numbers in standard form, use the laws of indices to simplify the power of 10.

Example 8

Giving your answer in standard form, work out
a　$12 \times 10^3 \times 3 \times 10^2$

$12 \times 10^3 \times 3 \times 10^2 = 12 \times 3 \times 10^3 \times 10^2$

> Rewrite the multiplication, grouping the numbers together and the powers of 10 together.

$= 36 \times 10^5$

> Multiply the numbers together. Use the laws of indices to combine the powers of 10.

$= 3.6 \times 10^6$

> If the number part is not between 1 and 10, rewrite the number in standard form.

b　$\dfrac{9 \times 10^4}{3 \times 10^2}$

$\dfrac{9 \times 10^4}{3 \times 10^2} = \dfrac{9}{3} \times \dfrac{10^4}{10^2}$

> Write the calculation as two fractions, grouping the numbers together and the powers of 10 together.

$= 3 \times 10^2$

> Divide the numbers and use the laws of indices for the powers of 10. Check that the answer is in standard form.

5 Giving your answers in standard form, work out

   a $3 \times 10^7 \times 7 \times 10^3$ b $5 \times 10^2 \times 4 \times 10^{11}$ c $4 \times 10^{-2} \times 3 \times 10^6$

   d $1.5 \times 10^7 \times 2 \times 10^{-1}$ e $3 \times 10^9 \times 3.2 \times 10^{-3}$ f $7 \times 10^{-6} \times 1.11 \times 10^{-5}$

6 Giving your answers in standard form, work out

   a $\dfrac{8 \times 10^{-7}}{2 \times 10^3}$ b $\dfrac{6 \times 10^{-9}}{2 \times 10^{-3}}$

   c $\dfrac{2 \times 10^{-10}}{4 \times 10^{-5}}$ d $(8.8 \times 10^8) \div (4 \times 10^3)$

   e $(9.01 \times 10^7) \div (1 \times 10^4)$ f $(2.4 \times 10^4) \div (3 \times 10^{-2})$

7 Use a calculator to work out these calculations.
   Give your answers to 3 significant figures.

   a $5.3 \times 10^7 \times 7.2 \times 10^3$ b $4.2 \times 10^7 \times 3.56 \times 10^{-2}$

   c $9.1 \times 10^{-4} \times 3.8 \times 10^{-6}$ d $(5.6 \times 10^3) \div (3.4 \times 10^8)$

   e $(3.2 \times 10^4) \div (5.02 \times 10^{-5})$ f $\dfrac{5.4 \times 10^{13}}{3.82 \times 10^{-5}}$

> **Q7 hint** To round a standard form number to 3 significant figures, only round the number part.

8 **STEM / Problem-solving** How many

   a kB in 1 GB

   b μg in 1 mg

   c μV in 1 GV?

> **Q8a hint** 1 kilobyte = 10^3 bytes, 1 gigabyte = 10^9 bytes

> **Q8b hint** 1 μg = 10^{-6} g, 1 mg = 10^{-3} g

9 **Reasoning / Problem-solving** A snail crawls 1×10^{-3} kilometres in one hour. How far will it travel in 3 days?

10 **Exam-style question**

   Work out

   a $8 \times 10^3 \times 5 \times 10^7$ **(2 marks)**

   b $\dfrac{1.8 \times 10^{-3}}{6 \times 10^{-5}}$ **(2 marks)**

   Give your answers in standard form.

> **Exam hint**
> Check that the number part of your answer is between 1 and 10.

Key point 9

To add and subtract numbers in standard form, write both numbers as ordinary numbers, add or subtract, and then convert back to standard form.

Example 9

Work out the value of

a $3.2 \times 10^7 + 1.9 \times 10^8$

$3.2 \times 10^7 = 32\,000\,000$ — Write both numbers as ordinary numbers.

$1.9 \times 10^8 = 190\,000\,000$

$32\,000\,000 + 190\,000\,000 = 222\,000\,000$ — Add the two numbers together.

$222\,000\,000 = 2.22 \times 10^8$ — Convert this number into standard form.

b $1.9 \times 10^{-4} - 3.4 \times 10^{-6}$

$1.9 \times 10^{-4} = 0.000\,19$ — Write both numbers as decimals.

$3.4 \times 10^{-6} = 0.000\,003\,4$

$0.000\,19 - 0.000\,003\,4 = 0.000\,186\,6$ — Subract the two numbers.

$0.000\,186\,6 = 1.866 \times 10^{-4}$ — Convert this number into standard form.

11 a Work out

    i $9 \times 10^{10} + 4 \times 10^{11}$

    ii $5 \times 10^8 + 7 \times 10^6$

    iii $9.2 \times 10^{-3} + 3.2 \times 10^{-2}$

    iv $5.4 \times 10^{-7} + 7.6 \times 10^{-5}$

    v $6.13 \times 10^7 + 7.2 \times 10^3$

    vi $5.05 \times 10^{21} + 5.05 \times 10^{20}$

  b Check your answers to part **a** using a calculator.

12 a Work out

    i $3 \times 10^9 - 7 \times 10^8$

    ii $6 \times 10^{-7} - 8 \times 10^{-9}$

    iii $5.3 \times 10^5 - 8.1 \times 10^3$

    iv $3.2 \times 10^{-4} - 4.7 \times 10^{-6}$

    v $4.7 \times 10^{-10} - 5.09 \times 10^{-12}$

    vi $9.99 \times 10^{-15} - 4.4 \times 10^{-18}$

  b Check your answers to part **a** using a calculator.

13 Use a calculator to work out

  a $3.2 \times 10^{-4} + 8.6 \times 10^{-8}$

  b $5.2 \times 10^{-7} - 4.11 \times 10^{-12}$

  c $3.2 \times 10^{11} + 5.004 \times 10^{14} - 4.19 \times 10^{13}$

14 Use a calculator to work out these calculations.
Give your answers to 3 significant figures.

  a $\dfrac{3.2 \times 10^4 - 7.6 \times 10^3}{4.1 \times 10^{-2}}$

  b $4.3 \times 10^8 \times 3.15 \times 10^6 - 6.1 \times 10^{14}$

  c $\dfrac{5.4 \times 10^{-3}}{2.5 \times 10^6 + 3.5 \times 10^{-8}}$

> **Q14a hint** Enter the calculation as a fraction.

15 **STEM / Problem-solving**
A litre of water contains approximately 3.35×10^{25} molecules of water.
What is the volume of one molecule of water?

> **Q15 hint**
> How many cm^3 in 1 litre?

16 **STEM** The distance between Earth and Mars is estimated to be 57.6 million km.

  a Write this distance in metres.

  b Write your distance from part **a** in standard form.

  c The speed of light is approximately 3×10^8 m/s.

  Use the formula time $= \dfrac{\text{distance}}{\text{speed}}$ to calculate an estimate for the time taken for light to travel from Mars to the Earth.

17 **STEM / Problem-solving** The closest distance from Venus to Earth is 3.8×10^7 km.
The highest speed a manned spacecraft ever achieved was 11.08 km/s, in 1969.
How long would it take a spacecraft travelling at this speed to reach Venus?

Give your answer in days to 3 significant figures.

18 Problem-solving: Space race

Objective • Interpret and calculate with numbers given in standard form.

You are the navigator of a space shuttle team which has entered the Fourth Great Solar Race.
The average speed of your spacecraft is $\frac{1}{10}$ of the speed of light, or 3×10^4 km/s.
Table 1 shows the distances between the planets in km.
The rules of the race are as follows:
- You must start on your home planet, Thales
- You get ten points for visiting each of the inner planets (Banneker, Boole and Pacioli)
- You get twenty points for visiting each of the outer planets (Cantor, Turing and Wiles)
- You cannot visit a planet more than once
- You must return to Thales at the end
- The time limit for the race is 36 hours

Plan your route by completing a journey table with the headings below. Make sure you follow all of
the rules and try to get as many points as possible.

From	To	Distance (km)	Time taken
Thales			

Hint How many seconds would it take
to travel from Thales to another planet?
Use the formula time $= \dfrac{\text{distance}}{\text{average speed}}$.
How long is this in hours and minutes?

What total distance did you travel?
What total time did it take you?
How many points did you score?
Compare with others in your class.
Reflect How did you decide which planets to visit, in which order?
Were you happy with your strategy? Could you improve it? If so, how?

Table 1

	Banneker	Boole	Cantor	Pacioli	Thales	Turing
Boole	6.6×10^8					
Cantor	1.3×10^9	1.5×10^9				
Pacioli	7.8×10^8	3.6×10^8	1.2×10^9			
Thales	1.8×10^8	5.2×10^8	1.5×10^9	7.2×10^8		
Turing	1.3×10^9	1.8×10^9	3×10^8	1.6×10^9	1.6×10^9	
Wiles	2.4×10^9	2.2×10^9	3.2×10^9	2.1×10^9	1.8×10^9	3.6×10^9

Hint To find the distance, for example, from Thales:
Move your finger across the Thales row. Stop at each number, and
look up the column. That is the distance from Thales to that planet.
Move your finger down the Thales column. Stop at each number, and
look across the row. That is the distance from Thales to that planet.

18 Check up

Reciprocals and fractions

1 Find the reciprocal of 0.4.

2 Work out
   a $\frac{4}{5} \times 1\frac{5}{7}$ b $4\frac{1}{4} \times \frac{2}{3}$

3 Work out
   a $1\frac{1}{2} \times 2\frac{2}{3}$ b $4\frac{1}{2} \div \frac{3}{4}$ c $3\frac{1}{3} \div 1\frac{5}{6}$

Indices

4 Write as a single power
   a $(5^2)^3$ b $(2^{-3})^4$ c $x^{-2} \times x^4$ d $\frac{3^{-8}}{3^5}$

5 Work out the value of
   a 4^{-1} b 2^{-3} c y^0 d $\left(\frac{3}{5}\right)^{-2}$ e $\left(\frac{3}{4}\right)^0$

Standard form

6 Write each number as an ordinary number.
   a 4×10^8 b 5.26×10^3 c 3.5×10^{-1} d 8.099×10^{-7}

7 Write each number in standard form.
   a 190 000   b 1 050 000 000   c 0.000 007   d 0.000 045 2

8 Giving your answers in standard form, work out
   a $4 \times 10^4 \times 2 \times 10^5$ b $5 \times 10^3 \times 3 \times 10^4$ c $2.2 \times 10^{-2} \times 3 \times 10^7$

9 Work out
   a $9 \times 10^8 \div 3 \times 10^2$ b $4 \times 10^4 \div 8 \times 10^{-3}$ c $(7.7 \times 10^{-8}) \div (2.2 \times 10^{-2})$

10 Work out
   a $3 \times 10^{12} + 2.4 \times 10^{11}$ b $9.1 \times 10^{-5} - 3.5 \times 10^{-7}$

11 Use a calculator to work out
   a $4.5 \times 10^{-3} \times 9.2 \times 10^{-6}$ b $\frac{4.5 \times 10^7}{2.3 \times 10^{-2} + 5.7}$

Give your answers in standard form to 3 significant figures.

12 How sure are you of your answers? Were you mostly

Just guessing   Feeling doubtful   Confident ?

What next? Use your results to decide whether to strengthen or extend your learning.

* Challenge

13 The answer to a question is 4^{-5}.
Write as many questions as possible that give this answer.
Try to use
   • laws of indices
   • standard form
   • reciprocals.

Reflect

18 Strengthen

Reciprocals and fractions

1   Work out the reciprocal of
    a   0.2   b   0.5   c   0.1   d   0.25   e   0.05

Q1a hint Reciprocal of 0.2 $= \dfrac{1}{0.2} = \square$

2   a   Write $3\frac{2}{5}$ as an improper fraction.
    b   Copy and complete the calculation.

$$3\frac{2}{5} \times \frac{2}{3} = \frac{\square}{5} \times \frac{\square}{\square}$$

$$= \frac{\square \times \square}{\square \times \square}$$

$$= \frac{\square}{\square} = \square\frac{\square}{\square}$$

3   Work out
    a $2\frac{1}{2} \times \frac{1}{3}$ b $3\frac{1}{5} \times \frac{2}{5}$ c $\frac{3}{7} \times 4\frac{1}{2}$

4   Copy and complete the calculation.

$$3\frac{1}{3} \times 1\frac{3}{4} = \frac{\square}{3} \times \frac{\square}{\square}$$

$$= \frac{\square \times \square}{\square \times \square}$$

$$= \frac{\square}{\square} = \square\frac{\square}{\square}$$

Q4 hint Divide the 10 and the 4 by 2.

5   Work out
    a $1\frac{1}{2} \times \frac{1}{3}$ b $2\frac{4}{5} \times \frac{5}{7}$ c $\frac{2}{3} \times 5\frac{1}{4}$

6   Copy and complete the calculation.

$$1\frac{1}{4} \times 2\frac{1}{3} = \frac{\square}{4} \times \frac{\square}{3}$$

$$= \frac{\square \times \square}{\square \times \square}$$

$$= \frac{\square}{\square} = \square\frac{\square}{\square}$$

7   Work out
    a $1\frac{3}{5} \times 1\frac{1}{3}$ b $2\frac{1}{3} \times 4\frac{1}{2}$ c $1\frac{1}{4} \times 1\frac{1}{3}$

8   Janice starts a problem.

$$2\frac{1}{3} \div \frac{2}{3} = \frac{7}{3} \div \frac{2}{3}$$
$$= \frac{7}{3} \times \ldots$$

Copy and complete the calculation.

9   Work out
    a $3\frac{1}{5} \div \frac{1}{4} = \frac{\square}{5} \div \frac{1}{4} =$ b $\frac{4}{5} \div 2\frac{1}{2}$ c $1\frac{1}{2} \div \frac{2}{3}$

Indices

1 a i Copy and complete $8^2 \div 8^2 = \dfrac{8^2}{8^2} = \dfrac{\square}{\square} = \square$.

   ii Copy and complete $8^2 \div 8^2 = 8^{\square - \square} = 8^{\square}$.

   iii Use your answers to parts **i** and **ii** to find the value of 8^0.

   b Repeat part **a** for $10^2 \div 10^2$ to find the value of $10^{\square}$.

   c Copy and complete the rule:

   When you write a number to the power 0, the answer is _____.

2 a i Copy and complete $7^2 \div 7^5 = 7^{\square}$.

   ii Copy and complete

   $7^2 \div 7^5 = \dfrac{7^2}{7^5} = \dfrac{\cancel{7} \times \cancel{7}}{\cancel{7} \times \cancel{7} \times 7 \times 7 \times 7} = \dfrac{1}{7^{\square}}$

> **Q2a hint** To divide powers of the same number, subtract the indices.

   iii Use your answers to parts **i** and **ii** to copy and complete $7^{-3} = \dfrac{1}{7^{\square}}$.

   b i Copy and complete $4^3 \div 4^5 = 4^{\square}$

   ii Copy and complete

   $4^3 \div 4^5 = \dfrac{4^3}{4^5} = \dfrac{\cancel{4} \times \cancel{4} \times \cancel{4}}{\cancel{4} \times \cancel{4} \times \cancel{4} \times 4 \times 4} = \dfrac{1}{4^{\square}}$

   iii Use your answers to parts **i** and **ii** to copy and complete $4^{-2} = \dfrac{1}{4^{\square}}$.

3 Work out

   a 8^{-1} **b** 5^{-1} **c** 9^0 **d** $\left(\dfrac{3}{4}\right)^0$ **e** $\left(\dfrac{1}{2}\right)^{-1}$

4 Write as a single power

   a $2^3 \times 2^{-4}$ **b** $4^{-2} \times 4^5$ **c** $8^{-9} \times 8^2$ **d** $19^{-7} \times 19^{-3}$

> **Q4a hint** $2^{(3 + -4)} = 2^{\square}$

5 Write as a single power

   a $7^{-2} \div 7^3$ **b** $12^8 \div 12^{-2}$ **c** $6^2 \div 6^{-5}$ **d** $2^{-7} \div 2^{-4}$

6 Write as a single power

   a $(3^2)^4 = 3^{(2 \times 4)} = 3^{\square}$ **b** $(5^2)^7$ **c** $(19^3)^{10}$ **d** $(6^2)^3$
   e $(5^{-2})^3$ **f** $(5^3)^{-4}$ **g** $(9^{-2})^{-3}$ **h** $(6^{-4})^{-2}$

> **Q6e hint** The rule is the same if one or both of the powers are negative.

7 Copy and complete

   a $3^{-1} = \dfrac{1}{3}$

   $3^{-2} = \left(\dfrac{1}{3}\right)^2 = \dfrac{1}{\square} \times \dfrac{1}{\square} = \dfrac{1}{\square}$

   $3^{-3} = \left(\dfrac{1}{3}\right)^3 = \dfrac{1}{\square} \times \dfrac{1}{\square} \times \dfrac{1}{\square} = \dfrac{1}{\square}$

   b $\left(\dfrac{2}{5}\right)^{-1} = \dfrac{5}{2}$

   $\left(\dfrac{2}{5}\right)^{-2} = \left(\dfrac{5}{2}\right)^2 = \dfrac{5}{2} \times \dfrac{5}{2} = \dfrac{\square}{\square}$

   $\left(\dfrac{2}{5}\right)^{-3} = \left(\dfrac{5}{2}\right)^3 = \dfrac{5}{2} \times \dfrac{5}{2} \times \dfrac{5}{2} = \dfrac{\square}{\square}$

8 Work out the value of

   a $\left(\dfrac{2}{3}\right)^{-2}$ **b** $\left(\dfrac{1}{4}\right)^{-3}$ **c** $\left(\dfrac{4}{7}\right)^{-2}$ **d** $\left(\dfrac{3}{10}\right)^{-3}$

Standard form

1. Copy and complete this table of powers of 10.

10^{-6}	10^{-5}	10^{-4}	10^{-3}	10^{-2}	10^{-1}	10^0	10^1	10^2	10^3	10^4	10^5	10^6
		0.0001			0.1		10		1000	10 000		

2. Write each number as an ordinary number.
   - a $3 \times 10^{-4} = 3 \times 0.0001 = \square$
   - b $5 \times 10^2 = 5 \times \square = \square$
   - c 9×10^{-4}
   - d 1.2×10^3
   - e 5.7×10^{-5}
   - f 1.12×10^2
   - g 9.03×10^{-4}
   - h 1.01×10^6

   > **Q2 hint** Use your values from the table in Q1.

3. A number written in standard form looks like this.

   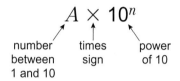

   $A \times 10^n$

   number between 1 and 10 → times sign → power of 10

   Write each number in standard form.
   - a 18 000
   - b 960 000
   - c 40 000
   - d 9 000 000
   - e 751 000
   - f 1 080 000

   > **Q3a hint** 1.8 lies between 1 and 10.
   > Multiply by how many 10s to get to 18 000?
   > 1.8
   > 18 000

4. Write each number in standard form.
   - a 0.0036
   - b 0.000 12
   - c 0.234
   - d 0.06
   - e 0.000 04
   - f 0.000 005 08

   > **Q4a hint** 3.6 lies between 1 and 10.
   > Divide by how many 10s to get to 0.0036?
   > 3.6
   > 0.0036

5. Copy and complete
   $$3 \times 10^7 \times 2 \times 10^3 = 3 \times 2 \times 10^{\square} \times 10^{\square}$$
   $$= \square \times 10^{\square}$$

6. Work out
   - a $5 \times 10^2 \times 1 \times 10^9$
   - b $2 \times 10^{-8} \times 2 \times 10^2$
   - c $3.4 \times 10^{-4} \times 2 \times 10^{-5}$

7. Giving your answers in standard form, work out these calculations. The first one has been done for you.
   - a $4 \times 10^5 \times 3 \times 10^{-2} = 4 \times 3 \times 10^{5-2} = 12 \times 10^3 = 1.2 \times 10^4$
   - b $9 \times 10^{-2} \times 8 \times 10^4$
   - c $7 \times 10^8 \times 5 \times 10^5$
   - d $2.1 \times 10^{11} \times 6 \times 10^{-15}$

8. Copy and complete
   $$\frac{9 \times 10^7}{3 \times 10^3} = \frac{9}{3} \times \frac{10^7}{10^3} = 3 \times 10^{\square}$$

9. Giving your answers in standard form, work out
   - a $\dfrac{8 \times 10^8}{2 \times 10^3}$
   - b $\dfrac{1 \times 10^5}{2 \times 10^{-2}}$
   - c $\dfrac{6.9 \times 10^5}{3 \times 10^{-2}}$
   - d $(8 \times 10^{-4}) \div (2 \times 10^3)$

   > **Q9d hint** Write the division as a two-line expression: $\dfrac{8 \times 10^{-4}}{2 \times 10^3}$

10. a Rewrite 3.5×10^7 as $\square \times 10^8$.
    b Use your answer to part **a** to work out
       i $5.2 \times 10^8 + 3.5 \times 10^7$
       ii $5.2 \times 10^8 - 3.5 \times 10^7$

    > **Q10a hint**
    >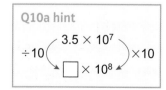
    > $\div 10 \, \big(\, 3.5 \times 10^7 \, \big) \times 10$
    > $\square \times 10^8$

11 a Rewrite 2.23×10^{-5} as $\boxed{} \times 10^{-3}$.

    b Use your answer to part **a** to work out

      i $2.23 \times 10^{-5} + 1.5 \times 10^{-3}$

      ii $1.5 \times 10^{-3} - 2.23 \times 10^{-5}$

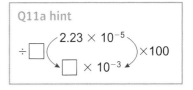

Q11a hint

12 Work out

    a $2.4 \times 10^6 + 3.1 \times 10^7$ b $7.6 \times 10^{-5} + 3.3 \times 10^{-6}$ c $5.04 \times 10^{-9} + 6.7 \times 10^{-7}$

    d $4.56 \times 10^{12} + 7.09 \times 10^{10}$ e $9.99 \times 10^{-9} + 1 \times 10^{-11}$ f $5.8 \times 10^9 - 5 \times 10^8$

    g $8.9 \times 10^{-3} - 4.7 \times 10^{-4}$ h $1.53 \times 10^{-11} - 7.2 \times 10^{-13}$ i $2.04 \times 10^{15} - 9.41 \times 10^{13}$

Q12 strategy hint Rewrite the number with the lower power of 10 as a number with the higher power of 10.

13 Use a calculator to work out

    a $6.47 \times 10^{-5} \times 1.55 \times 10^{12}$ b $7.904 \times 10^{11} - 3.9 \times 10^8$

    c $7.8 \times 10^5 \div (1.86 \times 10^{-5} + 9.79 \times 10^{-3})$ d $\dfrac{8.3 \times 10^{-8} + 1.33 \times 10^{-5}}{4.55 \times 10^5}$

Give your answers in standard form to 3 significant figures.

Q13 hint Make sure your calculator display shows the calculation correctly.

18 Extend

1 Write these numbers in order, smallest first.

    $\left(\frac{2}{3} \div 3\frac{4}{5}\right)$ 0.98 57% $\left(3\frac{1}{3} \div 1\frac{2}{3}\right)$ $\frac{7}{9}$

2 a $2^n \times 3 = 24$ b $a^{-1} \times 12 = 2$

      What is the value of n?   What is the value of a?

3 $a = -2$, $b = -3$ and $c = 4$.

    Work out the value of

    a $(10^a)^b$ b the reciprocal of a c c^b d $c^b \times c^c$

    e $c^a \div c^b$ f $12b^0$ g $\frac{1}{a} \times a$ h $(bc)^a$

4 **Reasoning**   a Write $(4^3)^8$ as a power of 2.

    b Write $(125^{-3})^9$ as a power of 5.

    c Write $(81^4)^{-5}$ as a power of the lowest possible integer.

5 Use a calculator to work out

    $\dfrac{0.3^{-7} + 8.2^{-4}}{3.02^{15} - 5.2^{-3}}$

Give your answer in standard form to 3 significant figures.

6 **Reasoning** A cuboid has sides of length $0.01\,\text{m}$, $0.03\,\text{m}$ and $0.03\,\text{m}$.

    Work out

    a the volume   b the surface area.

    Give your answers in standard form.

Q6 strategy hint

Sketch the cuboid and label the dimensions.

7 a How many cm in a km?

    b Copy and complete: $1\,\text{km} = 1 \times 10^{\square}\,\text{cm}$

    c Copy and complete

      i $1\,\text{km} = 10^{\square}\,\text{mm}$ ii $1\,\text{kg} = 10^{\square}\,\text{g}$ iii $1\,\text{ton} = 10^{\square}\,\text{g}$

8 Copy and complete these multiplication pyramids. Write all numbers in standard form. The number on each brick is the product of the numbers in the two bricks below.

a

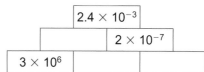

2×10^{-3}	
1×10^6	4×10^4

b

2.4×10^{-3}	
	2×10^{-7}
3×10^6	

9 **Reasoning / STEM** The distance from Earth to a distant comet (in kilometres) is recorded at five equally spaced intervals over a year.

3.2×10^9 3.91×10^9 4.201×10^9 3.99×10^9 4.29×10^9

Work out the mean distance between Earth and the comet.

10 **Reasoning / STEM / Real** A satellite travels 4×10^4 km in 1.5 hours.
Work out the speed of the satellite. Give your answer in standard form.

11 Write these numbers in order, largest first.

$2\frac{1}{3} \times 3\frac{3}{5}$ 6.5×10^2 $(3^2)^2$ $5.6 \times 10^4 - 3.2 \times 10^3$

12 **Modelling / Real** The diameter of the Earth is approximately 1.3×10^4 km.
Modelling the Earth as a sphere, estimate
   a   the surface area of the Earth
   b   the volume of the Earth.
Give your answers in standard form.
   c   About 70% of the Earth's surface is water. What is its land area?

> **Q12 hint** For a sphere of radius r, the surface area $= 4\pi r^2$ and the volume $= \frac{4}{3}\pi r^3$.
> Two significant figures is accurate enough for an estimate.

13 $a = 2 \times 10^{-3}$, $b = 3 \times 10^{-2}$ and $c = 5 \times 10^{-4}$.
Work out the value of
   a $2a$
   b $a + b$
   c ab
   d abc
   e $\dfrac{b}{c}$
   f $\dfrac{b}{a}$
   g   the reciprocal of a
   h c^2

14 **Reasoning / STEM** A bacterium cell has a surface area of 2×10^{-6} cm^2.
In a Petri dish there are ten cells.
The number of cells doubles each day.
What is the total surface area of the bacteria after
   a   1 day
   b   2 days
   c   10 days?
Give your answers in standard form.

> **Q14 hint** Assume that the cells do not join together.

15 **STEM / Problem-solving** A glass of water contains 240 grams of water.
1 mole of water weighs approximately 16 grams.
A mole contains 6.02×10^{23} molecules.
   a   How many molecules are there in the glass of water?
The ratio of the number of hydrogen atoms to oxygen atoms in water is $2 : 1$.
   b   How many hydrogen atoms are there in the glass of water?

16 **STEM / Problem-solving** Graphite and diamond are both forms of carbon.
The density of graphite is 2.27×10^3 kg/m^3.
The density of diamond is 3.52×10^3 kg/m^3.
1 g of carbon contains 5.01×10^{22} atoms.
How many carbon atoms are there in a 1 cm cube of
   a   graphite
   b   diamond?

17

> **Exam-style question**
>
> **a** Write down the value of 10^0. **(1 mark)**
>
> **b** Write 6.7×10^{-5} as an ordinary number. **(1 mark)**
>
> **c** Work out the value of $(3 \times 10^7) \times (9 \times 10^6)$.
>   Give your answer in standard form. **(2 marks)**
>
> *June 2012, Q17, 1MA0/1H*

Exam hint

Check that your answer is in standard form with the decimal point after the first digit.

18

> **Exam-style question**
>
> $$p^2 = \frac{x - y}{xy}$$
>
> $x = 8.5 \times 10^9$
>
> $y = 4 \times 10^8$
>
> Find the value of p.
>
> Give your answer in standard form correct to 2 significant figures. **(3 marks)**
>
> *June 2012, Q19, 1MA0/2H*

Exam hint

Check that your answer is the value of p and not the value of p^2.

18 Knowledge check

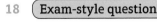

- When multiplying or dividing mixed numbers, change the mixed number to an improper fraction first. .... *Mastery lesson 18.1*

- To raise the power of a number to another power, multiply the indices. For example, $(5^3)^4 = 5^{3 \times 4} = 5^{12}$. .... *Mastery lesson 18.2*

- Any number to the power 0 is equal to 1. For example, $12^0 = 1$. .... *Mastery lesson 18.2*

- The reciprocal of a number is found by raising the number to the power −1. .... *Mastery lesson 18.2*

- The reciprocal of a number is 1 divided by the number. For example, $3^{-1} = \frac{1}{3}$. .... *Mastery lesson 18.2*

- To find the reciprocal of a fraction, turn the fraction upside down. For example, $\left(\frac{2}{3}\right)^{-1} = \frac{3}{2} = 1\frac{1}{2}$. .... *Mastery lesson 18.2*

- To find a negative power, find the reciprocal and then raise the value to the positive power. For example, $5^{-2} = \left(\frac{1}{5}\right)^2$. .... *Mastery lesson 18.2*

- **Standard form** is used to write very large or very small numbers. ...... *Mastery lesson 18.3*

- A number written in standard form is a value between 1 and 10 multiplied by a power of 10. For example, $3.5 \times 10^9 = 3\,500\,000\,000$. .... *Mastery lesson 18.3*

- To multiply or divide numbers in standard form, multiply or divide the number parts and combine the powers of 10 using the laws of indices. .... *Mastery lesson 18.5*

- To add or subtract numbers in standard form, write both numbers as ordinary numbers, add or subtract, and then convert back to standard form. .... *Mastery lesson 18.5*

Look back at the topics in this unit.

a Which one are you most confident that you have mastered?
What makes you feel confident?

b Which one are you least confident that you have mastered?
What makes you least confident?

c Discuss a question you feel least confident about with a classmate. How does discussing it make you feel?

18 Unit test

Log how you did on your Student Progression Chart.

1 Work out
   a $3\frac{1}{5} \times \frac{3}{4}$ *(2 marks)*
   b $4\frac{2}{3} \times 2\frac{1}{7}$ *(2 marks)*

2 Write as a single power
   a $3^{-7} \times 3^2$ *(1 mark)*
   b $\dfrac{10^8}{10^{-3}}$ *(1 mark)*
   c $(11^{-2})^5$ *(1 mark)*

3 Work out
   a $1\frac{1}{2} \div \frac{2}{5}$ *(2 marks)*
   b $3\frac{1}{3} \div 1\frac{3}{5}$ *(2 marks)*

4 Write each number as an ordinary number.
   a 3.04×10^7 *(1 mark)*
   b 2.1×10^{-3} *(1 mark)*

5 Write each number in standard form.
   a 9 070 000 000 *(1 mark)*
   b 0.000 031 4 *(1 mark)*

6 $a = 2$. Work out the value of
   a a^0 *(1 mark)*
   b a^1 *(1 mark)*
   c $(a^2)^3$ *(1 mark)*
   d a^{-2} *(1 mark)*

7 Work out
   a $\left(\dfrac{1}{8}\right)^2$ *(2 marks)*
   b $\left(\dfrac{3}{4}\right)^{-3}$ *(2 marks)*

8 Use a calculator to work out $7234 \times 69\,147$.
   a Write your answer to 3 significant figures. *(1 mark)*
   b Write your answer to part **a** in standard form. *(1 mark)*

ActiveLearn Homework, practice and support: Foundation 18 Unit test

9   Work out the reciprocal of 0.001. Give your answer in standard form.   *(2 marks)*

10 Write these numbers in order of size, smallest first.
    10^6 3.5×10^2 4.2×10^{-1} 0.5   *(3 marks)*

11 a   Work out $5.2^2 \times 8.5^{-8}$.
        Write down all the figures on your calculator display.   *(1 mark)*
    b   Write your answer to part **a** as an ordinary number to 2 significant figures.   *(1 mark)*

12 Work out
    a $3 \times 10^5 \times 2 \times 10^2$ *(2 marks)*
    b $\dfrac{4 \times 10^{-9}}{2 \times 10^4}$ *(2 marks)*
    c $8 \times 10^8 \times 7 \times 10^{-3}$ *(2 marks)*
    d $2 \times 10^3 \times 3.1 \times 10^{-7}$ *(2 marks)*
    Give your answers
    i   in standard form   ii   as ordinary numbers.

13 Giving your answers in standard form, work out
    a $8.1 \times 10^7 + 2 \times 10^5$ *(2 marks)*
    b $6.07 \times 10^{-8} - 4.2 \times 10^{-9}$ *(2 marks)*

14 **Reasoning** The mass of an atom of gold is 3.18×10^{-22} g.
    How many atoms of gold are there in 1 gram?
    Give your answer in standard form to 3 significant figures.   *(2 marks)*

15 **Reasoning** The table shows the results of weighing 100 feathers.

Mass (g)	Frequency
1.2×10^{-2}	22
1.3×10^{-2}	37
1.4×10^{-2}	35
1.5×10^{-2}	6

    Work out the mean mass of a feather.
    Give your answer as an ordinary number.   *(4 marks)*

Sample student answers

Which student gives the best answer and why?

> **Exam-style question**
>
> One sheet of paper is 9×10^{-3} cm thick.
> Mark wants to put 500 sheets of paper into the paper tray in his printer.
> The paper tray is 4 cm deep.
> Is the paper tray deep enough for 500 sheets of paper?
> You must explain your answer.
>
> *June 2013, Q15, 1MA0/1H*

Student A

$500 \times 9 \times 10^{-3} = 4.5$ cm

No, the paper tray is not deep enough.

Student B

$500 \div 9 \times 10^{-3} = 55\,555.56$

$55\,555.55 \div 500 = 111.11$

Student C

$4 \div (9 \times 10^{-3}) = 444.44$

Yes, the tray is deep enough.

19 CONGRUENCE, SIMILARITY AND VECTORS

A Russian doll set is made up of dolls that are a similar shape but different sizes. The smallest doll fits inside the next one and so on up to the largest.

The smallest doll in the photo is 2 cm tall. The largest is 16 cm. What is the scale factor between the two?

The third doll is an enlargement of the largest, with scale factor $\frac{1}{4}$. How tall is it?

19 Prior knowledge check

Numerical fluency

1   Simplify

   a $\frac{15}{20}$ b $\frac{9}{27}$ c $\frac{24}{30}$ d $\frac{36}{44}$

2   Work out

   a $\frac{2}{3}$ of 18   b $\frac{4}{5}$ of 30

   c $\frac{3}{4}$ of 36   d $\frac{5}{7}$ of 28

3   Work out

   a 3 − 7   b −2 + 8

   c 5 − −4   d −2 + −6

   e −2 × 5   f −3 × −4

Geometrical fluency

4   a Which shapes are enlargements of shape A? Give the scale factor for each one.

   b Which shapes are congruent to shape A?

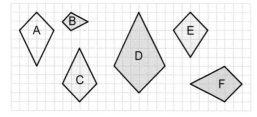

5   Name the angle that is

   a alternate to angle c

   b corresponding to angle b

   c vertically opposite to angle g.

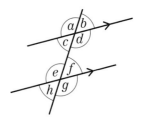

6   Find the size of

   a angle MOL

   b angle MLO.

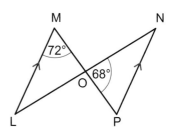

7   Write the column vector for the
    translation from
    a   A to B
    b   A to C
    c   A to D
    d   B to A
    e   C to A
    f   D to A.

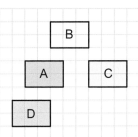

Algebraic fluency

8   Solve $\frac{x}{5} = \frac{3}{4}$.
    Write your answer as
    a   a mixed number
    b   a decimal.

*Challenge

9   In the diagram, the column vector from A
    to B is $\begin{pmatrix} 5 \\ -2 \end{pmatrix}$. The column vector from B
    to C is $\begin{pmatrix} 5 \\ 6 \end{pmatrix}$.

    B and D are on the same vertical line.
    C and D are on the same horizontal line.
    Find the column vector from A to D.

19.1 Similarity and enlargement

Objectives

- Understand similarity.
- Use similarity to solve angle problems.

Did you know?

Brunel's SS *Great Eastern* ship laid the first
successful transatlantic telegraph cable in
1866. The Science Museum has a model of the
ship at 1 : 96 scale.

Fluency

- When an enlargement makes a shape bigger, the scale factor is _____ than 1.
- When an enlargement makes a shape smaller, the scale factor is _____ than 1.

1   Which of these fractions are equivalent to $\frac{3}{4}$?
    $\frac{4}{5}$ $\frac{6}{8}$ $\frac{5}{6}$ $\frac{15}{20}$

2   Find the scale factor of the enlargement that maps shape A onto shape B.

a   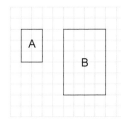   b   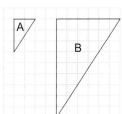   c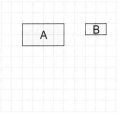

3   Find the scale factor of the enlargement that maps shape B onto shape A in each part of **Q2**.

Key point 1

When one shape is an enlargement of another, the shapes are **similar**.

Warm up

Questions in this unit are targeted at the steps indicated.

4 **Exam-style question**

A small photograph has a length of 4 cm and a width of 3 cm.
Shez enlarges the small photograph to make a large photograph.
The large photograph has a width of 15 cm.

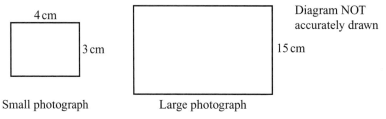

Diagram NOT
accurately drawn

Small photograph   Large photograph

The two photographs are similar rectangles.
Work out the length of the large photograph.   **(3 marks)**

June 2012, Q16, 5MB3F/01

Exam hint
Make sure
you show your
working. Most of
the marks will be
for your method.

5   What can you say about each pair of shapes in **Q2**?

6   **Communication** Identify the two similar triangles.
     Give a reason for your answer.

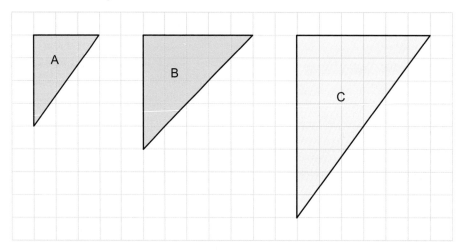

7   a   Measure the angles in the two similar triangles in **Q6**.
          What do you notice?
     b   Measure the angles in the other triangle.
          Discussion How can you use angles to decide if two shapes are similar?

Key point 2

These triangles are similar.
Corresponding sides are shown
in the same colour.
Corresponding angles are shown
in the same colour.

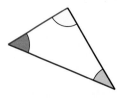
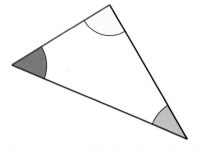

8   Triangle XYZ is similar to triangle ABC.

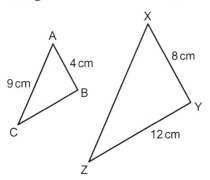

a   Which side corresponds to
    i   AB   ii   XZ   iii   BC?
b   What is the scale factor of the enlargement that maps
    i   triangle ABC to triangle XYZ   ii   triangle XYZ to triangle ABC?
c   Work out the length of
    i   XZ   ii   BC

> **Q8b hint** Compare the lengths of corresponding sides.

d   Copy and complete.
    i   ZX = 2 × ☐   ii   XY = 2 × ☐   iii   BC = ☐ × YZ
e   Which angle is the same as
    i   angle CAB   ii   angle XZY   iii   angle ABC?
f   Write $\dfrac{AB}{AC}$ as a fraction. Then write $\dfrac{XY}{XZ}$ as a fraction. What do you notice?

Key point 3

For similar shapes:
Corresponding sides are all in the same ratio.
Corresponding angles are equal.

---

9   Triangles ABC and PQR are similar. AC and PR are corresponding sides.
    a   Which side corresponds to BC?
    b   Work out the length of PQ.
    c   Work out the size of
        i   angle ACB   ii   angle ABC.

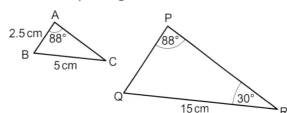

10   **Communication** a Are triangles PQR and UVW similar? Explain your answer.

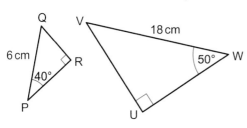

> **Q10a hint** Work out all the angles.

b   Which side corresponds to RQ?
c   Copy and complete.
    i   VW = ☐ × PQ   ii   UW = 3 × ☐   iii   QR = ☐ × ☐

19.2 **More similarity**

Objectives

- Find the scale factor of an enlargement.
- Use similarity to solve problems.

Did you know?

A fractal is a never-ending pattern. The Sierpinski triangle consists of a triangle containing similar triangles containing similar triangles reducing in size forever.

Fluency

- Work out $\frac{2}{3}$ of 24 $\frac{3}{4}$ of 32 $\frac{5}{12}$ of 36
- These triangles are similar. Find the values of x and y.

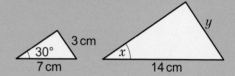

1   Find all the shapes similar to
    a   Shape A   b   Shape H   c   Shape E

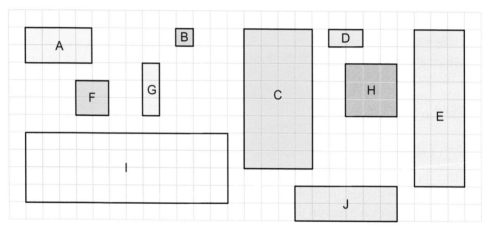

2   **Reasoning** Copy the diagram.
    Label all of the angles of 53°.
    Choose a reason from the box for each.

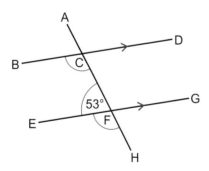

corresponding angle
vertically opposite angle
alternate angle

3   In the diagram, AB = 18 cm.
    The point C divides AB in the ratio 2 : 1.
    Work out the lengths of
    a   AC   b   CB

A   C   B

4

> **Exam-style question**
>
> Quadrilaterals *ABCD* and *LMNP* are mathematically similar.
> Angle *A* = angle *L*   Angle *B* = angle *M*
> Angle *C* = angle *N*   Angle *D* = angle *P*
>
> Diagram NOT accurately drawn
>
>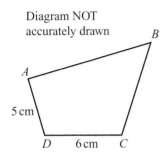
>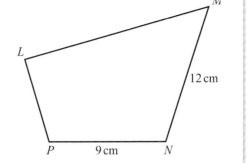
>
> **a** Work out the length of *LP*.   **(2 marks)**
> **b** Work out the length of *BC*.   **(2 marks)**
>
> *June 2014, Q17, 1MA0/2H*

Q4 strategy hint Start by working out the scale factor of enlargement using a pair of similar sides.

5 **a** **Communication** Explain how you know that these triangles are similar.
  b Write the scale factor of the enlargement that maps triangle A to triangle B.
  c Use the scale factor to find the value of x.

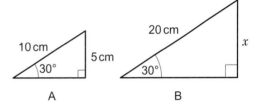

6 **a** Use the diagram in **Q5** to copy and complete $\dfrac{x}{20} = \dfrac{5}{\square}$.

Q5b hint You should get the same answer as you got in **Q5**.

  b Solve the equation in part **a** to find the value of x.

7 **a** **Communication** Explain how you know that this triangle is similar to the triangles in **Q5**.

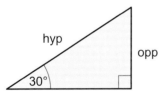

  b Write the value of $\dfrac{\text{opp}}{\text{hyp}}$.

Q7b hint The value of $\dfrac{\text{opp}}{\text{hyp}}$ is always the same when the angle is 30° because the triangles are similar.

  c The ratio $\dfrac{\text{opp}}{\text{hyp}}$ is called the sine of the angle.
  Check that $\sin 30° = \frac{1}{2}$.

8 Find the values of x and y.

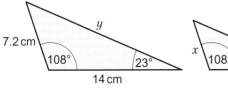
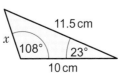

> **Example 1**

The lines AB and DE are parallel.

a   Show that triangles ABC and EDC are similar.

b   Work out angle DEC.

c   Write the scale factor of the enlargement that maps triangle ABC to triangle EDC.

d   Work out the length of AC.

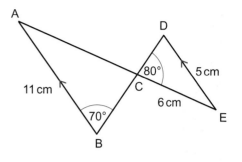

a   Angle ACB = 80°   (vertically opposite angles)

   Angle CDE = 70°   (corresponding angles)

   Triangles ABC and EDC are similar
   because they have the same angles.

> Two pairs of angles are the same, so the third pair has to be the same.

b   80° + 70° = 150°

   180° − 150° = 30°

> Angles in a triangle add to 180°.

   Angle DEC = 30°

c   AB corresponds to ED.

   The scale factor of the enlargement is $\frac{5}{11}$.

> Write the ratio $\frac{ED}{AB}$.

d   AC corresponds to EC.

   $\frac{AC}{6} = \frac{11}{5}$

> Corresponding sides are in the same ratio.
> Write the unknown length, AC, on the top of the fraction.

   $AC = 6 \times \frac{11}{5}$

   AC = 13.2 cm

> Solve the equation. Multiply both sides by 6.

9   **Reasoning** The lines PQ and ST are parallel.
   Triangles PQR and TSR are similar.

   a   Find the size of angle PRQ.

   b   Which side corresponds to PR?

   c   Write the scale factor of the enlargement that maps triangle PQR to triangle TSR.

   d   Work out the length of ST.

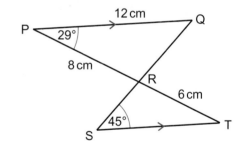

10   **Reasoning** The lines AB and DE are parallel.

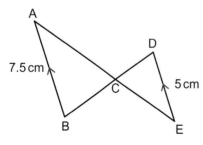

   a   Show that triangles ABC and EDC have the same angles, so they are similar.

   b   BD = 8 cm
      Work out the length of
      i   BC   ii   CD

> Q10a hint You don't need to know what the angles are.

11 **Reasoning** The lines DE and BC are parallel.
Triangles ABC and ADE are similar.

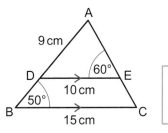

   a Work out the size of angle BAC.
   b Write the scale factor of the enlargement
      that maps triangle ADE to triangle ABC.
   c Work out the length of AB.
   d Work out the length of BD.

> **Q11d hint** Use your answer to part **c**.

12 **Problem-solving / Communication**
   a Show that triangle ABC and triangle ADE have the
      same angles, so they are similar.
   b Work out the scale factor of the enlargement
      that maps triangle ADE to triangle ABC.
   c Work out the length of BD.

> **Q12a hint** How do you know that DE and BC are parallel?

> **Q12b hint** Compare the sides AE and AC.

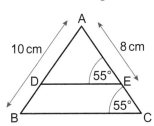

19.3 Using similarity

Objectives

- Understand the similarity of regular polygons.
- Calculate perimeters of similar shapes.

Did you know?

The Blackpool Tower (1894) was built to look 'similar' to the Eiffel Tower (1889) in Paris.

Fluency

Work out 3^2 0.5^2 2^3 0.1^3

Warm up

1 These rectangles are similar. Find the value of x.

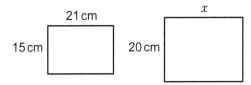

2 **Communication** These pentagons have the same angles.
   Are they similar?
   Explain your answer.

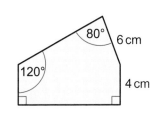
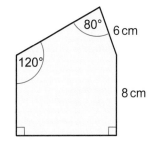

3   **Communication** These hexagons have sides in the same ratio. Are they similar?
     Explain your answer.

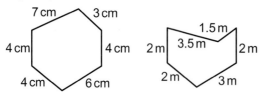

4   For each question, write 'Yes', 'No' or 'Need more information'.
     a   Two triangles share two identical angles. Are they similar?
     b   Two polygons share two identical angles. Are they similar?
     c   Two triangles have two pairs of sides in the same ratio. Are they similar?
     d   Two polygons have all sides in the same ratio. Are they similar?
     e   Two polygons share the same angles and have sides in the same ratio. Are they similar?

5   **Communication** Here are two regular pentagons. Are they similar?
     Explain your answer.

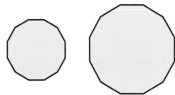

> **Q5 communication hint**
> A regular polygon has angles that are all the same size and sides that are all the same length.

6   **Communication** a Are all regular pentagons similar? Explain your answer.
     b   Are all regular hexagons similar? Explain your answer.

7   **Reasoning** Here are two regular dodecagons (12 sides).
     They are similar.

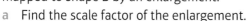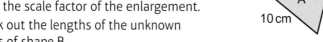

     Regular polygons with lots of sides start to look like circles.
     What does this suggest about the similarity of circles?

8   **Problem-solving** In the diagram, shape A is
     mapped to shape B by an enlargement.
     a   Find the scale factor of the enlargement.
     b   Work out the lengths of the unknown
         sides of shape B.
     c   Work out the perimeter of shape A.
     d   Work out the perimeter of shape B.
     e   Copy and complete.
         Perimeter of shape B = ☐ × perimeter of shape A
     f   Look at your answers to parts **a** and **e**. What do you notice?

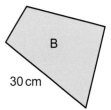

> **Key point 4**
>
> When a shape is enlarged, the perimeter of the shape is enlarged by the same scale factor.

9 **Problem-solving**
These two windows are
mathematically similar.
The perimeter of window
A is 8 m.
Work out the perimeter
of window B.

A  B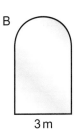

1.2 m   3 m

> **Q9 strategy hint** Work
> out the scale factor of the
> enlargement using the two
> corresponding sides given.

10 **Exam-style question**

The diagram shows two flower beds made in the shape of similar kites.

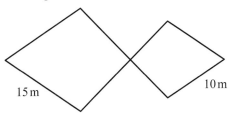

15 m   10 m

The perimeter of the larger flower bed is 48 m.
Work out the perimeter of the smaller flower bed.   **(2 marks)**

Exam hint

In questions
on similarity,
use the words
'scale factor'
in your answer.

19.4 **Congruence 1**

Objectives

- Recognise congruent shapes.
- Use congruence to work out unknown angles.

Why learn this?

Congruent triangles are used in the supports for a roof.

Fluency

Work out the size of angle x.

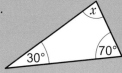

30°   70°

1   How many pairs of congruent triangles are there in this roof support?

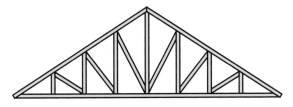

2   Which *two* of these triangles are congruent?

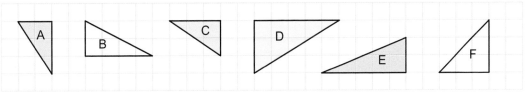

Warm up

3 a On squared paper, draw shapes congruent to these shapes:

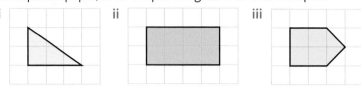

i   ii   iii

> **Q3 communication hint** Different **orientation** means facing another way, e.g.

b Draw another shape congruent to each shape in part **a**, in a different orientation.

Key point 5

Triangles are **congruent** if they have equivalent
- SSS (all three sides)
- SAS (two sides and the included angle)
- ASA (two angles and the included side)
- RHS (right angle, hypotenuse, side)

Triangles where all angles are the same (AAA) are similar, but might not be congruent.

---

4 Here are some pairs of congruent triangles. Work out the size of angles x and y.

a   b

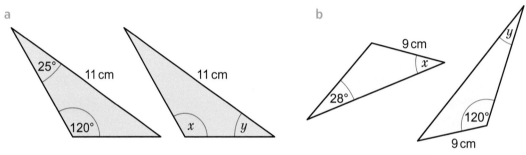

> **Q4a hint** The angle opposite the 11 cm side will be the same in each triangle. To find angle y, use the fact that angles in a triangle add to 180°.

5 Construct these triangles accurately using a ruler and compasses
Are your two triangles congruent?

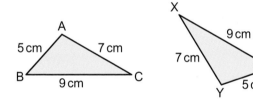

Key point 6

When two shapes are congruent, one can be rotated or reflected to fit exactly on the other.

Example 2

XYZ is an isosceles triangle. M is the midpoint of YZ.
a What can you say about triangles XMY and XMZ?
b Find the size of angle MXZ.

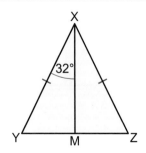

> Justify every statement.

a XM cuts the triangle in half, so triangle XMY is congruent to triangle XMZ

b Angle MXZ = angle MXY   (corresponding positions in congruent triangles)
    Angle MXZ = 32°

6 **Communication** ABCD is a rectangle.

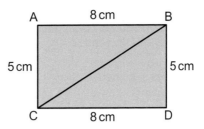

a What can you say about triangles DAB and BCD?

b Which angle is the same as angle DBA?

c Are triangles DAB and BCD congruent? If so give a reason for congruency (SSS, SAS etc).

7 **Problem-solving / Communication**
ABCD is an arrowhead.

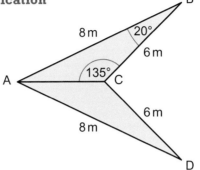

a Find the size of angle ADC.

b Find the size of angle CAD.

c Are triangles ABC and ACD congruent? If so give a reason for congruency.

> **Q7 hint**
> What can you say about triangles ABC and ACD?

8 **Problem-solving** The diagram shows a rhombus WXYZ.
Find the size of

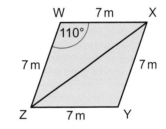

a angle XYZ

b angle ZXY.

9 **Problem-solving**
In the diagram, AB = BC and BD bisects angle ABC.

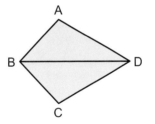

a Are triangles ABD and BCD congruent?

b Which angle is the same as angle BAD?

c Which side is the same as AD?

> **Q9 hint** Label your diagram with information about equal sides and angles.

10 **Reasoning** Draw these triangles accurately using a ruler and protractor. Are your triangles congruent?

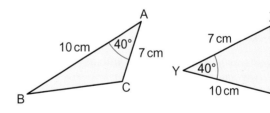

11 **Problem-solving / Communication**
In this diagram, C is the centre of the circle, radius 6 cm.

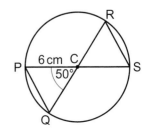

a Write down the size of angle RCS. Give a reason for your answer.

b Write the lengths of CQ, CR and CS. Give reasons for your answers.

c Are triangles PCQ and CRS congruent? Give a reason for your answer.

12

In the diagram, C is the centre of the circle.

a Are triangles ABC and CDE congruent?

b Work out the size of angle CAB.

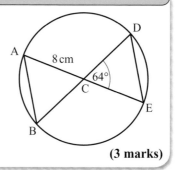

(3 marks)

Exam hint

For part **a**, mark equal lengths and angles on the diagram to help you decide.

19.5 Congruence 2

Objectives

• Use congruence to work out unknown sides.

Why learn this?

If you can prove a result then you know it must be true.

Fluency

Find two pairs of corresponding angles.
Find two pairs of alternate angles.

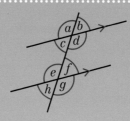

1 Find the values of x, y and z in each diagram.

a

b

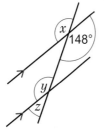
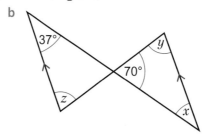

2 M is the midpoint of AB. M is also the midpoint of CD.

a Which length is the same as AM?

b Which length is the same as MD?

c Which angle is the same as angle AMC?

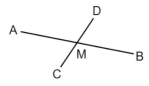

3 Draw these triangles accurately using a ruler and protractor.

a   b

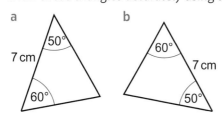

Are your triangles congruent?

4 Which triangle is congruent to triangle A?

Q4 hint Do they rotate or reflect to fit exactly on top of each other?

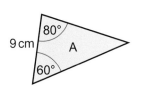
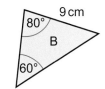
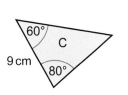
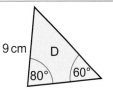

Example 3

In the diagram, R is the midpoint of QS.

a Are triangles PQR and TSR congruent?

b Find the length of RT.

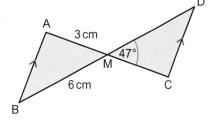

Mark angles and lengths on the diagram.

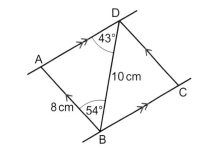

Vertically opposite angles.

$180° - 70° - 50° = 60°$

R is the midpoint of QS.

a Yes, triangle PQR is congruent to triangle TSR.

b RT = RP = 14 cm   (corresponding sides) — Use the fact that the triangles are congruent.

5 **Reasoning / Communication** In the diagram, M is the midpoint of AC. M is also the midpoint of BD.

a Copy the diagram. Label any sides or angles that are equal.

b Are triangles ABM and CDM congruent? Give a reason for your answer.

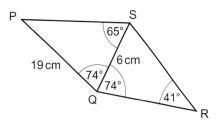

6 **Reasoning / Communication** In the diagram, ABCD is a parallelogram.

a Find the size of angle DBC.

b Find the size of angle BDC.

c Is triangle ABD congruent to triangle CDB? Give a reason for your answer.

d Find the length of CD.

7 **Reasoning / Communication**

a Work out the size of angle RSQ.

b Are triangles PQS and RQS congruent? Give a reason for your answer.

c PQ = 19 cm. Which other side has length 19 cm?

8   **Reasoning / Communication** Draw these pairs of right-angled triangles accurately using a ruler, protractor and compasses.
Are the triangles in each pair congruent? Give a reason for your answer.

a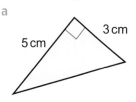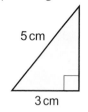

b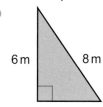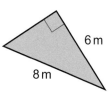

9   **Exam-style question**

   a Find the length of AC.
   b Find the length of *FD*.
   c Are triangles *ABC* and *DEF* congruent?
   d Find the length of *DE*.

   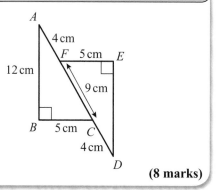

   (8 marks)

10   **Reasoning / Communication**
   a   Which angle in triangle NOL is 64°?
   b   Name a side equal in length to LM.
   c   Are triangles LMN and NOL congruent?
       Give a reason for your answer.

   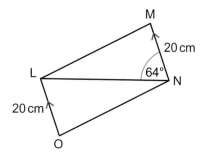

11   **Reflect** In your own words, write a definition for the terms 'congruent' and 'similar'.

19.6 **Vectors 1**

Objectives

- Add and subtract vectors.
- Find the resultant of two vectors.

Why learn this?

Column vectors may be used to solve geometrical problems in 2D or 3D.

Fluency

Work out   −2 + 4   3 − 5   6 − −4   −5 − 2

1   Here are some column vectors.

   $\begin{pmatrix} 4 \\ 2 \end{pmatrix}$ $\begin{pmatrix} 2 \\ 4 \end{pmatrix}$ $\begin{pmatrix} -2 \\ -4 \end{pmatrix}$ $\begin{pmatrix} -4 \\ -2 \end{pmatrix}$

   a   Which is the column vector from A to B?
   b   Which is the column vector from B to A?

2   Copy the grid and shapes. Translate

   a   shape A by $\begin{pmatrix} 3 \\ 3 \end{pmatrix}$

   b   shape B by $\begin{pmatrix} -6 \\ 4 \end{pmatrix}$

   c   shape C by $\begin{pmatrix} -3 \\ -5 \end{pmatrix}$.

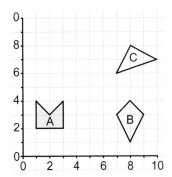

3   Describe the translation from shape A to shape B as a column vector.

   a   b

4   Describe each of these translations as
    column vectors.

   a   A to B
   b   B to C
   c   A to C

   Discussion What is the connection between
   your answers to parts **a**, **b** and **c**?

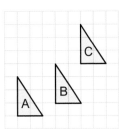

Key point 7

To add two column vectors, add the top numbers and add the bottom numbers.

$$2 + 1$$
$$\begin{pmatrix} 2 \\ 5 \end{pmatrix} + \begin{pmatrix} 1 \\ 3 \end{pmatrix} = \begin{pmatrix} 3 \\ 8 \end{pmatrix} \text{ and } \begin{pmatrix} -1 \\ 7 \end{pmatrix} + \begin{pmatrix} 5 \\ -2 \end{pmatrix} = \begin{pmatrix} 4 \\ 5 \end{pmatrix}$$
$$5 + 3 \qquad\qquad 7 + -2$$

5   Add these column vectors. Give your answers in the form of a single vector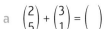

   a $\begin{pmatrix} 2 \\ 5 \end{pmatrix} + \begin{pmatrix} 3 \\ 1 \end{pmatrix} = (\quad)$ b $\begin{pmatrix} 4 \\ 1 \end{pmatrix} + \begin{pmatrix} 3 \\ 2 \end{pmatrix}$ c $\begin{pmatrix} 5 \\ 4 \end{pmatrix} + \begin{pmatrix} 2 \\ 0 \end{pmatrix}$

   d $\begin{pmatrix} 8 \\ -3 \end{pmatrix} + \begin{pmatrix} -1 \\ 2 \end{pmatrix}$ e $\begin{pmatrix} -5 \\ 4 \end{pmatrix} + \begin{pmatrix} 5 \\ -4 \end{pmatrix}$ f $\begin{pmatrix} -2 \\ -3 \end{pmatrix} + \begin{pmatrix} 2 \\ 4 \end{pmatrix}$

Key point 8

Two translations can be combined into a single translation by adding the column vectors.

6   Shape A is translated to shape B by the column vector $\begin{pmatrix} 3 \\ 1 \end{pmatrix}$.

   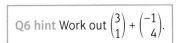
   Q6 hint Work out $\begin{pmatrix} 3 \\ 1 \end{pmatrix} + \begin{pmatrix} -1 \\ 4 \end{pmatrix}$.

   Shape B is translated to shape C by the column vector $\begin{pmatrix} -1 \\ 4 \end{pmatrix}$.

   Find the column vector of the single translation from shape A to shape C.

7   Shape X is translated to shape Y by the column vector $\begin{pmatrix} 4 \\ -3 \end{pmatrix}$.

   Shape Y is translated to shape Z by the column vector $\begin{pmatrix} 2 \\ -2 \end{pmatrix}$.

   Find the column vector of the single translation from shape X to shape Z.

The diagram shows a vector that starts at A and finishes at B.

You can write this vector as $\overrightarrow{AB} = \begin{pmatrix} 3 \\ 2 \end{pmatrix}$.

8   Write as column vectors

   a $\overrightarrow{PQ}$ b $\overrightarrow{XY}$ c $\overrightarrow{UV}$

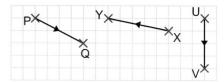

9   a   Write as column vectors

      i $\overrightarrow{AB}$ ii $\overrightarrow{BC}$ iii $\overrightarrow{AC}$

   b   **Communication**

      Show that $\overrightarrow{AB} + \overrightarrow{BC} = \overrightarrow{AC}$.

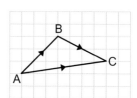

A to B followed by B to C is equivalent to A to C.

$\overrightarrow{AB} + \overrightarrow{BC} = \overrightarrow{AC}$

$\overrightarrow{AC}$ is the **resultant** of $\overrightarrow{AB}$ and $\overrightarrow{BC}$.

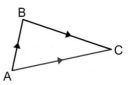

10   Copy and complete.

   a $\overrightarrow{NQ} + \overrightarrow{QT} = \square$ b $\overrightarrow{MP} + \square = \overrightarrow{MR}$ c $\square + \overrightarrow{TZ} = \overrightarrow{VZ}$

11   Find the resultant of

   a $\begin{pmatrix} -6 \\ 4 \end{pmatrix}$ and $\begin{pmatrix} 3 \\ 2 \end{pmatrix}$ b $\begin{pmatrix} -2 \\ 5 \end{pmatrix}$ and $\begin{pmatrix} -1 \\ 4 \end{pmatrix}$ c $\begin{pmatrix} 4 \\ -2 \end{pmatrix}$ and $\begin{pmatrix} 2 \\ 3 \end{pmatrix}$.

   Q11a hint Add the column vectors.

A single letter may be used to represent a vector. The letter is shown in **bold type**.

$\mathbf{a} = \begin{pmatrix} -2 \\ 5 \end{pmatrix}$

$-\mathbf{a}$ is the negative of $\mathbf{a}$ and points in the opposite direction.

$-\mathbf{a} = \begin{pmatrix} 2 \\ -5 \end{pmatrix}$

You can't do bold handwriting, so you should underline a letter that represents a vector. For **a**, write <u>a</u>.

12   Here are two vectors, **p** and **q**.

   a   Copy vector **p**.

   b   Draw vector **q** at one end of vector **p**.

   c   Draw in the third side of the triangle and label it **p** + **q**.

   d   Repeat the steps in parts **a–c**, but draw **q** at the other end of **p**.

   Discussion Does it matter which end of **p** you draw **q**?

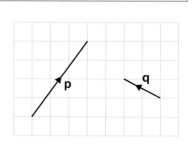

13 $\mathbf{p} = \begin{pmatrix} 4 \\ 2 \end{pmatrix}$ $\mathbf{q} = \begin{pmatrix} 1 \\ 3 \end{pmatrix}$ $\mathbf{r} = \begin{pmatrix} -3 \\ -1 \end{pmatrix}$

Copy the diagram and label the vectors using $\mathbf{p}$, $\mathbf{q}$, $\mathbf{r}$, $-\mathbf{p}$, $-\mathbf{q}$, $-\mathbf{r}$, $\mathbf{p} + \mathbf{q}$ and $\mathbf{q} + \mathbf{r}$.

19.7 **Vectors 2**

Objectives

- Subtract vectors.
- Find multiples of a vector.

Did you know?

Pilots use vectors to predict the position of their aircraft.

Fluency

Work out $-2 - 4$ $-3 - -7$ -2×5 -3×-4

1 Copy the diagram and label all of the vectors.

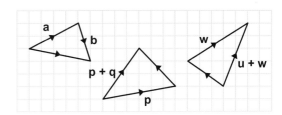

2 Find the resultant of $\begin{pmatrix} 4 \\ -1 \end{pmatrix}$ and $\begin{pmatrix} -3 \\ 5 \end{pmatrix}$.

Key point 12

$\mathbf{a} - \mathbf{b}$ is the same as $\mathbf{a} + (-\mathbf{b})$.

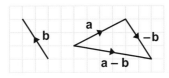

3 Copy the diagram and label all of the vectors.

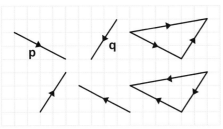

Key point 13

To subtract column vectors, subtract the top numbers and subtract the bottom numbers.

4 Subtract these column vectors.

a $\begin{pmatrix} 8 \\ 5 \end{pmatrix} - \begin{pmatrix} 3 \\ 4 \end{pmatrix}$ b $\begin{pmatrix} -2 \\ 3 \end{pmatrix} - \begin{pmatrix} 1 \\ 3 \end{pmatrix}$ c $\begin{pmatrix} 4 \\ -3 \end{pmatrix} - \begin{pmatrix} -1 \\ 2 \end{pmatrix}$

Warm up

5   $\mathbf{p} = \begin{pmatrix} 3 \\ 2 \end{pmatrix}$ $\mathbf{q} = \begin{pmatrix} -2 \\ 5 \end{pmatrix}$ $\mathbf{r} = \mathbf{p} - \mathbf{q}$

    a   Write **r** as a column vector.

    b   Show **p**, **−q** and **r** as a triangle on a grid.

6   $\mathbf{x} = \begin{pmatrix} 7 \\ -4 \end{pmatrix}$ $\mathbf{y} = \begin{pmatrix} 3 \\ -6 \end{pmatrix}$ $\mathbf{z} = \mathbf{x} - \mathbf{y}$

    a   Write **z** as a column vector.

    b   Show **x**, **−y** and **z** as a triangle on a grid.

Key point 14

You can multiply a vector by a number.

For example, if $\mathbf{a} = \begin{pmatrix} 4 \\ 3 \end{pmatrix}$ then $2\mathbf{a} = \begin{pmatrix} 8 \\ 6 \end{pmatrix}$ and $-3\mathbf{a} = \begin{pmatrix} -12 \\ -9 \end{pmatrix}$.

---

7   $\mathbf{p} = \begin{pmatrix} 2 \\ 5 \end{pmatrix}$ $\mathbf{q} = \begin{pmatrix} 3 \\ -2 \end{pmatrix}$

    Write as column vectors

    a   2**p**   b   3**q**   c   −4**p**   d   −5**q**

8   $\mathbf{r} = \begin{pmatrix} 3 \\ 0 \end{pmatrix}$ $\mathbf{t} = \begin{pmatrix} -2 \\ 3 \end{pmatrix}$

    Write as column vectors

    a   2**r**   b   −4**t**   c   2**t** + **r**   d   3**r** − **t**

9   $\mathbf{x} = \begin{pmatrix} 3 \\ 2 \end{pmatrix}$

    Show on a grid

    a   **x**   b   2**x**   c   −2**x**

    Discussion Compare the lengths and directions of these vectors. What do you notice?

10   $\mathbf{q} = \begin{pmatrix} -2 \\ 4 \end{pmatrix}$

    Write as column vectors

    a   a vector in the same direction as **q** but twice as long

    b   a vector the same length as **q** but in the opposite direction

    c   a vector 3 times as long as **q** but in the opposite direction

    d   a vector parallel to **q** but half as long.

> **Q10d hint** A **parallel vector** may have the same direction or the opposite direction. There are two possible answers.

11   $\mathbf{v} = \begin{pmatrix} 6 \\ -2 \end{pmatrix}$

    Choose from the vectors in the box to find a vector that is

    a   twice as long as **v**

    b   in the opposite direction to **v**

    c   half as long as **v**

    d   parallel to **v** but 3 times as long.

    You can only choose each vector once.

$$\begin{pmatrix} 18 \\ -6 \end{pmatrix} \quad \begin{pmatrix} 12 \\ -4 \end{pmatrix}$$
$$\begin{pmatrix} -3 \\ 1 \end{pmatrix} \quad \begin{pmatrix} -30 \\ 10 \end{pmatrix}$$

12   **Reflect** What is a vector? Write a definition in your own words.

19 Problem-solving

Objective • Use problem-solving strategies and then 'explain' or 'show that…'

> **Key point 15**
>
> There are many different problem-solving strategies. Here are some you can use:
> 1. pictures   2. smaller numbers   3. bar models   4. x for the unknown
> 5. flow diagrams   6. formulae   7. arrow diagrams

> **Example 4**
>
> These two shapes are mathematically similar.
>
> Perimeter = 46 cm
>
>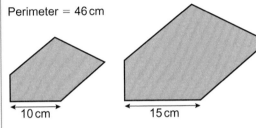
>
> 10 cm   15 cm
>
> Show that the large shape has a perimeter of 69 cm.
>
> | When asked to 'Explain' or 'Show that', you need to show some working. |
>
> | Decide on a problem-solving strategy. There is one missing number and you need to multiply or divide. An arrow diagram may help. Draw an arrow from a number you know to the number you need to find. |
>
>   side length   perimeter
> Small shape   ×1.5 ⌜ 10 cm   46 cm ⌝
> Large shape   ⌞ 15 cm   ☐ cm ⌟
>
>   side length   perimeter
> Small shape   ×1.5 ⌜ 10 cm   46 cm ⌝ ×1.5
> Large shape   ⌞ 15 cm   69 cm ⌟
>
> | When a shape is enlarged, the perimeter is enlarged by the same scale factor. So multiply 46 cm by 1.5. |
>
> The large shape has a perimeter of 69 cm. ——— | Write a sentence to answer the question. |

1   These two shapes are mathematically similar.

Perimeter = 32 cm

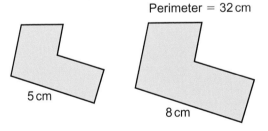

5 cm   8 cm

> **Q1 hint** What problem-solving strategy could help you?

The larger shape has a perimeter of 32 cm. Show that the perimeter of the smaller shape is 20 cm.

2   Patricia cuts out a rhombus. Then she cuts off one corner with a straight cut. Geoffrey says that now she has three corners, so she has made a triangle. Is Geoffrey correct? Explain.

3   David's age is x. Jesse is 3 years older than Lucy.
David is 5 years younger than Jesse.
The product of Lucy and David's age is 15.
a   Show that $x^2 + 2x - 15 = 0$.
b   Find Jesse's age.

> **Q3 hint** First write expressions for David's age, Jesse's age and Lucy's age.

4   Points X, Y, Z have coordinates (1, 2), (4, 6), (13, 18). Show that XZ is four times the length of XY.

5   Antony, Ross and Crista each invested money to start a business. Antony invested £5000, Ross invested £4000 and Crista invested £3000. In the first year, the business made a profit of £6000. Antony thinks he should get £2500, Ross thinks they should each get £2000. Who is correct? Explain.

6   Naomi is learning French. Every day she reads 12 pages of her French vocabulary book. There are 13 words on every page. After 9 days, she has 96 words left.
   a   Show that there are 1500 words in the French vocabulary book.
   b   Write an expression for the number of words in a vocabulary book when: someone reads p pages per day, there are q words on every page, and after r days, the person only has s words left.

7   **Reflect** Which problem-solving strategies did you use to answer these questions? How did they help you?

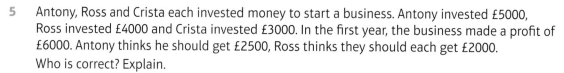

19 Check up

Log how you did on your Student Progression Chart.

Similarity and enlargement

1   Rectangle A is similar to rectangle B.
   a   Work out the scale factor of the enlargement that maps A to B.
   b   The perimeter of A is 48 cm. Work out the perimeter of B.

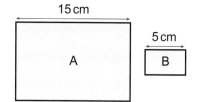

2   **Reasoning** Triangle ABC is similar to triangle EDC.
   a   Find the scale factor of the enlargement that maps triangle EDC to triangle ABC.
   b   Find the length of CD.

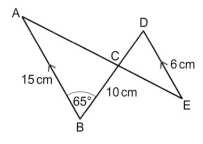

3   **Reasoning** Triangles ABC and PQR are similar. BC and QR are corresponding sides.

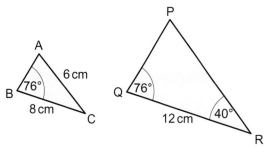

   a   Which side corresponds to AB?
   b   Find the length of PR.
   c   Find the size of
      i   angle ACB
      ii   angle BAC.

*Active*Learn   Homework, practice and support: Foundation 19 Check up

Congruence

4 **Communication** Decide whether each pair of triangles is congruent.
Give a reason for each answer.

a 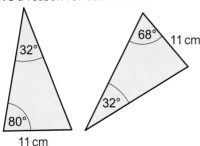 b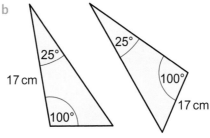

5 **Reasoning / Communication** In the diagram,
AB is parallel to CD.

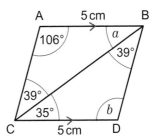

a Work out angles a and b.
Give reasons for your answers.

b Is triangle ABC congruent to triangle DCB?
Give a reason for your answer.

Vectors and translations

6 Shape A is translated to shape B by the column vector $\begin{pmatrix} 2 \\ 5 \end{pmatrix}$.

Shape B is translated to shape C by the column vector $\begin{pmatrix} -1 \\ 3 \end{pmatrix}$.

Find the column vector of the single translation from shape A to shape C.

7 Find the resultant of these vectors.

a $\overrightarrow{UV}$ and $\overrightarrow{VW}$ b $\begin{pmatrix} 4 \\ -3 \end{pmatrix}$ and $\begin{pmatrix} 2 \\ 3 \end{pmatrix}$

8 $\mathbf{a} = \begin{pmatrix} 3 \\ -1 \end{pmatrix}$ $\mathbf{b} = \begin{pmatrix} -2 \\ 4 \end{pmatrix}$

Write as column vectors

a $2\mathbf{a}$ b $\mathbf{a} - \mathbf{b}$ c $0.5\mathbf{b}$

9 Here are two vectors, **c** and **d**.
On squared paper, draw and label the vectors

a $-\mathbf{c}$ b $2\mathbf{d}$

c $\mathbf{c} + \mathbf{d}$ d $\mathbf{c} - \mathbf{d}$

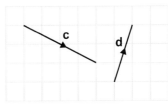

10 How sure are you of your answers? Were you mostly

Just guessing   Feeling doubtful   Confident

What next? Use your results to decide whether to strengthen or extend your learning.

* Challenge

11 ABCDEF is a regular hexagon.

$\overrightarrow{AB} = \mathbf{a}$ $\overrightarrow{BC} = \mathbf{b}$ $\overrightarrow{CD} = \mathbf{c}$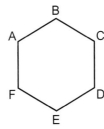

Find

a $\overrightarrow{ED}$ b $\overrightarrow{EF}$ c $\overrightarrow{FA}$

d $\overrightarrow{AC}$ e $\overrightarrow{DA}$

19 **Strengthen**

Similarity and enlargement

1 a   On squared paper, draw a rectangle 2 by 3. Label it A.
   b   Enlarge the rectangle by scale factor 2. Label it B.
   c   Work out   i   the perimeter of A   ii   the perimeter of B.
   d   Copy and complete.
      A to B, scale factor 2
      Perimeter of A to perimeter of B, scale factor ☐
   e   Enlarge rectangle A by scale factor 3. Label it C.
   f   Predict the perimeter of C. Work out the perimeter to check your prediction.

2   Triangle A is similar to triangle B.

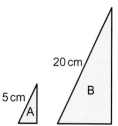

> **Q2b hint** To find lengths on triangle B, multiply lengths on triangle A by ☐.

   a   Work out the scale factor by comparing the given side lengths.
   b   The perimeter of triangle A is 12 cm. What is the perimeter of triangle B?

3   **Reasoning / Communication**
   The diagram shows two triangles.
   a   Explain why
      i $a = f$
      ii $b = e$
      iii $c = d$

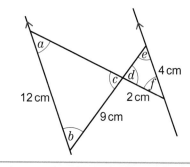

   Triangles with equal angles are similar.
   b   i   Trace the triangles and then sketch
          them the same way up.
      ii Label the measurements you know.
      iii Find the missing lengths.

> **Q3b iii hint** Use a pair of corresponding sides to work out the scale factor.

4   Triangles ABC and DEF are similar. AB and DE are corresponding sides.

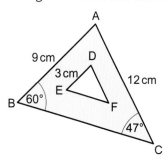

> **Q4c hint** Use the same scale factor to work out DF.

   a   Which side corresponds to BC?
   b   Work out the scale factor by comparing the sides AB and DE.
   c   Find the length of DF.
   d   Find the size of   i   angle DFE   ii   angle EDF.

> **Q4d hint** Find the matching angle in the larger triangle.

5 Triangles ABC and XYZ are similar. BC and YZ are corresponding sides.

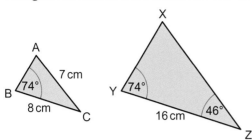

Q5 hint Use the same method as in **Q4**.

a Which side corresponds to AB?
b Find the length of XZ.
c Find the size of
  i angle ACB   ii angle BAC.

Congruence

1 Reasoning Draw diagrams rotating or reflecting the triangles to show they are identical.

a

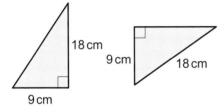

Q1a hint

b

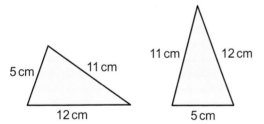

c

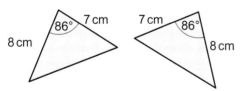

Q1c hint Draw a mirror line.

2 Communication
a Copy the diagram and label the missing angles.
b Draw each triangle accurately using a ruler and protractor.
  Are they congruent?
  Give a reason for your answer.
c Which side in triangle SVU has length 12 cm?

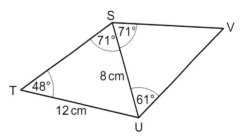

Vectors and translations

1 Write as a single column vector.

a $\begin{pmatrix} 4 \\ 3 \end{pmatrix} + \begin{pmatrix} 2 \\ 5 \end{pmatrix}$ b $\begin{pmatrix} -2 \\ 1 \end{pmatrix} + \begin{pmatrix} 1 \\ -3 \end{pmatrix}$

c $\begin{pmatrix} 5 \\ 1 \end{pmatrix} - \begin{pmatrix} 2 \\ 3 \end{pmatrix}$ d $\begin{pmatrix} 6 \\ -2 \end{pmatrix} - \begin{pmatrix} -3 \\ -4 \end{pmatrix}$

Q1a hint Add the two top numbers and add the two bottom numbers.

Q1b hint 1 + −3 is the same as 1 − 3.

2 **Reasoning** Point X is translated to point Y by the vector $\begin{pmatrix} 4 \\ -1 \end{pmatrix}$.

Point Y is translated to point Z by the vector $\begin{pmatrix} -3 \\ 5 \end{pmatrix}$.

a Find the vector of the translation X to Z.

b Work out $\begin{pmatrix} 4 \\ -1 \end{pmatrix} + \begin{pmatrix} -3 \\ 5 \end{pmatrix}$. What do you notice?

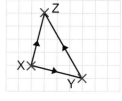

3 Point A is translated to point B by the vector $\begin{pmatrix} 2 \\ 5 \end{pmatrix}$.

Point B is translated to point C by the vector $\begin{pmatrix} -3 \\ 2 \end{pmatrix}$.

Find the vector of the single translation from point A to point C.

4 Find the resultant of these vectors.

$\overrightarrow{PR}$ and $\overrightarrow{RX}$

> **Q4 hint** Which vector starts at P and ends at X?

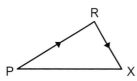

5 Copy and complete.

$\overrightarrow{GT} + \boxed{} = \overrightarrow{GX}$

> **Q5 hint** $\overrightarrow{GT}$, then another vector, start and end at the same point as $\overrightarrow{GX}$. Use the diagram to help you.

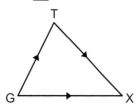

6 $\mathbf{p} = \begin{pmatrix} 4 \\ 1 \end{pmatrix}$ $\mathbf{q} = \begin{pmatrix} 2 \\ -6 \end{pmatrix}$

Write as column vectors.

a 3**p**   b −2**q**   c $\frac{1}{2}$**q**   d −4**p**

> **Q6a hint** Multiply each of the numbers in **p** by 3.

7 $\mathbf{r} = \begin{pmatrix} -2 \\ 5 \end{pmatrix}$ $\mathbf{s} = \begin{pmatrix} 3 \\ 4 \end{pmatrix}$

Write as column vectors.

a **r** + **s**   b **r** − **s**   c **s** − **r**   d 2**s** + **r**

8 **Reflect** The hint in **Q5** tells you to use the diagram to help you. Did you find the diagrams helpful? Explain why in your own words.

19 **Extend**

1 The football pitch at Britannia Stadium measures 100 m by 64 m.
Sam has made a scale model of the stadium using 1 cm to represent 1 m.
Work out the perimeter of the pitch on Sam's model.
Give your answer in metres and centimetres.

2 **Reasoning** X has coordinates (20, 15). Y has coordinates (40, 10).
A straight line is drawn through these points.
Find the coordinates of the point where this line crosses
 a the *y*-axis   b the *x*-axis.

Q2 hint Draw a diagram and look for similar triangles.

3 **Reasoning** Triangle ABC is similar to triangle DEC.
 a Find the scale factor of the enlargement that maps triangle DEC to triangle ABC.
 b Find the length of
  i BC   ii BE.

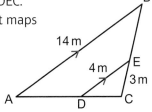

Q3b ii hint
BE = BC − EC

4 **Reasoning** In the diagram,
PT = 3.6 cm and TS = 5.4 cm.
PQ is parallel to TR.
Work out the length of PQ.

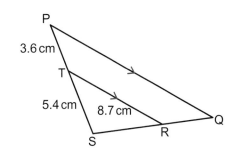

5 The diagram shows the positions of towns A, B and C on a map.
1 cm on the map represents 4.5 km.
 a How far is town A from town B?
 b The distance between towns A and C is 51.3 km. Find the value of *x*.
 c The bearing of town C from town A is 147°. Work out the size of angle BAC.

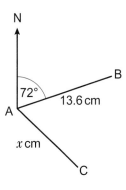

6 **Exam-style question**

ABC is a triangle.

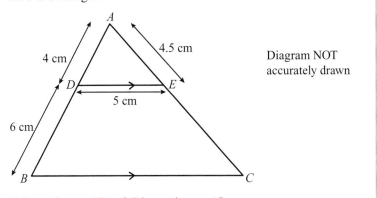

Diagram NOT accurately drawn

Exam hint
To get full marks, you will need to show how you worked out each of the missing lengths.

D is a point on *AB* and *E* is a point on *AC*.
DE is parallel to *BC*.
AD = 4 cm, *DB* = 6 cm, *DE* = 5 cm, *AE* = 4.5 cm.
Calculate the perimeter of the trapezium *DBCE*.   **(4 marks)**
Nov 2013, Q13, 5MB3H/01

7

Exam-style question

Here are some regular polygons.

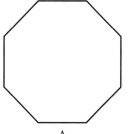

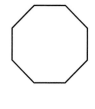

A   B   C

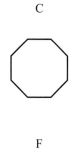

D   E   F

All these regular polygons have the same mathematical name.

a Write down this mathematical name.   **(1 mark)**

Two of these polygons are congruent.

b Write down the letters of these two polygons.   **(1 mark)**

The interior angles of this regular polygon add up to 1080°.

Diagram NOT
accurately drawn

c What is the size of one interior angle?   **(2 marks)**

June 2012, Q1, 5MB3F/01

8   Work out the missing lengths in these right-angled triangles.

i

ii

iii
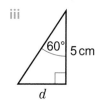

iv

v
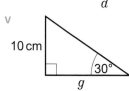

a   Which are congruent?

b   Which are similar?

9 Simplify $\overrightarrow{FJ} + \overrightarrow{JT} + \overrightarrow{TV}$

Q9 hint Simplify by combining the vectors into a single vector.

10 Write in terms of **a** and or **b**

a **v**   b **w**   c **x**   d **y**

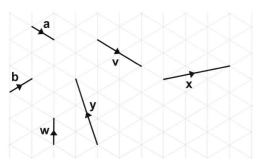

11 Write in terms of **a** and/or **b**

a $\overrightarrow{AO}$ b $\overrightarrow{AB}$ c $\overrightarrow{BA}$

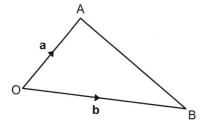

12 a Write **u** in terms of **a** and **b**.

b Write **v** in terms of **a** and **b**.

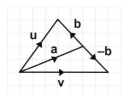

13 $\mathbf{v} = \begin{pmatrix} 6 \\ -3 \end{pmatrix}$.

$$\begin{pmatrix} 2 \\ -1 \end{pmatrix} \quad \begin{pmatrix} -6 \\ 3 \end{pmatrix} \quad \begin{pmatrix} -18 \\ 9 \end{pmatrix} \quad \begin{pmatrix} -3 \\ 6 \end{pmatrix}$$

Choose from the vectors in the box to find a vector that is

a in the opposite direction to **v**

b not parallel to **v**

c one-third as long as **v**

d parallel to **v** but 3 times as long.

You can only choose each vector once.

19 Knowledge check

- When one shape is an enlargement of another, the shapes are **similar**. In the diagram, corresponding sides and corresponding angles are shown in the same colour. .... *Mastery lesson 19.1*

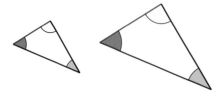

- For similar shapes, corresponding sides are in the same ratio and corresponding angles are equal. .... *Mastery lesson 19.1*

- When a shape is enlarged, the perimeter of the shape is enlarged by the same scale factor. .... *Mastery lesson 19.3*

- Triangles are congruent if they have equivalent: SSS, SAS, ASA or RHS. .... *Mastery lesson 19.4*

- When two shapes are congruent, one can be rotated or reflected to fit exactly on the other. .... *Mastery lesson 19.4*

- To add column vectors, add the top numbers and add the bottom numbers. .... *Mastery lesson 19.6*

$$\overset{2\,+\,1}{\begin{pmatrix} 2 \\ 5 \end{pmatrix} + \begin{pmatrix} 1 \\ 3 \end{pmatrix} = \begin{pmatrix} 3 \\ 8 \end{pmatrix}} \text{ and } \overset{-1\,+\,5}{\begin{pmatrix} -1 \\ 7 \end{pmatrix} + \begin{pmatrix} 5 \\ -2 \end{pmatrix} = \begin{pmatrix} 4 \\ 5 \end{pmatrix}}$$
$$\underset{5\,+\,3}{\qquad} \qquad \underset{7\,+\,-2}{\qquad}$$

- Two translations can be combined into a single translation by adding the vectors. .... *Mastery lesson 19.6*

- To find the **resultant** of two vectors, add them together.
  $\overrightarrow{AB} + \overrightarrow{BC} = \overrightarrow{AC}$. .... *Mastery lesson 19.6*

- A single letter may be used to represent a vector. The letter is shown in **bold type**. −**a** is the negative of **a** and points in the opposite direction. You can't write in bold, so you should underline a letter that represents a vector. .... *Mastery lesson 19.6*

- **a** − **b** is the same as **a** + (−**b**). .... *Mastery lesson 19.7*

- To subtract column vectors, subtract the top numbers and subtract the bottom numbers. .... *Mastery lesson 19.7*

- You can multiply a vector by a number. For example,
  if $\mathbf{a} = \begin{pmatrix} 4 \\ 3 \end{pmatrix}$ then $2\mathbf{a} = \begin{pmatrix} 8 \\ 6 \end{pmatrix}$ and $-3\mathbf{a} = \begin{pmatrix} -12 \\ -9 \end{pmatrix}$. .... *Mastery lesson 19.7*

'Notation' means symbols. Mathematics uses a lot of notation.

For example:

= means 'is equal to'   ° means degrees   └ means a right angle

Look back at this unit. Write a list of all the maths notation used.

Why do you think this notation is important?

Could you have answered the questions in this unit without understanding the maths notation?

19 Unit test

Log how you did on your Student Progression Chart.

1 **Reasoning** Triangle ADE is similar to triangle ABC.

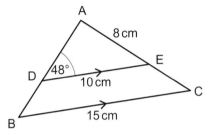

a Write the scale factor of the enlargement that maps triangle ADE to triangle ABC.

b Find the size of angle ABC.

c Work out the length of EC.

(5 marks)

2 **Reasoning** Shapes A and B are regular hexagons.
The height of B is 3 times the height of A.
The perimeter of A is 18 cm. Find the perimeter of B.

(2 marks)

3 These two triangles A and B are mathematically similar.

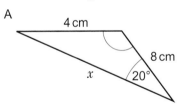

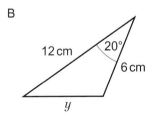

Find x and y.

(4 marks)

4 a Are triangles ABC and RQP congruent.
Give a reason for your answer.

b Write the length of AC.

c Work out the size of angle ACB.

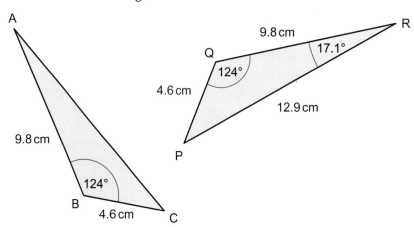

(4 marks)

5 **Communication / Reasoning** Write as a single column vector

a $\begin{pmatrix} 3 \\ 5 \end{pmatrix} + \begin{pmatrix} -4 \\ 2 \end{pmatrix}$

b $\begin{pmatrix} 3 \\ -2 \end{pmatrix} - \begin{pmatrix} -2 \\ 0 \end{pmatrix}$

(4 marks)

6 **Communication / Reasoning** ABCD is a parallelogram.

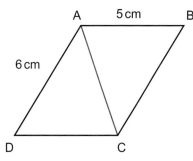

a Is triangle ABC congruent to triangle CDA?
Give a reason for your answer.

b Which angle is the same as angle ABC?

(4 marks)

7 Find the resultant of

a $\begin{pmatrix} 7 \\ -2 \end{pmatrix}$ and $\begin{pmatrix} -3 \\ -5 \end{pmatrix}$

b $\overrightarrow{LM}$ and $\overrightarrow{MN}$.

(4 marks)

8 Find a vector that is

a in the same direction as $\begin{pmatrix} 3 \\ -4 \end{pmatrix}$ but twice as long

b the same length as $\begin{pmatrix} -6 \\ 5 \end{pmatrix}$ but in the opposite direction

c parallel to $\begin{pmatrix} 4 \\ -8 \end{pmatrix}$ but half as long.

(6 marks)

9 Find the missing numbers.

$\begin{pmatrix} 3 \\ 2 \end{pmatrix} + \begin{pmatrix} \Box \\ \Box \end{pmatrix} = \begin{pmatrix} -2 \\ 7 \end{pmatrix}$

(3 marks)

Sample student answer

Describe two things the student has done to help them find the vector accurately.

Exam-style question

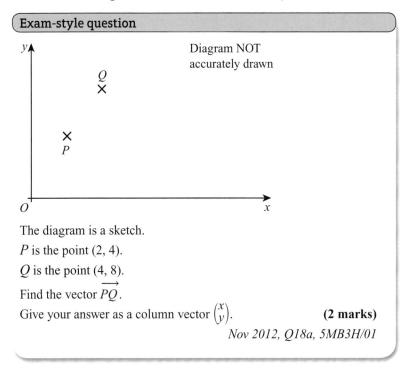

Diagram NOT accurately drawn

The diagram is a sketch.

P is the point (2, 4).

Q is the point (4, 8).

Find the vector $\overrightarrow{PQ}$.

Give your answer as a column vector $\begin{pmatrix} x \\ y \end{pmatrix}$.   **(2 marks)**

Nov 2012, Q18a, 5MB3H/01

Student answer

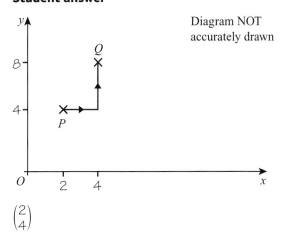

Diagram NOT accurately drawn

$\begin{pmatrix} 2 \\ 4 \end{pmatrix}$

20 MORE ALGEBRA

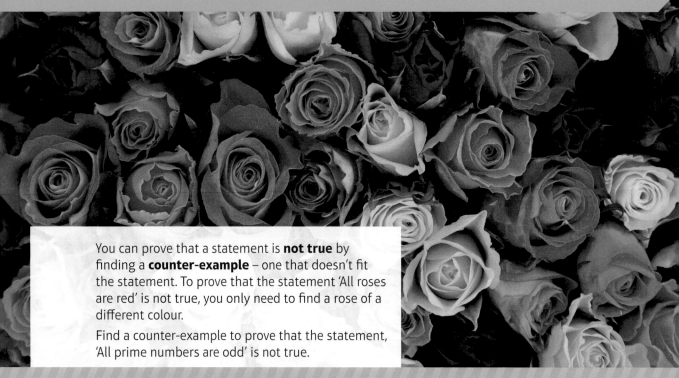

You can prove that a statement is **not true** by finding a **counter-example** – one that doesn't fit the statement. To prove that the statement 'All roses are red' is not true, you only need to find a rose of a different colour.

Find a counter-example to prove that the statement, 'All prime numbers are odd' is not true.

20 Prior knowledge check

Numerical fluency

1 Here are two numbers: 47 and 82. Work out
   a their sum   b their difference.

2 Work out
   a 2^3 b $3^3 + 4$
   c $2^2 + 3 \times 2 + 7$ d $(-5)^3$

3 Work out the reciprocal of each number. Write as an integer or decimal.
   a 2   b –10   c $\frac{1}{6}$ d 0.2

Algebraic fluency

4 a A bus can take n passengers.
     Write an expression for the number of passengers 4 buses can take.

   b A minibus can take x passengers.
     Write an expression for the number of passengers 7 minibuses can take.

5 Millie has x £2 coins. Altogether she has £50. Write an equation to show this information.

6 Solve these equations.
   a $5x = 15$ b $3x = 2$
   c $-2x = 8$ d $4x = -10$

7 Factorise.
   a $5x + 10$ b $8x + 4$
   c $x^2 + 4x$ d $6x - 8$

8 Make x the subject of each formula.
   a $y = x + 2$ b $m = x - 5$
   c $d = 4x$ d $f = \frac{1}{6}x$
   e $t = 2x + 5$ f $s = 3x - 4$

9 When $x = 4$, work out
   a x^3 b $x^3 + 3$
   c $x^3 + 2x$ d $(-x)^3$
   e $\frac{1}{x}$

10 Expand.
   a $4(x + 3)$ b $-2(x + 5)$
   c $x(x + 2)$ d $(x + 3)(x - 2)$
   e $(x - 1)^2$

Graphical fluency

11 a Copy and complete the table of values for $2x + 5y = 10$.

x	0	
y		0

b Draw the graph of $2x + 5y = 10$ on a grid with axes from −5 to + 5.

c Use your graph to find the value of y when $x = -2.5$

d Draw the graph of $y = 1$ on the same grid.

e Write the coordinates of the point where the two graphs intersect.

12 Write the gradient of the line $y = 4x - 1$, and the coordinates of the point where it crosses the y-axis.

* Challenge

13 In this puzzle, the rows and columns add to the totals given.
Work out the values of □, △ and ✻.

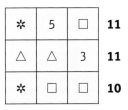

> **Q13 hint** Copy the grid and replace shapes with numbers as you work them out.

20.1 Graphs of cubic and reciprocal functions

Objectives

- Draw and interpret graphs of cubic functions.
- Draw and interpret the graph of $y = \frac{1}{x}$.

Why learn this?

You can estimate cube roots from a graph of the function $y = x^3$.

Fluency

What shape is
- a linear graph
- a quadratic graph?

1 Work out the reciprocal of

   a 4   b $\frac{1}{3}$ c −5   d $-\frac{1}{4}$

2 Copy and complete this table of values for $y = x^3$.

x	−3	−2	−1	0	1	2	3
y	−27					8	

> **Key point 1**
> A **cubic function** contains a term in x^3 but no higher power of x.
> It can also have terms in x^2 and x and number terms.

> Questions in this unit are targeted at the steps indicated.

3 a On graph paper, draw a coordinate grid with the x-axis from −3 to +3 and the y-axis from −30 to +30.

b Now plot the points from your table of values for $y = x^3$ in **Q2**.

c Join the points with a smooth curve. Label your graph with its equation.

4 a Use your graph from **Q3** to estimate

   i 2.3^3 ii $\sqrt[3]{15}$

  b Use a calculator to work out

   i 2.3^3 ii $\sqrt[3]{15}$

Discussion Which pair of answers is more accurate? Why?

5 a Make a table of $y = -x^3$ for values of x from -3 to $+3$.

   b Plot the graph of $y = -x^3$.

   Discussion What is the same and what is different about the graphs of $y = x^3$ and $y = -x^3$? What type of symmetry do they have?

> **Q5a hint** Draw a table like the one in **Q2**. Draw the axes from **Q3** again.

6 a Copy and complete this table of values for $y = x^3 + 2$.

x	-3	-2	-1	0	1	2	3
x^3		-8					27
$+2$	$+2$	$+2$	$+2$	$+2$	$+2$	$+2$	$+2$
y		-6					29

   b Draw a pair of axes on graph paper.
      Plot the graph of $y = x^3 + 2$.

   c Draw a table of values for $y = x^3 - 1$.

   d Plot the graphs of $y = x^3 + 2$ and $y = x^3 - 1$ on the same axes.

   e What is the same about your two graphs? What is different?

   Discussion What do you think the graph of $y = x^3 + 5$ would look like? What about $y = x^3 - \frac{1}{2}$?

> **Q6b hint** Make sure all the x- and y-values in the table are on your axes.

> **Q6c hint** Draw a table like the one in part **a**.

> ### Key point 2
>
> A cubic function can have one, two or three roots.

7 Use your graph of $y = x^3 + 2$ from **Q6** to find the solution of $x^3 + 2 = 0$.

> **Q7 hint** Read off the x-value where the graph of $y = x^3 + 2$ crosses the x-axis.

8 **Exam-style question**

   a Complete the table of values for $y = x^3 - 4x$ **(2 marks)**

x	-3	-2	-1	0	1	2	3
y			3	0			15

   b On the grid, draw the graph of $y = x^3 - 4x$ from $x = -3$ to $x = 3$ **(2 marks)**

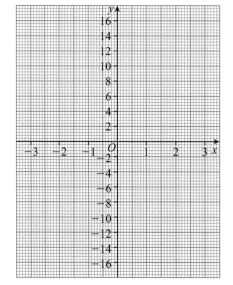

Nov 2013, Q17, 1MA0/2H

Exam hint

You might find it easier to draw your own table with a row for each part of the function

x	
x^3	
$-4x$	
y	

Plot the points using crosses. Join them with a smooth curve.
Always use a sharp pencil.

> **Key point 3**
> The x- and y-axes are **asymptotes** to the curve $y = \frac{1}{x}$. An asymptote is a line that the graph gets closer and closer to, but never actually touches.

9 a Copy and complete the table of values for $y = \frac{1}{x}$.

x	-4	-3	-2	-1	$-\frac{1}{2}$	$-\frac{1}{4}$	$\frac{1}{4}$	$\frac{1}{2}$	1	2	3	4
y	$-\frac{1}{4}$				-2			2	1			$\frac{1}{4}$

 b Plot the points. Join the two parts with smooth curves.

 c Label your graph $y = \frac{1}{x}$.

 Discussion Why can't you read the value of y when $x = 0$ from this graph?

10 **Reflect**

 a Sketch the graphs of $y = x^3$ and $y = \frac{1}{x}$. How can you remember which is which?

 b How can you find the graph of $y = -x^3$ from your sketch in part **a**?

> **Q10 hint**
> Look back at the graphs you drew in **Q3**, **Q5** and **Q9**.

11 **Reasoning** Match each equation to a graph.

 a $y = x^2$ b $y = x^3$ c $y = 4x$ d $y = \frac{1}{x}$ e $y = -x^3$ f $y = -2x$

i
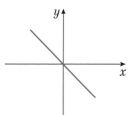

ii
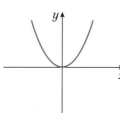

iii
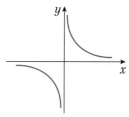

iv
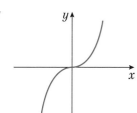

v
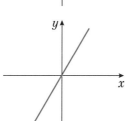

vi
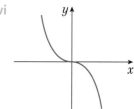

20.2 **Non-linear graphs**

Objectives

- Draw and interpret non-linear graphs to solve problems.

Did you know?

Scientists use non-linear graphs to model radioactive decay.

Fluency

Decide which of these relationships show *direct* proportion and which show *inverse* proportion:

- As x increases, y increases.
- As x increases, y decreases.
- $y = 3x$ $y = \frac{5}{2}x$ $y = \frac{2}{x}$

1 Match each graph to its equation: $y = x$ and $y = \frac{1}{x}$

a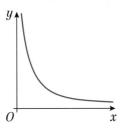

b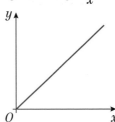

2 What is the volume of a cube with side

 a 2 cm b 5 cm c x cm?

3 Write 1.3 hours in hours and minutes.

4 **Reasoning** This solid is made from two cubes.
 a Write an equation for the volume, y, of the solid.
 b Draw the graph of your equation in **a**, for values of x from 0 to 4.
 c Estimate the volume of the solid when $x = 2.7$ cm
 d A solid like this has volume 40 cm³. Estimate the value of x.

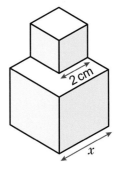

> **Q4a hint** Add the volumes of the two cubes.

> **Q4b hint** Make a table of values first.

Key point 4

When x and y are in *direct* proportion, $y = kx$

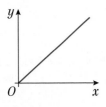

When x and y are *inversely* proportional, $y = \frac{k}{x}$

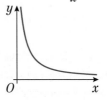

Example 1

This graph shows the number of days it takes different numbers of workers to build a swimming pool.

 a Describe the relationship between the number of workers and the number of days.

 b Estimate how long it would take 8 workers to build a swimming pool.

 c The job needs to be done in 4 days. How many workers will now be needed?

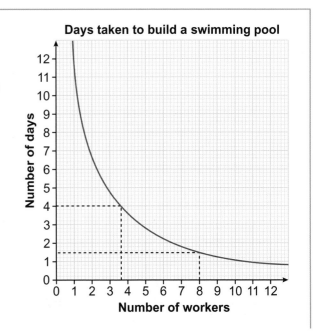

Days taken to build a swimming pool

a Number of workers and number of days are in inverse proportion.

> Does the graph show direct or inverse proportion?

b 8 workers will take 1.5 days.

> Use the graph to find the number of days for 8 workers.

c 4 days will need 4 workers.

> Use the graph to find the number of workers for 4 days. If the answer is a decimal then round up.

5 **Problem-solving** This graph shows how long it takes different numbers of cleaners to clean the bedrooms in a hotel.

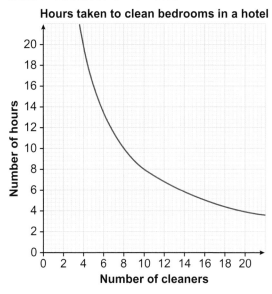

Hours taken to clean bedrooms in a hotel

a Describe the relationship between the number of cleaners and the number of hours.

b Estimate how long it would take 12 cleaners to clean the bedrooms.

c The bedrooms need to be cleaned between 10am and 4pm. How many cleaners will now be needed?

> **Q5c hint**
> 10am–4pm is ☐ hours

Discussion Why must the answer to part **b** be an estimate?

6 **STEM / Reasoning**
The graph shows the amount of water in a tank as it empties.

a How much water was there in the tank to start with?

b How much water was in the tank after 10 minutes?

c Estimate how many minutes it took for

  i half the water in the tank to empty out

  ii 75 litres to empty out.

d Show that the tank empties less quickly as the volume of water in it decreases.

> **Q6d hint** Compare the change in volume between 0 and 1 minutes with the change between 9 and 10 minutes.

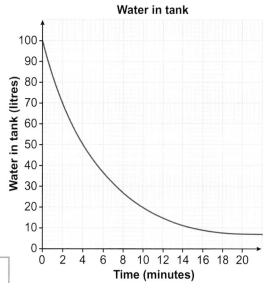

Water in tank

7 Exam-style question

The table shows the times taken to drive 120 km at different speeds:

Speed (km/h)	20	30	40	60	80
Time (hours)	6	4	3	2	1.5

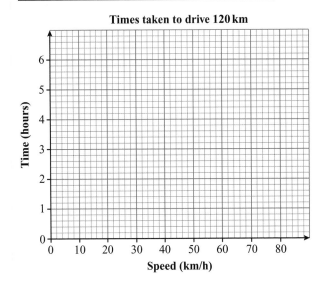

Times taken to drive 120 km

a Describe the relationship between speed and time. **(2 marks)**

b Estimate how long it would take to drive 120 km at 35 km/h. **(1 mark)**

c Liberty took $2\frac{1}{2}$ hours to drive 120 km.
Estimate her average speed for the journey. **(1 mark)**

Exam hint
If the exam question gives you a grid, use it to draw a graph even if the question does not ask you to.

Q7b hint Give times in hours and minutes.

8 **Reasoning / STEM** The graph shows the count rate against time for a radioactive material, phosphorus-32.

The count rate measures the number of radioactive emissions per second.

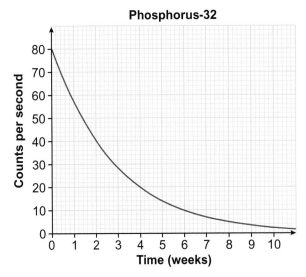

Phosphorus-32

a What is the count rate after 4 weeks?

b How many weeks does it take for the count rate to reduce to 10?

c The half-life of a radioactive material is the time it takes for the count rate to halve. What is the half-life of phosphorus-32?

Discussion Does the count rate ever reach zero?

9 **Modelling / STEM** The graph shows how the number of cells in a yeast sample grows over a 4-hour period.

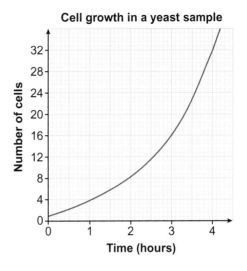

Cell growth in a yeast sample

a Copy and complete the table to show the numbers of cells.

Time (hours)	0	1	2	3	4
Number of cells	2				

b Describe the sequence of the number of cells.

> **Q9b hint** Give the first term and the term to term rule.

c Use the graph to estimate the number of cells after $3\frac{1}{4}$ hours.

d Estimate when the number of yeast cells reaches 28.

10 **Finance / Problem-solving**
The graph shows the value of an investment over a 5-year period.

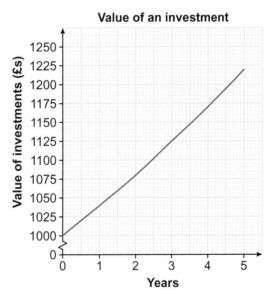

Value of an investment

a What was the initial value of the investment?

b Estimate the value of the investment after 5 years.

c By how much did the value increase in the first year?

d The rate of interest remained the same for the five years.
Work out the percentage interest rate.

> **Q10d hint** $\dfrac{\text{Actual change}}{\text{original amount}} \times 100.$

20.3 Solving simultaneous equations graphically

Objectives

- Solve simultaneous equations by drawing a graph.
- Write and solve simultaneous equations.

Why learn this?

Drawing a graph can help you find the solution to an equation. Drawing *two* graphs can help you find a solution that satisfies *two* equations.

Fluency

A coffee costs £2 and a lemonade £1.50. Write an expression for the price of
- x coffees
- y lemonades
- x coffees and y lemonades.

1 a Plot the graphs of $y = 3$ and $2x + y = 7$ on x- and y-axes from 0 to 10.
   b Write the coordinates of the point where the lines cross.

2 A cake costs 30p and a biscuit 15p.
   a Write an expression for the cost of x cakes and y biscuits.
   b x cakes and y biscuits cost 90p. Write an equation to show this.

> **Q2b hint** Write your expression from part **a** equal to 90.

Warm up

> **Key point 5**
>
> **Simultaneous equations** are equations that are both true for a pair of variables (letters).
> To find the solution to simultaneous equations:
> **1** draw the lines on a coordinate grid
> **2** find the point where the lines cross (the point of **intersection**).

3 **a** Draw the graph of $2x + y = 9$.

    b Draw the graph of $y = 1$ on the same axes.

    c Write the solution to the simultaneous equations $2x + y = 9$ and $y = 1$.

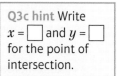

Q3c hint Write $x = \square$ and $y = \square$ for the point of intersection.

4 Draw graphs to solve these simultaneous equations
$3x + 2y = 9$ and $x - 2y = -1$.

Q3 communication hint
From two simultaneous equations you can find the values of two variables.

5 Solve these simultaneous equations graphically.
$4x + y = -5$ and $2x + 3y = 10$

6 Use these graphs to solve the pairs of simultaneous equations.

    a $x - y = 2$
       $x + y = 8$

    b $x - y = 2$
       $x + 4y = 11$

    c $x + y = 8$
       $x + 4y = 11$

Discussion Are there any values of x and y that satisfy all three equations simultaneously?

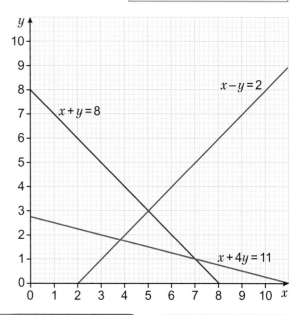

7

> **Exam-style question**
>
> The graph of $2y + x = 4$ is shown on this grid.
>
>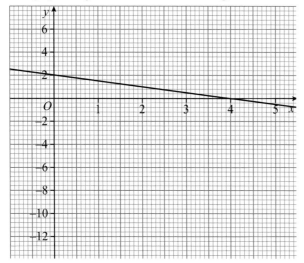

Exam hint
Plot points accurately with crosses. Join the points with a sharp pencil and ruler.
Write your solutions clearly: $x = \square$, $y = \square$

a   Copy and complete this table of values for $y = 3x - 11$ (**2 marks**)

x	0	2	4
y			

b   Copy the graph grid and the line $2y + x = 4$
Draw the line $y = 3x - 11$ on the grid.   (**2 marks**)

c   Use your graphs to solve the simultaneous equations:
$2y + x = 4$
$y = 3x - 11$ (**2 marks**)

8   **Reasoning / Communication**
By drawing their graphs, show that the simultaneous
equations $x + 2y = 4$ and $x + 2y = -7$ have no solution.

Q8 hint Draw the graphs. Write a sentence beginning: They do not have a solution because …

Example 2

The sum of two numbers is 4 and their difference is −2.
Find the two numbers.

$x + y = 4$
$x - y = -2$

Use x and y for the two numbers. Write two equations, one for the sum and one for the difference.

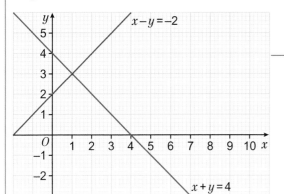

Draw the graphs of $x + y = 4$ and $x - y = -2$.
Read the solution from the point of intersection.

$x = 1$ and $y = 3$

---

9   **Reasoning** The sum of Abi's and Ben's ages is 18. The difference in their ages is 4 years.
Abi is older.

a   Write an equation for the sum of their ages.

b   Write an equation for the difference in their ages.

c   Draw the graph of each equation.

d   What are Abi's and Ben's ages?

Q9 hint Use x for Abi's age and y for Ben's age. Read the values for x and y from your graph.

10   **Problem-solving / Real** A Scout leader plans a trip to a theme park.
Scout group: 2 adult tickets, 7 child tickets, total cost £76.
Cub group: 4 adult tickets, 9 child tickets, total cost £112.

a   What do x and y stand for in this equation for the Scout group: $2x + 7y = 76$?

b   Write an equation for the Cub group.

c   Draw graphs to find the cost of an adult ticket and a child ticket.

Q10c hint Draw graphs of the equations from parts **a** and **b**. Write your answers in £s.

11 **Problem-solving / Modelling** A school trip for Year 10 uses 2 minibuses and 1 coach for 60 people.

A trip for Year 11 uses 1 minibus and 2 coaches for 84 people.

a Write equations for Years 10 and 11.

b Solve your simultaneous equations graphically to work out the number of people in a minibus and the number of people in a coach.

> **Q11a hint** Use x for the number of people in a minibus and y for the number in a coach.

Discussion The equations you wrote model the numbers of people in the minibuses and coaches. What assumptions did you make about what type of numbers to expect?

12 **Reflect** Look back at your answers in this lesson. How easy is it to read accurate answers from a graph?

Which set of simultaneous equations would be easier to solve accurately from a graph:

a $x + y = 7, 3x - y = 1$ or   b $x + y = 115, 2x - 3y = 5$?

Explain your answer.

20.4 **Solving simultaneous equations algebraically**

Objectives

- Solve simultaneous equations algebraically.

Why learn this?

Solving simultaneous equations algebraically can give more accurate solutions than solving them by drawing graphs.

Fluency

Simplify
- $2x - 2x$ • $2x - -2x$ • $-2x + 2x$ • $-2x - -2x$

If $y = 3x$, work out the value of y when
- $x = 2$ • $x = 1.5$

1 Copy and write in the missing + or −.

a $3y \boxed{} -3y = 0$ b $5x \boxed{} 5x = 0$ c $-6x \boxed{} 6x = 0$ d $-4y \boxed{} -4y = 0$

2 a Multiply all the terms in this equation by 2: $3x + 4y = 2$

b Multiply all the terms in this equation by 3: $x - 2y = 5$

3 Solve these equations.

a $3y = 18$ b $4x = 6$ c $2y = -8$ d $5x - 3 = 7$

4 Here are two simultaneous equations: $2x + y = 9$ and $y = 1$

a Substitute $y = 1$ into the equation $2x + y = 9$

b Solve your equation in x from part **a**.

c Write the solutions to the simultaneous equations.

d In Lesson 20.3 **Q3**, you drew graphs to solve these equations. Are your solutions the same?

> **Q4c hint** $x = \boxed{}, y = \boxed{}$

Key point 6

To solve simultaneous equations by the elimination method, add or subtract the equations to eliminate either the x or the y terms.

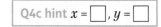
*Active*Learn Homework, practice and support: Foundation 20.4

Example 3

Solve these simultaneous equations algebraically: $3x + 2y = 9$ and $x - 2y = -1$

$3x + 2y = 9$ (1)

$\quad x - 2y = -1$ (2)

> Write the equations one above the other and number them. The y terms have coefficients 2 and –2. Add the two equations together to eliminate the y term.

$4x = 8$ (1) + (2)

$x = 2$ ——— Solve for x.

$3 \times 2 + 2y = 9$ (1)

$6 + 2y = 9$

> Substitute the x-value into one of the equations.

$2y = 3$

$\quad y = 1.5$

> Write both the solutions.

$x = 2, y = 1.5$

$x - 2y = 2 - 3 = -1$ ✓

> Check the solutions satisfy the other equation.

5   Solve these simultaneous equations algebraically: $x - y = 2$ and $x + y = 8$
   In Lesson 20.3 **Q6a**, you solved these equations using graphs. Are your solutions the same?

6   Solve these pairs of simultaneous equations.
   a $2x - y = 12$ and $-2x - 3y = -4$
   b $4x + y = 2$ and $8x + y = 3$
   c $-x + 2y = 10$ and $-x - 3y = -13$
   Discussion How do you decide whether to add or subtract the equations to eliminate terms?

> **Q6a hint** Eliminate the x terms.

> **Q6b hint** If adding the equations does not eliminate some terms, try subtracting.

7   **Reasoning** The sum of two numbers is 24 and their difference is 14. Write and solve a pair of simultaneous equations to find the two numbers.

> **Q7 hint** Write two equations; one for the sum and one for the difference. Use x and y for the two numbers.

8   Solve these simultaneous equations.
   $2x - y = -1$ (1)
   $5x - 2y = 4$ (2)
   Discussion Are equations (1) and (3) equivalent? How did multiplying by 2 help solve the equations?

> **Q8 hint** Multiply the first equation by 2 and label this equation (3). Now solve equations (2) and (3).

9   **Exam-style question**

   Solve the simultaneous equations
   $3x + y = 11$
   $2x - 3y = -11$ **(3 marks)**

> **Exam hint**
> Number the equations. Show clearly how you eliminate one of the letter terms. Write your answer $x = \boxed{}$, $y = \boxed{}$

10   Solve these simultaneous equations:
   $2x + 3y = 11$ (1)
   $3x + 4y = 15$ (2)
   a   First multiply equation (1) by 3. Label this equation (3).
   b   Now multiply equation (2) by 2. Label this equation (4).
   c   Solve the simultaneous equations (3) and (4).

11   Repeat **Q10**, but this time:
   a   Multiply equation (1) by 4. Label this equation (3).
   b   Multiply equation (2) by 3. Label this equation (4).
   c   Solve the simultaneous equations (3) and (4).
   Reflect Did you find **Q10** or **Q11** easier to solve these equations? Explain why.

12 **Problem-solving / Finance** Charlie is paid £55 for 6 hours' work plus 1 hour's overtime.
For 10 hours' work plus 2 hours' overtime he is paid £95.
How much is he paid for

a 1 hour's work?

b 1 hour's overtime?

> **Q12 hint** Write two simultaneous equations and solve them.

13 Follow the steps to find the equation of the line that passes through the points $A(2, 4)$ and $B(1, 10)$.

a Copy and complete these two equations for the line.
Use the x and y values from each coordinate pair in $y = mx + c$
At point A: $4 = \boxed{}m + c$
At point B: $\boxed{} = m + c$

b Solve the two simultaneous equations to find the values of m and c.

c Use your values of m and c to write the equation of the line.

> **Q13a hint** The points lie on the line, so their coordinates 'fit' the equation for the line.

> **Q13c hint** Substitute the values of m and c into $y = mx + c$

14 **Problem-solving** Find the equation of the line that passes through

a $C(2, 5)$ and $D(4, 7)$

b $E(2, 8)$ and $F(-4, -1)$

20.5 Rearranging formulae

Objectives

- Change the subject of a formula.

Did you know?

You can rearrange the formula mass = $\dfrac{\text{volume}}{\text{density}}$ to make volume or density the subject.

Fluency

What is the inverse operation for
- $+ x$ • $- z$ • $\times y$ • $\times -s$ • $\div m$ • n^2 • $\sqrt{t}$

1 Solve these equations.

a $5x - 4 = 11$ b $\dfrac{3x}{2} = 6$ c $x + 2 = 5x - 6$

2 a Which of these lines are parallel?

i $y = \frac{1}{2}x + 4$ ii $y = 4x + \frac{1}{2}$ iii $y = 2x + \frac{1}{2}$ iv $y = \frac{1}{2}x - \frac{1}{2}$

b Which of the lines have the same y-intercept?

Key point 7

The subject of a formula is the letter on its own on one side of the equals sign.

3 What is the subject of each formula?

a $C = 3d$ b $d = \dfrac{m}{v}$ c $V = IR$ d $A = lw$

Key point 8

You can change the subject of a formula using inverse operations.

ActiveLearn Homework, practice and support: Foundation 20.5

4   Copy and complete to make w the subject of each formula.

   a $A = 5w$

   $\dfrac{A}{\Box} = \Box$

   b $M = \dfrac{w}{2}$

   $\Box M = \Box$

5   Make the letter in brackets the subject of each formula.

   a $C = 3d$ [d]

   b $V = IR$ [R]

   c $s = \dfrac{d}{t}$ [d]

   d $p = \dfrac{4}{q}$ [q]

   e $d = \dfrac{m}{v}$ [m]

   f $P = \dfrac{F}{A}$ [A]

   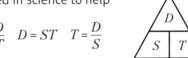 **Q5d hint** Rearrange $p = \dfrac{4}{q}$ to $pq = 4$ and then $q = \dfrac{\Box}{\Box}$

6   This is a distance, speed, time triangle used in science to help remember formulae.

   From the triangle you can see that $S = \dfrac{D}{T}$ $D = ST$ $T = \dfrac{D}{S}$

   Now rearrange $S = \dfrac{D}{T}$ to make

   a D the subject of the formula

   b T the subject of the formula.

   Reflect Do you get the same formulae?
   Which is easier to remember – the triangle or how to change the subject of $S = \dfrac{D}{T}$?

7   **STEM** The table shows the densities and masses of samples of different materials.

   Density $(d) = \dfrac{\text{mass } (m)}{\text{volume } (v)}$. Work out the volume of each block in m³, correct to 4 d.p.

Material	Density (kg/m³)	Mass (kg)
Copper	8960	3.46
Iron	7870	10.75
Aluminium	2700	92.4

   Q7 hint Change the subject of $d = \dfrac{m}{v}$

   Discussion How could you work out the volumes without changing the subject of the formula? Which is more efficient?

Example 4

   a   Rearrange to make a the subject of $v = u + at$

   b   Rearrange to make z the subject of $x = y - az$

   a $v = u + at$

   $v - u = at$ — Rearrange so the term including a is on its own on one side of the equals sign.

   $\dfrac{v - u}{t} = a$ — Use the inverse operation to make a the subject.

   b $x = y - az$

   $x - y = -az$

   $\dfrac{x - y}{-a} = z$ — Multiply top and bottom by -1 to remove the negative sign from the denominator. $\dfrac{-1 \times (x - y)}{-1 \times -a} = \dfrac{-x + y}{a} = \dfrac{y - x}{a}$

   $\dfrac{y - x}{a} = z$

---

8   Change the subject of each formula to the letter in brackets.

   a $v = u + at$ [t]

   b $x = -y + mz$ [z]

   c $l = q - rp$ [p]

   d $f = e - gh$ [h]

   Discussion How is changing the subject of a formula similar to solving equations?

9   a   Rearrange the equation of the line $6x + 2y = 5$ into the form $y = mx + c$

   b   What is the gradient and y-intercept of the line?

   Q9a hint Rearrange to make y the subject of the formula.

10 **Problem-solving** Which of these lines pass through the point (0, 5)?

> **Q10 hint** Rearrange to make y the subject of the formula

   a $y = 2x + 5$ b $2y - 5x = 5$ c $5y + 3x = 5$

   d $5y + 2x = 25$ e $5y + 3x = 10$

11 **Problem-solving** Jane wants to draw three circles with circumferences 15 cm, 20 cm and 25 cm.
How far should she open her compasses for each circle?
Round your answers to a sensible degree of accuracy.

> **Q12 hint** How accurately can you measure the gap between the point of your compasses and your pencil?

12 Make the letter in brackets the subject of each formula.

   a $C = 2\pi r$ [r]   b $V = lwh$ [l]   c $V = lwh$ [w]   d $A = 2\pi rh$ [h]

13 Make the letter in brackets the subject of each formula.

   a $I = \dfrac{PRT}{100}$ [T]   b $\dfrac{PV}{T} = k$ [T]   c $A = \dfrac{bh}{2}$ [b]   d $P = 2(l + w)$ [w]

   e $X = m(m + n)$ [n]   f $M = \dfrac{n + 2}{5}$ [n]   g $P = \dfrac{3(Q - t)}{2}$ [Q]   h $A = 2\pi r(r + h)$ [h]

14 a Copy and complete to make x the subject of $y = \dfrac{x + 1}{a} + 2$

$$y - \square = \dfrac{x + 1}{a}$$
$$\square(y - \square) = x + 1$$
$$\square(y - \square) - \square = x$$

> **Q14biii hint** First get all the x terms on one side of the equation and all the terms without x on the other.

   b Make x the subject.

     i $z = \dfrac{x - 1}{b} + 5$ ii $m = \dfrac{x + 3}{n} + p$ iii $4x - g = x + 6$ iv $2x + 7 = 5x - q$

Example 5

a Make x the subject of $y = ax^2$

b Make x the subject of $y = \sqrt{4x}$

a $y = ax^2$

  $\dfrac{y}{a} = x^2$

  $\sqrt{\dfrac{y}{a}} = x$ ——— Take the square root of both sides.

b $y = \sqrt{4x}$

  $y^2 = 4x$ ——— Square both sides.

  $\dfrac{y^2}{4} = x$

---

15 Make the letter in brackets the subject.

   a $y = x^2$ [x]   b $y = 5z^2$ [z]   c $y = \frac{1}{2}x^2$ [x]   d $A = \pi r^2$ [r]

   e $y = \sqrt{x}$ [x]   f $t = \sqrt{3s}$ [s]   g $P = \sqrt{t + r}$ [t]   h $V = \frac{4}{3}\pi r^3$ [r]

   i $A = 4\pi r^2$ [r]   j $V = \pi r^2 h$ [r]   k $V = \frac{1}{3}\pi r^2 h$ [h]   l $c^2 = a^2 + b^2$ [a]

16 **Problem-solving / STEM**
Each of these cylinders is made from 150 cm³ of steel.
Work out the missing dimensions to 1 d.p.

> **Communication hint**
> Dimensions are measurements.

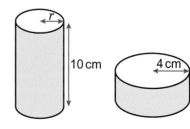

17

┌───┐
│ **Exam-style question** │
│ │
│ Make t the subject of the formula │
│ $2(d - t) = 4t + 7$ **(3 marks)** │
│ *June 2012 Q20 1MA0/2H* │
└───┘

Exam hint
This means you need to isolate all terms in t on one side of the equation.

18 **STEM** These formulae connect an object's final velocity (v) with its initial velocity (u), acceleration (a), distance travelled (s) and time (t).

a Make u the subject of $v^2 = u^2 + 2as$

b Make u the subject of $s = ut + \frac{1}{2}at^2$

c Make a the subject of $s = ut + \frac{1}{2}at^2$

20.6 **Proof**

Objectives

- Identify expressions, equations, formulae and identities.
- Prove results using algebra.

Did you know?

In the year 1637, Pierre de Fermat said that the equation $a^n + b^n = c^n$ has no integer solutions when $n > 2$. He forgot to say why – and it was to be his Last Theorem. In 1995, 358 years later, Andrew Wiles proved the theorem.

Fluency

Are these expressions, formulae or equations?

$2x + 3$ $5x = 20$ $C = 2\pi r$ $2v - 3u + 6$ $38 = \pi r^2$ $x^2 + 3x + 2 = 0$

1 Expand

a $x(x - 2)$

b $(x + 1)(x - 4)$

c $(x + 2)^2$

2 Factorise

a $3x + 3$

b $x^2 - 3x$

c $4x - 2$

3 Work out the area of this shape.

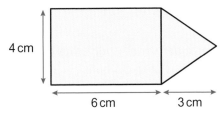

4 cm

6 cm   3 cm

┌──┐
│ **Key point 9** │
│ An **equation** has an equals sign. You can solve it to find the value of the letter. │
│ An **identity** is similar, but is true for *all* values of x and uses the symbol '≡'. │
└──┘

4 **Reasoning**

a Solve $3x + 5 = 26$

b Is $3x + 5 = 26$ true for all values of x?

c Is $3x + 5 = 26$ an equation or an identity?

Q4c hint Remember an identity is true for all values of x. An equation is only true for some values of x.

Warm up

5 **Reasoning** Decide if these are expressions, formulae, equations or identities.

  a $2x + 11 = 19$

  b $v = u + at$

  c $3x(x - 1) = 0$

  d $x^2 - 2x + 5$

Example 6

Show that $(x - 3)^2 - 2 \equiv x^2 - 6x + 7$ ————————— | $\equiv$ means 'is identically equal to' |

LHS $\equiv (x - 3)^2 - 2 \equiv (x - 3)(x - 3) - 2$

$\equiv x^2 - 6x + 9 - 2$ ————————— | Expand the brackets on the left hand side (LHS). |

$\equiv x^2 - 6x + 7 \equiv$ RHS

| Collect like terms to get the original expression from the right hand side (RHS) |

Key point 10

To show a statement is an identity, expand and simplify the expressions on one or both sides of the identity sign, until the two expressions are the same.

6 **Communication** Show that

  a $x(x^2 - 4) \equiv x^3 - 4x$

  b $(x - 2)^2 + 4x \equiv x^2 + 4$

  c $x^2 + 6x + 25 \equiv (x + 5)^2 - 4x$

  d $2x^3 - 5x^2 \equiv x^2(2x - 5)$

  e $(x + 4)(x - 4) \equiv x^2 - 16$

  f $x^2 - a^2 \equiv (x + a)(x - a)$

> **Q6c hint** Expand or factorise to prove the identity.

  Discussion How did you answer part **d**? Is expanding the brackets the only way to answer it?

7 **Communication** Follow the steps to show that $(x + 2)(x + 3) + x + 3 \equiv (x + 3)^2$

  a Multiply out the brackets on the LHS.

  b Simplify the LHS by collecting like terms.

  c Multiply out the brackets on the RHS.

  d Show that the LHS and RHS are the same.

8 **Communication** A rectangular card with length $x + 3$ and width $x + 1$ has a smaller rectangle cut out of the middle. The smaller rectangle has length $x - 1$ and width x.

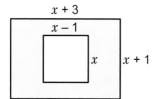

> **Q8c hint** Subtract your expression from part **b** from your expression from part **a**.

  a Write an expression for the total area of the rectangle before the middle was cut out.

  b Write an expression for the area of the rectangle that has been cut out.

  c Show that the area of the remaining card is $5x + 3$.

9 **Exam-style question**

  Show that the area of this pentagon can be written as $2x^2 + x - 1$ **(4 marks)**

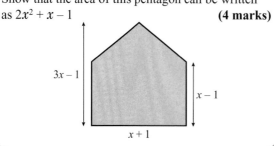

> **Exam hint**
> Write any other lengths you work out on the diagram.
> Simplify algebraic expressions by collecting like terms.
> End with the statement $= 2x^2 + x - 1$

10 **Modelling**

a Write any three consecutive integers.
Make the first one equal to n.
Write expressions for the other two numbers.

b Write a different set of three consecutive numbers.
Make the second one equal to n and write
expressions for the other two numbers.

> **Communication hint**
> **Consecutive integers** follow one
> after the other, like 1, 2, 3 or –5, –4, –3

> **Q10a hint** $n, n + \square, n + \square$

11 **Communication** The median of a set of five consecutive
integers is n.

a Write expressions for the five integers.

b Show that the range of the five integers is 4.

c Show that the mean of the five integers is n.

> **Q11a hint** The median is
> the middle one in the set.

> **Q11b hint** Range =
> highest value – lowest value.

12 **Communication** Follow these steps to show that the sum of any three consecutive
integers is a multiple of 3.

a Write expressions for the three integers as n, ___ , ___ .

b Work out the sum of these integers.

c Simplify the sum as much as possible, then factorise it.

d Copy and complete: The sum of three consecutive numbers is _____ , which is ____ × 3,
which is a multiple of 3.

13 **Communication** Show that the sum of any four
consecutive numbers is a multiple of 2.

> **Q13 hint** Follow the steps in **Q12**.

14 **Reasoning**

a Work out the first five terms of the sequence with general
(nth) term
  i $2n$ ii $2n - 1$.

b What is the general term for
  i an even number   ii an odd number?

Discussion What type of number is $2n + 1$?

> **Q14a hint** Substitute
> 1, 2, 3, … for n.

> **Q14b hint** Look at your
> sequences in part **a**.

> **Key point 11**

An even number is a multiple of 2.
$2m$ and $2n$ are both general terms for even numbers where m and n are integers.

15 **Reasoning / Communication** Copy and complete this proof to show that the sum of
two even numbers is even.
Two even numbers are $2m$ and $2n$.
$2m + 2n = \square(\square + \square)$
This is a multiple of $\square$, so it is an even number.

16 **Reasoning / Communication**
Show that the sum of two odd numbers is even.

> **Q16 hint** Use $2m + 1$ and
> $2n + 1$ as the odd numbers.

17 **Reasoning / Communication**
Given that $2(x - a) = x + 4$, where a is an integer,
show that x must be an even number.

> **Q17 hint** Expand the brackets.
> Rearrange to make x the subject.
> 2 × an integer = an even number.

20 Problem-solving: Breaking even

Objectives
· Be able to draw straight-line graphs arising from real life situations.
· Be able to solve simultaneous equations algebraically and graphically.

Company A

Income and expenses
A company makes commemorative plates for special occasions.

Income
Each plate sells for £16.

Expenses
Fixed costs (rent, utilities and wages): £3000
Variable costs: £4 to make each plate

Communication hint
Fixed costs are the same, however many goods are made or sold.
Variable costs vary with the number of goods made or sold.

Break-even point
When income equals expenses, this is called the break-even point.

Company B

Income and expenses
Company B sells digital photo frames.
Each photo frame costs £8 to make but sells for £30. The company has fixed running costs of £4400 per month.

Businesses can graph their income and expenses to model when they begin to make a profit.

1 a Draw a graph that shows the Company A's income from selling up to 500 plates per month.

> **Q1a hint** You could create a table of values to help you draw the graph.

Number of plates sold, x	0	100	200
Income, £ y			

   b Add a second graph to your axes to show the company's expenses.

> **Q1b hint** Use a different colour for the income and expenses lines.

   c Copy and complete these sentences with the word 'profit' or 'loss'.
      When income is less than expenses, the company is making a _____ .
      When income is more than expenses, the company is making a _____.
      When a company is at break-even point it is not making a ____ or a ____.

   d Using your graph, how many plates need to be sold for the Company A to reach break-even?

   e Construct two equations; one for income and one for expenses.
      Solve these simultaneous equations to show that your answer to **d** is correct.

> **Q1e hint** Both your equations will be in the form $y = mx + c$. For the income equation, c will be equal to 0.

   f How much profit does the company make if it sells 350 plates?

> **Q1f hint** Profit = Income − Expenses

   g How much money does the company lose if it sells 125 plates?

2 Find the break-even point for Company B.

> **Q2 hint** Draw graphs of income and expenses on the same axes. Check your break-even point using simultaneous equations.

20 Check up

Log how you did on your
Student Progression Chart.

Graphs

1 a Copy and complete this table of values for $y = x^3 - 2$

x	–3	–2	–1	0	1	2	3
y							

b Copy the grid. Draw the graph of $y = x^3 - 2$ from $x = -3$ to $x = +3$.

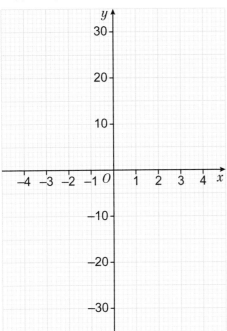

2 a Draw the graph of $y = \frac{1}{x}$ from $x = -3$ to $x = +3$

x	–3	–2	–1	–0.5	0.5	1	2	3
y								

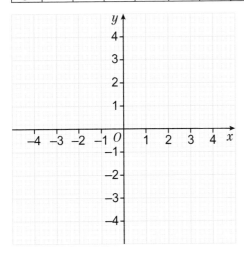

b Use your graph to find
   i the value of y when $x = -1.5$
   ii the value of x when $y = 4$

3 **Reasoning** The graph shows how the volume of a fixed amount of a gas varies as the pressure changes.

Estimate

a the volume when the pressure is 4 kilopascals.

b the pressure when the volume is 10 litres.

c the change in volume when the pressure is increased from 2 kilopascals to 6 kilopascals.

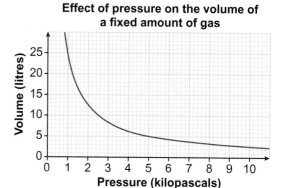

Effect of pressure on the volume of a fixed amount of gas

Simultaneous equations

4 A cup of coffee costs £y and a glass of juice costs £x. Three coffees and two juices cost £9. Write an equation to show this information.

5 Use the graph to find the solutions to the simultaneous equations $2x + y = 7$ and $x + 2y = 8$

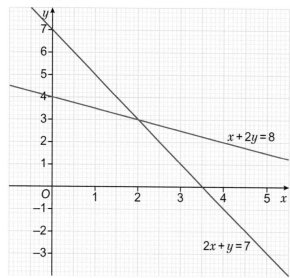

6 **Reasoning** Two numbers added together make 24 and their difference is 8.

a Write two equations for the two numbers. Use x and y.

b Solve your equations to find the two numbers.

7 Solve the simultaneous equations $2x + 3y = 14$ and $5x + y = 22$

Using algebra

8 Make the letter in brackets the subject of each formula.

a $m = pr + t$ [t]   b $z = ac + y$ [c]   c $p = 3(m + n)$ [m]   d $v = \dfrac{pq}{r}$ [r]

9 Make the letter in brackets the subject.

a $R = t^2$ [t]   b $n = \dfrac{pt}{x}$ [p]   c $y = 4 - \dfrac{x}{2} + z$ [x]   d $6x + 4 = y + x$ [x]

10 Show that $xy(x - y) \equiv x^2y - xy^2$ is an identity.

11 How sure are you of your answers? Were you mostly

Just guessing   Feeling doubtful   Confident 😊

What next? Use your results to decide whether to strengthen or extend your learning.

* Challenge

12 In this puzzle, the rows and columns add to the totals given.
Work out the value of ✿, Δ and □

Δ	✿	□	1	**10**
2	Δ	3	□	**11**
✿	□	Δ	✿	**12**
□	2	Δ	Δ	**13**
11	**11**	**14**	**10**	

> **Q12 hint** Look for rows or columns with only two shapes. Write and solve a pair of simultaneous equations.

---

20 Strengthen

Graphs

1 **Reasoning** Decide which graph is $y = x^3$ and which is $y = -x^3$.
Copy the graphs and label them with their equations.

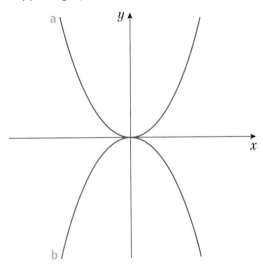

> **Q1 hint** When x is 1, is the value of y positive or negative?
> On the $y = x^3$ graph, when $x = 1$, $y = x^3 = 1$
> On the $y = -x^3$ graph, when $x = 1$, $y = -x^3 = -1$

2 **Reasoning** Shona makes a table of values and plots the graph of $y = x^3 - 1$

   a From the graph, which points do you think are incorrect?

> **Q2a hint** Which points do not fit the shape of an x^3 graph?

   b Find the incorrect points in Shona's table of values.
       Work out the correct values.

x	-3	-2	-1	0	1	2	3
y	-28	7	-2	-1	0	7	-26

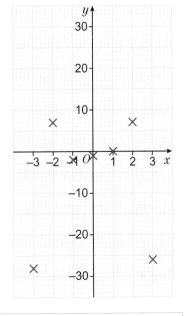

---

3 a Copy and complete this table of values for the graph of $y = x^3 + 1$.

x	−3	−2	−1	0	1	2	3
y	−26					9	

b Copy the coordinate grid from Q2 and then draw the graph of $x^3 + 1$ on the grid.

> **Q3b hint** First sketch the shape you expect the graph to be – is it like $y = x^3$ or like $y = -x^3$?
> Check that your points fit the shape before you join them.

4 Match the graphs to their equations.

$y = x$ $y = x^2$ $y = \dfrac{1}{x}$

a 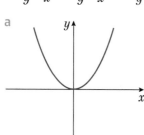   b 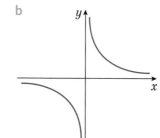   c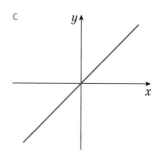

5 **Reasoning** Dan makes a table of values and plots the graph of $y = \dfrac{1}{x}$
   a From the graph, which points do you think are incorrect?

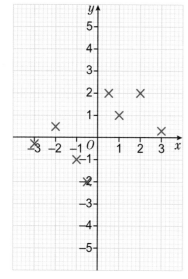

> **Q5a hint** The graph should be two smooth curves.

b Find the incorrect points in Dan's table of values. Work out the correct values.

x	−3	−2	−1	−0.5	0.5	1	2	3
y	−0.3	0.5	−1	−2	2	1	2	0.3

c Copy the coordinate grid and plot the correct points. Join them with smooth curves to draw the graph of $y = \dfrac{1}{x}$

6 **Reasoning** The graph shows the numbers of rats recorded in a colony.

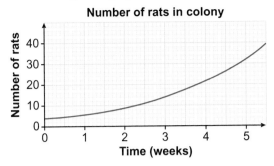

Number of rats in colony

a What does the horizontal axis show?
b What does the vertical axis show?
c Estimate the number of rats at
i 3 weeks   ii 5 weeks.
d Describe the change in the number of rats from week 3 to week 5.

> **Q6c hint** The number of rats must be a whole number.

> **Q6d hint** Is it an increase or decrease? By how much? Write a sentence beginning: The number of rats …

Simultaneous equations

1 Write an equation for each statement.
a 3 bananas and 5 apples cost £10.
b 6 adult tickets and 10 child tickets cost £70.
c 2 large bags and 4 small bags weigh 18 kg.
d 5 washers and a bolt measure 65 mm.

> **Q1 hint** To write an equation for the statement '2 pens and 3 notebooks cost £14': Call cost of pen x and cost of notebook y. Use a bar model.
>
> | x | x | y | y | y |
> |---|---|---|---|---|
>
> $2x + 3y = 14$

2 a What are the unknown values in these equations: $x + y = 8$, $x + 5y = 14$ and $-x + y = 2$?
b Use the graph to find the solutions to the simultaneous equations.
i $x + y = 8$ $x + 5y = 14$ ii $-x + y = 2$ $x + y = 8$ iii $x + 5y = 14$ $-x + y = 2$

> **Q2b hint** Find the intersection of the two graph lines with these equations.
> Make sure you write the solution for each unknown value in part **a**.

$-x + y = 2$

$x + y = 8$

$x + 5y = 14$

3 For each pair of these simultaneous equations, try (1) adding them and (2) subtracting them. One method will eliminate x or y. Use this method to find both x and y.
a $2x + y = 22$ $x + y = 12$
b $x + y = 20$ $x - y = 8$
c $2x - y = 11$ $x - y = 4$
d $-x + y = 3$ $x + 3y = 29$

> **Q3a hint** Adding – does this help here?
> $$\begin{array}{r} 2x + y = 22 \\ + \quad x + y = 12 \\ \hline 3x + 2y = 34 \end{array}$$

4 a Copy this pair of simultaneous equations
   $2x + 3y = 9$ (1)
   $x + 2y = 5$ (2)

   b Ring the coefficients of x. Is one a multiple of the other?

   c Multiply equation (2) to give $2x$, as in equation (1).
   $2x + 3y = 9$ (1)
   $\times \square \curvearrowright \begin{array}{l} x + 2y = 5 \quad (2) \\ 2x + \square = \square \quad (3) \end{array}$

   d Solve the simultaneous equations (1) and (3).

5 a Copy this pair of simultaneous equations
   $3x + y = 20$ (1)
   $2x - 4y = 4$ (2)

   b Ring the coefficients of x. Is one a multiple of the other?

   c Ring the coefficients of y. Is one a multiple of the other?

   d Multiply equation (1) to give $4y$, so the numbers in the coefficients of y are the same.

   e Solve the simultaneous equations (1) and (3).

> **Q5d hint** The numbers need to be the same, not the signs.
> $\times \square \curvearrowright \begin{array}{l} 3x + y = 20 \quad (1) \\ 2x - 4y = 4 \quad (2) \\ \square + 4y = \square \quad (3) \end{array}$

Using algebra

1 Rearrange each formula to make a the subject.
   a $v = 4x + a$
   b $p = qy - a$
   c $s = at + 3$
   d $y = 4a + b$

> **Q1a hint** Draw two function machines: one to show the formula and the other its inverse.

> **Q1c hint**
>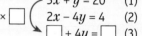
> $a = \dfrac{s - \square}{\square}$

2 Rearrange each formula to make d the subject.
   a $x = 2(d + 3)$ b $s = 2(d - t)$
   c $y = \dfrac{dt}{2}$ d $t = \dfrac{kr}{d}$

3 Rearrange each formula to make n the subject.
   a $n^2 = c^2$ b $n^2 = r$
   c $T = n^2$ d $6y = n^2$

> **Q3a hint** the inverse operation for 'square' is 'square root'.

4 a Solve the equation $\dfrac{3x}{9} = 6$ using the balancing method.

   b Use balancing to rearrange each formula to make m the subject.
   i $n = \dfrac{3m}{p}$ ii $k = \dfrac{mt}{2}$
   iii $r = \dfrac{mp}{q}$ iv $v = \dfrac{bm}{z}$

> **Q4bi hint**
> $\times p \left(\begin{array}{c} n = \dfrac{3m}{p} \\ pn = 3m \end{array} \right) \times p$
> $\div 3 \left(\begin{array}{c} pn = 3m \\ \dfrac{pn}{\square} = m \end{array} \right) \div 3$

5 a Solve $5x - 4 = 2 + 2x$ using the balancing method.

   b Use the balancing method to make x the subject of each formula
   i $2x - y = x + 6$ ii $5x - y = x + 7y$ iii $3x - 7 = 2x - y$

6 a Solve $5 = 3 + \dfrac{x}{2}$ using the balancing method.

   b Use the balancing method to make x the subject of each formula
   i $y = 5 + \dfrac{x}{3}$ ii $z = x + \dfrac{2}{a}$ iii $t = x + \dfrac{1}{b} + 4$

7 **Communication** Show that each statement is an identity.
   a $2a + 3b + 6a - 2b \equiv 8a + b$ b $x^2 + 3x - 2x + 4x^2 - 5 \equiv 5x^2 + x - 5$
   c $2(x + 4) \equiv 2x + 8$ d $x^2(y + 2) - 4y \equiv y(x^2 - 4) + 2x^2$

20 **Extend**

1 a **Reasoning** Plot these readings from a science experiment on a graph.

x	1	2	5	10	20	30
y	4.0	3.4	3.2	3.1	3.05	3.03

> **Q1a hint** You only need the y-axis to go from 3 to 4. Use a large scale so you can plot the points accurately.

b Are the two quantities in direct or inverse proportion?

2 **Problem-solving** Solve these simultaneous equations graphically: $3x + 5y = 5$ and $2x - y = 12$

> **Communication hint**
> Graphically means: by drawing graphs.

3 **Reasoning / Modelling** The graph shows the population of an ant colony.

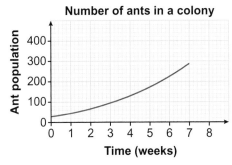

a Estimate the number of ants in week 3.

b Estimate the increase in their number from week 3 to week 5.

c Between which two weeks was the number of ants increasing fastest?

d When do you think there will be more than 400 ants in the colony?

4 a Draw the graph of $y = x^3 - 2x + 5$ for $-3 \leqslant x \leqslant 3$.

b Use your graph to find the solutions to the equation $x^3 - 2x + 5 = 0$ that lie between -3 and 3, by reading the x-value where the graph line crosses the x-axis.

5 Make the letter in brackets the subject:

a $m = \sqrt{t}$ [t] b $p = q\sqrt{r}$ [r] c $c = 3d^2$ [d] d $4a = b^2$ [b]

e $e = 9f^2$ [f] f $s^2 - wx = v$ [s] g $s^2 - wx = v$ [w]

6 **STEM** The table shows the masses of some copper samples.
The density of copper = 8.92 g/cm³.

Use the formula density = $\dfrac{\text{mass}}{\text{volume}}$, $d = \dfrac{m}{v}$, to calculate the volume of each sample.

Give your answers to 1 d.p.

Sample	Mass (m) in g	Volume (v) in cm³
a	25 g	
b	60 g	
c	90 g	
d	125 g	
e	240 g	
f	350 g	

> **Q6 hint** First rearrange the formula to make v the subject.

7 Make x the subject of each formula.

   a $m + n = p - rx$ b $a(x - b) = c$

   c $d(e + x) = fx$ d $g(h + x) = mx - n$

   e $p(x - q) = r(x + s)$ f $a(a + x) = cd$

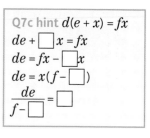

Q7c hint $d(e + x) = fx$

$de + \square x = fx$

$de = fx - \square x$

$de = x(f - \square)$

$\dfrac{de}{f - \square} = \square$

8 **Communication** In this grid, a 2 by 2 square is highlighted.

1	2	3	4	5
6	7	8	9	10
11	12	13	14	15
16	17	18	19	20

   a Show that adding the diagonal numbers 8 + 14 and 9 + 13 gives the same total.

   b Choose another 2 by 2 square on the grid.
      Add the diagonal numbers. Do they give the same total?

   c Use algebra to show that for any 2 × 2 square on this grid, the sum of both diagonals will always be equal.

Q8c hint Write expressions for the numbers in the grid:

8	9
n	$n + \square$
13	14
$n + \square$	$n + \square$

9 **Communication** Look at the 2 x 2 square highlighted in **Q8**.

   a Show that when you multiply the diagonal numbers 8 × 14 and 9 × 13, the answers have a difference of 5.

   b Choose another 2 by 2 square on the grid. Multiply the diagonal numbers. Do the answers have a difference of 5?

   c Use algebra to show that when you multiply the diagonal numbers in a 2 by 2 square in this grid, the answers always have a difference of 5.

10 **Exam-style question**

The diagram shows a cube and a cuboid.

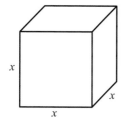

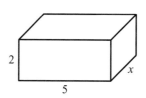

Diagram NOT accurately drawn

All the measurements are in cm.

The volume of the cube is 100 cm³ more than the volume of the cuboid.

Show that $x^3 - 10x = 100$ **(2 marks)**

Nov 2012 Q11a 1MA0/2H

Exam hint

When a question gives you a diagram, you need to use it.

Write an equation connecting the volumes of the two cuboids.

11 **Exam-style question**

Solve the simultaneous equations

$5x + 2y = 11$

$4x - 3y = 18$ **(4 marks)**

June 2012 Q20 1MA0/1H

Exam hint

Label the equations with numbers. Show clearly what you do to each one.

12 **Reasoning** Write the pair of simultaneous equations with solution $x = 4$, $y = 5$

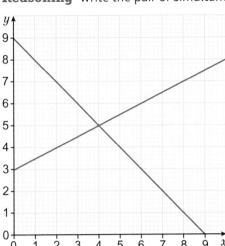

> Q12 hint The graphs show this solution.
> Find the equations of the two lines.

13 **Problem-solving / Communication** The diagram shows a rectangle and a square.
All the measurements are in centimetres.
The total area of the rectangle and the square is $35\,cm^2$.

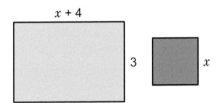

> Q13a hint Write an expression for the combined area and put it equal to 35.

a Show that $x^2 + 3x = 23$
b Show that the value of x is between 3 and 4.

> Q13b hint Try $x = 3$ and $x = 4$ in the equation in part **a**.

14 (Exam-style question)

    a Factorise $2t^2 + 5t + 2$
    b t is a positive whole number.
       The expression $2t^2 + 5t + 2$ can never have a value that is a prime number.
       Explain why.   **(3 marks)**
       Nov 2012 Q14c 1MA0/2H

Exam hint
Use your factorised form to help with your explanation.

15 **STEM / Reasoning** When a ball is thrown into the air, its height h metres above the ground after t seconds is given by the equation $h = kt - mt^2$
The ball was thrown into the air. At time $t = 2$, the height was 20 m.
At time $t = 3$, the height was 15 m.
Substitute these values into $h = kt - mt^2$ to get two equations in k and m.
Solve the equations to find k and m.

20 Knowledge check

- A **cubic function** contains a term in x^3 but no higher power of x. It can also have terms in x^2 and x and number terms. When a cubic function is equal to zero, it can have one, two or three solutions – the x-values where the graph crosses the x-axis. .... *Mastery lesson 20.1*

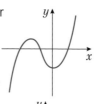

- For the reciprocal function $y = \dfrac{1}{x}$, the x- and y-axes are **asymptotes** to the curve. An asymptote is a line that the graph gets closer and closer to, but never actually touches. .... *Mastery lesson 20.1*

- When two quantities are in **direct proportion**, they are linked by an equation of the form $y = kx$ and their graph is a straight line through the origin. .... *Mastery lesson 20.2*

- When two quantities are in **inverse proportion**, they are linked by an equation of the form $y = \dfrac{k}{x}$ and their graph is a reciprocal graph. .... *Mastery lesson 20.2*

- The point where two (or more) lines cross is called the **point of intersection**. *Mastery lesson 20.3*

- **Simultaneous equations** are equations that are both true for a pair of variables (letters). .... *Mastery lesson 20.3*

- To find the solution to a pair of simultaneous equations graphically:
  - Draw the lines on a coordinate grid
  - Find the point of intersection. .... *Mastery lesson 20.3*

- To solve simultaneous equations by the elimination method, add or subtract the equations to eliminate either the x or the y terms. You may need to multiply one or both equations first, to get equal coefficients of x or y. .... *Mastery lesson 20.4*

- To find the equation of a line between two points, substitute the values of x and y from each pair of coordinates into $y = mx + c$ to get two simultaneous equations. Solve to find m and c. .... *Mastery lesson 20.4*

- The **subject** of a formula is the letter on its own on one side of the equals sign. *Mastery lesson 20.5*

- You can change the subject of a formula using inverse operations. .... *Mastery lesson 20.5*

- An **equation** has an equals sign. You can solve it to find the value of the letter. .... *Mastery lesson 20.6*

- An **identity** looks similar to an equation, but is true for *all* values of x. .... *Mastery lesson 20.6*

- To show a statement is an identity, expand and simplify the expressions on one or both sides until the two expressions are the same. .... *Mastery lesson 20.6*

- **Consecutive integers** follow one after the other, like 1, 2, 3 or −5, −4, −3 or n, $n + 1$, $n + 2$. .... *Mastery lesson 20.6*

- As an even number is a multiple of 2, $2m$ and $2n$ are both **general terms** for even numbers. .... *Mastery lesson 20.6*

Write down a word that describes how you feel

a before a maths test   b during a maths test   c after a maths test.

> **Hint** Here are some possible words: OK, worried, excited, happy, focused, panicked, calm.
>
> Beside each word, draw a face, 🙂 or 🙁, to show if it is a good or a bad feeling.
>
> Discuss with a classmate what you could do to change 🙁 feelings to 🙂 feelings.

Reflect

20 Unit test

1 Make x the subject of each formula.
   a $z = 4(x + y)$ b $t = \dfrac{rs}{x}$ *(2 marks)*

2 **Reasoning** Jim is 10 years older than Elise. The sum of their ages is 58.
   Write and solve a pair of simultaneous equations to work out
   a Jim's age.   b Elise's age.   *(3 marks)*

3 **Reasoning** The graph shows how atmospheric pressure varies with the height above sea level.

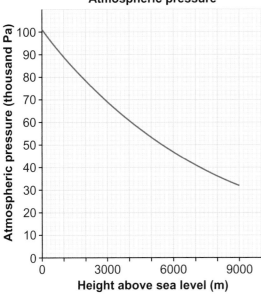

Atmospheric pressure

a Estimate
   i the atmospheric pressure at the summit of Everest, 8848 m above sea level.
   ii the fall in atmospheric pressure on the climb from Base Camp at 5380 m to the summit.
b Climbers carry oxygen to use at atmospheric pressures below 44 000 Pa. At what height
   above sea level should a climber start to use oxygen?
c Work out the percentage fall in atmospheric pressure for each extra 1000 m above
   sea level.   *(4 marks)*

4 Make r the subject of each formula
   a $N = \sqrt{r - s}$ b $F = \dfrac{2 + r}{x + 3}$ *(2 marks)*

5 Max plots this graph of his results from a science experiment.
   Describe the relationship between the two variables.   *(1 mark)*

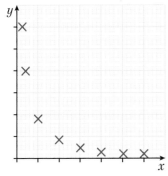

6 **Reasoning** a Use the graph to solve the simultaneous equations

$3x + 2y = 10$ $x - 2y = 2$ *(2 marks)*

b The lines $x + 5y = 9$ and $x - 2y = 2$ meet at point M.

Use algebra to find the coordinates of M. *(3 marks)*

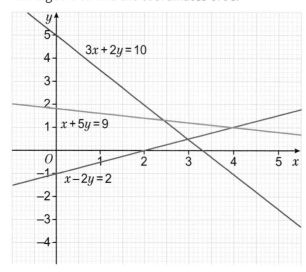

7 a Draw the graph of $y = x^3 + 3x - 7$ for $-2 \leqslant x \leqslant 3$

x						
y						

b Use your graph to find the solution to the equation $x^3 + 3x - 7 = 0$ that lies
between -2 and 3. *(4 marks)*

8 **Exam-style question**

Solve the simultaneous equations:

$3x + 4y = 5$ $2x - 3y = 9$ **(4 marks)**

Nov 2013 Q15 1MA0/1H

9 Show that $(n + 2)^2 \equiv (n + 1)^2 + 2n + 3$ *(4 marks)*

10 A taxi firm charges a fixed cost and an amount per km.
A 6 km journey costs £8.80 and a 10 km journey costs £12.
Work out the cost of a 15 km journey. *(4 marks)*

11 Make x the subject of

a $T = \sqrt{\dfrac{2s}{x}}$ *(2 marks)*

b $n = 3\sqrt{5 - 3x}$ *(3 marks)*

Sample student answer

1   One point is plotted incorrectly. Which one?

2   How has this error occurred?

Exam-style question

a Complete the table of values for $y = \dfrac{6}{x}$.

x	0.5	1	2	3	4	5	6
y		6	3		1.5		1

b On the grid, draw the graph of $y = \dfrac{6}{x}$ for $0.5 \leqslant x \leqslant 6$.

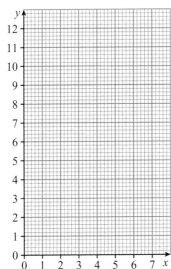

(4 marks)

Nov 2012, Q18, 1MA0/2H

Student answer

a

x	0.5	1	2	3	4	5	6
y	12	6	3	2	1.5	1.2	1

b

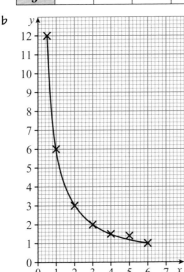

ANSWERS

UNIT 1

1 Prior knowledge check

1 a 20   b 2000   c 20 000
   d $\frac{2}{10}$ e $\frac{2}{1000}$ f $\frac{2}{100}$
2 4.61, 4.3, 4.12, 3.12, 3.09
3 a 2000   b 900   c 60   d 84
4 a 2500   b 1.4   c 3500
   d 1.386   e 4700 000   f 600
5 a 57   b 172   c 58   d 438
6 a 750   b 29 250
7 a 15, 27, 29, 49   b 4, 8, 12, 36
   c 12, 15, 27, 36   d 4, 36, 49
   e 8, 27
8 5, 7
9 1, 2, 3, 4, 6, 8, 12, 24
10 a 102   b 47   c 13.4
   d 18   e 106.5
11 a 81   b 8   c 64   d 4
12 a 2   b −2   c −5
   d −1   e 7   f 3
13 a −20   b −80
14 a −10   b −10   c 10
   d −4   e −4   f 4
15 a 8.37   b 7.22
16 a 0.8   b 3.5   c 0.6   d 0.4
17 a $5^2 = 25$ b $3^3 = 27$
18 a 400   b 216
19 a 9   b 4
20 a $1 < 7$ b $0.8 > 0.2$ c $-4 < 3$ d $-1 > -9$
21 49 cm^2
22 £11.45
23 2.82 kg
24 a 24 °C   b 75 °F
25 a 7 °C   b 15 °C   c 8 °C
26 32 °C
27 8 glasses
28 2 packs

1.1 Calculations

1 a $\frac{5}{2} = 2.5$ or $2\frac{1}{2}$ b 3.5   c 2.5
2 a 3   b 4, −8   c −6, −7
3 a $\frac{1}{4}$ b $\frac{1}{2}$ c $\frac{6}{5}$
4 a $+ \times + = +$ b $+ \times - = -$ c $- \times + = -$ d $- \times - = +$
   e $+ \div + = +$ f $+ \div - = -$ g $- \div + = -$ h $- \div - = +$
5 a $(4 - 1)^2 + 2 = 3^2 + 2 = 11$
   b 3   c 3   d −3
6 a −50   b 3   c 75   d 2
7 A and E
8 a ≠   b =   c ≠
   d ≠   e ≠
9 a $\frac{3 \times 8}{6} = \frac{3}{6} \times 8 = \frac{1}{2} \times 8 = 4$
   b $\frac{30 \times 20}{40} = 30 \times \frac{20}{40} = 30 \times \frac{1}{2} = 15$
   c $\frac{1}{6}$ d $\frac{5}{3}$ or $1\frac{2}{3}$ e $\frac{1}{4}$ f $\frac{1}{6}$
10 a −5   b ÷2   c ×3   d +4
11 a 966 − 579 = **387** or 966 − 387 = **578**
   b 89 + 598 = **687**   c 46 × 11 = **506**   d 3168 ÷ 12 = **264**
12 C
13 a 72   b 5   c 20
14 a 25   b 125   c 2025   d 11
15 a £64   b €125

1.2 Decimal numbers

1 a 1 cm = 10 mm   b 1 m = 100 cm
   c 1 km = 1000 m

2 a 79   b 171.5   c 52.75
3 a $\frac{3}{5} = \frac{30}{50}$ b $\frac{7}{10} = \frac{70}{100}$
4 a 3.5   b 0.5   c 12.1   d 9.0
5 a 4.03   b 16.17   c 0.13   d 11.90
6 a 8.462   b 22.806   c 9.106
7 a 3.5 cm   b 9.462 m   c 6.85 m   d 25.325 km
8 583 mm
9 £1.31 or 131 pence
10 £8.98 making a total of £35.92
11 a 0.8 × 10 = 8   b 18   c 150
12 a 6   b 0.6   c 0.6   d 0.06
13 a 25.5   b 4.96   c 21.3   d 293.22
14 £8.50
15 £29.75
16 0.064
17 e.g. 0.3 × 0.2
18 a 2   b 0.2   c 20   d 2
19 a 4.9   b 9.8   c 4.9
20 a 15.6   b 150   c 4000   d 130
21 16

1.3 Place value

1 a 1300   b 2600 000   c 34.5
2 a 16   b 7.5
3 7.6 × 2.1 = 15.96, so 15.96 ÷ 7.6 = 2.1 and 15.96 ÷ 2.1 = 7.6
4 Answer to 14.3 × 0.96 must be less than 14.3 as 0.96 is between 0 and 1
5 a 4 600 000   b 10 100 000   c 2 450 000   d 3 125 000
6 a 34.1 million   b 4.25 million
   c 58.42 million   d 16.325 million
7 a 11 550 000   b 10 750 000
   c 12 860 000   d 10 550 000
8 a 561.8   b 0.003 47   c 49 000
9 a 76 400   b 0.06   c 340 000   d 47.38
10 a 280 000   b 4   c 2 400 000   d 2000
11 a 300 × 400 = 120 000   b $\frac{900}{30} = 30$
   c $\frac{40 \times 500}{1000} = 20$ d $\frac{700 \times 2000}{400} = 3500$
   e $\frac{300 \times 10}{0.5} = 6000$ f $\frac{5 \times 4}{7} \approx 3$
   g $\frac{10 \times 3}{8} \approx 4$
12 a 115 775   b 30.206…   c 16.719…
   d 4234.09…   e 5647.23…   f 2.977….   g 4.163…
13 a 4   b ≈ 20 ÷ 8 = 5 ÷ 2 = 2.5
14 a 5.715   b 4.5   c 45
15 a 5920   b 5733
16 a 1.256258235   b 1.26
17 a 0.3680743774   b 0.37

1.4 Factors and multiples

1 2
2 7, 14, 21, 28, 35, 42, 49, 56
3 a factor   b multiple   c product   d prime
4 23, 29
5 53, 59, 61, 67
6 False, e.g. 3 × 5 = 15
7 1, 2, 3, 4, 6, 9, 12, 18, 36
8 a 1, 2, 4, 7, 14, 28   b 1, 2, 4, 8, 16, 32
   c 1, 2, 4, 5, 8, 10, 20, 40
9 a 8   b 7   c 12
   d 15   e 8
10 a i 3, 6, 9, 12, 15, 18, 21, 24, 27, 30
     ii 4, 8, 12, 16, 20, 24, 28, 32, 36, 40
   b 12, 24   c 12
11 a 35   b 12   c 24
   d 60   e 40

12 a HCF = 6; LCM = 36   b HCF = 20; LCM = 120
13 e.g. 16 and 32
14 e.g. 5 and 8
15 9
16 100 seconds or 1 minute 40 seconds
17 a 3 packs of bread rolls; 5 packs of sausages
   b 120 hot dogs
18 2.4 seconds

1.5 Squares, cubes and roots

1 a 13.3   b 4.63   c 0.848
2 a 52.4   b 110   c 500
3 a −12   b −12   c 12
4 a 29.16   b 8.9   c 226.981   d 4.82
5 a 36   b 81
6 a ±6   b ±9
7 a 125   b −125
8 a 5   b −5
9 a 17.64   b 50.7   c 21.9   d −8.26
10 1.6
11 $12.96\,m^2$
12 7 m
13 1^3 and 1; 2^3 and 8; 2^4 and 16; 2^5 and 32; 3^3 and 27; 4^3 and 64; 5^3 and 125
14 a 12   b 13   c 19
   d −17   e −3   f 400
15 a 10.665   b 198.207   c 7.24   d 1.22
16 9 and 10
17 8
18 15
19 a 3   b 5
20 a 3.648670991   b 3.6
21 9.476841579
22 9.48
23 a $2\sqrt{3}$ b 3.464101615
24 a $\sqrt{11}$ b $2\sqrt{2}$ c $3\sqrt{5}$
   d $2\sqrt{2}$ e $5\sqrt{3}$

1.6 Index notation

1 a −1   b −3   c 4   d 4
2 108
3 a 196   b 1331
4 a 6.5 cm   b 2.75 m
5 $10^1 = 10$
   $10^2 = 10 \times 10 = 100$
   $10^3 = 10 \times 10 \times 10 = 1000$
   $10^4 = 10 \times 10 \times 10 \times 10 = 10\,000$
   $10^5 = 10 \times 10 \times 10 \times 10 \times 10 = 100\,000$
   $10^6 = 10 \times 10 \times 10 \times 10 \times 10 \times 10 = 1\,000\,000$
6 a 3^6 b 5^3 c 4^5
7 a $2 \times 2 \times 2 \times 2 \times 2 \times 2$ b $5 \times 5 \times 5 \times 5$
   c $7 \times 7 \times 7 \times 7 \times 7$
8 a 4^5 b 6^3 c $2^3 \times 3^2$ d $2^2 \times 5^3$
9 a =   b =   c ≠   d ≠
10 $2^7 = 2^4 \times 2^3$; $2^6 = 2^4 \times 2^2$; $2^5 = 2^2 \times 2^3$
11 a 3^7 b 5^5 c 7^5 d 8^9
12 a 27   b 3^3 c $\frac{3^6}{3^3} = 3^3$
   d $\frac{3 \times 3 \times 3 \times 3 \times 3}{3 \times 3 \times 3} = 3^2$
13 a 4^2 b 5^3 c $7^1 = 7$ d 2^5
14 a 8   b 16   c 25   d 27
15 a 3^{10} b 4^{12} c 5^8 d 6^{15}
16 a 5^6 b 5^3 c 5^4 d 5^9
17 $10^3 = 1000$ $10^{-1} = \frac{1}{10} = 0.1$
   $10^2 = 100$ $10^{-2} = \frac{1}{100} = = 0.01$
   $10^1 = 10$ $10^{-3} = \frac{1}{1000} = \frac{1}{10^3} = 0.001$
   $10^0 = 1$
18 a 10^2 b 10^{-1} c 10^{-3} d 10^4
19 a 1 million $= 1\,000\,000 = 10^6$
   b 1 billion $= 1\,000\,000\,000 = 10^9$
   c 1 trillion $= 1\,000\,000\,000\,000 = 10^{12}$

20

Prefix	Letter	Power	Number
tera	T	10^{12}	1 000 000 000 000
giga	G	$\mathbf{10^9}$	1 000 000 000
mega	M	10^6	**1 000 000**
kilo	k	$\mathbf{10^3}$	1000
deci	d	10^{-1}	**0.1**
centi	c	$\mathbf{10^{-2}}$	0.01
milli	m	10^{-3}	**0.001**
micro	μ	$\mathbf{10^{-6}}$	0.000 001
nano	n	10^{-9}	**0.000 000 001**
pico	p	$\mathbf{10^{-12}}$	0.000 000 000 001

21 a 1000 g   b 1 000 000 B
   c 0.000 001 s   d 0.000 000 000 001 m
22 1000 gigabytes

1.7 Prime factors

1 $2^3 \times 3^4$
2 60
3 a 4   b 36
4 a

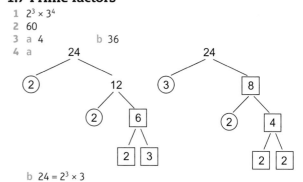

   b $24 = 2^3 \times 3$
5 a $2 \times 3 \times 3$ b $2 \times 3 \times 5$
   c $2 \times 2 \times 2 \times 7$ d $2 \times 2 \times 2 \times 3 \times 3$
6 False, e.g. $2 = 1 \times 2$, $3 = 1 \times 3$
7 a HCF = 12; LCM = 480   b HCF = 12; LCM = 216
   c HCF = 30; LCM = 600
8 D
9 D
10 a $3 \times 221 = 663$ b $2 \times 221 = 442$
11 a Yes, $24 = 2 \times 2 \times 2 \times 3$ b No, $50 = 2 \times 5 \times 5$
   c Yes, $90 = 2 \times 3 \times 3 \times 5$
12 15 chocolates per box
13 $\sqrt{324} = 2 \times 3^2$

1 Problem-solving

1 10 times
2 26 posts
3 1.3 m and 5 m
4 4 hot chocolates and 2 teas
5 8 lambs and 2 sheep
6 15
7 12
8 10 am

1 Check up

1 a 29   b 13   c −5
   d −54   e 15   f $\frac{3}{4}$
2 a $35 \times 24 = 840$ b $(32 + 7) \div 3 = 13$
3 a 15.96   b 5
4 a $500 \div 50 = 10$ b $\frac{800 \times 100}{40} = 2000$
5 a 77.608   b 178
6 a 0.035   b 32
7 a 18.93 570 946   b 18.9
8 15.822
9 a $5^3 = 125$ b $3^4 = 81$ c $\sqrt[3]{64} = 4$
   d $6 \times 6 \times 6 \times 6 \times 6 = 6^5$ e $10^3 = 1000$

10 a 72   b 5   c 4
11 a 7^7 b 7^2 c 7^8 d 7^4
12 31, 37
13 a 6   b 40
14 a $72 = 2^3 \times 3^2$, $96 = 2^5 \times 3$ b HCF = 24   c LCM = 288
16 a 9   b 12

1 Strengthen

Calculations

1 a 6.3   b 15.1   c 0.4   d 12.0
2 a 11.26   b 9.07   c 0.64   d 28.98
3 a 8.046   b 14.173   c 0.057   d 21.814
4 a **4**7.823; 40   b 0.00**5**72; $\frac{5}{1000}$
  c **4**32 650; 400 000   d 0.**6**718; $\frac{6}{10}$
5 a 50   b 500   c 6000   d 9000
6 a 14.1   b 7.2   c 0.0432   d 0.0061
7 a 11   b 28   c 17
  d 31   e 2   f 68
8 a 21   b 9   c $\frac{7}{2}$
  d $\frac{3}{5}$ e 1   f 2
9 a −12   b −32   c 12   d −3
  e 2   f 36   g −2   h 40
10 a 336 + 456   b 566 − 245   c 37 × 23   d 2976 ÷ 62
  e (16 − 5) × 2   f (2 + 3) × 5   g (50 + 2) ÷ 4   h 24 ÷ 3 × 4
11 a 37.12   b 13.11   c 2.272   d 7.3944
12 a 20   b 1.85   c 15.75
  d 120   e 90   f 650
13 a ≈ 500 × 400 = 200 000   b ≈ $\frac{600}{30}$ = 20
  c ≈ $\frac{300 \times 40}{60}$ = 200   d ≈ $\frac{50 \times 200}{0.4}$ = 25 000
14 a 226.8   b 2268   c 2.268   d 0.2268
15 a i $\frac{16.65}{3.7}$ = 4.5   ii $\frac{16.65}{4.5}$ = 3.7
  b i 45   ii 0.37   iii 0.37
16 a 4.486 089 611   b 4.49

Powers and roots

1 a $2^3 = 2 \times 2 \times 2 = 8$ b $4^3 = 4 \times 4 \times 4 = 64$
  c $2 \times 2 \times 2 \times 2 = 2^4$ d $2^2 \times 3^3 = 2 \times 2 \times 3 \times 3 \times 3$
2 a i $\sqrt[3]{64}$ = 4 because $4^3 = 4 \times 4 \times 4 = 64$
    ii $\sqrt[3]{1000}$ = 10 because $10^3 = 10 \times 10 \times 10 = 1000$
    iii $\sqrt[3]{8}$ = 2 because $2^3 = 8$
    iv The cube root of 27 is 3
  b i 5   ii 9   iii 1   iv 12
3 a i 16   ii 16
  b The two square roots of 16 are +4 and −4
  c i ±3   ii ±5   iii ±12
4 a 2^7 b 4^5 c 3^9
  d 5^5 e 6^9 f 10^{10}
5 a 7^3 b 4^6 c 3^3 d 6^3
6 a 6^6 b 2^{10} c 3^{12}
7 a 2   b 11   c 7
8 a 4   b 6   c 57

Factors, multiples and primes

1 a 1, 2, 4, 5, 10, 20   b 1, 2, 3, 5, 6, 10, 15, 30
  c 1, 2, 5, 10   d 10   e 14
2 a 6: 6, 12, 18, 24, 30, 36, 42, 48, 54, 60
  b 7: 7, 14, 21, 28, 35, 42, 49, 56, 63, 70
  c 42   d 40
3 a

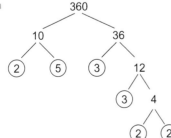

b $360 = 2^3 \times 3^2 \times 5$
c i $144 = 2^4 \times 3^2$ ii $396 = 2^2 \times 3^2 \times 11$
  iii $450 = 2 \times 3^2 \times 5^2$ iv $72 = 2^3 \times 3^2$
  v $84 = 2^2 \times 3 \times 7$
4 a $72 = 2^3 \times 3^2$; $84 = 2^2 \times 3 \times 7$

Prime factors of 72 | Prime factors of 84
2, 3, 2, 3 | 2, 7

b 12   c 504

1 Extend

1 $(1.6 + 3.8 \times 2.4) \times 4.2$
2 a 10.38 032 755   b 10.4
3 Zparts £525; CompParts £540; so Zparts cheaper
4 a 490 000 000   b 3 870 000
  c 16 000 000   d 353 000 000
5 16 children
6 a 40   b 30   c 50   d 30
7 1 pm
8 $2^3 = 2 \times 2 \times 2 = 8$ $2^{-1} = \frac{1}{2^1} = \frac{1}{2}$
  $2^2 = 2 \times 2 = 4$ $2^{-2} = \frac{1}{2^2} = \frac{1}{4}$
  $2^1 = 2$ $2^{-3} = \frac{1}{2^3} = \frac{1}{8}$
  $2^0 = 1$
9 6
10 3.2
11 0.17 lies between 0 and 1 so answer must be greater than 14.314
12 a 2.26541555   b 2
13 a 650 000 000 m   b 24 000 pm
  c 6000 mm
14 a False, e.g. 2 × 3 = 6   b False, e.g. 5 + 7 = 12
  c True
15 225
16 6
17 a $2^2 \times 135$ b $3^3 \times 20$ c $3^2 \times 240$
18 $n = 4$
19 a 64   b 3   c 12
20 a $2^3 \times 3^2 \times 5$ b $2^4 \times 3^4 \times 5^2$
21 a 48   b 6

1 Unit test

Sample student answer

Both methods give the correct answer. However, Student B has not written a sentence to make their answer clear, so Student A has given the better answer.

UNIT 2

2 Prior knowledge check

1 a −5   b −9   c −12   d −2
  e 20   f −2   g 5   h −60
  i 21   j 1
2 a 100   b 31   c 1
  d 12   e 9   f 6
3 a 3^5 b 7^4 c 4^3 d 9^3
4 a 4   b 3   c 6
5 a $15t$ b $12m$ c $-2x$ d $-5a$
6 £20
7 24 cm
8 Students' own answer, e.g. $\frac{100x}{4}$, $35x − 10x$, $20x + 5x$, $5x \times 5$

2.1 Algebraic expressions

1 a $6a$ b $7r$ c $-4x$ d $-s$
2 14 cm
3 a $9y + 12$ b $2a − 3b$ c $m^2 + 2$ d $9r^2 + r$
4 Ben is correct.
5 a $3m − 9$ b Same answer

6 a $3m$ b $40c$ c $8t$
   d cd e $\frac{h}{3}$ f $\frac{a}{b}$
7 a $3f$ b $6m$ c $4a + 6h$
8 a $x + 6$ b $x - 7$ c $12y$
   d $3m$ e $\frac{y}{2}$ f $\frac{d}{2}$
9 a $y - 2$ b $5y$ c $y + 3$ d $8y + 1$
10 a $2n$ b $4n$ c $10n$ d xn
11 $P = 2b + 2h$

2.2 Simplifying expressions

1 a $3y$ b lm c $\frac{h}{4}$ d $8n$
2 a 2^3 b $3^2 \times 4^3$ c 7^2 d 5^3
3 a 20 b -24 c -2 d -4
4 a 3^2 b x^2 c 4^3 d y^3
5 a 2^6 b x^6 c y^{11}
6 a $y \times y$ b $x^2 \times x$ c e.g. $x^4 \times x^3$
7 a 5^3 b y^3 c 9^5 d x^5
8 a a^4 b z^4 c x^3 d g^5
9 a $12ab$ b $14rs$ c $36abc$ d $18xyz$
   e $-42mn$ f $4a^2$ g $48rs^2$ h $-16a^2b$
10 Dave wrote it better. Algebraic terms should have numbers first, then letters in alphabetical order.
11 $-6m^2, -8mn, 10mn, 12mn, -15mn, -20n^2$
12 a $3b$ b $-4a$ c $2z$
   d $\frac{m}{2}$ e $\frac{p}{3}$ f $-2e$
   g $\frac{3t}{4}$ h $\frac{2f}{3}$ i $\frac{d}{2}$
13 Jessie is correct.
14 a $3a$ b $4c^2$ c $\frac{1}{4}$
   d $-5z$ e $-\frac{p}{6}$ f $\frac{3m^2}{4}$
15 a x b $2x$ c x^2
   d $2x$ e x^2

2.3 Substitution

1 a 2 b 42 c -7 d $-\frac{1}{4}$
2 a $\frac{x}{2}$ b $2x$ c $x - 2$ d $x + 2$
3 a 14 b 31 c 4 d 14
4 a $n - 20$ b $10n$ c $\frac{n}{5}$
5 a $4n + 5$ b $2n - 6$ c $\frac{3n}{4}$
   d $\frac{n}{2} + 4$ e $\frac{n+4}{2}$
6 a 7 b 3 c 10 d 20
   e 16 f 50 g -8 h -10
7 a -20 b -7 c -6 d 3
   e 50 f 7 g 22 h 13
8 a 23 b 14
9 a $c + p$ b 8
10 a $n - 1$ b $\frac{n-1}{2}$ c 2
11 a $3x$ b $2y$ c $3x + 2y$ d £2.85
12 a k represents each kg over the limit.
   b 59.50, 68.50, 86.50, 100

2.4 Formulae

1 a 9 b -4 c 41
2 a 19 b 15 c 100 d 34
3 $r + 4$
4 a $n + 10$ b $C = n + 10$
5 a $12x$ b $n = 12x$
6 a $C = x - 15$ b $C = 3x$ c $C = \frac{x}{2}$
7 a $20\,\text{N}$ b $45\,\text{N}$
8 a $27\,\text{cm}^2$ b $75\,\text{cm}^2$ c $300\,\text{cm}^2$
9 a 3 hours   b 4 hours   c 2 hours
10 a 25 mph   b 35 mph   c 60 mph   d 350 mph
11 35.4
12 a Expression   b Formula   c Expression
   d Expression   e Formula   f Formula
13 a $D = nr$ b £56   c $S = \frac{t}{2}$
   d £15   e £71
14 a $C = 3 + 4k$ b £23
15 a 5 b 25 c 4.8 d 12.3

2.5 Expanding brackets

1 a 45 b -24 c 2 d 6
2 a $-3y$ b $a - 3b$ c $3m^2 - 9$
3 a $C = 35 + 20h$ b £95
4 a 8 b 45 c 121
   d -8 e 100
5 a 27 b 27 c 14 d 14
6 a $3x + 6$ b $2a + 14$ c $10t - 60$
   d $15 - 5m$ e $6w + 3$ f $-2a - 10$
   g $-6b - 18$ h $-4a + 24$ i $-x - 4$
7 Benton is correct. Hannah has made the common mistake of forgetting that the negative sign is part of the $2d$ term.
8 a $x^2 + x$ b $r^2 + 4r$ c $3g^2 - 2g$
9 a $2(n + 3)$ b $4(n + 1)$ c $10(n + 5)$
10 a $5(3n + 1) = 15n + 5$ litres
   b 305 litres
11 a $3t + 14$ b $5m - 4$ c $a^2 - 6a$
   d $-2m + 1$ e $14 + 10x$ f $16e - 2f$
12 a $5x + 8$ b $5a + 6$ c $7d - 18$
13 a 100 b 400 c 1 d $\frac{4}{9}$
14 a $x + 2$ b $\frac{x+2}{2}$ c 8
15 a $12n + 25$ b $C = 12n + 25$
   c $E = 3(12n + 25)$ d £1155
16 a 70 represents the first 70 litres that are used; 5 represents the additional five minutes of water flow, and $x + 2$ represents the rate of water flow during these 5 minutes.
   b 110 litres

2.6 Factorising

1 a 4 b 3
2 a $3x + 12$ b $5x - 30$ c $x^2 + x$ d $2x^2 - 2x$
3 a 4 b 3
4 a $3b$ b 4 c $a + b$ d 12
5 a   i $1 \times 20, 2 \times 10, 4 \times 5$
    ii $1 \times 10t, 2 \times 5t, 5 \times 2t, 10 \times t$
    iii 10
   b   i Factor pairs for $24r$: $1 \times 24r, r \times 24, 2 \times 12r, 2r \times 12,$
     $3 \times 8r, 3r \times 8, 4 \times 6r, 4r \times 6$
     Factor pairs for 42: $1 \times 42, 2 \times 21, 3 \times 14, 6 \times 7$
    ii 6
   c   i 3 ii 4
6 a $8(y + 2)$ b $5(2m - 5)$ c $6(y + 4)$ d $7(m - 3)$
7 Phil is correct as he has taken out not just any factor, but the HCF out of the expression. Factorising an expression requires you to 'completely factorise'.
8 a $9(x + 2)$ b $3(w - 4)$ c $5(3a + 2)$ d $3(4 - 7t)$
9 Zhir is correct, Charlotte has not divided -5 by 5.
10 a d b a c b
   d $6y$ e $10a$ f $2x$
11 a $n(n + 1)$ b $x(4x + 3)$ c $2t(s + 2)$ d $b(5a - 3)$
12 a A   b B   c B   d A
13 a $x(x + 5)$ b $7x(x - 3)$ c $3(3x + 4y)$ d $2y(3y - 1)$
14 a $2(2x + 5y)$ b $x(x + 7)$
15 b $2t + 4 \equiv 2(t + 2)$ d $5t + 7 \equiv 7 + 5t$
16 a $4a + a \equiv 5a$ b $0.5a \equiv \frac{a}{2}$
   c $3(a + 4) \equiv 3a + 12$ d $a + 2 \equiv 2 + a$
17 a Formula   b Identity   c Expression
   d Identity   e Expression   f Formula
18 a $\neq$ b $\neq$ c $\equiv$ d $\equiv$

2.7 Using expressions and formulae

1 a 7 b -10 c 5 d -6
   e 9 f 16 g -2 h 2
2 a $3r$ b $r + 20$ c $C = 4r + 20$
3 a $P = 4x + 4$ b $P = 2a^2 + 2a^3$
   c $P = 8r + 4$ d $P = 2b^2 + 10b^3$
4 a $5n + 2$ b $2(n - 1)$ c $\frac{3n}{2}$
   d n^2 e n^3 f $\sqrt{n}$

5 a $3x$ b $4y$ c $3x + 4y$
  d $px + by$ e 22
6 a i $m - 20$ ii $2(m - 20)$
  b $4m - 60$ c 180
7 6 kg
8 a 60 m/s b 160 m/s
9 a 15 m b 40 m
10 10 m/s
11 a 59 °F b 25 °C = 77 °F. Monday was hotter.
12 a 120 minutes b $M = 40w + 20$
  c 180 minutes d 5.20 pm

2 Problem-solving

1 205
2 a 377.5 m b $22.5b + 25c$
3 a 89 b $rs - t$
4 a 56 b $q(q - 1)$
5 a 460 b $23p$
6 a 73 b $2n - 1$

2 Check up

1 a w b $5a - 5b$ c $m^2 - 8m$
  d $20mn$ e $4a^2b$ f $8x$ g $15a$
2 a 18 b 2 c 4 d -5
3 a $x + 5$ b $\frac{4x}{5}$
4 a $2a + 2$ b $15f^2 + 10f$ c $6y^2 - 2y$ d $-6a - 10$
5 a $12(3x + 1)$ b $4x(x + 4)$ c $3(3x + 7y)$ d $5y(3x - 1)$
6 a $\neq$ b $\equiv$ c $\equiv$ d $\neq$
7 a Formula b Identity c Expression
8 a 8 km/h b 25
9 a $T = b + p$ b 35
10 8
12 a $x^2 + 3x$, $2x^2 + 4x$, $x^2 - 2x$, $3x^2 - x$
  b $7x^2 + 4x$ c $x(7x + 4)$

2 Strengthen

Expressions and substitution

1 a $7x$ b $14b$ c $2h$ d $11y$
2 a $6d + 10e$ b $15x + 10y$ c $2r + 8s$ d $4p - 9q$
  e $-3v - 10w + 4$ f $2g + 4h + 10$
  g $10x^2$ h $8a^2 + 2a$
3 a $6a$ b $4n$ c $-2y$
  d $-4k$ e ab f fg
4 a $10a$ b $18y$ c $-40st$ d $-12pq$
  e $2a$ f $-6b$ g 4
5 a $n + 4$ b $n - 6$ c $2n$ d $\frac{n}{3}$
6 a a^2 b $2a^2$ c a^5 d q^5
7 a 4 b a c st
8 a x^3 b a^4 c $3x^4$ d $3f^4$
9 a 4 b 20 c 24
  d 8 e 12 f 75
  g 100 h 9 i 2
10 a 7 b 5 c 49 d 18
11 a 2 b 6 c 8
  d 16 e -4 f -12

Expanding and factorising

1 a $3b^2 + 4b$ b $5t + 10$ c $2d^2 - 5d$ d $10f - 5$
2 a $4p^2 + 4p$ b $2x^2 - 2x$ c $9r - 42$
  d $2b^2 + b$ e $12a + 29$ f $-2r^2 - r$
3 a 5 b 7 c 4 d 9
  e a^2 f a^2 g a h ab
4 a $3a(a - 3)$ b $4x(4x + 3)$ c $5a(a + 3b)$
  d $2q(q^2 + 4)$ e $12(7a - 1)$ f $a(5a + b - c)$
  g $y^2(y + 1)$
5 No. $8x$ is the HCF so the answer should be $8x(x + 2)$
6 a $\equiv$ b $\neq$ c $\equiv$ d $\neq$

Writing and using formulae

1 a 120 N b 30 N c -500 N

2 a $P = 2b + 2h$ b $P = 26$ cm
3 a i £100 ii £350
  b $T = ad$ c £76.50
4 a £30 b $hx + t$ c $E = hx + t$ d £70

2 Extend

1 a $2n + 6$
  b She could have doubled the number and then added 6.
2 a $10x^2 + 4x + 14$ b $-6x - 5xy + 3xz$
3 a $1.5x$ b $3x^2 + 3x$ c $0.25x$
4 a a^9 b $16a^{14}$ c a^{m+n} d $24a^{y+z}$
  e a^3 f $5a^2$ g a^{-5} h a^{m-n}
5 a ab b $\frac{c^4}{c^2} \times d = c^2d$ c xy^2 d m^3p^4
6 a 2250 m b 195 m
7 a 4 g/cm³ b 1.5 g/cm³
8 a 15 days b 8 ml
9 a $9x + 18y - 9z$
  b $-2r + 14t + 6s$
  c $9ac - 9bc$
  d $ps + pt - pr$
  e $5x^2 + 40xy - 45xz$
10 a $-7(t + 3)$ b $-a(a + 5)$
  c $-6f(f + 2)$ d $-2r(1 + 2r)$
11 a $10c + 15d$ b $2x - 7$
12 Alex is correct.
  A negative number raised to an even power is always positive because negative × negative is always positive. A negative number raised to an odd power is always negative because (negative × negative) × negative = positive × negative and is always negative.
13 a -120 b 9 c 5 d 5.4
14 Students' own answers, e.g. $\frac{qp}{2} = 12$, $\sqrt{prs} = 6$, $qp + s^2 = 28$, $\frac{r^2p}{s^2} = 27$
15 a $10a^2 + 9a - 10ab$ b $-4p^3 + 5p^2q + 2pq$
  c $2ab$

2 Unit test

Sample student answer

a It is a good idea to write out what you know first, then it is clear to see what the girls' individual expressions are.
b So that you don't forget to multiply both terms in the bracket by what is outside.
c You may still get marks for the method even if you make an error with the final answer.

UNIT 3

3 Prior knowledge check

1 a 108 b 180 c 12
  d 10 e 180 f 18
2 a

Flavour	Frequency
lemon	3
lime	9
orange	3
blackberry	5

  b 20 c 3
3 4 hours
4 a French b Ella c Penny
5 a 15 b Cat
6 a Clothes b Jess c 40%
7 a 7500 b 10 000
  c 2012 and 2013
  d 2011 and 2012
8 a 6 b ○ ○ ◖

9 a 5   b 9   c 16

10

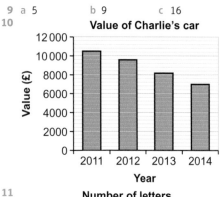
Value of Charlie's car

11

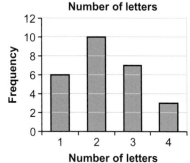
Number of letters

12 Accurate drawing of a circle with radius 3 cm
13 Accurate drawing of an arc of radius 4 cm with an angle of 120°
14 $\frac{1}{4}$ turn and 90°; $\frac{3}{4}$ turn and 270°
15 Missing angle is 130°

3.1 Frequency tables

1 a 122   b 140
2 Any suitable tally chart.
3 a 26   b 46
4

Shoe size	Tally	Frequency
2–4	Ⅲ Ⅱ	5
5–7	Ⅲ Ⅲ	10
8–10	IIII	4
11–13	I	1

5 a 25.5, 26   b 26, 26.5, 27
6 a

Age (years)	Tally	Frequency
$15 < y \leqslant 20$	Ⅲ	5
$20 < y \leqslant 25$	Ⅲ II	7
$25 < y \leqslant 30$	Ⅲ III	8
$30 < y \leqslant 35$	II	2

   b $30 < y \leqslant 35$ c 20
7 a Continuous
   b Any suitable grouped frequency table. e.g. Length split as $11 < y \leqslant 14$, $14 < y \leqslant 17$, $17 < y \leqslant 20$
8 Any suitable data collection sheet.
9 Any suitable data collection sheet.
10 Any suitable data collection sheet. e.g. Hours split as 11–14, 15–18, …

3.2 Two–way tables

1 a 130 minutes or 2 hours 10 minutes
   b 125 minutes or 2 hours 5 minutes
   c 11 hours 15 minutes
2 a 4 hours 15 minutes   b 13.15
3 a 23 minutes   b The 10.37
   c e.g. Yes, because the two trains both take 11 minutes.

4 a 42 miles   b 42 miles   c 45 miles
5 a 158 km   b 300 − 137 = 163 km
   c Manchester and Liverpool
6 5 + 26 + 24 = 55
7 34 − 9 = 25
8 Any suitable two-way table.
9 Any suitable two-way table.
10 a 10   b 8
   c

	Small	Medium	Large	Total
Pine	7	**12**	**4**	23
Oak	**10**	16	**8**	34
Yew	3	8	2	13
Total	20	**36**	14	**70**

11 25
12 35

3.3 Representing data

1 a i A   ii C   b 24 °C
2 a

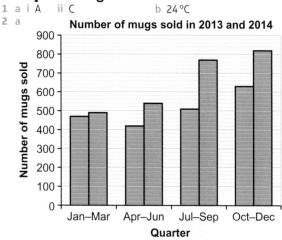
Number of mugs sold in 2013 and 2014

Key ▨ 2013 ▨ 2014

   b Oct–Dec 2014
   c e.g. More mugs were sold in 2014 than in 2013.
3 a

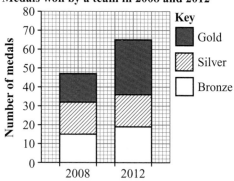
Medals won by a team in 2008 and 2012

Key: Gold, Silver, Bronze

   b e.g. In 2012 the team won more medals than in 2008. Or: The greatest improvement in 2012 was in the number of Gold medals won.
4

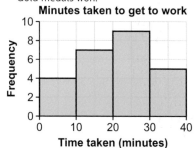
Minutes taken to get to work

5

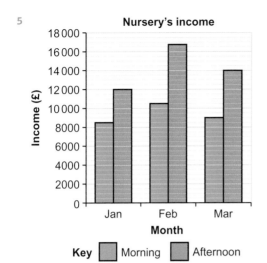

Nursery's income

6 a, b

Number of books and ebooks sold over 5 years

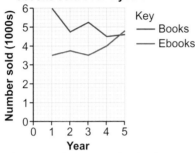

c Book sales are decreasing, ebook sales are increasing.

7

Income in first three months of 2014

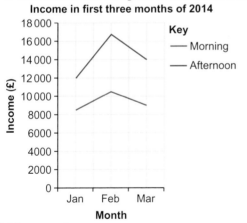

3.4 Time series

1

Amount of rainfall each day

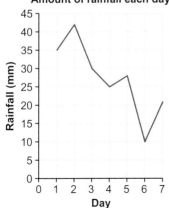

2 a 45 metres   b 32 seconds

3 Mandy is not correct. Her sales figures are increasing but NOT at a steady rate, i.e. the differences between the months are not the same.

4

Temperature one morning in January

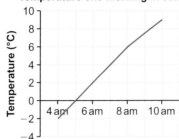

5 a

Month	March	April	May
Money (£)	5800	2600	4500

b

Money received each month

6 a

Number of overseas visitors to UK

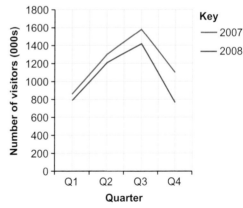

b e.g. The number of visitors increases from Q1 to Q3 then decreases in Q4.

7 a

Numbers taking A-level Physics and French

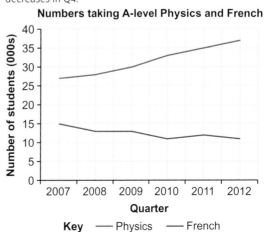

b e.g. More students have taken A level Physics than A Level French in every year.

c e.g. She is correct. From 2007 to 2012 Physics numbers have increased, so her prediction is based upon the data.

d e.g. Based on the data, it is likely that the number of A Level French candidates will continue to fall.

3.5 Stem and leaf diagrams

1 a 123, 123, 125, 125, 132, 135, 143, 146

b 1.4, 3.2, 3.9, 3.9, 4.1, 6.7, 7.5, 8.4

2 a 6   b 34 m

3 a 19   b 9   c 37 mm

4
0	8
1	0 2 3 5 7 8
2	0 1 2 2 2 3 3
3	1 3 4 5
4	4 5 6

Key
4 | 4 means 44 minutes

5
1	7 8 8 9
2	0 0 1 2 3 5 9
3	3 7 7
4	2

Key
3 | 3 means 33 years old

6
19	2 8
20	5
21	
22	0 0
23	0 5 8 8 8

Key
19 | 2 means 192 seconds

7
3	4 9 9 9
4	6 7 7 8 9
5	1

Key
5 | 1 means 5.1 metres

8 a 75 bpm   b 107 bpm   c 23 bpm

d e.g. More males than females had a heart rate of more than 100 bpm. Three males did (101, 103 and 107) whereas the highest female heart rate was 99 bpm.

9 a
Abbey Hotel		Balmoral Hotel
2	0	
9 7 5 1	1	
9 8 3 1	2	6
9 9 9 9 7 6 6 5 3 3 2	3	4 4 7
9 8 7 7 5 0	4	0 0 5 5 6 9
8	5	0 0 0 0 1 3 6 6 7
	6	2 3 3 4 5 7
	7	0 1 5

Key
For Abbey Hotel   For Balmoral Hotel
9 | 1 means 19 years old   7 | 0 means 70 years old

b e.g. She is correct. Balmoral Hotel has several residents over 50 years old, whereas Abbey Hotel only has one resident over the age of 50.

10
Boys		Girls
9	14	6 8
6 5 3	15	1 1 2 5 6 7
5 5 5 3	16	4 9
2 0	17	

Key
For boys   For girls
6 | 15 means 156 cm   14 | 6 means 146 cm

11 a
Dark		Light
9 9 8 7 6 5	1	
2 0	2	7 8
	3	5 8
	4	1 3 4 5

Key
For dark   For light
9 | 1 means 1.9 cm   3 | 8 means 3.8 cm

b e.g. They grew better in the light because those seedlings grew taller.

3.6 Pie charts

1 a, b Accurate drawing of a circle with an angle of 60° at the centre.

2 a $\frac{1}{2}$

b e.g. The percentage of householders who thought waste collection was poor has reduced from 50% in 2013 to 25% in 2014.

c e.g. Yes, because the proportion of householders who answered poor in 2014 has fallen and the proportion who answered good has increased.

3 a 25%

b 45° (allow 43–47°) which equates to 12.5% of the students (accept 12–13%).

c e.g. Twice the 'yes' vote is 50%. 'Don't know' is more than 50%, so Bethan is correct.

4 a Five games.

b No; Team A won 10 games and Team B won 14 games.

5 a 90

b 4°

c Hot chocolate 80°; milkshake 60°, coffee 100°, tea 120°

d Accurate pie chart drawn with hot chocolate 80°, milkshake 60°, coffee 100°, tea 120°.

6 Accurate pie chart drawn with France 144°, Spain 108°, Germany 45°, Italy 63°.

7 Accurate pie chart drawn with gym 144°, swimming 36°, squash 72°, aerobics 108°.

8 a 60

b i 12

ii 6

9 a 32

b Non-fiction

10 a

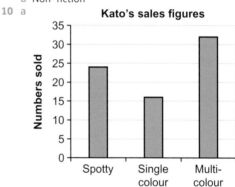

Kato's sales figures

b Accurate pie chart drawn with spotty 120°, single colour 80°, multicolour 160°.

3.7 Scatter graphs

1 a, b

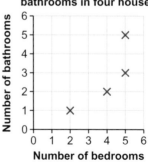

Number of bedrooms and bathrooms in four houses

2 a
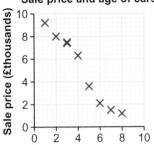
Sale price and age of cars

b decreases

3 a
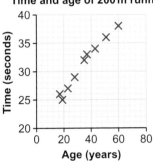
Time and age of 200 m runners

b increases

4 Positive correlation (Q1), negative correlation (Q2), positive correlation (Q3).

5 a B   b C   c A

6 a Negative correlation   b 2

7 a
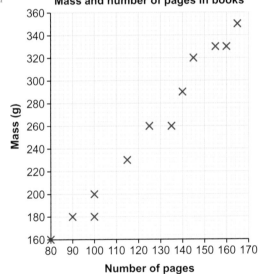
Mass and number of pages in books

b increases; positive

8 a
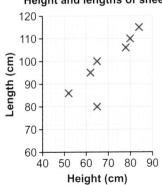
Height and lengths of sheep

b (65, 80)   c Positive correlation

9
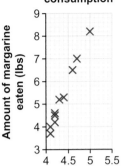
Divorces and margarine consumption

10 Arm length and leg length: positive correlation
Exercise and weight: negative correlation
Size of garden and running speed: no correlation
Hours of TV watched and shoe size: no correlation

3.8 Line of best fit

1 a 6
  b 2.5

2 a positive
  b negative

3 a, b
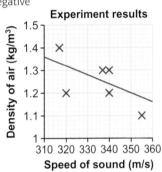
Experiment results

4 a
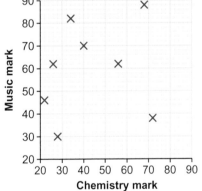
Chemistry and music exam results

b No correlation

c e.g. No, a line of best fit needs a correlation, either positive or negative.

5 a, b

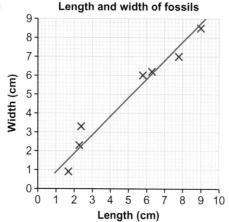

Length and width of fossils

c Allow 3.6 to 4.2 cm.

6 a

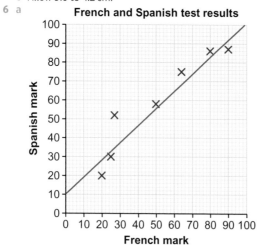

French and Spanish test results

b (27, 52) is an outlier. e.g. These marks do not fit the pattern as the Spanish mark is a lot higher than the French mark.

c As the French mark increases the Spanish mark increases. Positive correlation.

d Allow 60–68 marks

7 a, c

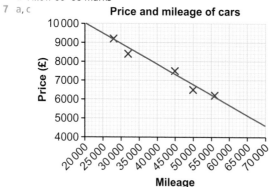

Price and mileage of cars

b Negative correlation

d Allow £5000 to £5400

8 a Positive

b Allow 83 to 87 cm

3 Problem-solving

1 a GDP per person

b Graph of GDP per person against infant mortality

c Students' own answers

2 a Graph of life expectancy against unemployment %

b Students' own answers

3 Students' own answers

3 Check up

1

Height (m)	Tally	Frequency
$1.2 \leqslant h < 1.3$	III	3
$1.3 \leqslant h < 1.4$	IIHI	5
$1.4 \leqslant h < 1.5$	IIII	4

2 a Huntingdon b 3 minutes c 10:05

3 a 19

b

	Budget seats	Standard seats	Luxury seats	Total
Adult	**15**	17	19	**51**
Child	24	**25**	30	**79**
Total	39	**42**	**49**	130

4 Accurate pie chart drawn with perch 50°, bream 115° and carp 195°.

5 a

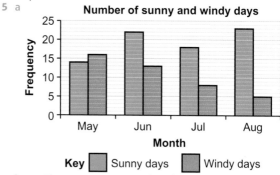

Number of sunny and windy days

Key Sunny days   Windy days

b e.g. There are more sunny days than windy days during the four-month period.

6 1 – Marc won 40 games and Caroline 39

7 e.g. After, because more people had heart rates over 80 bpm after walking up the stairs.

8 a 5 cm

b Allow 2 hours 35 minutes to 2 hours 45 minutes

9 e.g. As the temperature increases, ice cream sales increase. Positive correlation.

11 Any suitable data collection sheet.

3 Strengthen

Tables

1

Length, l (cm)	Tally	Frequency
$7 \leqslant l < 9$	II	2
$9 \leqslant l < 11$	II	2
$11 \leqslant l < 13$	IIHI	6
$13 \leqslant l < 15$	IIH	5

2 Any suitable data collection sheet.

3 7 hours 5 minutes

4 a 34 minutes b 10:29

5

	White	Black
Circle	3	4
Square	6	5

6 a 5   b 4   c Elephant

7 a 22   b 25

c

	Walk	Car	Bicycle	Total
Boys	15	**25**	14	54
Girls	**22**	8	16	**46**
Total	37	**33**	30	100

8

	London	York	Total
Boys	23	14	**37**
Girls	19	**24**	**43**
Total	**42**	**38**	80

Graphs and charts

1 a

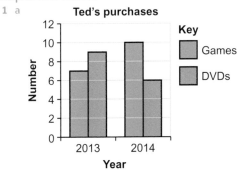

Ted's purchases

Key: Games, DVDs

b e.g. In 2013 he bought more DVDs than games but in 2014 he bought more games than DVDs. He bought the same number, 16, of games and DVDs combined in both years.

2 a

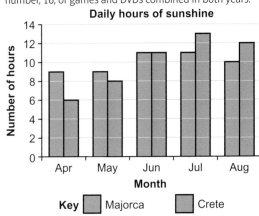

Daily hours of sunshine

Key: Majorca, Crete

b e.g. In April and May Majorca has more sunshine hours. In July and August Crete has more.

3 a 30   b 12°   c 120°
d Green 60°, yellow 72°, orange 108°
e Accurate pie chart drawn with red 120°, green 60°, yellow 72°, orange 108°. All labelled correctly.

4 a 103, 105, 108, 110, 112, 113, 114, 117, 119, 121, 123, 125, 125, 125, 127

b
```
10 | 3 5 8
11 | 0 2 3 4 7 9        Key
12 | 1 3 5 5 5 7        10 | 3 = 103 grams
```

Time series and scatter graphs

1 a

Temperature in a school greenhouse

b 2 pm and 5 pm
c 27°   d 9° (25° − 16°)
2 A–positive, B–negative, C–positive, D–no correlation, E–negative

3 Extend

1 a $3m$ b $2.5m$ c $1.5m$
2 14
3 Students' own answers
4 He is not correct. 50 + 110 + 220 + 140 = £520, $\frac{520}{12}$ = £43.33, so he needs to pay £43.33 per month.
5 Accurate pie chart drawn with Tennis 89°, Football 144°, Swimming 44°, Basketball 83°
6 a 58%
 b 54%
 c e.g. No, the chart tells us the about the proportion of mice in the two areas. It does not tell us anything about the numbers of mice.
7 a June
 b December
 c 18 hours.
8 Accurate bar chart or line graph.
9 e.g. Harrow Lane – Swipe Crescent 08:02 – 08:41
Swipe Crescent – Harrow Lane 15:16 – 15:49

3 Unit test

Sample student answers

a The bar chart is the best one to use. Line graphs can only be used for continuous data. Number of people is discrete data, not continuous.
b Both students forgot to label the vertical axis. They should have labelled it either 'Frequency' or 'Number of people'.

UNIT 4
4 Prior knowledge check

1 $\frac{1}{3}$
2 B, C
3 a $\frac{6}{8} = \frac{3}{4}$ b $\frac{1}{4} = \frac{2}{8}$ c $\frac{3}{6} = \frac{1}{2}$
4 a $\frac{4}{5}$ b $\frac{2}{4} = \frac{1}{2}$ c $\frac{2}{5}$ d $\frac{2}{4} = \frac{1}{2}$
5 $\frac{1}{9}, \frac{2}{9}, \frac{5}{9}, \frac{8}{9}$
6 a $\frac{3}{4} = \frac{6}{8}$ b $\frac{1}{5} = \frac{4}{20}$ c $\frac{6}{10} = \frac{3}{5}$
7 a $\frac{1}{3}$ b $\frac{3}{4}$ c $\frac{4}{5}$
8 $1\frac{5}{7}$
9 a $\frac{9}{2}$ b $\frac{27}{4}$
10 a 0.5   b 0.75   c 0.7   d 0.$\dot{6}$
11 a 50%   b 20%   c 75%   d 33.$\dot{3}$%
12 a $\frac{1}{4}$ b $\frac{2}{5}$ c $\frac{9}{10}$
13 a 20%   b 50%   c 75%
14 a $\frac{3}{10}$, 50%, 0.75 b 1, $\frac{4}{5}$, 60%, 0.5
15 a $\frac{5}{12}$ b $\frac{1}{4}$
16 a 2   b 150
17 a £18   b £6   c 16 cm
18 a $\frac{1}{2}$ of an hour b $\frac{1}{4}$ of an hour
19 $\frac{1}{5}$
20 $3\frac{1}{4}$, $\frac{17}{5}$, $\frac{17}{3} = 5\frac{2}{3}$, $6\frac{3}{4}$

4.1 Working with fractions

1 a 12   b 24   c 6
2 a >   b <
3 a $\frac{2}{3} = \frac{4}{6}$ b $\frac{4}{9} = \frac{12}{27}$ c $\frac{2}{5} = \frac{8}{20}$ d $\frac{2}{3} = \frac{4}{6} = \frac{8}{12}$
4 a $\frac{3}{4}$ b $\frac{2}{3}$ c $\frac{3}{5}$ d $\frac{2}{3}$
5 e.g.   a $\frac{2}{4}$ and $\frac{3}{4}$ b $\frac{7}{10}$ and $\frac{6}{10}$
6 Spinner A
7 Yes, e.g. $\frac{4}{9} = \frac{12}{27}$ $\frac{1}{3} = \frac{9}{27}$

8 $\frac{2}{3}$ is larger

9 a $\frac{3}{5}, \frac{2}{3}, \frac{3}{4}$ b $\frac{5}{6}, \frac{2}{3}, \frac{7}{12}$

10 a Students' own answers b $\frac{3}{10}$

11 a Students' own answers
 b e.g. $\frac{7}{8} = \frac{63}{72}, \frac{8}{9} = \frac{64}{72}. \frac{8}{9}$ is larger.

12 a $\frac{7}{8}$ b $\frac{11}{15}$ c $\frac{1}{4}$ d $\frac{5}{9}$
 e $\frac{1}{2}$ f $\frac{2}{3}$ g $\frac{1}{3}$ h $\frac{3}{5}$

13 a $\frac{8}{15}$ b $\frac{1}{6}$ c $\frac{39}{70}$

14 $\frac{5}{6}$

15 $\frac{7}{10}$

16 a $\frac{8}{15}$ b $\frac{1}{3} + \frac{1}{4}$ or $\frac{1}{2} + \frac{1}{12}$

17 a $\frac{13}{24}$ b $\frac{11}{12}$ c $\frac{7}{12}$
 d $\frac{3}{20}$ e $\frac{17}{20}$ f $\frac{11}{16}$

18 a $\frac{2}{3}$ b $\frac{2}{5}$ c $1\frac{3}{4}$ or $\frac{7}{4}$

4.2 Operations with fractions

1 a 500 cm b 3.5 m

2 a $\frac{1}{4}$ b $\frac{7}{24}$

3 a $3\frac{1}{2}$ b $1\frac{3}{8}$ c $8\frac{1}{3}$ d $2\frac{1}{12}$

4 a $\frac{5}{2}$ b $\frac{19}{4}$ c $\frac{43}{6}$ d $\frac{21}{2}$

5 a 4 b 28 c 64

6 a 150 kg b 72 cm

7 45

8 £30

9 e.g. 7 × 150 = 1050 cm = 10.5 m

10 e.g. $\frac{1}{10} + \frac{1}{5} = \frac{3}{10}$ $\frac{7}{10}$ are 17 year olds $\frac{7}{10}$ of 700 = 490

11 e.g. Total frequency = 30 + 15 + 5 + 10 = 60
 Goats = $\frac{10}{60} = \frac{1}{6}$

12 a 150 b Answers between 220 and 260

13 a $\frac{35}{56} + \frac{32}{56} = \frac{67}{56} = 1\frac{11}{56}$ b $\frac{47}{40} = 1\frac{7}{40}$
 c $\frac{41}{35} = 1\frac{6}{35}$ d $\frac{35}{24} = 1\frac{11}{24}$

14 a $5\frac{3}{4}$ b $13\frac{1}{4}$ c $11\frac{3}{4}$
 d $16\frac{1}{4}$ e $3\frac{3}{5}$ f $6\frac{2}{3}$

15 a $1\frac{5}{12}$ b $4\frac{3}{4}$ c $3\frac{1}{10}$
 d $\frac{3}{4}$ e $1\frac{1}{2}$ f $2\frac{1}{2}$

16

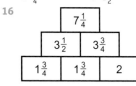

4.3 Multiplying fractions

1 a $\frac{5}{6}$ b $\frac{6}{7}$ c $\frac{10}{15} = \frac{2}{3}$ d $\frac{14}{16} = \frac{7}{8}$

2 a $7\frac{1}{2}$ b $\frac{18}{5}$

3 a $\frac{7}{2} = 3\frac{1}{2}$ b $\frac{35}{2} = 17\frac{1}{2}$

4 a 5 b $7\frac{1}{2}$ c $16\frac{4}{5}$
 d 10 e $9\frac{3}{4}$

5 $5 \times \frac{3}{4} = \frac{15}{4} = 3\frac{3}{4}$

6 £345

7 a 4 b 4 c 3 d $1\frac{1}{2}$

8 6 kg

9 a $\frac{1}{8}$ b $\frac{1}{15}$ c $\frac{3}{8}$ d $\frac{6}{49}$
 e $\frac{12}{35}$ f $\frac{8}{63}$ g $\frac{1}{5}$ h $\frac{1}{2}$

10 £187 500

11 a $\frac{2}{5}$ b $\frac{7}{30}$ c $\frac{5}{14}$
 d $\frac{2}{9}$ e $\frac{2}{9}$ f $\frac{13}{28}$

12 $\frac{1}{6}$

13 e.g. $\frac{3}{4} \times \frac{2}{3} = \frac{1}{2}$ kg = 500 g. No, he doesn't have enough.

14 39 litres

15 a $5\frac{1}{4}$ b $12\frac{6}{7}$ c $17\frac{1}{2}$ d $35\frac{1}{5}$

16 $7\frac{1}{2}$ hours

17 $6 \times 1\frac{3}{4} = 10\frac{1}{2}$ m. No, she doesn't have enough.

18 Option 1: 22 000 + 37 000 = 59 000   59 000 × $3\frac{1}{2}$ = 206 500
 Option 2: 222 000 + 33 000 = 255 000
 $\frac{1}{4}$ million = 250 000
 Only option 2 is suitable.

4.4 Dividing fractions

1 a 2 b 4

2 $\frac{15}{4}$

3 $6\frac{1}{3}$

4 a $\frac{10}{3} = 3\frac{1}{3}$ b $\frac{6}{11}$

5 a $\frac{4}{3}$ b 5 c $\frac{1}{4}$

6 a 4 × 3 = 12 b 25 c 96 d $5\frac{1}{3}$

7 2

8 a $\frac{10}{3} = 3\frac{1}{3}$ b 9 c $\frac{70}{18} = 3\frac{8}{9}$

9 a $\frac{14}{5} = 2\frac{4}{5}$ b $\frac{50}{12} = 4\frac{1}{6}$ c $1\frac{1}{2}$

10 a $\frac{1}{6}$ b $\frac{1}{9}$ c $\frac{1}{5}$ d $\frac{15}{32}$

11 1375 m

12 $\frac{4}{5}$

13 25

14 a $\frac{3}{2} = 1\frac{1}{2}$ b $\frac{15}{4} = 3\frac{3}{4}$ c $\frac{14}{10} = 1\frac{2}{5}$

4.5 Fractions and decimals

1 a 0.7 b $\frac{2}{5}$

2 a 1.25 b $4.1\dot{6}$

3 a $\frac{7}{12}$ b $\frac{2}{3}$

4 0.28

5
Fraction	$\frac{1}{1000}$	$\frac{1}{100}$	$\frac{1}{10}$	$\frac{1}{8}$	$\frac{1}{5}$	$\frac{1}{4}$	$\frac{1}{2}$
Decimal	0.001	0.01	0.1	0.125	0.2	0.25	0.5

6 a 0.375 b 0.4 c 1.25 d 3.5

7 a 2.5 b 12 c 25 d 150 kg

8 $\frac{7}{20}$

9 a $\frac{61}{100}$ b $\frac{39}{50}$ c $\frac{229}{1000}$
 d $\frac{9}{20}$ e $\frac{12}{125}$

10 a 0.6, $\frac{5}{8}$, 0.628, $\frac{2}{3}$
 b $\frac{1}{2}$, 0.51, $\frac{3}{5}$, 0.605
 c $-4.5, \frac{-17}{4}, \frac{5}{2}, 2.8$

11 $\frac{1}{7} \times 250 = 35.71$

12 a Ed pays £133.33; Sam pays £66.67
 b Yes

13 $2\frac{1}{3}$ hours = 2 hours 20 minutes

14 $\frac{150}{360} = \frac{15}{36}$

15 $1\frac{1}{6}$

16 $1\frac{1}{2}$

17 $\frac{3}{10}$

4.6 Fractions and percentages

1 a 0.875 b 0.4

2 a 20% b 35% c 80% d 55%

3 $\frac{1}{3}$

4 a $\frac{2}{25}$ b $\frac{6}{25}$ c $\frac{13}{20}$
 d $\frac{16}{25}$ e $\frac{24}{25}$

5 $\frac{7}{20}$

6 e.g. $\frac{1}{10} = 10\%$

7 a 40%   b 60%

8 $\frac{1}{4}$

9 e.g. 30% $= \frac{3}{10}$

10 a $\frac{35}{75} = 46.7\%$ b e.g. half of 35 is 17.5; Kia + Audi is 20

11 a 30%   b 46.7%   c 20%   d 6.25%

12 $\frac{70}{80}$ is 87.5%; Harry achieved the higher score.

13 English

14 35%

15 6.5%

4.7 Calculating percentages 1

1 a $\frac{3}{10}$ b $\frac{3}{5}$ c $\frac{1}{3}$

2 a 0.2   b 0.004   c 0.9

3 a 0.65   b 0.42

4 a Fraction–decimal–percentage triangle showing $\frac{1}{4}$, 25%, 0.25
  b Fraction–decimal–percentage triangle showing $\frac{4}{5}$, 80%, 0.8
  c Fraction–decimal–percentage triangle showing $\frac{3}{5}$, 60%, 0.6
  d Fraction–decimal–percentage triangle showing $\frac{2}{3}$, 66.6%, 0.6̇

5 a 65%, 0.6, $\frac{11}{20}$, $\frac{1}{2}$ b 0.5, $\frac{26}{100}$, $\frac{1}{100}$, 0.5%

6 0.033

7 a $\frac{1}{5}$ b 300

8 a 360   b 210   c 270

9 a 0.2   b 60

10 a 225   b 210   c 14

11 £8.75

12 15% off is £6.08 off, which is more than £6

13 105

14 75% of 1050.60 = 787.95   87.50 × 12 = 1050
   The council are NOT charging correctly.

15 £13

16 a 1.3   b 1.2   c 3.5

17 150%

18 a £120   b 250 ml   c £1750

19 £10 400

20 £2.53

21 a £82.50   b £24.38   c £132

22 £2310

23 130% of 240 is 312. Her wage was only £300.
   Wage has not increased with the cost of living

24 £8330

25 £7800

4.8 Calculating percentages 2

1 a 4   b 132   c 175

2 a 39.6   b 532   c 2002

3 £41.20

4 a 36   b 330   c 4970

5 £180

6 23.75 minutes = 23 minutes 45 seconds

7 a £4.50   b £22.50   c £22.50
  d Same calculations.

8 a 1.2   b 1.75   c 1.8725

9 a 144   b 525   c 9375   d 3.6

10 a 0.6   b 0.65   c 0.94   d 0.988

11 a 120   b 325   c 122.2   d 3952

12 £54.60

13 Paint R Us

14 £1788

15 Quote 2 is £600 including VAT. Quote 1 is more expensive

16 £717.60

17 6 winners

18 £40

19 253

4 Problem–solving

1 150

2 a 4   b 10

3 $\frac{1}{9}$

4 £1400

5 $3x + 13$, where x is the number of balls Ed gets in.

6 Bar chart showing flip 9, aerial 6, grind 3, slide 6

4 Check up

1 a $\frac{8}{15}$ b $\frac{5}{8}$ c $\frac{1}{4}$ d $\frac{19}{24}$

2 a $\frac{1}{6}$ b $\frac{6}{25}$ c $\frac{4}{5}$

3 a $\frac{32}{3} = 10\frac{2}{3}$ b $\frac{3}{32}$ c $\frac{3}{2} = 1\frac{1}{2}$

4 a $1\frac{1}{24}$ b $5\frac{7}{8}$ c $2\frac{1}{4}$ d $1\frac{3}{4}$

5 0.375

6 a $\frac{7}{1000}$ b $\frac{13}{40}$

7 37.5

8 a $\frac{1}{20}$ b $\frac{7}{25}$ c $\frac{3}{2} = 1\frac{1}{2}$

9 a 12%   b 37.5%

10 58%, $\frac{3}{5}$, 0.62, $\frac{2}{3}$

11 a 13%   b Greater

12 a 18 cm   b 18 km   c £106.60

13 £3.30

14 £78

15 £325

16 £10 200

17 £43.75

19 Ole is wrong because $2\frac{1}{5}$ is 2.2 as a decimal.

4 Strengthen

Operations with fractions

1 a $\frac{3}{4}$ b $\frac{5}{8}$ c $\frac{5}{9}$ d $\frac{1}{6}$

2 a $\frac{7}{24}$ b $\frac{13}{40}$ c $\frac{1}{6}$ d $\frac{7}{30}$

3 a $\frac{7}{12}$ b $\frac{19}{20}$ c $\frac{17}{24}$ d $\frac{3}{20}$

4 a $1\frac{1}{4}$ b $1\frac{1}{9}$ c $1\frac{1}{20}$ d $1\frac{5}{14}$

5 a $7\frac{7}{8}$ b $8\frac{1}{4}$ c $8\frac{3}{8}$ d $6\frac{3}{8}$

6 b $1\frac{1}{2}$ c $2\frac{6}{7}$ d 9

7 a $3\frac{1}{4}$ b $3\frac{3}{7}$ c $1\frac{3}{4}$

8 a $\frac{1}{6}$ b $\frac{1}{15}$ c $\frac{4}{15}$ d $\frac{2}{5}$
  e $\frac{9}{28}$ f $\frac{28}{55}$ g $\frac{16}{25}$

9 b $\frac{2}{7}$ c $\frac{3}{5}$ d $\frac{4}{7}$
  e $\frac{1}{2}$ f $\frac{3}{5}$ g $\frac{1}{2}$

10 a $4\frac{1}{2}$ b 16   c 6   d $10\frac{1}{2}$

11 b $\frac{1}{12}$ c $\frac{1}{15}$ d $\frac{1}{14}$

12 a $1\frac{7}{8}$ b $1\frac{3}{4}$ c $\frac{8}{15}$ d $1\frac{1}{2}$

Percentages, decimals and fractions

1 a 0.625   b 0.375   c 0.6

2 a $\frac{3}{5}$ b $\frac{3}{100}$ c $\frac{1}{4}$
  d $\frac{1}{200}$ e $\frac{33}{1000}$ f $\frac{1}{8}$

3 a $\frac{1}{50}$ b $\frac{2}{25}$ c $\frac{12}{25}$
  d $\frac{39}{50}$ e $1\frac{3}{5}$ f $1\frac{4}{5}$

4 a 28%   b 82%   c 12%

5 a 0.8 = 80%   b 0.4 = 40%   c 0.375 = 37.5%

6 a 223%, 2.35, $2\frac{3}{4}$
  b 0.335, 35%, 0.357, $\frac{3}{8}$
  c 12.5%, 0.127, 55%, $\frac{5}{8}$
  d 0.07, 72.3%, 73.2%, $\frac{3}{4}$

7 a $\frac{3}{8}$ b $\frac{17}{25}$

8 a 48%   b 16%   c 48%   d 25%
9 a 0.4   b 0.7   c 0.02   d 0.025

Calculating percentages

1 a 24   b 14   c 132
   d 3   e 50   f 937.5
2 a 5.52   b 34.5   c 329   d 111.6
3 a 44   b 33   c 30
   d 324   e £3150
4 a 63   b 54   c 72
   d 16   e £950
5 a £900   b £1440   c £38.40
6 a £26.25   b £350   c £112 500

4 Extend

1 3750 square miles
2 a $\frac{1}{3}$ b 28.6%
3 £3468
4 Designer Bargains
5 35%
6 Bethany saves $\frac{1}{8}$ = 12.5%. Bethany saves more
7 Investment A
8 He saved £2160
9 No she hasn't. March was her best month.
10 £3492
11 £152.40
12 £224.70
13 a £720   b No
14 £4800
15 Yes; 10% of 8.9 is 0.89; 8.9 + 0.89 = 9.79

4 Unit test

Sample student answer
The student must state their decision about which shop to buy from and why, using the final prices as evidence. They must also put the correct units (e.g. £59 not just 59) in their statement. e.g. 'Samantha should buy her trainers from Edexcel Sports as they only cost £56 instead of £59 or £60 from the other shops.'

UNIT 5

5 Prior knowledge check

1 a $<$ b $>$ c $>$ d $<$
2 a 7   b 9   c 3   d 3
3 a 5   b 3   c −2   d −8
   e 81   f −9   g −32   h 8
4 a 7 squares   b 11   c +2 to previous term
5 a 22, 26 (+4)   b 12, 8 (−4)   c 15, 17 (+2)   d 7, 2 (−5)
6 a 8, 10   b 10, 13   c 9, 7   d 29, 24
7 a 70   b 31
8 a 32, 64 (×2)
   b 100 000, 1 000 000 (×10)
   c 486, 1458 (×3)
   d 50, 25 (÷2)
9 a 7, 8, 9, 10, 11   b 3, 5, 7, 9, 11
10 a $5x$ b $3a - 10$ c $5b + 16$ d $6x - 2$
11 $8c + 4p$
12 a $a = 1$ b $c = 10$ c $g = 11$
   d $h = 2$ e $k = 17$ f $l = 21$
13 a $h = 10$ b $m = 5$
14 a $2a + 16$ b $44 + 4c$ c $45 - 5f$
   d $10b + 6$ e $16 - 16x$ f $-2x - 6$
15 a 7   b 12   c 24   d 13
16 a $v = 13$ b $s = 40$
17

$7a + 6b$	$2a + b$	$3a + 8b$
$7b$	$4a + 5b$	$8a + 3b$
$5a + 2b$	$6a + 9b$	$a + 4b$

5.1 Solving equations 1

1 a 6   b 10   c 3
   d 2   e 13   f 3
2 $y = 8 \times 7$, $y = 56$
3 a 90   b 12   c 20
   d 12   e 64   f 12
4 a −2, $y = 2$ b +4, $a = 7$ c ÷3, $b = 4$ d ×4, $c = 12$
5 a 2   b 18   c 5   d 4
6 a 4   b 3   c 3   d 4
7 a $9a = 180°$ b $a = 20°$
8 a $6c = 30$ b $c = 5$ c 10 cm
9 $9b = 360$; $b = 40$; 200°
10 $10v = 60$, $v = 6$
   Length = 24 cm, Width = 6 cm
11 a $4s = 36$ b £9
12 a $n - 11 = 18$ b 29

5.2 Solving equations 2

1 a 6, 10, 14, 18, 22   b −1, 2, 5, 8, 11
2 a $4d$ b $8f$ c $3p + 12$ d $b - 1$
3 a −4, ÷3   b +2, ×5
4 a 2
   b i $k = 3$ ii $w = 5$ iii $w = 4.5$
5 a 3, 20, 10
   b 4, 18, 9
6 a $a = 2$ b $a = 3$ c $a = 2$
   d $a = -\frac{1}{3}$ e $f = 3$ f $c = 2$
   g $a = \frac{7}{3}$ h $p = -\frac{1}{2}$ i $t = -\frac{5}{8}$
7 a $y = 7$ b $x = 12$ c $w = 3.2$
8 9
9 a $3a + 120° = 180°$ b $a = 20°$
10 $a = 30°$
11 $y = 4$
12 15 cm
13 a $c = 5$ b $d = 24$ c $e = 6$ d $f = -3$
14 a $a = 12$ b $b = 30$ c $c = 12$
   d $d = 12$ e $e = 16$ f $f = 27$
15 a $x = 16$ b $x = 120$ c $x = 12$
   d $x = 30$ e $x = \frac{9}{2}$ f $x = \frac{10}{7}$

5.3 Solving equations with brackets

1 a $4a + 6$ b $30p - 20$ c $18 - 21c$ d $-12 + 8x$
2 a $e = 6$ b $a = -1$ c $b = 3$
   d $y = -2\frac{2}{7}$ e $x = 24$ f $x = -9$
3 a $a = 19$ b $b = 0$ c $d = 10$
   d $d = 7$ e $m = 7$ f $b = 12$
   g $c = -1$ h $e = 3$ i $f = 6$
4 a $c = \frac{2}{3}$ b $g = 5$ c $g = 3.4$
5 a $k = 1$ b $a = 14$ c $c = 5$ d $p = 6$
6 a $a = -4$ b $b = 1$ c $d = 12$
   d $v = 4\frac{2}{3}$ e $e = 4.5$ f $h = 1\frac{1}{3}$
7 a $x = 13$ b $x = 1$ c $x = 8$
   d $x = -1$ e $x = -2$ f $x = -\frac{2}{3}$
8 $x = 6$
9 Steve is 33 years old and Jenson is 3 years old.
10 a $s = 28$ b $f = -2$ c $x = -8$
   d $m = 13$ e $y = -20$ f $t = 3$
11 a Width = $1\frac{2}{3}$ cm, Length = $10\frac{1}{6}$ cm
   b $23\frac{2}{3}$ cm
12 a $5(b + 5) = 45$, $b = 4$ b 28 m

5.4 Introducing inequalities

1 a $<$ b $>$ c $>$
   d $>$ e $>$ f $=$
2 a
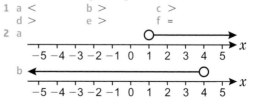
   b

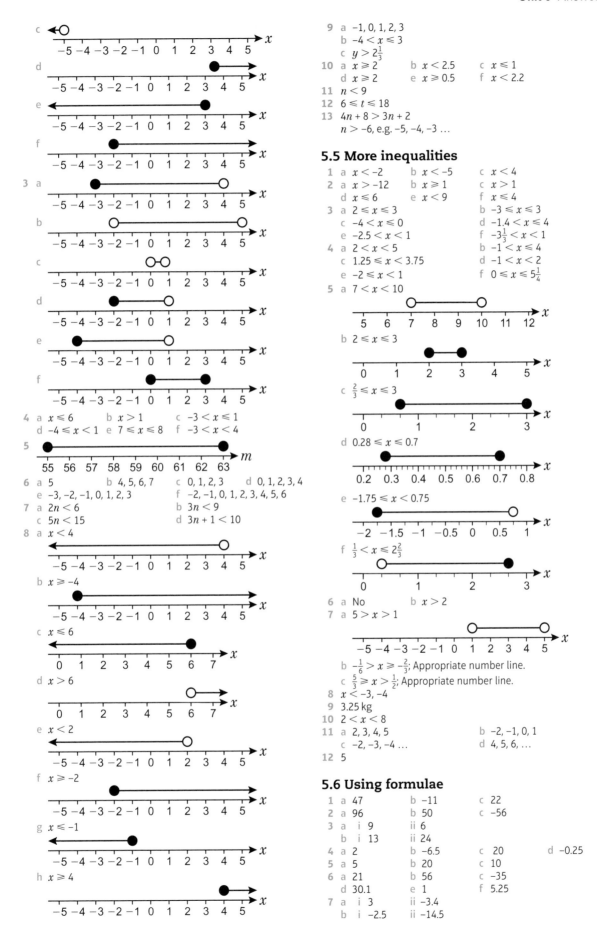

c
d
e
f

3 a
b
c
d
e
f

4 a $x \leqslant 6$ b $x > 1$ c $-3 < x \leqslant 1$
  d $-4 \leqslant x < 1$ e $7 \leqslant x \leqslant 8$ f $-3 < x < 4$

5

6 a 5   b 4, 5, 6, 7   c 0, 1, 2, 3   d 0, 1, 2, 3, 4
  e $-3, -2, -1, 0, 1, 2, 3$ f $-2, -1, 0, 1, 2, 3, 4, 5, 6$

7 a $2n < 6$ b $3n < 9$
  c $5n < 15$ d $3n + 1 < 10$

8 a $x < 4$

b $x \geqslant -4$

c $x \leqslant 6$

d $x > 6$

e $x < 2$

f $x \geqslant -2$

g $x \leqslant -1$

h $x \geqslant 4$

9 a $-1, 0, 1, 2, 3$
  b $-4 < x \leqslant 3$
  c $y \geqslant 2\frac{1}{3}$
10 a $x \geqslant 2$ b $x < 2.5$ c $x \leqslant 1$
  d $x \geqslant 2$ e $x \geqslant 0.5$ f $x < 2.2$
11 $n < 9$
12 $6 \leqslant t \leqslant 18$
13 $4n + 8 > 3n + 2$
  $n > -6$, e.g. $-5, -4, -3 \ldots$

5.5 More inequalities

1 a $x < -2$ b $x < -5$ c $x < 4$
2 a $x > -12$ b $x \geqslant 1$ c $x > 1$
  d $x \leqslant 6$ e $x < 9$ f $x \leqslant 4$
3 a $2 \leqslant x \leqslant 3$ b $-3 \leqslant x \leqslant 3$
  c $-4 < x \leqslant 0$ d $-1.4 < x \leqslant 4$
  e $-2.5 < x < 1$ f $-3\frac{1}{3} < x < 1$
4 a $2 < x < 5$ b $-1 < x \leqslant 4$
  c $1.25 \leqslant x < 3.75$ d $-1 < x < 2$
  e $-2 \leqslant x < 1$ f $0 \leqslant x \leqslant 5\frac{1}{4}$
5 a $7 < x < 10$

  b $2 \leqslant x \leqslant 3$

  c $\frac{2}{3} \leqslant x \leqslant 3$

  d $0.28 \leqslant x \leqslant 0.7$

  e $-1.75 \leqslant x < 0.75$

  f $\frac{1}{3} < x \leqslant 2\frac{2}{3}$

6 a No   b $x > 2$
7 a $5 > x > 1$

  b $-\frac{1}{6} > x \geqslant -\frac{2}{3}$; Appropriate number line.
  c $\frac{5}{3} \geqslant x > \frac{1}{2}$; Appropriate number line.
8 $x < -3, -4$
9 3.25 kg
10 $2 < x < 8$
11 a 2, 3, 4, 5   b $-2, -1, 0, 1$
  c $-2, -3, -4 \ldots$ d 4, 5, 6, …
12 5

5.6 Using formulae

1 a 47   b -11 c 22
2 a 96   b 50   c -56
3 a i 9   ii 6
  b i 13   ii 24
4 a 2   b -6.5 c 20   d -0.25
5 a 5   b 20   c 10
6 a 21   b 56   c -35
  d 30.1   e 1   f 5.25
7 a i 3   ii -3.4
  b i -2.5 ii -14.5

8 a i 200 miles ii 357.5 miles
  b i 4 hours ii 4 hours
9 a 35.4 b 5.3
10 30 m
11 5 m/s^2
12 64.8 g
13 a Equation b Formula c Expression
  d Formula e Equation f Formula
14 a $x = y - 4$ b $x = y + 7$ c $I = \dfrac{P}{V}$

  d $d = \dfrac{P}{5}$ e $x = \dfrac{y-3}{5}$ f $x = \dfrac{y+3}{4}$

  g $N = \dfrac{M+5}{7}$ h $W = 3V$ i $V = \dfrac{D}{T}$

5.7 Generating sequences

1 a 25, 21 b 5, 2
2 a 37 (+4) b 39 (+4) c 40 (+4)
3 a 3.5, 4 b $\frac{2}{3}$, 1 c 0.3, −0.5
  d −5.5, −6.5 e $-2\frac{3}{5}, -3\frac{2}{5}$ f −7.8, −7.1
4 a 3, 3.4, 3.8, 4.2, 4.6 b 10, 9.8, 9.6, 9.4, 9.2
  c 7, 10, 13, 16, 19 d 7, 9, 11, 13, 15
  e −3, −1, 1, 3, 5, f −7, −12, −17, −22, −27
5 a 8, 13, 21 b 24, 39, 63 c 40, 65, 105
6 a i

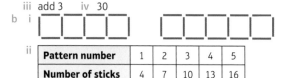

  ii

Pattern number	1	2	3	4	5
Number of sticks	3	6	9	12	15

  iii add 3 iv 30
  b i

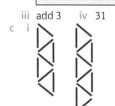

  ii

Pattern number	1	2	3	4	5
Number of sticks	4	7	10	13	16

  iii add 3 iv 31
  c i

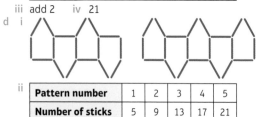

  ii

Pattern number	1	2	3	4	5
Number of sticks	3	5	7	9	11

  iii add 2 iv 21
  d i

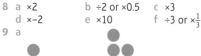

  ii

Pattern number	1	2	3	4	5
Number of sticks	5	9	13	17	21

  iii add 4 iv 41
7 a 50 000, 500 000 b $\frac{1}{8}$, $\frac{1}{16}$ or 0.125, 0.0625
  c 0.01, 0.001 d 512, 2048
8 a ×2 b ÷2 or ×0.5 c ×3
  d ×−2 e ×10 f ÷3 or ×$\frac{1}{3}$
9 a

b 1, 3, 6, 10, 15
c +2, +3, +4, +5…
d 55
10 a

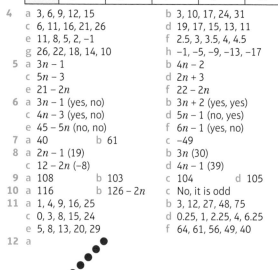

b 64
11 a

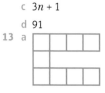

b 12 c 11

5.8 Using the nth term of a sequence

1 a 50 b 11
2 a $x = 2$ b $x = 11$ c $x = 4.5$
3

n	1	2	3	4	5
Term	$2 \times 1 + 1$ $= 3$	$2 \times 2 + 1$ $= 5$	$2 \times 3 + 1$ $= 7$	$2 \times 4 + 1$ $= 9$	$2 \times 5 + 1$ $= 11$

4 a 3, 6, 9, 12, 15 b 3, 10, 17, 24, 31
  c 6, 11, 16, 21, 26 d 19, 17, 15, 13, 11
  e 11, 8, 5, 2, −1 f 2.5, 3, 3.5, 4, 4.5
  g 26, 22, 18, 14, 10 h −1, −5, −9, −13, −17
5 a $3n - 1$ b $4n - 2$
  c $5n - 3$ d $2n + 3$
  e $21 - 2n$ f $22 - 2n$
6 a $3n - 1$ (yes, no) b $3n + 2$ (yes, yes)
  c $4n - 3$ (yes, no) d $5n - 1$ (no, yes)
  e $45 - 5n$ (no, no) f $6n - 1$ (yes, no)
7 a 40 b 61 c −49
8 a $2n - 1$ (19) b $3n$ (30)
  c $12 - 2n$ (−8) d $4n - 1$ (39)
9 a 108 b 103 c 104 d 105
10 a 116 b $126 - 2n$ c No, it is odd
11 a 1, 4, 9, 16, 25 b 3, 12, 27, 48, 75
  c 0, 3, 8, 15, 24 d 0.25, 1, 2.25, 4, 6.25
  e 5, 8, 13, 20, 29 f 64, 61, 56, 49, 40
12 a

  b

Pattern number	1	2	3	4	5	6
Number of dots	4	7	10	13	16	19

  c $3n + 1$

  d 91
13 a

  b i Yes ii No
  c 17
14 a

Pattern number	1	2	3	4	5
White tiles	4	5	6	7	8
Blue tiles	2	4	6	8	10

  b $2n$ c $n + 3$ d 40
  e 33 f 25

5 Problem-solving

1 a, b

Name	Clark's	Young's
Alex	116 mg	120 mg
Brian	90 mg	75 mg
Clara	64 mg	60 mg
Dilip	104 mg	111 mg
Emily	48 mg	43 mg
Farhana	116 mg	136 mg

2 a Jon 4 years old, Fiona 6 years old, Jill 14 years old, Wayne 11 years old

5 Check up

1 a $c = 1$ b $f = -2$ c $x = 14$
2 $a = \frac{7}{3}$
3 a $x = 3$ b $x = -5$
4 $I = \frac{1}{2}$
5 $l = \frac{19}{2}$
6 a $M = 4P$ b $x = \frac{y+5}{3}$
7 a $a = 3$ b $t = \frac{13}{4}$
8 a $8x - 20$ b 33, 33, 2
9 a

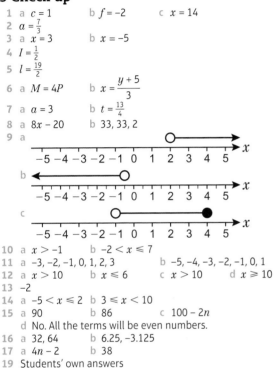

b

c

10 a $x > -1$ b $-2 < x \leq 7$
11 a $-3, -2, -1, 0, 1, 2, 3$ b $-5, -4, -3, -2, -1, 0, 1$
12 a $x > 10$ b $x \leq 6$ c $x > 10$ d $x \geq 10$
13 -2
14 a $-5 < x \leq 2$ b $3 \leq x < 10$
15 a 90   b 86   c $100 - 2n$
   d No. All the terms will be even numbers.
16 a 32, 64   b 6.25, -3.125
17 a $4n - 2$ b 38
19 Students' own answers

5 Strengthen

Equations and formulae

1 a -7 b $+10$ c $\div 2$
2 a $x = 3$ b $x = 4$
3 a $x = 5$ b $y = 10$ c $s = 6$
   d $p = 6$ e $s = 4$
4 a $F = 30$ b $m = 3$
5 a $v = 20$ b $t = 5$ c $a = 2$
6 a $x = 5$ b $x = 10$ c $x = 6$
7 a $a = 6$
8 a $a = 4$ b $p = 2$ c $y = 5$ d $b = -2$
9 a $x = -3$ b $q = \frac{19}{2}$ c $a = \frac{1}{3}$ d $z = \frac{1}{2}$
10 a $-x$ b $+y$ c $\div a$ d $\times p$
11 A: y, B: A, C: q, D: F, E: m
12 a $x = M + y$
   b A: $x = y + z$, B: $x = A - h$, C: $x = \frac{q}{r}$, D: $x = F + ab$, E: $x = 2m$

Inequalities

1 a 7, 9   b 1, 0, -1, -2, -5, -8
   c 7, 9   d 7, 9   e -5, -8 f -5, -8
2 a $-3, -2, -1, 0, 1, 2, 3, 4, 5, 6$
   b, c $-3, -2, -1, 0$
3 a 3, 4, 5, 6, 7   b $-7, -6, -5, -4$
   c $-5, -6, -7 \ldots$ d 4

4 a F   b B   c C
   d D   e E   f A
5 $n < -1$

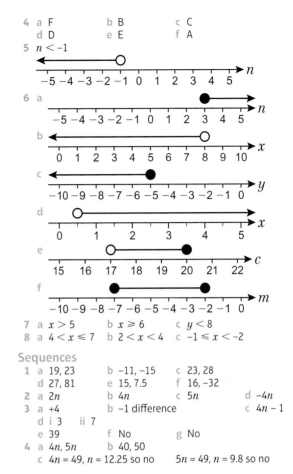

6 a
   b
   c
   d
   e
   f

7 a $x > 5$ b $x \geq 6$ c $y < 8$
8 a $4 < x \leq 7$ b $2 < x < 4$ c $-1 \leq x < -2$

Sequences

1 a 19, 23   b -11, -15 c 23, 28
   d 27, 81   e 15, 7.5   f 16, -32
2 a $2n$ b $4n$ c $5n$ d $-4n$
3 a $+4$ b -1 difference   c $4n - 1$
   d i 3   ii 7
   e 39   f No   g No
4 a $4n, 5n$ b 40, 50
   c $4n = 49, n = 12.25$ so no $5n = 49, n = 9.8$ so no

5 Extend

1 a 3.35   b 2.63   c 0.33
2 a 2   b 1   c 16.8
   d $\frac{2}{3}$ e $-\frac{5}{6}$
3 a 29
   b i 45   ii nth term is $4n + 1$
4 a 6   b 102
5 a $+1$ b -2 c $\times x$ d $+4$
   e add the last 2 terms together   f $\times 2$
6 $p = 1$
7 a $127 - 7n$ b 19th term, -6
8 $x = 1\frac{1}{2}$
9 $210°$
10 a $x = \frac{1}{2}$ b $x = \frac{1}{3}$ c $x = \frac{3}{5}$
11 2.5 litres
12 a $2x + 18 = 4x + 6$ b 30 cm
13 a 4, 5, 6, 7, 8
   b 2
   c $-15, -14, -13, -12, -11, -10, -9, -8, -7$
   d 2, 3, 4, 5   e None
14 a $15x + 10$ b 7, 49, 59
15 The nth term is $6n + 4$ and n doesn't equal an integer when 555 is a term.

5 Unit test

Sample student answer

$x = 11$ is correct but is not the final answer. The student was asked to find the perimeter of the rectangle. They must use this answer to substitute into the sides to find the perimeter of the rectangle. They must remember to check they have answered the actual question fully.

UNIT 6

Note: for some questions in this chapter there is more than one way of reaching the correct answer.

6 Prior knowledge check

1. a 20   b 130   c 60
   d 90   e 540   f 180
   g 25   h 80   i 95
2. a 75   b 100   c 75
3. a 76   b 252   c 36   d 540
4. a i $x + 90$ ii $270 - x$ b i 130 ii 230
5. a 30   b 90
6. a e.g. CD and HI, AB and EF   b perpendicular
7. a Acute   b Reflex
   c Right-angled   d Obtuse
8. a Obtuse   b Acute   c Reflex
   d Acute   e Obtuse   f Reflex
9. a Students' estimates   b $a = 95°$, $b = 45°$
10. $a = 90°$; Angles on a straight line add up to 180°; $b = 130°$; Vertically opposite angles are equal or Angles around a point add up to 360°
11. a YZ   b Angle XYZ, Angle Y, $X\hat{Y}Z$, ∠XYZ
12. a i Isosceles ii Right-angled iii Scalene iv Equilateral
    b
    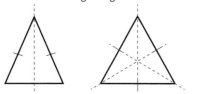
    c i Order 1 ii Order 1 iii Order 1 iv Order 3
13. a Shape A: 6, Shape B: 7
    b Shape A: order 6, Shape B: order 7
14. a Trapezium   b Kite
    c Parallelogram   d Rhombus
15. a
    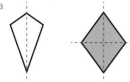
    b Shape a: Order 1, Shape b: Order 1, Shape c: Order 2, Shape d: Order 2
16.
    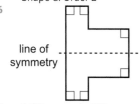
    line of symmetry
17. a 360°   b 70°
18.
    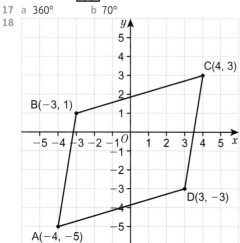

19. a AB = BC = CD = DA, angle A = angle C, angle B = angle D, sum of angles = 360°
    b At least 2 sides have different lengths, and angle A and B sum to 180° and angle C and D sum to 180° or angle B and C sum to 180° and angle D and A sum to 180°

6.1 Properties of shapes

1. a Accurate drawing of a 65° angle
   b Accurate drawing of a 128° angle
2. a i 360°
      ii 360°
      iii 360°
   b 360°
3. A and D, C and F, E and H
4. a
   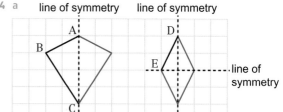
   b Kite: 83°, 126°, 83°, 68°
   Rhombus: 127°, 53°, 127°, 53°
5. a A rhombus; B kite; C parallelogram; D rectangle; E trapezium
   b i A, C, D
     ii A, B
     iii A, C, D
6. a Rhombus, square
   b Square
   c Kite
   d Isosceles trapezium
7.
   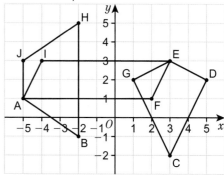

8. a 144°, 36°, 144°
   b 65°, 115°, 65°
   c 110°, 82°, 55°
9. $a = 90°$, $b = 99°$, $c = 81°$, $d = 107°$, $e = 73°$, $f = 114°$, $g = 66°$
10. PSR = 47° (angles on a straight line add to 180°)
    PQR = 133° (angles in a quadrilateral add to 360°)
11. a, b, c
    i
    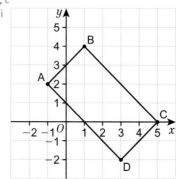

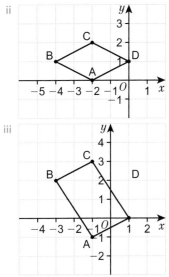

d i Rectangle   ii Rhombus   iii Parallelogram
12 Angle b = Angle c = 65°
Angle a = Angle d = 115° (angles on a straight line)
65° + 115° + 65° + 115° = 360°
13 38°

6.2 Angles in parallel lines

1 Acute: a, b, d, e, h, j
Obtuse: c, g, i
2 q and r, p and s
3 a a = 60°   b p = 80°, q = 100°   c x = 120°, y = 60°
4 a a = 83°   b b = 111°   c c = 99°, d = 81°
5 s = 50° Angles on a straight line add up to 180°;
angles s and 50°are alternate.
t = 95° Angles t and 95° are corresponding.
u = 85° Angles on a straight line add up to 180°.
v = 80° Angles on a straight line add up to 180°;
angles v and 80° are alternate.
w = 105° Angles on a straight line add up to 180°;
angles w and 105° are corresponding.
6 $a = c = i = h = g = e$ Corresponding angles, alternate angles
and vertically opposite angles.
$b = d = f$ Corresponding angles and vertically opposite angles.
7 $x + y$ = 180° because of angles on a straight line. $z = y$
because of corresponding angles.
8 x = 115° Angle BEF = 65° (corresponding angles); Angle
ABE = 65° (alternate angles); x = 180° – 65° (angles on a
straight line)
9 a = 66° co-interior angles
b = 27° vertically opposite angles and lies on straight line
with a and 87°
c = 65° corresponding angles and lies on straight line with
115°
d = 30° vertically opposite angles and lies on straight line
with c and 85°
e = 81° corresponding
f = 27° lies on straight line with 72° and 81°
g = 108° corresponding with f + 81°
h = 39° corresponding angles and angles on a straight line
i = 116° co-interior angles
j = 108° corresponding angles
10 68°
11 x = 100°; y = 130°

6.3 Angles in triangles

1 180°
2 a 30   b 66   c 72.5
3 a = 30° Reason 2; b = 90° Reason 2; c = 50° Reason 1;
d = 45° Reason 2

4 a a and b are equal.
b d, e and f are all equal.
5 a a = 55°   b b = 53°   c c = 84°; d = 48°
d e = 54°; f = 54°
6 a 75°
b   i No; 45° with vertical
ii No; 30° with vertical
7 a = 40°; b = 100°
8 $d + b + e$ = 180°
$d = a$ Angles d and a are alternate.
$e = c$ Angles e and c are alternate.
So $d + b + e = a + b + c$ = 180°
This proves that the angles in a triangle sum to 180°.
9 a = 23°; b = 157°; c = 44°; d = 116°
10 $x + y$ = 180° – z because the angles in a triangle add up to
180°.
w = 180° – z because the angles on a straight line add up
to 180°
11 C
12 Students' own answers, e.g. a = 55°, b = 45°, c = 100°
13 a $a = b = c$ = 60°; $d = e$ = 45°; f = 33°; g = 135°;
b w = 165°, x = 40°, y = 74°
14 110°

6.4 Exterior and interior angles

1 a 36   b 30   c 20
2 a 90°   b 60°
3 a 103   b 43   c 87
d 143   e 167
4 Students' own answers
5 a Regular   b Irregular   c Irregular
d Irregular   e Regular
6 a   i 120°
ii 72°
iii 90°
b They all equal 360°.
7 a 15   b 24°
8 a 12 sides   b 8 sides   c 20 sides
9 360°
10 a a = 44°   b b = 54°
11 No; this would result in exterior angles being equal which
is incorrect as the shape is irregular.
12 a 18   b 180°   c 180°
d Angles lie on a straight line.
13 a = 112°; b = 97°; c = 85°; d = 148°

14

Regular polygon	Exterior angle	Interior angle
Pentagon	72°	108°
Hexagon	60°	120°
Decagon	36°	144°

15 Exterior angle = 36°, 10 sides
16 54°
17 112.5°

6.5 More exterior and interior angles

1 a 720   b 180   c 1440
2 a Interior   b Interior   c Exterior
b a = 135°
b = 94°
c = 72°
3 Students' own answers
4 No. The interior angles are 108° and they don't add to
360°.
5 Students' own answers
6 This shape has 5 sides.
It is made of 3 triangles.
The interior angles in a triangle add up to 180°.
So the sum of the interior angles in this shape is
3 × 180° = 540°

7 a 720°   b $n - 2$ c $S = 180° \times (n - 2)$

8 a i 360°   b i 540°   c i 720°
    ii 156°   ii 142°   ii 146°

9 a i 720°   b i 540°   c i 2880°
    ii 120°   ii 108°   ii 160°

10 a 11 sides   b 14 sides   c 17 sides   d 24 sides

11 a $p = 108°$ Exterior and interior angles add up to 180°
    $q = 36°$ The triangle is isosceles
    $r = 72°$ Exterior and interior angles add up to 180°
   b $s = 18°$ The triangle is isosceles
    $t = 54°$ Exterior and interior angles add up to 180°
   c $w = 45°$ Exterior angle or an octagon
    $u = 135°$ Exterior and interior angles add up to 180°
    $v = 22.5°$ The triangle is isosceles

12 30°

13 $x = 54°$

6.6 Geometrical problems

1 a $x = 36$ b $x = 50$ c $x = 105$ d $x = 82.5$

2 a $x + 60$ b $x + x + 60 = 2x + 60$

3 a $4x = 180$ b $x = 45$ c 45°, 135°

4 a $x = 20°$ b $x = 40°$ c $x = 30°$ d $x = 22.5°$

5 a i Vertically opposite angles are equal.
    ii $x = 50$ iii 90°
   b 165°

6 a $x = 35, y = 55$ b $y = 90°, y = 80°$

7 a $4x = 140$ b $x = 35$ c 35°, 105°, 40°

8 80°, 80°, 20°

9 a 20°, 20°, 140°: $2a + b = 180°$
    60°, 60°, 60°: $3a = 180°$
    30°, 50°, 100°: $a + b + c = 180°$
    90°, 45°, 45°: $2a = 90°$
   b $100 + 20 + 70 \neq 180$

10 $3x + 75 = 180, x = 35$, so angles are 60°, 60°, 60°,
    equilateral because all angles = 60°

11 $x, 3x, 5x$: Sometimes
    x, x, x: Sometimes
    $x + 60°, x - 60°, 60° - 2x$: Never
    $x + 60°, x - 60°, x + 180°$: Never
    $60° - x, x + 90°, 30°$: Always

12 a $y = x + 50 + 2x - 10$ (exterior angle of triangle = sum of
    opposite interior angles, simplifies to $y = 3x + 40$)
    or $\angle PBA = 180 - (x + 50) - (2x - 10) = 140 - 3x$ (angles in
    a triangle add up to 180°)
    $y = 180 - \angle PBA = 180 - 140 + 3x$ (y and $\angle PBA$ lie on a
    straight line), simplifies to $y = 3x + 40$
   b i $x = 35$ ii 85°

13 $\angle CBD = \angle BCD = (180 - 2(x - 30)) \div 2 = 120 - x$
    $120 - x + x + \angle BCA = 180°$, $\angle BCA = 60°$ and $\angle ABC = 60°$
    $\angle EAC = 180 - 90 - x = 90 - x$,
    $\angle BAC = 180 - (x + 30) - (90 - x) = 60°$
    $\angle ABC = \angle ACB = \angle BAC = 60°$ means ABC is equilateral

14 $\angle ABD = 25°$

6 Problem-solving

1 a 120°   b Isosceles

2 £37

3 £5

4 8 cm

5 Students' own drawing of pie chart, UK = 10, France = 30,
   Other = 5, Spain = 45

6 Check up

1 a 55°   b 88°

2 a $a = 77°$ Corresponding angles; $b = 77°$ Alternate angles
   b $c = 68°$ Alternate angles; $d = 68°$ Corresponding angles;
    $e = 84°$ Corresponding angles; $f = 28°$ Angles on a
    straight line
   c $g = 90°$ Angles on a straight line and corresponding
    angles; $h = 104°$ Co-interior angles.

3 a 120°   b 108°   c 43°

4 $a = 108°$ reason 4; $b = 95°$ reasons 4 and 1 or 4 and 2;
   $c = 13°$ reasons 3 and 4

5 a $4x = 180°$ b $x = 45°$ c 45°, 105°, 30°

6 a $p = 72°$ (p and 72° are alternate or opposite angles
    in a parallelogram are equal); $q = 108°$ (angles in a
    quadrilateral equal 360°)
   b $r = 91°$ (angles on a straight line add up to 180°);
    $s = 104°$ (angles in a quadrilateral equal 360°)

7 a 24 sides   b 20°

8 a $x = 108°$ b $y = 60°; z = 120°$

9 1440°

10 $a = 145°$

12 Students' own answers

6 Strengthen

Angles between parallel lines

1 a Alternate   b Corresponding
   c Corresponding   d Vertically opposite

2 a $p = 49°$ Alternate angles
   b $q = 88°$ Corresponding angles
   c $r = 37°$ Alternate angles, $s = 37°$ vertically opposite
    angles
   d $t = 76°$ Vertically opposite angles; $u = 76°$
    Corresponding angles; $v = 104°$ Angles lie on a straight
    line

3 a $a = 63°$ Vertically opposite angles; $b = 63°$ a and b are
    corresponding angles; $c = 117°$ c and b lie on a straight line
   b $d = 57°$ d and 123° lie on a straight line; $e = 57°$ d and e
    are alternate; $f = 57°$ f and d are vertically opposite;
    $g = 65°$ g and 65° are corresponding
   c $h = 108°$ h and 72° lie on a straight line; $i = 95°$ i and
    85° lie on a straight line.

Triangles and quadrilaterals

1 a 180°   b 360°

2 a $a = 28°$ b $b = 70°$ c $c = 55°$ d $d = 95°$

3 a b and c b e and f c h and i

4 a 152°   b 76°

5 $a = 36°; b = 34°; c = 48°$

6 $\angle BAC = 30°$ Reason 3; $\angle ABC + \angle ACB = 150°$ Reason 1;
   $\angle ABC = \angle ACB = 75°$ Reason 2

7 a $9x = 180°$ b $x = 20°$ c 20°, 60°, 100°

8 c $3x = 180$ d 60°

Interior and exterior angles

1 a 360°   b 18 sides   c 12°

2 a $x = 60°$ b $y = 120°$

3 a iii 540°
    iv 101°
   b iii 720°
    iv 155°

4 d 1080°   e 135°

6 Extend

1 90°, 30°, 60°

2 Yes $129 + 51 = 180$ and $52 + 77 + 51 = 180$

3 155°, 155°, 25°

4 $\angle BFE = 114°$ alternate with $\angle ABF$; $\angle CFE = 48°$ vertically
   opposite angles are equal; $x = \angle BFE - \angle CFE = 66°$

5 126°

6 52°

7 a 900°, $(7 - 2) \times 180°$
   b 100°
   c 100°, 100°, 110°, 175°, 150°, 100°, 165°
   d Sum of angles = 900°

8 a Isosceles   b $2x - 180$
   c i $\angle CDB = 180 - 120 - (2x - 180) = 240 - 2x$
    ii $\angle BCD = 180 - x - (240 - 2x) = x - 60$
    iii $(x - 60) + y = 180$, so $x = 240 - y$
    iv $2(240 - y) - 180 = 300 - 2y$

9 No because 500 is not divisible by 180.
10 $x = 120°$, $y = 150°$
11 48°
12 $x = y = 60°$

6 Unit test

Sample student answers

Student A's answer is better. When giving reasons for angles, students must state the angle rule they have used, in a sentence and not just the calculations done.

UNIT 7

7 Prior knowledge check

1 a 41.2   b 0.248   c 7.25   d 2.3
2 a 20.5   b 23   c 14.9   d 23.3
3 a 0.8   b 12.5   c 8.8   d 21.0
4 a 2.57   b 0.55   c 0.05   d 0.50
5 a 12   b 4.8   c 4.5   d 1.8
6 3, 5.9, 6.2, 6.3, 7, 8
7 a 14.33   b 4.13
8 25
9 30.5
10 a Qualitative   b Quantitative
   c Quantitative   d Quantitative
11 a 13   b 5   c 13
12 125
13 a 3   b 4   c 15
14 a 3.9   b 3.8

7.1 Mean and range

1 a 7.2   b 0.68   c 3.73
2 a Red   b 118
3 35.2 books
4 10.75 km
5 a 8 °C   b 13 °C   c 13.9 °C
6 a 7 minutes   b 4 minutes
7 a

Waiting time, w (mins)	Frequency, f	$w \times f$
7	6	42
8	4	32
9	3	27

   b 140 minutes   c 20
   d 7 minutes
8 a 5.57 portions   b 3 portions
9 a

Score, s	Frequency, f	$s \times f$
1	1	1
2	2	4
3	4	12
4	3	12
5	6	30
6	5	30
7	2	14
8	3	24
9	3	27
10	1	10
Total	30	164

   b Mean = 5.5; range = 9
   c Any suitable frequency table.
   d Mean = 5.267; range = 4
   e close, 11W, 11Y, spread
10 a Tess 44.5; Jo 49.25   b Tess 28; Jo 96   c Tess
   d Tess, because she always scores consistently

11 7
12 1.6 children

7.2 Mode, median and range

1

14	8 9	
15	2 5 7 8	
16	0 2 3 4 5 7	**Key**
17	2 3 5 6 7	14 | 9 means 149 bpm
18	0 1 6	

2 a Median = 6; range = 8
   b Median = 23.5; range = 12
3 a 5, 6, 7, 8, 9   b 0, 1, 2
4 a 28   b 63 000 miles
5 Most of the values occur just once but several occur twice.
6 a 7.0 cm   b 7.0 cm   c 2.8 cm
7 a 21   b 17 years   c 40 years
8 a 87 points   b 76 points   c 42 points
9 a i 23   ii 195   iii 2.76
   b i Range = 32, range without outlier = 14
     ii Range = 83, range without outlier = 21
     iii Range = 6.09, range without outlier = 2.46
10 a Outlier = 86, range without outlier = 14
   b Outlier = 2.9, range without outlier = 0.6
11 9 volts
12 50 points
13 a e.g. 25 cm (anything in the range $20 < h \leqslant 25$ is acceptable)
   b e.g. 0.1 cm (anything in the range $0 < h \leqslant 5$ is acceptable)
   c 25 cm
14 25 years

7.3 Types of average

1 59
2 30
3 20
4 a 25   b 132   c Set a Yes and Set b No
5 a 51.6   b 45.2   c Set a No and set b Yes
6 a 18   b 1.9
   c Yes because **a** is a good spread of results and **b** has outliers.
7 a Median = 40.5; outlier so use median
   b Median = 1.9; good spread of results so use median
   c Mean = 48; similar results so mean
   d Mode = blue; qualitative so use mode
8 a 31   b 31   c 31
9

Average	Advantages	Disadvantages
Mean	Every value makes a difference	Affected by extreme values
Median	Not affected by extreme values	May not change if a data value changes
Mode	Easy to find; not affected by extreme values; can be used with non-numerical data	May not be one

10 40–49
11 $7750 \leqslant p < 8000$
12 3
13 a 2 people   b 1 person   c 1.8367 = 1.8 (1 d.p.)
14 151–175
15 a $220 \leqslant d < 240$ b $240 \leqslant d < 260$ c 14 times

7.4 Estimate of the mean

1 3.92
2 12.4
3 a 3 hours 44 minutes
   b 2 hours 50 minutes
   c 4 hours 39 minutes

4 8.666 … = 9 people as an estimate
5 21 minutes
6 3 hours 41 minutes
7

Temp, t (°C)	Frequency, f	Midpoint, m	$m \times f$
$16 \leqslant t < 18$	2	17	34
$18 \leqslant t < 20$	5	19	95
$20 \leqslant t < 22$	5	21	105
$22 \leqslant t < 24$	10	23	230
$24 \leqslant t < 26$	8	25	200

  a 22.1 °C   b $22 \leqslant t < 24$
  c $22 \leqslant t < 24$ d 10 °C
8 a 28.8 cm
  b Loss of raw data once the data is grouped

7.5 Sampling

1 C Putting all the names in a hat and choosing one without looking
2 a People in the Midlands
  b Too many people to survey in total   c 1000
3 a His local area
  b His office is in the local area with working age people, so this is not representative of everyone
4 a All in one street so not a full representation
  b Ask people in different streets too
5 No. The people in the shopping centre could live near the centre and do not drive very far. The people in the motorway service area could be travelling long distances. This survey would be more accurate if it is taken in a wide variety of places.
6 a Biased   b Ask students in different classes too
7 a Yes – all students have the same chance of selection
  b No – not all students have the same chance of being selected
8 a e.g. 03 83 41 33 81 39 22 09
  b Random numbers could give the 3rd, 33rd, 39th, 22nd and 9th people
9 Method 1 – put all room numbers into a hat and pick out 20 without looking.
  Method 2 – use a random number list to select 20 numbers between 1 and 200.
10 a Landon   b Roosevelt

7 Problem-solving: Watching statistics

1 Any suitable summary written.
2 e.g. if a linear trend continues, series 3 (excluding the Christmas Special) should achieve a mean rating of about 8.2 million and a mean appreciation of about 75%

7 Check up

1 4.3 people
2 a 40.35714 = 40.4 (1 d.p.) minutes
  b 50 minutes
3 a 1.96   b 4   c 2
4 a Jade scored the most runs. Ria's total = 149; Jade's total 32.4 × 5 = 162 runs
  b Jade is more consistent as her range is smaller, showing that her scores are less spread out. Jade's range = 28; Ria's range = 53 − 11 = 42
5 a 3.4 kg   b 4.3 kg   c 2.1 kg
6 a Median
  b Mean would be affected by large break value and mode is the two lowest values in the data set so it doesn't give a good picture of the data
7 a 27.1129 = 27 emails   b 39 emails
  c 31–40   d 31–40
8 Random sample, e.g.names in a hat and pick out 3 without looking.

9 Monday morning – people are at school or work so not representative. Shopping centre – needs to include a more varied area so all people can be included in sample.
11 a

Age	Frequency, f	Class midpoint, m	$f \times m$
18–24	32	**21**	672
25–**29**	**41**	27	1107
30–40	70	**35**	**2450**
41–59	**53**	**50**	2650
60–70	14	**65**	**910**

  b 37.1   c 30–40
  d 30–40, unequal class sizes.

7 Strengthen

Averages and range

1 a Students' own answer.
  b 8   c 192   d 24 seconds
2 14.2 °C
3 88.83333…=£88.83
4 2.125 = 2.1 goals (1 d.p.)
5 a Sam   b Ellie = 5, Sam = 62
  c Ellie
  d Ellie (Ellie's mean = 37, Sam's mean = 32)
  e Ellie as she scores more consistently
6 a Kyle (Kyle's total = 166, George's total = 158)
  b George as he has the smaller range and is therefore more consistent.
7 a i 8   b i 5
     ii 8   ii 10
  c

Number of cards	Frequency	Number of cards × frequency
1	8	**8**
2	5	**10**
3	3	**9**
4	2	**8**
Total	**18**	**35**

  d 18   e 35   f 1.944…
8 a 39 cars   b 89 tyres
  c 2.28205…= 2.3 tyres (1 d.p.)
9 a 80   b 32   c 2.5   d 4
  e i 32   ii 16.5th value   iii 2
10 a Liz = 28.345, Katie = 27.578
  b Liz = 28.4, Katie = 29.8
  c 20.47 m
  d Median is insensitive to extreme values
  e Median
  f Students' own answer
11 a 2.8 kg   b 3.9 kg   c 2.8 kg

Averages and range for grouped data

1 a

Puzzles completed	Frequency, f	Midpoint of class, m	$m \times f$
1–5	11	3	33
6–10	15	**8**	**120**
11–15	23	**13**	**299**
16–20	16	**18**	**288**
Total	**65**	**Total**	**740**

  b 65   c 740
  d Raw data is lost once it has been grouped.
  e 11.3846…= 11.4 (1 d.p.)
  f 11–15   g 33rd data value   h 11–15

2 a

Score	Frequency, f	Midpoint of class, m	$m \times f$
0–20	9	10	$10 \times 9 = 90$
21–40	6	30.5	**183**
41–60	2	**50.5**	**101**
61–80	2	**70.5**	**141**
Total	**19**	**Total**	**515**

  b 19   c 515
  d Raw data is lost once it has been grouped.
  e 27.10526…=27   f 80

3 a

Distance, d (miles)	Frequency, f	Midpoint of class, m	$m \times f$
$0 < d \leq 2$	9	1	9
$2 < d \leq 4$	7	3	21
$4 < d \leq 6$	8	**5**	**40**
$6 < d \leq 8$	6	**7**	**42**
Total	**30**	**Total**	**112**

  b 30   c 112   d 3.733… = 4

Sampling

1 a Checking the whole population would take too long and slow down the process.
   b 10
   c If the fault occurs later in the batch it will remain unnoticed.
   d A random sample would be best.
2 It would be better to take the survey on several days of the week. Saturday morning may not provide a representative sample of people. It should also be considered that people who have gone to the town centre may be less likely to do their weekly shopping online than the people who have stayed at home.

7 Extend

1 11
2 4
3 a $\frac{73}{120}$ b $\frac{11}{24}$
4 $\frac{4x + 3}{3}$
5 a Mode   b Mean   c Mode   d Median
6 Boys range = 26 cm, boys median = 180.5 cm.
   Girls range = 26 cm, girls median = 168.5 cm.
   Boys are taller on average but same spread in data.
7 a 15   b 14.5   c 25
8 a £10 222 (nearest £1)   b $9000 < p \leq 12\,000$
   c $9000 < p \leq 12\,000$ d £13 000
9 a 2 minutes 6 seconds   b 8 seconds
10 a 180
   b Use a random sample, e.g. use random numbers to select the number on 180 of the tickets at random
11 a $\frac{3}{5}$ b 9
13 50

7 Unit test
Sample student answer

a The student will get 1 mark for showing the working to find out the range for boys and girls (8 – 2 and 6 – 1). They will get 1 mark for finding the actual range for boys and girls (6 and 5).

b The student has to say Adele is correct and they must use the working to describe why she is correct, i.e. state that as the girls' range is 6 and the boys' range is 5, then the range of the girls' marks is 1 more.

UNIT 8
8 Prior knowledge check

1 a 7   b 3
2 a 8   b 6   c 9
   d 4   e 25
3 a 5.3   b 15.4   c 2.0   d 21.1
4 a 6.38   b 0.79   c 80.02   d 2.20
5 a 1.238   b 78.027   c 0.932   d 60.690
6 a 13 656   b 14 000   c 13 656.0   d 13 700
7 $4 \times 20 \div 5 = 16$ or close estimate
8 a 2.8 m   b 4.4 m   c 5.8 m
   d 150 kg   e 325 kg
9 No. The mass shown is closer to 10 than 20, so a better estimate would be 12 or 13 g.
10 a 3 cm, 30 mm   b 4.2 cm, 42 mm
11 a No, better estimate is around 50 g
   b No, better estimate is around 450 ml
   c Yes   d Yes
12 a 6000 g   b 3600 cm   c 1750 mm   d 0.305 km
13 a 0.89 litres   b 3600 ml
14

15 a Sphere   b Square-based pyramid
   c Cuboid   d Triangular prism
   e Triangle-based pyramid (or tetrahedron)
16 a 24 cm²   b 12.5 cm²
17 a 9.5 cm² (accept 9 or 10 cm²)
   b 28 cm² (accept 29 or 30 cm²)
18 a Perimeter 16 cm, area 15 cm²
   b Perimeter 17.6 cm, area 17.9 cm² (1 dp)
19 a 22 cm   b 4 cm
20 a $h = 4$ b $m = 20$
21 a 7   b 12   c 12   d 21
22 a $t = 4$ b $p = 3$ c $r = 8$
23 Students' own answers.

8.1 Rectangles, parallelograms and triangles

1 a 16 cm   b 26 m   c 17 m   d 250 mm
2 a Area 25 cm², perimeter 20 cm
   b Area 370 mm², perimeter 94 mm
3 a i $l \times w$ ii $2l + 2w$
   b $A = l \times w$ c $P = 2l + 2w$
4 a 72 cm²   b 63 cm³   c 2432 mm²   d 14.64 cm²
5 a Perimeter = 40 cm, area = 45 cm²
   b Perimeter = 23.6 cm, area = 27.6 cm²
   The two measurements for base and height have to be perpendicular.
6 a 40 cm²   b Students' drawing
   c Half   d 20 cm²
7 a 13.5 cm²   b 20 cm²   c 17.15 cm²   d 294 mm²
8 a Perimeter 26 cm, area 30 cm²
   b Perimeter 19.8 cm, area 13.86 cm²
9 16 cm²
10 a 4800 cm²   b 1440 cm
11 3 triangles sketched and labelled, with area 18 cm²
12 a 20 cm²   b 9 cm²   c 6 cm²
13 a $a = 10$ cm   b $b = 4$ cm   c $c = 5$ cm
14 a Sketch of two rectangles 3.7 m × 2.6 m, two rectangles 2.9 m × 2.6 m
   b Estimate $2 \times 4 \times 3 + 2 \times 3 \times 3 = 42$ m²
   c $7 \times £12 = £84$

8.2 Trapezia and changing units

1 a 8   b 6   c 30
2 a $h = 4$ b $x = 3$
3 81.3 cm²
4 a 56 cm²   b 57.8 cm²   c 31 cm²   d 46.4 cm²
5 Area 16 cm², perimeter 12.8 cm

6 a $40 = \frac{1}{2} \times (6 + 10) \times h$ b $40 = 8h$ c $h = 5\,cm$

7 7 mm

8 a 27.5 cm²   b 140 mm²

9 a 4.41 cm²   b 441 mm²

10 a 1 m² = 10 000 cm²

   b

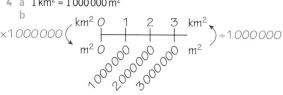

11 a 22 000 cm²   b 0.5 m²

12 Students' own conversions to m²

13 a 1.05 m²   b 10 500 cm²   c 10 500 cm²

   d Students' own answers.

14 a 2.8 cm²   b 20 250 cm²   c 11.2 cm²

15 a 13.2 cm²   b 1320 mm²

16 Slide A has 0.5 bacteria per mm². Slide B has 0.6 bacteria per mm².

   So slide B has more bacteria per square millimetre.

8.3 Area of compound shapes

1 a Area 24 cm², perimeter 20 cm

   b Area 30 cm², perimeter 30 cm

2 a 50 000 cm²   b 22 000 cm²   c 2 m²   d 0.72 m²

3 a 80 000 m²   b 35 000 m²   c 4 ha   d 22.5 ha

4 a 1 km² = 1 000 000 m²

   b

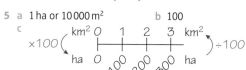

5 a 1 ha or 10 000 m²   b 100

   c

6 a Windermere   b 983 ha or 9.83 km²

7 a Area C 60 cm²; Area D 40 cm²

   b 100 cm²   c 44 cm²

8 a Area 71 cm², perimeter 40 cm

   b Area 57 m², perimeter 38.6 m

   c Area 64.5 cm², perimeter 39 cm

9 a $12 \times 8 - 8 = 88\,cm^2$

    $4 \times 8 + 6 \times 4 + 4 \times 8 = 88\,cm^2$

   b Students' own answers

10 a Area = 295 cm², perimeter = 82 cm

   b Area = 106.62 m², perimeter = 63.6 m

   c Area = 130 cm², perimeter = 58 cm

11 a 35 cm²   b 630 mm²   c 216.5 cm²

12 a Students' own answers   b Students' own answers

   c 153 cm²

   d Split it into rectangle and triangles and find areas separately.

13 36 cm²

14 360 cm²

8.4 Surface area of 3D solids

1 Net of cuboid, e.g.

2 a 12 cm²   b 24 cm²   c 18 cm²

3 164 cm²

4 a i 36 cm²   b i 18 cm²

     ii bottom   ii back

   c i 8 cm²

     ii left hand side (or opposite side)

   d 2 × 36 + 2 × 18 + 2 × 8 = 124 cm²

5 a 160 cm²   b 15.5 m²   c 1350 mm²

   d 6 × area of one face

6 a Net of triangular prism, e.g.

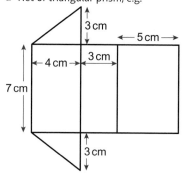

   b i 35 cm², 21 cm², 28 cm², 6 cm², 6 cm²

     ii 96 cm²

7 a 660 cm²   b 2864 cm²

8 a 7.5 m²   b 2

9 a Cuboid 7.46 m², prism 3.51 m², total 10.97 m²

   b Estimate 11 × 13 = £143

10 8 cm

11 a 96 cm²   b 520

8.5 Volume of prisms

1 a 12 cm²   b 18 mm²   c 100 cm²

2 a 64   b 210   c 72

3 a 2.4 m   b 0.52 m   c 430 cm   d 70 cm

4 9 cm³

5 a 4 cm³   b 8 cm³   c 12 cm³

6 a 48 cm³   b 3.6 m³   c 7500 mm³   d 125 cm³

    4³ or 4 × 4 × 4 = 64 cm³

7 12 cm³

8 a 40 cm³   b 72 cm³   c 300 cm³

9 58.88 cm³

10 a 1.3 m³   b £90.20

11 7 cm

12 Student's sketches of 3 cuboids with volume 24 cm³

8.6 More volume and surface area

1 Teaspoon 5 ml, drink can 330 ml, bucket 5 litres, juice carton 1 litre

2 a $h = 9$ b $h = 3$

3 a 496.6 cm²   b 421.5 cm³

4 a 24 000 mm³   b 30 cm³   c 5700 mm³   d 4.81 cm³

5 a 10.416 cm³ and 10 416 mm³   b 10 416 mm³

6 a 1 000 000   b 1 m³ = 1 000 000 cm³

7 a 3 500 000 cm³   b 6 870 000 cm³

   c 8.2 m³   d 3.159 m³

8 a 500 cm³   b 4.9 litres   c 33 000 000 cm³

   d 2000 litres   e 3400 litres   f 4 m³

   g 2400 cm³   h 1 m³ = 1000 litres

9 200 cm³

10 80 litres

11 a 1280 ml   b 1.28 litres

12 a 350 m³ = 350 000 litres

   b 105 m²

     No, already calculated to use in the volume calculation in part a.

13 a 2.55 cm³

   b Volume of block = 300 cm³, 300 ÷ 2.55 = 117.6, which rounds down to 117.

14 a 2 cm   b 8 cm
15 500 cm²

8 Problem-solving

1 105°
2 56%
3 12 cm
4 Capacity of trough = 103 125 cm³ = 103.125 litres; 26 min
5 Students' own answers.

8 Check up

1 a i 30 cm   b i 20 cm
     ii 48 cm²   ii 25 cm²
2 9 cm
3 5 cm
4 a 34 cm   b 57 cm²
5 260 mm²
6 13.75 m²
7 402 cm²
8 a 376 cm²   b 480 cm³
9 48 cm³
10 a 1500 mm²   b 5.47 m²   c 12 000 cm²   d 9.8 cm²
11 a 24 000 litres b 450 ml   c 200 cm³
     d 8.4 cm³   e 0.549 m³   f 1 600 000 cm³
13 a 8 cm, 6 cm, 48 cm³
     b 24 cm³
     c No, because shorter side will have length 0

8 Strengthen

2D shapes

1 a 18 cm   b 16 cm   c 22 cm   d 10.5 cm
2 a 6 cm, 4 cm   b 2 cm, 3 cm
3 a 18 cm   b 30 cm
4 a Sketch of L-shape divided into two rectangles
     b 15 cm², 4 cm²   c 19 cm²
5 a 18 cm²   b 64 cm²
6 Students' drawings showing lines perpendicular to each other, e.g.

7 a 9 cm²   b 9 cm²
8 a 35 cm²   b 8 cm²   c 70 cm²
9 a 12 cm², 4 cm²; 16 cm²   b 28 cm², 7 cm², 35 cm²
10 a 300 cm²   b 150 cm²   c 150 cm²
11 a 40 cm²   b 36 cm²
12 a $h = 3$ cm   b $b = 5$ cm

3D solids

1 a

front 5 cm   side 5 cm   top 4 cm
3 cm   4 cm   3 cm

     b Labelled sketches showing measurements: back (3 cm by 5 cm), other side (4 cm by 5 cm), bottom (3 cm by 4 cm)
     c Top and bottom, two sides, front and back
     d 15 cm², 20 cm², 12 cm²   e 94 cm²
2 a

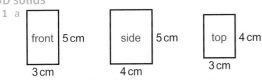

b Triangles 6 cm², rectangles 45 cm², 36 cm², 27 cm²
     c 120 cm²
3 a 36 cm³   b 140 cm³   c 2250 cm³

Measures

1 a i 1 cm²   ii 2 cm²   iii 3 cm²
     b i 100 mm²   ii 200 mm²   iii 300 mm²
     c

×100 ⟋ mm² 0   100   200   300   400 ⟍ ÷100
         cm² 0   1   2   3   4

2 a i 1 m², 10 000 cm²   ii 2 m², 20 000 cm²
       iii 3 m², 30 000 cm²
     b

×10 000 ⟋ cm² 0   10 000   20 000   30 000   40 000   50 000 ⟍ ÷10 000
             m² 0   1   2   3   4   5

3 a 500 mm²   b 3 cm²
     c 60 000 cm²   d 8 m²
4 a 3000 mm³   b 9 cm³   c 4720 mm³
     d 2 000 000 cm³   e 1 200 000 cm³
5 a 9 m³   b 3.2 m³   c 5000 ml = 5 litres
     d 875 cm³ = 875 ml = 0.875 litres
     e 2300 ml = 2300 cm³
     f 1345 ml = 1345 cm³

8 Extend

1 3.9 m²
2 240
3 Yes, Area of wall = 150 m²
4 Area of floor 72 + 42 = 114 m²
     Needs 10 tins polish at a cost of £133.
     No, he doesn't have enough money
5 a $14x$ b $x = 2$ cm   c 36 cm²
6 260 cm²
7 a $6x + x = 35$ b $x = 5$
8 a 2 m   b 9 m   c $\sqrt{50} = 7.1$ m
9 a 88 m²
     b Rectangle sketch with area 88, e.g. 11 × 8 m
     c, d, e Students' own answers. Smallest possible perimeter is for a square enclosure, 37.5 m
10 $50 = \frac{1}{2}(a + 6) \times 10$
     $50 = \frac{1}{2} \times 10 \times (a + 6)$
     $50 = 5(a + 6)$
     $10 = a + 6$
     $a = 4$ cm
11 a xy mm²   b $100xy$ mm²   c $\frac{xyz}{10}$ cm³   d $50rst$ cm³
12 50 cm³
13 a 4000 cm³ = 0.004 m³   b 103 200
14 4000 cm³
15 60
16 12.4 cm
17 682 (682.666… round down)
18 24 ha
19 a 750 cm³   b 600 cm²
20 192 cm³

8 Unit test

Sample student answer

a e.g. the student could have divided the shape into smaller shapes, and numbered them to show their working more clearly, e.g.

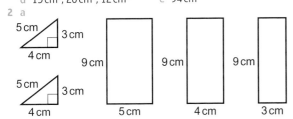

     (1) Triangle = $\frac{1}{2} \times 8 \times 12 = 48$ cm²
     (2) Square = $12 \times 12 = 144$ cm²
     (3) Rectangle = $10 \times 9 = 90$ cm²
     Total = 48 + 144 + 90 = 282 cm²
     So Ellen needs 282 cm² of material.
b She has missed a bit of the perimeter that wasn't labelled. She needs to add on 3 cm.

UNIT 9

9 Prior knowledge check

1 a i 7   b i +3
    ii −5   ii ×3
2 a 10 am   b 11 am   c 12.30 pm   d 1.45 pm
3 a A (3, 2), B (4, 0), C (−3, 1), D (−4, −3), E (0, −2),
   b i H
     ii F
     iii G
4 a Coordinate grid with the points (1, 80), (2, 160), (3, 240),
    (4, 320) and (5, 400) correctly plotted and joined with a
    straight line
   b 280 pence
   c 480 pence

5 a
Number of miles	0	5	20	50	100
Number of km	0	8	32	80	160

   b Coordinate grid with the points (0, 0), (5, 8), (20, 32),
    (50, 80) and (100, 160) correctly plotted and joined with
    a straight line
   c 112 km
6 a −3   b 6   c 10   d 10
7 a $x = 2$ b $x = 6$
8 $D = (1, 1)$
9 Students' own answers

9.1 Coordinates

1 a −2   b −5   c 1   d 4
2 a 5   b 6   c 3.5
   d 1   e 3   f 2
3 a (1, 5), (1, 1), (1, 0), (1, 4), (1, 2)
   b Coordinate grid with the points (1, 5), (1, 1), (1, 0), (1, 4),
    (1, 2) correctly plotted
    All points are on a vertical line through $x = 1$.
   c Vertical line through $x = 3$
   d Coordinate grid with the points (3, 0), (3, −2), (3, 4) and
    (3, 2) correctly plotted
4 a (4, −1), (4, 0), (4, 1), (4, 2), (4, 3), (4, 4), (4, 5)
   b All have x-coordinate 4
   c i $x = 4$ ii $x = -2$ iii $y = 3$
5 Line P is $x = -4$
   Line Q is $x = 5$
   Line R is $y = -3$
   Line S is $y = 1$
6

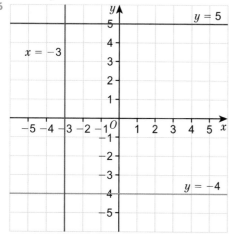

   d $y = 0$
7 a (−5, −5), (−4, −4), (−3, −3), (−2, −2), (−1, −1), (0, 0), (1, 1),
    (2, 2), (3, 3), (4, 4), (5, 5)
   b x and y coordinates are the same
   c (6, 6), (−8, −8)
   d $y = x$

8

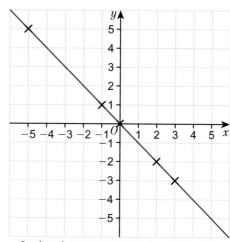

   c Students' own answers, e.g. (4, −4), (1, −1), (5, −5)

9
Line segment	Start point	End point	Midpoint
AB	(4, 1)	(4, 3)	(4, 2)
CD	(1, 4)	(5, 4)	(3, 4)
EF	(2, −5)	(5, 1)	(3.5, −2)
GH	(1, −4)	(3, −2)	(2, −3)
JK	(−3, −3)	(1, −3)	(−1, −3)
LM	(−4, 1)	(−4, 5)	(−4, 3)
PQ	(−2, 4)	(1, −2)	(−0.5, 1)
RS	(−1, 5)	(2, 0)	(0.5, 2.5)

10 (4, 5.5)
11 a (5, 7)   b (3.5, 8.5)   c (−1, 5)   d (−1, −1)

9.2 Linear graphs

1 a 1   b 20   c −18
2 a i 8   b i −3   c i $3\frac{1}{2}$
   ii −1   ii 3   ii $2\frac{3}{4}$
3 a 4.2   b 2.5
4 a (−3, 2), (0, 5), (1, 6)
   b, c, d, g

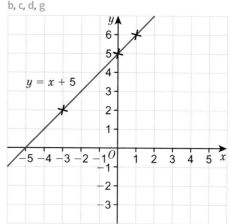

   e Students' own answers, e.g. (−1, 4), (−2, 3)
   f $y = x + 5$

5 a (–1, –1), (0, 1), (2, 5)
  b, c, d, g

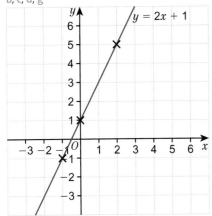

  e Students' own answers, e.g. (–2, –3), (1, 3)
  f $y = 2x + 1$

6 a i

x	–3	–2	–1	0	1	2	3
$y = x + 1$	–2	–1	0	1	2	3	4

  ii

x	–3	–2	–1	0	1	2	3
$y = 2x - 3$	–9	–7	–5	–3	–1	1	3

  b

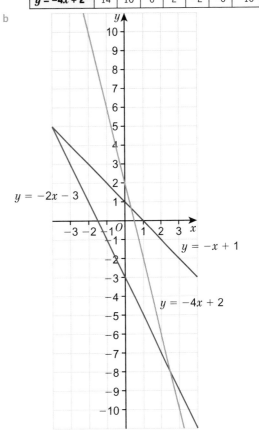

7

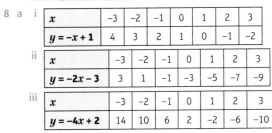

8 a i

x	–3	–2	–1	0	1	2	3
$y = -x + 1$	4	3	2	1	0	–1	–2

  ii

x	–3	–2	–1	0	1	2	3
$y = -2x - 3$	3	1	–1	–3	–5	–7	–9

  iii

x	–3	–2	–1	0	1	2	3
$y = -4x + 2$	14	10	6	2	–2	–6	–10

  b

9 a, b, c, d

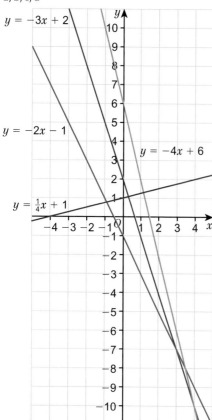

$y = -3x + 2$

$y = -2x - 1$

$y = -4x + 6$

$y = \frac{1}{4}x + 1$

e i Upwards
  ii Downwards

10 a

x	−2	−1	0	1	2	3	4
y	−2	0	2	4	6	8	10

b

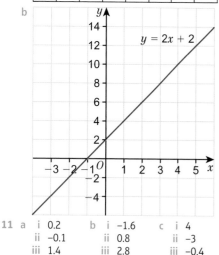

$y = 2x + 2$

11 a i 0.2   b i −1.6   c i 4
   ii −0.1   ii 0.8   ii −3
   iii 1.4   iii 2.8   iii −0.4

9.3 Gradient

1 a 5   b −2   c $1\frac{1}{3}$ d $\frac{1}{2}$
2 Line B
3 a i 3   ii 6   iii 9
  b i 2   ii 4   iii 6
4 a Parallel   b 2   c Gradient

5 Gradient line A = 1
  Gradient line B = −2
  Gradient line C = −3
  Gradient line D = −$\frac{1}{2}$
  Gradient line E = 2
  Gradient line F = $\frac{1}{2}$

6   a   b   c   d   e

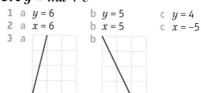

7 1.5
8 a 0.12   b 0.07   c 0.0625
9 a i

x	−3	−2	−1	0	1	2	3
$y = x + 1$	−2	−1	0	1	2	3	4

  ii

x	−3	−2	−1	0	1	2	3
$y = 2x − 3$	9	7	5	3	1	−1	−3

b i Straight-line graph passing through (−1, 0) and (3, 4), labelled $y = x + 1$
  ii Straight-line graph passing through (0, 3) and (3, −3), labelled $y = −2x + 3$
c Gradients are 1, −2
10 a $y = 5x + 2$
   b $y = 3x + 2$ and $y = 3x$
   c $y = −3x + 4$
11 $\frac{1}{2}$

9.4 $y = mx + c$

1 a $y = 6$ b $y = 5$ c $y = 4$
2 a $x = 6$ b $x = 5$ c $x = −5$
3 a   b

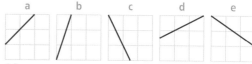

4 a Number = y-intercept   b 4   c 1
5 Line A $y = x + 1$
  Line B $y = 2x − 1$
  Line C $y = −5x + 1$
  Line D $y = −\frac{1}{2}x + 3$
  Line E $y = \frac{1}{2}x − 3$
6 a students' own answers, e.g. $y = 5x − 1$
7 a $y = 3x, y = −3x$ b $y = 3x − 3, y = x − 3$
8 a $y = 3x − 5$ b $y = 2x − 2$ c $y = 4x + 6$
  d $y = \frac{1}{2}x + 11$ e $y = 3x − 27$
9 a $y = 2x − 1$ b $y = \frac{1}{2}x + 3$
  c $y = 2x − 6$ d $y = −x + 2$
10 a Straight-line graph passing through (0, −7) and (1, −5)
   b Straight-line graph passing through (0, 5) and (1, 2)
   c Straight-line graph passing through (0, −4) and (2, −3)
   d Straight-line graph passing through (0, 6) and (1, 5)
11 a Straight-line graph passing through (0, 8) and (8, 0)
   b Straight-line graph passing through (0, 6) and (6, 0)
   c Straight-line graph passing through (0, 4) and (4, 0)
12 a Straight-line graph passing through (0, 8) and (4, 0)
   b Straight-line graph passing through (0, 3) and (2, 0)
   c Straight-line graph passing through (0, 2) and (1, 0)
   d Straight-line graph passing through (0, $1\frac{1}{2}$) and (3, 0)

9.5 Real-life graphs

1 a £3.50
  b £120
2 Straight-line graph passing through (5, −3) and (0, −4)

3 a

Fuel used (litres)	0	1	2	3	4	5
Distance travelled (km)	0	10	20	30	40	50

b Straight-line graph passing through (0, 0) and (5, 50)
c 6 litres   d 17 litres

4 a

Water used (m³)	0	1	2	3	4	5	6	7	8	9	10
Cost (£)	30	32	34	36	38	40	42	44	46	48	50

b Straight-line graph passing through (0, 30) and (10, 50)
c $5\frac{1}{2}$ cubic metres
d 5500 litres

5 a i £25   ii £2.50
b 2.5

6 a Line 1 = B   Line 2 = A   Line 3 = C
b Higher or larger

7 a

Hours worked	0	1	2	3	4	5
Total cost (£)	60	90	120	150	180	210

b Straight-line graph passing through (0, 60) and (5, 210)
c $y = 30x + 60$
d Gradient is hourly cost
e y-intercept is callout fee

8 a Line 1 = A Line 2 = B
b Plumber A is cheaper at £520; plumber B would cost £540

9 a Current rate = £25.60; offer rate = £25.30 – he is 30p better off if he changes
b Draw a graph. You will see that the cost is equal at 75 minutes. The original rate is better below 75 minutes; the offer rate is better above 75 minutes.

10 a Option A – one off joining fee of £600 then no cost per game
Option B – flat rate of £20 per game but no joining fee
b Option A is cheaper for more than 30 games

11 a Straight-line graph passing through (0, 600) and (20 000, 2600)
b £600
c Yes, on average he would be £200 better off

12 a Lines from top to bottom: C, A, B
b £110
c C is cheapest. It charges £240, A charges £260, B charges £320.

9.6 Distance–time graphs

1 a 30   b 20   c 12   d 10
2 a 6 × 10 = 60   b 5 × 12 = 60
c 30 × 2 = 60   d 2 × 30 = 60
3 a

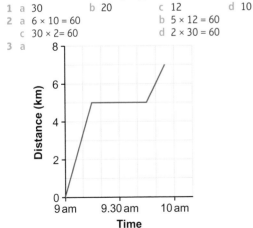

b First part = 20 km/hour, last part = 12 km/hour
c Last part is uphill as speed is slower
4 a 40 km   b 10 am   c 2 hours   d 20 km/h
e 30 minutes   f 3 hours
g 100 km   h 33.33… km/h

5 a 60 km/h   b 45 minutes   c 30 km/h   d 30 km
6 a 1.6 km   b 20 minutes   c 15 minutes
d Walking to surgery at 4.8 km/h – other parts 3.6 km/h and 4 km/h

7 a

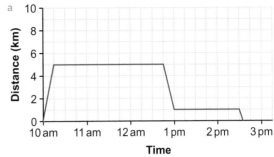

b First leg is fastest at 20 km/h, second leg 16 km/h, last leg 12 km/h

8 a Red line
b Southampton to Winchester arrives at 11:20 and Winchester to Southampton arrives at 11:40.
c Southampton to Winchester is faster.

9 a 3000 feet   b 3 hours 15 minutes
c 1 hour 30 minutes   d 9:30 am

10 a

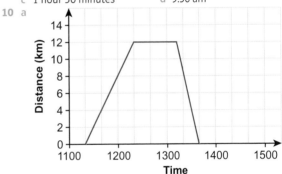

b 11:20   c 12 km
11 Debbie was faster at 75 km/h
12 a iii   b iv   c i   d ii
13 a 5 m/s   b 5 m/s²   c $y = 5x + 5$

9.7 More real-life graphs

1 a Positive   b Negative   c None
2 $y = 5x + 10$
3 a B   b A is Line 2 , B is Line 1, C is Line 3
4 A is Line 1, B is Line 2
5 a

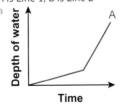 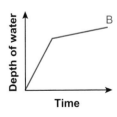

b A is Line 1, B is Line 2
6 a 20 °C   b 54 cm
c The graph is linear.   d 0.29 to 2 d.p.
7 a −4   b Water is flowing out at a rate of 4 cm/h
8 a Positive correlation: the more hours of study the better the test score.
b $y = 9x + 34$ (answer from the line of best fit may vary)
c 74 or 75   d It would be negative hours

9 a, b

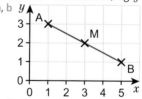

c Students' own answers for line of best fit e.g. $y = 0.9x - 2$
d ≈ 52 e ≈ 11.5

10 a Between 2.1 and 2.2
b Results outside the range of the data are always less reliable. Assume that the decrease continues at the same rate.

9 Problem solving

1, 2, 3

4 T rex, Allosaurus and Microceratus were walking. Velociraptor was running.
5 2.8 m
6 If the stride length of the Velociraptor was reduced by 10% it would be trotting, not running.

9 Check up

1 a

x	−2	−1	0	1	2
$y = 2x - 1$	−5	−3	−1	1	3

b Straight-line graph passing through (−2, −5) and (2, 3)
2 (4, 4)
3 A $x = -5$ B $y = 2$
C $y = x$ D $y = -\frac{1}{2}x + 3$
4 a $y = 3x + 4$ and $y = 3x - 7$
b $y = x + 3$ and $y = 4x + 3$
c $y = 4x + 3$
5 a 5 minutes
b Between 10 minutes and 16 minutes
c 2.4 km/h
6 a $y = x + 10$ b ≈ 58
7 a £80   b £40   c £60
9 Students' own answers

9 Strengthen

Algebraic straight-line graphs

1 a

x	0	1	2	3	4
$y = 3x$	0	3	6	9	12

b (0, 0) (1, 3) (2, 6) (3, 9) (4, 12)
c, d, e Straight-line graph passing through (0, 0) and (4, 12) labelled $y = 3x$

2 a

x	0	1	2	3	4
$y = 4x - 3$	−3	1	5	9	13

b, c, d Straight-line graph passing through (0, −3) and (4, 13) labelled $y = 4x - 3$
e (0,−3)

3 a Four points plotted with the x-coordinate 3.
b The line crosses the x-axis at (3, 0)
c (3, 4) (3, 2)
d x value = 4, y value = 3

4 a Straight-line graph passing through (1, −5) and (1, 5)
b Straight-line graph passing through (−5, 2) and (5, 2)
c Straight-line graph passing through (−4, −5) and (−4, 5)

5 a Positive   b Negative

6 a Line A positive; Line B positive; Line C negative; Line D positive
b Line A 3; Line B 2; Line C −3; Line D 2
c Lines B and D are parallel

7 a $\frac{1}{2}$ b $\frac{1}{3}$ c $-\frac{1}{4}$

8 a Gradient always = 2
b

Line	y-intercept
$y = 2x + 3$	3
$y = 2x + 1$	1
$y = 2x$	0
$y = 2x - 2$	−2
$y = 2x - 4$	−4

c −5

9 a Line A gradient = 4; Line B gradient −2
b Line A intercept 2; Line B intercept 1
c Line A $y = 4x + 2$; Line B $y = -2x + 1$

10 Line A $y = 3x + 2$, Line B $y = 4x - 3$, Line C $y = \frac{1}{2}x - 1$, Line D $y = -2x + 4$

11 a i 5   ii 3   iii $-\frac{1}{2}$ iv 3   v $\frac{1}{2}$
b Lines ii and iv are parallel

12 a Students' own answers, e.g. $y = 4x$
b Students' own answers, e.g. $y = -2x$
c Students' own answers, e.g. $y = \frac{1}{2}x$

13 a, b

c (3, 2)

14 a (5, 6)   b (2, 5)   c (1, 2)   d $(2\frac{1}{2}, 6\frac{1}{2})$

Distance–time graphs and scatter graphs

1 a Horizontal line
b 20 minutes
c 18 minutes
d 10 miles
e 12 minutes
f Return home 50 miles per hour
g 8:20 pm
h 30 miles per hour

2 a i 4   ii 4   iii $y = 4x + 4$
b 112 ice creams

Real-life graphs

1 a

	Translator A	Translator B
Fixed fee	£50	£150
Cost for 1000 words	£140	£190
Rate per word	9p	4p
Cost for 1500 words	£185	£210
Cost for 2500 words	£275	£250

b Translator A
c Translator B

2 diagram A; description 1; graph W
diagram B; description 5; graph V
diagram C; description 2; graph Y
diagram D; description 4; graph Z
diagram E; description 3; graph X

9 Extend

1 a 10 m/s   b Particle B
2 Gradient A = 6.66…%, gradient B = 6.25%, gradient C = 6% so A is steepest.
3 a 5 cm   b 8 kg   c $y = 1.5x + 5$
   d $l = 1.5m + 5$
   e 23 cm
   f Unreliable as the weight is outside the plotted points
   g The elasticity of the spring

4 a

Width in cm (x)	1	2	3	4	5	6
Perimeter in cm (y)	10	14	18	22	26	30

   b Straight-line graph passing through (1, 10) and (6, 30)
   c $y = 4x + 6$
   d 3.5 cm

5 Approx. 2030 to 2040
6 15 square units
7 8 square units
8 a

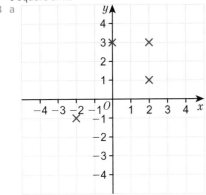

b Kite
c

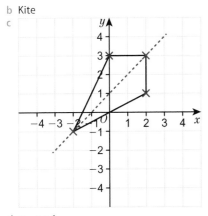

d $y = x + 1$

9 a

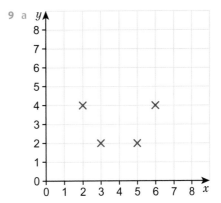

b Trapezium
c

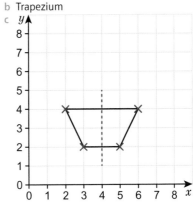

d $x = 4$

10 a   i $v = 0$
     ii $v = 20$
   b

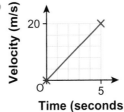

c

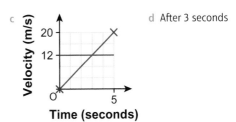

d After 3 seconds

11 $y = -2x + 13$
12 a $y = -x + 7$ b Gradient $= -1$, intercept $= 7$
13 a $y = 4x$ (or: $v = 4t$)   b 40 m/s   c 400 m
14 a 1991
   b Yes; 14.28…% increase, or 14% increase is only 3192 thousand
15 a 30 minutes   b 20 km
   c Plot (5:30, 20) and (7:00, 0) and join with straight lines
16 Straight-line graph from (−2, 4) to (4, 7) passing though (0, 5)

9 Unit test

Sample student answer
a The student forgot to join the points with a straight line.
b e.g. Use the equation to find the y-intercept (0, −1) and the gradient (2).

UNIT 10

10 Prior knowledge check

1 a 3 b 6 c $3\frac{1}{2}$ d 12
2 a 24 cm b 48 cm
3 a 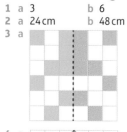 b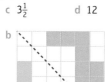

4 a 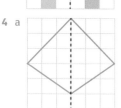 b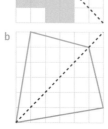

5 a Incorrect b Incorrect
6 a   b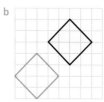

7 a $\frac{1}{4}$ turn clockwise
  b $\frac{1}{4}$ turn anticlockwise
  c $\frac{1}{4}$ turn anticlockwise

8 a   b   c

  d

9 & 10 a b c d

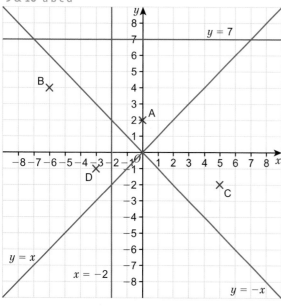

11 a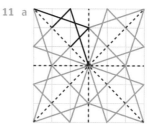

  b Students' own answers

10.1 Translation

1 a B b C c C d A
2 a 2 right, 4 up
  b 2 right, 2 down c 7 right, 5 up
3 a 3 right, 3 up b 5 left, 2 up c 6 right, 1 down
4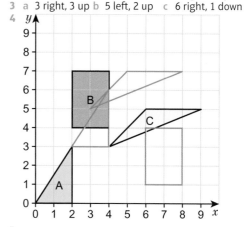

5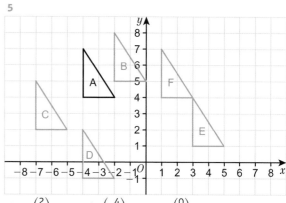

6 a $\begin{pmatrix} 2 \\ 4 \end{pmatrix}$ b $\begin{pmatrix} 4 \\ -4 \end{pmatrix}$ c $\begin{pmatrix} 0 \\ 4 \end{pmatrix}$

7 a $\begin{pmatrix} 6 \\ -1 \end{pmatrix}$ b $\begin{pmatrix} 0 \\ -7 \end{pmatrix}$ c $\begin{pmatrix} 6 \\ -9 \end{pmatrix}$ d $\begin{pmatrix} 10 \\ -7 \end{pmatrix}$

  e $\begin{pmatrix} 0 \\ -8 \end{pmatrix}$ f $\begin{pmatrix} 6 \\ -2 \end{pmatrix}$ g $\begin{pmatrix} -6 \\ 2 \end{pmatrix}$

8 Translation $\begin{pmatrix} 6 \\ -1 \end{pmatrix}$

9 a, b

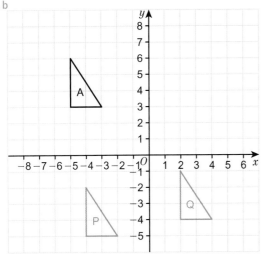

c Translation $\begin{pmatrix} 7 \\ -7 \end{pmatrix}$

10.2 Reflection

1 B

2 a

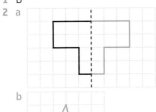

b

c

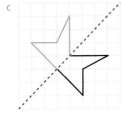

3 a, b, c

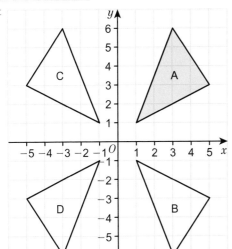

4 a, b, c, d

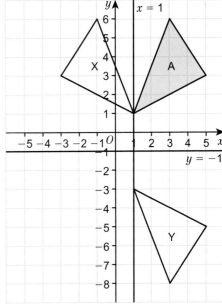

5 a, b

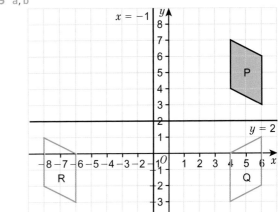

c Yes

6

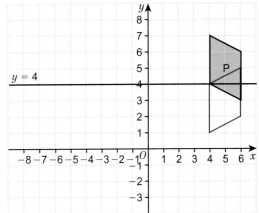

7 a, b

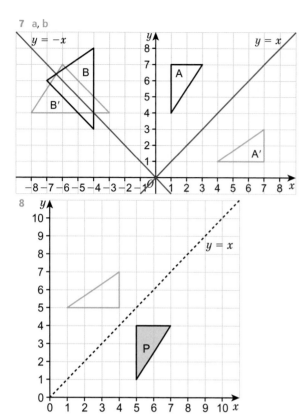

8

9 e.g. Draw the mirror line.
Reflect each vertex in the mirror line.
Join the vertices with a ruler.

10 a Reflection in the line $y = 0$ or x-axis
b Reflection in the line $y = -\frac{1}{2}$
c Reflection in the line $y = x$
d Reflection in the line $y = -x$

11 Reflection in the line $y = x$

12 The type of transformation, in this case a reflection and the equation of the mirror line

10.3 Rotation

1 a 270° clockwise
b 180° anticlockwise
c 90° clockwise

2 a   b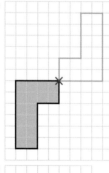

c 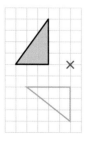 d

3

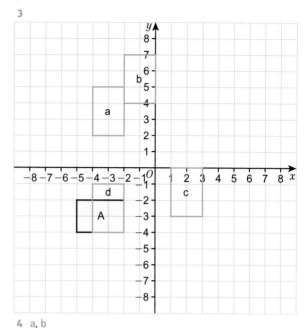

4 a, b

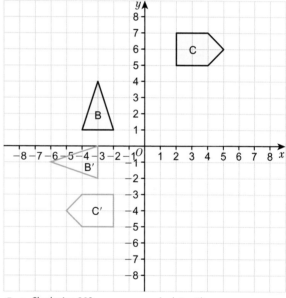

5 a Clockwise 90° b (−1, −1)
c Rotation of 90° clockwise around (−1, −1)

6 a 90° anti-clockwise around (0, 0)
b 180° around (0, 2) c 180° around (0, −1)
d 90° anti-clockwise (−2, 2)

7 Rotation of 180° (clockwise or anticlockwise) around (3, 3)

10.4 Enlargement

1 A = 1 and B = 9

2 a 6   b 3   c 20
   d 9   e 2   f 25

3 a
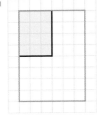

   b

   c   d

4 a
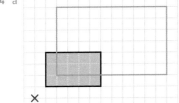

   b
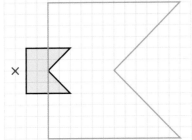

   c

   d
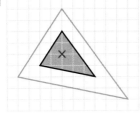

5
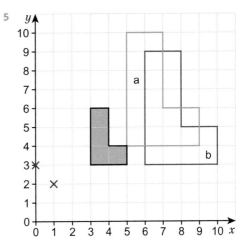

7 a
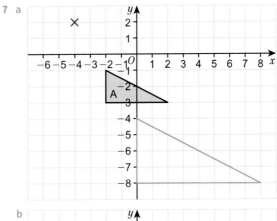

   b
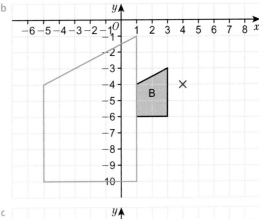

   c
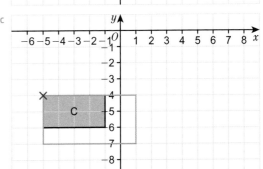

8 a Length = 24 cm and width = 18 cm
   b Length = 16 cm and width = 12 cm

10.5 Describing enlargements

1 a 2   b $\frac{1}{2}$ c $1\frac{1}{2}$

  d 3   e $\frac{1}{4}$

2 a 2   b $\frac{1}{2}$

3 a 2   b 10 cm

4 a 3   b (2, 2)

  c Enlarged by a scale factor of 3 from centre (2, 2)

5 a Enlarged by a scale factor of 3 from centre (–6, –3)

  b Enlarged by a scale factor of 2 from centre (–6, 6)

  c Enlarged by a scale factor of 2 from centre (1, 1)

6 Enlarged by a scale factor of 2.5 from centre (0, 0)

7 a i Enlarged by a scale factor of 1.5 from centre (6, –6)

    ii Enlarged by a scale factor of $\frac{2}{3}$ from centre (6, –6)

  b i Enlarged by a scale factor of 3 from centre (2, –3)

    ii Enlarged by a scale factor of $\frac{1}{3}$ from centre (2, –3)

10.6 Combining transformations

1 a Translation   b Reflection

  c Rotation   d Enlargement

3 a Direction of rotation   b Centre of enlargement

  c Translation vector

  d Equation of the mirror line

4 a, b

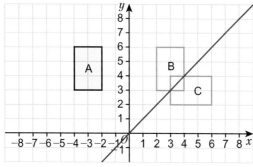

  c Rotation of 90° clockwise around (0, 0)

5 a, b

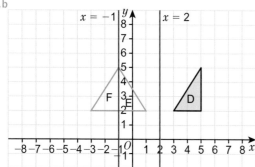

  c Translate with the vector $\begin{pmatrix} -6 \\ 0 \end{pmatrix}$

6 a, b

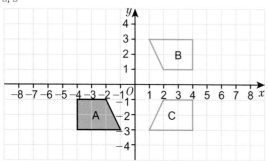

  c Reflection in the y-axis

7 a, b

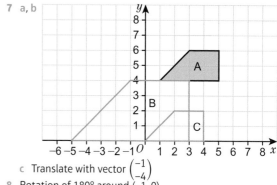

  c Translate with vector $\begin{pmatrix} -1 \\ -4 \end{pmatrix}$

8 Rotation of 180° around (–1, 0)

9 a Translation   b Rotation   c Translation

10 Problem-solving: Exploding shapes

1 a, b, c, d

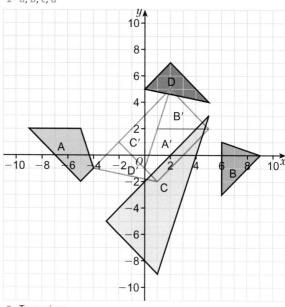

2 Trapezium

3 Students' own answers

10 Check up

1 a, b, c, d

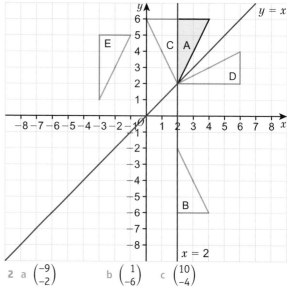

2 a $\begin{pmatrix} -9 \\ -2 \end{pmatrix}$ b $\begin{pmatrix} 1 \\ -6 \end{pmatrix}$ c $\begin{pmatrix} 10 \\ -4 \end{pmatrix}$

3 a $x = 0$ or y-axis **b** $y = -1$ **c** $y = -x$

4

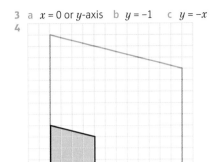

5 a, b

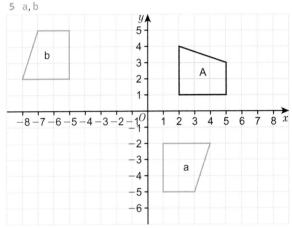

6 a Rotation of 90° anticlockwise around (0, 0)
   b Rotation of 90° clockwise around (−0.5, −0.5)

7

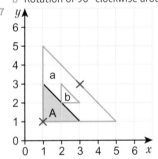

8 a Enlargement by a scale factor of 3 from centre (−6, 7)
   b Enlargement by a scale factor of 0.5 from centre (6, −6)
10 Students' own answers

10 Strengthen

Translations and reflections

1 a 3 right, 1 up   **b** 3 left and 4 down

2 a $\begin{pmatrix} 3 \\ 2 \end{pmatrix}$ **b** $\begin{pmatrix} -2 \\ -1 \end{pmatrix}$

   c $\begin{pmatrix} 4 \\ -3 \end{pmatrix}$ **d** $\begin{pmatrix} -6 \\ 5 \end{pmatrix}$

3 a, b

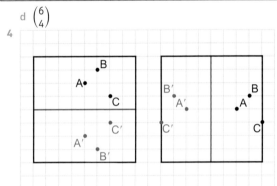

c

Coordinates of vertex	P	Q	R	S
Shape A	(−1, 0)	(−1, −3)	(−5, −3)	(−5, 0)
Shape D	(5, 4)	(5, 1)	(1, 1)	(1, 4)

d $\begin{pmatrix} 6 \\ 4 \end{pmatrix}$

4

5 a, b

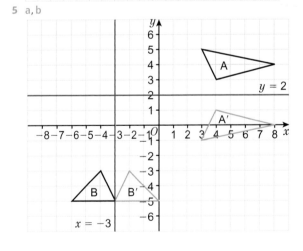

6 a

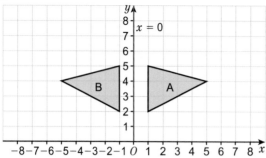

b Reflection in the *y*-axis or *x* = 0

7

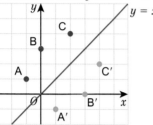

Enlargements and rotations

1 a   b

c

2 a, b, c, d – See below
3 a, b – See below

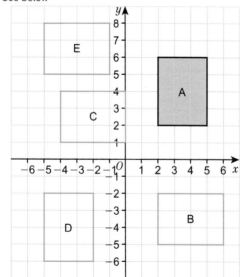

4 b (0, 0)   c Rotation 90° anticlockwise around (0, 0)
  d i Rotation of 90°clockwise around (1.5, −1.5)
     ii Rotation of 180° around (0, −2)
5 a B, E and F
  b Scale factor for B is 2, scale factor for E is 3 and scale
    factor for F is 0.5
  c i 16 cm   ii 8 cm

6 a

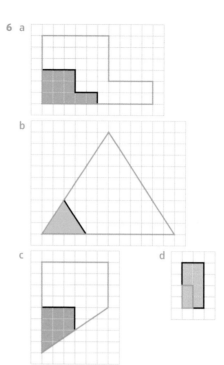

b

c   d

7 a, b, c

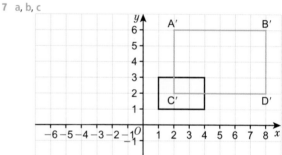

8 a, b

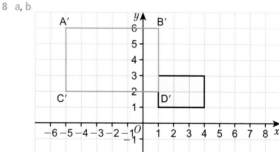

9 a 3   b (0, 5)
  c Enlargement by a scale factor of 3 from a centre of
    enlargement at (0, 5)

10 Extend

1 a Reflection about *x* = 3.5

b **Triangle ABC**	A(1, 1)	B(1, 5)	C(3, 4)
Triangle DEF	D(6, 1)	E(6, 5)	F(4, 4)
Triangle GHI	G(10, 1)	H(10, 5)	I(12, 4)

  c J (14, 1), K (14, 5), L (12, 4)
2 a 2   b 3   c 1.5

d **Rectangle**	A	B	C
Perimeter (cm)	10	20	30

  e Original Perimeter × Scale Factor = New Perimeter
    18 cm × 4 = 72 cm

3 $24n$
4 a 2.5   b Length = 12 cm and width = 18 cm
5 Enlargement by a scale factor of 2 from a centre of enlargement at (6, −5)
6 a Irregular pentagon   b 60.5
   c

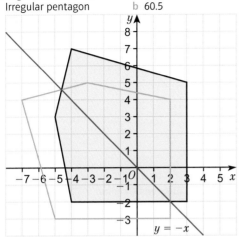

   d Yes, reflection does not change the area of a shape.
7 a $\begin{pmatrix} -x \\ -y \end{pmatrix}$ b $\begin{pmatrix} -6 \\ 1 \end{pmatrix}$
8 Reflection in the line $y = -x$
   Translation $\begin{pmatrix} -4 \\ -4 \end{pmatrix}$
   Rotation of 180° around (0, 0)
9 Rotation of 60° clockwise around O
10 $\begin{pmatrix} a + c \\ b + d \end{pmatrix}$
11 a 8 and 6
    b i 6 cm²   ii Length = 6 cm and width = 4 cm
       iii 24 cm²
12 36 cm²
13 1.5 cm²

10 Unit test

Sample student answers
a Student B has the correct answer. You must count the squares from one corner to the corresponding corner of the second shape, not just the nearest corner.
b e.g. Draw a dot onto 2 corresponding corners so they make sure they start and finish in the right place.

UNIT 11
11 Prior knowledge check
1 a 6   b 100
2 a 5   b 8
3 a 8 kg   b 9 mm   c 70p   d 60 g
4 a 3.5   b 32.4
5 a 24   b 72
6 a $\frac{1}{4}$ b $\frac{3}{5}$
7 a 12   b 4
8 a 1000   b 1000
9 2
10 a 3   b $y = 3x$
11 Students' own answers

11.1 Writing ratios
1 a 45   b Canoeing
2 a 1:2   b 2:5
3 a

b

4 a 6   b 24
5 a 1:3   b 2:1   c 3:1   d 1:6
   e 3:4   f 9:4   g 3:8   h 4:5
6 a 1:3
   b The coach is wrong, they won three times as many other medals as gold medals.
7 Yes, 4:32 is 1:8
8 A, B, and D are equivalent
   C and E are equivalent
9 a 4:5:3   b 6:4:5   c 8:6:5   d 2:5:7
10 All the ratios simplify to 2:3:5.
11 25:11:36
12 5:11

11.2 Using ratios 1
1 a 50   b 25   c 75   d 50
2 a 15   b 245   c 37.1   d 937
3 a 6   b 7   c 9   d 16
4 10
5 320 m
6 a 1:15   b 5:6   c 10:3   d 16:29
7 a 5:62   b 5:1   c 9:10   d 3:7
8 a Old, ratio simplifies to 4:3
   b New, ratio simplifies to 16:9
9 37:3
10 675 g
11 No, he will use 300 g.
12 £780
13 240

11.3 Ratios and measures
1 a 9   b i $\frac{7}{9}$ ii $\frac{2}{9}$
2 35 cm²
3 8 cm³
4 a 1:4   b 2:5   c 240:1
5 1:4
6 a 2500   b 34   c 3800   d 9.5
7 a 8 km   b 10 miles   c 41.9504 km
8 a 32 oz   b 800 g
9 7
10 $840
11 Three answers from this table

Currency	£200	£350	£475	£690
Dollars	336	588	798	1159.20
Euros	246	430.50	584.25	848.70
Rupees	19 824	34 692	47 082	68 392.80
Yen	34 440	60 270	81 795	118 818
Swiss francs	300	525	712.50	1035

12 a £200   b 49 560 rupees
13 a 1:3   b 1:3   c 1:9
14 a 3:5   b 9:25   c 27:125
15 a 11:3   b $\frac{11}{14}$ c $\frac{3}{14}$
16 a 3:1   b 24   c 32
17 a 3:2   b 18 is not a multiple of 5.

11.4 Using ratios 2
1 a 5   b 2.5   c 1.5   d 0.7
2 a 12.47   b 4.97   c 3.158   d 0.012
3 a 5:2   b 2:7   c 3:1   d 3:4
4 a £12, £6   b £6, £36   c £12, £15
   d 14 kg, 21 kg e 25 m, 35 m   f 3 litres, 4.5 litres
5 a 4.875 g   b 1.625 g
6 a 12.5 litres b 7.5 litres
7 5 litres, 10 litres, 20 litres
8 No, Talil only has 15 kg of cement and he needs 20 kg.
9 a £16, £24, £32   b 20 g, 30 g, 50 g
   c 90 ml, 120 ml, 150 ml

d £3.12 or £3.13, £9.38 or £9.37, £12.50. Ensure it adds to £25.

10 a £22.86, £57.14   b 15.556 litres, 54.444 litres

11 a 3000 : 1500   b 2 : 1

    c Bob gets £12 000, Phil gets £6000

12 Andrea gets £190, Penny gets £285

13 6 and 18

14 a Rob £35 and Simon £15   b £20

15 £30

11.5 Comparing using ratios

1 a 6   b 5   c 5   d 4

2 a 6   b $\frac{7}{9}$ c 75   d 8

3 a 4 : 5 : 3   b 11 : 9 : 6   c 6 : 1   d 7 : 10

4 2 : 5

5

Fraction of group that are girls	Ratio of girls : boys
$\frac{5}{9}$	5 : 4
$\frac{3}{5}$	3 : 2
$\frac{7}{10}$	7 : 3
$\frac{4}{11}$	4 : 7

6 7 : 3 : 2

7 a $\frac{3}{4}$ b $\frac{1}{4}$

8 a $\frac{4}{7}$ b $\frac{3}{7}$

9 a $\frac{21}{25}$ b $\frac{4}{25}$

10 No, there are 7 parts so $\frac{2}{7}$ of the people are office staff.

11 a i $\frac{17}{20}$ ii $\frac{1}{20}$

    b Copper has a mass of 85 g. Zinc, tin and lead each have a mass of 5 g.

    c 1200 g

12 a $\frac{7}{20}$

    b Milk 200 g, chocolate 175 g, butter 75 g, syrup 50 g

    c 1 kg using all the syrup

13 a 1.67 : 1   b 1 : 3.5

14 a 0.4 : 1   b 1.75 : 1   c 1.78 : 1   d 0.14 : 1

    e 0.21 : 1   f 0.20 : 1

15 Anna's squash is stronger as it has a squash-to-water ratio of 1 : 17. Jeevan's squash has a ratio of 1 : 18.5.

16 Dexter's paint is darker as it has a red-to-blue ratio of 3.75 : 1. Josh's has a ratio of 4 : 1.

17 Raj makes concrete in the ratio 7.28 : 1 and Sunil makes concrete in the ratio 5.34 : 1. Therefore Sunil's concrete has a higher proportion of cement.

18 Both have the same proportion of cement, 4 : 1.

19 Yes, gold : silver is 2 : 3. 3 is one and a half times 2.

11.6 Using proportion

1 a 120   b 4050   c 13.5   d 1.6

    e 2   f 2   g 5   h 7.2

2 a 1 : 4.5 or 0.22 : 1   b 1 : 0.45 or 2.2 : 1

    c 1 : 2.5 or 0.4 : 1   d 1 : 0.44 or 2.25 : 1

3 a 50   b 200

4 a 1.8p per gram   b 3.25 ml for 1p

5 a 12   b 3   c 9   d 15

6 £450

7 £44.41 or £44.42

8 a £56.79   b £164.06   c £189.30

9 £1347.30

10 a i 0.29545… kg per £1   ii 0.433333… kg per £1

    b 2.6 kg size

11 300 g

12 2 litres

13 Mary

14 Office Deals, Office Deals = £32.40 and Paper World = £33.60

11.7 Proportion and graphs

1 a 40   b 72   c 15

2 a 4   b $y = 4x$

3 Yes

4 a

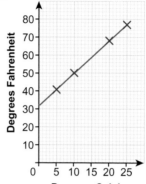

    b No, line does not go through origin.   c 32 °F

5 a

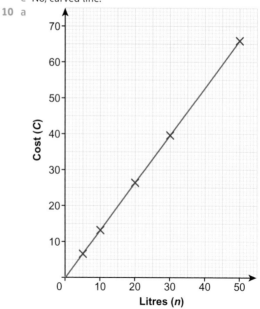

    b Yes, straight line through origin

6 a Yes   b $\frac{4}{1000}$ or 0.004   c £4

7 a Yes   b No   c No   d Yes

8 No

9 a No, not through origin.

    b Yes, straight line through origin.

    c No, curved line and not through origin.

    d Yes, straight line through origin.

    e No, curved line.

10 a

    b $y = 1.32x$ c Yes   d $C = 1.32n$ e £108.24

11 a 1:8   b 8   c $P = 8G$

12 a

Miles	0	5	10
Kilometres	0	8	16

b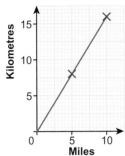

c 1:1.6   d $y = 1.6x$

13 No, when $C = 0$, $K = 273.15$

11.8 Proportion problems

1 a £2.50   b £7.20   c £7.35   d £17.70
2 a 5 hours 15 minutes   b 3 hours 20 minutes
   c 2 hours 45 minutes   d 4 hours 20 minutes
3 a 21 hours   b 4 hours 12 minutes
4 a 2 hours 30 minutes   b 3 hours 20 minutes
5 £112.50
6 7.5 so 8 people
7 4 minutes 30 seconds
8 1 hour 20 minutes
9 $16\frac{2}{3}$ days
10 4 kg
11 15 people
12 18 hours
13 2 hours 30 minutes
14 Students' own answers
15 2.5 days

11 Problem solving

1 40 musicians
2 Sketch of triangle with angles of 20°, 60° and 100°
3 Hilary pays £590, Ruth pays £295.
4 £6
5 £150
6 45 cows
7 3 lengths

11 Check up

1 a 300 g   b 75 g   c 225 g
2 300 ml
3 a 6   b 18
4 a 3:4   b 3:2:6   c 5:7
5 a 4200   b 50   c 246   d 15.625
6 15 m
7 20 tins
8 a Naadim £24, Bal £36
   b $\frac{2}{5}$ c 24 + 36 = 60
9 a 1.6:1 or 1:0.63   b 0.37:1 or 1:2.71
10 a 18   b 30
11 Both the same 1:9.5

12 a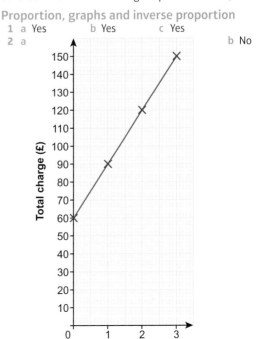

b Yes   c $y = 2x$ or $B = 2U$
13 a 12 days   b 2 days
14 1.5 minutes
16 Students' own answers

11 Strengthen

Simple proportion and best buys
1 a £6   b £15
2 a £64   b £16   c £80
3 21
4 300 g

Ratio and proprtion
1 a 4   b 8
2 a 1:3   b 2:3   c 2:3   d 2:3:5
   e 5:12   f 3:2
3 a 380   b 2500   c 738   d 3
4 15 tins
5 a 480 cm   b 4.8 m
6 3 oz
7 a Isabel £8, Freya £12   b £8 + £12 = £20
8 a 30 cm, 60 cm, 150 cm   b 30 + 60 + 150 = 240
9 a 27   b 45
10 a $\frac{7}{10}$ b $\frac{3}{10}$
11 a 3:1   b 9   c 36
12 a 1:6.5   b 12.5:1   c 2.6:1   d 1:3.14
13 Amar has made the stronger squash. Amar 1:9, Ben 1:10

Proportion, graphs and inverse proportion
1 a Yes   b Yes   c Yes
2 a   b No

3 Yes

4 $k = 2.2p$

5

Number of men	Longer or shorter	Hours
3	–	4
1	Longer	12
2	Longer	6
6	Shorter	2

6 8 hours

11 Extend

1 No, he has enough for 100 m².

2 10 adults

3 Small bottle

4 7.5 cm

5 a 400 ml b 12

6 a 9 hours b £648

7 40°, 60°, 80°

8 a Loam = $8\frac{3}{4}$ litres; peat = $3\frac{3}{4}$ litres
   b 15 litres

9 £200

10 Ollie = £51.30, Sam = £44.55, Peter = £33.75

11 4.15 pm

12 30

13 140°

14 She should import it from the USA.

15 273 kg

16 £358.90

11 Unit test

Sample student answers

Student A has the best answer. They have shown the working clearly, and have written a sentence answering the question.
Student B has shown the correct working, but has not written a sentence answering the question.
Student C has shown the correct working, but has written the wrong conclusion in answer to the question.

UNIT 12

12 Prior knowledge check

1 a 2.0 b 18.5

2 a 12.3 b 0.365

3 a 25 b 121 c 61
   d 121 e 6 f 5

4 a $2\sqrt{2}$, 2.83 b $\sqrt{5}$, 2.24 c $\sqrt{11}$, 3.32 d $\sqrt{58}$, 7.62

5 a 8.9 b 3.9 c 20

6 A, C, E, F

7 a AC = BC
   b Angle CAB = angle ABC
   c BC
   d Isosceles

8 a 32 mm b 55 mm c 41 mm

9 Accurate drawings of 30° and 47° angles.

10 a 25 b 5

11 $\frac{6}{9} = \frac{2}{3}$

12 a 3 b 2.4 c 4

13 a C
   b Longest side < short side + other short side

12.1 Pythagoras' theorem 1

1 a 625 b 25

2 a 8.06 b 18.6

3 a Students' accurate drawings of triangles
   b i 5 cm, 13 cm, 10 cm
     ii 5 cm, 13 cm, 10 cm
   c Both largest and opposite each other

4 a 6.5 cm b 34 m c 29 cm

5 b 9 cm, 16 cm, 25 cm c $5^2 = 3^2 + 4^2$

6 Same, squares

7 a 10.1 cm b 15.7 cm c 149 m
   d 173 m e 16.9 km

8 a 9.35 m b 10.19 m c 23.32 m

9 12.04 m

10 3.52 m

11 Diagonal of letter box is 22.56 cm, so yes it will fit.

12.2 Pythagoras' theorem 2

1 a AC = c b 11.3 m

2 a 12 b 4 c 5 d 6

3 a No b No c Yes
   d Yes e No f Yes

4 a $\sqrt{5}$ b $\sqrt{13}$ c $\sqrt{21}$

5 4.47 cm

6 5

7 a 5 b 13 c 17

8 13.86 cm

9 a 5.8 cm to 1 d.p. b 7.89 m to 2 d.p.

10 50.3 cm

11 a $\sqrt{13}$ cm b $\sqrt{17}$ cm c $\sqrt{51}$ cm

12 b $10^2 = 6^2 + 8^2$
   c $6^2 \neq 4^2 + 5^2$

12.3 Trigonometry: the sine ratio 1

1 a 0.5 b 0.4 c 0.7

2 a 6.7 b 1.239 c 6.5

3 a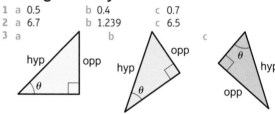

4 a Students' accurate drawings of triangles.
   b
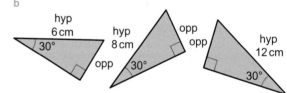

   c 3 cm, 4 cm, 6 cm
   d i 0.5
     ii They are all the same.
     iii $\frac{1}{2}$
     iv Students' own answers

5 a 0.766 b 0.616 c 0.729

6 a $\frac{9}{14}$ b $\frac{7}{10}$ c $\frac{3}{5}$

7 a 6.5 cm b 8.7 cm c 7.5 cm
   d 15.1 cm e 18.2 cm

8 2.78 m to 3 s.f. or to 2 d.p.

9 27.7 m

10 No; rope will only reach 6.9 m.

12.4 Trigonometry: the sine ratio 2

1 $\frac{5}{13}$

2 a 0.342 b 0.707 c 0.536
   d 1 e 0 f 0.125

3 a 40° b 70° c 10°

4 a 65° b 25° c 15°

5 Students' own checks to the answers to **Q3** and **Q4**

6 a 20.2° b 55.5° c 60.1°

7 a 66.4° b 38.7° c 38.8°

8 b 0.6427... sin⁻¹(0.6427...) c 25°... sin⁻¹(0.4226...)

9 a 41.8° b 53.8° c 40.8°

10 7.7°

11 58.2°

12.5 Trigonometry: the cosine ratio

1 a 9.6   b 10   c 7.63   d 40
2 a $\frac{5}{13}$ b $\frac{3}{5}$
3 a Students' accurate drawings of triangles.
   b

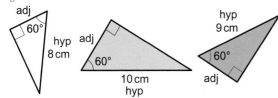

   c 4 cm, 5 cm, 4.5 cm
   d i 0.5
      ii They are the same.
4 a 0.799   b 0.139   c 0.995
   d 1   e 0   f 0.707
5 a $\frac{15}{17}$ b $\frac{7}{25}$ c $\frac{9}{41}$
6 cos (60°) = $\frac{adj}{hyp}$ = $\frac{1}{2}$; sin (30°) = $\frac{opp}{hyp}$ = $\frac{1}{2}$; so cos 60° = sin 30°
7 a 13.6   b 5.6   c 12.0
8 26.5 km
9 a 68.8°   b 55.0°   c 47.7°
10 a 51.3°   b 16.3°   c 51.1°
11 b 0.8829…then cos⁻¹(0.8829…)
   c cos 65° then cos⁻¹(0.42261…) then 65°
12 a 66.9°   b 62.9°   c 47.5°
13 48.2°

12.6 Trigonometry: the tangent ratio

1 a $\frac{35}{12}$ b $\frac{36}{77}$
2 a Students' accurate drawings of triangles.
   b

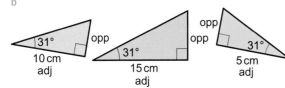

   c 6 cm, 9 cm, 3 cm
   d i 0.6   ii They are the same.
3 a 0.466   b 1.192   c 3.732
   d 0   e 57.290   f 1
4 a $\frac{8}{15}$ b $\frac{24}{7}$ c $\frac{40}{9}$
5 a 11.4   b 35.2   c 29.6
6 4.77 m
7 a is angle of depression; b is angle of elevation.
8 a   b 73 m

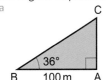

9 71 m
10 a 40.5°   b 59.9°   c 41.2°   d 60.5°
11 b 2.35585… then tan⁻¹(2.35585…)
   c tan⁻¹(1.920…) then 62.5°
12 a 51.3°   b 48.4°   c 21.9°
13 66.7°

12.7 Finding lengths and angles using trigonometry

1 sin θ = $\frac{12}{37}$, cos θ = $\frac{35}{37}$, tan θ = $\frac{12}{35}$
2 a 10.9   b 22.6   c 303
3 a $\sqrt{10}$ b $\sqrt{8}$ = $2\sqrt{2}$ c $\sqrt{29}$
4 a 32.1   b 17.9   c 50.9   d 14.0
5 a 66.0°   b 39.6°   c 54.1°
6 Students' own answers.

7 47.2°
8 10.0°
9 3.62 km
10 a 30°   b 60°   c 30°   d 45°
11 a 4   b 5   c 2
12 Isosceles (a = b = 30°; c = d = 60°)
13 45°

12 Problem-solving

1 18 mm
2 4 cm
3 9.98 cm²
4 12 noon
5 64 cm³
6 a 48.2°, 48.2°, 83.6°   b No
7 35.9 cm

12 Check up

1 a 65
   b 18
2 10.6 cm
3 Yes; 12.5² = 3.5² + 12²
4 a 10.2   b 32.8   c 87.5
5 3.83 m
6 111.96 m
7 a 41.6°   b 40.2°
8 a 1   b $\frac{1}{2}$ c $\frac{\sqrt{3}}{2}$ d 0
10 tan θ = $\frac{5}{5}$ = 1

12 Strengthen

Pythagoras' theorem

1 a 15 cm   b 20 cm   c 19.5 m
2 a Hypotenuse
   b 5 cm × 5 cm = 25 cm²
   c 6 cm × 6 cm = 36 cm²
   d 25 cm² + 36 cm² = 61 cm²
3 136 cm²
4 17 km
5 a 26 m   b 8.5 km
6 7.2 cm
7 12.5 cm
8 a 6.93 cm   b 8.06 cm
9 7.5² + 4² = 72.25; 8.5² = 72.25, so triangle is right-angled.

Finding lengths using trigonometry

1

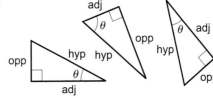

2 a i $\frac{55}{48}$ b i $\frac{55}{73}$ c i $\frac{48}{73}$
      ii $\frac{8}{6}$ = $\frac{4}{3}$ ii $\frac{8}{10}$ = $\frac{4}{5}$ ii $\frac{6}{10}$ = $\frac{3}{5}$
3 a 27.3 cm   b 13.2 cm   c 77.6 m
4 b tan   c 99 m
5 a cos   b cos 57 = $\frac{8.47}{AC}$
   c AC = $\frac{8.47}{\cos 57}$ d 15.55 m

Finding angles using trigonometry

1 a 11.5°   b 44.4°   c 64.6°
   d 78.5°   e 33.7°   f 26.6°
2 a 31.5°   b 34.4°
3 a 8 km labelled hyp; $4\sqrt{3}$ km labelled adj
   b cos   c cos θ = $\frac{adj}{hyp}$ = $\frac{4\sqrt{3}}{8}$
   d θ = cos⁻¹$\left(\frac{4\sqrt{3}}{8}\right)$ e 30°

4

Angle	0°	30°	45°	60°	90°
sin	0	$\frac{1}{2}$	$\frac{1}{\sqrt{2}}$	$\frac{\sqrt{3}}{2}$	1
cos	1	$\frac{\sqrt{3}}{2}$	$\frac{1}{\sqrt{2}}$	$\frac{1}{2}$	0
tan	0	$\frac{1}{\sqrt{3}}$	1	$\sqrt{3}$	

12 Extend

1 a 6.32 m b Wall is vertical, ground is horizontal.
2 130 km
3 7.5 m^2
4 11.2 cm
5 a $\sqrt{2}$ b 2
6 a 5 b 3 c 12
d $7\sqrt{2}$ e $6\sqrt{3}$ f $3\sqrt{5}$
7 Yes; $6^2 = (3\sqrt{2})^2 + (3\sqrt{2})^2$
8 a $AD^2 + BD^2 = (4\sqrt{5})^2 + (2\sqrt{5})^2 = 100$
$AB^2 = 100$
So the triangle is right angled.
b $\sqrt{5}$ c $5\sqrt{5}$
d $AB^2 + BC^2 = 10^2 + 5^2 = 125$
$AC^2 = (5\sqrt{5})^2 = 125$
So the triangle is right angled.
9 $x = 5$, $y = 5\sqrt{3}$ cm
10 $x = 8$, $y = 8\sqrt{2}$
11 a Increases b Decreases
12 6.6 m
13 a $37^2 = 35^2 + 12^2$ b 71.1°
14 a 28.1° b 920 cm^2
15 43.5 cm^2
16 60°
17 67.98°; it is safe as below 75°

12 Unit test

Sample student answers

a Student A
b The diagram has helped because the student can match the correct sides to the lengths given, and therefore put the correct numbers into the formula. Labelling the sides a, b and c reminds the student that it is the LONG side which is by itself in the formula. To improve the answer further, Student A should have written $a = \sqrt{33.75} = 5.809... = 5.8$ m after $a^2 = 33.75$ to ensure all method marks in case they made a careless error with rounding.

UNIT 13

13 Prior knowledge check

1 a $\frac{1}{2}$ b $\frac{3}{4}$ c $\frac{16}{25}$ d $\frac{3}{5}$
2 a $\frac{3}{4}$ b 0.23 c 80%
d 0.68 e $\frac{13}{25}$ f 31%
3 $\frac{16}{25}$
4 $\frac{1}{3}, \frac{7}{20}, \frac{2}{5}, \frac{3}{7}$
5 a 54 b 72% c 42 d 60%
6 a 4.5 b $\frac{1}{2}$ c 0.08 d 45
7 a Even chance b Impossible c Likely d Unlikely
8 a Marked at half way b Marked at left end
c Marked close to right end d Marked close to left end
9 a $\frac{1}{2}$ b $\frac{1}{4}$ c 0
10 Two red sectors, one yellow sector, the other two could be any other colour.

13.1 Calculating probability

1 a 1 b $\frac{1}{2}$ c $\frac{3}{4}$ d 0.3
2 a Yes – both are 0.4 b Red
3 a $\frac{3}{7}$ b $\frac{5}{7}$ c 0 d Green
4 a $\frac{1}{11}$ b $\frac{2}{11}$ c $\frac{2}{11}$
5 a $\frac{5}{12}$ b $\frac{3}{12}$ c $\frac{9}{12}$
6 $\frac{3}{7}$
7 a $\frac{17}{30}$ b $\frac{23}{30}$ c 0
8 a $\frac{2}{18}$ b 2
9 a Yes b No c No
10 a No b No c Yes
d i $\frac{1}{2}$ ii $\frac{1}{2}$ iii 1
11 $\frac{1}{2}$
12 a 0.3 b i 80% ii $\frac{8}{10}$
c No d 50
13 a 100 b $\frac{1}{100}$ c $\frac{1}{20}$

13.2 Two events

1 $\frac{1}{2}$
2 a The numbers from 1 to 6; 6 outcomes
b Yes
c i $\frac{1}{3}$ ii $\frac{1}{2}$ iii $\frac{1}{2}$
3 a Any suitable two-way table.
b 4 c $\frac{1}{4}$ d $\frac{1}{2}$ e $\frac{3}{4}$
4 a

	Red	Green	Blue	Yellow
Red	R, R	R, G	R, B	R, Y
Green	G, R	G, G	G, B	G, Y
Blue	B, R	B, G	B, B	B, Y

b $\frac{1}{12}$ c $\frac{1}{2}$ d 0
e Mischa is wrong because you have to look at the outcomes not the individual letters. There are 12 possible outcomes and 6 of them contain at least one blue so the probability of getting at least one blue is $\frac{6}{24}$.
5 a

	A	B	C
A	A, A	A, B	A, C
B	B, A	B, B	B, C
C	C, A	C, B	C, C

b $\frac{3}{9}$ or $\frac{1}{3}$ c 10
6 a (1, head) (1 , tail) (2, head) (2, tail) (3, head) (3, tail) (4, head) (4, tail) (5, head) (5, tail)
b $\frac{1}{10}$
7 a Any suitable sample space diagram. b 9
c i $\frac{1}{9}$ ii $\frac{1}{9}$ iii $\frac{5}{9}$
8 a

Dice 2						
11	13	15	17	19	21	23
9	11	13	15	17	19	21
7	9	11	13	15	17	19
5	7	9	11	13	15	17
3	5	7	9	11	13	15
1	3	5	7	9	11	13
	2	4	6	8	10	12

Dice 1

b i $\frac{3}{36}$ ii $\frac{21}{36}$ iii 0 iv $\frac{12}{36}$
c Multiple of 3 more likely: P(multiple of 3) = $\frac{12}{36}$; P(7) = $\frac{3}{36}$.
9 $\frac{3}{8}$
10 No, losing is more likely. There are 12 possible outcomes but only 4 are odd (1, 3, 3, 9) so the probability of losing is $\frac{8}{12}$
11 $\frac{5}{9}$

13.3 Experimental probability

1 a 60 b 26 c 15
2 a 20 b 70 c $\frac{8}{25}$

3 a

Outcome	1	2	3	4	5	6
Frequency	2	4	5	3	2	4

b Students' own answers

4 $\frac{25}{60}$

5 a 18   b $\frac{7}{18}$

6 a 100   b $\frac{72}{100}$ or 0.72   c 160

d $\frac{120}{160}$ or 0.75   e $\frac{192}{260}$ or 0.74 (2 d.p.)

f 0.74 is the best estimate as it uses the most trials.

7 She could get a more accurate measure using more trials.

8 a $\frac{18}{60}$ b $\frac{1}{4}$

9 a $\frac{42}{120}$ b $\frac{1}{6}$ c 20

d Students' own answers, e.g. The dice are biased as $\frac{42}{120}$ is more than twice as high as one would predict using the theoretical probability.

10 a £30   b $\frac{49}{160}$

11 a $\frac{30}{100}$ b 175

12 a i $\frac{175}{328}$ or 0.53 (2 d.p.)   ii $\frac{65}{328}$ or 0.20 (2 d.p.)

b $\frac{116}{175}$ or 0.66 (2 d.p.)

13 a i $\frac{5}{70}$ or $\frac{1}{14}$ ii $\frac{5}{70}$ or $\frac{1}{14}$

b Silver van   c Bus   d Car   e White

13.4 Venn diagrams

1 a i $\frac{4}{20}$ ii $\frac{8}{20}$ iii $\frac{10}{20}$ iv $\frac{4}{20}$

b 0.25

2 a A = {1, 3, 5, 7, 9}   B = {1, 4, 9}

b i False   ii True   iii False

c 1, 9   d 3, 5, 7

e 1, 3, 4, 5, 7, 9   f 2, 6, 8

3 a X = {2, 4, 6, 8, 10, 12}

Y = {3, 6, 9, 12}

ξ = {1, 2, 3, 4, 5, 6, 7, 8, 9, 10, 11, 12}

b ξ = {integers 1 to 12}

X = {multiples of 2 up to 12}

Y = {multiples of 3 up to 12}

4

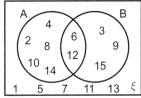

5 a X ∩ Y = {6, 12}

b X ∪ Y = {2, 3, 4, 6, 8, 9, 10, 12}

c X' = {1, 3, 5, 7, 9, 11}

d Y' = { 1, 2, 4, 5, 7, 8, 10, 11}

e X' ∩ Y = {3, 9}

6 a A ∩ B = {6, 12}

b A ∪ B = {2, 3, 4, 6, 8, 9, 10, 12, 14, 15}

c A' = {1, 3, 5, 7, 9, 11, 13, 15}

d B' = {1, 2, 4, 5, 7, 8, 10, 11, 13, 14}

e A' ∩ B = {3, 9, 15}

f A ∩ B' = {2, 4, 8, 10, 14}

7 a 12   b 24   c $\frac{8}{25}$ d $\frac{4}{25}$

e She has missed out the people who have mobile phones and tablets. 20 people have mobile phones.

8 a Venn diagram with 2 in rock, 3 in pop, 17 in intersect and 8 outside.

b $\frac{3}{30}$ c $\frac{19}{30}$

9 a $\frac{10}{32}$ b P(X' ∪ Y') = $\frac{3}{32}$

13.5 Tree diagrams

1 RR, RA, RG, AR, AA, AG, GR, GA, GG

2 a $\frac{7}{10}$ b $\frac{3}{10}$

3 a $\frac{2}{6}$ b $\frac{4}{6}$ c 1

4 a $\frac{5}{7}$ b 6   c 4   d $\frac{4}{6}$

5 A, B, C

6 a

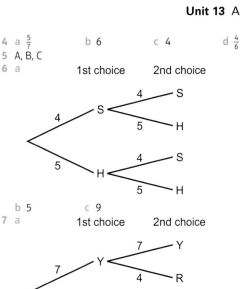

b 5   c 9

7 a

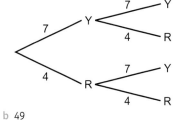

b 49

8

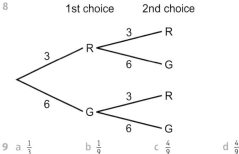

9 a $\frac{1}{3}$ b $\frac{1}{9}$ c $\frac{4}{9}$ d $\frac{4}{9}$

10 a $\frac{3}{4}$

b

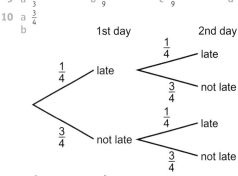

c $\frac{9}{16}$ d $\frac{3}{16}$

11 a

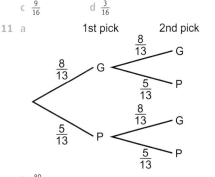

b $\frac{80}{169}$

12 a

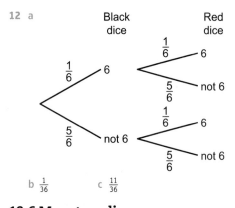

Black dice   Red dice

b $\frac{1}{36}$ c $\frac{11}{36}$

13.6 More tree diagrams

1 a

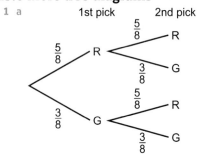

1st pick   2nd pick

b $\frac{9}{64}$

2 a $\frac{4}{10}$ b 9   c 3

   d i $\frac{3}{9}$ ii $\frac{6}{9}$

3 a Dependent   b Dependent

   c Independent   d Independent

   e Students' own answers

4

1st choice   2nd choice

5 a

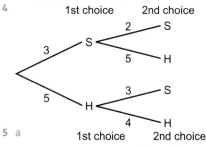

1st choice   2nd choice

b 42

6 a $\frac{42}{90}$ b $\frac{42}{90}$

7 a

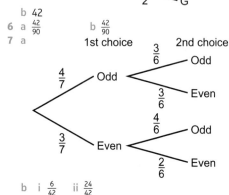

1st choice   2nd choice

b i $\frac{6}{42}$ ii $\frac{24}{42}$

8 a

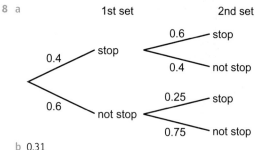

1st set   2nd set

b 0.31

9 a

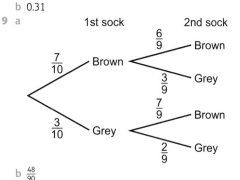

1st sock   2nd sock

b $\frac{48}{90}$

13 Problem-solving: Ciphers, language and probability

1 a i 0.04   ii 0.39

   b J, Q, X and Z

2 a $\frac{9}{135}$ = 0.07 (2 d.p.); q has replaced I

   b

a	b	c	d	e	f	g	h	i	j	k	l	m	n	o	p	q	r	s	t	u	v	w	x	y	z
M	A	T	H	S	C	O	D	E	F	B	G	–	–	L	N	I	–	–	R	–	Y	K	W	V	U

   I WILL MEET THE DOCTOR NEAR THE ROAD BRIDGE. HE HAS CASH AND I HAVE THE DOCUMENTS. IF YOU SEE ANYONE ELSE, RUN! DON'T TAKE A TRAIN OR BOAT. YOUR REAL MISSION: TELL THEM NOTHING.

3 A longer cipher would provide a bigger sample

4 Letters are not independent, for instance q is almost always followed by u.

13 Check up

1 $\frac{3}{4}$

2 a $\frac{3}{10}$ b $\frac{3}{10}$ c $\frac{4}{10}$

3 0.5

4 a $\frac{25}{110}$ b $\frac{15}{70}$ c Pierre's: more trials

5 a $\frac{80}{200}$ b $\frac{1}{2}$ c 100

   d Possibly; 80 is fewer than the 100 heads predicted.

6 a Any suitable sample space diagram.

   b $\frac{1}{4}$ c $\frac{1}{2}$

7 a Any suitable sample space diagram.

   b 16

   c i $\frac{3}{16}$ ii $\frac{11}{16}$

8 a 30   b 20   c $\frac{13}{30}$

   d i $\frac{26}{30}$ ii $\frac{7}{30}$

9 a

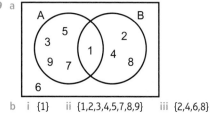

b i {1}   ii {1,2,3,4,5,7,8,9}   iii {2,4,6,8}

10 a

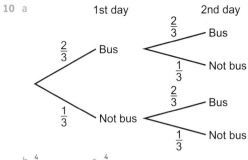

1st day   2nd day

b $\frac{4}{9}$ c $\frac{4}{9}$

11 a

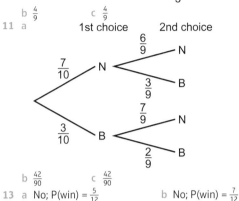

1st choice   2nd choice

b $\frac{42}{90}$ c $\frac{42}{90}$

13 a No; P(win) = $\frac{5}{12}$ b No; P(win) = $\frac{7}{12}$

13 Strengthen

Calculating probabilities

1 a 11
   b i $\frac{2}{11}$ ii $\frac{1}{11}$ iii $\frac{4}{11}$ iv $\frac{9}{11}$
2 a $\frac{7}{13}$ b $\frac{12}{13}$ c $\frac{8}{13}$
3 a $\frac{3}{5}$ b $\frac{2}{5}$ c $\frac{2}{5}$ d $\frac{3}{5}$
4 $\frac{5}{6}$
5 0.3

Experimental probability

1 a $\frac{1}{6}$
   b i $\frac{12}{60}$ or $\frac{1}{5}$ ii $\frac{30}{90}$ or $\frac{1}{3}$
   c Freya's dice is biased as $\frac{1}{3}$ is very different from the theoretical probability of $\frac{1}{6}$.
2 a 14
   b i $\frac{14}{120}$ ii $\frac{6}{120}$ iii $\frac{25}{120}$
3 a The experimental probability is $\frac{7}{10}$, which is a lot higher than the theoretical probability of $\frac{1}{2}$.
   b The experimental probability is $\frac{102}{200}$, which is close to the theoretical probability of $\frac{1}{2}$.
   c His second estimate is more accurate as more trials give a more accurate result.

Probability diagrams

1

	2	4	6
3	(2, 3)	(4, 3)	(6, 3)
6	(2, 6)	(4, 6)	(6, 6)
9	(2, 9)	(4, 9)	(6, 9)

b 9
c i $\frac{1}{9}$
   ii $\frac{4}{9}$
   iii $\frac{3}{9}$
d $\frac{6}{9}$

2 a

	1	2	3	4	5	6
1	1	2	3	4	5	6
2	2	4	6	8	10	12
3	3	6	9	12	15	18
4	4	8	12	16	20	24
5	5	10	15	20	25	30
6	6	12	18	24	30	36

b 36
c i $\frac{1}{36}$ ii $\frac{6}{36}$ iii $\frac{8}{36}$

3 a

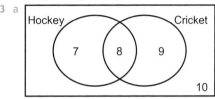

b 34
c i $\frac{15}{34}$ ii $\frac{8}{34}$ iii $\frac{7}{34}$ iv $\frac{10}{34}$
4 a $\frac{22}{25}$ b $\frac{6}{25}$ c $\frac{15}{25}$
5 a 1, 4, 5, 6, 8, 9
   b 1, 2, 3, 4, 10
   c 1, 2, 3, 4, 5, 6, 8, 9, 10
   d 1, 4
   e 1, 2, 3, 4, 5, 6, 7, 8, 9, 10
6 a 0.16   b i 0.24   ii 0.48
7 a $\frac{2}{5}$
   b

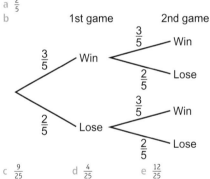

1st game   2nd game

c $\frac{9}{25}$ d $\frac{4}{25}$ e $\frac{12}{25}$

Dependent events

1 a After she has eaten one jelly, there are 15 sweets remaining, of which 7 − 1 = 6 are jellies. Therefore the new probability of picking a jelly is $\frac{6}{15}$.
   b After she has eaten one boiled, there are 15 sweets remaining, of which 9 − 1 = 8 are boiled. Therefore the new probability of picking a boiled is $\frac{8}{15}$.
   c

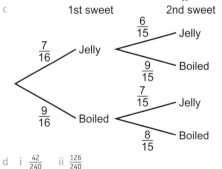

1st sweet   2nd sweet

d i $\frac{42}{240}$ ii $\frac{126}{240}$

13 Extend

1 0.74 = 74%
2 a $\frac{8}{15}$ b 9

3 a $\frac{1}{6}$

b

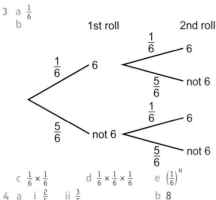

1st roll   2nd roll

c $\frac{1}{6} \times \frac{1}{6}$ d $\frac{1}{6} \times \frac{1}{6} \times \frac{1}{6}$ e $\left(\frac{1}{6}\right)^n$

4 a i $\frac{2}{5}$ ii $\frac{3}{5}$ b 8

5 a P(green) = 0.15 = $\frac{3}{20}$. You could not get this with 10 counters.

b 20

6 a

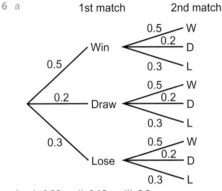

1st match   2nd match

b i 0.25   ii 0.15   iii 0.2

7 a 30   b 19   c 4

d i $\frac{19}{30}$ ii $\frac{3}{30}$ iii $\frac{2}{30}$ iv $\frac{1}{30}$ v $\frac{4}{30}$

8 Greater than 7 and less than 7 are equally likely. 15 outcomes are less than 7, 15 are greater than 7 and 6 are equal to 7.

9 58%

10 a

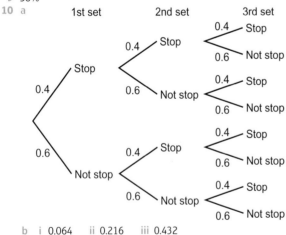

1st set   2nd set   3rd set

b i 0.064   ii 0.216   iii 0.432

c 0.288

11 a

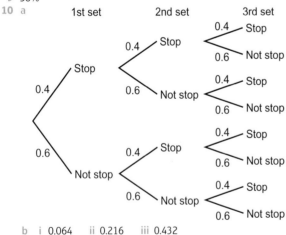

b $\frac{2}{10}$ c $\frac{3}{20}$ d $\frac{7}{20}$

12 a 3x   b $\frac{3}{4}$

13 a

First game   Second game

b p^2 c $2p(1-p)$

13 Unit test

Sample student answer

a He added the 0.4 as a hundredth instead of a tenth.

b Set out the numbers in a column with the decimal points lined up to ensure they have the correct place value.

UNIT 14

14 Prior knowledge check

1 a 0.2   b 0.175   c 0.0825   d 1.45

2 a 15   b 22

3 a 2:5   b 2:7   c 2:7   d 2:9

4 a 1:4 or 0.25:1   b 1:1.5 or 0.67:1

    c 1:0.42 or 2.4:1   d 1:4.25 or 0.24:1

5 a Length   b Area   c Mass

    d Length   e Volume   f Volume

6 a 5.4 kg   b 6.235 m   c 0.5 litres

    d 12 minutes 30 seconds   e 5 hours 40 minutes

7 a 2.25 hours   b 3.5 hours   c 1.75 hours

8 a 1:6   b 9:100   c 5:3   d 24:1

9 a 70 g   b 68 mph   c 43 km/h

10 64 cm^2

11 36 cm^3

12 a 2   b 9   c −1   d −12

13 a $x = 6$ b $x = 20$ c $x = 2$

14 a $x = \frac{y}{4}$ b $x = 2y$ c $x = \frac{3}{y}$

15 a 3   b $y = 3x + 1$

16 C

17 The higher the rate of euros per pound, the steeper the gradient. Mar, Jan, Feb/Apr, May, Jun, Aug, Jul

14.1 Percentages

1 a 40%   b 25%   c 20%

2 a £420   b £330   c £756

3 a £42.50   b £72.25   c £52.70

4 a 115%   b 75%   c 103.2%   d 98.25%

5 £520

6 £20

7 £2 450 000

8 £500

9 a £390   b Paul saves £7.50 more than Matt.

10 £635.58

11 Actual change = £3334.40 − £3200 = £134.40

    Percentage change = $\frac{134.4}{3200} \times 100 = 4.2\%$

12 3.2%

13 6.8%

14

Item	Cost price	Selling price	Actual profit	Percentage profit
ring	£5	£8	**£3**	**60%**
bracelet	£12	£18	**£6**	**50%**
necklace	£20	£30	**£10**	**50%**
watch	£18	£25	**£7**	**38.9%**

15 7.3%
16 9.04%
17 35.4%
18 a 21.6%
   b Percentage decrease
   c 8850 children
   d 2.6% decrease
19 Students' own answers

14.2 Growth and decay

1 a 60%   b 1.5%   c 130%
2 a 1.4   b 0.88   c 1.064   d 0.985
3 a 0.7   b £4200   c 0.9
   d £3780   e 0.63
4 a 1.05575   b £26393.75
5 No; 1.12 × 1.2 = 1.344; this is an increase of 34.4%
6 2% then 1.5% is the better offer: 1.02 × 1.015 = 1.0353; this is a 3.53% increase after two years.
7 £204
8 £3276.08

9

Year	Amount at start of year	Total amount at end of year
1	£500	£520
2	£520	£540.80
3	£540.80	**£562.43**
4	**£562.43**	**£584.93**
5	**£584.93**	**£608.33**

10 a 1.520875   b 0.4096   c 1.071
   d 0.66   e 1.026
11 512 000 bacteria
12 838.2 counts per second
13 3750 counts per minute
14 253 or 254 outlets
15 a e.g. The reduction in value will not be 100%.
   b $0.8^2 = 0.64$

14.3 Compound measures

1 a 5   b −24   c −0.125
2 a £24   b 12.5 g
3 24 m³
4 39 cm²
5 150 cm³
6 a 0.35 cm³   b 0.054 m³
7 a $x = \dfrac{y}{3}$ b $x = 4y$ c $x = \dfrac{c}{y}$
8 £352.50
9 a i 1 litre
     ii 2.5 litres
   b 40 hours
10 a 36 000 cm³
   b 7.2 minutes or 7 minutes 12 seconds
11 a 15 km/litre   b 4.3 litres
12 8.3 g/cm³

13 2.4 g/cm³
14 686 cm³
15 a Gold 19.32 g/cm³, platinum 21.45 g/cm³
   b Platinum is more dense as it has a higher density.
16 a 5115.5 g   b 5.1155 kg
17 a 120 cm³   b 2.316 kg
18 a 17.3 N/m²   b 90 N

19

Force (N)	Area (m²)	Pressure (N/m²)
60	2.5	**24**
72	4.8	15
100	**8.33…**	12

14.4 Distance, speed and time

1 a 1.8 m   b 4700 m   c 0.28 km   d 54.6 km
2 a $d = 3$ b $s = 5$ c $t = 3$
3 a 0.5 hours
   b 0.333…. hours
   c 1.25 hours
4 a 12 minutes
   b 1 hour 45 minutes
   c 3 hours 24 minutes

5

Distance (km)	Speed (km/h)	Time
280	**80**	3 hours 30 minutes
132	48	2 hours 45 minutes
350	60	**5 hours 50 minutes**

6 a 30 mph   b 12 km/h   c 95.2 mph
7 a 2000 miles   b 17.5 km   c 2.8 km
8 a 40 seconds
   b 45.5 seconds
   c 200 seconds
9 a 3600 m/h   b 43200 m/h   c 28800 m/h
10 44.4 km/h
11 1.4 m/s
12 a 120 miles
   b Tuesday (drives 120 miles on Monday and $200 \times \frac{5}{8} = 125$ miles on Tuesday)
13 a 18 km/h   b 64.8 km/h   c 108 km/h
14 a 15 m/s   b 20 m/s   c 2.5 m/s
15 900 km/h
16 The peregrine falcon is faster. 108 m/s = 388.8 km/h, which is faster than 350 km/h.
17 a 11   b 9
18 a 2   b 1
19 a 3   b 0.76

20

s (m)	u (m/s)	a (m/s²)	t (s)
10	**9**	2	1
8	2	**12**	1
15	**4.5**	3	2
12	4	**2**	2

21

v (m/s)	**10**	6	9	5	7
u (m/s)	8	**5.10**	4	**1**	3
a (m/s²)	2	1	**4.64**	3	**3.33**
s (s)	9	5	7	4	6

22 a 45 m/s   b Students' own answers

14.5 Direct and inverse proportion

1 a 1:4   b 3:1   c 5.5:1   d $\frac{3}{5}$:1
2 a

kg	0	5	10
pounds	**0**	**11**	**22**

b

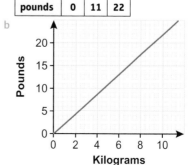

c Yes; straight-line graph through the origin
d $y = 2.2x$

3 a 4:1   10:2.5   16:4   20:5
  b They are all 4:1.   c $F = 4P$
4 a $x = 3y$ b $t = 6s$ c $q = 3.5p$
5 a 2.5:1 $F = 2.5m$
  b 1:1.25 $y = 1.25x$
  c 1:1.75 $s = 1.75r$
6 a 8:10   16:20   24:30   32:40   40:50
    All are equivalent to 1:1.25 or 0.8:1
  b d is in direct proportion to t as the ratio $d:t$ is the same for all values.
  c $\frac{d}{t} = 0.8$ or $d = 0.8t$ or $t = 1.25d$
  d 20 miles
7 a i $F \propto m$ b i $F = km$
    ii $V \propto x$ ii $V = kx$
8 a $x \propto t; x = kt$ b $k = 1.5$ c 15
9 a $r = 1.25s$ b 8.75
10 a $P = 1.25E$ b £100

11
A	10	20	15	**2.5**	**12.5**
B	15	**7.5**	**10**	60	12

12 a Direct   b Inverse   c Neither
   d Inverse   e Direct
13 a i $y \propto \frac{1}{x}$ ii $P \propto \frac{1}{A}$
   b i $y = \frac{k}{x}$ ii $P = \frac{k}{A}$
14 a $y \propto \frac{1}{x}$, $y = \frac{k}{x}$
   b $k = 60$ c 4
15 $s = \frac{k}{t}$; $20 = \frac{k}{0.4}$, so $k = 20 \times 0.4 = 8$; when $t = 0.5$,
   $s = 8 \div 0.5 = 16$.
16 15 amperes

14 Problem-solving

The graph for Q1–4 will look like this by the end of the lesson.

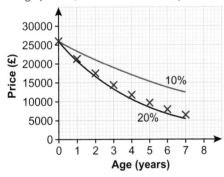

1 See graph
2 The price drops by roughly £4500 in the first year but then the drop gets smaller each year. In Year 7 the price drop is only £1450. Nick's theory is not true.

3 a
Age (years)	Price (£)
0 (new)	26 065
1	23 459
2	21 113
3	19 001
4	17 101
5	15 391
6	13 852
7	12 467

  b See graph
  c Yes

4 An annual price decrease of 18% is the best fit to the values collected.

14 Check up

1 82.5%
2 16.7%
3 £2678.06
4 5643 bees
5 £405.45
6 50 kg
7 2.8 N/cm²
8 a 47 km/h   b 3 hours 7.5 minutes
9 6.25 m/s²
10 Usain Bolt is faster: 13.4 m/s = 48.24 km/h, which is faster than 40 km/h.

11
Pounds (£)	Euros (€)
150	189
400	504
320	**403.20**
280	352.80

  a Yes; the ratio of $\frac{\text{pounds}}{\text{euros}}$ is constant so the number of pounds and the number of euros increase at the same rate.
  b $e = 1.26p$
12 a $k = 4$ b $f = 6$
13 a $y = \frac{45}{x}$ b $y = 3$
15 a
Year	Amount at start of year	Multiplier in index form	Total amount at end of year
1	£650	1.034	**£672.10**
2	**£672.10**	1.034^2	**£694.95**
3	**£694.95**	1.034^3	**£718.58**
4	**£718.58**	1.034^4	**£743.01**
5	**£743.01**	1.034^5	**£768.27**

  b 1.034^{10}
  c 1.034^n
  d Total amount at the end of year $n = P \times \left(1 + \frac{r}{100}\right)^n$

14 Strengthen

Percentages

1 a 25%   b 30%   c 85.7%
2 a 25%   b 20%   c 30%
3 500%
4 £50
5 £300 000
6 a £60   b £93.33   c £160
7 a 400 g   b 300 km   c 400 litres
8
Year	Amount at start of year	Interest at end of year	Total at end of year
1	£700	$0.038 \times 700 = $ **26.6**	**£726.60**
2	**£726.60**	$0.038 \times 726.60 = 27.6108$	**£754.21**

9 a £815.89   b £65.89
10 a Increase = 1.15, decrease = 0.85
   b Increase = 1.08, decrease = 0.92
   c Increase = 1.026, decrease = 0.974
   d Increase = 1.21, decrease = 0.79
   e Increase = 1.07, decrease = 0.93
   f Increase = 1.045, decrease = 0.955
11 $15 \times 1.2^{11} = 111$ (answer must be a whole number)
12 13.48 million square metres

Compound measures

1 £293.70

2

Metal	Mass (g)	Volume (cm³)	Density (g/cm³)
aluminium	**27**	10	2.70
copper	448	**50**	8.96
zinc	427.8	60	**7.13**

3

Force (N)	Area (cm²)	Pressure (N/cm²)
220	20	11
60	15	**4**
45	**5**	9

Distance, speed and time

1 a Higher   b Lower   c Lower

2

Distance (miles)	Time (hours)	Speed (mph)
180	4	45
145	2.5	58
150	3	**50**
120	1.25	**96**
45	**1.5**	30
154	**2.75**	56

3

km/h	m/h	m/min	m/s
9	**9 000**	**150**	**2.5**
18	**18 000**	**300**	5
12	**12 000**	**200**	**3.3**
28.8	**28 800**	480	8

4 a $s = ut + \frac{1}{2}at^2$
   b $v^2 = u^2 + 2as$
   c $v = u + at$

5

s (cm)	t (s)	u (cm/s)	v (cm/s)	a (cm/s²)
30	3	5		**3.3**
12		**4**	8	2
	7	4	**25**	3
15	2	**6.5**		1
8		3	**6.4**	2
9		2	8	**3.3**
	1	**6**	10	4
	2	2	12	**5**

Direct and inverse proportion

1 $W = 14$; $X = 24$, $Y = 28$, $Z = 20$
2 $W = 6$; $X = 4$, $Y = 3$, $Z = 4$
3 y is proportional to x; $y = kx$; $y \propto x$
   y is inversely proportional to x; $y = \dfrac{k}{x}$; $y \propto \dfrac{1}{x}$
4 a $y \propto x$ b $y = kx$ c $k = 3$
   d $y = 3x$ e 12

14 Extend

1 53.3 km/h
2 12.2 m/s
3 International Bank amount after 2 years = £2000 × 1.04 × 1.01 = £2100.80
Friendly Bank amount after 2 years = £2000 × 1.05 × 1.005 = £2110.50
She should invest in The Friendly Bank.

4 £7680
5 2.7 g/cm³
6 745.2 N
7 The car is travelling below the speed limit:
15 m/s = 54 km/h or 64 km/h = 17.8 m/s
8 a 13.9 m/s   b 50.1 km/h
9 a $m = 4.2r$ b $m = 50.4$ c $r = 5$
10 a $E = 0.048m$ b 28.8 mm   c 750 g
11 a 6   b 8
12 a $V = \dfrac{3000}{P}$ b 10 m³   c 400 N/m³
13 a i £440   ii £281.60
   b £250

14 Unit test

Sample student answer

Student B has the correct answer.
Student A has worked out 12% of the value of the house after the increase. This is not the same as 12% of the original value and is not the correct method to calculate the original value.
Original value × 1.12 = £168 000

UNIT 15

15 Prior knowledge check

1 a 1:2   b 1:8
2 a 2:1   b 2:5
3 a Compass points correctly drawn
   b 135°
4 a 6 cm   b 5.2 cm
5 Accurate drawing of an 8.7 cm line
6 a Accurate drawing of an angle of 54°
   b Accurate drawing of an angle of 163°
7 Sketch of net of 4 cm × 5 cm × 6 cm cuboid
8 a $\frac{1}{4}$ turn clockwise
   b $\frac{3}{4}$ turn anticlockwise
9 Shapes A, C and E.
10 a Accurate drawing of a circle with diameter 6 cm
   b Accurate drawing of a circle with radius 4.2 cm
11 100° and 80°; 120° and 60°

15.1 3D solids

1 a Accurate drawing of a square with side length 3 cm
   b Accurate drawing of a rectangle with length 6.5 cm and width 4 cm
2 a 6   b 12
3 a 6   b 12   c 8
4 a Rectangle
   b 10 cm × 4 cm, 10 cm × 3 cm, 3 cm × 4 cm
5 a Cube   b Cuboid   c Cylinder
   d Sphere   e Cone
6 a Equilateral triangle   b Rectangle
   c

	Number of faces	Number of edges	Number of vertices
Cube	6	12	8
Tetrahedron	4	**6**	**4**
Square-based pyramid	**5**	**8**	**5**
Triangular prism	**5**	**9**	**6**
Hexagonal prism	**8**	**18**	**12**

   d Number of faces + number of vertices − number of edges = 2
7 No, she is not correct. There are 14 faces and 36 edges.

15.2 Plans and elevations

1 2D are triangle, rectangle, square, pentagon circle, 3D are cube, sphere, cuboid, tetrahedron

2 2, 0, 1, 6

3 a

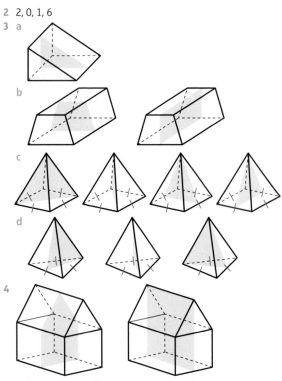

b

c

d

4

5 Infinite number of ways, assuming top loop is same size as bottom loop, or 2 ways assuming top loop has smaller size.

6 a

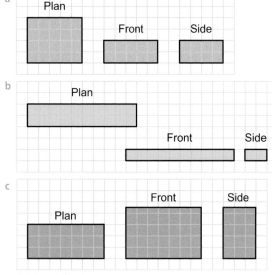

Plan

Front   Side

b

Plan

Front   Side

c

Front   Side

Plan

7 a Cube or cuboid
  b Cylinder or sphere
  c Triangular prism
  d Cone

8

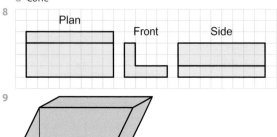

Plan

Front   Side

9

←front

10

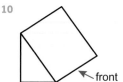

←front

11 a Accurate drawing of rectangle 6.2 cm × 3 cm
  b Accurate drawing of rectangle 6.2 cm × 4.5 cm

12

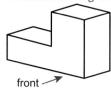

front →

13 Sketch of rectangle labelled 3 m by 2 m

15.3 Accurate drawings 1

1 a Accurate drawing of a 7.5 cm line
  b Accurate drawing of a 47 mm line
2 a Accurate drawing of a circle with radius 6 cm
  b Accurate drawing of half a circle with radius 4.5 cm
3 a Accurate drawing of an angle of 75°
  b Accurate drawing of an angle of 48°
  c Accurate drawing of an angle of 112°
4 Accurate drawing of triangle PQR, with PQ = 6.8 cm, angle RPQ = 75°, angle PQR = 48°
5 Accurate drawing of triangle XYZ, with XY = 5.6 cm, angle XYZ = 63°, YZ = 7 cm
6 a Accurate drawing of triangle labelled ABC, with BC = 7.3 cm, angle C = 38°, AC = 8 cm
  b 63°
7 a ASA triangles A and D, SAS triangles B, C and E
  b Accurate drawings of all triangles
  c A and D, C and E
8 Accurate drawing of triangle TUV, with angles 54°, 44°, 81°
9 Accurate drawing of triangle ABC, with angles 93°, 39°, 49°
10 Accurate drawing of an equilateral triangle with side length 5 cm
11 a Accurate drawing of triangle VWX, with side lengths 5 cm, 8.5 cm and 6.9 cm
  b VW = 6.9 cm
12 Both statements always true

15.4 Scale drawings and maps

1 a 1:2   **b** 1:25   **c** 1:0.2
2 a 12.5 mm   **b** 125 m
3 a 60   **b** 4
4 a **i** 2 cm   **b** **i** 10 m
    ii 4.5 cm   **ii** 13.6 m
5 Accurate scale diagram, with 3 m, 2 m, 8.5 m, 3.5 m represented by 6 cm, 4 cm, 17 cm, 7 cm
6 a 6 m   **b** 16 cm
7 a 0.5   **b** 1:50
  c **i** 4.5 m × 3.5 m
    ii 0.75 m × 1.5 m
    iii 15.75 m²
  d **i** 0.8 cm × 1 cm
    ii 0.2 cm × 3 cm
8 a 5 : 1.25   **b** 1:25
9 a **i** 135 km **ii** 248 km **iii** 192 km
  b Sligo
10 a 10 cm × 35 cm   **b** 83 cm (= 830 mm)
11 a The dimensions of the rooms are incorrectly shown. Living room drawn 2.5 m wide, hall 0.75 m x 2.5 m, kitchen 1.75 m × 2.5 m, bathroom drawn 2 m wide and the doors are drawn 0.75 m wide.
  b Diagram drawn correctly to scale.
12 B
13 a 2500 m   **b** 3.5 km   **c** 25 cm

14 a 48 cm   b 8 km
15 a Map drawn to scale: AB = 7.5 cm, AC = 12 cm
    b AD = 4.8 cm
    c   i   BD = 11.4 cm = 22.8 km
       ii   DC = 15.8 cm = 31.6 km

15.5 Accurate drawings 2

1 a Cuboid   b Tetrahedron
2 There are 11 different correct nets for a cube including the one given.
3 a Sketch of triangle
    b Accurate triangle drawn to scale
    c 2.5 cm or 7.2 cm (2 triangles possible)
4 Accurate triangle JKL drawn (angles 58°, 25°, 97°)
5 a Accurate drawing of quadrilateral
    b 6.9 cm   c 88°
6 Diagram of regular hexagon circumscribed by circle
7 a

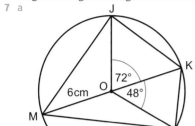

    b JOM = 108˚; KLM = 90˚.
      Triangle JOM is isosceles, check angles OJM and OMJ = 36°
      Check angles LKM and LMK add to 90°

8

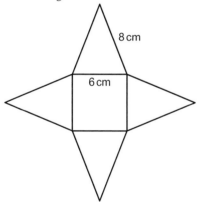

9 Accurate scale diagram drawn with scale 1:5

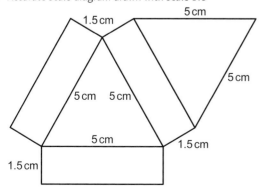

15.6 Constructions

1 a Accurate drawing of 5.5 cm line
    b Accurate drawing of 47 mm line

2 60°
3 a Perpendicular bisector of 7 cm line accurately constructed
4 Perpendicular line accurately constructed 4 cm from a line 9 cm long
5 Students' own answers
6 a Diagram showing Main road, with perpendicular line to the Sports Centre
    b 1.25 km
7 Perpendicular line at point P accurately constructed
8 Accurate 70° angle drawn and bisected, showing construction arcs
9 Accurate copies of triangles with angle Q bisected, showing construction arcs
10 Accurate copies of angle with angle bisected, showing construction arcs
11 Accurate 85° angle drawn and bisected, showing construction arcs
12 Construct 90° by drawing a straight line and constructing the perpendicular bisector at an end point.
    Construct 45° by first constructing 90° and then constructing its bisection.
13 Accurate equilateral triangle with one angle bisected, showing construction arcs.

15.7 Loci and regions

1 Accurate drawing of a circle with radius 6 cm
2 Accurate line 76 mm drawn with accurate construction of its perpendicular bisector
3 A circle of radius 5 cm
4

5

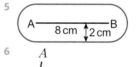

6

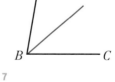

7

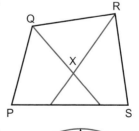

8

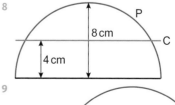

9

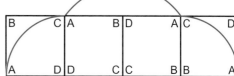

10 Statement B is true

11

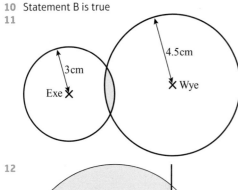

12

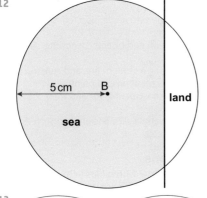

13

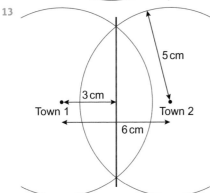

14

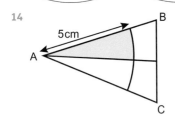

15.8 Bearings

1 a 80°   b 100°   c 260°
2 a 045°   b 125°   c 206°   d 200°
3 260°
4 a 240°   b 310°   c 200°
5 a 222°   b 243°
6 Accurate diagram, showing Venice–Rome = 8.4 cm on a bearing of 356° and Naples–Rome = 3.5 cm on a bearing of 128°
7 a Accurate diagram
   b 6.3 km   c 128°   d 308°

8 a

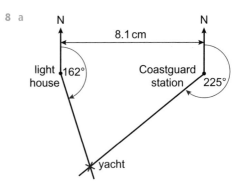

   b 045°   c Closer to lighthouse by 22 km
9 a Accurate diagram   b 308°   c 13.4 miles
10 a 166°   b 289°
11 a, b, c

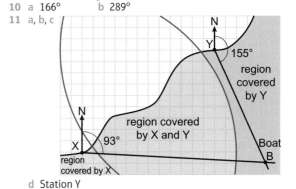

   d Station Y

15 Problem-solving

1 1 : 50 000
2 Car A (Car A travels 83 miles on 1 gallon of petrol; Car B travels 80.4 miles on 1 gallon of petrol).
3 a 525 g box (525 g box gives 500 g for £1; 720 g box gives 450 g for £1)
   b 720 g box (2 × 525 g boxes gives 500 g for £1; 2 × 720 g boxes gives 600 g for £1)
4 87.52 Euros
5 Students' own answers
6 21.5 cm
7 Students' own answers

15 Check up

1 a 6   b 9   c 5
2

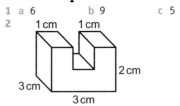

3

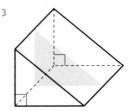

4 Accurate drawing of a 10 cm line and construction of its perpendicular bisector
5 Accurate drawing of a 110° angle and construction of its bisector
6 Accurate drawing of triangle PQR with angle QPR = 50°, QP = 10 cm, PR = 13 cm
7 Accurate construction of triangle with side lengths 4 cm,

8 cm, 7 cm
8 a Circle with radius 4 cm drawn and centre the fixed point
   b Circle in part a shaded in
9 Accurate drawing of two points 10 cm apart with perpendicular bisector of the line joining these points accurately constructed
10

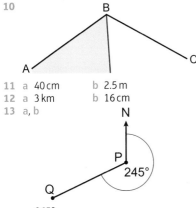

11 a 40 cm   b 2.5 m
12 a 3 km   b 16 cm
13 a, b

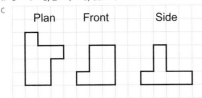

   c 065°
14 a 255°   b 075°   c 135°
16 a 3 × 4 × 1, 2 × 6 × 1, 12 × 1 × 1   b 2 × 2 × 3
   c

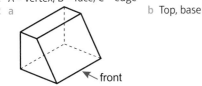

15 Strengthen

3D solids
1 A = vertex, B = face, C = edge
2 a   b Top, base

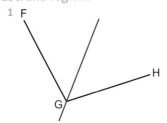

Constructions
1 Accurate drawing of line 10 cm long, with its perpendicular bisector accurately constructed
2 Accurate drawing of line 12.5 cm, with its perpendicular bisector accurately constructed
3 Accurate drawing of 50° angle, with its bisector accurately constructed
4 Accurate drawing of 120° angle, with its bisector accurately constructed
5 Accurate construction of triangle with side lengths 5 cm, 6 cm, 7 cm
6 Accurate construction of a right-angled triangle with base 5 cm, hypotenuse 9 cm
7 Accurate construction of triangle ABC with angle BAC = 42°, AC = AB = 8 cm

Loci and regions
1

2

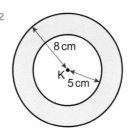

Scale drawings and bearings
1 a 4 cm   b 8 m by 4 m
   c Yes. Work out the area of the café, work out the area covered by 8 tables and compare the two answers.
2 a 10 km   b 9 cm   c 4.5 cm
3 a 027°   b 230°   c 134°   d 318°
4 a 237°
   b Accurate bearing drawn
   c Accurate bearing drawn
   d 305°
5 220°

15 Extend

1 Yes, e.g. the farmer could fit 6 bays on each of the barn with 3 m between them.
2 Yes. 8 cm represents 160 km, 3.2 gallons needed, car contains 5 gallons
3 a Accurate drawings to a suitable scale of a rectangle 6 m long × 2.5 m wide labelled Plan and a rectangle with length 2.5 m, height 0.48 m and a triangle with base 6 m, height 0.48 m labelled Front and Side in either order

   b

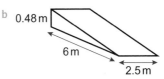

4 Accurate drawing of isosceles trapezium drawn, interior angles 123°, 57°, 57°, 123°, side lengths 2.5 cm, 1.85 cm, 4.5 cm, 1.85 cm
5 Accurate net of tetrahedron, side length 4 cm
6 a-f Students' drawings showing locus equidistant from point and line is a parabola
   g Parabola
7 a The locus of points reached by each transmitter
   b The region which is reached by both transmitters
   c 35 miles
8

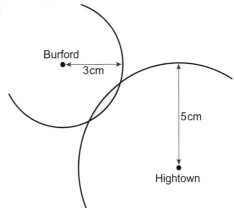

Scale: 1 cm represents 10 km

9 a Accurate construction of triangle DEF drawn, with e.g. scale 1 cm represents 10 km, DE = 8.6 cm, EF = 3.7 cm

b 337°
c 94 km
10 a Accurate construction of equilateral triangle with side 9 cm
   b 150°, 210°

15 Unit test

Sample student answer

The student didn't use a compass to draw the arc of the circle with centre C accurately.
The shaded area is closer to AD than AB.
The student didn't use a ruler for the line parrallel to DC.
The horizontal line is not 3 cm from DC.

UNIT 16

16 Prior knowledge check

1 a 9   b 49
2 a +7 or −7   b +4 or −4
3 a 4   b −8   c −1   d −13
   e −12   f 21   g −2   h 3
4 a i 1 and 9, −1 and − 9, 3 and 3, −3 and −3
     ii −3 and −3
   b i 1 and − 4, − 1 and 4, 2 and −2
     ii −1 and 4
5 a $8x$ b $−x$ c $x^2 + 9x$
   d $4x^2 + 7x + 3$ e $7x$ f $−5x$
   g x^2 h 16   i −9
6 a 16   b 6   c 12   d 21
7 a $2x + 6$ b $5a − 10$ c $x^2 + 4x$ d $d^2 − 9d$
8 a $5(2t + 1)$
   b $h(12h + 1)$
   c $7(2y − 1)$
   d $d(c − 1)$
9 a $x = 2$ b $x = 2$ c $x = 7$ d $x = 6$
10 a 28 cm², 22 cm
11 a 3   b 6
12 $y = 1$
13 Students' graphs of $y = x + 1$, with coordinate points
   (−2, −1), (−1,0), (0,1), (1,2), (2,3)
14 a The point where the line crosses the y-axis
   b i 3
     ii −5
     iii 0
     iv 4
15 $x^2 − 6x, 6x − x^2, 3x^2 − 18x, 2x^2 − 12x$ (any order)

16.1 Expanding double brackets

1 a $−9a$ b $−12b$ c $−3z^2$
   d $18a$ e $8y$ f $−3t$
   g $m^2 + 9m + 4$ h $y^2 − 20y − 3$
2 a $2a + 4$ b $3x + 12$ c $m^2 + 8m$ d $2b^2 − 5b$
3 a a^2 b $5a$ c $4a$
   d 20   e $(a + 5)(a + 4)$
4 a $(n + 4)(n + 2) = n^2 + 4n + 2n + 8 = n^2 + 6n + 8$
   b $(z + 6)(z + 3) = z^2 + 6z + 3z + 18 = z^2 + 9z + 18$
   c $(a + 1)(a + 7) = a^2 + a + 7a + 7 = a^2 + 8a + 7$
   d $(p + 12)(p + 5) = p^2 + 12p + 5p + 60 = p^2 + 17p + 60$
5 a $x^2 + 3x + 2$ b $t^2 + 7t + 12$
   c $q^2 + 15q + 54$ d $z^2 + 13z + 12$
   e $m^2 + 19m + 88$ f $y^2 + 17y + 70$
6 a $z^2 − z − 2$ b $m^2 + m − 30$
   c $a^2 − 5a − 36$ d $n^2 − 3n − 70$
   e $x^2 − 5x + 6$ f $y^2 − 7y + 6$
7 5
8 Multiplying both ways gives $x^2 − 4x − 21$ so Isabella is correct.
9 No, the answer is $a^2 − 11a + 28$. Rex has made mistakes with the negative terms.

10 a $6 + 3t$ b $6x^2 + 15x$ c $m^2 + 13m + 30$
11 a, c, e and f. A quadratic expession always contains a squared term as its highest power.
12 a $x^2 + 4x + 4$ b $a^2 + 10a + 25$
   c $y^2 − 18y + 81$ d $m^2 − 8m + 16$
13 a $x^2 + 12x + 36$ b $n^2 + 24n + 144$
   c $q^2 − 8q + 16$ d $t^2 − 20t + 100$
14 A and H, B and G, C and E, D and F
15 a $x^2 + 8x + 16$
   b $10(x^2 + 8x + 16) = 10x^2 + 80x + 160$
16 $(n + 6)(n + 3) = n^2 + 9n + 18$
17 $(x + 4)(x − 4) = x^2 − 16$. Answer is a quadratic expression as x^2 is the highest power. Both x^2 and 16 are squares.

16.2 Plotting quadratic graphs

1 $x = 3$
2 a 4   b 2   c 35   d 15
3 a

x	−3	−2	−1	0	1	2	3
y	−1	0	1	2	3	4	5

b Graph of $y = x + 2$
c (−1, 1)

4 a

x	−4	−3	−2	−1	0	1	2	3	4
y	16	9	4	1	0	1	4	9	16

b
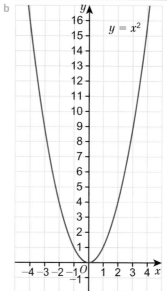

5 a

x	−4	−3	−2	−1	0	1	2
x^2	16	9	4	1	0	1	4
−3	−3	−3	−3	−3	−3	−3	−3
y	13	6	1	−2	−3	−2	1

b
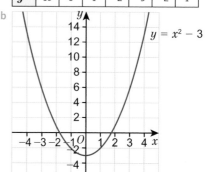

6 a

x	−4	−3	−2	−1	0	1	2
x^2	16	9	4	1	0	1	4
+2x	−8	−6	−4	−2	0	+2	+4
−4	−4	−4	−4	−4	−4	−4	−4
y	4	−1	−4	−5	−4	−1	4

b

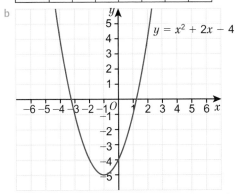

c The y-coordinates repeat before and after the lowest y value.

7 a i $x = -2$
   ii $(-2, -16)$
   iii −12
   b i $x = 1$
   ii $(1, 4)$
   iii 5

8

x	−3	−2	−1	0	1	2	3
y	−9	−4	−1	0	−1	−4	−9

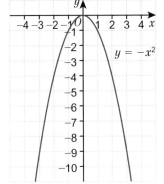

9 b and d are quadratic functions
10 a i 17 metres
   ii 2 seconds
   b 1 second, 3 seconds
   c 4 seconds
11 a i 1 cm² **ii** 16 cm² **iii** 6.25 cm²
   b Square. Area is always the square of the length.
12 a

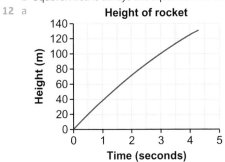

b 0.5 seconds
c Answers between 85 and 90 m

16.3 Using quadratic graphs

1 a

x	−3	−2	−1	0	1	2	3
y	10	4	0	−2	−2	0	4

b

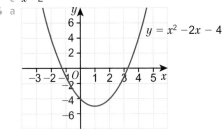

2 a (−1, 0) and (2, 0) **b** $x = -1$ and $x = 2$
3 a $x = -2$ or $x = 3$
   b $x = -5$ or $x = 2$
   c $x = 2$
4 a

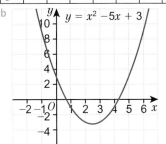

b $x = -1.2$ or $x = 3.2$

5

x	−1	0	1	2	3	4	5
y	9	3	−1	−3	−3	−1	3

b

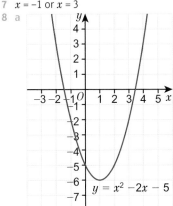

c $x = 0.7$ or $x = 4.3$
6 $x = 1$ or $x = -1$
7 $x = -1$ or $x = 3$
8 a

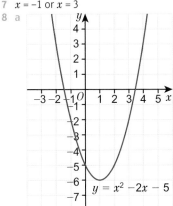

b $x = -1.4$ or $x = 3.4$
c $x = -2$ or $x = 4$
9 a $x = 0$ or $x = 3$ **b** $x = -0.6$ or $x = 3.6$

10 a

x	-2	-1	0	1	2	3	4
y	8	3	0	-1	0	3	8

b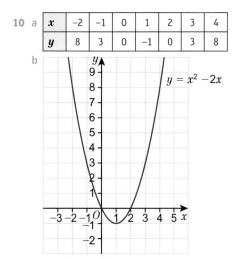

$y = x^2 - 2x$

c $x = -1$ or $x = 3$

11 a

x	-2	-1	0	1	2	3	4
y	7	2	-1	-2	-1	2	7

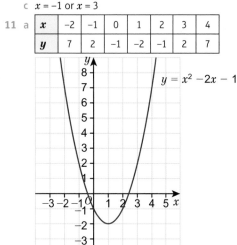

$y = x^2 - 2x - 1$

b $x = -1$ or $x = 3$
c The curve and the line do not intersect.
12 -5

16.4 Factorising quadratic expressions

1 a $x^2 + 7x + 6$ b $x^2 + 3x - 40$
   c $x^2 - 3x + 2$ d $x^2 - 9$
2 a −1 and −2
   b −1 and −4, −2 and −2
   c −1 and −6, −2 and −3
3 a 2 and 4   b 3 and 4
   c 3 and −2   d 1 and −10
4 a x b 7   c 5   d $x, 4$
5 a $(x + 4)(x + 3)$ b $(x + 2)(x + 4)$
   c $(x + 1)(x + 3)$ d $(x + 5)(x + 2)$
6 a $(x + 3)(x - 2)$ b $(x - 2)(x + 5)$
   c $(x + 1)(x - 3)$ d $(x - 5)(x + 4)$
   e $(x - 2)(x - 3)$ f $(x - 2)(x - 4)$
7 $(x - 2)(x + 5)$
8 a $y + 9$ b 144 cm²
9 a $(x + 5)(x + 6)$
   b $(x - 2)(x + 7)$
   c $(x - 4)(x + 1)$
10 $(x + 12)$ and $(x - 2)$
11 a $x^2 - 4$ b $x^2 - 9$
   c $x^2 - 16$ d $x^2 - 100$
12 $x^2 - 4$; $x^2 - 36$; $x^2 - y^2$
13 a $z^2 - 9$
   b Difference of 2 squares
14 a $p^2 + 5p - 36$ b $(x - 7)(x + 7)$

16.5 Solving quadratic equations algebraically

1 a $(x + 2)(x + 5)$
   b $(x - 3)(x + 4)$
   c $(x + 3)(x - 3)$
2 a $d = 18$ b $x = -1$ c $p = 4$
3 $x - 9 + 9 = 0 + 9$
   $x^2 = 9$, $x = 3$ or $x = -3$
4 a $x = 5$ or $x = -5$ b $y = 1$ or $y = -1$
5 a $x = 6$ or $x = -6$
   b $x = 7$ or $x = -7$
   c $x = 9$ or $x = -9$
6 a $x = -10$ or $x = 2$
   b $x = -5$ or $x = 3$
   c $x = 1$ or $x = 5$
7 a $x = -1$ or $x = -4$ b $x = -3$ or $x = -6$
   c $x = -3$ or $x = 2$ d $x = 4$ or $x = -1$
   e $x = 5$ or $x = 1$ f $x = 2$ or $x = 8$
8 A and E, B and G, C and F, D and H
9 a   i $(x - 9)(x - 3)$
     ii $x = 3$ or $x = 9$
   b $(y + 10)(y - 10)$
10 a $x = -6$ or $x = 6$ b $z = -9$ or $z = 9$
   c $p = 8$ or $p = -8$ d $d = 11$ or $d = -11$
11 $x^2 - 7x + 12 = 0$
12 a

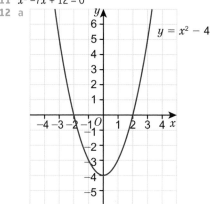

$y = x^2 - 4$

b, c $x = 2$ or $x = -2$

16 Problem-solving

1 5.5 bar
2 5 m
3 £14.70
4 Answers in the range 1.6–1.8 km
5 Answers in the range 0.60–0.70 seconds, and 1.10–1.15 seconds

16 Check up

1 Graphs a and c

2 a

x	-1	0	1	2	3	4	5	6	7
y	7	0	-5	-8	-9	-8	-5	0	7

   b $x = 3$ c $(3, -9)$
   d $(0, 0)$ e $x = 0$ or $x = 6$
3 a $x = -4.2$ or $x = -0.8$ b $x = -5.7$ or $x = 0.7$
4 a 8.5 m   b 1 s   c 0 s and 2 s   d 2.05 s
5 a $t^2 + 10t + 24$ b $f^2 + 2f - 35$
   c $a^2 - 1$ d $n^2 - 14n + 49$
6 $(x + 7)(x - 7)$
7 a $x = -3$ or $x = 3$ b $x = -6$ or $x = 6$
8 a $(x + 7)(x + 8)$ b $(x - 9)(x + 11)$
9 a $(x + 10)(x + 3)$, $x = -10$ or $x = -3$
   b $(x - 6)(x - 1)$, $x = 6$ or $x = 1$
11 a $(x + 3)^2$ $(x - 3)^2$ $(x + 2)^2$ $(x - 2)^2$
   $(x + 3)(x - 3)$ $(x + 2)(x - 2)$ $(x + 3)(x + 2)$
   $(x + 3)(x - 2)$ $(x + 2)(x - 3)$ $(x - 3)(x - 2)$

16 Strengthen
Quadratic graphs

1 a, b

x	−3	−2	−1	0	1	2	3
x^2	9	4	1	0	1	4	9
+1	+1	+1	+1	+1	+1	+1	+1
y	10	5	2	1	2	5	10

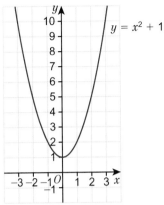

$y = x^2 + 1$

2 a, b

x	−4	−3	−2	−1	0	1	2
x^2	16	9	4	1	0	1	4
+2x	−8	−6	−4	−2	0	2	4
+3	+3	+3	+3	+3	+3	+3	+3
y	11	6	3	2	3	6	11

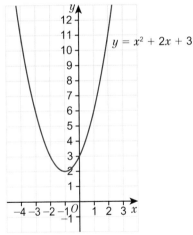

$y = x^2 + 2x + 3$

3 a, b

x	−3	−2	−1	0	1	2
x^2	9	4	1	0	1	4
+x	−3	−2	−1	0	1	2
y	6	2	0	0	2	6

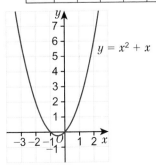

$y = x^2 + x$

4 a Roots   b y-intercept
   c Turning point   d Line of symmetry
5 a $y = 8$ b (3, −1)
   c $x = 3$ d $x = 2$ or $x = 4$
6 a, b

x	−5	−4	−3	−2	−1	0
x^2	25	16	9	4	1	0
+5x	−25	−20	−15	−10	−5	0
+5	+5	+5	+5	+5	+5	+5
y	5	1	−1	−1	1	5

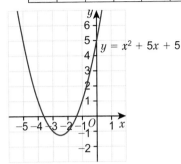

$y = x^2 + 5x + 5$

   c i Line $y = 1$ drawn
     ii $x = −4$ or $x = −1$
   d $x = −3.6$ and $x = −1.4$

7 a, b

x	−1	0	1	2	3	4	5
x^2	1	0	1	4	9	16	25
−2x	2	0	−2	−4	−6	−8	−10
y	3	0	−1	0	3	8	15

b

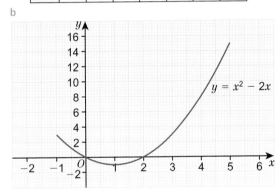

$y = x^2 − 2x$

   c $x = 1$, (1, −1)   d $x = 0$ or $x = 2$
8 Graphs a and d

Quadratic equations
1 a^2, 6, 2a, 3a, $a^2 + 6 + 2a + 3a$
2 a $a^2 + 5a + 6$ b $t^2 + 5t + 4$
   c $x^2 + 8x + 7$ d $y^2 + 15y + 50$
3 a $z^2 − z − 2$ b $f^2 + 2f − 63$ c $x^2 − 1$
4 a $m^2 + 8m + 16$
   b $x^2 + 16x + 64$
   c $g^2 − 20g + 100$
5 a $x = 2$ or $x = −2$
   b $a = 3$ or $a = 3$
   c $t = 7$ or $t = −7$
6 $x^2 − 9 = (x − 3)(x + 3)$
   $x^2 − 49 = (x + 7)(x − 7)$
   $x^2 − 16 = (x + 4)(x − 4)$
   $x^2 − 4 = (x − 2)(x + 2)$
7 a $(x + 4)(x − 4)$ b $(p + 1)(p − 1)$
   c $(y + 9)(y − 9)$ d $(k + 10)(k − 10)$
8 a $(x + 4)(x + 3)$ b $(x + 2)(x + 6)$
   c $(x + 6)(x + 1)$

9 a $(x + 4)(x + 2)$, $x = -4$ or $x = -2$
  b $(x + 5)(x + 1)$, $x = -5$ or $x = -1$
  c $(x + 6)(x - 3)$, $x = -6$ or $x = 3$
  d $(x - 4)(x - 3)$, $x = 4$ or $x = 3$

16 Extend

1 a

x	−2	−1	0	1	2	3	4	5
$-x^2$	−4	−1	0	−1	−4	−9	−16	−25
$+3x$	−6	−3	0	3	6	9	12	15
$+4$	+4	+4	+4	+4	+4	+4	+4	+4
y	−6	0	4	6	6	4	0	−6

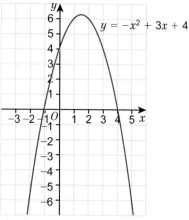

$y = -x^2 + 3x + 4$

  b i y-intercept = 4
    ii $x = 1.5$
    iii (1.5, 6.25)
    iv $x = -1$ or $x = 4$
2 a $3(2 + 3x)$
  b $(y - 4)(y + 4)$
3 a

x	0	1	2	3	4	5
y	10	6	4	4	6	10

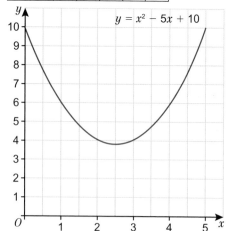

$y = x^2 - 5x + 10$

  b i 10 m   ii 3.75 m   iii 5 m
4 a

w	0	4	8	10	12	16	20
$20w$	0	80	160	200	240	320	400
$-w^2$	−0	−16	−64	−100	−144	−256	−400
A	0	64	96	100	96	64	0

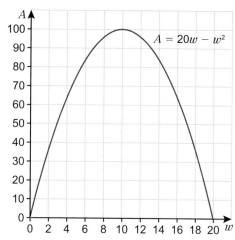

$A = 20w - w^2$

  b 100 m²   c 10 m   d 10 m
5 $3x^2 - 12 + 12 = 0 + 12$
  $3x^2 = 12$
  $x^2 = 4$
  $x = 2$ or $x = -2$
6 a $x = 2$ or $x = -2$
  b $y = 7$ or $y = -7$
  c $x = 3$ or $x = -3$
  d $x = 4$ or $x = -4$
7 a

x	0	0.5	1	1.5	2
y	1	0.25	0	0.25	1

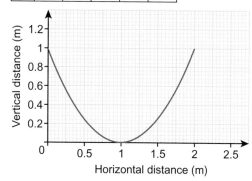

  b i 1 m   ii 2 m   iii 0 m
8 a 5 m   b 0 m
  c It has landed on the ground.
9 a $x^2 - 4x + 4$ b $12(x + 1)$
10 a $x^2 = 3x$ b $x = 3$ c 9 cm²
11 a $x = -1.3$ or $x = 2.8$
  b $x = -0.7$ or $x = 2.2$

16 Unit test

Sample student answer

Draw a smooth curve to join the points, not straight lines.
Label the graph with its equation.
Label the x- and y-axes.
Draw smaller crosses.
Draw thinner lines with a sharp pencil.

UNIT 17

17 Prior knowledge check

1 a 64   b 27   c 7
2 a 3   b 2.53   c 2 000 000
3 a $1\,cm^3$ b $1000\,cm^3$ c $1000\,cm^3$
4 a 15   b 12   c 6   d 3
5 a Cylinder   b Cone
   c Square-based pyramid   d Sphere
6

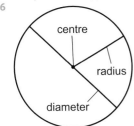

7 a $20\,cm^2$ b $15\,cm^2$
8 a $24\,cm^3$ b $52\,cm^2$
9 $142\,cm^3$
10 Students' own answers, e.g.
   Assuming that the 6 faces can be cut out separately and stuck together, could make a cube with side length 280.3 mm giving a volume of $22\,022\,636\,mm^2$.

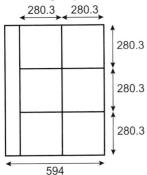

Or, assuming that a net must be drawn, the largest cube would have side length 198 mm (594 ÷ 3)

17.1 Circumference of a circle 1

1 $x = 5y$
2 4.19
3 a 120 m   b 5 inches   c $r = \dfrac{d}{2}$
   d $d = 2r$ e 2.6 inches
4 a All 3.14   b All 3.14   c $C = 3.14 \times d$ d 57.5 m
5 a 16.7 cm   b 1947.8 mm   c 10.0 m
6 a 42.1 cm   b 2450.4 mm   c 15.1 m
7 28.3 cm
8 a 64.8 cm   b 38
9 6 lengths
10 234

17.2 Circumference of a circle 2

1 a $d = 25$ b $d = 22.28$ c $r = 22.5$ d $C = 15.71$
2 2, 3, 4, 5, 6, 7, 8
3 a $l = \dfrac{A}{w}$ b 7.3
4 a i 7.5 cm
     ii 19.5 km
     iii 37.5 mm
   b i Any number smaller than 8.5 cm but bigger than 8 cm
     ii Any number smaller than 20.5 km but bigger than 20 km
     iii Any number smaller than 38.5 mm but bigger than 38 mm
5 14.5 cm
6 a 1.625 m   b 1.635 m   c $1.625 \le h < 1.635$

7 $25\,250 \le n < 25\,350$
8 a $4.355 \le n < 4.365$ b $715 \le n < 725$
   c $15.65 \le n < 15.75$ d $445 \le n < 455$
9 Height of book could be up to 265 mm, minimum height of shelf is 259.5 mm
10 a 3.14159   b 3.142   c 15.71 cm
11 a 15.1 m   b 19 mm   c 4715.5 cm
12 a 179π feet
   b Estimates between 500 and 600 feet
   c 562 feet
13 Mary is correct, Paul forgot to double the radius
14 a 111.4 cm   b 0.9 m
15 a 107 m   b 477 m
16 a $27.5\,cm \le$ circumference $< 28.5\,cm$
   b 4.4 cm or 4.5 cm

17.3 Area of a circle

1 a $p = 40$ b $t = 108$ c $k = 200$
2 a 3.2   b $\frac{4}{3}$ or 1.3   c 1.8
3 a $y = 8$ b $y = 16.6$
4 a $201\,cm^2$ (nearest cm)
   b $132.7\,m^2$ (1 d.p.)
   c $138\,540\,cm^2$ (nearest 10 cm)
   d $234\,826.4\,m^2$ (1 d.p.)
5 a i $27\,mm^2$
     ii 150 square feet
     iii Estimates between $18\,km^2$ and $27\,km^2$
   b i $9\pi\,mm^2$
     ii 49π square feet
     iii $6.25\pi\,km^2$
   c i $28\,mm^2$
     ii 150 square feet
     iii $20\,km^2$
6 Rita is correct; Jack forgot to halve the diameter.
7 a $63.6\,m^2$ b 3180 g (3 s.f.)
8 a $7.1\,m^2$ b $\approx 20\,000$ slugs
9 a i 6.3 m   b i 2.1 cm   c i 1.2 cm
     ii 12.6 m   ii 4.2 cm   ii 2.4 cm
10 11 m or 11.0 m
11 120 cm
12 $130.4\,m^2$
13 $867\,cm^2$
14 Area of lawn: $(17 - 2.8) \times 9.5 = 134.9\,m^2$;
   Area of pond: $\pi \times (3.8 \div 2)^2 = 11.34\,m^2$;
   $134.9 - 2 \times 11.34 = 112.2\,m^2$
   $112.2 \div 25 = 4.48\ldots$ so he needs 5 boxes of fertiliser.

17.4 Semicircles and sectors

1 a $\frac{1}{2}$ b $\frac{1}{4}$ c $\frac{1}{4}$
2 a 90   b 36   c 72
3 Area = $102.1\,cm^2$; circumference = 35.8 cm
4 a Chord   b Tangent   c Segment
   d Arc   e Sector
5 a i $11.52\pi\,cm^2$ b i $19.22\pi\,cm^2$
     ii $36.2\,cm^2$ ii $60.4\,cm^2$
6 a i $2.56\pi\,cm^2$ b i $90.25\pi\,mm^2$
     ii $8.04\,cm^2$ ii $284\,mm^2$
7 a 37.7 cm   b 18.8 cm   c 30.8 cm
8 $9\pi + 18$ cm
9 16.4 cm
10 a 44 cm   b 176 cm   c 22 cm
   d 11 cm   e 77 cm
11 a 24π cm   b 6   c $\frac{1}{6}$
12 a 12 and $\frac{1}{12}$ b 10 and $\frac{1}{10}$ c 3 and $\frac{1}{3}$
13 a 1.0 cm (1 d.p.)   b 11.1 cm (1 d.p.)
   c 101 mm (3 s.f.)
14 a $403\,mm^2$ (3 s.f.)   b $47.1\,km^2$ (3 s.f.)
   c $22.0\,m^2$ (3 s.f.)
15 a 32.3 cm   b 38.3 m
16 $58.9\,cm^2$

17.5 Composite 2D shapes and cylinders

1 One rectangle and two circles
2 a 24 cm²   b 4.5π cm²   c 4π cm²
3 a 22 cm   b 3π + 6 cm   c 2π + 8 cm
4 a Perimeter = 47.8 cm; area = 70.5 cm²
   b Perimeter = 28.9 cm, area = 43.6 cm²
5 a 60 m   b 7627 m²   c 348 m   d 3
6 8.79 cm²
7 a Circle   b $A = \pi r^2$ c $V = \pi r^2 h$
8 a 113 cm³   b 628 cm³   c 1130 cm³
9 a 21.2 m³   b 1060
10 Cylinder B (volume of A 277 cm³; volume of B = 362 cm³)
11 134.8 to 135 litres
12 a 1 256 637 cm³   b 5403.5 kg
13 a 242.8 cm²   b 294.1 cm²
14 a 750π ft³   b 300π ft²
15 243.1 cm²

17.6 Pyramids and cones

1 a 25 cm²   b 13 cm²
   c 60 cm²   d 63.6 cm²
2 a 6π   b 45π   c 80π
3 a 60.5 cm³   b 30 cm³   c 96 cm³
4 21 509 583 cubic yards
5 a 144 cm²   b 800 cm²
6 a 64 cm³   b 1280 cm³
7 a 49 000 cm³ b 10 500 cm²
8 a Circle   b $A = \pi r^2$ c $V = \frac{1}{3}\pi r^2 h$
9 a 288 cm³   b 1018 cm³   c 1571 mm³
10 a 4 cm   b 168 cm³   c 29
11 a i 144π m²   b i 224π cm² c i 50π cm²
      ii 452 m²   ii 704 cm²   ii 157 cm²
12 a i 128π m²   b i 392π cm² c i 40π cm²
      ii 402 m²   ii 1230 cm³   ii 126 cm³
13 3.4 or 4 litres

17.7 Spheres and composite solids

1 a 8   b 32   c 8   d 144
2 a 8   b $10\frac{2}{3}$
3 a 4.19   b 84.95
4 5 cm
5 a 2.304π cm³ b 7 cm³
6 75 766 mm³
7 Moira; Jill has used the formula πr^2.
8 18 cm²
9 300 000 000 km²
10 a 108π m²   b 339.29 m²
11 84 m³
12 a 56.25π m³
   b 36 000π cm³
   c 9π m³
13 a $3^2 + 4^2 = 25$
      Slant height, $l = \sqrt{25} = 5$ cm
   b 15π cm²   c 9π cm²   d 24π cm²   e 75.4 cm²
14 a i 576π mm²   ii   1809.6 mm²
   b i 297π cm²
      ii 933.1 cm²
15 a i 235π cm²
      ii 738 cm²
   b i 64π cm²
      ii 201 cm²
   c i 297π cm²
      ii 933 cm²
16 905 m³

17 Problem solving

e.g. Total surface area: $2\pi r^2 + 2\pi rh = 473$ cm²
      Cost of leather: 473 × 0.8 = 378p
      Cost of sewing 2 curves = 22p
      Cost of sewing 2 straight edges = 14p

Cost of 20 cm zip = 15p
Total cost = 378 + 22 + 14 + 15 = 429p = £4.29

17 Check up

1 a 10.5 ⩽ length < 11.5   b 5.315 ⩽ length < 5.325
2 Marker pen could be up to 145 mm but minimum length of
   pencil case is 141.5 mm
3 a 12π cm   b 37.7 cm
4 14 cm or 14.0 cm
5 78.5 cm²
6 7 m
7 a 82.3 cm   b 402 cm²
8 30 954 mm²
9 a 19.2 cm²   b 5.5 cm   c 19.5 cm
10 a i 54π cm³
      ii 170 cm³
   b i 40 500π cm³
      ii 127 000 cm³
   c i 2304π cm³
      ii 7240 cm³
11 a i 54π cm²
      ii 170 cm²
   b i 5400π cm²
      ii 170 000 cm²
   c i 576π cm²
      ii 1810 cm²
12 Volume = 23 cm³; surface area = 60 cm²
14 116.67 times

17 Strengthen

Accuracy

1 a 3.5 cm
   b i 3.5 ⩽ length < 4.5
      ii 2.45 ⩽ length < 2.55
      iii 5.455 ⩽ length < 5.465

Circles and sectors

1 a 12π cm   b 37.7 cm
2 a 8π cm   b 25.1 cm
3 a $C = \pi d$ b 21 cm
4 a 10 cm   b $A = \pi \times 10^2$ cm²   c 314.2 cm²
5 a 5 m   b 10 m
6 a $\frac{1}{2}$ b $\frac{1}{4}$ c $\frac{1}{6}$ d $\frac{13}{36}$
7 a Area = 314.2 cm²; circumference = 62.8 cm
   b i 50π cm²
      ii 10π cm
      iii 20 cm
      iv (10π + 20) cm
8 a 630 cm²   b 176.7 cm²   c 453.3 cm²
9 a 18.8 cm,   b 2.4 cm
10 a 144π cm²   b 24π cm²   c 24π cm
   d 4π cm   e (4π + 24) cm

Volumes and surface areas

1 a

   b 28.3 cm²   c 339 cm³

2 a
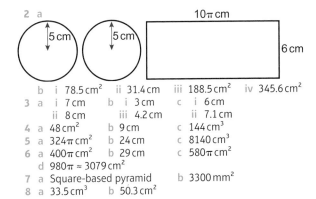

  b i 78.5 cm² ii 31.4 cm iii 188.5 cm² iv 345.6 cm²
3 a i 7 cm b i 3 cm c i 6 cm
  ii 8 cm iii 4.2 cm ii 7.1 cm
4 a 48 cm² b 9 cm c 144 cm³
5 a 324π cm² b 24 cm c 8140 cm³
6 a 400π cm² b 29 cm c 580π cm²
  d 980π ≈ 3079 cm²
7 a Square-based pyramid b 3300 mm²
8 a 33.5 cm³ b 50.3 cm²

17 Extend

1 a Maximum mass 187 g; Minimum mass 153 g
  b Maximum number 105; Minimum number 95
2 £104.70
3 $x = 3$
4 50.3 square units
5 $36 - 9\pi$ cm²
6 6.87 cm²
7 13.3 cm
8 No. Volume of tank = $35^2 \times 140 \times \pi$ = 538 783.14…cm³ =
  538.78…litres
  Takes 538.78 ÷ 0.4 = 1346.95…seconds to fill = 22.4
  minutes, which is less than 1 hour.
9 a 216π cm² b 448π cm³
10 2490 cm²
11 1:6
12 Yes, density of 0.96… g/cm³ < 1 g/cm³
13 10
14 2.6 cm²
15 25 cm

17 Unit test

Student A gives the better answer. Student A has found the
perimeter of each shape, but student B has found the area of
each shape.

UNIT 18

18 Prior knowledge check

1 a −6 b −5 c 8 d −11
2 a $\frac{1}{10}$ b $\frac{1}{100}$ c $\frac{1}{1000}$
3 a 9000 b 705
4 a 3.6 b 0.00081
5 a $\frac{1}{5}$ b $\frac{2}{5}$ c $\frac{2}{5}$
6 $3\frac{4}{5}$
7 $\frac{14}{3}$
8 a $\frac{3}{10}$ b $\frac{2}{21}$
9 a $\frac{3}{8}$ b 8
10 a 20 b 25
11 a 4^3 b 9^6
12 a 16 b 125 c 64
13 a 3^5 b 5^4 c 6^5 d 8^8
14 a $\frac{4}{1} \times \frac{3}{2}$ or $\frac{4}{2} \times \frac{3}{1}$ b $\frac{4}{1} \div \frac{2}{3}$ or $\frac{4}{2} \div \frac{1}{3}$
  c $3^4 + 2^1$ d $3^4 - 1^2$

18.1 Multiplying and dividing fractions

1 a $\frac{1}{11}$ b $\frac{7}{3}$ c $\frac{2}{7}$ d $\frac{5}{14}$
2 a $7\frac{1}{2}$ b 16 c $\frac{12}{35}$ d $\frac{1}{4}$
3 a $\frac{5}{12}$ b $\frac{2}{3}$ c $26\frac{2}{3}$ d $3\frac{1}{2}$
4 a 10 b 2 c 1.25 d 0.4

5 a $1\frac{4}{35}$ b $2\frac{2}{15}$ c $\frac{13}{20}$
  d $1\frac{11}{24}$ e $1\frac{1}{9}$ f $\frac{14}{15}$
6 a $\frac{3}{14}$ b $1\frac{3}{11}$ c $\frac{2}{3}$
  d $\frac{1}{2}$ e $2\frac{14}{15}$ f $\frac{54}{77}$
7 a $4\frac{4}{25}$ b $4\frac{4}{9}$ c $3\frac{1}{21}$
  d $1\frac{7}{9}$ e 18 f $3\frac{27}{35}$
8 a $3\frac{1}{5}$ b $1\frac{11}{16}$ c $3\frac{5}{7}$
  d 18 e $\frac{4}{7}$ f $\frac{2}{3}$
9 a $\frac{20}{39}$ b $1\frac{1}{9}$ c $\frac{4}{9}$
  d $1\frac{1}{13}$ e $1\frac{2}{3}$ f $\frac{9}{22}$
10 a $3\frac{33}{35}$ b $267\frac{2}{9}$ c $7\frac{4}{13}$ d $1\frac{97}{185}$
11 a $24\frac{4}{9}$
  b Students' own answers
12 26
13 $3\frac{3}{8}$ square inches
14 a 1 b 3 c 3 d 2
15 $6\frac{2}{3} \times 5\frac{4}{5}$
16 £7.40
17 $4\frac{4}{7}$ litres

18.2 The laws of indices

1 a 5 b 20 c 2.5 d 0.4
2 a 2^6 b 7 or 7^1 c 11 or 11^1 d 5^4
3 a $\frac{1}{36}$ b $\frac{1}{16}$ c $\frac{8}{125}$ c $\frac{81}{10000}$
4 a i 2 ii 5 iii 100
  iv 2.5 v $-\frac{1}{3}$ vi $-\frac{2}{5}$
  b The product is 1.
5 a $3^4 \times 3^4 \times 3^4 \times 3^4 \times 3^4 = 3^{20}$
  b $4^6 \times 4^6 \times 4^6 \times 4^6 = 4^{24}$
6 a 5^8 b 6^6 c 7^{16} d 11^{25}
7 a 3^6 b 3^{12} c 3^6 d 3^5
  e 3^5 f 3^4 g 3^{12} h 3^6
8 $2^3 = 8$
  $2^2 = 4$
  $2^1 = 2$
  $2^0 = 1$
  $2^{-1} = \frac{1}{2^1} = \frac{1}{2}$
  $2^{-2} = \frac{1}{2^2} = \frac{1}{4}$
  $2^{-3} = \frac{1}{2^3} = \frac{1}{8}$
9 a 1 b 1 c 1
10 a $\frac{1}{4}$ b $\frac{1}{7}$ c $\frac{1}{6}$ d $\frac{1}{x}$
11 a $\frac{1}{8}$ b $\frac{1}{25}$ c $\frac{1}{1000}$
  d 8 e 25 f $20\frac{1}{4}$
12 a 2^3 b 2^{-3} c 8^0
  d 6^0 e 11^{14} f 3^{-8}
13 a a^3 b $\frac{1}{b^3}$ c $\frac{1}{m^5}$
  d n^5 e $\frac{1}{x^3}$ f c^0 or 1
14 100^{-2}, 10^{-3}, 1^4 and 1^{-5}, 10^3, 100^2
15 a 2 b 30 c 0.18 d 18
16 a 0.004 b 9.57 c 762 d 7480
17 a 27 b $\frac{1}{7}$ c $2\frac{1}{4}$
18 5^{-4}, 5^{-1}, 0.5^2, 0.5
19 a 9.849382716 b 9.85

18.3 Writing large numbers in standard form

1 a 500 b 3500 c 87.5 d 903 000
2 a 1 000 000 b 1 000 000 000
3 a 1000 b 13 400
4 a 300 b 5 000 000 c 70 000
  d 900 000 000 000
5 a 5×10^2 b 3×10^5 c 9×10^9 d 7×10^6

6 a 47 000   b 9 210 000   c 830 000 000 000
   d 923 000   e 630 000 000 000
   f 9 050 000   g 6 702 000 000
   h 407 000 000

7 a D   b B   c A   d H
   e G   f F   g E   h C

8 a 4.5×10^5 b 3.2×10^4 c 1.5×10^5 d 7.25×10^6
   e 6.291×10^6 f 1.5×10^6 g 7.03×10^8 h 7.6×10^{10}

9 44 368 000 000

10 a 3.34×10^4 b 2.7×10^6 c 4.2×10^3
   d 4×10^2 e 8.7×10^5 f 5.05×10^5

11

Planet	Average distance from Sun (km)
Mercury	5.8×10^7
Venus	1.08×10^8
Earth	1.5×10^8
Mars	2.28×10^8
Jupiter	7.78×10^8
Saturn	1.433×10^9
Uranus	2.871×10^9
Neptune	4.503×10^9

12 a 1 000 000 000 000   b 1×10^{12}

13 a 8×10^{12} bytes   b 4.6×10^6 metres
   c 1.77×10^{10} litres   d 9.5×10^2 grams

14 a <   b >   c <
   d <   e >   f >

15 3.2×10^9, 3.2×10^8, 3.2×10^5, 3.2×10^4, 320

16 a 7.05×10^8 b 34 500 000

18.4 Writing small numbers in standard form

1 a 0.0001   b 0.000 001
   c 0.01   d 0.000 000 0001

2 10^{-8}, $\frac{1}{1000}$, 10^{-2}, 0.1

3 a 0.000 03   b 0.08
   c 0.4   d 0.000 000 000 07

4 a 3×10^{-2} b 5×10^{-3} c 1×10^{-4} d 3×10^{-7}

5 a 0.000045   b 0.000 003 8
   c 0.000 000 008 34   d 0.1401

6 a 5.2×10^{-2} b 7.1×10^{-4} c 5.69×10^{-4}
   d 2.41×10^{-3} e 1.4×10^{-5} f 1.09×10^{-3}
   g 3.04×10^{-5} h 6.102×10^{-1}

7 a D   b A   c F   d B
   e H   f C   g E   h G

8 Lucy: The number should be between 1 and 9; she must divide by a further power of 10.
Ali: The zero between 9 and 7 is a place holder and should be included.
Sam: The power of 10 should be '−4'. otherwise you are making the number larger.
   b 9.07×10^{-4}

9 a 7×10^{-12} grams   b 1.4×10^{-6} seconds
   c 5.93×10^1 metres   d 1.05×10^{-8} volts
   e 3.8×10^{-4} amps   f 9.9×10^{-2} litres

10 a

Element	Radius (m)
Hydrogen	2.5×10^{-11}
Lithium	1.45×10^{-10}
Sodium	1.8×10^{-10}
Phosphorus	1×10^{-10}
Nitrogen	6.5×10^{-11}
Chromium	1.4×10^{-10}
Tin	1.45×10^{-10}

   b Hydrogen, Nitrogen, Phosphorus, Chromium, Lithium/Tin, Sodium

11 a 7.05×10^{-7} b 0.00032

18.5 Calculating with standard form

1 a 3.2×10^5 b 1.8×10^9
   c 9×10^3 d 5.96×10^{-5}

2 a 10^9 b 10   c 10^4 d 10

3 a 6×10^5 b 7.5×10^{-7} c 6×10^4 d 1.8×10^{-5}
   e 4×10^{-3} f 2×10^3 g 5×10^7 h 2×10^{-12}

4 5×10^4

5 a 2.1×10^{11} b 2×10^{14} c 1.2×10^5
   d 3×10^6 e 9.6×10^6 f 7.77×10^{-11}

6 a 4×10^{-10} b 3×10^{-6} c 5×10^{-6}
   d 2.2×10^5 e 9.01×10^3 f 8×10^5

7 a 3.82×10^{11} b 1.50×10^6 c 3.46×10^{-9}
   d 1.65×10^{-5} e 6.37×10^8 f 1.41×10^{18}

8 a 10^6 b 10^3 c 10^{15}

9 0.072 kilometres

10 a 4×10^{11} b 3×10^1

11 a, b i 4.9×10^{11} ii 5.07×10^8 iii 4.12×10^{-2}
     iv 7.654×10^{-5} v $6.130 72 \times 10^7$ vi 5.555×10^{21}

12 a, b i 2.3×10^9 ii 5.92×10^{-7} iii 5.219×10^5
     iv 3.153×10^{-4} v 4.6491×10^{-10} vi 9.9856×10^{-15}

13 a $3.200 86 \times 10^{-4}$ b $5.199 958 9 \times 10^{-7}$
   c 4.5882×10^{14}

14 a 5.95×10^5 b 7.45×10^{14} c 2.16×10^{-9}

15 2.99×10^{-23} cm^3

16 a 57 600 000 000 m   b 5.76×10^{10} m
   c 192 seconds

17 39.7 days

18 Problem-solving

One possible route would be (times taken have been rounded up):

From	To	Distance (km)	Time taken
Thales	Banneker	1.8×10^8	1 hour 40 minutes
Banneker	Turing	1.3×10^9	12 hours 3 minutes
Turing	Cantor	3×10^8	2 hours 47 minutes
Cantor	Pacioli	1.2×10^9	11 hours 7 minutes
Pacioli	Boole	3.6×10^8	3 hours 20 minutes
Boole	Thales	5.2×10^8	4 hours 49 minutes

This route would take 35 hours and 46 minutes and be awarded a total of 70 points.

18 Check up

1 $\frac{5}{2}$

2 a $1\frac{13}{35}$ b $2\frac{5}{6}$

3 a 4   b 6   c $1\frac{9}{11}$

4 a 5^6 b 2^{-12} c x^2 d 3^{-13}

5 a $\frac{1}{4}$ b $\frac{1}{8}$ c 1
   d $2\frac{7}{9}$ e 1

6 a 400 000 000
   b 5260
   c 0.35
   d 0.000 000 809 9

7 a 1.9×10^5
   b 1.05×10^9
   c 7×10^{-6}
   d 4.52×10^{-5}

8 a 8×10^9 b 1.5×10^8 c 6.6×10^5

9 a 3×10^6 b 5×10^6 c 3.5×10^{-6}

10 a 3.24×10^{12} b 9.065×10^{-5}

11 a 4.14×10^{-8} b 7.86×10^6

13 Students' own answers e.g. $4^{-2} \times 4^{-3}$ or $\left(\frac{1}{4}\right)^5$

18 Strengthen

Reciprocals and fractions

1 a 5   b 2   c 10
   d 4   e 20

2 a $\frac{17}{5}$ b $\frac{17}{5} \times \frac{2}{3} = \frac{17 \times 2}{5 \times 3} = \frac{34}{15} = 2\frac{4}{15}$

3 a $\frac{5}{6}$ b $1\frac{7}{25}$ c $1\frac{13}{14}$

4 $\frac{10}{3} \times \frac{7}{4} = \frac{5}{3} \times \frac{7}{2} = \frac{35}{6} = 5\frac{5}{6}$

5 a $\frac{1}{2}$ b 2   c $3\frac{1}{2}$

6 $\frac{5}{4} \times \frac{7}{3} = \frac{5 \times 7}{4 \times 3} = \frac{35}{12} = 2\frac{11}{12}$

7 a $2\frac{2}{15}$ b $10\frac{1}{2}$ c $1\frac{2}{3}$

8 $\frac{7}{3} \times \frac{3}{2} = \frac{7}{2} = 3\frac{1}{2}$

9 a $\frac{16}{5} \div \frac{1}{4} = \frac{16}{5} \times \frac{4}{1} = \frac{64}{5} = 12\frac{4}{5}$ b $\frac{8}{25}$ c $2\frac{1}{4}$

Indices:

1 a i $\frac{64}{64} = 1$
     ii $8^{2-2} = 8^0$
     iii 1
   b 1
   c 1

2 a i 7^{-3}
     ii $\frac{1}{7^3}$
     iii $7^{-3} = \frac{1}{7^3}$
   b i 4^{-2}
     ii $\frac{1}{4^2}$
     iii $4^{-2} = \frac{1}{4^2}$

3 a $\frac{1}{8}$ b $\frac{1}{5}$ c 1
   d 1   e 2

4 a 2^{-1} b 4^3 c 8^{-7} d 19^{-10}

5 a 7^{-5} b 12^{10} c 6^7 d 2^{-3}

6 a 3^8 b 5^{14} c 19^{30} d 6^6
   e 5^{-6} f 5^{-12} g 9^6 h 6^8

7 a $\frac{1}{3} \times \frac{1}{3} = \frac{1}{9}$; $\frac{1}{3} \times \frac{1}{3} \times \frac{1}{3} = \frac{1}{27}$ b $\frac{25}{4}, \frac{125}{8}$

8 a $\frac{9}{4}$ b 64   c $\frac{49}{16}$ d $\frac{1000}{27}$

Standard form

1

10^{-6}	0.000001
10^{-5}	0.00001
10^{-4}	0.0001
10^{-3}	0.001
10^{-2}	0.01
10^{-1}	0.1
10^0	1
10^1	10
10^2	100
10^3	1000
10^4	10000
10^5	100000
10^6	1000000

2 a 0.0003   b 500   c 0.0009   d 1200
   e 0.000057   f 112   g 0.000903   h 1010000

3 a 1.8×10^4 b 9.6×10^5 c 4×10^4
   d 9×10^6 e 7.51×10^5 f 1.08×10^6

4 a 3.6×10^{-3} b 1.2×10^{-4} c 2.34×10^{-1}
   d 6×10^{-2} e 4×10^{-5} f 5.08×10^{-6}

5 $3 \times 2 \times 10^7 \times 10^3 = 6 \times 10^{10}$

6 a 5×10^{11} b 4×10^{-6} c 6.8×10^{-9}

7 b 7.2×10^3 c 3.5×10^{14} d 1.26×10^{-3}

8 3×10^4

9 a 4×10^5 b 5×10^6 c 2.3×10^7 d 4×10^{-7}

10 a 0.35×10^8
   b i 5.55×10^8
     ii 4.85×10^8

11 a 0.0223×10^{-3}
   b i 1.5223×10^{-3}
     ii 1.4777×10^{-3}

12 a 3.34×10^7 b 7.93×10^{-5}
   c 6.7504×10^{-7} d 4.6309×10^{12}
   e 1×10^{-8} f 5.3×10^9
   g 8.43×10^{-3} h 1.458×10^{-11}
   i 1.9459×10^{15}

13 a 1.00×10^8 b 7.90×10^{11}
   c 7.95×10^7 d 2.94×10^{-11}

18 Extend

1 $\left(\frac{2}{3} \div 3\frac{4}{5}\right)$, 57%, $\frac{7}{9}$, 0.98, $\left(3\frac{1}{3} \div 1\frac{2}{3}\right)$

2 a $n = 3$ b $a = 6$

3 a 1000000   b $-\frac{1}{2}$ c $\frac{1}{64}$ d 4
   e 4   f 12   g 1   h $\frac{1}{144}$

4 a 2^{48} b 5^{-81} c 3^{-80}

5 2.88×10^{-4}

6 a $9 \times 10^{-6}\,m^3$ b $3 \times 10^{-3}\,m^2$

7 a 100000   b $1\,km = 1 \times 10^5\,cm$
   c i $1\,km = 1 \times 10^6\,mm$
     ii $1\,kg = 1 \times 10^3\,g$
     iii $1\,ton = 1 \times 10^6\,g$

8 a

	8×10^7	
2×10^{-3}		4×10^{10}

| 2×10^{-9} | 1×10^6 | 4×10^4 |

b

	2.4×10^{-3}	
1.2×10^4		2×10^{-7}

| 3×10^6 | 4×10^{-3} | 5×10^{-5} |

9 $3.9182 \times 10^9\,km$

10 $2.67 \times 10^4\,km/h$

11 $5.6 \times 10^4 - 3.2 \times 10^3$, 6.5×10^2, $(3^2)^2$, $2\frac{1}{3} \times 3\frac{3}{5}$

12 a $5.3 \times 10^8\,km^2$ (2 s.f.)
   b $1.2 \times 10^{12}\,km^3$ (2 s.f.)
   c $1.6 \times 10^8\,km^2$ (2 s.f.)

13 a 4×10^{-3} b 3.2×10^{-2} c 6×10^{-5} d 3×10^{-8}
   e 60   f 15   g 500   h 2.5×10^{-7}

14 a $4 \times 10^{-5}\,cm^2$ b $8 \times 10^{-5}\,cm^2$ c $2.048 \times 10^{-2}\,cm^2$

15 a 9.03×10^{24} b 6.02×10^{24}

16 a 1.13727×10^{23} b 1.76352×10^{23}

17 a 1   b 0.000067   c 2.7×10^{14}

18 $p = 4.9 \times 10^{-5}$

18 Unit test

Sample student answers

Student A gives the best answer. There are 500 sheets with a thickness of 9×10^{-4} each. To find the total thickness multiply the thickness of each sheet by the number of sheets.

UNIT 19

19 Prior knowledge check

1 a $\frac{3}{4}$ b $\frac{1}{3}$ c $\frac{4}{5}$ d $\frac{9}{11}$

2 a 12   b 24   c 27   d 20

3 a -4 b 6   c 9
   d -8 e -10 f 12

4 a B scale factor 0.5, C scale factor 1, D scale factor 1.5, F scale factor 1
   b C, F

5 a f b f c e

6 a 68°   b 40°

7 a $\begin{pmatrix} 2 \\ 3 \end{pmatrix}$ b $\begin{pmatrix} 5 \\ 0 \end{pmatrix}$ c $\begin{pmatrix} -1 \\ -3 \end{pmatrix}$

    d $\begin{pmatrix} -2 \\ -3 \end{pmatrix}$ e $\begin{pmatrix} -5 \\ 0 \end{pmatrix}$ f $\begin{pmatrix} 1 \\ 3 \end{pmatrix}$

8 a $3\frac{3}{4}$ b 3.75

9 $\begin{pmatrix} 5 \\ 4 \end{pmatrix}$

19.1 Similarity and enlargement

1 $\frac{6}{8}, \frac{15}{20}$

2 a 2   b 3   c $\frac{1}{2}$

3 a $\frac{1}{2}$ b $\frac{1}{3}$ c 2

4 20 cm

5 They are similar.

6 A and C. Triangle A is an enlargement of triangle C.

7 a 37°, 53°, 90°. Both triangles have the same angles

    b 45°, 45°, 90°. Similar triangles have the same angles.

8 a i XY   b i 2   c i 18 cm

      ii AC   ii $\frac{1}{2}$ ii 6 cm

      iii YZ

    d i CA   e i angle ZXY

      ii AB   ii angle ACB

      iii $\frac{1}{2}$ iii angle XYZ

    f $\frac{4}{9}, \frac{8}{18}$. The fractions are equivalent.

9 a QR   b 7.5 cm

    c i 30°   ii 62°

10 a Yes. They have the same angles.

    b UW

    c i 3   ii QR   iii $\frac{1}{3}$, UW

19.2 More similarity

1 a C, D   b B, F   c G, I, J

2

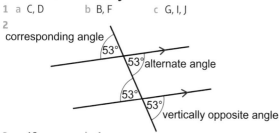

corresponding angle

53°

53° alternate angle

53°

53° vertically opposite angle

3 a 12 cm   b 6 cm

4 a 7.5 cm   b 8 cm

5 a They have the same angles.

    b 2   c 10 cm

6 a 10   b $x = 10$

7 a It has the same angles.   b $\frac{1}{2}$

8 $x = 5.1$ cm; $y = 16.1$ cm

9 a 106°   b TR   c $\frac{3}{4}$ d 9 cm

10 a Angle ACB = angle DCE (vertically opposite)

      Angle BAC = angle DEC (alternate angles)

      Angle ABC = angle CDE (alternate angles)

    b i 4.8 cm

      ii 3.2 cm

11 a 70°   b 1.5   c 13.5 cm   d 4.5 cm

12 a Angle A is common to both triangles

      Angle AED = angle ACB

      Since two angles are the same, the third angle in each triangle must be the same.

    b 1.25

    c 2 cm

19.3 Using similarity

1 28 cm

2 No. Corresponding sides are not in the same ratio.

3 No. They don't have the same angles.

4 a Yes

    b Need more information

    c Need more information

    d Need more information

    e Yes

5 Yes. They have the same angles and corresponding sides are in the same ratio.

6 a, b Yes. They have the same angles and corresponding sides are in the same ratio.

7 All circles are similar.

8 a 3   b 18 cm, 24 cm, 24 cm   c 32 cm

    d 96 cm   e 3

    f The scale factor of the perimeters is the same as the scale factor of the lengths.

9 20 m

10 32 m

19.4 Congruence 1

1 10

2 A, C

3 Students' own answers

4 a $x = 120°$, $y = 35°$ b $x = 32°$, $y = 28°$

5 Accurate drawings of triangles ABC and YZX. The triangles are congruent.

6 a They are congruent.   b Angle BDC

    c Yes; SSS.

7 a 20°   b 25°   c Yes; SSS.

8 a 110°   b 35°

9 a Yes   b Angle BCD   c CD

10 Accurate drawings of triangles ABC and YZX. Yes

11 a 50° (vertically opposite)

    b All 6 cm (radii)   c Yes; SAS

12 a Yes   b 58°

19.5 Congruence 2

1 a $x = 148°$, $y = 148°$, $z = 32°$ b $x = 37°$, $y = 73°$, $z = 73°$

2 a MB   b MC   c Angle BMD

3 Accurate drawings of triangles a and b. The triangles are congruent.

4 C

5 a AM = MC, BM = MD, AB = CD, angle AMB = 47° = angle CMD, angle BAM = angle DCM, angle ABM = angle CDM

    b Yes (SSS or SAS)

6 a 43°   b 54°   c Yes (ASA)   d 8 cm

7 a 65°   b Yes (ASA)   c RQ

8 a Not congruent   b Congruent (RHS)

9 a 13 cm   b 13 cm   c Yes   d 12 cm

10 a angle NLO   b NO   c Yes; SAS

19.6 Vectors 1

1 a $\begin{pmatrix} 4 \\ 2 \end{pmatrix}$ b $\begin{pmatrix} -4 \\ -2 \end{pmatrix}$

2

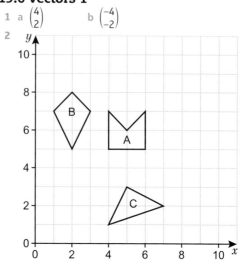

3 a $\begin{pmatrix} 3 \\ -1 \end{pmatrix}$ b $\begin{pmatrix} -3 \\ 0 \end{pmatrix}$

4 a $\begin{pmatrix} 3 \\ 1 \end{pmatrix}$ b $\begin{pmatrix} 2 \\ 3 \end{pmatrix}$ c $\begin{pmatrix} 5 \\ 4 \end{pmatrix}$

5 a $\begin{pmatrix} 5 \\ 6 \end{pmatrix}$ b $\begin{pmatrix} 7 \\ 3 \end{pmatrix}$ c $\begin{pmatrix} 7 \\ 4 \end{pmatrix}$

   d $\begin{pmatrix} 7 \\ -1 \end{pmatrix}$ e $\begin{pmatrix} 0 \\ 0 \end{pmatrix}$ f $\begin{pmatrix} 0 \\ 1 \end{pmatrix}$

6 $\begin{pmatrix} 2 \\ 5 \end{pmatrix}$

7 $\begin{pmatrix} 6 \\ -5 \end{pmatrix}$

8 a $\begin{pmatrix} 4 \\ -2 \end{pmatrix}$ b $\begin{pmatrix} -5 \\ 1 \end{pmatrix}$ c $\begin{pmatrix} 0 \\ -4 \end{pmatrix}$

9 a i $\begin{pmatrix} 3 \\ 3 \end{pmatrix}$ ii $\begin{pmatrix} 4 \\ -2 \end{pmatrix}$ iii $\begin{pmatrix} 7 \\ 1 \end{pmatrix}$

   b $\begin{pmatrix} 3 \\ 3 \end{pmatrix} + \begin{pmatrix} 4 \\ -2 \end{pmatrix} = \begin{pmatrix} 7 \\ 1 \end{pmatrix}$

10 a $\overrightarrow{NT}$ b $\overrightarrow{PR}$ c $\overrightarrow{VT}$

11 a $\begin{pmatrix} -3 \\ 6 \end{pmatrix}$ b $\begin{pmatrix} -3 \\ 9 \end{pmatrix}$ c $\begin{pmatrix} 6 \\ 1 \end{pmatrix}$

12 Students' own answers

13

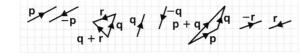

19.7 Vectors 2

1

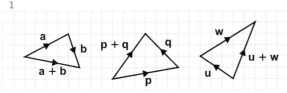

2 $\begin{pmatrix} 1 \\ 4 \end{pmatrix}$

3

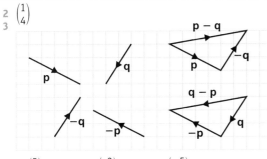

4 a $\begin{pmatrix} 5 \\ 1 \end{pmatrix}$ b $\begin{pmatrix} -3 \\ 0 \end{pmatrix}$ c $\begin{pmatrix} 5 \\ -5 \end{pmatrix}$

5 a $r = \begin{pmatrix} 5 \\ -3 \end{pmatrix}$

   b

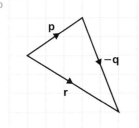

6 a $z = \begin{pmatrix} 4 \\ 2 \end{pmatrix}$ b

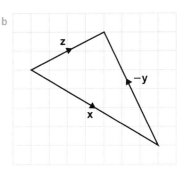

7 a $\begin{pmatrix} 4 \\ 10 \end{pmatrix}$ b $\begin{pmatrix} 9 \\ -6 \end{pmatrix}$ c $\begin{pmatrix} -8 \\ -20 \end{pmatrix}$ d $\begin{pmatrix} -15 \\ 10 \end{pmatrix}$

8 a $\begin{pmatrix} 6 \\ 0 \end{pmatrix}$ b $\begin{pmatrix} 8 \\ -12 \end{pmatrix}$ c $\begin{pmatrix} -1 \\ 6 \end{pmatrix}$ d $\begin{pmatrix} 11 \\ -3 \end{pmatrix}$

9

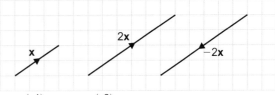

10 a $\begin{pmatrix} -4 \\ 8 \end{pmatrix}$ b $\begin{pmatrix} 2 \\ -4 \end{pmatrix}$

   c $\begin{pmatrix} 6 \\ -12 \end{pmatrix}$ d $\begin{pmatrix} -1 \\ 2 \end{pmatrix}$ or $\begin{pmatrix} 1 \\ -2 \end{pmatrix}$

11 a $\begin{pmatrix} 12 \\ -4 \end{pmatrix}$ b $\begin{pmatrix} -30 \\ 10 \end{pmatrix}$ c $\begin{pmatrix} -3 \\ 1 \end{pmatrix}$ d $\begin{pmatrix} 18 \\ -6 \end{pmatrix}$

19 Problem-solving

1 Scale factor $= \frac{5}{8}$; $\frac{5}{8} \times 32 \, cm = 20 \, cm$

2 Yes. A shape with three corners is always a triangle.

3 a David $= x$; Jesse $= x + 5$; Lucy $= x + 2$;
    David × Lucy = 15
    $x(x + 2) = 15$
    $x^2 + 2x = 15$
    Therefore $x^2 + 2x - 15 = 0$

   b $x^2 + 2x - 15 = (x + 5)(x - 3) = 0$
    Therefore $x = -5$ (not a valid age) or $x = 3$.
    Jesse $= x + 5 = 8$ years old

4 $XY = \sqrt{(4 - 1)^2 + (6 - 2)^2}$

   $XZ = \sqrt{(13 - 1)^2 + (18 - 2)^2}$

   $20 = 4 \times 5$, so $XZ = 4XY$

5 Anthony is correct. Anthony, Ross and Crista have invested in the company in the ratio of 5 : 4 : 3 and they will get a share of the £6000 profit in the same ratio. Therefore, Anthony gets £2500, Ross gets £2000 and Christa gets £1500.

6 a $12 \times 13 \times 9 = x - 96$; $1500 = x$

   b $pqr + s$

7 Students' own answers

19 Check up

1 a $\frac{1}{3}$ b 16 cm

2 a 2.5   b 4 cm

3 a PQ   b 9 cm   c i 40°   ii 64°

4 a Yes. ASA (work out the missing angles)
   b No. The 17 cm side is not between the 25° and 100° angles in the second triangle.

5 a $a = 35°$ alternate angle; $b = 106°$ interior or supplementary angle or angles in a triangle sum to 180°
   b Yes (ASA)

6 $\begin{pmatrix} 1 \\ 8 \end{pmatrix}$

7 a $\overrightarrow{UW}$ b $\begin{pmatrix} 6 \\ 0 \end{pmatrix}$

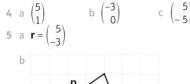

8 a $\begin{pmatrix} 6 \\ -2 \end{pmatrix}$ b $\begin{pmatrix} 5 \\ -5 \end{pmatrix}$ c $\begin{pmatrix} -1 \\ 2 \end{pmatrix}$

9

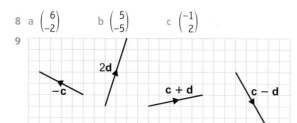

11 a **a**
   b −**b**
   c −**c**
   d **a** + **b**
   e −**a** − **b** − **c**

19 Strengthen

Similarity and enlargement

1 a, b, e

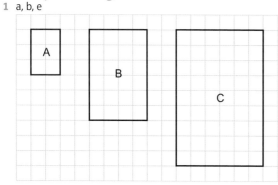

   c i 10   ii 20
   d 2   f 30
2 a 4   b 48 cm
3 a i alternate angles
     ii alternate angles
     iii vertically opposite angles
   b

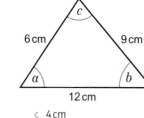

4 a EF   b 3   c 4 cm
   d i 47°
     ii 73°
5 a XY   b 14 cm
   c i 46°
     ii 60°

Congruence

1 Students' own answers
2 a

   b Accurate drawings of traingles STU and SVU.
     They are congruent (ASA).
   c UV

Vectors and translations

1 a $\begin{pmatrix} 6 \\ 8 \end{pmatrix}$ b $\begin{pmatrix} -1 \\ -2 \end{pmatrix}$ c $\begin{pmatrix} 3 \\ -2 \end{pmatrix}$ d $\begin{pmatrix} 9 \\ 2 \end{pmatrix}$

2 a $\begin{pmatrix} 1 \\ 4 \end{pmatrix}$ b $\begin{pmatrix} 1 \\ 4 \end{pmatrix}$ They are the same.

3 $\begin{pmatrix} -1 \\ 7 \end{pmatrix}$

4 $\overrightarrow{PX}$

5 $\overrightarrow{TX}$

6 a $\begin{pmatrix} 12 \\ 3 \end{pmatrix}$ b $\begin{pmatrix} -4 \\ 12 \end{pmatrix}$ c $\begin{pmatrix} 1 \\ -3 \end{pmatrix}$ d $\begin{pmatrix} -16 \\ -4 \end{pmatrix}$

7 a $\begin{pmatrix} 1 \\ 9 \end{pmatrix}$ b $\begin{pmatrix} -5 \\ 1 \end{pmatrix}$ c $\begin{pmatrix} 5 \\ -1 \end{pmatrix}$ d $\begin{pmatrix} 4 \\ 13 \end{pmatrix}$

19 Extend

1 3 m 28 cm
2 a (0, 20)   b (80, 0)
3 a 3.5   b i 10.5 m   ii 7.5 m
4 14.5 cm
5 a 61.2 km   b 11.4   c 75°
6 30.25 cm
7 a Octagon   b C, D   c 135°
8   i a = 10 cm, b = 8.7 cm
    ii c = 8.7 cm
    iii d = 8.7 cm
    iv e = 5 cm, f = 8.7 cm
     v g = 17.3 cm
   a i, ii, iii, iv
   b v is similar to each of the other triangles.
9 $\overrightarrow{FV}$
10 a 2**a**   b **b** − **a**   c **a** + 2**b**   d 2**b** − 3**a**
11 a −**a**   b **b** − **a**   c **a** − **b**
12 a **a** + **b**   b **a** − **b**
13 a $\begin{pmatrix} -6 \\ 3 \end{pmatrix}$ b $\begin{pmatrix} -3 \\ 6 \end{pmatrix}$ c $\begin{pmatrix} 2 \\ -1 \end{pmatrix}$ d $\begin{pmatrix} -18 \\ 9 \end{pmatrix}$

19 Unit test

Sample student answer
The student has drawn the direction lines from P to Q to help show that the directions are positive.
The student has labelled the axes with the coordinates known, to more clearly see how far they are counting, which is more difficult without grid lines.

UNIT 20

20 Prior knowledge check

1 a 129   b 35
2 a 8   b 31   c 17   d −125
3 a 0.5   b −0.1   c 6   d 5
4 a $4n$ b $7x$
5 $2x = 50$
6 a $x = 3$ b $x = \frac{2}{3}$ c $x = -4$ d $x = -\frac{5}{2}$
7 a $5(x + 2)$ b $4(2x + 1)$ c $x(x + 4)$ d $2(3x - 4)$
8 a $x = y - 2$ b $x = m + 5$ c $x = \frac{d}{4}$ d $x = 6f$
   e $x = \frac{t - 5}{2}$ f $x = \frac{s + 4}{3}$
9 a 64   b 67   c 72
   d −64   e $\frac{1}{4}$ or 0.25
10 a $4x + 12$ b $-2x - 10$ c $x^2 + 2x$
   d $x^2 + x - 6$ e $x^2 - 2x + 1$
11 a

x	0	5
y	2	0

b, d

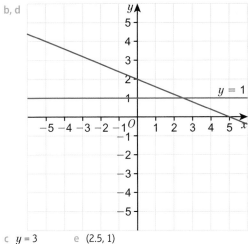

c $y = 3$ e (2.5, 1)

12 Gradient 4, y-intercept (0, −1)

13 □ = 4, Δ = 4 and ✿ = 2

20.1 Graphs of cubic and reciprocal functions

1 a $\frac{1}{4}$ b 3   c $-\frac{1}{5}$ d −4

2

x	−3	−2	−1	0	1	2	3
y	−27	**−8**	**−1**	**0**	**1**	8	**27**

3 Students' own graphs of $y = x^3$

4 a Students' estimates from their graphs
   b i 12.167   ii 2.47 (2 d.p.)

5 a

x	−3	−2	−1	0	1	2	3
y	27	8	1	0	−1	−8	−27

   b Students' own graphs of $y = -x^3$

6 a

x	−3	−2	−1	0	1	2	3
x^3	**−27**	−8	**−1**	**0**	**1**	**8**	27
+2	+2	+2	+2	+2	+2	+2	+2
y	**−25**	−6	**1**	**2**	**3**	**10**	29

b, d

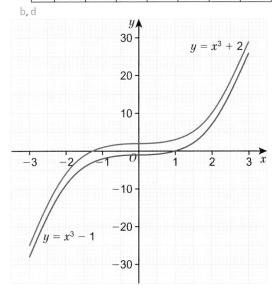

c

x	−3	−2	−1	0	1	2	3
x^3	−27	−8	−1	0	1	8	27
−1	−1	−1	−1	−1	−1	−1	−1
y	−28	−9	−2	−1	0	7	26

   e Same shape, cross the x- and y-axes at different places.

7 Answers between −1.1 and −1.5

8 a

x	−3	−2	−1	0	1	2	3
y	**−15**	**0**	3	0	**−3**	**0**	15

   b Students' own graphs correctly plotted from the table.

9 a

x	−4	−3	−2	−1	$-\frac{1}{2}$	$-\frac{1}{4}$	$\frac{1}{4}$	$\frac{1}{2}$	1	2	3	4
y	$-\frac{1}{4}$	$-\frac{1}{3}$	$-\frac{1}{2}$	−1	−2	−4	4	2	1	$\frac{1}{2}$	$\frac{1}{3}$	$\frac{1}{4}$

   b, c Students' own graphs correctly plotted from the table.

10 Students' own answers

11 a ii   b iv   c v
    d iii   e vi   f i

20.2 Non−linear graphs

1 a $y = \frac{1}{x}$ b $y = x$

2 a $8\,cm^3$ b $125\,cm^3$ c $x^3\,cm^3$

3 1 hour 18 minutes

4 a $y = x^3 + 8$
   b

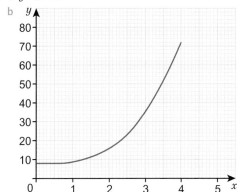

   c Answers between 26 and 29
   d Answers between 3.1 and 3.3

5 a Inverse proportion
   b Answers between 6.4 and 7.2 hours
   c 14

6 a 100 litres   b 20 litres
   c i 4 minutes
      ii Answers between 8 and 9 minutes
   d Between 0 and 2 minutes, 30 litres emptied out.
      Between 8 and 10 minutes, 7 litres emptied out. This
      shows that the tank emptied less quickly as time passed
      and more water emptied out.

7 a Speed is inversely proportional to time.
   b Estimates between 3 hours 20 minutes and
      3 hours 30 minutes
   c Estimate between 47 km/h and 49 km/h

8 a 20 counts per second   b 6 weeks   c 2 weeks

9 a

Time (hours)	0	1	2	3	4
Number of cells	2	4	8	16	32

   b First term is 2; multiply by 2 to get next term
   c 18 or 19
   d Answers between 3 hours 40 minutes and 3 hours 50
      minutes

10 a £1000   b Answers between £1215 and £1225
    c £40   d 4%

20.3 Solving simultaneous equations graphically

1. a Straight-line graph passing through y-coordinates = 3, labelled $y = 3$
   Straight-line graph passing through (0, 7) and (2, 3), labelled $2x + y = 7$
   b (2, 3)
2. a $30x + 15y$
   b $30x + 15y = 90$
3. a, b Straight-line graph passing through y-coordinates = 1, labelled $y = 1$
   Straight-line graph passing through (0, 9) and (2, 5), labelled $2x + y = 9$
   c $x = 4, y = 1$
4. $x = 2, y = 1.5$
5. $x = -2.5, y = 5$
6. a $x = 5, y = 3$
   b $x = 3.8, y = 1.8$
   c $x = 7, y = 1$
7. a

x	0	2	4
y	−11	−5	1

   b Straight-line graph passing through (0, −11) and (2, −5), labelled $y = 3x − 11$
   c $x = 3.6, y = 0.2$
8. They do not have a solution because the graphs are parallel lines, so do not intersect.
9. a $x + y = 18$ b $x − y = 4$
   c Straight-line graphs of $x + y = 18$ (passes through (0, 18) and (18, 0)) and $x − y = 4$ (passes through (10, 6) and (4, 0))
   d Abi is 11 and Ben is 7
10. a x cost of adult ticket; y cost of child ticket
    b $4x + 9y = 112$
    c $x = £10, y = £8$
11. a $2x + y = 60; x + 2y = 84$
    b minibus 12, coach 36
12. Set b would be harder to solve from a graph because to fit 115 and 5 on a graph you would need a very small scale, which can be difficult to read accurately.

20.4 Solving simultaneous equations algebraically

1. a +   b −   c +   d −
2. a $6x + 8y = 4$ b $3x − 6y = 15$
3. a $y = 6$ b $x = 1.5$ c $y = −4$ d $x = 2$
4. a $2x + 1 = 9$ b $x = 4$ c $x = 4, y = 1$
5. a $x = 5, y = 3$
6. a $x = 5, y = −2$
   b $x = \frac{1}{4}, y = 1$
   c $x = −\frac{4}{5}, y = 4\frac{3}{5}$
7. $x + y = 24$ and $x − y = 14$
   The two numbers are 5 and 19
8. $x = 6, y = 13$
9. $x = 2, y = 5$
10. a $6x + 9y = 33$
    b $6x + 8y = 30$
    c $x = 1, y = 3$
11. a $8x + 12y = 44$
    b $9x + 12y = 45$
    c $x = 1, y = 3$
12. a £7.50   b £10
13. a At point A: $4 = 2m + c$
       At point B: $10 = m + c$
    b $m = −6, c = 16$
    c $y = −6x + 16$
14. a $y = x + 3$ b $y = \frac{3}{2}x + 5$

20.5 Rearranging formulae

1. a $x = 3$ b $x = 4$ c $x = 2$
2. a i and iv   b ii and iii
3. a C b d c V d A
4. a $\frac{A}{5} = w$ b $2M = w$
5. a $d = \frac{C}{3}$ b $R = \frac{V}{I}$ c $d = st$
   d $q = \frac{4}{p}$ e $m = dv$ h $A = \frac{F}{P}$
6. a $D = ST$ b $T = \frac{D}{S}$
7. Copper 0.0004 m³
   Iron 0.0014 m³
   Aluminium 0.0342 m³
8. a $t = \frac{v − u}{a}$ b $z = \frac{x + y}{m}$ c $p = \frac{q − l}{r}$ d $h = \frac{e − f}{g}$
9. a $y = −3x + 2.5$
   b Gradient = −3; y-intercept = 2.5
10. a and d
11. 2.4 cm   15 cm circumference, radius 2.4 cm
    3.2 cm   20 cm circumference, radius 3.2 cm
    4.0 cm   25 cm circumference, radius 4.0 cm
12. a $r = \frac{C}{2\pi}$ b $l = \frac{V}{wh}$ c $w = \frac{V}{lh}$ d $h = \frac{A}{2\pi r}$
13. a $T = \frac{100I}{PR}$ b $T = \frac{PV}{k}$ c $b = \frac{2A}{h}$
    d $w = \frac{P − 2l}{2}$ or $w = \frac{P}{2} − lw$
    e $n = \frac{X − m^2}{m}$ or $n = \frac{X}{m} − m$
    f $n = 5M − 2$
    g $Q = \frac{2P}{3} + t$
    h $h = \frac{A}{2\pi r} − r$
14. a $y − 2 = \frac{x + 1}{a}$
       $a(y − 2) = x + 1$
       $a(y − 2) − 1 = x$
    b i $x = b(z − 5) + 1$ ii $x = n(m − p) − 3$
       iii $x = 2 + \frac{g}{3}$ iv $x = \frac{q + 7}{3}$
15. a $x = \sqrt{y}$ b $z = \sqrt{\frac{y}{5}}$ c $x = \sqrt{2y}$
    d $r = \sqrt{\frac{A}{\pi}}$ e $x = y^2$ f $s = \frac{t^2}{3}$
    g $t = P^2 − r$ h $r = \sqrt[3]{\frac{3V}{4\pi}}$ i $r = \sqrt{\frac{A}{4\pi}}$
    j $r = \sqrt{\frac{V}{\pi h}}$ k $h = \frac{3V}{\pi r^2}$ l $a = \sqrt{c^2 − b^2}$
16. $r = 2.2$ cm; $h = 3.0$ cm
17. $t = \frac{2d − 7}{6}$
18. a $u = \sqrt{v^2 − 2as}$
    b $u = \frac{s}{t} − \frac{at}{2}$ or $u = \frac{s − \frac{1}{2}at^2}{t}$
    c $a = \frac{2(s − ut)}{t^2}$ or $a = \frac{s − ut}{\frac{1}{2}t^2}$

20.6 Proof

1. a $x^2 − 2x$ b $x^2 − 3x − 4$ c $x^2 + 4x + 4$
2. a $3(x + 1)$ b $x(x − 3)$ c $2(2x − 1)$
3. 30 cm²
4. a $x = 7$
   b No, only true for $x = 7$
   c Equation
5. a Equation   b Formula
   c Equation   d Expression

6 Expand or factorise to prove the identities.

7 a $x^2 + 5x + 6 + x + 3$
  b $x^2 + 6x + 9$
  c $x^2 + 6x + 9$
  d Statement in part **b** = statement in part **c**

8 a $(x + 3)(x + 1) = x^2 + 4x + 3$
  b $x^2 - x$
  c $x^2 + 4x + 3 - (x^2 - x) = 5x + 3$

9 Area of rectangle = $(x - 1)(x + 1) = x^2 - 1$
  Height of triangle = $3x - 1 - (x - 1) = 2x$
  Area of triangle = $\frac{1}{2} \times 2x(x + 1) = x^2 + x$
  Total area = $x^2 - 1 + x^2 + x = 2x^2 + x - 1$

10 a $n, n + 1, n + 2$
   b $n - 1, n, n + 1$

11 a $n - 2, n - 1, n, n + 1, n + 2$
   b range = $n + 2 - (n - 2) = 4$
   c mean = $\dfrac{n - 2 + n - 1 + n + n + 1 + n + 2}{5} = \dfrac{5n}{5} = n$

12 a $n, n + 1, n + 2$
   b $3n + 3$
   c $3(n + 1)$
   d $3(n + 1)$, which is $(n + 1) \times 3$

13 Four consecutive numbers: $n, n + 1, n + 2, n + 3$
   Sum = $4n + 6 = 2(2n + 3)$
   2 is a factor of $2(2n + 3)$ so it is even.

14 a i 2, 4, 6, 8, 10
      ii 1, 3, 5, 7, 9
   b i $2n$
      ii $2n - 1$ or $2n + 1$

15 $2m + 2n = 2(m + n)$
   This is a multiple of 2, so it is an even number.

16 $2m + 1 + 2n + 1 = 2m + 2n + 2 = 2(m + n + 1)$.
   This is a multiple of 2, so it is an even number.

17 $2(x - a) = x + 4$
   $2x - 2a = x + 4$
   $x - 2a = 4$
   $x = 2a + 4 = 2(a + 2)$
   Since a is an integer, $a + 2$ is an integer, and 2 × an integer is a multiple of 2, or an even number.

20 Problem solving

1 a, b

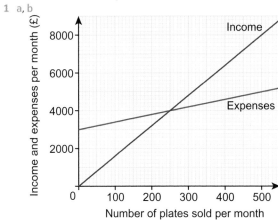

  c When income is less than expenses, the company is making a loss.
    When income is more than expenses, the company is making a profit.
    When a company is at break-even point it is not making a profit or a loss.
  d 250 plates
  e $y = 16x$ and $y = 4x + 3000$; $x = 250$; $y = £4000$
  f £1200
  g £1500

2 200 frames sold

20 Check up

1 a

x	−3	−2	−1	0	1	2	3
y	−29	−10	−3	−2	−1	6	25

  b

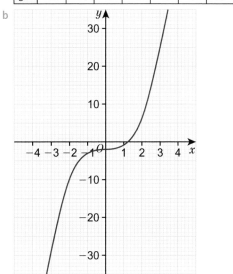

2 a

x	−3	−2	−1	−0.5	0.5	1	2	3
y	$-\frac{1}{3}$	$-\frac{1}{2}$	−1	−2	2	1	$\frac{1}{2}$	$\frac{1}{3}$

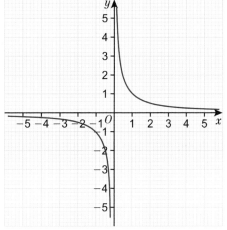

  b i Answers between $y = -0.6$ and -0.8
    ii Answers between $x = 0.2$ and 0.4

3 a Answers between 6 and 6.4 litres
  b 2.5 kilopascals
  c Answers between 7.5 and 9 litres

4 $3y + 2x = 9$

5 $x = 2, y = 3$

6 a $x + y = 24$; $x - y = 8$ b 16 and 8

7 $x = 4, y = 2$

8 a $t = m - pr$ b $c = \dfrac{z - y}{a}$

  c $m = \dfrac{p}{3} - n$ d $r = \dfrac{pq}{v}$

9 a $t = \sqrt{R}$ b $p = \dfrac{nx}{t}$

  c $x = 8 - 2y + 2z$ d $x = \dfrac{y - 4}{5}$

10 Expanding the brackets: $xy(x - y) = x^2y - xy^2$

12 ☼ = 3, Δ = 5 and ☐ = 1

20 Strengthen

Graphs

1 a $y = -x^3$ b $y = x^3$
2 a (−2, 7) and (3, −26)   b (−2, −9) and (3, 26)
3 a

x	−3	−2	−1	0	1	2	3
y	−26	−7	0	1	2	9	28

b

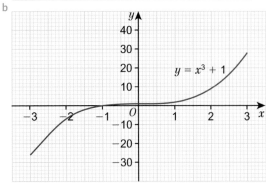

4 a $y = x^2$ b $y = \frac{1}{x}$ c $y = x$
5 a (2, 2) and (−2, 0.5)
   b (2, 0.5) and (−2, −0.5)
   c

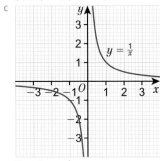

6 a Time in weeks
   b Number of rats
   c   i 14
     ii 32
   d The number of rats increased by 18.

Simultaneous equations

1 a $3x + 5y = 10$ b $6x + 10y = 70$
   c $2x + 4y = 18$ d $5x + y = 65$
2 a x and y
   b   i $x = 6.5, y = 1.5$
     ii $x = 3, y = 5$
     iii $x = 0.6–0.8, y = 2.6–2.8$
3 a $x = 10, y = 2$ b $x = 14, y = 6$
   c $x = 7, y = 3$ d $x = 5, y = 8$
4 b $2x$ is a multiple of x
   c $2x + 4y = 10$
   d $x = 3, y = 1$
5 b No   c $-4y$ is a multiple of y
   d $12x + 4y = 80$ e $x = 6, y = 2$

Using algebra

1 a $a = v - 4x$ b $a = qy - p$
   c $a = \frac{s-3}{t}$ d $a = \frac{y-b}{4}$
2 a $d = \frac{x}{2} - 3$ b $d = \frac{s}{2} + t$
   c $d = \frac{2y}{t}$ d $d = \frac{kr}{t}$
3 a $n = c$ b $n = \sqrt{r}$ c $n = \sqrt{T}$ d $n = \sqrt{6y}$

4 a $x = 18$
   b   i $m = \frac{np}{3}$
     ii $m = \frac{2k}{t}$
     iii $m = \frac{qr}{p}$
     iv $m = \frac{vz}{b}$
5 a $x = 2$
   b   i $x = y + 6$
     ii $x = 2y$
     iii $x = 7 - y$
6 a $x = 4$
   b   i $x = 3y - 15$
     ii $x = z - \frac{2}{a}$
     iii $x = t - 4 - \frac{1}{b}$
7 a LHS simplify to RHS $8a + b$
   b LHS simplify to RHS $5x^2 + x - 5$
   c LHS simplify to RHS $2x + 8$
   d Both sides simplify to $x^2y + 2x^2 - 4y$

20 Extend

1 a Students' own graphs correctly plotted from the table.
   b Inverse proportion
2 $x = 5, y = -2$
3 a Estimates between 90 and 100
   b Estimates between 70 and 80
   c Between week 6 and week 7
   d Week 8
4 a

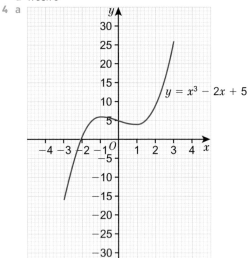

   b −2.1
5 a $t = m^2$ b $r = \left(\frac{p}{q}\right)^2$ c $d = \sqrt{\frac{c}{3}}$ d $b = 2\sqrt{a}$
   e $f = \frac{\sqrt{e}}{3}$ f $s = \sqrt{v + wx}$ g $w = \frac{s^2 - v}{x}$

6

Sample	Mass (g)	Volume (cm³)
a	25g	2.8
b	60g	6.7
c	90g	10.1
d	125g	14.0
e	240g	26.9
f	350g	39.2

7 a $x = \dfrac{p - m - n}{r}$ b $x = \dfrac{c + ab}{a}$

   c $x = \dfrac{de}{f - d}$ d $x = \dfrac{gh + n}{m - g}$

   e $x = \dfrac{rs + pq}{p - r}$ f $x = \dfrac{cd - a^2}{a}$

8 a $8 + 14 = 9 + 13 = 22$
   b Students' own answers
   c $n + n + 6 = n + 1 + n + 5 = 2n + 6$

9 a $9 \times 13 - 8 \times 14 = 5$
   b Students' own answers
   c $(n + 1)(n + 5) - n(n + 6) = n^2 + 6n + 5 - n^2 - 6n = 5$

10 Volume of cube = x^3; Volume of cuboid = $10x$.
   So $x^3 = 10x + 100$.
   Rearrange to give $x^3 - 10x = 100$.

11 $x = 3, y = -2$

12 $y = \frac{1}{2}x + 3$ (or $2y - x = 6$) and $x + y = 9$

13 a $x^2 + 3x + 12 = 35$, so $x^2 + 3x = 23$
   b When $x = 3$, $x^2 + 3x = 18$
     When $x = 4$, $x^2 + 3x = 28$
     So value of x between 3 and 4 gives value of
     $x^2 + 3x = 23$

14 a $(2t + 1)(t + 2)$
   b This is always a product of two whole numbers, each of
     which is greater than 1.

15 $20 = 2k - 4m$ and $15 = 3k - 9m$
   $k = 20, m = 5$

20 Unit test

Sample student answer
a $(5, 1.2)$
b The student has forgotten that the y-axis scale on the graph
   paper goes up in 0.2 instead of 0.1.

Index

702